# Principles of Wood Science and Technology

## I
### Solid Wood

Franz F. P. Kollmann   Wilfred A. Côté, Jr.

Springer-Verlag New York Inc. 1968

Franz F. P. Kollmann

Professor an der Universität München
Institut für Holzforschung und Holztechnik

Wilfred A. Côté, Jr.

Professor at the State University of New York
College of Forestry, Syracuse, N. Y.

All rights reserved. No part of this book may be translated or reproduced in any form
without written permission from
Springer-Verlag

ISBN-13: 978-3-642-87930-2    e-ISBN-13: 978-3-642-87928-9
DOI: 10.1007/978-3-642-87928-9

© by Springer-Verlag, Berlin · Heidelberg 1968
Softcover reprint of the hardcover 1st edition 1968

Library of Congress Catalog Card Number 67—29614

The use of general descriptive names, trade names, trade marks, etc. in this publication, even if the former are not especially
identified, is not to be taken as a sign that such names, as understood by the Trade Marks and Merchandise Marks
Act, may accordingly be used freely by anyone

Title-No. 1301

# PREFACE

Modern forest products research had its start hardly fifty years ago. Today we are in a position to apply the title "wood science" to the field of wood technology that is based on scientific investigation, theoretical as well as experimental. It is this research that fosters new uses for wood as a raw material and that creates the foundation for new industries for the manufacture of wood-base materials such as plywood, laminated products, particle and fiber board and sandwich construction.

Wood technology in its broadest sense combines the disciplines of wood anatomy, biology, chemistry, physics and mechanical technology. It is through this interdisciplinary approach that progress has been made in wood seasoning, wood preservation methods, wood machining, surfacing and gluing, and in the many other processes applied in its utilization.

In 1936 the senior author published a book entitled, "Technologie des Holzes", which was a first approach to a universal reference book on wood technology. The first edition of Volume I of the Textbook of Wood Technology, co-authored by H. P. BROWN, A. J. PANSHIN, and C. C. FORSAITH, was published in 1948. An indication of the rapid development of this field can be gained from the fact that the second edition of "Technologie des Holzes und der Holzwerkstoffe", completely revised, was needed by 1951. It contains 2233 pages compared with the 764 pages of the 1936 edition. Similarly, the many changes in wood science called for a revision of the Textbook of Wood Technology which was accomplished by A. J. PANSHIN and C. H. DE ZEEUW in 1964.

In spite of the wide acceptance of the above works, there appeared to be a need for a completely new and concise book on the fundamentals of wood science. The authors undertook to produce such a volume, agreeing to write it in English so that wider use might be made of it. The object of this volume is to provide a ready reference for the technologist who deals with wood as a structural material or who uses wood as a raw material in manufacturing improved wood products. It is essential that the fundamentals of wood structure be presented in this type of manual, but the identifying features of many individual species of wood, as well as lengthy keys, cannot be included if conciseness is to be maintained.

In the original planning for this book, a chapter on Adhesion and Adhesives was included. After noting that all of the other chapters dealt with "solid wood" this chapter was deleted. Instead it will appear in Volume II of this reference book, a more logical arrangement since this volume will be concerned with wood base and composite materials. Although solid wood is glued to make furniture and other products, it seemed more appropriate to place this topic in the context of other composite materials such as plywood, fiberboard, particleboard, corestock and laminated products. A chapter on the fundamentals of adhesion would then have immediate relevance to the other chapters of that volume.

We hope that we have achieved our objectives in this joint endeavor and that this volume does prove to be a useful one for the wood scientist. Our task has been made much easier by the aid and advice of a number of specialists. To them we extend our sincere appreciation. Included are Prof. Dr. GÜNTHER BECKER, Prof. Dr. ELLIS B. COWLING, Prof. Dr. JOHN B. SIMEONE, and Prof. Dr. TORE E. TIMELL.

Mr. A. C. Day offered invaluable assistance in the preparation of many of the photographs.

The senior author wishes to express his personal gratitude to Prof. Dr. Arno P. Schniewind and Prof. Skaar who reviewed part of the typescript and gave valuable linguistic advice. Mr. P. Kisseloff thoroughly studied and summarized the newest scientific publications on wood science, particularly the literature on wood research in the USA. Miss R. Preissler diligently assisted in the preparation of the manuscript. Mr. K. A. Sorg has been a capable and hard-working editorial assistant.

It is a pleasant duty to express sincere thanks to Springer-Verlag for their interest in this project and for permission to reproduce many illustrations from various volumes of the journal "Holz als Roh- und Werkstoff".

München/Germany,
Syracuse, N.Y., USA, 1967 **Franz F. P. Kollmann   W. A. Côté, Jr.**

# CONTENTS

**1. The Structure of Wood and the Wood Cell Wall.** By WILFRED A. CÔTÉ, JR. . . . 1

  1.0 Introduction . . . . . . . . . . . . . . . . . . . . . . . . . . . . 1

  1.1 Gross Structure of Wood . . . . . . . . . . . . . . . . . . . . . 1
    1.1.1 Cellular Composition . . . . . . . . . . . . . . . . . . . 2
    1.1.2 Wood Rays . . . . . . . . . . . . . . . . . . . . . . . . . 2
    1.1.3 Planes of Wood . . . . . . . . . . . . . . . . . . . . . . 3
    1.1.4 Sapwood and Heartwood . . . . . . . . . . . . . . . . . . 4
    1.1.5 Growth Increments . . . . . . . . . . . . . . . . . . . . 4
    1.1.6 Axial Parenchyma . . . . . . . . . . . . . . . . . . . . . 6
    1.1.7 Intercellular Canals . . . . . . . . . . . . . . . . . . . . 8
    1.1.8 Other Gross Features . . . . . . . . . . . . . . . . . . . 9

  1.2 Microscopic Structure of Wood . . . . . . . . . . . . . . . . . . 9
    1.2.1 Major Cell Types . . . . . . . . . . . . . . . . . . . . . 10
    1.2.2 Cell Sorting and Arrangement . . . . . . . . . . . . . . . 12
    1.2.3 Cell Inclusions . . . . . . . . . . . . . . . . . . . . . . . 15
      1.2.3.1 Tyloses . . . . . . . . . . . . . . . . . . . . . . 15
      1.2.3.2 Crystals . . . . . . . . . . . . . . . . . . . . . . 16
      1.2.3.3 Oil cells . . . . . . . . . . . . . . . . . . . . . . 17
      1.2.3.4 Gums and Resins . . . . . . . . . . . . . . . . . 18

  1.3 Organization of the Cell Wall . . . . . . . . . . . . . . . . . . . 18
    1.3.1 The Microfibril . . . . . . . . . . . . . . . . . . . . . . . 18
    1.3.2 Generalized Structure and Terminology . . . . . . . . . . 20
    1.3.3 Microfibrillar Orientation . . . . . . . . . . . . . . . . . 21
    1.3.4 Cell Wall Sculpturing . . . . . . . . . . . . . . . . . . . 27
      1.3.4.1 Pit Structure . . . . . . . . . . . . . . . . . . . 27
      1.3.4.2 Helical Thickenings . . . . . . . . . . . . . . . . 37
      1.3.4.3 Wart Structure . . . . . . . . . . . . . . . . . . 39
      1.3.4.4 Dentate Ray Tracheid Thickenings . . . . . . . . 42

  1.4 Reaction Wood Anatomy and Ultrastructure . . . . . . . . . . . 43
    1.4.1 Compression Wood . . . . . . . . . . . . . . . . . . . . 44
    1.4.2 Tension Wood . . . . . . . . . . . . . . . . . . . . . . . 48

  Literature Cited . . . . . . . . . . . . . . . . . . . . . . . . . . 52

**2. Chemical Composition of Wood.** By WILFRED A. CÔTÉ, JR. . . . . . . . . . . 55

  2.0 Introduction . . . . . . . . . . . . . . . . . . . . . . . . . . . 55

  2.1 Chemical Constituents of Wood and their Determination . . . . . . . . . 56

  2.2 Characteristics of the Principal Wood Constituents . . . . . . . . 58
    2.2.1 Cellulose . . . . . . . . . . . . . . . . . . . . . . . . . . 58
      2.2.1.1 Isolation from Wood . . . . . . . . . . . . . . . 58
      2.2.1.2 Structure . . . . . . . . . . . . . . . . . . . . . 58
      2.2.1.3 Properties . . . . . . . . . . . . . . . . . . . . . 60
    2.2.2 Hemicellulose . . . . . . . . . . . . . . . . . . . . . . . 61
      2.2.2.1 Hemicelluloses of Hardwoods . . . . . . . . . . . 61
      2.2.2.2 Hemicelluloses of Softwoods . . . . . . . . . . . 63
    2.2.3 Other Wood Polysaccharides . . . . . . . . . . . . . . . 64
    2.2.4 Lignin . . . . . . . . . . . . . . . . . . . . . . . . . . . 64
      2.2.4.1 Isolation from Wood . . . . . . . . . . . . . . . 67

# Contents

|  |  |
|---|---|
| 2.2.4.2 Structure | 67 |
| 2.2.4.3 Properties | 67 |
| 2.3 Wood Extractives | 70 |
| 2.4 Distribution of Chemical Constituents in Wood | 72 |
| Literature Cited | 75 |

**3. Defects and Abnormalities of Wood.** By WILFRED A. CÔTÉ, JR. . . . . . . . . . 79

| | |
|---|---|
| 3.0 Introduction | 79 |
| 3.1 Natural Defects | 79 |
| 3.1.1 Knots | 79 |
| 3.1.2 Reaction Wood | 81 |
| 3.1.2.1 Compression Wood | 83 |
| 3.1.2.2 Tension Wood | 83 |
| 3.1.3 Cross Grain | 84 |
| 3.1.4 Variations in Log Form | 85 |
| 3.1.5 Shake | 86 |
| 3.1.6 Miscellaneous Natural Defects | 87 |
| 3.2 Defects due to Processing | 90 |
| 3.2.1 Manufacturing Defects | 90 |
| 3.2.2 Seasoning Defects | 90 |
| 3.2.2.1 Checks | 91 |
| 3.2.2.2 Warp | 92 |
| 3.2.2.3 Casehardening | 92 |
| 3.2.2.4 Collapse | 93 |
| 3.2.2.5 Honeycomb | 94 |
| 3.2.2.6 Washboarding | 94 |
| 3.2.2.7 Miscellaneous Seasoning Defects | 94 |
| 3.2.3 Raised Grain | 94 |
| 3.2.4 Loosened Grain | 95 |
| Literature Cited | 95 |

**4. Biological Deterioration of Wood.** By WILFRED A. CÔTÉ, JR. . . . . . . . . . 97

| | |
|---|---|
| 4.0 Introduction | 97 |
| 4.1 Fungi causing Wood Deterioration | 97 |
| 4.1.1 Characteristics of Wood-destroying Fungi | 98 |
| 4.1.1.1 Comparison of Brown Rot and White Rot | 98 |
| 4.1.1.2 Soft Rot | 104 |
| 4.1.2 Characteristics of Wood-staining Fungi | 105 |
| 4.1.3 Physiological Requirements of Wood-destroying and Wood-inhabiting Fungi | 107 |
| 4.1.3.1 Temperature | 107 |
| 4.1.3.2 Oxygen | 107 |
| 4.1.3.3 Moisture | 108 |
| 4.1.3.4 Nutrients | 109 |
| 4.1.3.5 Hydrogen Ion Concentration | 109 |
| 4.1.3.6 Natural Durability | 110 |
| 4.1.3.7 Relationship of Wood Preservation to Physiological Requirements | 110 |
| 4.1.4 Mechanism of Wood Decay | 110 |
| 4.1.5 Influence of Decay on Mechanical Properties | 111 |
| 4.2 Wood-boring Insects | 112 |
| 4.2.1 Termites | 112 |
| 4.2.1.1 Characteristics | 113 |
| 4.2.1.2 Control | 114 |
| 4.2.2 Powder-post Beetles | 115 |
| 4.2.2.1 Lyctidae | 115 |

| | | |
|---|---|---|
| 4.2.2.2 | Wood-feeding Anobiidae | 118 |
| 4.2.2.3 | Cerambycidae. Long-horned Beetles or Round-headed Borers | 119 |
| 4.2.2.4 | Bostrichidae. Auger or Shot-hole Borers | 121 |
| 4.2.2.5 | Control Measures | 121 |
| 4.2.3 | Carpenter Ants | 124 |
| 4.2.4 | Carpenter Bees. Order Hymenoptera | 126 |
| 4.2.5 | Horntails. (Siricidae) | 126 |
| 4.3 | Marine Borers | 128 |
| 4.3.1 | Molluscan Borers | 129 |
| 4.3.2 | Crustacean Borers | 130 |
| 4.3.3 | Protection against Marine Wood Borers | 132 |

Literature Cited . . . . . . . . . . . . . . . . . . . . . . . . . . . . . . . . 133

## 5. Wood Preservation. By Wilfred A. Côté, Jr. . . . . . . . . . . . . . . . . 136

5.0 Introduction . . . . . . . . . . . . . . . . . . . . . . . . . . . . . . . . 136

5.1 General Considerations . . . . . . . . . . . . . . . . . . . . . . . . . . 136

    5.1.1 Effect of Structure on Treatment . . . . . . . . . . . . . . . . . . 137

    5.1.2 Timber Preparation . . . . . . . . . . . . . . . . . . . . . . . . . . 140

5.2 Wood Preservation Processes . . . . . . . . . . . . . . . . . . . . . . . 140

    5.2.1 Non-pressure Processes . . . . . . . . . . . . . . . . . . . . . . . . 141

        5.2.1.1 Brushing or Spraying . . . . . . . . . . . . . . . . . . . . 141

        5.2.1.2 Dipping . . . . . . . . . . . . . . . . . . . . . . . . . . . 141

        5.2.1.3 Steeping and Cold Soaking . . . . . . . . . . . . . . . . . 142

        5.2.1.4 Hot-and-Cold Bath . . . . . . . . . . . . . . . . . . . . . 142

        5.2.1.5 Diffusion Method . . . . . . . . . . . . . . . . . . . . . . 143

    5.2.2 Pressure Processes . . . . . . . . . . . . . . . . . . . . . . . . . . . 144

        5.2.2.1 Full-cell Processes . . . . . . . . . . . . . . . . . . . . . . 145

        5.2.2.2 Empty-cell Processes . . . . . . . . . . . . . . . . . . . . 146

    5.2.3 Miscellaneous Processes . . . . . . . . . . . . . . . . . . . . . . . 146

5.3 Wood Preservatives . . . . . . . . . . . . . . . . . . . . . . . . . . . . 147

    5.3.1 Characteristics of Preservatives . . . . . . . . . . . . . . . . . . . 147

    5.3.2 Preservative Materials Toxic to Insects, Fungi and Marine Borers . . . . . 148

5.4 Fire Retardant Treatment . . . . . . . . . . . . . . . . . . . . . . . . . 149

    5.4.1 General Remarks about the Combustibility of Wood . . . . . . . . 149

    5.4.2 Developed Heat and Strength . . . . . . . . . . . . . . . . . . . . 150

    5.4.3 Course of Temperature and Chemical Phenomena in Combustion of Wood . 151

    5.4.4 Effects and Properties of Fire Retardants . . . . . . . . . . . . . . 152

        5.4.4.1 Water Soluble Salts . . . . . . . . . . . . . . . . . . . . . 153

        5.4.4.2 Alkali Silicates . . . . . . . . . . . . . . . . . . . . . . . 154

        5.4.4.3 Foam Forming Organic Compounds . . . . . . . . . . . . 154

        5.4.4.4 Other Fire Retardants . . . . . . . . . . . . . . . . . . . 154

    5.4.5 Testing of Fire Retardants . . . . . . . . . . . . . . . . . . . . . 155

5.5 Dimensional Stabilization . . . . . . . . . . . . . . . . . . . . . . . . . 155

    5.5.1 Theory . . . . . . . . . . . . . . . . . . . . . . . . . . . . . . . . 156

    5.5.2 Methods . . . . . . . . . . . . . . . . . . . . . . . . . . . . . . . 156

Literature Cited . . . . . . . . . . . . . . . . . . . . . . . . . . . . . . . . 157

## 6. Physics of Wood. By Franz F. P. Kollmann . . . . . . . . . . . . . . . . 160

6.1 Density and Specific Gravity . . . . . . . . . . . . . . . . . . . . . . . . 160

    6.1.1 Density, Porosity, Specific Gravity of Wood Substance and of Wood Constituents . . . . . . . . . . . . . . . . . . . . . . . . . . . . . . . . 160

    6.1.2 Effect of Moisture Content in Wood on its Density . . . . . . . . . 164

    6.1.3 Density of Green Wood . . . . . . . . . . . . . . . . . . . . . . . 165

    6.1.4 Variations in Density . . . . . . . . . . . . . . . . . . . . . . . . . 168

## Contents

6.1.5 Density of Springwood and Summerwood, Correlation with Width of Annual Rings ... 173

6.1.6 Content of Solid Matter in Piles of Wood and Wood Residues ... 179

6.2 Wood-Liquid Relations ... 180

6.2.1 Moisture Content, Definition ... 180

6.2.2 Determination of Moisture Content ... 181

6.2.2.1 Oven-drying Method ... 181

6.2.2.2 Distillation Method ... 181

6.2.2.3 Titration According to K. Fischer (1935); Eberius (1952, 1958) ... 183

6.2.2.4 Hygrometric Methods ... 183

6.2.2.5 Electrical Moisture Meters ... 184

6.2.3 Sorption and Equilibrium Moisture Content ... 189

6.2.4 Recommended Moisture Content for Wood in Service ... 195

6.2.5 Fiber Saturation Point, Maximum Moisture Content of Wood ... 198

6.2.6 Thermodynamics of Sorption ... 201

6.2.7 Shrinkage and Swelling ... 204

6.2.7.1 Maximum Volumetric Shrinkage and Swelling, Influence of Drying Temperature ... 204

6.2.7.2 Anisotropy of Shrinkage and Swelling ... 205

6.2.7.3 Super-position of the Components of Swelling, Restrained Swelling 214

6.2.7.4 Swelling in Aqueous Solutions and Organic Liquids ... 216

6.2.7.5 Dimensional Stabilization of Wood ... 218

6.3 Capillary Movement and Diffusion in Wood ... 219

6.3.0 General Considerations on the Movement of Water in Wood Above and Below Fiber Saturation Point ... 219

6.3.1 Capillary Movement of Water in Wood ... 221

6.3.2 Diffusion of Water in Wood ... 224

6.3.3 Drying of Wood as a Diffusion Problem ... 225

6.3.3.1 Analogy to Fourier's Analysis for Heat Conduction ... 225

6.3.3.2 Approximated Calculation of the Drying Time ... 226

6.3.3.3 Stamm's Theoretical Drying Diffusion Coefficients ... 232

6.4 Physical Aspects of Wood Impregnation ... 235

6.4.1 Nonpressure Processes ... 235

6.4.2 Pressure Processes ... 236

6.4.2.0 General Considerations ... 236

6.4.2.1 Theory of Pressure Treatment of Wood ... 236

6.5 Thermal Properties of Wood ... 240

6.5.1 Thermal Expansion ... 240

6.5.2 Specific Heat of Wood ... 245

6.5.3 Thermal Conductivity of Wood ... 246

6.5.3.0 General Considerations ... 246

6.5.3.1 Influences of Structure and Density, Moisture Content and Temperature on the Thermal Conductivity of Wood ... 246

6.5.4 Diffusivity of Wood, Change of Temperature in Heated Wood ... 250

6.5.5 Radiation of Heat with Respect to Wood ... 256

6.6 Electrical Properties of Wood ... 257

6.6.1 Direct-current Properties: Electrical Resistance and Electrical Conductivity 257

6.6.2 Alternating-Current Properties of Wood ... 262

6.6.2.1 Resistivity ... 262

6.6.2.2 Dielectric Constant ... 263

6.6.2.3 Power Factor ... 267

6.6.3 Magnetic Properties of Wood and Wood Constituents ... 271

6.6.4 Piezoelectric Properties of Wood ... 271

6.7 Acoustical Properties of Wood ... 274

6.7.0 General Considerations ... 274

## Contents

IX

6.7.1 Sound Transmission in Wood . . . . . . . . . . . . . . . . . . . . . . 276
  6.7.1.1 Sound Velocity . . . . . . . . . . . . . . . . . . . . . . . 276
  6.7.1.2 Sound Wave Resistance, Damping of Sound Radiation and Internal Friction . . . . . . . . . . . . . . . . . . . . . . . . . 279
6.7.2 Acoustics of Buildings . . . . . . . . . . . . . . . . . . . . . . . 281
  6.7.2.1 Sound Energy . . . . . . . . . . . . . . . . . . . . . . . . 281
  6.7.2.2 Sound Transmission Loss for Various Types of Construction . . . 282
  6.7.2.3 Sound Absorption . . . . . . . . . . . . . . . . . . . . . . 284

Literature Cited . . . . . . . . . . . . . . . . . . . . . . . . . . . . . . . 285

## 7. Mechanics and Rheology of Wood. By Franz F. P. Kollmann . . . . . . . . . 292

7.1 Elasticity, Plasticity, and Creep . . . . . . . . . . . . . . . . . . . . . 292
  7.1.1 Hooke's Law, Modulus of Elasticity . . . . . . . . . . . . . . . . 292
  7.1.2 Rhombic Symmetry of Wood, Systems of Elastic Constants . . . . . . 293
  7.1.3 Poisson's Ratios . . . . . . . . . . . . . . . . . . . . . . . . 297
  7.1.4 Compressibility (Bulk Modulus) . . . . . . . . . . . . . . . . . . 299
  7.1.5 Determination of Elastic Constants . . . . . . . . . . . . . . . . 300
    7.1.5.1 Determination by Static Tests . . . . . . . . . . . . . . . 300
    7.1.5.2 Determination by Dynamic Tests . . . . . . . . . . . . . . . 301
  7.1.6 Influences Affecting the Elastic Properties of Wood . . . . . . . . 302
    7.1.6.1 Grain Angle . . . . . . . . . . . . . . . . . . . . . . . . 302
    7.1.6.2 Density . . . . . . . . . . . . . . . . . . . . . . . . . . 305
    7.1.6.3 Moisture Content . . . . . . . . . . . . . . . . . . . . . . 309
    7.1.6.4 Temperature . . . . . . . . . . . . . . . . . . . . . . . . 311
    7.1.6.5 Knots and Notches . . . . . . . . . . . . . . . . . . . . . 313
  7.1.7 Plasticity and Creep . . . . . . . . . . . . . . . . . . . . . . 315
    7.1.7.1 Stress-strain Behavior . . . . . . . . . . . . . . . . . . . 315
    7.1.7.2 Creep and Creep Recovery . . . . . . . . . . . . . . . . . . 317
    7.1.7.3 Rheological Models and Mathematical Considerations . . . . . 318

7.2 Tensile Strength . . . . . . . . . . . . . . . . . . . . . . . . . . . . 321
  7.2.1 Tensile Strength of Cellulose Molecules, of Single Wood Fibers, and Breaking Length . . . . . . . . . . . . . . . . . . . . . . . . . . . . . . . 321
  7.2.2 Determination of Tensile Strength Along the Grain . . . . . . . . . 324
  7.2.3 Factors Affecting the Tensile Strength Along the Grain . . . . . . . 326
    7.2.3.1 Grain Angle . . . . . . . . . . . . . . . . . . . . . . . . 326
    7.2.3.2 Density . . . . . . . . . . . . . . . . . . . . . . . . . . 326
    7.2.3.3 Moisture Content . . . . . . . . . . . . . . . . . . . . . . 327
    7.2.3.4 Temperature . . . . . . . . . . . . . . . . . . . . . . . . 327
    7.2.3.5 Knots and Notches . . . . . . . . . . . . . . . . . . . . . 328
  7.2.4 Determination of Tensile Strength Perpendicular to the Grain, Cleavage . 330
  7.2.5 Fatigue in Tension Parallel to the Grain . . . . . . . . . . . . . 334

7.3 Maximum Crushing Strength and Stresses in Wood Columns . . . . . . . . . . 335
  7.3.0 General Considerations . . . . . . . . . . . . . . . . . . . . . . 335
  7.3.1 Testing in Compression Parallel to Grain . . . . . . . . . . . . . 336
  7.3.2 Testing in Compression Perpendicular to Grain . . . . . . . . . . . 339
  7.3.3 Influences Affecting the Crushing Strength . . . . . . . . . . . . 341
    7.3.3.1 Grain Angle . . . . . . . . . . . . . . . . . . . . . . . . 341
    7.3.3.2 Density . . . . . . . . . . . . . . . . . . . . . . . . . . 342
    7.3.3.3 Moisture Content . . . . . . . . . . . . . . . . . . . . . . 346
    7.3.3.4 Temperature . . . . . . . . . . . . . . . . . . . . . . . . 349
    7.3.3.5 Knots and Notches . . . . . . . . . . . . . . . . . . . . . 353
    7.3.3.6 Chemical Constituents . . . . . . . . . . . . . . . . . . . 355
  7.3.4 Fatigue in Compression Parallel to the Grain . . . . . . . . . . . 355
  7.3.5 Stresses in Solid Wood Columns . . . . . . . . . . . . . . . . . . 356

7.4 Bending Strength (Modulus of Rupture) . . . . . . . . . . . . . . . . . . 359
  7.4.0 General Considerations . . . . . . . . . . . . . . . . . . . . . . 359
  7.4.1 Testing of Small Wooden Beams under Static Center Loading . . . . . 363

# Contents

7.4.2 Influences Affecting the Bending Strength (Modulus of Rupture) . . . . . 365
    7.4.2.1 Grain Angle . . . . . . . . . . . . . . . . . . . . . . 365
    7.4.2.2 Density . . . . . . . . . . . . . . . . . . . . . . . . 367
    7.4.2.3 Moisture content . . . . . . . . . . . . . . . . . . . 368
    7.4.2.4 Temperature . . . . . . . . . . . . . . . . . . . . . 370
    7.4.2.5 Shape and Size of Beams, Knots and Notches . . . . . . . . 370
    7.4.2.6 Fatigue in Bending . . . . . . . . . . . . . . . . . . 376

7.5 Shock Resistance or Toughness . . . . . . . . . . . . . . . . . . 379

    7.5.0 General Considerations . . . . . . . . . . . . . . . . . . . 379
    7.5.1 Determination of Shock Resistance . . . . . . . . . . . . . . 380
        7.5.1.1 Single Blow Impact Test . . . . . . . . . . . . . . . 380
        7.5.1.2 The Hatt-Turner Test (Successive Blows Impact Test) . . . . . . 381
    7.5.2. Comparison of Impact Test Results . . . . . . . . . . . . . . 383
    7.5.3 Influences Affecting the Shock Resistance . . . . . . . . . . . . 383
        7.5.3.1 Shape and Size of Beams, Notches (Izod-test) . . . . . . . . . 383
        7.5.3.2 Grain Angle . . . . . . . . . . . . . . . . . . . . . 385
        7.5.3.3 Density . . . . . . . . . . . . . . . . . . . . . . . 386
        7.5.3.4 Moisture Content . . . . . . . . . . . . . . . . . . . 388
        7.5.3.5 Temperature . . . . . . . . . . . . . . . . . . . . . 389
        7.5.3.6 Anatomical Properties, Chemical Constituents, Decay . . . . . . 390
        7.5.3.7 Types and Phenomena of Failures in Impact Bending . . . . . . 393

7.6 Torsional Properties and Shear Strength . . . . . . . . . . . . . . 394

    7.6.0 General Considerations . . . . . . . . . . . . . . . . . . . 394
    7.6.1 Determination of Torsional Strength . . . . . . . . . . . . . 395
    7.6.2 Determination of Shearing Strength Parallel to Grain . . . . . . . 397

7.7 Hardness and Abrasion Resistance . . . . . . . . . . . . . . . . . 403

    7.7.0 General Considerations . . . . . . . . . . . . . . . . . . . 403
    7.7.1 Hardness Tests . . . . . . . . . . . . . . . . . . . . . . 403
    7.7.2 Factors Influencing the Hardness of Wood . . . . . . . . . . . 406
    7.7.3 Abrasion Resistance . . . . . . . . . . . . . . . . . . . . 409
    7.7.4 Some Aspects of Nondestructive Testing of Wood and Timber Grading . . 413

Literature Cited . . . . . . . . . . . . . . . . . . . . . . . . . . 414

## 8. Steaming and Seasoning of Wood. By FRANZ F. P. KOLLMANN . . . . . . . 420

8.0 General Considerations . . . . . . . . . . . . . . . . . . . . 420

8.1 Air-drying . . . . . . . . . . . . . . . . . . . . . . . . . . 420

    8.1.1 Moisture Content of Green Wood . . . . . . . . . . . . . . . 420
    8.1.2 Course of Air-drying . . . . . . . . . . . . . . . . . . . 422
    8.1.3 Yard Seasoning . . . . . . . . . . . . . . . . . . . . . 424
        8.1.3.1 Lumberyard Layout . . . . . . . . . . . . . . . . . 424
        8.1.3.2 Seasoning Periods . . . . . . . . . . . . . . . . . . 427
    8.1.4 Accelerated Air-drying, Predrying . . . . . . . . . . . . . . 429
        8.1.4 1 Fan Air-drying . . . . . . . . . . . . . . . . . . . 429
        8.1.4.2 Air-drying by Means of Swings or Centrifuges . . . . . . . . 432
        8.1.4.3 Air-drying by Solar Heat . . . . . . . . . . . . . . . 433
        8.1.4.4 Predriers . . . . . . . . . . . . . . . . . . . . . 433

8.2 Steaming . . . . . . . . . . . . . . . . . . . . . . . . . . . 434

    8.2.1 Reasons for Steaming . . . . . . . . . . . . . . . . . . . 434
    8.2.2 Methods of Steaming and Heat Consumption . . . . . . . . . . 434
    8.2.3 Effects of Steaming on Wood . . . . . . . . . . . . . . . . 436

8.3 Kiln Drying . . . . . . . . . . . . . . . . . . . . . . . . . 438

    8.3.0 General Considerations . . . . . . . . . . . . . . . . . . . 438
    8.3.1 Fundamental Drying Factors . . . . . . . . . . . . . . . . 438
    8.3.2 Defects in Wood due to Kiln Drying . . . . . . . . . . . . . 451

| Contents | XI |

8.3.2.0 General Considerations . . . . . . . . . . . . . . . . . . . . 451
8.3.2.1 Staining . . . . . . . . . . . . . . . . . . . . . . . . . . 451
8.3.2.2 Deformations (Warping, Twisting, Cupping) . . . . . . . . . 452
8.3.2.3 Casehardening . . . . . . . . . . . . . . . . . . . . . . . 452
8.3.2.4 Collapse . . . . . . . . . . . . . . . . . . . . . . . . . . 453
8.3.3 Types of Kilns and Instruments . . . . . . . . . . . . . . . . . 456

8.4 Special Seasoning Methods . . . . . . . . . . . . . . . . . . . . . . 460
8.4.1 High Temperature Drying . . . . . . . . . . . . . . . . . . . . 460
8.4.2 Drying by Boiling in Oily Liquids . . . . . . . . . . . . . . . . 464
8.4.3 Solvent Seasoning . . . . . . . . . . . . . . . . . . . . . . . . 465
8.4.4 Vapor Drying . . . . . . . . . . . . . . . . . . . . . . . . . . 466
8.4.5 Vacuum Drying . . . . . . . . . . . . . . . . . . . . . . . . . 467
8.4.6 Chemical Seasoning . . . . . . . . . . . . . . . . . . . . . . . 468
8.4.7 Drying by Direct Application of Electricity . . . . . . . . . . . . 469
8.4.7.1 Drying by Joule's Heat . . . . . . . . . . . . . . . . . . 469
8.4.7.2 High-frequency Dielectric Drying . . . . . . . . . . . . . 470
8.4.8 Drying by Infrared Radiation . . . . . . . . . . . . . . . . . . 471

Literature Cited . . . . . . . . . . . . . . . . . . . . . . . . . . . . . . 471

## 9. Wood Machining. By Franz F. P. Kollmann . . . . . . . . . . . . . . 475

9.1 Introduction . . . . . . . . . . . . . . . . . . . . . . . . . . . . . 475

9.2 Technology of Sawing . . . . . . . . . . . . . . . . . . . . . . . . . 475
9.2.1 Sash Gang Sawing . . . . . . . . . . . . . . . . . . . . . . . . 475
9.2.1.1 Cutting Velocity . . . . . . . . . . . . . . . . . . . . . . 475
9.2.1.2 Chip Thickness and Average Cutting Resistance . . . . . . . 476
9.2.1.3 Consumption of Energy . . . . . . . . . . . . . . . . . . 478
9.2.1.4 Effects of Tooth Geometry, Tooth Height and Pitch . . . . . 479
9.2.1.5 Influence of Setting . . . . . . . . . . . . . . . . . . . . 481
9.2.1.6 Strain and Stresses in Gang Saw Blades; Thermal Effects . . . 482
9.2.1.7 Surface Quality . . . . . . . . . . . . . . . . . . . . . . 483
9.2.1.8 Yield . . . . . . . . . . . . . . . . . . . . . . . . . . . 484

9.2.2 Band Sawing . . . . . . . . . . . . . . . . . . . . . . . . . . . 486
9.2.2.1 General Considerations, Saw Blade Dimensions . . . . . . . 486
9.2.2.2 Cutting Velocity and Cutting Resistance . . . . . . . . . . 487
9.2.2.3 Influence of Feed Speed . . . . . . . . . . . . . . . . . . 488
9.2.2.4 Effect of Depth of Timber Cut and of Grain Orientation . . . 489
9.2.2.5 Effect of Tooth Geometry and Pitch . . . . . . . . . . . . 489
9.2.2.6 Band Tension and Stability . . . . . . . . . . . . . . . . . 490

9.2.3 Circular Sawing . . . . . . . . . . . . . . . . . . . . . . . . . 490
9.2.3.1 Introduction, Saw Blade Geometry, Kinematics . . . . . . . 490
9.2.3.2 Effect of Cutting Velocity on the Cutting Resistance . . . . . 492
9.2.3.3 Cutting Force and Cutting Power, Effect of Feed Rate or Feed per Tooth . . . . . . . . . . . . . . . . . . . . . . . . . . 494
9.2.3.4 Specific Cutting Energy . . . . . . . . . . . . . . . . . . 496
9.2.3.5 Effect of Depth of Timber Cut and of Grain Orientation . . . 497
9.2.3.6 Effect of Blade Diameter and Blade Thickness . . . . . . . . 499
9.2.3.7 Effect of Tooth Geometry and Pitch . . . . . . . . . . . . 500
9.2.3.8 Chip Formation . . . . . . . . . . . . . . . . . . . . . . 505
9.2.3.9 Thermal Effects, Stresses, and Stability of Circular Saw Blades . . 506
9.2.3.10 Special Types of Circular Saw Blades . . . . . . . . . . . 508

9.2.4 Chain sawing . . . . . . . . . . . . . . . . . . . . . . . . . . . 510
9.2.4.1 Introduction . . . . . . . . . . . . . . . . . . . . . . . . 510
9.2.4.2 Machine Types . . . . . . . . . . . . . . . . . . . . . . . 511
9.2.4.3 Chip Formation, Power Requirements . . . . . . . . . . . . 511

## Contents

9.3 Proposed Methods of Chipless Wood-Cutting . . . . . . . . . . . . . . . . . 513
    9.3.1 Peeling and Slicing . . . . . . . . . . . . . . . . . . . . . . . . . . . 513
    9.3.2 Cutting with Vibration Cutters . . . . . . . . . . . . . . . . . . . . 513
    9.3.3 Cutting with High-energy Jets . . . . . . . . . . . . . . . . . . . . 514
    9.3.4 Cutting with the Laser . . . . . . . . . . . . . . . . . . . . . . . . 517

9.4 Technology of Jointing, Planing, Moulding and Shaping . . . . . . . . . . 517
    9.4.1 General Considerations . . . . . . . . . . . . . . . . . . . . . . . . 517
    9.4.2 Geometry of Cutterhead-knives . . . . . . . . . . . . . . . . . . . 518
    9.4.3 Cutting Velocity and Cutting Force . . . . . . . . . . . . . . . . . 519
        9.4.3.1 Effect of Cutting Velocity on the Cutting Force . . . . . . . . 519
        9.4.3.2 Effect of Cutting-Circle Diameter, Feed Speed, and Number of Knives . . . . . . . . . . . . . . . . . . . . . . . . . . . . . 519
        9.4.3.3 Effect of Grain Orientation, Inclination of the Cutting Edge, and Chip Thicknes . . . . . . . . . . . . . . . . . . . . . . . . . 520
        9.4.3.4 Effect of Wood Species, Moisture Content and Temperature . . . 524
        9.4.3.5 Effect of Cutter Materials . . . . . . . . . . . . . . . . . . . 525
        9.4.3.6 Effect of Cutting Depth . . . . . . . . . . . . . . . . . . . . 525
        9.4.3.7 The Blunting of Cutter Head Knives . . . . . . . . . . . . . . 526
    9.4.4. Formation of Chips through Knife-cutting . . . . . . . . . . . . . . 527
        9.4.4.1 Influence of Wood Moisture Content on Chip Formation . . . . . 527
        9.4.4.2 Influence of Knife Geometry on Chip Formation . . . . . . . . 527
        9.4.4.3 Other Cutting Factors and their Effect on Chip Formation and Quality . . . . . . . . . . . . . . . . . . . . . . . . . . . . . 527

9.5 Sanding . . . . . . . . . . . . . . . . . . . . . . . . . . . . . . . . . . . . 528
    9.5.1 General Considerations . . . . . . . . . . . . . . . . . . . . . . . . 528
    9.5.2 Abrasives . . . . . . . . . . . . . . . . . . . . . . . . . . . . . . . 528
    9.5.3 Technology of Sanding Process . . . . . . . . . . . . . . . . . . . . 529

9.6 Turning . . . . . . . . . . . . . . . . . . . . . . . . . . . . . . . . . . . . 534
    9.6.1 General Considerations . . . . . . . . . . . . . . . . . . . . . . . . 534
    9.6.2 Effects on the Turning of Wood . . . . . . . . . . . . . . . . . . . 534
    9.6.3 Quality of Turned Surfaces . . . . . . . . . . . . . . . . . . . . . . 537

9.7 Tenoning, Mortising and Boring . . . . . . . . . . . . . . . . . . . . . . . 537

9.8 Bending of Solid Wood . . . . . . . . . . . . . . . . . . . . . . . . . . . . 541
    9.8.1 General Considerations . . . . . . . . . . . . . . . . . . . . . . . . 541
    9.8.2 Strains and Stresses in Wood Bending . . . . . . . . . . . . . . . . 542
    9.8.3 Pretreatment of the Wood Prior to Bending . . . . . . . . . . . . . 545
    9.8.4 Methods and Machines for Wood Bending . . . . . . . . . . . . . . 546
    9.8.5 Properties of Bent Wood . . . . . . . . . . . . . . . . . . . . . . . 548
        9.8.5.1 Sorption Properties . . . . . . . . . . . . . . . . . . . . . . 548
        9.8.5.2 Mechanical Properties . . . . . . . . . . . . . . . . . . . . . 549

9.9 Laminated Bending . . . . . . . . . . . . . . . . . . . . . . . . . . . . . . 549

Literature Cited . . . . . . . . . . . . . . . . . . . . . . . . . . . . . . . . . . 551

**Author Index** . . . . . . . . . . . . . . . . . . . . . . . . . . . . . . . . . . 555

**Subject Index** . . . . . . . . . . . . . . . . . . . . . . . . . . . . . . . . . 560

# 1. THE STRUCTURE OF WOOD AND THE WOOD CELL WALL

## 1.0 Introduction

The principal sources of commercial timber are the trees of the Coniferales (Gymnosperms) and of the Dicotyledons (Angiosperms). Softwood lumber is derived from coniferous trees while hardwood is a product of broad-leaved species, the Dicotyledons. Another class of Angiosperms, the Monocotyledons, is an important source of structural material used primarily for local construction "in the round". Its vascular tissue occurs as scattered bundles surrounded by ground tissue (Fig. 1.1), thus making it difficult to saw into boards. Palm and bamboo are examples of widely used Monocotyledons. In trees of the Dicotyledoneae and of the Coniferales, the secondary vascular tissue is continuous in the stem (Fig. 1.2) and this homogeneity allows ready conversion into lumber. There are many structural differences between the softwoods and the hardwoods. These are brought out in the descriptions of anatomy and ultrastructure when applicable.

A mature tree, of either the softwood or the hardwood type, generally consists of a single stem which is covered with a layer of bark. This central trunk is the principal source of woody material for the manufacture of lumber and other products. While there is a trend toward the utilization of a greater portion of the tree, the conversion of tree tops and branches into usable material is done only where it is economically feasible. These parts can be converted into chips for pulping or for the manufacture of chipboard, but in many timber producing regions of the world they are considered to be waste and are left in the forest during a logging operation.

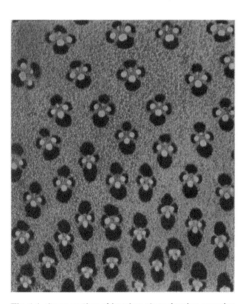

Fig. 1.1. Cross section of bamboo stem showing vascular bundles surrounded by ground tissue; a monocotyledon (10×)

## 1.1 Gross Structure of Wood

The gross features of wood are those that are visible to the naked eye or with the aid of a hand lens. Characteristics such as growth increments, sapwood heartwood differences, wood rays and cell distribution patterns can be recognized at this level.

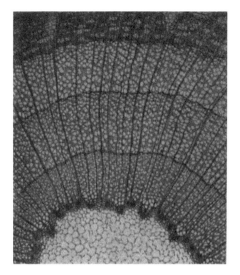 

Fig. 1.2a. Cross section of dicotyledonous stem, *Liriodendron tulipifera* L., showing pith, three concentric layers of xylem, and phloem at top. Rays run from pith into phloem (27 ×)

Fig. 1.2b. Cross section of gymnosperm stem, *Picea mariana* (Mill.) B. S. P., with pith, concentric growth increments of xylem, and rays radiating from pith (11 ×)

### 1.1.1 Cellular Composition

A tree trunk is composed of millions of individual woody cells. These cells differ in size and shape, depending upon their physiological role in the tree, most of them being many times longer than broad. They are arranged in recognizable patterns of distribution within the wood, the organization varying with the species. The long cells which are arranged longitudinally make up the bulk of the wood and provide "grain" to the material.

Cells of the xylem or wood portion of the tree are of two general types, *parenchyma* and *prosenchyma*. Parenchymatous cells are food storage elements and must, therefore, remain alive for a longer period than prosenchymatous cells which lose their protoplasm the year in which they are formed. Parenchymatous cells are found in wood rays (Section 1.1.2), as longitudinal or axial parenchyma (Section 1.1.6), and as epithelial cells surrounding resin canals (Section 1.1.7). The term prosenchyma may be applied to all of the other types of cells in mature wood, whether their major physiological role in the living tree may have been conductive or supportive.

### 1.1.2 Wood Rays

There are also shorter cells in the tree, cells which are oriented perpendicular to the longitudinal elements and organized into bands of tissue called *wood rays*. The cells found in wood rays are predominantly parenchyma cells, specialized for food storage, but the rays of some conifers contain prosenchymatous cells (Fig. 1.3). The presence and structure of this type of element (ray tracheid), the height, the width and the composition of the ray are features frequently useful for the identification of a wood species.

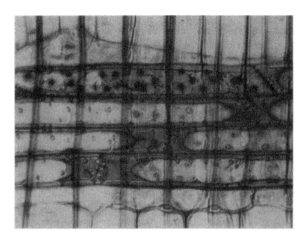

Fig. 1.3. Wood ray in *Tsuga heterophylla* (Raf.) Sarg. Parenchyma cells make up body of the ray. Ray tracheids form the upper and lower margins (390 ×)

Rays extend radially from the pith, at the center of the stem, to the cambium at the outer periphery of the xylem (wood), and continue into the phloem (bark). Depending on the manner in which a log is cut, these ribbons of tissue contribute to the natural figure of lumber and veneer. Wood splits easiest along the grain and particularly in the plane of the wood rays.

### 1.1.3 Planes of Wood

When discussing structural elements and features of wood, it is convenient to specify the aspect or viewpoint with respect to three planes, cross or transverse, radial, and tangential. By cutting across the stem perpendicularly, a surface is exposed which is called a *cross* or *transverse section*. A *radial section* results from cutting longitudinally in the plane of the wood rays, from the pith to the bark. The plane which is perpendicular to the rays and tangent to the bark is called the *tangential section*. These planes and some of the gross features of wood are illustrated in Fig. 1.4.

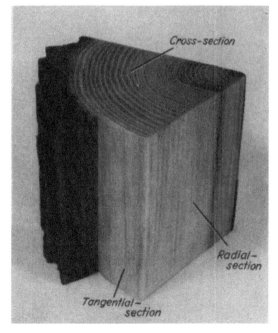

Fig. 1.4. Segment of hardwood stem showing cross-, radial and tangential sections

End grain and rays are visible on the cross section. This is also the aspect for viewing growth increments as concentric rings. The extent of heartwood development can also be observed in this plane, but both features can be seen on a radial section as well. Ends of rays appear on the tangential section and the pattern formed by them on this surface provides diagnostic evidence in some instances. Rays on the radial section give the appearance of broken ribbons and, depending on their size and contrast, may produce distinctive patterns.

The vascular cambium is located at the xylem-phloem interface and is the initiating layer for both of these tissue systems. At the gross level this layer can barely be detected since it is but a few cells wide.

### 1.1.4 Sapwood and Heartwood

The sapwood-heartwood pattern is one of the most obvious features that can be observed on the cross or radial section of a mature tree trunk. Although not as pronounced in all species, most trees have an inner core of dark colored wood, heartwood, and an outer shell of light colored tissue called sapwood (Fig. 1.5). This contrast in color has physiological significance in a general way, but it is not strictly correct to designate the core as heartwood only on the basis of its darker color. A more accurate criterion for heartwood determination is the absence of living cells within the zone, and in particular, cells of ray parenchyma which remain alive far longer than the neighboring prosenchymatous elements (FREY-WYSSLING and BOSSHARD, 1959). For commercial purposes, however, color is the determining factor used for the separation of sapwood from heartwood.

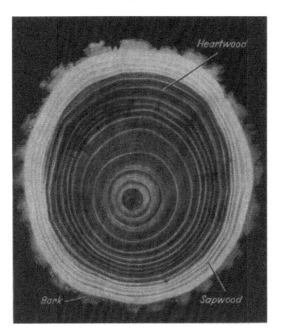

Fig. 1.5. Cross section of hardwood stem with dark colored heartwood at center, lighter colored sapwood surrounding the core, and bark on the outside

Certain characteristic differences can be found between the sapwood and heartwood of the same tree. For example, weight, durability and permeability are often quite different and can be related to the changes which accompany sapwood-heartwood transformation. Reference to permeability differences will be made in the section dealing with pit membranes (1.3.4.1).

### 1.1.5 Growth Increments

Another feature which is readily observed on the cross section is the growth increment. The boundaries of these increments are usually related to annual growth in trees grown in temperate climates. In tropical regions, however, growth increments can be the result of wet and dry seasons and the term "annual ring" often used for timbers from temperate zones would not be strictly applicable.

The nature of the growth layer can be a helpful feature in the identification of wood. In temperate zone hardwoods, for example, patterns can often be

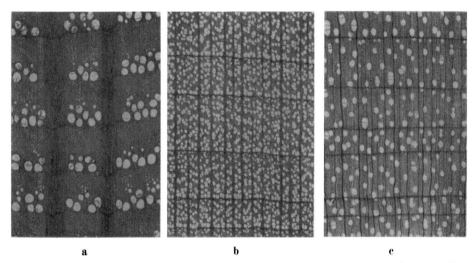

Fig. 1.6a—c. a) Ring porous, b) semi-ring porous, and c) diffuse porous hardwoods as seen in cross section. Distribution of vessels within growth increment determines its category (12×)

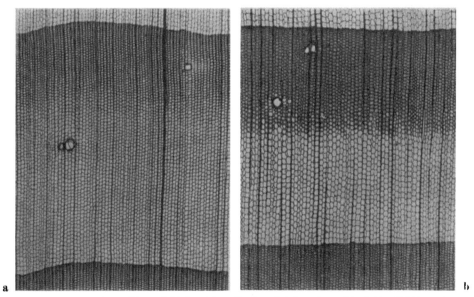

Fig. 1.7a, b. a) Gradual transition and b) abrupt transition in radial dimension and tracheid wall thickness from earlywood to latewood. a) *Picea jezoensis*, b) *Larix dahurica*. (27×)

observed with the unaided eye, if one examines the tissue within one annual increment. These patterns can be attributed to the size and distribution of pores, the term given to vessel openings as seen in cross section. When the pores in the earlywood are much larger than those formed later in the season, and when the size transition between earlywood and latewood pores is abrupt, the wood is classed as *ring porous*. If little or no transition in pore size exits between early- and latewood, the term *diffuse porous* is applied. *Semi-ring porous* or *semi-diffuse porous* are used to describe wood in which the pore patterns are not distinctly of the ring porous nor of the diffuse porous type. The three conditions are illustrated in Fig. 1.6.

A transition pattern can also be observed as a macroscopic feature in coniferous wood (Fig. 1.7). In this case it is associated with a reduction in the radial dimension of the tracheids as well as a thickening of the cell walls. Earlywood tracheids are thin-walled and Polygonal in cross section while latewood cells have thicker walls and are flattened radially, giving a darker tone to the summerwood portion of the annual ring. Some species exhibit an abrupt transition from the springwood to the summerwood (earlywood to the latewood) portion of the annual increment while in others the transition is very gradual.

When using transition features as an aid in the identification of coniferous wood, caution should be used because the presence of compression wood can alter the normal pattern (CORE *et al.*, 1961). For instance, woods that normally have an abrupt transition between earlywood and latewood may show a gradual transition in this reaction wood.

### 1.1.6 Axial Parenchyma

Axial or longitudinal parenchyma cells are found in various concentrations and distribution patterns in the growth increments of both softwoods and hardwoods. Parenchyma cells are generally short, thin-walled elements with simple pitting. The ray parenchyma described earlier are found in the wood rays while the longitudinal parenchyma occur in strands along the grain. On the transverse section they appear as dots, bands or lines when viewed with a hand lens.

The patterns of parenchyma arrangement are helpful in the description of wood anatomy and in the identification of wood species. The two major categories of parenchyma arrangement in hardwoods are the *apotracheal* (independent of pores or vessels) and *paratracheal* (associated with the vessels or vascular tracheids). Apotracheal parenchyma may be *terminal*, if found singly or as a more or less continuous layer at the end of a season's growth; *diffuse*, if distributed as individual cells without any particular pattern; or *banded*, if arranged in concentric lines or bands as seen in cross section (Fig. 1.8).

Paratracheal parenchyma arrangement may be qualified further by the term *vasicentric*, to describe parenchyma which forms a complete sheath around a vessel; *aliform*, where the parenchyma extends laterally from the vessels in winglike projections; *confluent*, when the aliform arrangement is so extensive as to form irregular tangential or diagonal bands; and *scanty*, when the parenchyma arrangement does not form a complete sheath around the vessels (Fig. 1.8). There are other more precise and detailed descriptive terms included in the Multilingual Glossary of Terms Used in Wood Anatomy (I. A. W. A., 1964), but these are the basic definitions.

In softwoods axial parenchyma occurs only in *diffuse, zonate* or *terminal* arrangements since vessels are lacking (Fig. 1.9).

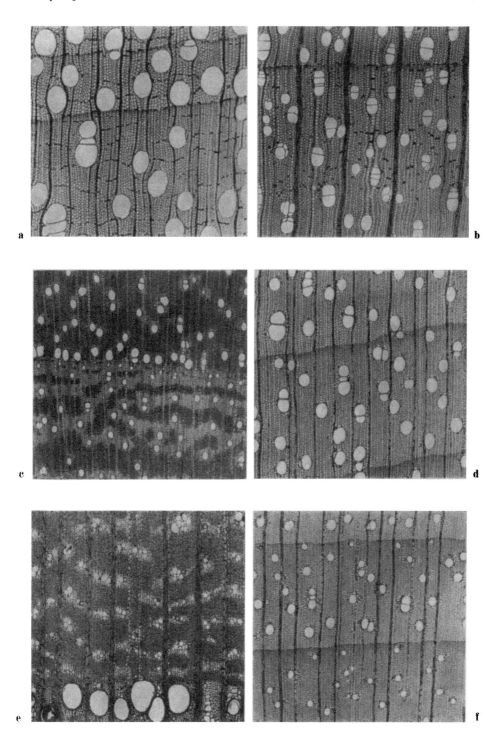

Fig. 1.8 a—f. Arrangement of axial parenchyma in hardwoods (27 ×): a) apotracheal-banded *(Pterocarya stenoptera)*, b) apotracheal-diffuse and terminal *(Betula insignis)*, c) paratracheal confluent *(Pteroceltis tartarinowie)*, d) paratracheal-scanty *(Betula luminifera)*, e) paratracheal-aliform and confluent *(Zelkova formosana)*, f) paratracheal-vasicentric and aliform *(Cinnamomum chingii)*

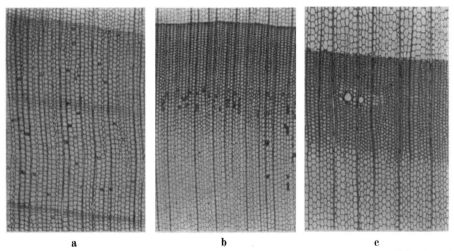

Fig. 1.9a–c. Arrangement of axial parenchyma in softwoods (27×): a) diffuse *(Podocarpus neriifolia)*, b) zonate *(Abies nephrolepis)*, c) terminal *(Pseudotsuga wilsoniana)*

## 1.1.7 Intercellular Canals

A number of coniferous woods (*e. g. Larix, Picea, Pinus*) have *resin canals* as normal anatomical features. These are long, tubular channels or cavities oriented longitudinally in the wood and surrounded by epithelial tissue which secretes resin into the canal. The epithelium may be composed of one or more layers of specialized parenchyma cells which are thin walled in some genera, thick walled

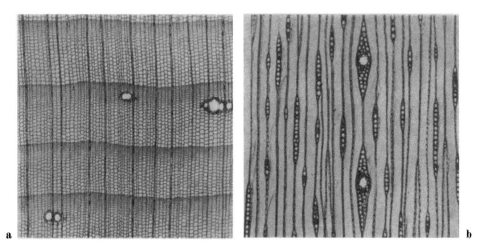

Fig. 1.10a–e. Resin canals in coniferous wood: a) normal canals with thick-walled epithelial cells *(Picea brachytyla* var. *complanata)* (27×), b) transverse canals, same species, in fusiform rays (67×) (Fig. 10c–e see opposite page)

in others (Fig. 1.10). In cross section, resin canals appear to the naked eye as dots or flecks. Their size, location and arrangement are often of diagnostic value. Normal resin canals also occur in the rays (called fusiform rays) of certain genera, but when resin ducts are traumatic (caused by injury), they are usually restricted to axial occurrence.

In dicotyledonous woods, intercellular canals are generally called *gum ducts* (Fig. 1.11). Their contents may be oily, resinous or mucilaginous. Gum ducts are either of lysigenous or of schizogenous origin. In the first instance, the canal is

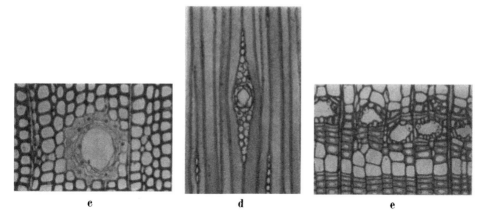

Fig. 1.10 c – e. c) normal canal with thin-walled epithelium *(Pinus strobus* L.*)* (270 ×), d) transverse canal, same species as c) (270 ×), e) Traumatic resin canals *(Larix laricina* (Du Roi) K. Koch*)* (270 ×)

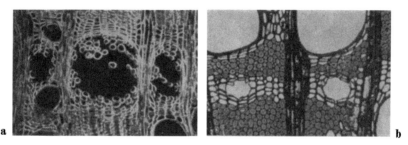

Fig. 1.11 a, b. Gum ducts in dicotyledonous wood: a) lysigenous canal *(Lovoa* sp.*)* (215 ×), b) schizogenous canals *(Parashorea stellata)* (270 ×)

formed by a dissolving or gummosis of the axial parenchyma cells. Schizogenous canals are formed by a separation or delamination of adjoining cell walls. There is no true epithelium in lysigenous canals.

### 1.1.8 Other Gross Features

Besides those already discussed, other macroscopic features of taxonomic significance or of importance in the selection of a wood for a particular application include texture, color and luster. It would be inaccurate to class odor, taste, hardness or weight of wood as macroscopic features, but these are helpful characteristics for describing and identifying woods.

## 1.2 Microscopic Structure of Wood

Most of the gross structural features of wood considered in the previous sections have their origin in the sorting and arrangement of the various cell types found in wood. When thin sections of wood are examined with a light microscope, cellular composition can readily be observed. The cells are held together by inter-cellular substance, but should this middle lamella be dissolved through chemical treatment, the cellular composite is separated into individual elements. This, in fact, is the commercial process known as pulping which is employed to produce fibers for paper manufacture. Microscopic examination of pulp samples of hardwood and of softwood reveals that there are significant differences in the sizes and shapes of cells from these two sources.

### 1.2.1 Major Cell Types

As was mentioned earlier, wood contains parenchyma and prosenchyma. Parenchyma cells in softwood do not differ greatly from those found in hardwood. They are generally short, thin-walled cells with simple pits. Their shape can be quite variable, but in the wood rays, where most parenchyma is located, they are predominantly brick-shaped, particularly in coniferous wood. In softwoods, parenchyma forms but a small portion of the total woody tissue, less than five percent of the total volume in some species, probably averaging 7 to 8% for all conifers. In some hardwoods, parenchyma may form as much as one-third of the wood volume. This occurs in woods having aggregate or oak-type rays, but the average for hardwoods is less than 20%, or more than double that of the softwoods.

The tissue of softwoods is simpler from the standpoint of variety of cell types. Besides the parenchyma in the wood rays, resin canals and longitudinal strands, only one other type of cell can be found, the *tracheid*. Longitudinal tracheids make up as much as 95% of the total wood volume of some coniferous species. These are long, imperforate cells having walls which may be thick or thin, depending on location of the cell in the growth increment. In cross section, tracheids are polygonal in shape, more or less square in earlywood and flattened radially in the latewood. In compression wood, tracheids are more or less rounded in transverse section.

Compared with the prosenchymatous elements of hardwood, coniferous tracheids are very long (Fig. 1.12a). In some species they are 7 mm or more in length. This feature makes wood of the conifers more desirable for paper manufacturing because it contributes to greater paper strength.

*Ray tracheids* appear in the wood of some conifers, usually at the upper or lower margins of the wood rays, but in some cases the ray may consist solely of this type of cell. These prosenchymatous elements have bordered pits that are smaller than those found in the longitudinal tracheids and which distinguish them from the ray parenchyma cells which have only simple pits (Fig. 1.3). In the hard pines, ray tracheids contain dentate thickenings (Section 1.3.4.4) on their inner walls (C. f. Fig. 1.49). Ray tracheids do not occur in woods of the angiosperms.

The *vessel* is the distinctive tracheary structure of the hardwoods. As viewed in cross section, where it is called a *pore*, it has a larger lumen than other cell types. Individual vessel elements are tubular, open-ended cells which, when joined end-to-end, form the vessel. They are relatively thin-walled cells specialized for conduction longitudinally in the tree. Extensive pitting is typical of most vessel walls, these openings providing for lateral conduction to neighboring cells. As viewed on the cross section, pores may be arranged in clusters, chains, multiples or as solitary openings. Grouping is sometimes of diagnostic significance.

Vessel elements vary in size and shape depending upon the wood species and upon location in the growth increment. Vessels of the springwood in ring-porous woods are of large diameter while those of the summerwood are quite small. In diffuse-porous woods, vessel diameters are nearly uniform across the growth ring. The variability in size and shape of vessel elements is illustrated in Fig. 1.12 b, d and e.

A cell type of completely different appearance than the vessel element is the *libriform fiber*. This type of element is thick-walled, narrow-lumened and elongated

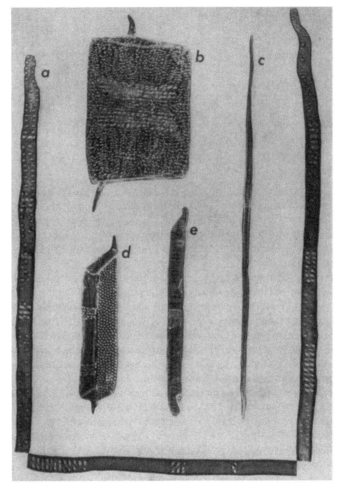

Fig. 1.12a—e. Comparative size and shape of hardwood and softwood elements: a) conifer tracheid. b), d). and e) vessel elements, c) libriform fiber

(Fig. 1.12c). It is adapted for strength and support rather than conduction since it has imperforate tapered ends and small, slit-like, simple pits.

In addition to these two specialized hardwood prosenchymatous elements, there are others which have features that would indicate a dual role in the living tree, involving both conduction and support. *Tracheids* or *fiber-tracheids*, similar in appearance to coniferous tracheids though shorter in length (*ca.* 1.5 mm), are relatively thick-walled with bordered pits. *Vascular tracheids* resemble small vessel elements, but the ends of the cells are imperforate. *Vasicentric tracheids*

12        1. The Structure of Wood and the Wood Cell Wall        [Ref. p. 52

are more fiber-like, but are nevertheless short, thin-walled and irregular-shaped. They also have bordered pits.

On a weight basis, coniferous tracheids make up 97.8% of the wood of *Pinus sylvestris* (PERILA and HEITTO, 1959) and an equal amount in *Picea abies* (PERILA and SEPPA, 1960). Therefore, the parenchyma constitute a small percentage of the weight of softwood and only about 7 % of the volume.

In hardwoods there is wide variation in the proportions of cell types. PERILA (1962) found that the fibers of *Betula verrucosa* make up 86.1% of the weight of the wood, the vessels 9.3%, and the parenchyma cells 4.6%. Volumetric data for this species are not available, but in *Betula lutea*, the fibers make up 63.8%, the vessels 21.4% and the parenchyma 12.8%. In a more porous species, *Tilia americana* L., the vessels form as much as 55.6% of the volume of the wood, the fibers 36.1%, and the total parenchyma volume is only 8.3%, a small amount for a hardwood. At the other extreme, the vessels of *Hicoria ovata* (now *Carya ovata*) constitute only 6.5% of the total wood volume, while parenchyma totals 28% and fibers 66.5% (FRENCH, 1923).

### 1.2.2 Cell Sorting and Arrangement

It is apparent from some of the discussion and illustrations in Section 1.1 that distinctive arrangement of cells characterizes the major wood groups. This is further illustrated in Fig. 1.13a and Fig. 1.13b, photomicrographic reconstructions of coniferous and hardwood microscopic structure, respectively. These allow a direct comparison of the longitudinal and radial tissue systems of representative woods of the two lumber producing groups.

In conifers the tracheids are aligned in nearly perfect, radial rows from the pith to the cambium. In hardwoods, the cells appear, at first glance, to be randomized. However, a little concentration, and the comparison of a large number of specimens reveals typical generic or family designs, repetitive features that are useful in their identification. Consideration of a combination of several different features is often necessary to arrive at a correct identification and this frequently includes microscopic examination as well as hand lens study. Some of the more useful features or patterns will be described.

Mention has been made of axial parenchyma arrangement (Section 1.1.6) with special regard to relationship of these cells with vessels or to location in the growth increment. In hardwoods, parenchyma arrangement is of particular diagnostic significance because of the diversity of arrangements that can be found. In addition, combinations of two or more arrangements occur in some woods. If considered along with ray structure, vessel arrangement, presence or absence of gum ducts, and other anatomical evidence, reliable diagnoses are often possible. Though of less importance in softwoods, parenchyma arrangement can be used as a microscopic feature in separating certain woods.

Vessel arrangement has also been referred to, in the discussion of ring- and diffuse-porosity. Typical vessel arrangements can be recognized within or in addition to those categories. For example, woods of the Ulmaceae have the vessels of the summerwood arranged in continuous, wavy, concentric bands (Fig. 1.14). The numbers of bands may vary from species to species, between genera in the family, or this feature can be influenced by rate of growth, but the wavy character of the summerwood pore arrangement is typical.

KRIBS (1959) employs eight different categories of vessel arrangement in describing the anatomy of hardwoods in his keysort system. Included are: pores

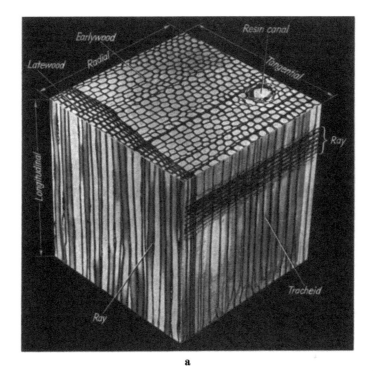

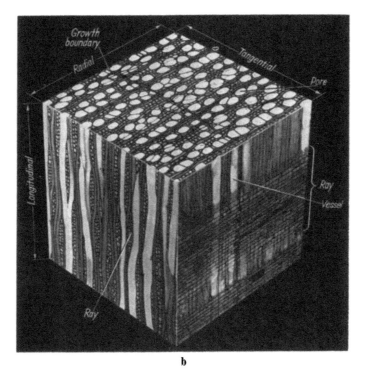

Fig. 1.13a, b. Photomicrographic model of softwood a) and hardwood b) structure, showing growth increments, longitudinal elements, ray structure, resin canal

evenly distributed, pores in echelon, pores and pore groups in radial lines, pores solitary, pores and pore groups in straight to wavy tangential bands (ulmiform) and pores in radial flame-like clusters. Descriptions that have been used by others vary somewhat from the above and might include such terms as pores in nestlike groups and pores crowded.

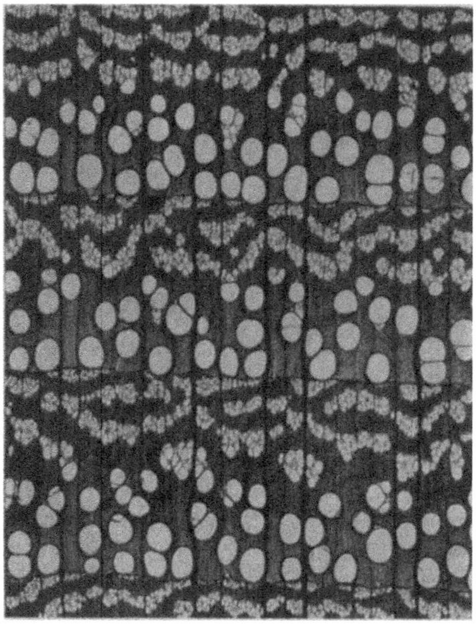

More restrictive anatomical descriptions can be written by considering pore diameters, vessel inclusions (gums and tyloses), and pitting, but these features are not in the category of pore arrangement.

The composition of wood rays is another characteristic that is used by wood anatomists. Wood rays are either *uniseriate* or *multiseriate*, that is, one cell wide or two or more cells wide, as viewed in tangential section. In conifers, rays are *homogeneous* or *homocellular* if made up only of ray parenchyma and *heterogeneous* or *heterocellular* if ray tracheids occur with the parenchyma. A third category, the *fusiform ray*, is a spindle-shaped ray as viewed in tangential section. This conformation is caused by the inclusion of a transverse resin canal within the body of the ray (Fig. 1.10).

Fig. 1.14. Cross section of wood from family Ulmaceae with typical wavy bands of summerwood vessels: *Celtis occidentalis* L. (15×)

Since there are no tracheids in the rays of hardwoods, the shape of the parenchyma cells is used as the criterion for classification. If the ray is composed only of procumbent (horizontally elongate) parenchyma, it is classed as homogeneous. A heterogeneous ray is made up wholly or in part of square or upright (vertically elongate) cells. Frequently the upright cells are located at the upper or lower margins of the ray (Fig. 1.15), or as sheath cells around the procumbent parenchyma of a multiseriate ray.

Description becomes more involved when the ray is partly uniseriate and partly multiseriate since the multiseriate portion may be composed of procumbent cells while the uniseriate portion is made up of vertically elongate cells. The marginal cells can also form single or multiple rows. KRIBS (1959) has devised a classification system for categorizing

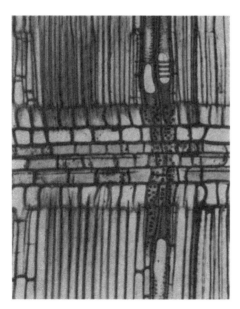

Fig. 1.15. Heterogeneous or heterocellular ray in hardwood, *Sassafras albidum* (Nutt.) Nees. Note that the cells on the upper and lower margins differ from the parenchyma making up the body of the ray (120×)

hardwood heterogeneous rays as types I, II, or III. The I. A. W. A. recommendation, in its International Glossary of Terms Used in Wood Anatomy (1957), is that each individual ray be described in full. However, this was deleted from the 1964 multilingual edition.

It must not be assumed from the above that hardwoods lack uniseriate rays. They do occur in woods of some Angiosperms, but they are not as common as multiseriate arrangements. In addition, some woods may exhibit more than one type of ray with various gradations between them.

### 1.2.3 Cell Inclusions

Wood cells often contain inclusions which may be stored food materials, products of cell metabolism or elaborations of cells. In the case of tyloses, there may be a proliferation of part of the protoplast of one cell into another. Among the inclusions are such materials as starch, proteins, oils, fats, tannins and inorganic substances recognizable in the form of crystals.

**1.2.3.1 Tyloses.** Tyloses are structures of common occurrence in vessels of hardwoods. They may appear membranous, thick-walled or sclerosed (Fig. 1.16). In some species they resemble soap suds, being frothy and shiny. In others, they take the form of ladder-like cross walls as viewed in longitudinal section. Though supposedly related to heartwood formation, they are found frequently in growth increments within the sapwood (Fig. 1.17).

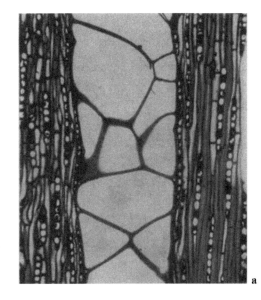

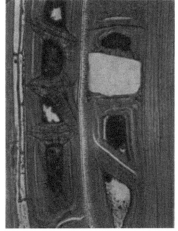

Fig. 1.16a–d. Tyloses in hardwood vessels: a) tangential section of *Quercus alba* L. with membranous tyloses (130×), b) cross section of *Ocotea rodiaei* (R. Schomb.) Mez. with sclerosed tyloses (40×), c) longitudinal section of same species as b) showing vessels completely plugged with sclerosed tyloses (170×), d) isolated vessel from same species as (b) and (c) filled with sclerosed tyloses (36×). Photos (c) and (d): C. KEITH.

A tylosis has its origin in a living ray or axial parenchyma cell. It is an outgrowth from such a cell, through a pit pair opening into an adjoining vessel. Though described by CHATTAWAY (1949) and others as a ballooning or stretching of the pit membrane due to increased pressures in the parenchyma cell, KORAN (1964) reports that in reality the membrane is ruptured. The tylosis develops from protoplasmic material that proliferates into the vessel (Fig. 1.18). Growth

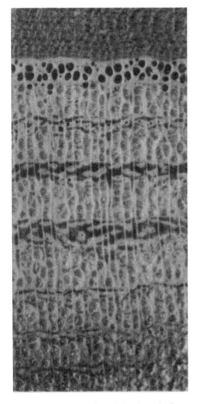
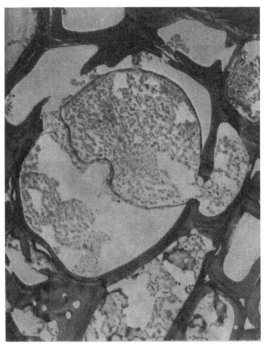

Fig. 1.17. Cross section of hardwood showing vessels of all but last-formed growth increment, including sapwood, containing tyloses (5×)

Fig. 1.18. Electron micrograph of cross section through vessel containing a developing tylosis which originated in the parenchyma cell at right. *Quercus alba* L. (1500×). Photo: Z. KORAN

takes place in a manner that resembles ordinary cell development. Primary wall structure is evident, as is pitting, secondary-wall type of layering, and encrustation.

Since they effectively plug the vessels, tyloses are not only of taxonomic significance but are of importance in wood processing and utilization as well (GERRY, 1914). The obstructions prevent the movement of wood preservative or pulping liquor into the wood through the normal tracheary channels. On the other hand, tyloses may be beneficial. White oak, which has tyloses, can be used for tight cooperage while red oak, which lacks them, cannot.

*Tylosoids* resemble the tyloses of hardwoods, but they are proliferations of epithelial cells into a resin or intercellular canal. They do not pass through a pit opening.

**1.2.3.2 Crystals.** A variety of crystals occur as inclusions in wood cells. They are found in both softwoods and hardwoods but are smaller and less common in the former. Crystals are nearly always located in parenchyma cells, either in the wood rays or in longitudinal strand parenchyma. Most commonly, these inorganic deposits are calcium salts, especially calcium oxalate. Only in some instances is the presence of crystals of diagnostic value (CHATTAWAY, 1955).

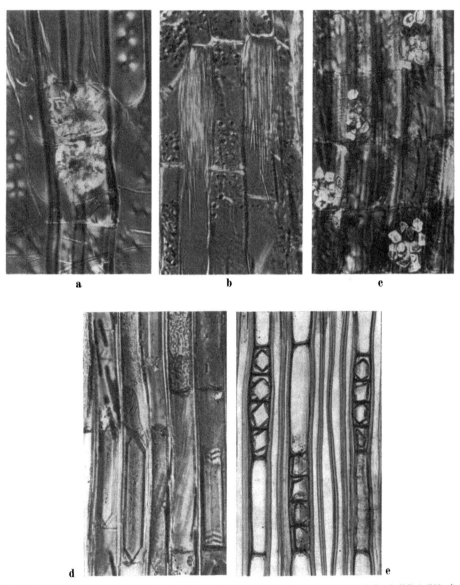

Fig. 1.19a—e. Crystals in wood: a) Druses *(Ginkgo biloba)* (310×). b) Raphides *(Morinda citrifolia)* (310×) c) Crystal sand *(Neocauclea calycina)* (310×), d) Styloid crystals *(Glochidion* sp.*)* (310×), e) Crystals in wood parenchyma strand *(Diospyros virginiana* L.*)* (180×)

Among the recognized crystal types found in wood are: druses, raphides, crystal sand, acicular and styloid crystals (Fig. 1.19).

**1.2.3.3 Oil cells.** Oil cells are rounded and enlarged specialized parenchyma containing oil. They occur in xylary (wood) rays or in longitudinal strand parenchyma of woody dicotyledons. A distinction should be made between oil cells and those containing true gums since these two cell types have rather different properties. The family Lauraceae contains many wood species famous for their aromatic oils which are contained in oil cells. Among them are Bois de Rose (*Aniba rosaedora* Ducke), sassafras (*Sassafras albidum* (Nutt.) Nees), and camphorwood (*Cinnamomum camphora* (L.) Nees & Eberm.) (RECORD and HESS, 1943).

**1.2.3.4 Gums and Resins.** True gums either dissolve in water or, when placed in water, absorb moisture and swell to form a mucilaginous or jelly-like material (HOWES, 1949). In practice, the term "gum" is used rather loosely and is often applied to gumlike substances which are insoluble in water. Probably the polyphenols are classed as gums in wood as often as any of the cell inclusions (CÔTÉ and MARTON, 1962).

Intercellular gum ducts contain gums but since the ducts are channels lined with secretory cells, the gum is perhaps not strictly a cell inclusion. The same situation exists in the case of resin canals which are lined with epithelial cells; the resin is stored in the cavity.

There are evidently few, if any, gums produced in commercial quantities by the wood portion of trees. Most of the gums are derived from bark, flowers or fruits, many by tapping or wounding of the tree. The resins are also collected from bark blisters, by tapping or similar harvesting procedures. Many of the resins are valued for medicinal purposes. Probably gum and resin production by the xylem is too limited and harvesting too costly to be feasible when compared with other sources.

## 1.3 Organization of the Cell Wall

The walls of wood cells are composed of three groups of structural substances which Wardrop (1964 a) has classed as framework, matrix and encrusting materials. The framework substance is cellulose which occurs in the form of microfibrils. As will be discussed in Section 1.3.1, cellulose is closely associated with the matrix and encrusting substances in the paracrystalline regions of the microfibrils. The microcapillaries surrounding the cell wall framework are also filled with these amorphous substances.

Hemicelluloses and other carbohydrate materials (excluding cellulose) are incorporated into the cell wall as the matrix substances. Encrustation is not initiated with the start of cell wall formation but occurs as the final phase of differentiation (lignification).

The lignin skeleton that remains following the enzymatic removal of carbohydrates from wood provides clear evidence of the microcapillary system in the cell wall (MEIER, 1955). Similarly, acid removal of the framework and matrix substances creates a pattern of lignin distribution (SACHS *et al.*, 1963).

The physical and chemical properties of these cell wall constituents, their isolation from wood and their distribution in wood are discussed in Chapter 2. The present section is concerned with the organization of the structural substances in the cell wall.

### 1.3.1 The Microfibril

The crystalline nature of cellulose in wood has been known for several decades, based on evidence provided by studies with X-ray diffraction and polarization microscopy (Section 2.2.1.2). That the cellulose molecular chains were organized into strands could not be clearly demonstrated until application of electron microscopy. The term *fibril* had wide use and was applied to wall components that could be seen with the light microscope. On the basis of present knowledge, it seems logical to continue using the general term fibril and to modify it as required.

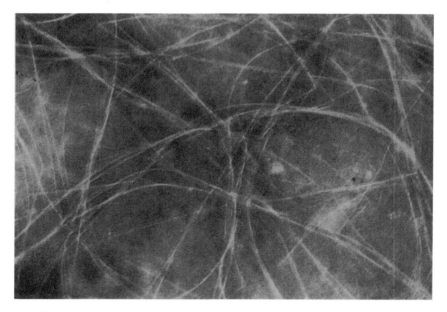

Fig. 1.20. Electron micrograph, negative contrast technique, of elementary fibrils of cellulose from onion root cell walls. Elementary fibril diameter 35 Å. (85,500×). Photo: K. MÜHLETHALER

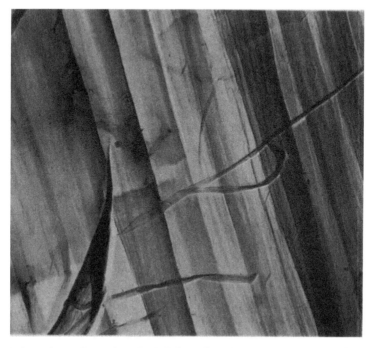

Fig. 1.21. Electron micrograph, pseudo-replica technique, of lamellae composed of cellulose microfibrils making up cell walls in *Tilia americana* L. (16,200×)

*Elementary fibrils* are presumably the cellulosic strands of smallest possible diameter. MÜHLETHALER (1960; 1965) applies this term to fibrils with a diameter of approximately 35Å (Angstrom units) which can be resolved using "negative contrast" preparation technique for electron microscopy (Fig. 1.20). An elementary fibril of this cross-sectional dimension could contain about forty cellulose chains.

Aggregates of elementary fibrils are classed as *microfibrils* and occur in nature in a broad spectrum of sizes, depending on the source of cellulose. They are probably 100 to 300 Å wide in wood and evidently of indefinite length. Microfibrils aggregate into larger units which may be called *macrofibrils* and these are joined into lamellae (Fig. 1.21) that are organized into cell wall layers (FREI *et al.*, 1957).

Within the elementary fibrils are zones in which the cellulose chains are oriented to such a high order that X-ray diffraction patterns can be produced as with true crystals. These regions are termed *crystallites* or *micelles*. The crystalline regions are interrupted by *amorphous zones* along the elementary fibril. In these portions the cellulose molecules are simply not as perfectly aligned.

Since the elementary fibrils are in reality components of the larger microfibril, producible by ultra-sonic treatment of the major unit, these units are repeated in lateral order. There is considerable evidence that a paracrystalline phase exists between elementary fibrils and within the microfibril. In these regions the cellulose chains are again less highly ordered than within the crystalline portion of the elementary fibril, and further, may be in close association with matrix substances and lignin (FREY-WYSSLING, 1959).

## 1.3.2 Generalized Structure and Terminology

Early workers in the field of cell wall research were able to detect evidence of its apparent lamellar organization. The chemical swelling of woody tissue served as one means of magnifying this detail when the structure was otherwise too fine to be resolved by the light microscope. Individual wood elements were also swollen chemically and the ballooning phenomenon which resulted from this treatment brought out indications of the reversal of orientation in the layers. When sections of wood were examined with the aid of a polarizing microscope, layering was again indicated by the variable birefringence in different portions of the cell wall.

On the basis of the above evidence and some additional experiments, BAILEY and KERR (1935) proposed a generalized structure for the secondary cell walls of most wood cells. According to their concept, the secondary wall of normal wood cells consists of three layers: a relatively narrow outer layer, a narrow or thin inner layer and a middle layer of variable thickness. They noted that the cellulosic constituents of the cell wall, the fibrils, are oriented approximately parallel to the long axis of the cell in the middle layer while the orientation in the inner and outer layers is close to 90 degrees to the longitudinal axis.

At the same time, they attempted to clear up the confusing terminology which had developed in the field during the last half of the nineteenth century. They suggested that the term *middle lamella* be used in referring to the intercellular substance, the isotropic substance which separates the walls of adjoining cells. The term *primary wall* should be applied only to the original wall of the cell, formed at the cambium, while the term *secondary wall* should be applied only to the layers of secondary thickening which are formed inside the primary wall. With a few exceptions, their concept and terminology were accepted, although until the development of the electron microscope there was no convincing direct method available for refuting their proposal.

### 1.3.3 Microfibrillar Orientation

Although one should not make sweeping generalizations about the organization of the walls of all wood cells without the use of statistical sampling

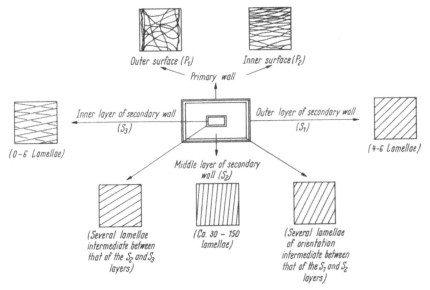

Fig. 1.22. Cell wall organization represented diagrammatically for typical coniferous tracheid and some hardwood elements. Adapted from WARDROP and HARADA (1963) as shown in WARDROP (1964a)

techniques, there have been enough investigations of a large number of species by many different workers to permit reporting on "typical" orientation. The current concept of microfibrillar orientation in layers of wood cell walls is based largely on evidence from electron microscope observations supported by X-ray diffraction and polarized light microscopy.

In Fig. 1.22, textures to be found in cell wall layers of typical xylary elements are shown diagrammatically. The thin primary wall (P) consists of a loose aggregation of microfibrils (Fig. 1.23) randomly arranged on the outer surface and oriented more or less transverse to the cell axis on the inner surface (Fig. 1.24). At the cell corners the primary wall is often thickened into a

Fig. 1.23. Primary wall from a cell of *Eucalyptus deglupta*. Direct carbon replica; electron micrograph. (32,000×)

rib-like structure that extends along the length of the cell. The actual cellulose content of the mature tracheid primary wall has been estimated to be about 25% of the total dry weight (WARDROP and DADSWELL, 1953). Although the intercellular layer (middle lamella) contains most of the lignin found in wood, the primary wall is the most highly lignified of the cell layers. Between 60% (hardwoods) and 90% (softwoods) of all the lignin in wood is concentrated in the middle lamella and primary wall (MEIER, 1964).

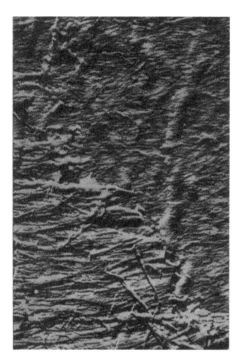

Fig. 1.24. Primary wall ($P_2$) upper right, and $S_1$, lower left, from a springwood tracheid of *Picea jezoensis*. (7,760×). Photo: H. HARADA

The three layers of the secondary wall, designated as $S_1$ (outer layer), $S_2$ (middle layer), and $S_3$ (inner layer), are organized in a plywood-type of construction. A photomicrograph of a thin cross section of normal coniferous wood taken with a polarizing microscope indicates that the microfibrils of the inner and outer layers of the secondary wall are oriented more or less perpendicular to those of the middle layer (Fig. 1.25). The $S_1$ orientation generally ranges between 50° to 70° from the cell axis, the $S_2$ varies between 10° to 30° depending on the source of the cell, earlywood or latewood, and the $S_3$ is oriented between 60° to 90° from the cell axis.

The layers themselves are composed of lamellae of microfibrils with varying amounts of shift in orientation visible in electron micrographs. In the $S_1$, there may be a crossed fibrillar texture (WARDROP, 1964a), but the predominating orientation is approximately perpendicular to the $S_2$ direction (Fig. 1.26). The $S_1$ is invariably a thin layer, consisting of but a few lamellae, with the microfibrils wound helically around the main axis of the cell. JAYME and FENGEL (1961) report that in earlywood spruce tracheids the $S_1$ is 0.12 to 0.35 μm (micrometer = micron) in thickness. Others have reported similar values for coniferous species.

Although the main body of $S_2$ exhibits an orientation nearly parallel to the long axis of the cell, there are transition lamellae on its inner and outer faces. The several lamellae in these regions show a gradual shift from $S_1$ to main $S_2$ alignment and from main $S_2$ to $S_3$ orientation (Fig. 1.27) (HARADA et al., 1958; HARADA, 1965a). The middle layer of the secondary wall, which can reach a thickness of five microns or more, contributes most to the bulk of the cell wall material as well as to its physical properties. It is a compact region in which a high degree of parallelism of the microfibrils exists except in the transition lamellae. The number of lamellae composing the $S_2$ may vary from thirty or forty in thin-walled or earlywood cells to 150 or more in latewood elements. FREI et al. (1957) demonstrated, with electron micrographs of ultrathin sections, that there is a slight alternation of direction in the orientation of the lamellae.

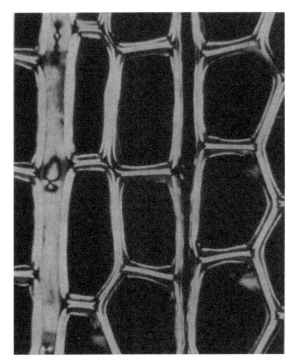

Fig. 1.25. Polarized light photomicrograph of normal wood cross section of *Larix occidentalis* Nutt. Note the bright lines at the inner and outer layers of the cell wall, and the dark middle layer, $S_2$, between them. At the left side of micrograph is a ray. (640×)

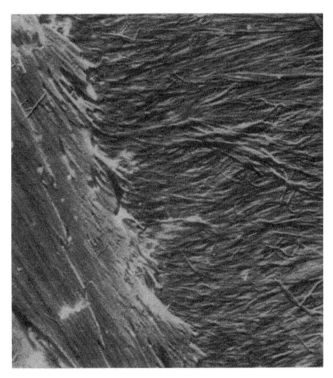

Fig. 1.26. Electron micrograph of replica of tracheid wall from *Pseudotsuga menziesii* (Mirb.) Franco, with $S_1$ layer at right, and $S_2$ at left. Note the relative orientation of the two layers. (7,200×)

The $S_3$, when present, is a thin layer of helically arranged microfibrils like the $S_1$, probably not exceeding five or six lamellae in thickness. LIESE (1963) estimates that in softwoods this layer is about 700 to 800 Å thick, but that in hardwoods it is thinner. As opposed to the highly oriented $S_2$, the inner layer of the secondary wall is loose-textured. The microfibrils criss-cross at small angles of about 20° to 30° (Fig. 1.28). The $S_3$, birefringent to a somewhat lesser degree then the $S_1$, is poorly developed or absent in some genera, *e. g.* *Picea*. The $S_3$ is also lacking in compression wood tracheids.

Fig. 1.27. The $S_2$–$S_3$ transition lamellae in a tracheid wall of *Picea jezoensis*. $S_2$ lower left and $S_3$ upper right. Electron micrograph. (9,500) 10,000×. Photo: H. HARADA

Caution is necessary in determining $S_3$ development by means of polarizing microscopy because birefringence as observed on cross sections is influenced by microfibrillar orientation. It is maximum when the microfibrils are arranged perpendicular to the longitudinal axis of the cell. Since this angle has been found to be as great as 20° to 30° from this cell axis, intensity would be quite low in such cases and might then be interpreted erroneously as being due to poor development of $S_3$.

Microfibrillar orientation in the walls of specialized cells such as vessel elements cannot be described quite as simply as for coniferous tracheids and hardwood fibers to which the above conditions apply. Vessels are heavily pitted and the straight-line arrangement of microfibrils is often disrupted by these openings in the cell wall. WARDROP (1964a) reports that this pitting apparently interferes with the interpretation of cell wall layering when examined with the polarizing microscope. In vessels of *Eugenia* that he examined, the walls appeared uniformly birefringent, but thin sections studied with the electron microscope clearly showed a three-layered wall. Replicas of vessel wall surfaces reveal how variable the layering can be in these elements (Fig. 1.29).

There has been relatively little work reported on the fine structure of ray parenchyma cells. This is unfortunate because rays probably influence certain wood properties out of proportion to the volume of wood they constitute. HARADA and WARDROP (1960) have investigated the cell wall structure of ray parenchyma in *Cryptomeria japonica*. In this species, the cell walls are remarkably similar to those of longitudinal elements. There are three major layers inside the primary wall. The $S_1$ and $S_3$ are thicker than the $S_2$, a condition that differs from the vertical tracheids. However, they are helically arranged and have a microfibrillar

Fig. 1.28. Electron micrograph, direct carbon replica technique, of the $S_3$ layer in a tracheid of *Pinus strobus* L. Note the relatively loose texture and criss-crossing of microfibrils. (12,000 ×)

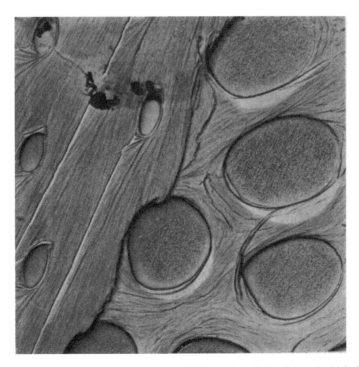

Fig. 1.29. Wall layers and fibrillar organization in vessel of *Tilia americana* L. The $S_2$ layer is at left, $S_1$ surrounds the pit membranes at right. Primary wall structure is visible on the membrane surfaces. Electron micrograph, replica technique. (5,600 ×)

orientation at an angle to the major axis of the cell ranging from 30° to 60°. The middle layer, $S_2$, nearly parallels the long axis of the parenchyma cell.

A layer which was designated as the $S_4$ lines the cell lumen. It is composed of randomly arranged microfibrils and is quite thin (Fig. 1.30). No warts were found on the lumen walls of the ray parenchyma cells although they do occur in the longitudinal tracheids of this species.

In light microscopic studies of decayed hardwood ray cells, WARDROP and DADSWELL (1952) found evidence of helical organization in the secondary walls. Complex lamellation was detected in cells of sound material by using polarizing microscopy.

Fig. 1.30. Direct carbon replica of lumen lining of ray parenchyma cell of *Pinus lambertiana* Dougl. showing organization of the $S_4$ layer. Material slightly delignified. (21,500×). Photo: H. HARADA

At the ultrastructural level, hardwood ray parenchyma have been examined by several investigators, but there has been no comprehensive study of a large number of species. HARADA (1965b) reports briefly on the organization of ray parenchyma in *Fagus crenata*. Electron micrographs of this species reveal little deviation from the basic wall organization of *Cryptomeria* ray parenchyma, except that in *Fagus* the $S_2$ is thicker than the $S_1$ and $S_3$. PREUSSER et al. (1961) studied another species of beech, *Fagus sylvatica* L., but primarily from the cytological rather than the structural approach. Lamellation of the ray cell walls is clearly illustrated in their electron micrographs of ultra-thin sections in this species.

## 1.3.4 Cell Wall Sculpturing

Perhaps the microfibrillar orientation of wood cell walls described above is oversimplified. Certainly there are no cells in which the organization is not interrupted by pitting, by the presence of specialized thickenings, by a coating of cytoplasmic débris, or by some other type of sculpturing. In many cases these deviations from a smooth, helically-oriented structure are features that characterize families, genera, or species, and they can, therefore, be helpful in wood identification. In other instances, these structures are so variable in form, size, or distribution as to be of little diagnostic value.

**1.3.4.1 Pit Structure.** A pit is a recess in the secondary wall of a cell, together with its external closing membrane (*which is actually primary wall*); it is open internally to the lumen (I. A. W. A., 1964). In the living tree, pits provide intercell communication for the translocation of liquid, so there must be a complementary pit in the adjacent cell. Connecting pits are correctly termed *pit pair*. However, if a pit of one cell has no matching pit in the contiguous element, it is called a blind pit. Blind pits often lead to intercellular spaces.

Pit pairs may be simple, bordered, or half-bordered, depending on the types of cells which they connect. In general, simple pits occur in parenchyma and bordered pits in prosenchyma cells. Thus, simple pit pairs link contiguous ray parenchyma cells, bordered pit pairs connect neighboring vessels or longitudinal tracheids, and half-bordered pit pairs occur between parenchyma and prosenchyma. Some prefer to base pit-pair nomenclature on cell function rather than on morphological features. In this case, a bordered pit-pair between two vessels would be designated as intervascular and a simple pit-pair between parenchyma cells would be called interparenchymatous (SCHMID, 1965). Since there are certain exceptions to the pit type found normally in a given cell (*e. g.* libriform fibers may have simple pits), the latter system of designation may have some merit.

Simple pits are mere gaps in the secondary wall. The pit cavity is of nearly constant width from the pit aperture at the cell lumen to the membrane at the cell's outer envelope. The pit cavity of bordered pits is overarched by the secondary wall, forming a pit chamber. In thin-walled cells, such as earlywood coniferous tracheids, the pit aperture is centered in a domed area, the pit border, which bulges into the cell lumen. In thick-walled cells, especially in hardwoods, the pit border may be quite thick and the dome is either absent or is reduced in size. The inner aperture of the pit, at the lumen, then opens into a pit canal leading to the outer aperture at the pit chamber. The canal can be circular or lenticular in cross section. When flattened, its long axis generally parallels the microfibrillar orientation of the $S_2$ wall layer.

It follows from these descriptions that the membrane of a pit pair consists of the external (primary wall) membranes of two pits with middle lamella sandwiched between them. In fact, this is the structure that appears in electron micrographs of simple pit pair membranes in both softwoods and hardwoods. In bordered pit pairs of hardwoods, the same basic membrane organization exists. Only in coniferous bordered pit pairs is there a distinctly different type of membrane.

*1.3.4.1.1 Coniferous Bordered Pits.* Before secondary wall formation begins in tracheids, the pit membrane is of a primary wall structure, *i. e.* more or less randomly arranged microfibrils. It is at this stage of differentiation, before the $S_1$ layer is deposited, that a pit border is initiated. Microfibrils are arranged concentrically, delineating the pit area. Then, as the border is built up during secondary wall formation, the primary wall structure of the membrane is modified through

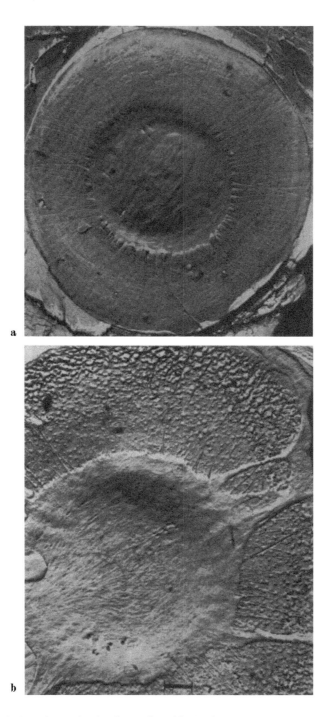

Fig. 1.31 a, b. Electron micrographs of surface replicas of bordered pit membranes of conifers: a) *Pseudotsuga macrocarpa* (Vasey) Mayr (4,300×), b) *Tsuga canadensis* (L.) Carr. (7,200×). From KRAHMER and CÔTÉ (1963). By permission TAPPI

Ref. p. 52]   1.3 Organization of the Cell Wall   29

the addition of a central thickening, a torus. The torus is composed of microfibrils also arranged circularly. Meanwhile there is a rearrangement of the microfibrils surrounding the torus so that supporting strands radiate from the torus

Fig. 1.32a, b. Bordered pit membrane of *Thuja plicata* Donn. sapwood. a) surface replica (8,500 ×) and b) section (7,000 ×). This species does not have a torus thickening on the membrane, but there is a denser region at the center where microfibrils cross. Some encrustation may take place even in sapwood. From KRAHMER and CÔTÉ (1963). By permission TAPPI

to the periphery of the pit, leaving openings of varying sizes between the strands (Fig. 1.31) (FREY-WYSSLING et al., 1956a; WARDROP, 1958). Direct communication with the neighboring cell is accomplished through the dissolution of substances surrounding the primary wall microfibrils and forming the intercellular layer. The resultant pit membrane openings in species of the Pinaceae range up to 0.2 μm, based on filtration experiments (LIESE, 1954). In the normal position, the torus appears as a disk suspended by a system of fine strands, often called the margo, but when aspirated, the torus is pressed against the pit border (cf. Fig. 1.34).

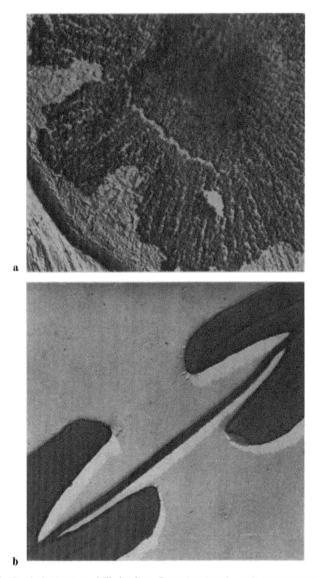

Fig. 1.33a, b. Bordered pit structure of *Thuja plicata* Donn. heartwood. a) surface replica showing heavy encrustation of microfibrillar strands (7,700×) b) section indicating reduced porosity of membrane (8,800×). Compare with Fig. 1.32. (a) From KRAHMER and CÔTÉ (1963), by permission TAPPI. (b) Photo by D. LANTICAN

Although this is the completed pit structure found in many coniferous genera, particularly in the Pinaceae, it is not a universal pattern for softwoods. Some genera have more highly elaborated pit tori with scalloped (*Cedrus*) or strap-like

(*Tsuga*) extensions of the central thickening (Fig. 1.31). Many genera have no additional torus thickening, but only a thicker or denser region in the center of the membrane where the re-oriented microfibrils cross. This is the case in bordered pit pairs of *Thuja plicata* Donn. (Fig. 1.32) (KRAHMER and CÔTÉ, 1963). With the formation of heartwood in this species, there is often an accumulation of encrusting substances, particularly near the center of the membrane (Fig. 1.33). The lack of an additional thickening does not prevent the membrane from sealing the aperture as it does in aspirated pits of species with a true torus thickening (Fig. 1.34).

Investigation of representative species from more than forty-five genera of the Gymnosperms by LIESE (1965a) revealed that a distinct torus formed by secondary apposition occurs only in the Pinaceae and the Cephalotaxaceae. In all of the other conifers, the membrane consists of more densely packed microfibrils, radially arranged, but lacking a thickening composed of cellulose microfibrils. Some appear to have matrix substances at the center of the membrane while others have no additional encrustation.

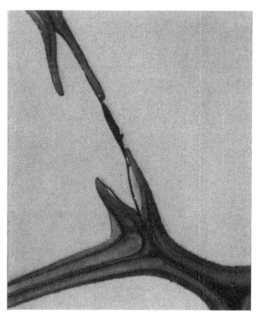

Fig. 1.34. Aspirated bordered pits in coniferous tracheids: Bordered pit pair in sapwood of redwood, *Sequoia sempervirens* (D. Don) Endl. (2,250 ×)

The secondary wall layers that surround the pit chamber are composed of cellulose microfibrils with associated matrix and encrusting substances. The microfibrils follow the orientation of their respective lamellae, but where they are interrupted by the pit area, they sweep around it in streamline fashion (Fig. 1.35). Therefore, a flattened aperture reflects the microfibrillar orientation of the main $S_2$ direction. The dome enclosing the pit chamber consists of $S_1$, $S_2$ and $S_3$ wall layers. The initial pit border of circularly oriented microfibrils lines the pit chamber.

*1.3.4.1.2 Hardwood Bordered Pits.* Knowledge of the fine structure of hardwood bordered pits is based on relatively few studies compared with those of the conifers. A limited number of species has been examined at electron microscopic level and many of those were probably not investigated comprehensively. Nevertheless, it appears that the bordered pits of the dicotyledonous woods bear considerable similarity to those of the softwoods. For example, the shape of the pit pair between fiber tracheids, in a sectional view, is much the same as the structure found between coniferous tracheids (Fig. 1.36). The membrane is overarched by pit borders of approximately the same configuration, though perhaps somewhat shorter. There are certain other notable differences, first in the membrane and, second, in the construction of the pit border.

There is no torus or secondary thickening in the center of Angiosperm bordered pit membranes. In effect, the external membrane consists of unmodified primary wall. No openings can be seen even at magnifications of 100,000 times normal size. The pit pair membrane does not become aspirated in the sense of sealing

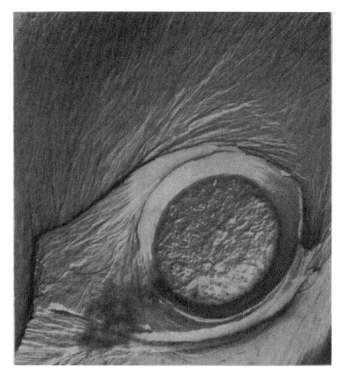

Fig. 1.35. Organization of the cell wall around coniferous bordered pit. *Pseudotsuga menziesii* (Mirb.) Franco (5,400 ×)

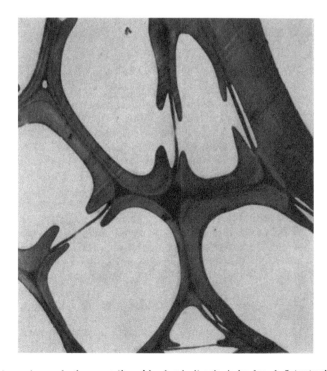

Fig. 1.36. Electron micrograph of cross section of bordered pit pairs in hardwood: Inter-tracheid pit pairs in *Quercus rubra* L. (3,400 ×)

against one of the borders. However, some reduction in movement of liquid through the membrane is unavoidable, since ultra-fine pores left by the removal

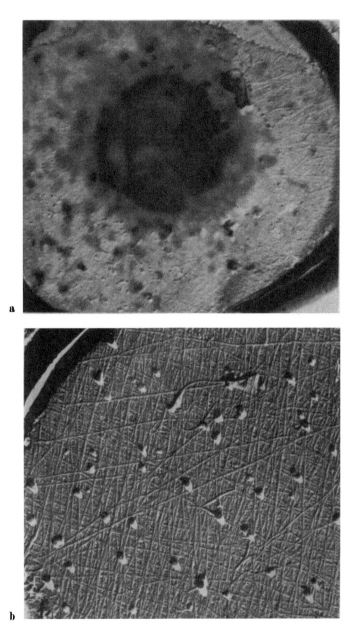

Fig. 1.37a, b. Electron micrographs of surface replicas of bordered pit pair membranes in hardwoods: a) Encrusted membrane in *Ulmus americana* L. (14,800×), b) Unencrusted membrane in *Tilia americana* L. showing the microfibrillar organization of the primary wall of which it is composed. (24,000×)

of plasmodesmata must allow diffusion across the membrane even though clear openings are lacking. Otherwise, there would be no accumulation of materials near the center of the membrane, opposite the pit apertures (Fig. 1.37a).

3 Kollmann/Côté, Solid Wood

That there is an encrustation of the membranes is clear from Fig. 1.38 which shows portions of two bordered pits in a vessel of *Tilia americana* L. The primary wall texture is evident on the surfaces of the membranes, but where the membrane at the right has been folded over, fortuitously, its underside is free of masking substances. The microfibrils are sharply defined and their original position in the intercellular substance can be traced (CÔTÉ, 1958).

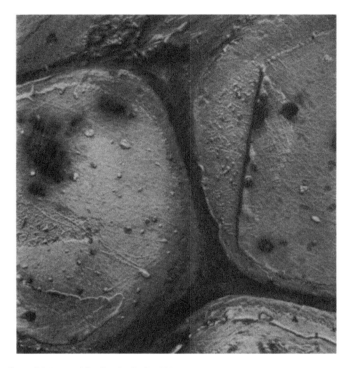

Fig. 1.38. Portions of intervessel bordered pits in *Tilia americana* L. Note the encrustation of the membrane surfaces, but in the pit at the right, edge flap turned over, the microfibrils are more distinct. Note also the impression of the microfibrils in the intercellular layer. (14,000 ×). From CÔTÉ (1958) by permission Forest Products Journal

The pit border is composed primarily of cell wall material of the $S_2$ layer, and the $S_1$ does not extend beyond the pit cavity. According to WARDROP (1964a), the lining of the chamber in fiber tracheids consists of circularly arranged microfibrils much as in the initial pit border of softwood tracheids. While this may be true in some species, in *Quercus rubra* L. (Fig. 1.39), it appears more likely that the $S_3$ layer from the cell lumen extends into the pit chamber. This is unlike the condition in conifers where the $S_3$ stops at the inner aperture of the pit, that is, at the cell lumen. Apparently variations from a basic pattern are possible. Admittedly, interpretation is sometimes difficult, especially if a warty layer covers the $S_3$. Terminology is also subject to misinterpretation and if a surface replica of a fiber tracheid bordered pit were to exhibit circular organization, the term *initial pit border* would be an appropriate one.

This is the condition found in some vessel bordered pits (Fig. 1.40), although the structure proposed by WARDROP for these does not include an initial pit border. The $S_3$ again terminates at the aperture, the $S_1$ at the cavity, and the $S_2$ makes up the bulk of the pit border. Again, it must be emphasized that wider ranging

*1.3.4.1.3 Vestured Pits.* Vestured pits are hardwood bordered pits containing outgrowths or processes which extend from the cell wall into the pit chamber and overhang the membrane. In some cases the vestures project from the pit aperture into the cell lumen. They occur most commonly in the vessels of some of the more specialized families of dicotyledonous woods, particularly in the Leguminosae. When the pitting in hardwoods is vestured, the interpretation of pit structure can be more complex because vestures mask much of the underlying organization. However, it is known that the pit membrane of vestured pits has the same structure as other hardwood bordered pits.

The term "vestured pit" is of quite recent origin, having been suggested by BAILEY in 1933. Until that time these structures of recognized diagnostic significance were called cribriform (sieve-like) membranes. The vestures are at the limit of resolution of the light microscope and, to an average microscopist, solid structures of this small size could easily be interpreted as openings in a membrane. However, Bailey's skill with a microscope was phenomenal and his interpretations of pit vesture structure remain fundamentally unchanged even though the electron microscope has been employed in later studies (CÔTÉ and DAY, 1962b; WARDROP et al., 1963; SCHMID and MACHADO, 1964).

Pit vestures can be separated into two broad categorries on the basis of their shape: branched or unbranched. They may occur in the form of dense growths or as simple projections from the cell wall. Occasionally vestures may be found on the

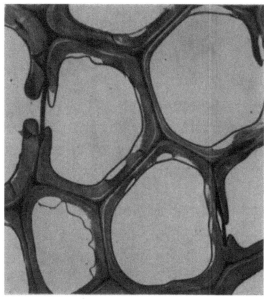

Fig. 1.39. Cross section of partially decayed wood of *Quercus rubra* L. showing isolated inner layer of cell wall $S_3$. Note that the $S_3$ in the bordered pit pair at left also lines the pit cavity. (6,500 ×)

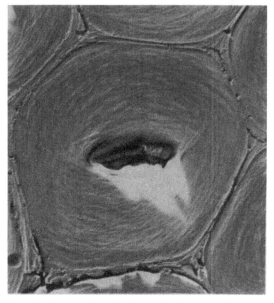

Fig. 1.40. Inner surface of intervessel pit border in basswood. *Tilia americana* L. Note circular arrangement of microfibrils. Electron micrograph of surface replica (9,200 ×)

pit membrane, but this is apparently quite rare and it is difficult to determine in such cases whether the structure originates at the wall of the pit chamber or at the membrane.

From electron micrographs (Fig. 1.41), it is apparent that some vestures resemble warts (Section 1.3.4.3). They were believed to be of similar origin (CÔTÉ and DAY, 1962b) on the basis of structural resemblance and similarity of electron density. SCHMID and MACHADO (1964), however, investigating their development, concluded that vestures are formed by living protoplasm, whereas warts have

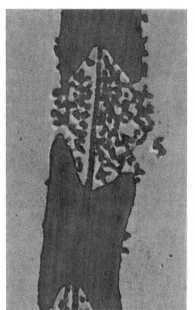
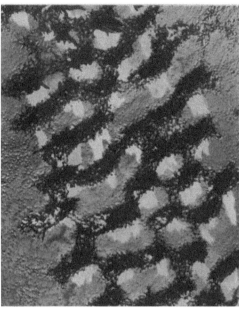

Fig. 1.41a, b. Vestures in the bordered pits of dicotyledonous woods: a) Section of *Parashorea plicata* showing branched nature of vestures and similarity to warts on vessel walls (6,700×), b) surface replica of lumen lining of vessel of *Calycophyllum spruceanum* Benth. showing vestures projecting from pit apertures into lumen. Note warty layer on vessel wall. (2,800×). From CÔTÉ and DAY (1962b) by permission TAPPI.

been considered to be the remnants of the dead protoplast. Whatever the relationship, it is perhaps significant that in all of the species investigated, woods with vestured pitting also have warty membranes.

*1.3.4.1.4 Simple and Half-bordered Pit Pairs.* The term *simple pit* implies that it is only a recess in the secondary wall, an area where the wall layers are not elaborated into an overhanging border. The pit cavity is approximately of the same width from cell lumen to membrane. Simple pits occur in ray or axial parenchyma cells and in libriform wood fibers. It follows that a simple pit pair can be found where cells of these categories make contact. Its membrane, whether in Gymnosperms or in Angiosperms, has the same structure as hardwood bordered pit membranes. The surface is covered with more or less randomly oriented microfibrils and there are no visible openings on the surface, even at electron microscopic levels of magnification. However, in thin cross sections it is sometimes possible to see either plasmodesmata or fine channels perforating the membrane (Fig. 1.42). There is no evidence of an initial pit border or of any particular modification of the layers in the wall surrounding a simple pit. Neither does the $S_3$ layer line the pit cavity.

Where parenchyma and prosenchyma are joined, a simple pit is aligned with a bordered pit and this combination is designated as a half-bordered pit pair. This is a very frequent occurrence. Since in conifers each tracheid is in contact with at least one ray, but usually with several, and since each individual ray parenchyma cell contains one or more pits at each contact area, there may be hundreds of such pit pairs formed. The same applies to hardwoods where rays contact longitudinal elements. Similar pairs are found between ray parenchyma and ray tracheids, and between axial parenchyma and adjoining prosenchyma.

The membrane of a half-bordered pit pair has the same surface texture as that of a simple pit, but it may be considerably thicker in some cases (Fig. 1.43a). The parenchyma side of the membrane is usually encrusted to such a degree that it is difficult to see its surface organization (Fig. 1.43b). This condition is especially true of conifers.

The bordered side of the pit pair is variable in structure but in a pattern consistent enough to be of diagnostic value. For example, the soft pines, such as *Pinus strobus* L., have a large window-like or fenestriform pit on the tracheid side of the crossfield. The hard pines have smaller pinoid pits, variable in size and shape, and narrow bordered or simple. Other conifers have cupressoid, taxodioid or piceoid pit pairs (Fig. 1.44).

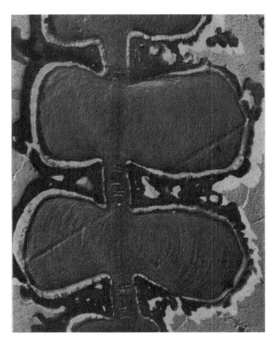

Fig. 1.42. Simple pit pairs between parenchyma cells of *Fagus grandifolia* Ehrh. heartwood. Note the fine channels across the pit membrane, caused by plasmodesmata. Polyphenolic substances can be seen in the pit canals. (11,700×). Photo: Z. KORAN

**1.3.4.2 Helical Thickenings.** These are ridges of aggregated microfibrils arranged like a coil spring inside the cell lumen of certain conifer tracheids and Angiosperm elements (Fig. 1.45). Called spiral thickenings in some of the literature, helical thickenings are characteristic features of many woods and occur sporadically in others. The thickenings often occur only in the tips of some cells, but where they are found in the main body of the cell, they often taper off and disappear some distance from the cell tip.

In electron micrographs of surface replicas of helical thickenings, the characteristic $S_3$ criss-crossing texture is sometimes evident. In other cases, the microfibrils are more closely packed in parallel strands. In a section of a *Tilia americana* L. vessel, the thickening appears to be part of the $S_3$ (Fig. 1.46). WERGIN and CASPERSON (1961) found this to be true in *Taxus baccata* L. However, WARDROP (1964a) states that in this same species, it may clearly be part of an additional layer. Though it may be too soon to assign them to a particular cell wall layer, it is obvious that in many species, for instance *Pseudotsuga menziesii* (Mirb.) FRANCO, the helical thickenings blend into the last-formed layer of the wall

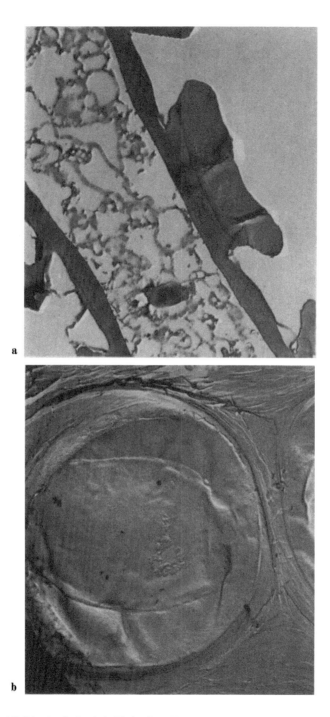

Fig. 1.43 a, b. a) Half-bordered pit pair in *Thuja plicata* Donn. Electron micrograph of cross section. (6,400×). From KRAHMER and CÔTÉ (1963) by permission TAPPI, b) Half-bordered pit pair membrane, surface replica, in *Pinus strobus* L. This is a window-like or fenestriform type of pit. (4,000×). From CÔTÉ (1958) by permission Forest Products Journal

(Fig. 1.47). This is true even though the thickenings have an orientation somewhat different from the microfibrils of the underlying layer.

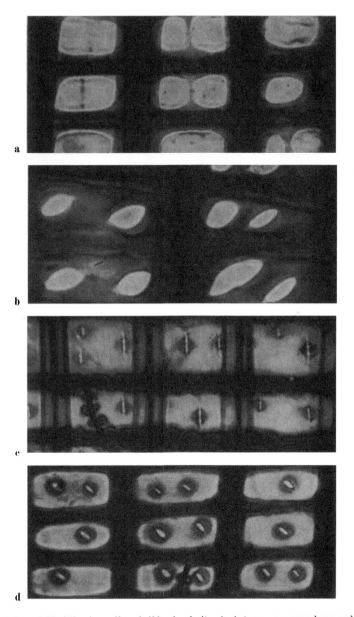

Fig. 1.44a–d. Cross field pitting in conifers; half-bordered pit pairs between ray parenchyma and longitudinal tracheids (600×): a) window-like or fenestriform (soft pines), b) pinoid (hard pines), c) piceoid *(Picea, Larix, Pseudotsuga)*, d) cupressoid *(Chamaecyparis, Libocedrus, Juniperus, Tsuga)*

**1.3.4.3 Wart Structure.** The warty layer has the distinction of being the only major structural feature of wood cells to have been "discovered" by means of electron microscopy. Although these small protuberances on the lumen lining are large enough to be seen with the light microscope, under optimum optical conditions, their identity was not established until 1951 when KOBAYASHI and UTSUMI described them as "particle structure" and LIESE called them "kleine Erhebungen" (small raised structures). Subsequently, HARADA and MIYAZAKI (1952), HARADA

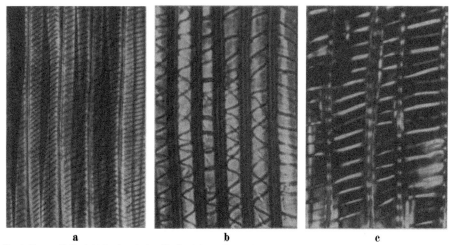

Fig. 1.45a–c. Helical thickenings in longitudinal elements of softwoods and hardwood: a) tracheids of *Pseudotsuga menziesii* (Mirb.) Franco, polarized light photomicrograph (135×), b) tracheids of *Taxus brevifolia* Nutt., bright field photomicrograph (335×). c) vessels of *Magnolia grandiflora* L., polarized light photomicrograph (335×)

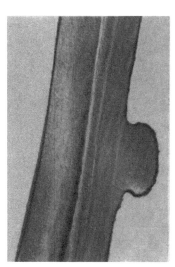
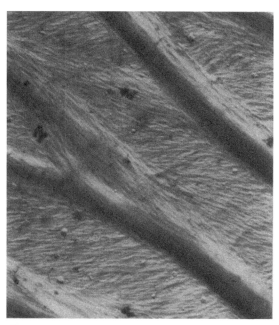

Fig. 1.46. Section across a helical thickening in a basswood vessel, *Tilia americana* L. The thickening appears to form part of the $S_3$ cell wall layer. (9,300×)

Fig. 1.47. Helical thickenings in *Pseudotsuga menziesii* (Mirb.) Franco. Note the manner in which the microfibrils of the thickenings merge with the background layer, the $S_3$. Electron micrograph, replica technique (4,500×)

(1956), LIESE and co-workers (1953, 1954, 1956a & b, 1957) and FREY-WYSSLING et al. (1955, 1956) carried out systematic studies which helped to determine the variation, distribution and structure of what LIESE then called "warzenähnlich" (wart-like) structures. Later the term "Warzenstruktur" or wart structure was introduced and, more recently (LIESE, 1963), warty layer or "Warzenschicht" has been adopted as a more specific name.

It is of historical interest to note that warts appeared in some of the earliest electron micrographs of wood (FISCHBEIN, 1950) and perhaps in many light microscopy studies where they were not recognized as consistent features of most woods.

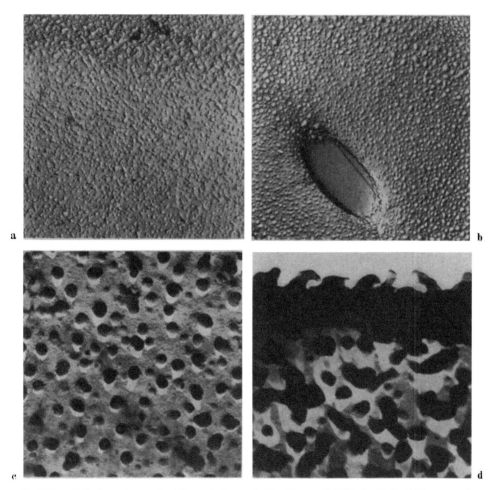

Fig. 1.48a—d. Electron micrographs of the warty layer: a) *Abies balsamea* (L.) Mill. (6,000×), b) *Chamaecyparis lawsoniana* (A. Murr.) Parl. (4,700×), c) *Sequoia sempervirens* (D. Don) Endl. (22,700×), d) *Fagus grandifolia* Ehrh. (17,000×)

Now that their structure is known, it is possible to find warts in many woods with only a light microscope. The smaller warts are near the limit of resolution of the light microscope and their index of refraction of approximately 1.52 renders them virtually invisible in the usual mounting media such as Canada balsam. However, if the sections are mounted in water, all but the smallest warts can be distinguished with careful microscopy (LIESE, 1965b).

Warts can be found in hardwoods and in softwoods, in fibers, tracheids, fiber tracheids and vessels. The warty layer covers the $S_3$, or the lumen lining ($S_2$) when the $S_3$ is absent, such as in compression wood tracheids. Even pit chambers and pit canals can be coated with this terminal structure. On the basis of its wide distribution and frequent occurrence, LIESE (1965b) suggests that it be regarded as a general structural feature of wood cells.

The size and shape of warts is extremely variable between tree species (Fig. 1.48) but rather consistent within species. Some warts are as small as 0.01 μm while others are as large as one micrometer in diameter. The height of warts may be as great as one micrometer, but usually they are much shorter than that. Warts may be present in great numbers or very sparsely distributed. LIESE (1965b) examined representative species from fifty genera of the Gymnosperms and classified them according to the degree of development of the warty layer. He found that, though the warts are rare in many genera, apparently all conifers have some warts.

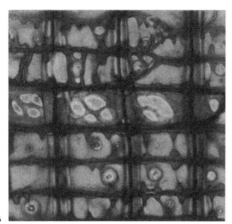
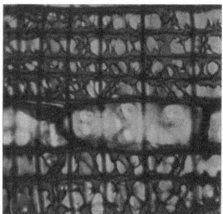

Fig. 1.49a, b. Dentate ray tracheids in (a) earlywood and (b) latewood of *Pinus taeda* L. These are typical of the hard pines (350×)

The origin of the warts has been subject to considerable speculation. It is clear that this layer is different and distinct from the $S_3$. One point seems to be settled: the warty layer is formed coincident with the death of the protoplast. One suggestion (WARDROP, 1964a) is that various cytoplasmic organelles are trapped between the plasmalemma and the tonoplast as the cell becomes vacuolated. The sandwiched particles are adpressed to the cell wall under the confining tonoplast which finally coats the cell lumen. Since there can be encrustations or protuberances on the cell wall before this terminal process, the final wart structure could conceivably consist of both, one superimposed upon the other. There are still many aspects of this question yet to be resolved.

**1.3.4.4 Dentate Ray Tracheid Thickenings.** Ray tracheids are prosenchymatous cells found in the rays of some coniferous species. They have small bordered pits in contrast to the simple pitting of the ray parenchyma with which they are associated in the radially oriented bands of tissue. Most ray tracheids have smooth lumen walls, but some of the *Pinus* species have localized tooth-like projections lining the cell cavity (Fig. 1.49). The hard pines are characterized by these dentate or reticulate structures.

The fine structure of dentate ray tracheids was investigated only by MEIER (1960) who selected *Pinus silvestris* L. for the study. The dentations are not comparable to helical thickenings but appear to be enlarged regions of the usual secondary cell wall layers.

# 1.4 Reaction Wood Anatomy and Ultrastructure

The normal growth pattern of forest trees produces erect, vertical stems. When trees are forced out of this pattern, caused to lean or bend, by wind or by gravitational forces, abnormal woody tissue is formed in certain parts of the tree. This wood is called reaction wood. In conifers, it forms on the lower side of a leaning stem or branch and is therefore termed compression wood. In hardwoods, reaction tissue occurs on the upper side of a leaning or bent tree and on the upper part of a branch, and it is called tension wood. In both cases the tree stems are generally eccentric in cross section in the area of the reaction wood. Because reaction wood behaves quite differently from normal or unaffected wood, and because some of its properties are not desirable for construction or processing applications, it is classed as a defect if found in pronounced form or in extensive areas in a wood product.

The gross characteristics of reaction wood are described in Chapter 3 which is devoted to defects and abnormalities of wood. Table 1.1 in this chapter summarizes the comparative features, properties, and structure of compression wood and of tension wood. Since reaction wood contains certain deviations from the normal chemical composition of wood, these are detailed in Chapter 2. The anatomical and ultrastructural characteristics of reaction wood, with interpretations of their effect on behavior and properties, are discussed below.

Table 1.1. *Comparison of Tension Wood and Compression Wood Characteristics*

| | Tension Wood | Compression Wood |
|---|---|---|
| Gross and physical characteristics; mechanical properties | Eccentricity of stem cross section: "upper side" <br> Dry, dressed lumber: silvery sheen of tension wood zones in many species; darker than normal in certain tropical and Australian species <br> Green sawn boards woolly on surface <br> Longitudinal shrinkage up to $1+\%$ <br> Particularly high tensile strength in dry tension wood; lower than normal in green condition | Eccentricity of stem cross section: "lower side" <br> Non-lustrous, "dead" appearance <br> Darker colored ("Rotholz") than normal wood <br> Longitudinal shrinkage up to $6-7\%$ <br> Modulus of elasticity, impact strength, tensile strength: low for its density |
| Anatomical characteristics | Gelatinous (tension wood) fibers present though may be lacking in some species <br> Vessels reduced in size and number in tension wood zones <br> Ray and vertical parenchyma apparently unmodified | Rounded tracheids <br> Inter-cellular spaces <br> Transition pattern, springwood-summerwood, altered: more gradual than in normal wood |
| Micro-structure | G-layer present; convoluted or not <br> Slip planes and compression failures in tension wood fiber walls <br> G-layer in secondary wall of gelatinous fibers in 3 types of arrangements: $S_1 + S_2 + S_3 + G$ or $S_G$ <br> $S_1 + S_2 + G$ or $S_G$ <br> $S_1 + G$ or $S_G$ | Helical checks or cavities in $S_2$ <br> Slip planes and compression failures generally absent |

Continuation Table 1.1.

|  | Tension Wood | Compression Wood |
|---|---|---|
| Ultra-structure | Primary wall appears normal<br>$S_1$ may be thinner than normal<br>Microfibrillar orientation of G-layer nearly parallels fiber axis; high parallelism within G-layer | $S_3$ layer absent<br>$S_1$ may be thicker than normal<br>$S_2$ microfibrillar orientation approaches 45°<br>Ribs of cellulose parallel to direction of microfibril orientation; cellulose lamellae parallel with wall surface |
| Chemical composition | Lignification of tension wood fibers variable; G-layer only slightly lignified<br>Abnormally high *cellulose* content<br>Abnormally low *lignin* content<br>More *galactan* than normal<br>Less *xylan* than normal | "Extra" lignin deposited as a layer between $S_1$ and $S_2$<br>Abnormally low *cellulose* content<br>Abnormally high *lignin* content<br>More *galactan* than normal<br>Less *galactoglucomannan* than normal |

### 1.4.1 Compression Wood

In normal wood, coniferous tracheids generally appear as square, rectangular or polygonal cells in cross section, but in compression wood the summerwood tracheids are circular in shape. As a consequence, inter-cellular spaces are formed at the junction of four tracheids (Fig. 1.50). This is a feature that is quite consistent

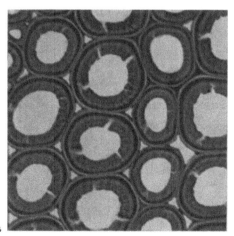
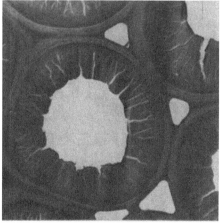

Fig. 1.50a. Cross section of compression wood of *Larix laricina* (Du Roi) K. KOCH showing rounded tracheids, intercellular spaces, and checks in the secondary cell walls. Photomicrograph (1,050×)
Fig. 1.50b. Cross section of same species, electron micrograph, with greater resolution of wall layers and checks (1,900×)

in compression wood anatomy. In cases where a wood exhibits abrupt transition in cell wall thickness from normal earlywood to latewood, it appears more gradual in compression wood because the earlywood cells have thicker walls than usual.

The secondary cell walls of compression wood tracheids invariably contain helical "checking" and these cavities generally extend radially through the $S_2$ layer, as seen on the cross section (Fig. 1.50) (CORE et al., 1961). Fibrillar alignment also differs from the normal in coniferous reaction wood tracheids. This can be detected by comparing the angle of the pit orifices in the latewood tracheids of normal and of compression wood, or through the use of the electron microscope.

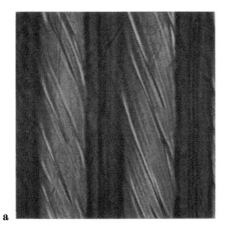

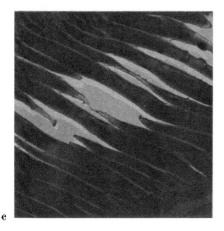

Fig. 1.51a—c. Longitudinal sections of compression wood tracheids showing nature of helical checking: a) *Sequoia sempervirens* (D. Don) Endl., Photomicrograph (900×), b) *Larix laricina* (Du Roi) K. Koch Photomicrograph, (900×), c) Same species as (b), electron micrograph (2,250×)

The "checks" in the $S_2$ layer follow the general fibrillar alignment (Fig. 1.51).

The above comparison shows that the microfibrils in the $S_2$ layer of compression wood tracheids are arranged in a flatter helix than in normal tracheids. This has been suggested (WARDROP and DADSWELL, 1950) as a source of the abnormally high longitudinal shrinkage of compression wood. These authors also state that this flatter helix is related to tracheid length and that the tracheids of compression wood are substantially shorter than those of normal wood.

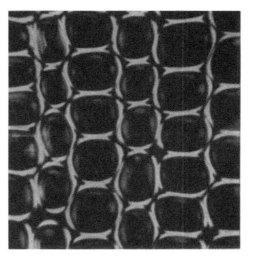

Fig. 1.52. Polarized light photomicrograph of cross section of *Larix laricina* (Du Roi) K. Koch compression wood. Compare with Fig. 1.25 and note that the $S_3$ layer is absent in this case; the bright inner line is missing. (1,300×)

When thin cross sections of normal and compression wood are examined with the polarizing microscope, a striking difference in cell wall organization becomes

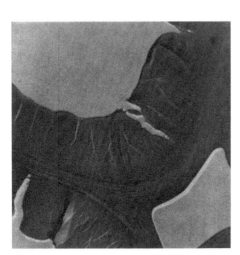 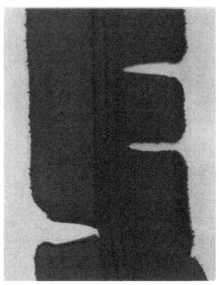

Fig. 1.53. Portions of compression wood tracheids in *Pinus* sp. illustrating small and large helical checks restricted to the $S_2$ wall layer. Note absence of $S_3$. Electron micrograph (2,600 ×)

Fig. 1.54. Cross section of compression wood tracheid wall from *Sequoia sempervirens* (D. Don) Endl. showing extent of checking through $S_2$ to $S_1$, as well as lack of $S_3$. Note also the warts shown in profile along the lumen and in the checks. Electron micrograph (6,300 ×)

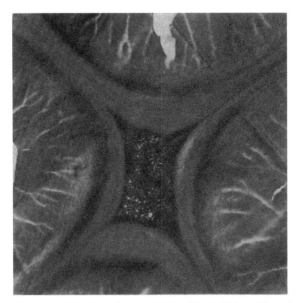

Fig. 1.55. Unshadowed cross section of portions of compression wood tracheids from *Picea rubens* Sarg. There is an electron-dense zone between the thickened $S_1$ and the checked $S_2$ layers, interpreted to be lignin deposit. Electron micrograph (3,900 ×)

apparent. Normal tracheids have three secondary wall layers, as a rule. In a photomicrograph of such a wall, made with a polarizing microscope (Fig. 1.25), the $S_2$ layer appears as a wide, dark band bounded by two bright, but narrower bands, the $S_1$ and $S_3$ layers. In compression wood one of these bright bands is lacking; namely, the $S_3$, which lines the cell lumen. (Fig. 1.52). The absence of an $S_3$ or inner layer of the secondary wall in compression wood tracheids can be confirmed by electron micrographs of ultra-thin cross sections or of surface replicas of the lumen lining (Fig. 1.53).

WARDROP and DAVIES (1964) have shown that the so-called "checks" are not due to drying. They are morphological features found in compression wood tracheids in the green condition and can even be seen in differentiating tracheids of reaction xylem. Since prominent warts appear within these helical openings (HARADA et al., 1958), there can be no question but that the cavities are formed before the death of the cell. The wart structure is known to be terminal in nature, forming on the lumen lining when the cell dies. Warts can be seen in ultra-thin sections of compression wood tracheids or in surface replicas of the cell lumen lining (Fig. 1.54).

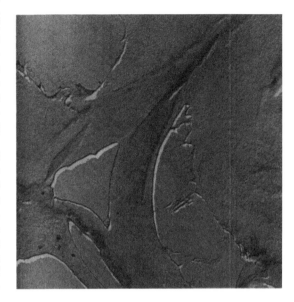

Fig. 1.56. Cross section of portions of compression wood tracheids from same source as Fig. 1.55, delignified to show separation of wall at the $S_1$–$S_2$ interface. Electron micrograph (7,600×)

In effect, the $S_2$ layer of the secondary wall of compression wood tracheids is composed of ribs of cellulose extending toward the lumen between the helical cavities. Within the ribs, the lamellae parallel the wall surface (WERGIN and CASPERSON, 1961; WARDROP and DAVIES, 1964). The helical cavities are believed to be contributory to the high longitudinal shrinkage of coniferous reaction wood (WARDROP, 1964b).

The chemical composition of compression wood is presented in detail in Chapter 2. However, it can be summarized briefly here. The cellulose content is lower than in normal wood, and there is an abnormally high lignin content. A galactan, so far not encountered in normal wood, is present in considerable amounts, while the galactoglucomannan level is lower than in normal wood. In Fig. 1.55, which is an electron micrograph of an unshadowed ultra-thin cross section of spruce compression wood, a zone of greater electron density can be seen between the $S_1$ and $S_2$ layers. This has been suggested by DADSWELL, WARDROP and WATSON (1961) as the location of the "excess" lignin found in compression wood tracheids. Partial delignification of wood from the same source results in a preferential separation of $S_1$ and $S_2$, demonstrating that this is undoubtedly the region where the extra lignin is concentrated (Fig. 1.56) (TIMELL and CÔTÉ, 1964).

## 1.4.2 Tension Wood

Probably the most readily recognized anatomical feature of angiosperm reaction wood is the gelatinous fiber. In the majority of cases, the occurrence of a large number of gelatinous fibers concentrated in an area of the wood cross section constitutes clear evidence of the presence of tension wood. However, the I. A. W. A. (1964) definition for tension wood states "... it is characterized anatomically by lack of cell wall lignification and often by the presence of an internal gelatinous layer in the fibers".

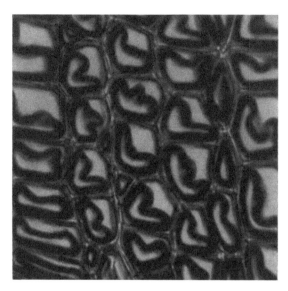

Fig. 1.57. Cross section of tension wood fibers in *Populus tremuloides* Michx. showing G-layers displaced in same direction by microtoming. Photomicrograph (700×)

Undoubtedly, this criterion is more accurate than the presence of gelatinous fibers alone because ONAKA (1949) found that certain genera lack gelatinous fibers in portions of the tree that otherwise exhibit properties of tension wood. In cases where tension wood characteristics are present without gelatinous fibers, WARDROP (1964b) suggests that the phloem in the eccentric portion of the stem be checked for possible modifications in its structure. It was JACCARD (1938) who noted that, though the usual tension wood fibers are lacking in *Tilia*, the bark is greatly modified. Earlier, METZGER (1908) recognized that changes in the phloem accompany the formation of reaction xylem. DADSWELL and WARDROP (1955) reported that the phloem fibers are also unlignified and have much thicker walls than normal bark fibers. SCURFIELD and WARDROP (1962) found no deviation from the normal anatomical pattern in reaction phloem, but electron micrographs confirmed the presence of a G-layer in the $S_2$ layer of the bark fibers.

There have been reports of some exceptions to the list of genera lacking gelatinous fibers prepared by ONAKA (1949). One of them was made by BAREFOOT (1963) who found gelatinous fibers in the wood of *Liriodendron tulipifera* L., a member of a genus which ONAKA listed as being without them. Perhaps it is significant, however, that they are not numerous in this species.

Tension wood fibers contain a gelatinous or G-layer as part of the secondary wall. This layer may appear thick or swollen in cross section, sometimes nearly filling the fiber lumen. In some fibers the G-layer is convoluted, giving the impression that it is attached only loosely to the remainder of the secondary wall. During microtoming, the G-layers in many of the cells are displaced in the same direction, identifying this as a cutting artifact and, at the same time, indicating that it is a loose attachment (Fig. 1.57).

SCURFIELD and WARDROP (1962) found that these convolutions are representative of various developmental stages of cell wall formation, though in some species convoluted G-layers appear in mature cells. If a tension wood fiber is examined before it has completely developed, convolutions may be prominent. In most species, as cell wall formation continues, these convolutions are reduced or disappear altogether and the layers flatten out against the existing wall.

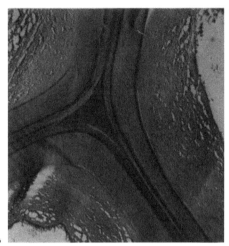 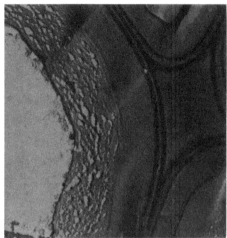

Fig. 1.58a, b. Cell wall layering of gelatinous fibers in *Celtis occidentalis* L. (7,700×): a) $S_1$, $S_2$, $S_3$ and $G$ layers present, b) $S_1$, $S_2$ and $G$ layers present. From CÔTÉ and DAY (1962a) by permission Forest Products Journal

The wall structure of tension wood fibers can be quite variable. Three variations of layering are possible, according to DADSWELL and WARDROP (1955). The first type contains the usual three secondary wall layers, $S_1$, $S_2$, and $S_3$, plus an additional layer designated as $G$ or $S_G$ (Fig. 1.58). This arrangement is not as common as the following two types, however. In the second, the $S_3$ is lacking and the G-layer is found adjacent to the $S_2$. In the third variation, both the $S_2$ and $S_3$ are missing, the G-layer occurring immediately after the $S_1$. Apparently there is no relationship between species and type of gelatinous fiber formed. There may be more than one type formed in a single specimen.

Besides having an almost unlignified, thick layer forming the inner part of the secondary wall, gelatinous fibers have another characteristic feature that can best be observed with a polarizing microscope. If longitudinal sections of tension wood are examined, slip planes and minute compression failures can be seen in the fiber walls (Fig. 1.59). Although these were first believed to result from machining or cutting stresses, they can be found even in macerated fibers prepared with careful handling (WARDROP and DADSWELL, 1947).

Deviations from the normal in size, arrangement and number of cells can also be found in reaction xylem of hardwoods. Like all tension wood characteristics, these modifications are influenced by the degree of reaction wood development, whether mild or pronounced. Variations can also occur between species and genera. The ray and longitudinal parenchyma are apparently unmodified by the formation of reaction wood, but the size and number of vessels may be reduced. In this case, the number of fibers would generally be greater than in normal wood (WARDROP, 1964b). In a wood such as *Tilia*, which shows little structural modification, perhaps reduced lignification is the only change that is related to reaction wood formation.

Studies on the ultrastructure of tension wood have largely been limited to the fibers since they undergo the greatest modification. As mentioned above, presence of a thick, unlignified G-layer in the fibers is the salient feature of tension wood anatomy. Consequently, this layer has been investigated with staining reactions, UV absorption, X-ray diffraction and electron microscopy. Incidental observations such as the behavior of the G-layer in methacrylate embedding, and its effect on the equilibrium moisture content of tension wood have also been reported (WARDROP and DADSWELL, 1955; PRESTON and RANGANATHAN, 1947; CÔTÉ and DAY 1962a).

Fig. 1.59. Polarized light photomicrograph of longitudinal section of tension wood in *Populus* sp. showing slip planes and minute compression failures in the cell walls (640×)

When sections of tension wood fibers are stained with a combination of safranin and fast green or light green, the unlignified regions stain green and the lignified layers stain red (JUTTE, 1956). Although staining reactions cannot always be considered reliable evidence, in this instance they have been supported by adequate physical evidence. The G-layer invariably stains green since it is composed of highly crystalline cellulose and is virtually unlignified. X-ray diffraction patterns of fibers containing a well-developed G-layer are sharp compared with the diffuse diagrams of normal wood. Ultra-violet photomicrographs of tension wood show that the G-layer absorbs very weakly in the UV region while the heavily lignified middle lamella absorbs strongly.

Fig. 1.60. Longitudinal section through G-layer of tension wood fiber from *Acer saccharum* Marsh. The microfibrils in this layer are oriented almost parallel to cell axis. Electron micrograph (5,450×). From CÔTÉ and DAY (1962a) by permission Forest Products Journal

In electron micrographs of longitudinal sections through tension wood fibers, cellulose microfibrils can be seen very clearly if the specimens are embedded in methacrylate. The gelatinous layer expands during polymerization, but the remainder of the cell wall does not (Fig. 1.60). This artifact has been interpreted as an indication of relatively weak lateral bonding between the microfibrils. The lack of encrusting substances revealed in the micrographs may be a further explanation.

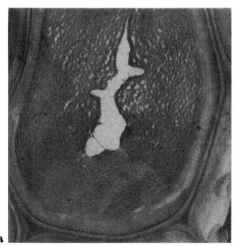 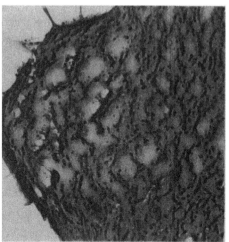

Fig. 1.61a. Cross section of tension wood fiber in *Acer saccharum* Marsh. $S_3$ layer replaced by G-layer which is expanded by methacrylate embedding. Electron micrograph (5,200×)
Fig. 1.61b. Portion of (a) enlarged to show individual isolated microfibrils, 200 to 400 Ångström units in diameter. Electron micrograph (23,300×). From CÔTÉ and DAY (1962a) by permission Forest Products Journal

It is clear from Fig. 1.60, and from X-ray diffraction evidence, that the microfibrils are oriented nearly parallel to the longitudinal axis of the cell. Lamellation within the layer can be seen in cross sections of a swollen wall (Fig. 1.61a) and, upon further enlargement, isolated microfibrils are visible (Fig. 1.61b). In this figure the microfibrils range in diameter from 200 to 400 Å units, approximately.

Methacrylate-expanded G-layers serve to isolate zones that are resistant to this swelling phenomenon. A dense, thin layer can be seen in some tension wood fibers. It lines the very narrow cell lumen and appears to be analogous to the terminal lamella of normal wood elements (Fig. 1.62).

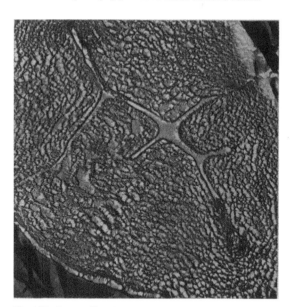

Fig. 1.62. Cross section of tension wood fiber from *Populus* sp. with methacrylate-expanded G-layer demonstrating lamellated structure. Note thin, dense terminal lamella bordering the narrowed lumen. Electron micrograph (5,440×). From CÔTÉ and DAY (1962a) by permission Forest Products Journal

Although its chemical composition has not been determined, it is possible that it is lignin-like because its UV absorption is very similar to that of the middle lamella (WARDROP, 1964b).

The chemical composition of tension wood differs from normal wood. With a large number of gelatinous fibers, abnormally low lignin levels can be expected and usually typify tension wood. The cellulose content is much higher than normal, and a new type of galactan is present, unknown in normal hardwood and different from that occuring in compression wood. In tension wood the proportion of xylan is lower than in normal wood of the same species.

## Literature Cited

BAILEY, I. W., (1933) The cambium and its derivative tissues. VIII. Structure, distribution, and diagnostic significance of vestured pits in dicotyledons. J. Arnold Arboretum **14**: 259—273.

— and KERR, T., (1935) The visible structure of the secondary wall and its significance in physical and chemical investigations of tracheary cells and fibers. J. Arnold Arboretum **16**: 273—300.

BAREFOOT, A. C., Jr., (1963) Abnormal wood in yellow-poplar (*Liriodendron tulipifera* L.) For. Prod. J. **XIII**: 16—22.

CHATTAWAY, M. M., (1949) The development of tyloses and secretion of gum in heartwood formation. Australian J. Sci. Res. Ser. B., Biol. Sci. **2**: 227—240.

—, (1955) Crystals in woody tissues. Part I. Trop. Woods **102**: 55—74.

CORE, H. A., Côté, W. A., Jr. and DAY, A. C., (1961) Characteristics of compression wood in some native conifers. For. Prod. J. **XI**: 356—362.

CÔTÉ, W. A., Jr., (1958) Electron microscope studies of pit membrane structure. For. Prod. J. **VIII**: 296—301.

— and DAY, A. C., (1962a) The G-layer in gelatinous fibers — electron microscopic studies. For. Prod. J. **XII**: 333—338.

— and—, (1962b) Vestured pits — fine structure and apparent relationship with warts. TAPPI **45**: 906—910.

— and MARTON, R., (1962) Brightness of high yield pulps IV. Electron microscopy of white birch heartwood. TAPPI **45**: 46—53.

DADSWELL, H. E. and WARDROP, A. B., (1955) The structure and properties of tension wood. Holzforschung **9**: 97—104.

—, — and WATSON, A. J., (1961) The morphology, chemistry and pulp characteristics of reaction wood. In: Fundamentals of Papermaking Fibers, Francis Bolam, Editor. Tech. Section of British Paper and Board Maker's Asso., Inc. London. pp. 187—219.

FISCHBEIN, I. W., (1950) A study of the cell wall of coniferous tracheids by means of the electron microscope. Unpublished doctoral dissertation. Biophysics, grad. div. Univ. of California, Berkeley.

FREI, E., PRESTON, R. D. and RIPLEY, G. W., (1957) The fine structure of the walls of conifer tracheids. VI. Electron microscope investigations of sections. J. Exp. Bot. **8**: 139—146.

FRENCH, G. E. (1923) The effect of the internal organization of the North American hardwoods upon their more important mechanical properties. Unpublished M. F. thesis, New York State College of Forestry, Syracuse, N. Y.

FREY-WYSSLING, A., (1959) Die Pflanzliche Zellwand. Springer-Verlag. Berlin/Göttingen/Heidelberg, 367 pp.

— and BOSSHARD, H. H., (1959) Cytology of the ray cells in sapwood and heartwood. Holzforschung **13**: 129—137.

—, — and MÜHLETALER, K., (1956a) Die submikroskopische Entwicklung der Hoftüpfel. Planta **47**: 115—126.

—, MÜHLETHALRE, K. and BOSSHARD, H. H., (1955) Das Elektronenmikroskop im Dienste der Bestimmung von Pinusarten. Holz Roh- Werkstoff **13**: 245—249.

—, — and —, (1956b) Nachtrag zu: Das Elektronenmikroskop im Dienste der Bestimmung von Pinusarten. Holz Roh- Werkstoff **14**: 161—162.

GERRY, E. J., (1914) Tyloses: Their occurrence and practical significance in some American woods. J. Agric. Res. **1**: 445—469.

HARADA, H., (1956) The electron microscopic investigation of wood. On the fine structure of the wart-like structure and of the pit membrane. Transactions 65th Mtg. Japanese Forestry Soc.

—, (1965a) Ultrastructure and organization of gymnosperm cell walls. In: Cellular Ultrastructure of Woody Plants, W. A. CÔTÉ, Jr., Editor. Syracuse University Press, Syracuse, N. Y.

# Literature Cited

—, (1965b) Ultrastructure of angiosperm vessels and ray parenchyma. In: Cellular Ultrastructure of Woody Plants, W. A. CÔTÉ, Jr., Editor. Syracuse University Press, Syracuse, N. Y.

— and MIYAZAKI, Y., (1952) Electron microscopic observation of compression wood. Bulletin of Govt. For. Exp. Sta. No. 54: 101—108.

—, — and WAKASHIMA, T., (1958) Electron-microscopic investigation of the cell wall structure of wood. Bull. For. Exp. Sta., Meguro, Tokyo, Japan. No. 104. 115 pp.

— and WARDROP, A. B., (1960) Cell wall structure of ray parenchyma cells of a softwood. J. Japan Wood Research Society (Mokuzai Gakkaishi) 6: 34—41.

HOWES, F. N., (1949) Vegetable Gums and Resins. Chronica Botanica Co., Waltham, Mass. 188 pp.

I. A. W. A., (1957) International glossary of terms used in wood anatomy. International Association of Wood Anatomists, Commitee on Nomenclature. Tropical Woods 107: 1—36.

—, (1964) Multilingual glossary of terms used in wood anatomy. International Association of Wood Anatomists, Committee on Nomenclature. Verlagsanstalt Buchdruckerei Konkordia Winterthur, Switzerland. 186 pp.

JACCARD, P., (1938) Exzentrisches Dickenwachstum und Anatomisch-Histologische Differenzierung des Holzes. Ber. Schweiz. Botan. Ges. 48: 491.

JAYME, G. and FENGEL, D., (1961) Beitrag zur Kenntnis des Feinbaus der Frühholztracheiden. Beobachtungen an Ultradünnschichten von Fichtenholz. Holz Roh-Werkstoff. 19: 50—55.

JUTTE, S. M., (1956) Tension wood in Wane (*Ocotea rubra* Mez). Holzforschung 10: 33—35.

KOBAYASHI, K. and UTSUMI, N., (1951) Cell wall structure of coniferous tracheids. (In Japanese) Electron Microscope Committee Note No. 56: 93—94.

KORAN, Z., (1964) Ultrastructure of tyloses and a theory of their growth mechanism. Unpublished doctoral dissertation. State Univ. College of Forestry at Syracuse University, Syracuse, N. Y. 95 pp.

KRAHMER, R. L. and CÔTÉ, W. A., Jr., (1963) Changes in coniferous wood cells associated with heartwood formation. TAPPI 46: 42—49.

KRIBS, D. A., (1959) Commercial Foreign Woods on the American Market. Edwards Brothers. Inc. Ann Arbor, Mich. 203 pp.

LIESE, W., (1951) Demonstration elektronenmikroskopischer Aufnahmen von Nadelholztüpfeln. Ber. Dtsch. Bot. Ges. 64: 31—32.

—, (1954) Der Feinbau der Hoftüpfel im Holz der Koniferen. Proceedings International Conf. Electron Microscopy, London. pp. 550—554.

—, (1956a) Elektronenmikroskopische Beobachtungen über die Warzenstruktur bei den Koniferen. Proc. Stockholm Conference on Electron Microscopy. pp. 276—279.

—, (1956b) Zur systematischen Bedeutung der submikroskopischen Warzenstruktur bei der Gattung *Pinus* L. Holz Roh- Werkstoff 14: 417—424.

—, (1957) Beitrag zur Warzenstruktur der Koniferentracheiden unter besonderer Berücksichtigung der Cupressaceae. Ber. Dtsch. Bot. Ges. 70: 21—30.

— (1963) Tertiary wall and warty layer in wood cells. J. Poly. Sci. Part C, No. 2: 213—229.

— (1965a) The fine structure of bordered pits in softwoods. In: "Cellular Ultrastructure of Woody Plants", W. A. CÔTÉ, Jr., Editor. Syracuse Univ. Press, Syracuse, New York.

— (1965b) The warty layer. In: „Cellular Ultrastructure of Woody Plants", W. A. CÔTÉ, Jr., Editor. Syracuse Univ. Press, Syracuse, New York.

— and HARTMANN-FAHNENBROCK, M., 1953 Elektronenmikroskopische Untersuchungen über die Hoftüpfel der Nadelhölzer. Biochem. et Biophys. Acta 11: 190—198.

— and JOHANN, I. (1954) Elektronenmikroskopische Beobachtungen über eine besondere Feinstruktur der verholzten Zellwand bei einigen Koniferen. Planta 44: 269—285.

MEIER, H., (1955) Über den Zellwandabbau durch Holzvermorschungspilze und die submikroskopische Struktur von Fichtentracheiden und Birkenholzfasern. Holz Roh- Werkstoff 13: 232—338.

— (1960) Über die Feinstruktur der Markstrahltracheiden von *Pinus silvestris* L. Beih. Z. Schweiz. Forstv. 30: 40—53.

— (1964) General chemistry of cell walls and distribution of the chemical constituents across the walls. In: The Formation of Wood in Forest Trees, M. H. ZIMMERMANN, Editor. Academic Press, New York pp. 137—151.

METZGER, K., (1908) Über das Konstruktionsprinzip des sekundären Holzkörpers. Naturw. Z. Forst- u. Landwirtsch. 6: 249.

MÜHLETHALER, K., (1960) Die Feinstruktur der Zellulosemikrofibrillen. Beih. Zeit. Schweiz. Forstv. 30: 55—65.

—, (1965) The fine structure of the cellulose microfibril. In: Cellular Ultrastructure of Woody Plants, W. A. CÔTÉ, Jr., Editor. Syracuse Univ. Press, Syracuse, New York.

ONAKA, F., (1949) Studies on compression and tension wood. Wood Research Bulletin of the Wood Research Institute, Kyoto Univ., Japan. **1**: 1—83.

PERILÄ, O., (1962) The chemical compositions of wood cells. III. Carbohydrates of birch cells. Suomen Kemistilethti **B35**: 176—178.

— and HEITTO, H., (1959) The chemical compositions of wood cells. I. Carbohydrates of pine cells. Suomen Kemistilethti **B32**: 76—80.

— and SEPPA, T., (1960) The chemical compositions of wood cells. II. Carbohydrates of spruce cells. Suomen Kemistilethti **B33**: 114—116.

PRESTON, R. D. and RANGANATHAN, S., (1947) The fine structure of the fibres of normal and tension wood in beech (*Fagus sylvatica* L.) Forestry **21**: 92—98.

PREUSSER, H. J., DIETRICHS, H. H. and GOTTWALD, H., (1961) Elektronenmikroskopische Untersuchungen an Ultradünnschnitten des Markstrahlparenchyms der Rotbuche — *Fagus sylvatica* L. Holzforschung **15**: 65—75.

RECORD, S. J. and HESS, R. W., (1943) Timbers of the New World. Yale Univ. Press. 640 pp.

SACHS, I. B., CLARK, I. T. and PEW, J. C., (1963) Investigation of lignin distribution in the cell wall of certain woods. J. Poly. Sci., Part C, No. **2**: 203—212.

SCHMID, R., (1965) The fine structure of pits in hardwoods. In: Cellular Ultrastructure of Woody Plants, W. A. CÔTÉ, Jr., Editor. Syracuse University Press, Syracuse, N. Y.

— and MACHADO, R. D., (1964) Zur Entstehung und Feinstruktur skulpturierter Hoftüpfel bei Leguminosen. Planta **60**: 612—626.

SCURFIELD, G. and WARDROP, A. B., (1962) The nature of reaction wood VI. The reaction anatomy of seedlings of woody perennials. Australian J. Bot. **10**: 93—105.

TIMELL, T. E. and CÔTÉ, W. A., Jr., (1964) Unpublished results.

WARDROP, A. B., (1958) Organization of the primary wall in differentiating conifer tracheids. Australian J. Bot. **6**: 299—305.

—, (1964a) The structure and formation of the cell wall in xylem. In: The Formation of Wood in Forest Trees, M. H. ZIMMERMANN, Editor. Academic Press, Inc. New York. pp. 87—134.

—, (1964b) The reaction anatomy of arborescent angiosperms. In: Formation of Wood in Forest Trees, M. H. ZIMMERMANN, Editor. Academic Press, Inc. New York. pp. 405—456.

— and DADSWELL, H. E., (1947) Contributions to the study of the cell wall. 5. The occurrence, structure and properties of certain cell wall deformations. CSIRO (Australia) Bulletin No. **221**.

— and —, (1950) The nature of reaction wood II. The cell wall organization of compression wood tracheids. Australian J. Sci. Res. B **3**: 1—13.

— and —, (1952) The cell wall structure of xylem parenchyma. Australian J. Sci. Res. Ser. B, Biol. Sci. **5**: 223—236.

— and —, (1953) The development of the conifer tracheid. Holzforschung **7**: 33—39.

— and —, (1955) The nature of reaction wood IV. Variations in cell wall organization of tension wood fibres. Australian J. Botany **3**: 177—189.

— and DAVIES, G. W., (1964) The nature of reaction wood VIII. The structure and differentiation of compression wood. Australian J. Bot. **12**: 24—38.

— and HARADA, H., (1963) As shown in Wardrop, A. B., (1964a).

—, INGLE, H. D. and DAVIES, G. W., (1963) Nature of vestured pits in angiosperms. Nature (London) **197**: 202—203.

WERGIN, W. and CASPERSON, G., (1961) Über Entstehung und Aufbau von Reaktionsholzzellen. 2. Mitt. Morphologie der Druckholzzellen von *Taxus baccata* L. Holzforschung **15**: 44—49.

# 2. CHEMICAL COMPOSITION OF WOOD

## 2.0 Introduction

The walls of wood cells are composed of three principal chemical materials, cellulose, hemicelluloses and lignin, all of which are polymeric. With the initiation of daughter cells through cell division at the cambium, new walls are formed, the primary wall which encloses the new unit, and the middle lamella or intercellular layer which separates adjoining cells. Both of these regions are rich in pectic material. During the phase of cell wall thickening, cellulose and hemicelluloses are synthesized within the cell and deposited onto the primary wall, forming a secondary wall. The formation of lignin begins before this phase is complete (cell elongation may still be going on), starting at the cell corners and spreading along the primary wall and intercellular layer. Finally, lignification proceeds to the secondary wall after which the cell dies and the remaining cytoplasmic débris is deposited on the lumen walls in the form of a terminal lamella or a warty membrane.

Cellulose, as the skeletal substance, contributes its high tensile strength to the complex of wood structure. The function of hemicelluloses is less clear, there existing some possibility that they serve as a temporary matrix before lignification. The role of lignin is to provide rigidity to the tree, making upright growth possible. It also adds toxicity, thus making wood durable. Its concentration in the middle lamella serves the purpose of cementing individual cells together.

Reaction wood, classed by some as abnormal wood, follows a somewhat different pattern of cell wall development. Compression wood, the reaction wood formed on the lower side of leaning coniferous stems, is characterized by the absence of an $S_3$ layer in the secondary wall, by the deposition of an extra layer of lignin between two layers of the secondary wall, $S_1$ and $S_2$, and by the presence of a galactan which is not formed in normal wood. A number of other morphological features of reaction wood are described in Chapter 3.

Cells containing unique "gelatinous" layers (G-layers) are formed in tension wood, the reaction wood of hardwoods. These zones are generally found on the upper side of a branch or of a leaning stem in angiosperms. The thick, inner cell layer consists largely of cellulose. In addition, tension wood contains a newly discovered type of galactan.

The only fully differentiated cells to remain alive in the sapwood portion of a tree are the parenchyma. In the sapwood-heartwood transition zone a certain portion of these cells give rise to polyphenolic substances, many of them polymeric, which permeate and encrust the surrounding tissues. As a result, the heart wood core of most tree species has a darker color than the sapwood. Just as the prosenchymatous elements die after lignification, with heartwood formation the parenchyma cell nuclei disintegrate. In members of the genus *Larix*, heartwood formation is associated with the deposition of an arabinogalactan. The physical properties of heartwood that are considerably different from sapwood can be related, in many instances, to the chemical changes associated with this transformation.

In the discussion that follows, the position taken by the authors has been that wood technologists are primarily concerned with the chemical composition of wood. The chemistry of wood has been ably covered by HÄGGLUND (1951), WISE and JAHN (1952), BRAUNS and BRAUNS (1960), TREIBER (1957), HILLIS (1962) and BROWNING (1963). The biochemical, morphological and physiological aspects can be found in KRATZL and BILLEK (1959), NORTHCOTE (1958) FREY-WYSSLING (1959), ROELOFSEN (1959) and ZIMMERMANN (1964).

## 2.1 Chemical Constituents of Wood and their Determination

In considering the overall chemical composition of wood, it must be emphasized that wood is not a homogeneous material. Just as cell dimensions vary in wood taken from various parts of the tree, so does its chemical composition. For example,

Table 2.1. *Chemical Composition of Wood from five Angiosperms* (TIMELL, unpublished)
(All values in percent of extractive-free wood)

| Component | Acer rubrum L. | Betula papyrifera Marsh. | Fagus grandifolia Ehrh. | Populus tremuloides Michx. | Ulmus americana L. |
|---|---|---|---|---|---|
| Cellulose | 45 | 42 | 45 | 48 | 51 |
| Lignin | 24 | 19 | 22 | 21 | 24 |
| O-Acetyl-4-O-methyl-glucurono-xylan | 25 | 35 | 26 | 24 | 19 |
| Glucomannan | 4 | 3 | 3 | 3 | 4 |
| Pectin, starch, ash, etc. | 2 | 1 | 4 | 4 | 2 |

Table 2.2. *Chemical Composition of Wood from five Conifers* (TIMELL, unpublished)
(All values in percent of extractive-free wood)

| Component | Abies balsamea (L.) Mill | Picea glauca (Moench) Voss | Pinus strobus L. | Tsuga canadensis (L.) Carr. | Thuja occidentalis L. |
|---|---|---|---|---|---|
| Cellulose | 42 | 41 | 41 | 41 | 41 |
| Lignin | 29 | 27 | 29 | 33 | 31 |
| Arabino-4-O-methyl-glucurono-xylan | 9 | 13 | 9 | 7 | 14 |
| O-Acetyl-4-O-methyl-glucurono-xylan | 18 | 18 | 18 | 16 | 12 |
| Pectin, starch, ash, etc. | 2 | 1 | 3 | 3 | 2 |

variation is found within a single tree from the center of the stem to the bark, from the stump to the crown, between springwood and summerwood, and between sapwood and heartwood. The occurrence of zones of reaction wood also affects the overall chemical composition. It is therefore not surprising to find that the chemical composition of wood varies widely, even in trees of the same species. In spite of these limitations, it is possible to make certain generalizations concerning the gross chemical composition of wood, as long as statistical sampling is employed in selecting the material for analysis.

Wood is often analyzed by a series of so-called summative analyses, standard methods which are easy to apply and which account for the entire material. Such an analysis includes estimation of extractives, lignin, ash, acetyl, uronic anhydride, and residues of galactose, glucose, mannose, arabinose and xylose. The first three of these are determined by standard methods developed by the Technical Association of the Pulp and Paper Industry (TAPPI), New York, N. Y. (1962). A number of special methods are available for the determination of the other constituents.

Both lignin and the various polysaccharides can be more or less quantitatively isolated from wood by a variety of extraction procedures. These will be referred to later.

Angiosperms from the temperate zones contain less lignin and more xylan than tropical hardwoods, the lignin content of which often exceeds that of many conifers. The softwoods differ from the temperate zone hardwoods in their higher lignin and mannose and lower xylose contents. The chemical composition of five hardwoods (Table 2.1) and five softwoods (Table 2.2) is given in terms of the major polymeric constituents, namely cellulose, lignin, hemicellulose and other polysaccharides. The carbohydrate values have been obtained from summative data and from a knowledge of the composition of each polysaccharide. Values resulting from a direct isolation of cellulose and hemicelluloses agree well with these indirect data.

Normal hardwoods and softwoods both usually contain $42 \pm 2\%$ of cellulose. The lignin content of hardwoods varies between 18 and 25%, while in softwoods the range is 25 to 35%. The most characteristic chemical feature of hardwoods is their high

Table 2.3. *Composition of Normal and Compression Wood from Sapwood Portion of Larix laricina* (Du Roi) K. Koch (Timell, unpublished)

| Component | Normal Wood | Compression Wood |
|---|---|---|
| | Percent of extractive-free wood | |
| Lignin | 27 | 38 |
| Pentosan | 8.7 | 9.6 |
| Uronic anhydride | 2.8 | 3.1 |
| Residues of : | Percent of total neutral carbohydrates | |
| Galactose | 3.6 | 17.7 |
| Glucose | 68.3 | 59.0 |
| Mannose | 18.2 | 9.2 |
| Arabinose | 2.1 | 2.1 |
| Xylose | 7.8 | 12.0 |

Table 2.4. *Composition of Normal and Tension Wood from Eucalyptus goniocalyx* F. Muell. (Schwerin, 1958) (All values in percent of extractive-free wood)

| | Cellulose | Lignin | Pentosan | Acetyl | Galactose residue |
|---|---|---|---|---|---|
| Normal Wood | 44.0 | 29.5 | 15.1 | 3.0 | 2.5 |
| Tension Wood | 57.3 | 13.8 | 11.0 | 1.9 | 7.4 |

content of a partly acetylated, acidic xylan which accounts for 20 to 35% of the wood. A second hemicellulose, a glucomannan, occurs in only small amounts. A partly acetylated galactoglucomannan makes up almost 20% of coniferous wood, but the xylan, in this case, comprises only 10% of the total wood substance.

Table 2.5. *Relative Sugar Composition of Normal and Tension Wood from Betula pubescens* Ehrh. (Gustafsson et al., 1952) *and Fagus sylvatica* L. (Meier, 1962) (All values in percent)

| Sugar Residue | Betula pubescens | | Fagus sylvatica | |
|---|---|---|---|---|
| | Normal | Tension | Normal | Tension |
| Galactose | 2.6 | 11.6 | 1.3 | 4.9 |
| Glucose | 57.6 | 73.5 | 57.6 | 73.3 |
| Mannose | 0.9 | 0.4 | 4.3 | Trace |
| Arabinose | 0.9 | Trace | 1.0 | 2.0 |
| Xylose | 38.0 | 14.5 | 35.8 | 18.8 |

In addition to these major polysaccharides, both hardwoods and softwoods contain small amounts of pectic material, starch, and as yet unknown hemicelluloses. The ash content seldom exceeds 0.5%. The amount and nature of the extractives removed with water and organic solvents vary widely from one species to another, but they are always more abundant in the heartwood than in the sapwood.

58 2. Chemical Composition of Wood [Ref. p. 75

Compression wood (Table 2.3) can contain as much as 40% lignin and 10% galactose residues, with corresponding lesser quantities of cellulose and gluco-mannan. Tension wood (Tables 2.4 and 2.5) contains less lignin and xylan, but has much more cellulose and galactose residues than normal hardwoods.

## 2.2 Characteristics of the Principal Wood Constituents

The physical and chemical characteristics of the principal constituents of wood are of particular significance when wood is to be treated or utilized as the raw material for a wood-based product. The behavior of the various components, individually, and in association with others, is a primary consideration of many wood processors.

### 2.2.1 Cellulose

Cellulose is one of the most abundant organic substances on earth, occurring throughout the plant kingdom. Wherever it occurs in nature, it takes the forms of microfibrils, long strands composed of cellulose molecules arranged more or less parallel to each other. Cotton fibers are almost pure cellulose, but usually cellulose is found in association with hemicelluloses and lignin in land plants. Regardless of its source, cellulose is always the same chemically even if the nature and arrangement of microfibrils varies somewhat.

**2.2.1.1 Isolation from Wood.** For wood, the necessity of attaining a quantitative isolation of the cellulose without causing some depolymerization poses a real problem. Wood is too heavily lignified to allow a direct extraction of the cellulose with a suitable solvent (JORGENSEN, 1950). If the lignin is first removed, for example by the chlorine-ethanolamine (TAPPI Standard T 9 m-54) or by the acid chlorite (WISE, et al., 1946) methods, the hemicelluloses can be extracted with aqueous alkali from the resulting holocellulose, leaving the entire cellulose portion as an insoluble residue. This cellulose has been severely degraded, however, first by the delignifying agents and then by the alkali. Fortunately, most woods are susceptible to direct nitration with a non-degrading acid mixture (ALEXANDER and MITCHELL, 1949), which is also capable of swelling the wood, thus giving an undegraded cellulose nitrate in quantitative yield (TIMELL, 1957). This nitrate is suitable for molecular weight determinations.

Like all natural polymers, cellulose is polymolecular; that is, it contains molecules of different sizes. It is also a linear macromolecule of considerable length, a fact which has given rise to difficulties when attempts have been made to determine its average molecular weight by methods such as sedimentation-diffusion or light scattering. However, results obtained by light scattering measurements of cellulose nitrates, of wood, seed hairs of cotton, kapok, milkweed floss, and bast fibers of flax, hemp, jute and ramie indicate that native cellulose have an average molecular weight value $(P_w)$ close to 10,000 (GORING and TIMELL, 1962).

**2.2.1.2 Structure.** Cellulose is a $(1 \rightarrow 4)$ — linked ß-D-glucoglycan, consisting of $\beta$-D-glucopyranose residues which are linked together into straight chains by $(1 \rightarrow 4)$ - glycosidic bonds (OTT, et al., 1954; HERMANS, 1949; HONEYMAN, 1959). The glucose residues in cellulose are present in the C1-Chair conformation with the three hydroxyl groups all equatorial (Fig. 2.1).

Cellulose is partly crystalline and gives an X-ray diagram quantitative evaluation of which allows the dimensions of the crystallographic unit cell to be

## 2.2 Characteristics of the Principal Wood Constituents

calculated. A monoclinic unit cell (Cellulose I) based on this evidence is shown in Fig. 2.2. It was first suggested by MEYER and MARK (1929) and later modified by

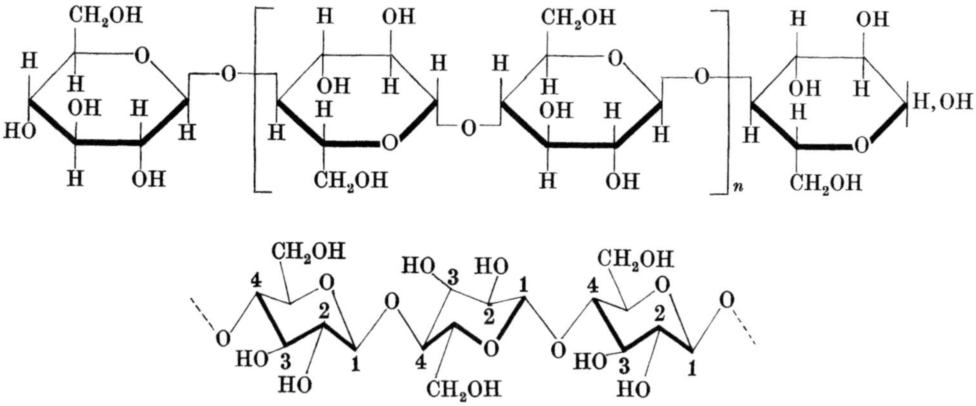

Fig. 2.1. Structural formula of cellulose

MEYER and MISCH (1937). Each unit cell contains four glucose residues. The two sets of chains possess a twofold screw axis, making a residue of cellobiose the repeating unit, and are tentatively assumed to run in opposite directions. In the $a$–$b$ plane the glucose residues are kept together by numerous and thus fairly strong, hydrogen bonds, in the $a$–$c$ plane by much weaker VAN DER WAAL's forces, and in the $b$–$c$ plane by covalent bonds. The cellulose lattice is

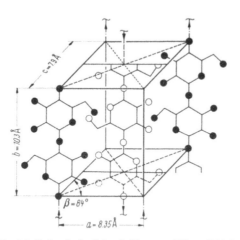

Fig. 2.2. Unit cell of cellulose I (MEYER and MISCH [1937])

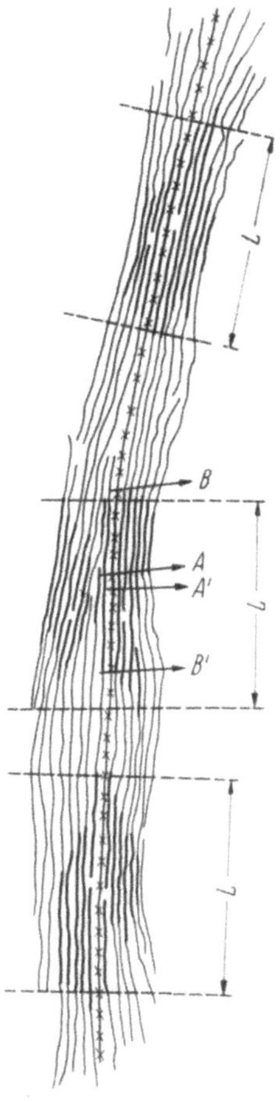

Fig. 2.3. Crystallized and amorphous regions of cellulose (MARK [1940]). $A$, $A'$, $B'$ are chain-ends within the crystallized region; $B$ is a chain-end outside the crystallized region; $L$ is the length of the crystallized region

thus at once a chain lattice and a layer lattice. It should be noted that certain X-ray and electron reflections cannot be explained with this model.

One of cellulose's most characteristic properties is the presence of both crystalline and amorphous regions within its microfibrils. According to the so-called *fringe micellar theory*, a single cellulose chain may run through several crystalline regions (called crystallites or micelles) as well as through amorphous zones in between. This is illustrated in Fig. 2.3. It has been shown that cellulose acetate can form single crystals in which the cellulose chains are arranged in a direction perpendicular to the crystal surface and folded (MANLEY, 1963). This interesting observation cannot be entirely reconciled with the presently accepted fringe micellar theory.

X-ray and small-angle X-ray scattering indicate that the crystalline regions have a diameter of 50 to 100 Å. For determination of the relative crystallinity of cellulose, numerous methods have been used. Actually, many of them do not measure the crystallinity but rather the accessibility of cellulose. In other words, what is measured is the relative proportion of glucose residues located in the amorphous regions and on the surface of the crystallites. Chemical methods are all of very limited usefulness in this connection. Among the best physical methods are X-ray diffraction (HERMANS, 1951), deuterium exchange, easily followed with the aid of infrared spectroscopy (MANN and MARRINAN, 1956), the sensitive tritium exchange method (SEPALL and MASON, 1961), and moisture regain (HOWSMON, 1949). Most physical techniques indicate 70 to 80% crystallinity for cotton cellulose, 60 to 70% for wood pulp cellulose and 40 to 50% for regenerated cellulose (rayon).

It has become quite clear, with the wide application of electron microcopy, that cellulose occurs throughout nature in the form of microfibrils. In certain algae, mannans and xylans have been reported to form microfibrils (FREI and PRESTON, 1961). However, in land plants only one polysaccharide besides cellulose has been reported to form microfibrils. MEIER (1958) found these in the amorphous mannan $B$ in vegetable ivory nuts.

Much controversy has existed concerning the true dimensions of the cellulose microfibrils. According to COLVIN (1963) their width is 150 to 200 Å. However, the use of "negative contrast" preparation technique has made it possible to show that the smallest structural unit, an *elementary fibril*, has a diameter of 35 Å (see Fig. 1.20). A microfibril of this size would contain approximately 40 cellulose chains (MÜHLETHALER, 1960; 1963). Such elementary fibrils can easily aggregate to form microfibrils of greater diameter and further combination can then lead to the formation of macrofibrils, lamellae and cell walls. See Fig. 1.23 which is an electron micrograph of cellulose microfibrils.

**2.2.1.3 Properties.** Cellulose, unlike starch, is not soluble in water. This is due to its regularity and large number of hydrogen bonds. The only agents that will dissolve cellulose are those which are capable of complexing with it such as aqueous cuprammonium hydroxide, cupriethylenediamine, and cadmium triethylenediamine (cadoxen). The last solvent is the best presently available, giving colorless, stable solutions of any cellulose (HENLEY, 1960; REIMER, 1962).

Aqueous alkalis swell native cellulose but do not dissolve it. Alkali will degrade cellulose in several ways. At low temperatures and in the presence of oxygen, the glycosidic bonds are broken by a chain reaction involving free radicals (ENTWISTLE, et al., 1949; WHISTLER and BEMILLER, 1958). At high temperatures, alkali alone can attack glycosidic bonds.

Cellulose dissolves in several strong acids such as 72% sulfuric acid, 41% hydrochloric acid, and 85% phosphoric acid. Degradation sets in rapidly, however, especially in the first two cases.

## 2.2.2 Hemicellulose

Hemicelluloses are relatively low-molecular-weight, non–cellulosic polysaccharides which occur in plant cell walls together with lignin and cellulose. They are isolated by extraction of untreated or of delignified wood with water or, more frequently, with aqueous alkali. Many hemicelluloses have numerous, short side chains, but they are otherwise essentially linear. One exception is the heavily branched arabinogalactan found in larch wood.

Hemicelluloses in land plants are built up of D-xylose, D-mannose, D-glucose, D-galactose, L-arabinose, 4-O-methyl-D-glucuronic acid and, to a lesser extent, of D-glucuronic acid, L-rhamnose, L-fucose, and various methylated neutral sugars. Hemicelluloses seldom have more than 150 to 200 sugar residues in their molecular backbone and are therefore quite small macromolecules compared to cellulose.

Although related, the hemicelluloses in the wood of angiosperms and gymnosperms are not the same. The polysaccharides of the gymnosperms are more complex, both with respect to the number of hemicelluloses present and with respect to their structure.

The chemical and physical properties of all wood hemicelluloses have been critically discussed in considerable detail by TIMELL (1964) and the methods used in polysaccharide chemistry have been reviewed by BOUVENG and LINDBERG (1960).

**2.2.2.1 Hemicelluloses of Hardwoods.** There are two principal hemicelluloses found in the wood of angiosperms, O-Acetyl-4-O-methylglucurono-xylan and glucomannan. The xylan is the predominant hemicellulose of all hardwoods and forms 20 to 35% of the extractive-free material. The glucomannan makes up 3 to 5% of extractive-free wood.

*2.2.2.1.1 O-Acetyl-4-O-methylglucurono-xylan.* This polysaccharide can be isolated in a yield of 80 to 90% by direct extraction of the wood with aqueous potassium hydroxide, sodium hydroxide being less selective (HAMILTON and QUIMBY, 1957). The 10 to 20% of the xylan that remains in the wood is probably not able to diffuse out of the cell wall (NELSON and SCHUERCH, 1956, 1957).

All hardwoods have so far been found to contain the same type of xylan, the basic structure of which is presented in Fig. 2.4. A simplified, but complete,

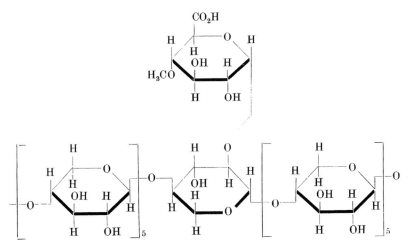

Fig. 2.4. Structure of 4-O-Methylglucuronoxylan

structure is shown in Fig. 2.5. The polysaccharide consists of a framework of (1→4)-linked β-D-xylopyranose residues. Some carry as a directly attached side chain a 4-O-methyl-a-D-glucuronic acid residue at their 2-positions.

Most hardwoods contain a xylan with ten xylose residues per acid side chain. Exceptions so far reported include *Malus pumila*, *Prunus serotina* and *Ulmus americana*, the xylans of which have one acid group per 6 to 7 xylose residues.

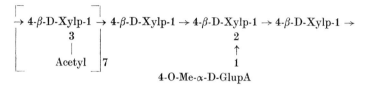

Fig. 2.5. Repeating unit of O-Acetyl-4-O-Methylglucuronoxylan

Unlike cellulose, 4-O-methylglucuronoxylans have a three-fold screwaxis, as indicated in Fig. 2.6 (MARCHESSAULT and LIANG, 1962). Like cellulose, they are partly crystalline and can also be obtained in the form of single crystals (MARCHESSAULT, et al., 1961). The native, acetylated xylan in the wood is probably entirely amorphous, but it might be somewhat oriented in the direction of the fiber axis (LIANG, et al., 1960). It is soluble in water, but the deacetylated product obtained on extraction with alkali is not. Summary data on this xylan are included in Table 2.6.

*2.2.2.1.2 Glucomannan.* After removal of the xylan by extraction of a hardwood holocellulose with potassium hydroxide, the remaining glucomannan can be separated from the cellulose by treatment with an aqueous solution of sodium hydroxide containing borate, in which it is soluble (JONES, et al., 1956; TIMELL, 1960). This polysaccharide (Fig. 2.7) consists of probably randomly dis-

→4-β-D-Manp-1 → 4-β-D-Glup-1 →4-β-D-Manp-1 →4-β-D-Manp-1 →

Fig. 2.7. Structural formula of hardwood glucomannan

tributed β-D-glucopyranose and β-D-mannopyranose residues linked together by (1→4)-glycosidic bonds. The ratio between the glucose and mannose residues is usually close to 1:2, but seems to be 1:1 in wood of the genus *Betula*. It is not known whether hardwood glucomannans are branched or not, nor have the molecular properties of the native polymer been studied. Glucomannans are more readily depolymerized by acids than cellulose and they are also easily degraded by alkali.

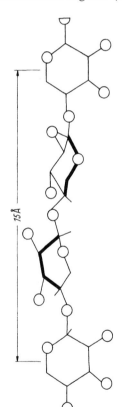

Fig. 2.6. Conformation of xylan backbone (MARCHESSAULT and LIANG[1962])

*2.2.2.1.3 Tension Wood Galactan.* MEIER (1962b) has isolated a galactan from tension wood of beech. The constitution of this polysaccharide, which might also contain galacturonic and glucuronic acid residues, is not as yet clear, but it is evident that both (1→4) and (1→6)-linked galactose units are present.

**2.2.2.2 Hemicelluloses of Softwoods.** Unlike hardwoods, softwoods, when treated with alkali, remain practically unchanged, probably because of their higher concentration of lignin in the secondary wall. For isolation of softwood hemicellulose, the lignin must first be eliminated. Usually this is done with acid chlorite. The resulting holocellulose, on extraction with potassium hydroxide, gives a mixture of two hemicelluloses, an acidic arabinoxylan and a galactoglucomannan (TIMELL, 1961).

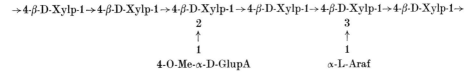

→4-β-D-Xylp-1→4-β-D-Xylp-1→4-β-D-Xylp-1→4-β-D-Xylp-1→4-β-D-Xylp-1→4-β-D-Xylp-1→
             2                                              3
             ↑                                              ↑
             1                                              1
      4-O-Me-α-D-GlupA                                  α-L-Araf

Fig. 2.8. Structural formula of arabino 4-O-Methylglucuronoxylan

*2.2.2.2.1 Arabino-4-O-methylglucurono-xylan.* The constitution of this xylan is shown in Fig. 2.8. Its backbone and the nature and attachment of the acid side chains are the same as those of the hardwood xylan. In addition, however, α-L-arabinofuranose residues are also attached to C-3 of the xylose residues (ASPINALL and ROSS, 1963), probably randomly distributed along the chain (GOLDSCHMID and PERLIN, 1963). Unlike the hardwood xylan, the softwood xylan is not acetylated in its native state (MEIER, 1961a), but it is nevertheless soluble in water because of its numerous 4-O-methylglucuronic acid and arabinose side chains.

*2.2.2.2.2 Galactoglucomannans.* The predominant softwood hemicelluloses are the galactoglucomannans, a family of closely related polysaccharides containing various amounts of galactose residues. They were the last major wood hemicelluloses to be discovered (HAMILTON, et al., 1957; 1960). The major portion consists of a galactoglucomannan which contains galactose, glucose and mannose residues in a ratio of 0.1:1:3. The general structure of all galactoglucomannans is shown in Fig. 2.9.

→4-β-D-Manp-1→4-β-D-Glup-1→4-β-D-Manp-1→4-β-D-Manp-1→4-β-D-Glup-1→4-β-D-Manp-1→
             6                        2 or 3
             ↑                           |
             1                        Acetyl
          β-D-Galp

Fig. 2.9. Structural formula of O-Acetyl-galacto-glucomannan

The hexosan backbone consists of (1→4)-linked β-D-glucopyranose and β-D-mannopyranose residues which are probably randomly distributed. Some glucose and mannose residues carry a varying number of α-D-galactopyranose residues, directly attached to their 6-position. The glucomannan framework is probably slightly branched (CROON, et al., 1959) and contains at least 150 hexose residues. This polysaccharide is quite labile to alkali which degrades it from the reducing end group.

*2.2.2.2.3 Compression Wood Galactan.* A polysaccharide has been isolated from spruce compression wood, containing mostly galactose, but probably also uronic acid residues (BOUVENG and MEIER, 1959). The neutral part of this galactan consists of (1→4)-linked β-D-galactopyranose residues. This galactan is thus different from that present in tension wood.

*2.2.2.2.4 Larch Arabinogalactan.* When wood from conifers other than larches is extracted with cold water, small amounts of several polysaccharides are obtained, one of which is an acidic arabinogalactan. Members of the genus *Larix*, on direct extraction of their wood with water, yield considerable quantities of similar polysaccharide (BOUVENG, 1961). Many larches contain approximately 10% by weight of arabinogalactan, and wood from *Larix occidentalis* Nutt. and *L. dahurica* may contain as much as 20 to 25 percent. Larch arabinogalactan

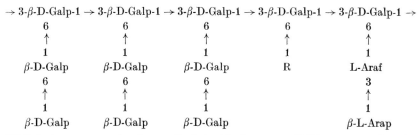

Fig. 2.10. Structural formula for larch arabinogalactan

occurs almost exclusively in the heartwood. The product present in western larch is now recovered by extraction of the wood with water and is used commercially.

All species of larch contain the same arabinogalactan, a simplified formula of which is shown in Fig. 2.10. The basic framework is composed of $(1\rightarrow 3)$-linked $\beta$-D-galactopyranose residues, each of which carries various side chains attached to the 6-position. A summary of the occurrence, isolation and properties of larch arabinogalactans has been prepared by ADAMS and DOUGLAS (1963).

### 2.2.3 Other Wood Polysaccharides

In addition to cellulose and hemicelluloses, wood contains other polysaccharides, chief among them being pectin and starch. Pectin (HIRST and JONES, 1946; WHISTLER and SMART, 1953) is more abundant in bark than in wood where it is formed only during the earlier stages of cell development. On hydrolysis, pectin usually gives galacturonic acid and minor quantities of arabinose and galactose. Its structure is still largely unknown.

Starch, the main reserve polysaccharide of a tree, consists of two components, amylose and amylopectin (GREENWOOD, 1956; MANNERS, 1962; WHELAN, 1959), both of very high molecular weight, in the case of amylopectin considerably higher than that of cellulose. In normal starch the proportion between amylose and amylopectin is usually 1:2.

A summary of the relative abundance and properties of the principal wood polysaccharides is given in Table 2.6.

### 2.2.4 Lignin

Lignin, the encrusting substance in wood, is a three-dimensional polymer of phenylpropane units (I), linked together by C—O—C and C—C linkages. In softwoods, each unit carries one phenolic oxygen and one methoxyl group (II),

[Ref. p. 75]     2.2 Characteristics of the Principal Wood Constituents     65

Table 2.6. *Summarizing Data for Wood Polysaccharides* (TIMELL, 1964, Unpublished)

| Polysaccharide | Occurrence | Per cent of extractive–free wood | Composition | Parts | Linkages |
|---|---|---|---|---|---|
| 1. Cellulose | All woods | $42 \pm 2$ | $\beta$-D-Glup | All | $1 \to 4$ |
| 2. *O*-Acetyl-4-*O*-methylglucurono-xylan | Hardwoods | $20-35$ | $\beta$-D-Xylp | 10 | $1 \to 4$ |
| | | | 4-*O*-Me-$\alpha$-D-GlupA | 1 | $1 \to 2$ |
| | | | *O*-Acetyl | 7 | |
| 3. Glucomannan | Hardwoods | $3-5$ | $\beta$-D-Manp | $1-2$ | $1 \to 4$ |
| | | | $\beta$-D-Glup | 1 | $1 \to 4$ |
| 4. Arabino-4-*O*-methylglucurono-xylan | Softwoods | $10-15$ | $\beta$-D-Xylp | 10 | $1 \to 4$ |
| | | | 4-*O*-Me-$\alpha$-D-GlupA | 2 | $1 \to 2$ |
| | | | L-Araf | 1,3 | $1 \to 3$ |
| 5. Galactoglucomannan (Water-soluble) | Softwoods | $5-10$ | $\beta$-D-Manp | 3 | $1 \to 4$ |
| | | | $\beta$-D-Glup | 1 | $1 \to 4$ |
| | | | $\alpha$-D-Galp | 1 | $1 \to 6$ |
| | | | *O*-Acetyl | 0.24 | |
| 6. Galactoglucomannan (Alkali-soluble) | Softwoods | $10-15$ | $\beta$-D-Manp | 3 | $1 \to 4$ |
| | | | $\beta$-D-Glup | 1 | $1 \to 4$ |
| | | | $\alpha$-D-Galp | 0.1 | $1 \to 6$ |
| | | | *O*-Acetyl | 0.24 | |
| 7. Arabinogalactan | Larch wood | $10-20$ | $\beta$-D-Galp | 6 | $1 \to 3, 1 \to 6$ |
| | | | L-Araf | 2/3 | $1 \to 6$ |
| | | | $\beta$-L-Arap | 1/3 | $1 \to 3$ |
| | | | $\beta$-D-GlupA | Few | $1 \to 6$ |
| 8. Pectin | All woods | 1 | $\alpha$-D-GalpA | Most | $1 \to 4$ |
| | | | D-Galp | Few | |
| | | | L-Araf | Few | |
| 9. Starch | All woods | Variable, but 5% | | | |
| Amylose | | | $\alpha$-D-Glup | All | $1 \to 4$ |
| Amylopectin | | | $\alpha$-D-Glup | All | $1 \to 4, 1 \to 6$ |

Table 2.6.
(Continued)

| Polysaccharide | Linear or branched | Specific rotation, degrees | Solvent | $\overline{P}_n$ | $\overline{P}_w$ |
|---|---|---|---|---|---|
| 1. Cellulose | Linear | $-2$ | Water | — | 10,000 |
| 2. *O*-Acethyl-4-*O*-methylglucurono-xylan | Undecided | $-80 \pm 5$ | Water Alkali | $150-200$ | $180-250$ |
| 3. Glucomannan | Undecided | $-30 \pm 2$ | Alkali | 70 | 120 |
| 4. Arabino-4-*O*-methylglucurono-xylan | Undecided | $-37 \pm 2$ | Water | 120 | — |
| 5. Galactoglucomannan (Water-soluble) | Slightly branched | $-7$ to $-8$ | Water | 100 | 150 |
| 6. Galactoglucomannan (Alkali-soluble) | Slightly branched | $-35 \pm 5$ | Alkali | 100 | 150 |
| 7. Arabinogalactan | Heavily branched | $+10 \pm 2$ | Water | 220 | 600 and 100 |
| 8. Pectin | Essentially linear | $+250$ $+200$ | Water Water | — | 3,000 to 5,000 |
| 9. Starch | Linear | | | — | 15,000 to 20,000 |
| Amylose | Heavily | | | — | 30,000 to 60,000 |
| Amylopectin | branched | | | | |

5  Kollmann/Côté, Solid Wood

while in hardwoods approximately half of the units contain an additional methoxyl group (III). High yields of degradation products are formed on oxidation of lignin with nitrobenzene with alkali at high temperature, when softwoods give vanillin (IV) and hardwoods a mixture of vanillin and syringaldehyde (V), in addition to minor amounts of $p$-hydroxybenzaldehyde (VI).

Softwood lignin has a methoxyl content of $15-16\%$ while that of hardwoods has $21\%$. The fact that lignin, unlike polysaccharides, cannot be hydrolyzed by acids, that it is difficult to remove from the wood in an unchanged state, and that it does not contain a simple repeating unit, all have made the structural chemistry of lignin an unusually difficult subject. The structure and properties of lignin are now fairly well known, however.

Ref. p. 75]     2.2 Characteristics of the Principal Wood Constituents     67

**2.2.4.1 Isolation from Wood.** All methods available for isolation of lignin from wood cause changes in the native polymer in one way or another. The two mildest methods both employ neutral solvents for extraction. *Braun's native lignin* (BRAUNS, 1939) is obtained by extracting wood with ethanol, when a few percentages of the lignin dissolve. A more representative preparation is obtained from wood if it is ground in a special vibratory ball mill and subsequently extracted with suitable organic solvents. As much as 30 to 50% of the lignin can then be isolated in a state which is, of course, changed physically, but which has undergone very minor, if any, chemical changes. This so-called *Björkman lignin*, or *milled wood lignin* (MWL) (BJÖRKMAN, 1956, 1957; BJÖRKMAN and PERSON, 1957) is closer to native lignin than any other preparation, and is presently the preferred material in lignin research. Treatment of wood with organic solvents in the presence of small amounts of mineral acids gives better yields than alcohol alone, but the lignin preparations obtained are changed chemically. The same also applies to the "periodate lignin" (WALD *et al.*, 1947) which has lost methoxyl groups.

**2.2.4.2 Structure.** The analytical determination of the functional groups in lignin is a difficult problem. It has been studied by many investigators, most recently by ADLER (1961) and GIERER (1958). The many ingenious methods developed in this connection will not be reviewed here, only the results so far obtained. Most investigations have been concerned with softwood lignin, and since in this polymer each $C_6-C_3$ (guaiacyl, VII) unit carries one methoxyl group, analytical data are usually expressed in numbers per one or per 100 (%) methoxyl groups.

Of the total guaiacyl units, 30% carry a free (VIII) and 70% an etherified (IX) phenolic group. Of the former, 16% are "uncondensed" (X) and 14% are "condensed" in the ortho-position (XI) to the phenolic group. Condensation occurs mainly (12%) by a diphenyl linkage (XII), while the remaining units assume structure (XIII). These figures probably apply to the entire macromolecule, not only to those units which carry a free phenolic group.

Carbonyl groups are present in 20 out of 100 guaiacylpropane units, distributed over the $\alpha$-, $\beta$- and $\gamma$-carbon atoms thus: (XIV) 1%, (XV) 3%, (XVI) 1%, (XVII) 6%, and (XVIII) 10%. Structure (XIV) is responsible for the majority of the known lignin color reactions.

A benzyl alcohol or benzyl ether group occurs in 43% of the units, distributed in the following way: (XIX) 5%, (XX) 15%, and (XXI) 23% (LINDGREN, 1952). These groups are sulfonated in the technical sulfite process.

Two ways in which the guaicyl propane units of lignin can be linked to each other are seen in (XII) and (XIII), together representing $15-16\%$ of the total units. Another 8% are united to *phenylcoumaran* dimers (XXII) while a few are condensed to the likewise cyclic *pinoresinol* system (XXIII). A much larger proportion (25 to 30%) of the interunit linkages in lignin consist of the monocyclic *guaiacylglycerol-$\beta$-coniferyl* ether type (XXIV).

A summary (but *not* a structural formula) of the functional groups and interunit linkages in lignin, as it is now known, and constructed by ADLER (1961), is given in Fig. 2.11.

**2.2.4.3 Properties.** Lignin, as it is deposited in wood, is a colorless, completely amorphous substance. Its ultraviolet absorption spectrum has a characteristic band at 280 m$\mu$. The molecular weight of the lignin network in wood is undetermined. Molecular weights of soluble lignosulfonic acids, produced in the sulfite cooking process, have been found to vary within the enormous range of 260 to 50 million (GORING, 1962). BJÖRKMAN lignin has a molecular weight of only 11,000.

5*

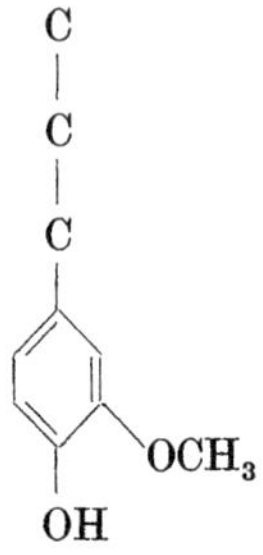

C
|
C
|
C

OCH₃

OH

(X)

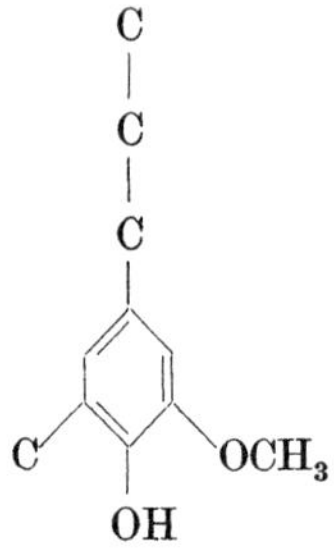

C
|
C
|
C

C′        OCH₃

OH

(XI)

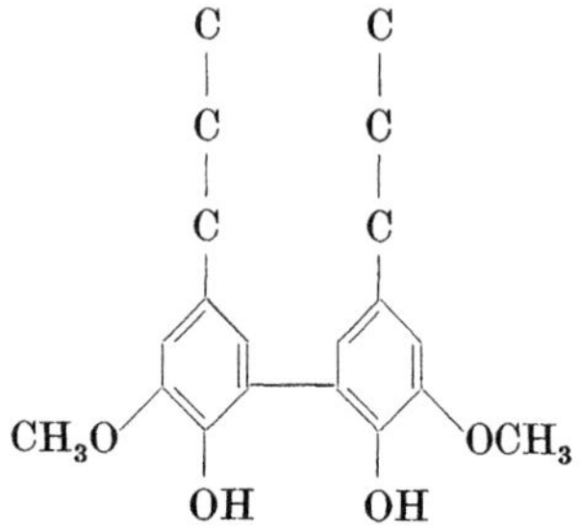

C        C
|        |
C        C
|        |
C        C

CH₃O        OCH₃

OH        OH

(XII)

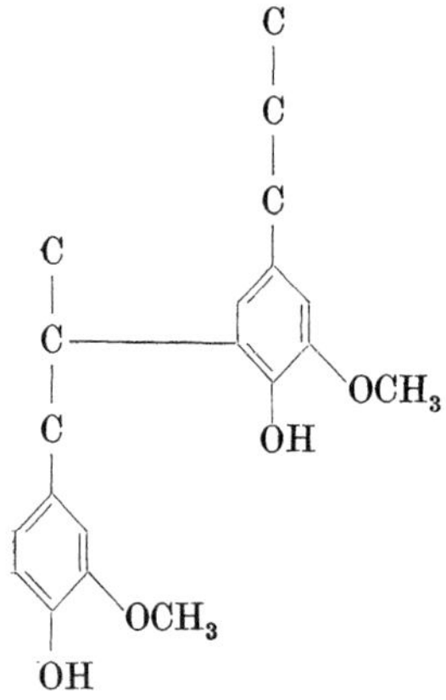

C
|
C
|
C

C
|
C
|
C        OCH₃

OH

OCH₃

OH

(XIII)

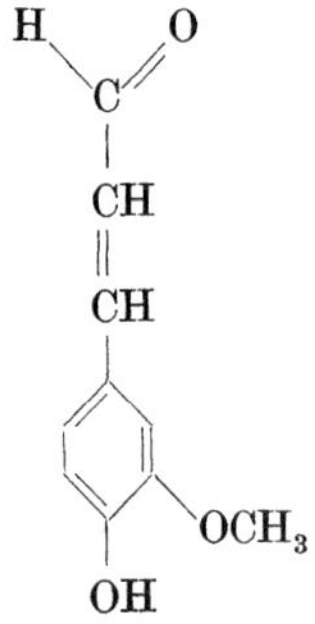

H        O
\\      /
C

CH
‖
CH

OCH₃

OH

(XIV)

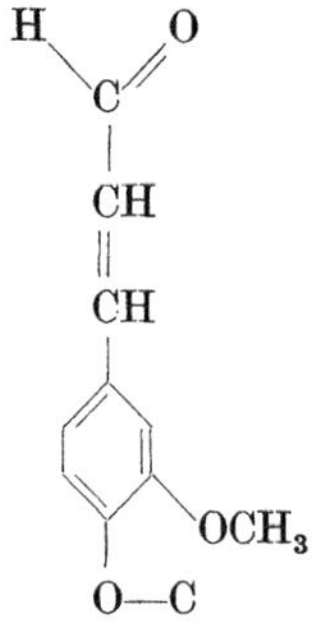

H        O
\\      /
C

CH
‖
CH

OCH₃

O—C

(XV)

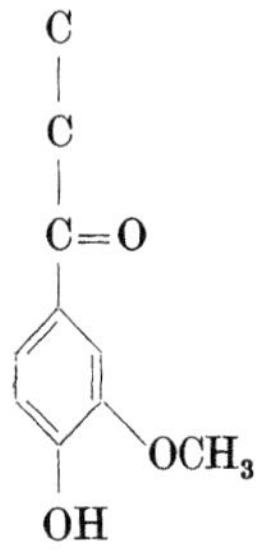

C
|
C
|
C=O

OCH₃

OH

(XVI)

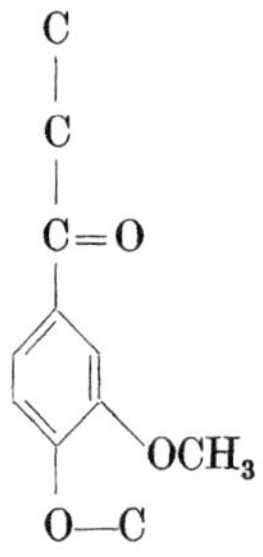

C
|
C
|
C=O

OCH₃

O—C

(XVII)

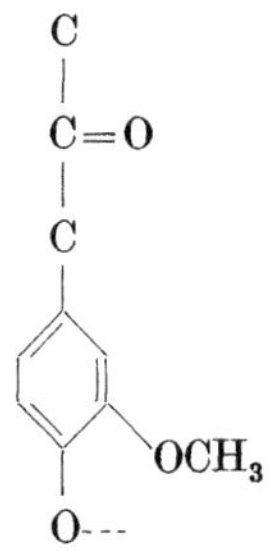

C
|
C=O
|
C

OCH₃

O---

(XVIII)

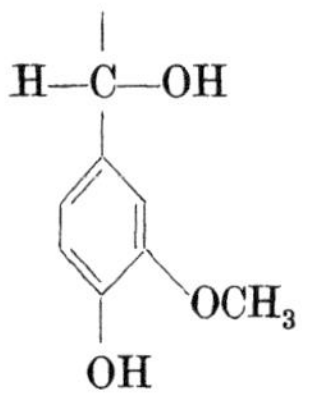

H—C—OH

OCH₃

OH

(XIX)

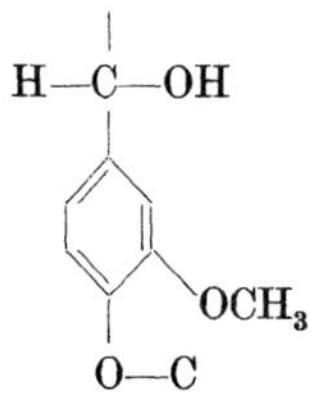

H—C—OH

OCH₃

O—C

(XX)

H—C—O—C

OCH₃

O---

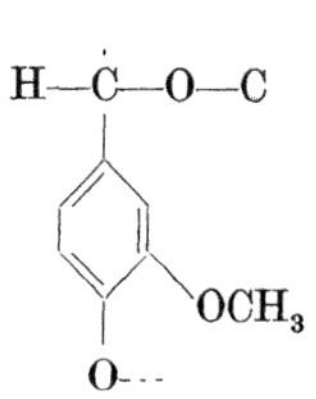

(XXI)

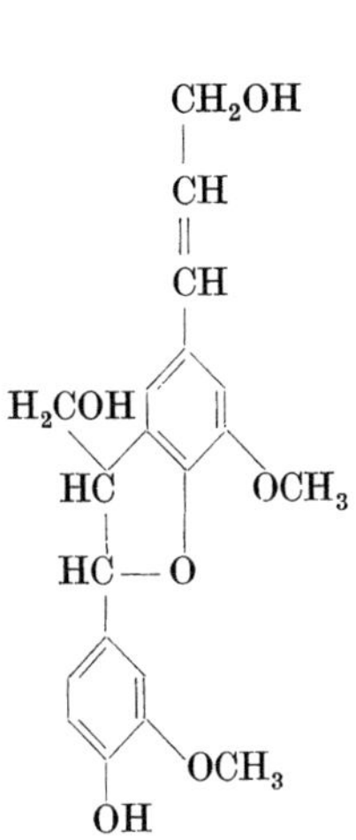

# 2.2 Characteristics of the Principal Wood Constituents

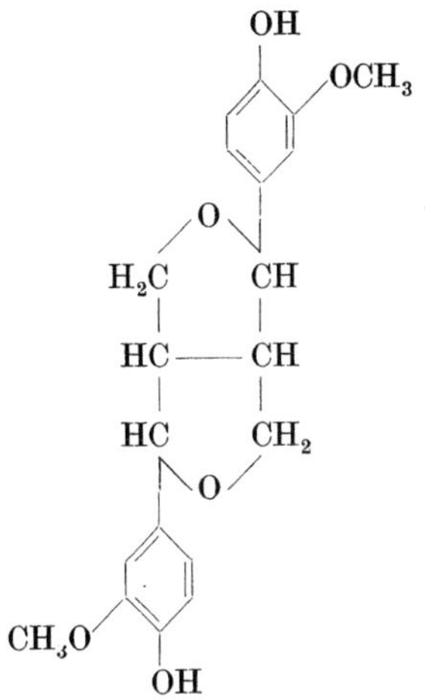

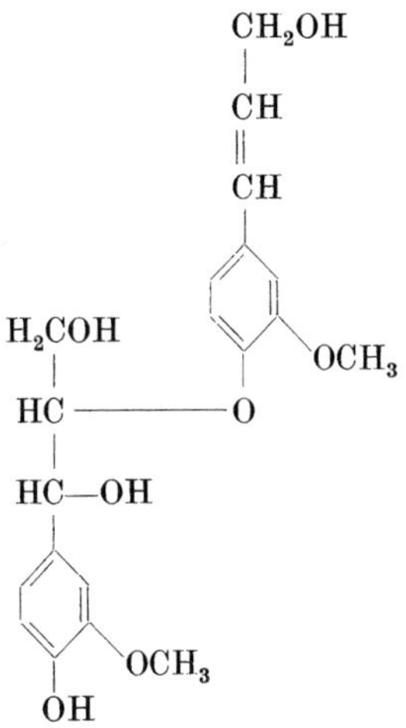

69

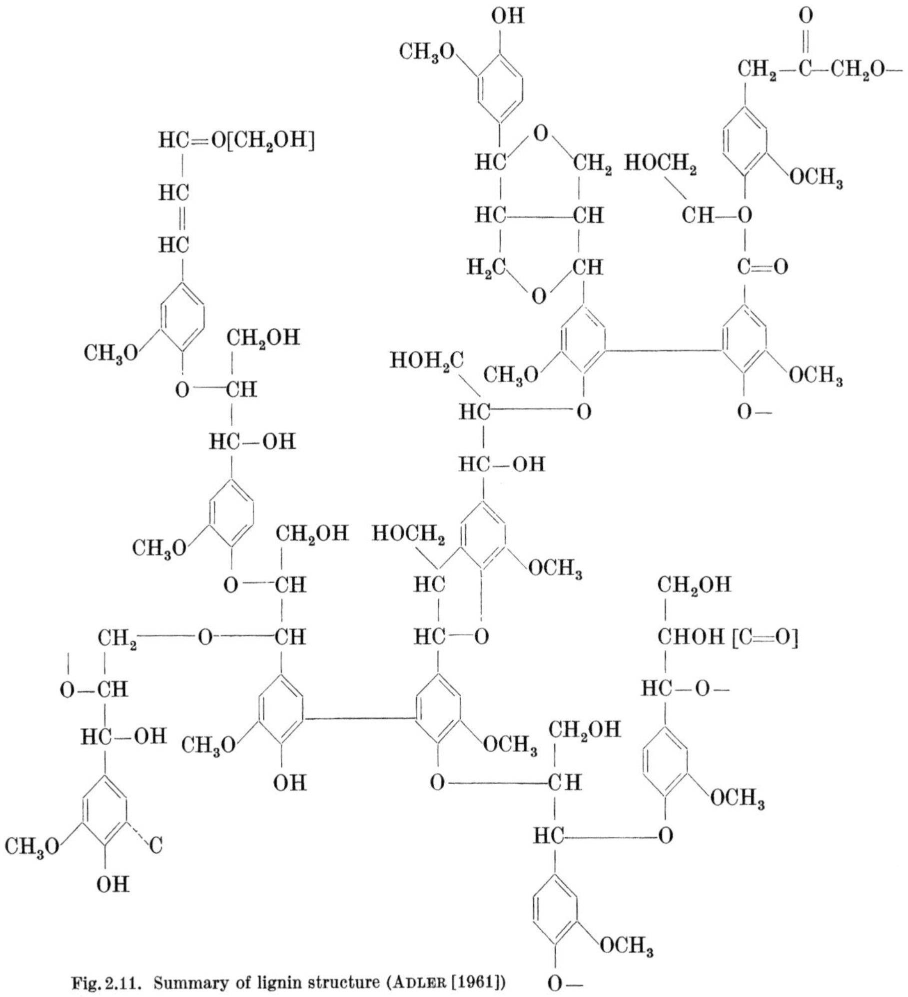

Fig. 2.11. Summary of lignin structure (ADLER [1961])

70 2. Chemical Composition of Wood [Ref. p. 75

The possibility of a chemical bond between lignin and the carbohydrates in wood, and especially the hemicelluloses, has been a question which has been much discussed for many years (MEREWETHER, 1957). Much of the evidence in favor of the occurrence of such a bond can actually just as well be explained from physical encrustation. The problem is, however, far from settled and has attracted much interest (BJÖRKMAN, 1957; LINDGREN, 1958).

Almost all experimental evidence so far discussed has been concerned with softwood lignins. In the more complicated hardwood lignins, the ratio between guaiacyl and syringyl units is usually 1:1. Many angiosperms, and especially the monocotyledons, contain units of p-hydroxybenzaldehyde in addition. It is not known whether one single copolymer or a mixture of two polymers are present.

The presence of a softwood (guaiacyl) or a hardwood (guaiacyl-syringyl) lignin can easily be ascertained with the aid of the Mäule reaction which produces a brown color in wood containing the former and a brilliant red in hardwoods.

## 2.3 Wood Extractives

In addition to its major, structural components, cellulose, hemicelluloses, and lignin, wood also contains an exceedingly large number of other compounds, both low- and high-molecular. Among the more important ones are the *terpenes*, and the wood resins, both of which are composed of isoprene units, *polyphenols*, such as flavonols, anthocyanins, quinones, stilbenes, lignans, and tannins, *tropolones*, *glycosides*, *sugars*, *fatty acids*, and *inorganic constituents*.

Most of the extractives are located in the heartwood, the presence of some of them being the source of general darkening of this portion of the tree. When the extractive materials are particularly toxic, the heartwood portion of the tree is relatively resistant to attack by biological deteriorating agencies (see Chapter 4).

The nature and amount of the terpenes can vary considerably from one species to another. ERDTMAN (1959) has been able to create a chemical taxonomy for the genus *Pinus*, based on the nature of the heartwood polyphenols. Excellent reviews dealing with a number of the principal wood extractives are available (LINDBERG, 1957; ERDTMAN, 1959; KURTH, 1952; SIMONSEN, 1947; PINDER, 1960; HARRIS, 1952; MUTTON, 1962 and HATHWAY, 1962).

The most common *terpenes* (PINDER, 1960; SIMONSEN, 1947) in the softwoods, and especially in the pines, are $\alpha$-pinene (XXV), $\beta$-pinene (XXVI), limonene (XXVII), and $\Delta_3$-carene (XXVIII). Bornyl acetate (XXIX) occurs in firs and spruces, while camphor (XXX) is abundant in the camphor tree.

Among the *resin acids*, abietic acid (XXXI) represents a common type (HARRIS, 1952; MUTTON, 1962).

One of the most important *polyphenols* is pinosylvin (XXXII). It is strongly toxic and protects the pine heartwood where it is found. Its occurrence prevents pine heartwood from being digested in sulfite pulping.

*Lignans* (HATHWAY, 1962) are formed from two phenylpropane units and bear a close relationship to lignin. Lariciresinol (XXXIII) occurs in larches, pineresinol (XXIII) in spruce and pine, and conidendrin (XXXIV) in spruce and hemlock.

*Tannins* (HATHWAY, 1962; JURD, 1962) can be divided into three classes, namely gallotannins, ellagtannins, and condensed tannins. Gallotannins are polymeric esters of gallic acid (XXXV) or digallic acid with sugars, such as glucose. The ellagtannins give the insoluble ellagic acid (XXXVI) on hydrolysis.

*Tropolones* (GARDNER, 1962) occur in members of the Cupressales. They are quite toxic and responsible for the unusual durability of cedar wood. They constitute one of the relatively rare examples of a natural product with a ring of seven carbon atoms. Examples are α-, β-, and γ-thujaplicin (XXXVII).

Glucose, fructose and sucrose are present together with starch in the living ray cells. The glucosides coniferin (XXXVIII) and syringin (XXXIX) are widely distributed in all woods and are of special interest in connection with the bio-synthesis of lignin.

The ash content of wood is usually 0.2—0.3%. The most common constituents, in this order, are calcium, potassium and magnesium; carbonates, phosphates, silicates and sulfates.

The biogenesis of wood constituents is now at least partly known. The bio-synthesis of cellulose, hemicelluloses, pectin, starch and lignin has been discussed in several recent publications (GREATHOUSE, 1959; ELBEIN et al., 1964; NEUFELD and HASSID, 1963; NEISH, 1959, 1960; WHELAN, 1963; NORD and SCHUBERT, 1959; KRATZL and BILLEK, 1957; FREUDENBERG, 1959, 1960, 1962; KRATZL, 1959, 1960, 1961; ADLER, 1957, 1959).

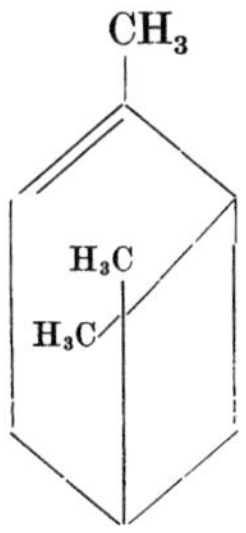

(XXV)

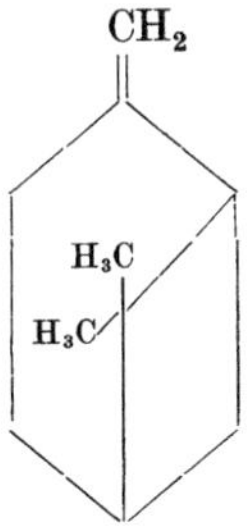

(XXVI)

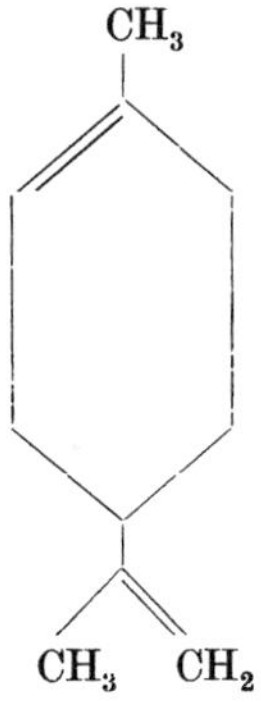

(XXVII)

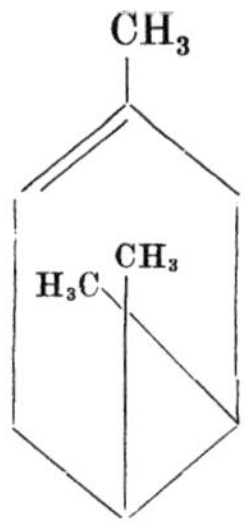

(XXVIII)

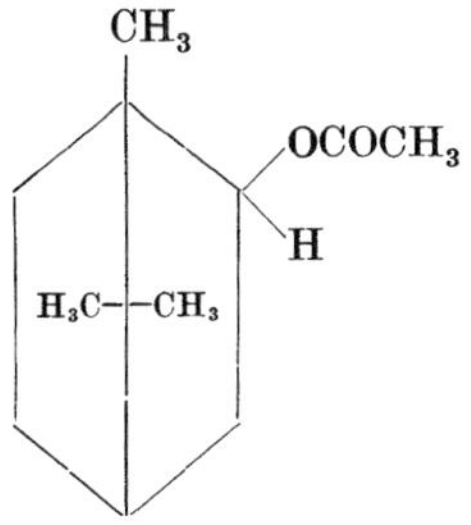

(XXIX)

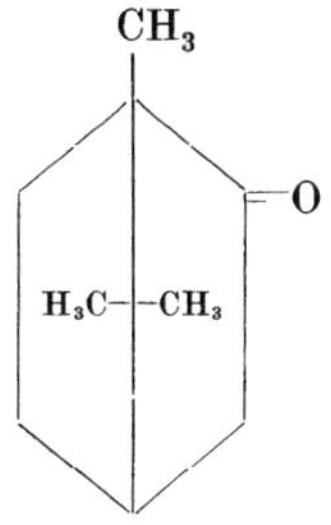

(XXX)

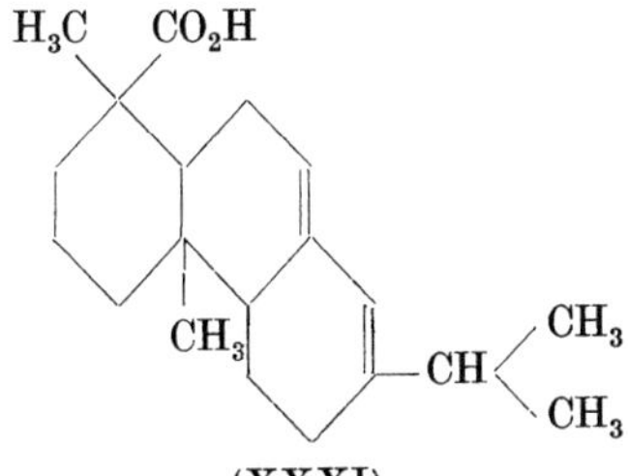

(XXXI)

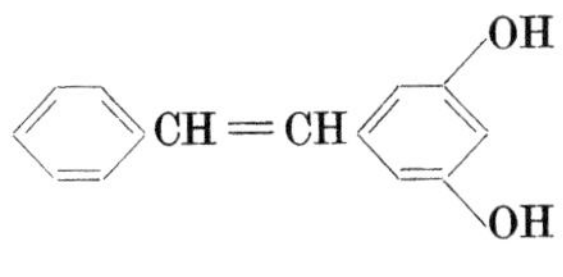

(XXXII)

## 2.4 Distribution of Chemical Constituents in Wood

It has already been pointed out that wood is not a homogeneous raw material. Variations in chemical composition between its various portions are actually quite pronounced. Heartwood has a different chemical composition from sapwood, the cambial zone and the mature xylem are unlike, and springwood and summerwood are not only morphologically but also chemically dissimilar. Prosenchymatous elements differ chemically from parenchyma cells. The constituents of the middle lamella are entirely different from those of the cell wall. The distribution of polysaccharides and lignin over the cell wall is not uniform, the primary wall and the three layers of the secondary wall each being different in this respect. The outer layer of $S_2$ appears to have a chemical composition unlike that of the inner portion. Considerable progress has been made towards a better understanding of these differences, particularly since the early 1950's.

The major event associated with the transformation of sapwood into heartwood is the death of the living ray parenchyma cells in the sapwood. Simultaneously, several other changes usually occur, namely deposition of polyphenols,

Ref. p. 75]          2.4 Distribution of Chemical Constituents in Wood          73

aspiration of the bordered pits of softwoods, encrustation of pit membranes, formation of tyloses in hardwoods, and depletion of stored starch (HILLIS, 1962; KRAHMER and CÔTÉ, 1963). Often, the acidity of the wood increases toward the center of the tree because of a gradual hydrolysis of acetyl groups with concomitant formation of acetic acid (STEWART et al., 1961). The total amount of polyphenols often increases from the center of the tree towards the sapwood-heartwood boundary. The inner portion of the heartwood tends to contain more high-molecular-weight polyphenols than the outer. In wood from larches, considerable quantities of an arabinogalactan are deposited in the heartwood (CÔTÉ and TIMELL, 1964), especially in the ray cells and in the epithelial cells lining the resin canals (ZAITSEVA, 1959). It has been claimed (THORNBER and NORTHCOTE, 1961, 1962) that the sapwood-heartwood transition involves formation of glucose and mannose residues in softwoods and xylose in hardwoods.

The chemical composition of cambial tissue in differentiating xylem and phloem has been reported (THORNBER and NORTHCOTE, 1961, 1962). Pectic substances are reported to be lost during development of a cambial cell. Secondary thickening is accompanied by formation of cellulose and hemicelluloses. The xylan in a differentiated cell contained more acid side chains than xylan from the cambial region.

It has long been known that ray cells have a higher pentosan content than the total wood. PERILÄ and his co-workers have studied the carbohydrate composition of ray cells and prosenchyma cells in *Pinus sylvestris*, *Picea abies*, and *Betula verrucosa* (PERILÄ, 1961, 1962; PERILÄ and HEITTO, 1959; PERILÄ and SEPPÄ, 1960). Because of their lesser size, the ray cells could be separated from the other cells of the wood by screening of the defibered holocellulose. Although their total hemicellulose content was the same, both the pine and the spruce ray cells contained more xylose and less mannose residues than the tracheids. In birch wood, the ray cells contained as much as 77% xylose units compared to only 36% for the longitudinal elements. All xylose originated from a 4-$O$-methyl-glucuronoxylan.

As long ago as 1936, BAILEY succeeded in isolating and analyzing the middle lamella in Douglas-fir wood and found that it contained 70% lignin and 14% pentosan. LANGE (1954) used a microspectrographic technique, obtaining a similar result. However, when ASUNMAA and LANGE (1954) attempted to determine the location of the cellulose and hemicelluloses, difficulties were encountered.

The distribution of lignin over the cell wall is now fairly well known, due in part to the increased use of electron microscopy. Ultra-thin sections of wood from which the carbohydrates have been previously removed by either acid hydrolysis (SACHS et al., 1963) or by the use of suitable microorganisms (MEIER, 1955) reveal that a larger portion of the lignin is located in the secondary wall in the softwoods than in the hardwoods. The lignin network in the $S_2$-layer seems to be more loosely organized in the hardwoods. The $S_3$-layer appears to be more heavily lignified than the $S_2$.

During the early stages of the growing season, cell wall division is rapid, and the resulting springwood cells have thin walls compared to the more compact summerwood elements. Since the middle lamella remains of uniform thickness throughout a growth ring, the fibers in the springwood contain more lignin, proportionately, than those in the summerwood. The G-layer in tension wood is free of lignin (WARDROP and DADSWELL, 1948, 1955). The "extra" lignin formed in compression wood is deposited as a separate layer between $S_1$ and $S_2$ (WARDROP and DADSWELL, 1950, 1952). Often the middle lamella does not completely surround the tracheids in compression wood, intercellular spaces being left at the junction of four cells due to a change from the normal square or polygonal configuration

74                            2. Chemical Composition of Wood                    [Ref. p. 75

of the tracheids to a rounded shape as viewed in cross-section. The lignin in compression wood, but not in tension wood, differs chemically from normal lignin to some degree (BLAND, 1958, 1961; CROON, 1961).

The orientation of the cellulose microfibrils is quite different in the various layers of the cell wall, as was brought out in Chapter I. There are apparently no microfibrils in the middle lamella. In the primary wall, the microfibrils form a thin, random network in a matrix which consists largely of lignin, pectin, fats and waxes. The $S_1$-layer may have a crossed fibrillar structure, with the microfibrils arranged in two helices which form an angle of about 50° with the fiber axis. The typical feature of the thick $S_2$-layer is that all microfibrils run in the same direction and nearly parallel to the fiber axis. In the $S_3$-layer, finally, the cellulose microfibrils are deposited in a flat helix, but they have a tendency to criss-cross each

Table 2.7. *Relative Percentages of Polysaccharides in the Different Layers of the Cell Wall* (MEIER and WILKIE, 1959; MEIER, 1961 b, 1962 a)

| Polysaccharide | $M + P^a$ | $S_1$ | $S_2$ outer | $S_2$ inner + $S_3$ |
|---|---|---|---|---|
| *Betula verrucosa* | | | | |
| Silver birch | | | | |
| Galactan | 16.9 | 1.2 | 0.7 | Nil |
| Cellulose | 41.4 | 49.8 | 48.0 | 60.0 |
| Glucomannan | 3.1 | 2.8 | 2.1 | 5.1 |
| Arabinan | 13.4 | 1.9 | 1.5 | Nil |
| Glucuronoxylan | 25.2 | 44.1 | 47.7 | 35.1 |
| *Picea abies* | | | | |
| Norway spruce | | | | |
| Galactan | 16.4 | 8.0 | Nil | Nil |
| Cellulose | 33.4 | 55.2 | 64.3 | 63.6 |
| Glucomannan | 7.9 | 18.1 | 24.4 | 23.7 |
| Arabinan | 29.3 | 1.1 | 0.8 | Nil |
| Arabinoglucuronoxylan | 13.0 | 17.6 | 10.7 | 12.7 |
| *Pinus sylvestris* | | | | |
| Scots pine | | | | |
| Galactan | 20.1 | 5.2 | 1.6 | 3.2 |
| Cellulose | 35.5 | 61.5 | 66.5 | 47.5 |
| Glucomannan | 7.7 | 16.9 | 24.6 | 27.2 |
| Arabinan | 29.4 | 0.6 | Nil | 2.4 |
| Arabinoglucuronoxylan | 7.3 | 15.7 | 7.4 | 19.4 |

[a] Contains also a high percentage of pectic acid.

other on the surface (at the lumen lining). The cellulose in the tension wood G-layer is organized in microfibrils which are oriented almost parallel to the fiber axis and which appear to be highly oriented (CÔTÉ and DAY, 1962). Amorphous lignin surrounds the cellulose microfibrils in normal cell wall layers, but it probably does not penetrate the strands to any great extent.

Using an ingenious summative technique, MEIER and WILKIE (1959) and MEIER (1961 b, 1962 a), were able to determine the distribution of the polysaccharides over the cell wall of three woods. Fibers in various stages of development were isolated and analyzed. From the analytical data, the relative weight of the various cell wall layers in the mature fiber, and from the known composition of the major wood polysaccharides, the data summarized in Table 2.7 were obtained.

As might be expected from earlier results, the middle lamella and the primary wall were both found to be rich in pectic material. The concentration of cellulose was highest in the inner portion of the secondary wall in the birch, but in the outer region in the pine and spruce. The content of arabino-glucurono-xylan was quite high in the $S_3$-layer of the softwoods in accordance with previous observations (MEIER and YLLNER, 1956; BUCHER, 1960).

The relative amount of glucomannan showed a gradual increase from the outer to the inner portions of the cell wall, reaching its highest value in the central part of $S_2$. This layer constitutes a larger portion of the xylem in summerwood than in springwood. Summerwood could thus be expected to contain more glucomannan and less xylan than springwood.

Meier's technique has furnished very valuable information which could hardly have been obtained in any other way. A direct isolation and analysis of each cell wall layer, although perhaps not entirely impossible, would be a most difficult task.

## Literature Cited

ADAMS, M. F. and DOUGLAS, C., (1963) Arabinogalactan — A review of the literature. Tappi 46: 544—548.

ALEXANDER, W. J. and MITCHELL, R. L., (1949) Rapid measurement of cellulose viscosity by the nitration method. Anal. Chem. 21: 1497—1500.

ADLER, E., (1957) Newer views on lignin formation. Tappi 40: 294—301.

—, (1959) Chinoide Strukturen und Benzolalkoholgruppierungen in der Chemie und Biochemie des Lignins. In Kratzl, K. and BILLEK, G., 137—153.

—, (1961) Über den Stand der Ligninforschung. Papier 15: 604—609.

ASPINAL, G. O. and ROSS, K. M., (1963) The degradation of two periodate-oxidized arabinoxylans. J. Chem. Soc. 1681—1686.

ASUNMAA, S. and LANGE, P. W., (1954) The distribution of "cellulose" and "hemicellulose" in the cell wall of spruce, birch and cotton. Svensk Papperstid. 57: 501—516.

BAILEY, A. J., (1936) Lignin in Douglas-fir. Composition of the middle lamella. Ind. Eng. Chem. Anal. Ed. 8: 52—55.

—, (1936) Lignin in Douglas-fir. The pentosan content of the middle lamella. Ind. Eng. Chem. Anal. Ed. 8: 389—391.

BJÖRKMAN, A., (1956) Studies on finely divided wood. Part I. Extraction of lignin with neutral solvents. Svensk Papperstid. 59: 477—485.

—, (1957) Studies on finely divided wood. Part 2. The properties of lignins extracted with neutral solvents from softwoods and hardwoods. Svensk Papperstid. 60: 158—169.

—, (1957) Studies on finely divided wood. Part 3. Extraction of lignin-carbohydrate complexes with neutral solvents. Svensk Papperstid. 60: 243—251.

—, (1957) Studies on finely divided wood. Part 5. The effect of milling. Svensk Papperstid. 60: 329—335.

—, (1957) Lignin and lignin-carbohydrate complexes — Extraction from wood meal with neutral solvents. Ind. Eng. Chem. 49: 1395—1398.

— and PERSON, B., (1957) Studies on finely divided wood. Part 4. Some reactions of the lignin extracted by neutral solvents from *Picea abies*. Svensk Papperstid. 60: 285—292.

BLAND, D. E., (1958) The chemistry of reaction wood. Part I. The lignins of *Eucalyptus goniocalyx* and *Pinus radiata*. Holzforschung 12: 36—43.

—, (1961) The chemistry of reaction wood. Part 3. The milled wood lignins of *Eucalyptus goniocalyx* and *Pinus radiata*. Holzforschung 15: 102—106.

BOUVENG, H. O., (1961) Studies on some wood polysaccharides. Svensk Kem. Tidskr. 73: 115—131.

— and LINDBERG, B. G., (1960) Methods in structural polysaccharide chemistry. Advan. Carbohydrate Chem. 15: 53—89.

— and MEIER, H., (1959) Studies on a galactan from Norwegian spruce compression wood (*Picea abies* Karst.). Acta Chem. Scand. 13: 1884—1889.

BRAUNS, F. E., (1939) Native lignin I. Its isolation and methylation. J. Am. Chem. Soc. 61: 2120—2127.

— and BRAUNS, D. A., (1960) The Chemistry of Lignin. Academic Press, New York, N. Y.

BROWNING, B. L., Editor (1963) The Chemistry of Wood. John Wiley and Sons — Interscience Publishers, New York, N. Y.

BUCHER, H., (1960) Zur Topochemie des Holzaufschlusses. Papier 14: 542—549.

COLVIN, J. R., (1963) The size of the cellulose microfibril. J. Cell Biol. 17: 105—109.

# Literature Cited

CÔTÉ, W. A., Jr. and DAY, A. C., (1962) The G-layer in gelatinous fibers — electron microscopic studies. Forest Prod. J. XII (7): 333—339.
— and TIMELL, T. E., (1964) Unpublished results.
CROON, I., (1961) Tryckved och dragved, morfologi och kemisk sammansättning. Svensk Papperstid. **64**: 175—180.
—, LINDBERG, B. and MEIER, H., (1959) Structure of a glucomannan from *Pinus silvestris* L. Acta Chem. Scand. **13**: 1299—1304.
ELBEIN, A. D., BARBER, G. A. and HASSID, W. Z., (1964) The synthesis of cellulose by an enzyme system from a higher plant. J. Am. Chem. Soc. **86**: 309—310.
ENTWISTLE, D., COLE, E. H. and WOODING, N. S., (1949) The autoxidation of alkali cellulose I. An experimental study of the kinetics of the reaction. Textile Res. J. **19**: 527—546.
ERDTMAN, H., (1959) Conifer chemistry and taxonomy of conifers. In KRATZL, K. and BILLEK, G., 1—28.
FREUDENBERG, K., (1959) Biochemische Vorgänge bei der Holzbildung. In KRATZL, K. and BILLEK, G., 121—136.
—, (1960) Principles of lignin growth. J. Polymer Sci. **48**: 371—377.
—, (1962) Biogenesis and constitution of lignin. Pure Appl. Chem. **5**: 9—20.
FREY-WYSSLING, A., (1959) Die Pflanzliche Zellwand. Springer-Verlag, Berlin.
GARDNER, J. A. F., (1962) The tropolones. In HILLIS, W. E., 317—330.
GIERER, J., (1958) Über die Isolierung, Struktur und Biosynthese des Lignins. Holz Rohwerkstoff **16**: 251—262.
GOLDSCHMID, H. R. and PERLIN, A. S., (1963) Interbranch sequences in the wheat arabinoxylan. Can. J. Chem. **41**: 2272—2277.
GORING, D. A. I., (1962) The physical chemistry of lignin. Pure Appl. Chem. **5**: 233—254.
— and TIMELL, T. E., (1962) Molecular weight of native celluloses. Tappi **45**: 454—460.
GREATHOUSE, G. A., (1959) On the enzymic polysaccharide synthesis. In KRATZL, K. and BILLEK, G., 76—81.
GREENWOOD, C. T., (1956) Aspects of the physical chemistry of starch. Advan. Carbohydrate Chem. **11**: 335—393.
GUSTAFSSON, C., OLLIMAA, P. J. and SAARNIO, J., (1952) The carbohydrates in birch wood. Acta Chem. Scand. **6**: 1299—1300.
HÄGGLUND, E., (1951) Chemistry of Wood. Academic Press, New York, N. Y.
HAMILTON, J. K., PARTLOW, E. V. and THOMPSON, N. S., (1960) The nature of a galactoglucomannan associated with wood cellulose from southern pine. J. Am. Chem. Soc. **82**: 451—457.
— and QUIMBY, G. R., (1957) The extractive power of lithium, sodium, and potassium hydroxide solutions for the hemicelluloses associated with wood cellulose and holocellulose from western hemlock. Tappi **40**: 781—786.
HARRIS, G. C., (1952) Wood resins. In Wise, L. E. and JAHN, E. C., 590—617.
HATHWAY, D. E., (1962) The condensed tannins. In HILLIS, W. E., 191—228.
—, (1962) The lignins. In HILLIS, W. E., 159—190.
HENLEY, D., (1960) The cellulose solvent cadoxen, a preparation, and a viscometric relationship with cupriethylenediamine. Svensk Papperstid. **60**: 143—146.
HERMANS, P. H., (1949) Physics and Chemistry of Cellulose Fibres. Elsevier Publishing Co., New York, N. Y.
—, (1951) X-ray investigations on the crystallinity of cellulose. Makromol. Chem. **6**: 25—29.
HILLIS, W. E., Editor (1962) Wood Extractives. Academic Press, New York, N. Y.
—, (1962) The distribution and formation of polyphenols within the tree. In HILLIS, W. E., 59—131.
HIRST, E. L. and JONES, J. K. N., (1946) The chemistry of pectic materials. Advan. Carbohydrate Chem. **2**: 235—251.
HONEYMAN, J., Editor (1959) Recent Advances in the Chemistry of Cellulose and Starch. Heywood and Company, London.
HOWSMON, J. A., (1949) Water sorption and the polyphase structure of cellulose. Textile Res. J. **19**: 152—162.
JONES, J. K. N., WISE, L. E. and JAPPE, J., (1956) The action of alkali containing metaborates on wood cellulose. Tappi **39**: 139—141.
JORGENSEN, L., (1950) Studies on the Partial Hydrolysis of Cellulose. Emil Moestue A/S, Oslo.
JURD, L., (1962) The hydrolyzable tannins. In HILLIS, W. E., 229—260.
KRAHMER, R. L. and CÔTÉ, W. A., Jr., (1963) Changes in coniferous wood cells associated with heartwood formation. Tappi **46**: 42—49.
KRATZL, K., (1959) Biochemie des Holzes. Ein Bericht über Symposium II. In KRATZL, K. and BILLEK, G., 247—285.
—, (1960) On the biosynthesis of gymnospermae and angiospermae lignins. Tappi **43**: 650—653.
—, (1961) Zur Biogenese des Lignins. Holz Roh- Werkstoff **19**: 219—232.

## Literature Cited

KRATZL, K., and BILLEK, G., (1957) Synthesis and testing of lignin precursors. Tappi **40**: 269—285.

—, Editors (1959) Biochemistry of Wood. Pergamon Press, London.

KURTH, E. F., (1952) The volatile oils. In WISE, L. E. and JAHN, E. C., 548—589.

LANGE, P. W., (1954) The distribution of lignin in the cell wall of normal and reaction wood from spruce and a few hardwoods. Svensk Papperstid. **57**: 525—532.

LIANG, C. Y., BASSETT, K. H., McGINNES, E. A. and MARCHESSAULT, R. H., (1960) Infrared spectra of crystalline polysaccharides VII. Thin wood sections. Tappi **43**: 1017—1024.

LINDBERG, B., (1957) Die Chemie der übrigen Wandsubstanzen. In TREIBER, E., 386—397.

LINDGREN, B. O., (1958) The lignin-carbohydrate linkage. Acta Chem. Scand. **12**: 447—452.

MANLEY, R. ST. J., (1963) Growth and morphology of single crystals of cellulose triacetate. J. Polymer Sci. Part A **1**: 1875—1892.

—, (1963) Hydrolylis of cellulose triacetate crystals. J. Polymer Sci. Part A **1**: 1893—1899.

MANN, J. and MARRINAN, H. J., (1956) Reaction between cellulose and heavy water I. Qualitative study by infrared spectroscopy. Trans. Faraday Soc. **52**: 481—487. II. Measurement of absolute accessibility and crystallinity. Trans. Faraday Soc. **52**: 487—492. III. Quantitative study by infrared spectroscopy. Trans. Faraday Soc. **52**: 492—497.

MANNERS, D. J., (1962) Enzymic synthesis and degradation of starch and glycogen. Advan. Carbohydrate Chem. **17**: 371—430.

MARCHESSAULT, R. H. and LIANG, C. Y.. (1962) The infrared spectra of crystalline polysaccharides VIII. Xylans. J. Polymer Sci. **59**: 357—378.

—, MOREHEAD, F. F., WALTERS, N. M., GLAUDEMANS, C. P. J., and TIMELL, T. E., (1961) Morphology of xylan single crystals. J. Polymer Sci. **51**: S 66—S 68.

MARK, H., (1940) Intermicellar hole and tube system in fiber structure. J. Phys. Chem. **44**: 764—788.

MEIER, H., (1955) Über den Zellwandabbau durch Holzvermorschungspilze und die submikroskopische Struktur von Fichtentracheiden und Birkenholzfasern. Holz Roh- Werkstoff **13**: 323—338.

—, (1958) On the structure of cell walls and cell wall mannans from ivory nuts and from dates. Biochim. Biophys. Acta **28**: 229—240.

—, (1961 a) Isolation and characterisation of an acetylated glucomannan from pine (*Pinus silvestris* L.). Acta Chem. Scand. **15**: 1381—1385.

—, (1961 b) The distribution of polysaccharides in wood fibers. J. Polymer Sci. **51**: 11—18.

—, (1962 a) Chemical and morphological aspects of the fine structure of wood. Pure App. Chem. **5**: 37—52.

—, (1962 b) Studies on a galactan from tension wood of beech (*Fagus silvatica* L.) Acta Chem. Scand. **16**: 2275—2283.

— and WILKIE, K. C. B., (1959) The distribution of polysaccharides in the cell wall of tracheids of pine (*Pinus silvestris* L.) Holzforschung **13**: 177—182.

— and YLLNER, S., (1956) Die Tertiarwand in Fichtenzellstoff-Tracheiden. Svensk Papperstid. **59**: 395—401.

MEREWETHER, J. W. T., (1957) Lignin-carbohydrate complex in wood. Holzforschung **11**: 65—80.

MEYER, K. H. and MARK, H., (1929) Über den Bau des kristallisierten Anteils der Cellulose II. Z. physik. Chem. B **2**: 115—145.

— and MISCH, L., (1937) Position des atomes dans le nouveau modèle spatial de la cellulose. Helv. Chim. Acta **20**: 232—244.

MÜHLETHALER, K., (1960) Die Feinstruktur der Zellulosemikrofibrillen. Beih. Zeit. Schweiz. Forstv. **30**: 55—65.

—, (1963) Feinstruktur der Cellulosefaser. Papier **17**: 546—550.

MUTTON, D. B., (1962) Wood resins. In HILLIS, W. E., 331—363.

NEISH, A. C., (1959) Biosynthesis of hemicelluloses. In KRATZL, K. and BILLEK, G., 82—91.

—, (1960) Biosynthetic pathways of aromatic compounds. Ann. Rev. Plant Physiol. **11**: 55—80.

NELSON, R. and SCHUERCH, C., (1956) Factors influencing the removal of pentosan from birch wood: evidence on the lignin-carbohydrate bond. J. Polymer Sci. **22**: 435—448.

— and —, (1957) The extraction of pentosans from woody tissues II. Tappi **40**: 419—426.

NEUFELD, E. F. and HASSID, W. Z., (1963) Biosynthesis of saccharides from glycopyranosyl esters of nucleotides ("Sugar nucleotides"). Advan. Carbohydrate Chem. **18**: 309—356.

NORD, F. F. and SCHUBERT, W. J., (1959) Lignification. In KRATZL, K. and BILLEK, G., 189—206.

NORTHCOTE, D. H., (1958) The cell walls of higher plants: their composition, structure and growth. Biol Reviews **33**: 53—102.

OTT, E., SPURLIN, H. M. and GRAFFLIN, M. W., Editors (1954) Cellulose and Cellulose Derivatives. Interscience Publishers, New York, N. Y.

# Literature Cited

PERILÄ, O., (1961) The chemical composition of carbohydrates of wood cells. J. Polymer Sci. **51**: 19—26.

—, (1962) The chemical composition of wood cells. III. Carbohydrates of birch cells. Suomen Kemistilehti B **35**: 176—178.

— and HEITTO, P., (1959) The chemical composition of wood cells I. Carbohydrates of pine cells. Soumen Kemistilehti B **32**: 76—80.

— and SEPPÄ, T., (1960) The chemical composition of wood cells II. Carbohydrates of spruce cells. Suomen Kemistilehte B **33**: 114—116.

PINDER, A. R., (1960) Chemistry of the Terpenes. JOHN WILEY and Sons. New York, N.Y.

PRESTON, R. D. and FREI, E., (1961) Variants in the structural polysaccharides of algal cell walls. Nature **192**: 939—943.

REIMER, H., (1962) Ein vereinfachtes Verfahren zur Herstellung von Triäthylendiamin-Cadmium (II)-hydroxid. Papier **16**: 566—568.

ROELOFSEN, P. A., (1959) The Plant Cell Wall. Gebrüder Borntraeger, Berlin-Nikolassee.

SACHS, I. B., CLARK, I. T. and PEW, J. C., (1963) Investigation of lignin distribution in the cell wall of certain woods. J. Polymer Sci., Part C, No. **2**: 203—212.

SCHWERIN, G., (1958) The chemistry of reaction wood II. The polysaccharides of *Eucalyptus goniocalyx* and *Pinus radiata*. Holzforschung **12**: 43—48.

SEPALL, O. and MASON, S. G., (1961) Hydrogen exchange between cellulose and water I. Measurement and accessibility. Can. J. Chem. **39**: 1934—1943. II.Interconversion of accessible and inaccessible regions. Can. J. Chem. **39**: 1944—1955.

SIMONSEN, J. L., (1947) The Terpenes. Cambridge University Press, 1947—1957.

STEWART, C. M., KOTTEK, J. F., DADSWELL, H. E. and WATSON, A. J., (1961) The process of fiber separation III. Hydrolytic degradation within living trees and its effect on the mechanical pulping and other properties of wood. Tappi **44**: 798—813.

TAPPI, (1954) Holocellulose in Wood. Standard Method T 9 m-54.

THORNBER, J. P. and NORTHCOTE, D. H., (1961) Changes in the chemical composition of a cambial cell during its differentiation into xylem and phloem tissue in trees 1. Main components. Biochem. J. **81**: 449—455.

— and —, (1962) 2. Carbohydrate constituents of each main component. Biochem. J. **81**: 455—464. 3. Xylan, glucomannan and α-cellulose fractions. Biochem. J. **82**: 340—346.

TIMELL, T. E., (1957) Molecular properties of seven wood celluloses. Tappi **40**: 25—29.

—, (1960) Isolation of hardwood glucomannans. Svensk Papperstid. **63**: 472—476.

—, (1960) Isolation and properties of a glucomannan from the wood of white birch (*Betula papyrifera* Marsh.) Tappi **43**: 844—888.

—, (1961) Isolation of galactoglucomannans from the wood of gymnosperms. Tappi **44**: 88—96.

—, (1964, 1965) Wood Hemicelluloses. Advan. Carbohydrate Chem. Vols. 19 and 20.

TREIBER, E., Editor (1957) Die Chemie der Pflanzenzellwand. Springer-Verlag, Berlin.

WALD, W. J., RITCHIE, P. F. and PURVES, C. B., (1947) The elementary composition of lignin in northern pine and black spruce woods, and of the isolated Klason and periodate lignins. J. Am. Chem. Soc. **69**: 1371—1377.

WARDROP, A. B. and DADSWELL, H. E., (1948) The nature of reaction wood. I. The structure and properties of tension wood fibres. Australian J. Sci. Res. B **1**: 3—16.

— and —, (1950) The nature of reaction wood. II. The cell wall organization of compression wood tracheids. Australian J. Sci. Res. B **3**: 1—13.

— and —, (1952) The nature of reaction wood. III. Cell division and cell wall formation in conifer stems. Australian J. Sci. Res. B **5**: 385—398.

— and —, (1955) The nature of reaction wood. IV. Variations in cell wall organization of tension wood fibres. Australian J. Bot. **3**: 177—189.

WHELAN, W. J., (1959) The enzymic synthesis and degradation of cellulose and starch. In Honeyman, J., 307—336.

—, (1963) Recent advances in starch metabolism. Stärke **15**: 247—251.

WHISTLER, R. L. and BeMILLER, J. N., (1958) Alkaline degradation of polysaccharides. Advan. Carbohydrate Chem. **13**: 289—329.

— and SMART, C. L., (1953) Polysaccharide chemistry. Academic Press, Inc., New York, N.Y. pp. 161—197.

WISE, L. E. and JAHN, E. C., (1952) Wood Chemistry. Reinhold Publ. Corp., New York, N.Y.

—, MURPHY, M. and D'ADDIECO, A. A., (1946) Chlorite holocellulose, its fractionation and bearing on summative wood analysis and on studies on the hemicelluloses. Paper Trade J. 122 (**2**): 35—43.

ZAÏTSEVA, A. F., (1959) The localization of arabinogalactan in the cell walls of *Larix dahurica*. Trudy Inst. Lesa, Aka. Nauk S. S. S. R., Izuchenie Khim. Sostava Drevesiny Daursk Listvennitsy **45**: 50—60. (Chem. Abstracts **53**: 4439 (1959).)

ZIMMERMANN, M. H., Editor (1964) The Formation of Wood in Forest Trees. Academic Press, New York, N.Y.

# 3. DEFECTS AND ABNORMALITIES OF WOOD

## 3.0 Introduction

From the point of view of economics, a defect in wood is any feature that lowers its value on the market. It may be an abnormality that decreases the strength of the wood or a characteristic that limits its use for a particular purpose. There is a certain amount of risk involved in classifying an abnormality as a defect because what is judged to be definitely unsuitable for one application may prove to be ideal for a different or special use.

Good examples of this are the popularity of "knotty pine", "pecky cypress", "wormy chestnut" and similar woods containing natural growth defects or evidence of degrade from fungal or insect attack. Basically, wood of this kind is undesirable for applications where strength or soundness are required, but its novelty occasionally creates a demand in certain localized areas or for a limited period of years.

Many so-called defects are not abnormalities in a strict sense, but are simply the product of natural growth. Variations from the normal form of a tree trunk, knots, and reaction wood are natural defects whose formation man can control only within certain limits. Closely allied to these, but directly attributable to environmental factors of wind, heavy snow load, severe cold, heat or lightning, are other natural defects which the forest manager can do little about. Section 3.1 of this chapter is devoted to the nature and causes of natural defects.

The second part of the chapter is concerned with defects that arise during the conversion of logs into lumber, as a result of seasoning procedures or through faulty machining technique.

Other common sources of defect are insects and fungi which can attack living trees, logs, lumber or processed wood products. Biological deterioration of wood from these causes is taken up in Chapter 4.

## 3.1 Natural Defects

### 3.1.1 Knots

Undoubtedly the most common natural defect is one over which the forester has limited control: knots. A knot is a branch that is included in the wood of a tree stem by growth around its base. As long as the branch is living, the cambium of the stem and branch are continuous and the resulting knot is *intergrown* or *tight*. Once the branch has died, the continuity of the cambium is interrupted and the knot produced is *encased* or *loose*.

The appearance of a knot in a piece of lumber depends upon the direction of cut through the included branch. When the cut is made along the axis of the original branch, the knot appears as a spike knot on the surface of the wood (Fig. 3.1). When a transverse cut is made through the branch, the knot appears round or oval (Fig. 3.2).

If a branch is pruned naturally after dying, or if a living branch is artificially pruned, the branch end is grown over and any wood produced subsequently in this area of the stem will be knot free. When the lower branches are pruned early in the life of a tree, there is a larger proportion of clear wood produced than if they are allowed to remain for a longer period (Fig. 3.3). Smaller knots also result from this approach to forest management since branch diameter increases as stem diameter increases. Some tree species produce small knots and relatively larger amounts of clear wood because of inherent growth characteristics or because of the nature of the stands in which they grow (Fig. 3.4).

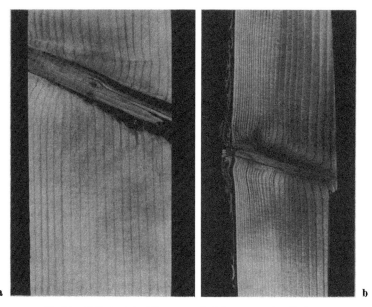

Fig. 3.1a. Spike knot in *Abies* sp., partly intergrown and partly loose,
Fig. 3.1b. Spike knot in Douglas-fir, *Pseudotsuga menziesii* (Mirb.) Franco, partly intergrown and partly loose

For structural purposes there is no question but that knots lower the grade of lumber because the grain is distorted or deflected around knots leaving local areas of cross grain which affect mechanical properties. The location or distribution of the knots is a critical consideration which must be related to the limiting strength property in a specific application (GRAF, 1938; SIIMES, 1944). Chapter 7 contains a comprehensive coverage of this topic.

The effect of knots on lumber grade depends upon their size, number, whether encased or intergrown, and whether the knots are decayed or sound. Grading rules or standards are established by lumber manufacturers or national industrial standards associations depending on the practice in the country concerned. Knots are the principal defect considered in grading, but other defects taken up in this chapter are also included.

The market value of lumber is, of course, related to lumber grade, clear wood bringing the highest price. Lumber containing large and numerous knots is normally graded quite low. However, as mentioned earlier, "knotty pine" has a particularly high value for its use in paneling where its decorative character influences this architectural application. In Switzerland and Tyrolia this wood is produced by trees of *Pinus cembra* L. and in northeastern United States by *Pinus strobus* L.

### 3.1.2 Reaction Wood

Reaction wood is the name applied to woody tissues produced in certain parts of leaning trees and at the upper or lower sides of branches. The evidence indicates that this type of wood develops as a reaction to lean or to the influence of prevailing wind forces. It has also been demonstrated that forcing a tree out of its

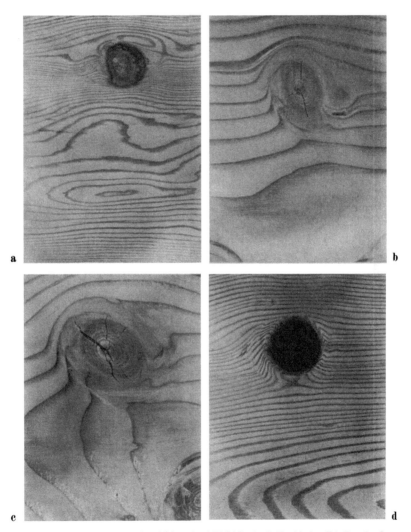

Fig. 3.2a—d. a) Loose knot, b) Tight knot, c) Tight knot, d) Knot hole, all in Douglas-fir

normal growth pattern initiates reaction wood formation (SINNOTT, 1952). The factor of lean in coniferous stems is one which is of actual proven statistical significance in the development of compression wood in plantation-grown *Pinus resinosa* Ait. (CORE, 1962).

Compression wood is the reaction wood of conifers. It develops on the underside of leaning trees and branches and on the leeward side of trees frequently bent by wind forces. In hardwood trees reaction wood is called tension wood and it forms on the upper side of leaning stems. In both cases, stems usually show eccentric growth on the transverse section (HARTMANN, 1932) (Fig. 3.5).

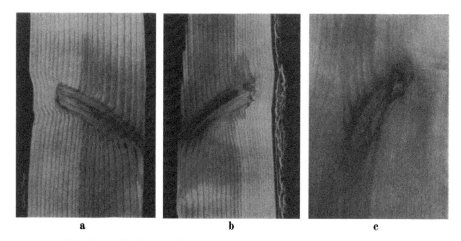

Fig. 3.3a. Artificially pruned branch, now a spike knot grown over, in Douglas-fir
Fig. 3.3b. Spike knot from naturally pruned branch, grown over, in Douglas-fir
Fig. 3.3c. Decayed knot (spike), grown over, in western paper birch, *Betula papyrifera* var. *commutata* (Reg.) Fern

Fig. 3.4. Pin knots in Parana pine, *Araucaria angustifolia* (Bert.) O. KTZE

The frequency of occurrence of reaction wood has undoubtedly been underestimated in spite of its obvious abnormal behavior in seasoning and machining. Perhaps this is due to the fact that only *pronounced* reaction wood is readily distinguishable from normal wood unless some special effort is made to recognize mild cases by learning the characteristic deviations from normal wood. In general, reaction wood is not classed as a defect in lumber grading unless it is present in sufficiently large areas to cause degrade.

The gross physical characteristics, anatomical and microscopic features, ultrastructure and chemical composition of tension wood and compression wood are summarized in Chapter 1, Table 1.1, where a direct comparison of these abnormal woods can be made. The anatomical and ultrastructural features of reaction wood are discussed in Chapter 1, Sections 1.4.1 and 1.4.2. The chemical composition of reaction wood and its deviation from the normal can be found in Chapter 2. Only the general physical characteristics of reaction wood as they relate to defects are discussed below.

**3.1.2.1 Compression Wood.** Reaction wood in conifers is recognizable in logs through the presence of eccentric growth rings and a more gradual transition between springwood and summerwood than in normal wood, made apparent because of lower contrast between these zones. Another obvious characteristic is the non-lustrous, lifeless appearance of this wood compared with adjacent normal wood. Compression wood also has a considerably higher density than normal wood, but it is nevertheless lower in impact and tensile strength, on a density basis. When compression wood breaks, there is a characteristic lack of splintering that usually accompanies normal wood failure.

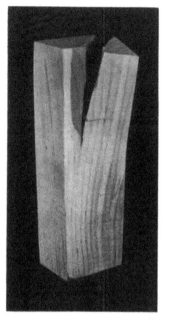

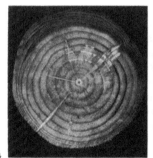 

a b

Fig. 3.5a. Compression wood (dark area in eccentric portion of stem cross section) in plantation grown red pine, *Pinus resinosa* Ait. (Material supplied by H. A. CORE)
Fig. 3.5b. Tension wood (eccentric portion of stem cross section) in basswood, *Tilia americana* L

Fig. 3.6. Differential longitudinal shrinkage in eastern white pine, *Pinus strobus* L., containing compression wood

Upon seasoning, compression wood zones generally shrink more longitudinally and less transversely than normal wood. Longitudinal shrinkage in normal wood is practically negligible (0.1···0.2%) while it has been known to reach 6 or 7% in compression wood. Consequently, where bands of compression wood occur adjacent to normal rings, there is a great tendency for bending, twisting and splitting due to differential shrinkage (Fig. 3.6) (TRENDELENBURG, 1932; PILLOW and LUXFORD, 1937).

**3.1.2.2 Tension Wood.** As mentioned earlier, tension wood occurs on the upper side of leaning trees and branches of angiosperms. The presence of tension wood can be indicated by eccentric growth as seen on the end of a log, wider rings occurring in the region containing tension wood (Fig. 3.5). In dressed lumber, tension wood stands out because it is more reflective than adjacent normal tissue. This silvery sheen may be difficult to detect in sapwood of certain species unless the angle of reflected light is suitable. In some Australian species, tension wood zones are darker than normal wood areas (DADSWELL and WARDROP, 1949).

When green logs containing tension wood tissue are sawn into lumber, the surfaces will often be woolly in the tension wood zones (Fig. 3.7). The same condition is found in peeling rotary veneer from logs containing tension wood. When seasoned material is sawn, there is considerably less woolliness. However, the denser tension wood becomes more difficult to cut when dry.

Though not as great as in compression wood, longitudinal shrinkage of tension wood may be as high as 1%, which is considerably greater than the negligible longitudinal shrinkage of normal hardwood. PILLOW (1950) found that in tension wood of mahogany, longitudinal shrinkage was as high as 0.64%. Lumber containing extensive areas of tension wood warps and twists, creating a serious problem in the hardwood industry. Collapse is another industry problem associated with the seasoning of tension wood from the green condition (DADSWELL and WARDROP, 1955). Some species have a greater tendency for collapse than others, but the presence of tension wood aggravates this condition.

Tension wood tested in the green condition is particularly low in tensile strength; when air dried, however, its tensile strength is higher than that of normal wood. KLAUDITZ and STOLLEY (1955) attribute this to increased lateral bonding, G-layer to remainder of the secondary wall, as shrinkage takes place. This can be more readily understood from the discussion on the ultrastructure of tension wood fibers in Chapter 1 and from the chemical composition of reaction wood of hardwoods presented in Chapter 2.

### 3.1.3 Cross Grain

Fig. 3.7. Woolly surface of board of mahogany, *Swietenia mahagoni* Jacq., caused by sawing green log containing tension wood

The term cross grain is applied to any condition in which fiber alignment is not parallel to the long axis of the piece of wood. When this condition is serious enough to effectively reduce the strength of the wood used structurally, it must be considered to be a defect. This occurs naturally only in cases of spiral or diagonal grain although the term cross grain applies to wavy, curly and interlocked grain as well. In these latter instances, the value of the cross grain stock is sometimes substantially greater than straight-grain material due to the demand for unusually figured woods produced from such material. Some of the variations of cross grain produced naturally are illustrated in Fig. 3.8.

Spiral grain lumber is produced from trees in which the fibers grow in a left- or a right-hand helix around the stem. The cause of this phenomenon is, to a large extent, genetically controlled according to NOSKOWIAK (1963) who reviewed this topic from the point of view of detection, measurement, growth patterns and causal factors. The condition can be observed by stripping the bark from the tree or log (Fig. 3.9). It is difficult to adjust for this defect at the sawmill except in cases of mild spiral grain.

Diagonal grain, on the other hand, can result from poor cutting practice at the sawmill. When there is considerable taper in a log and the cut is made parallel to the pith rather than to the bark, the resulting lumber exhibits diagonal grain. This particular defect is related to variations in log form, discussed in the following section.

### 3.1.4 Variations in Log Form

Conventional sawmilling is geared to the production of lumber from straight logs of average form. Logs from trees having abnormal butt swell, crook, sweep or excessive taper can yield lumber of lower grade due primarily to so-called short grain or cross grain defect. depending upon the degree of control that the sawyer has with a particular sawmill.

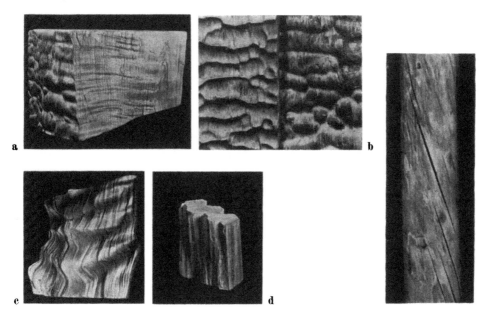

Fig. 3.8a. Wavy (quilted or blistered) grain in Oregon maple, *Acer macrophyllum* Pursh. Fig. 3.9. Spiral grain
Fig. 3.8b. Inner bark (left) and outer portion of tree (right) tracing the origin of wavy (quilted) grain in Oregon maple to cambium
Fig. 3.8c. Wavy grain in American beech, *Fagus grandifolia* Ehrh.
Fig. 3.8d. Interlocked grain in a tropical hardwood species

When logs with butt swell are sawn there is, of course, considerable wood loss in the form of slabs, the bark and outer portion of wood cut from the log. In addition, the wood produced by the tree near its base contains cross grain in varying degrees so that the lumber cut from this portion will contain this defect.

Excessive taper in a log also leads to greater waste than normal and, to cross grain or diagonal grain defect in the lumber unless taper sawing can be done; that is, sawing parallel to the bark. Full length boards of clear, straight-grained lumber are produced from the outside of the log by this method, but tapered lumber is often permitted only in hardwoods. An added advantage of taper sawing is that the waste in logs of excessive taper is confined to the center portion which contains the lowest grade of wood because of pith and many knots (MALCOLM, 1956). Of course, in gang sawmill operations this degree of control is not possible.

Lumber produced from logs exhibiting crook or sweep also contains cross grain. Furthermore, short material and excessive waste result from cutting logs having such abnormal form.

### 3.1.5 Shake

Shake is a defect which is classed as natural because it occurs in the standing tree. The splitting and complete separation of portions of the stem occur as a result of differential stresses of various origins. Ruptures which follow the growth rings are called ring or wind shakes (Fig. 3.10) while those that cross the annual rings, extending radially from the pith, are called heart checks, heart shake or rift cracks. The cause of shakes has been attributed to wind, causing the tree trunks to sway; severe frost, with contraction of the outer portion of the stem followed by expansion upon warming; shrinkage of the heartwood; and unbalanced growth stresses. The amount of loss in a log containing shake depends upon the location and size of the ruptures.

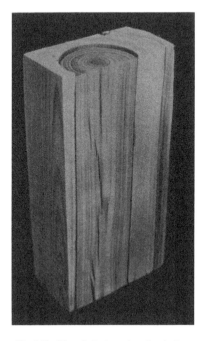

Fig. 3.10. Ring shake in eastern hemlock, *Tsuga canadensis* (L.) Carr

Fig. 3.11a. Frost crack on outside of tree
From KNUCHEL (1947)

Fig. 3.11b. Frost crack as seen in transverse section

### 3.1.6 Miscellaneous Natural Defects

There are many other less important natural defects which arise in the growing tree. These defects may affect trees only in certain localized regions, or they may be limited to a single species. They may also be due to unidentifiable causes.

In areas where very low temperatures are reached, frost injuries in the form of cracks or splits along the grain, or as frost rings following the general pattern of the growth rings, can be of economic importance (Fig. 3.11). Damage of somewhat similar nature can occur in areas of drought or wherever excessive drying occurs (Fig. 3.12).

Pitch pockets and pitch streaks in softwoods can be attributable to insect damage in some instances, but in other cases, cracks or other breaks of unknown origin at the cambium are filled with resin. These pockets of resin are subsequently grown over (Fig. 3.13).

Bark pockets is another defect which is of unknown origin in some cases, but traceable to insects or birds in others. Patches of bark are included in the wood, grown over after damage (Fig. 3.14). When found in western hemlock, *Tsuga heterophylla* (Raf.) Sarg., this defect is called black check.

[Fig. 3.12a. Drying cracks on outside of tree trunk
From KNUCHEL (1947)

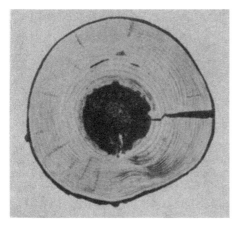

Fig. 3.12b. Drying crack as seen in transverse section     Fig. 3.13. Pitch pocket in wood of Douglas-fir

Mineral streak or mineral stain causes considerable degrade in hardwood lumber. This defect is reported to be due to an unusual concentration of minerals with the resultant discoloration of the wood. It differs from chemical staining, such as brown kiln stain which develops during the seasoning or storage of certain soft pines. Although it is different from the type of staining that appears to be due to

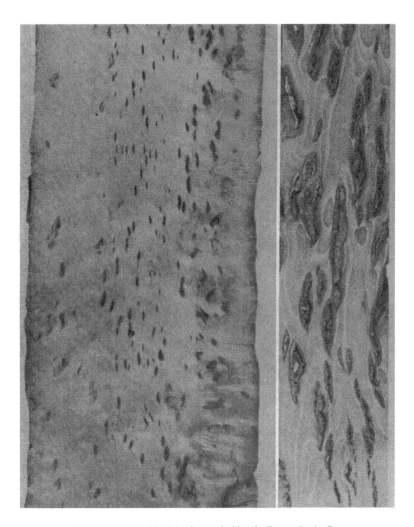

Fig. 3.14. Included bark in the wood of beech, *Fagus silvatica* L.
From KNUCHEL (1947)

the oxidation of cell contents, some mineral streak has been attributed to boring and the resulting chemical action. It is not associated with fungi or other microorganisms (LORENZ, 1944). ROTH (1950) concludes that it is due to oxidation. Mineral streak frequently appears in wood from trees damaged by bird pecks (Fig. 3.15), but other openings into the tree may also be involved (Fig. 3.16). Lumber produced from maple trees that have been tapped for maple syrup production often has mineral streak above and below the bore holes.

Fire or logging injury to the living tree can be the source of defect in lumber. Even superficial wounding is sufficient to permit entry of insects or fungi before callus growth covers the opening (Fig. 3.17).

Fig. 3.15. Mineral streak in hickory, *Carya* sp., caused by bird peck damage to living tree

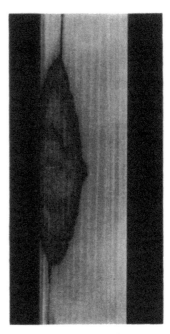

Fig. 3.16. Mineral streak associate with compression failure in basswood *Tilia americana* L.

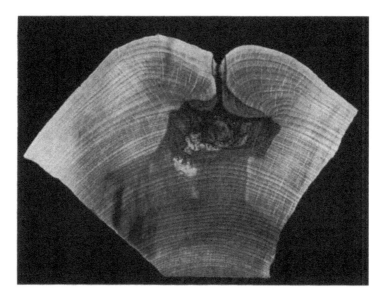

Fig. 3.17. Logging injury and resultant decay in yellow birch, *Betula alleghaniensis* Britton

Compression failures represent one type of defect which can form in the living tree, during felling or logging operations or after the wood is placed in service. The failure zones are regions where the cells have been permanently deformed or fractured by excessive compression forces parallel to the grain. This defect is particularly serious in wood for structural applications where it might break suddenly under relatively small load. Compression failures are occasionally very visible (Fig. 3.18) or it may require high magnification microscopy to see them. Methods for detecting compression failures in wood have been reported by ANDERSON (1944) and LIMBACH (1945).

Fig. 3.18. Compression failure in wood of pine, *Pinus* sp

## 3.2 Defects due to Processing

The defects that arise in the process of converting logs into finished lumber can be attributed to two major phases of the operation: manufacturing and seasoning.

### 3.2.1 Manufacturing Defects

In the sawmilling operation, defects of slope of grain, wane, or excessive variation in dimension may be due to inaccuracy of equipment or poor judgment and lack of skill of the operators.

As mentioned before, should a log exhibiting considerable taper be sawn parallel to the pith rather than to the bark, cross grain defect can result in the lumber.

Thickness variation in rough, green lumber is permitted within certain limits. When it exceeds these limits or when the thickness varies greatly within one piece, this is considered as a defect. In dressed lumber, oversized material may be planed to standard finished dimensions after seasoning, but undersized material may be unsalvageable.

Inaccurate edging can be the source of wane, which is bark or lack of wood from any cause, on the edge or corner of a piece of lumber. This type of defect can generally be eliminated by remanufacture, but at some loss of material.

Defects such as shakes, splits or checks which are exposed when the log is broken down are classed as natural defects and discussed in Section 3.1.

### 3.2.2 Seasoning Defects

The drying or seasoning of green lumber is a critical step in the conversion of logs into finished lumber because it is a potential source of much degrade. The subject of wood-liquid relations is fundamental to a study of this area and is discussed in another chapter, but one point bears repeating. The removal of free water, the water contained in the cell lumens, does not initiate shrinkage. Shrinkage begins when drying proceeds below the fiber saturation point, when the absorbed water is removed from the cell walls. Depending on species, the moisture content of the wood at the fiber saturation point, based on oven-dry weight, ranges between 20% and 35%.

Many of the common seasoning defects can be related to shrinkage and the internal stresses that accompany it. This is largely due to the fact that shrinkage and swelling in the tangential direction is greater than in the radial direction; in some species tangential shrinkage is more than double radial shrinkage. The heterogeneity of wood is further emphasized when longitudinal shrinkage is determined. Wood from the center of the tree will often exhibit high longitudinal shrinkage. Except in such juvenile wood, and in the case of reaction wood, mentioned in Section 3.1.2, shrinkage along this axis is negligible.

There is a difference in longitudinal shrinkage evident between earlywood and latewood from the same growth ring and this undoubtedly contributes to some of the defects discussed in subsequent sections.

**3.2.2.1 Checks.** Checks are longitudinal openings at weak points in the wood occasioned by the stresses produced in differential shrinkage. They usually extend across the growth rings, but may occasionally develop between them, as in shake. However, shake is generally more extensive and develops in the living tree, being exposed when the log is sawn.

End checks are very common seasoning defects which develop because of the more rapid loss of moisture from exposed ends of fibers than from inner portions of the wood. The cells at the ends of boards attempt to shrink in response to their loss of moisture below the fiber saturation point, but they are restrained by the adjoining fibers which have a higher moisture content. This results in tension stresses near the ends of the lumber and compression stresses further in. Wood is weakest in tension across the grain so the end checks develop. End checks tend to close as the wood dries and stresses are relieved, the final end checks being narrower than the initial ones. Painting the ends of green lumber with a moisture-impermeable coating prevents too rapid drying and reduces trim losses (RASMUSSEN, 1961). Further reference to seasoning techniques is made in Chapter 8.

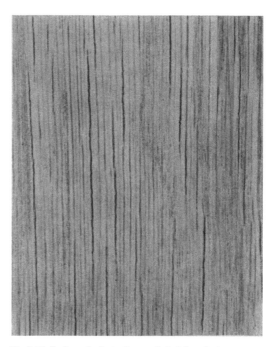

Fig. 3.19. Surface checks in the wood of white oak, *Quercus alba* L. Note that the checks follow the orientation of the rays

Surface checks frequently develop along the rays and are observable on the tangential surfaces (Fig. 3.19). Surface checking is sometimes indicative of a stage of casehardening, a seasoning defect that is discussed below.

The presence of compression wood in lumber can lead to cross checks or tension failures which are ruptures across the grain. They can occur if a band or zone of compression wood is located adjacent to, or between, two normal wood areas. The compression wood exhibits abnormally high longitudinal shrinkage, as indicated earlier, and when restrained from shrinking by the normal wood, will often check across the fiber axis (Fig. 3.20) (CORE et al., 1961).

**3.2.2.2 Warp.** In general, warp in lumber is any deviation from a true or plane surface. A variety of terms is used to describe special forms of warp such as bow, cup, twist, crook and diamond. All of these distortions can be traced to differential shrinkage. There may be differences in longitudinal shrinkage on two faces of a piece, or differences between edges related to variable radial or tangential orientation, or a combination of such factors to create warp in the seasoned lumber. It is possible to predict, to a certain degree, how a piece of green lumber will shrink and warp in drying by examining the orientation of the annual rings.

Fig. 3.20. Cross fracture in the wood of mountain hemlock, *Tsuga mertensiana* (Bong.) Carr. The checks are located in compression wood zones where longitudinal shrinkage is greater than in adjoining normal wood

**3.2.2.3 Casehardening.** Though the term casehardening is not accurately descriptive of this defect, is has been used to such an extent that changing it would only serve to confuse. A piece of seasoned, casehardened lumber is one in which the internal stresses cause the piece to crook if rip sawn or to cup and bow if re-sawn. When cross grain is present, twist can accompany cup.

The cause of this stress pattern is too rapid drying in which the zones near the surface of the board lead the inner portion in passing the fiber saturation point and consequently, in shrinking. The outer cells dry in a somewhat stretched condition and, in a sense, lose their capacity to shrink further. The term "set" has been applied to this phenomenon. At this stage of seasoning, the shell of the wood is in tension while the core is in compression. Surface checking related to casehardening can occur at this point.

The core then dries below the fiber saturation point initiating shrinkage which creates tension stresses greater than those in the shell. This causes a reversal in stress pattern with the shell in compression and the core in tension (Fig. 3.21).

Conditioning procedure for relieving the stresses of case-hardening consists of exposing the lumber to high temperature and high relative humidity. The wood substance becomes more plastic and the moisture content difference between the shell and the core is reduced. Care must be exercised in steaming to prevent the development of reverse casehardening by excessive humidification.

Fig. 3.22 illustrates the "prong test" devised by TIEMANN (1942) to determine the stress condition of a board at any time during seasoning. Samples are taken from various parts of a lumber pile and can be used for moisture content determinations as well as for stress analysis. This enables the kiln operator to vary the seasoning schedule according to changes in moisture content and the development of stress. It also serves as a check on the effectiveness of conditioning procedures to relieve casehardening.

**3.2.2.4 Collapse.** Collapse is a defect which occurs in the seasoning of wood, usually when excessively high dry bulb temperatures are employed in the early stages of kiln drying. The green lumber becomes somewhat plasticized, lowering its compressive strength and allowing internal crushing under high stress. The outside of the collapsed wood is generally distorted, suggesting aggravated warp (Fig. 3.23). Honeycomb and washboard defects are frequently associated with collapse. The cells in the collapsed areas are severely distorted or flattened.

There are apparently two major sources of stress which contribute to the collapse of wood cells. The first is the tension produced by capillary forces in the cells which are partially filled with water at this stage.

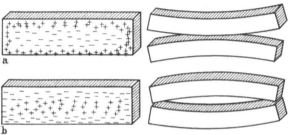

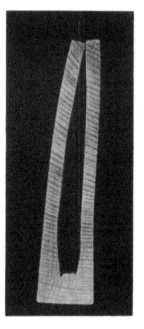

Fig. 3.21 a, b. Stress conditions in drying wood leading to casehardening defect

Fig. 3.22. Prong test for casehardening defect

The second is the drying stress due to the moisture content gradient across a piece of lumber during seasoning. One or the other of these two stresses, or a combination of both, may be involved when the resistance to stress applied in compression perpendicular to the grain is exceeded and the cells collapse.

Fig. 3.23. Cross section through redwood board that collapsed during kiln drying

Certain woods are particularly susceptible to collapse. The Australian *Eucalyptus* species, American baldcypress, western redcedar, redwood, redgum, bottomland oak and cottonwood all show varying tendencies to collapse. Any portion of the wood can collapse, but it is believed to be initiated in wet pockets. In Pacific madrone, *Arbutus menziesii* Pursh, BRYAN (1960) finds that the red heartwood is more susceptible. In aspen, KEMP (1950) reports that the wetwood zones are the areas in which collapse in initiated, but in no case are completely filled fibers associated with it. This last point is of particular significance since most reports and theories on collapse emphasize the fact that full or saturated cells are necessary to produce the conditions leading to collapse.

The reconditioning of collapsed lumber is possible when no actual ruptures of the wood have developed. A large percentage of collapsed material shows some response to a steaming treatment (100 °C, 100% relative humidity) if it is carried out when the wood is seasoned to about 15% moisture content. Since there is no free water present in the cells below the fiber saturation point, the high temperature cannot duplicate the conditions leading to collapse, but there is evidently enough plasticization to permit recovery to normal shape, or nearly so (TIEMANN, 1942).

**3.2.2.5 Honeycomb.** Honeycomb, also called hollow-horning, is serious internal checking brought about by the stresses generally associated with casehardening and collapse (Fig. 3.24). It is a defect which is not always detectable externally,

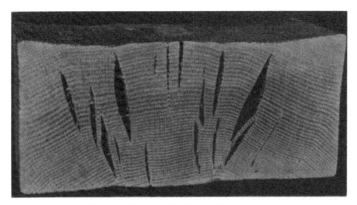

Fig. 3.24. Honeycomb defect in the wood of oak

but is serious when re-sawing operations are required. The use of mild kiln drying schedules recommended to avoid collapse can be expected to reduce losses from honeycomb defect.

**3.2.2.6 Washboarding.** Washboarding is an external symptom of collapse in edge-grain lumber showing the variable depth of internal crushing across the growth rings.

**3.2.2.7 Miscellaneous Seasoning Defects.** Kiln brown stain is one type of seasoning degrade that is common in the soft pines. The stain is believed to be caused by the oxidation, or other chemical reaction, of water soluble materials in green lumber. Its presence does not impair the strength of the lumber, but it does reduce its value where natural finishes are to be used.

Sticker stain or sticker marking is the cross streaking of lumber with chemical stain at the point of contact of the stickers separating the boards in a pile of drying lumber. Just as in brown stain, it is considered a defect for certain applications.

### 3.2.3 Raised Grain

Raised grain is a defect which is the product of inadequate seasoning followed by machining. In the planing of plain sawn lumber having a moisture content greater than about 12%, the thin-walled springwood cells are frequently crushed by the denser summerwood bands. The compressed springwood has a tendency to swell more than the summerwood with changes in the moisture content and upon recovery from compression, will push the summerwood bands above the surface (Fig. 3.25) (KOEHLER, 1929).

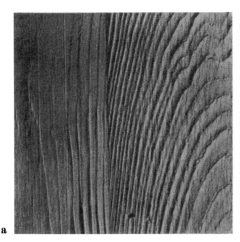 

Fig. 3.25a. Raised grain in western white pine, *Pinus monticola* Dougl
Fig. 3.25b. Raised grain in redwood, *Sequoia sempervirens* (D. Don) Endl

### 3.2.4 Loosened Grain

Loosened grain is the loosening and raising of tops or edges of annual rings on the surface of wood to give the impression of shelling off. It can develop directly from raised grain defect in plain sawn lumber. There is an actual separation of the growth increments as a result of repeated shrinking and swelling and the summerwood layers tend to lift from the surface in large splinters. Loosening generally occurs on the pith side only. The use of too great pressure in sanding or planing can lead to this defect (KOEHLER, 1940).

## Literature Cited

ANDERSON, E. A., (1944) Detection of compression failures in wood. U. S. For. Prod. Lab. Report No. 1588.
BRYAN, E. L., (1960) Collapse and its removal in Pacific Madrone. For. Prod. J. X: 598—604.
CORE, H. A., (1962) Variables affecting the formation of compression wood in plantation red pine. Unpublished doctoral dissertation, State Univ. College of Forestry at Syracuse Univ., Syracuse, N. Y.
—, CÔTÉ, W. A., Jr. and DAY, A. C., (1961) Characteristics of compression wood in some native conifers. For. Prod. J. XI: 356—362.
DADSWELL, H. E. and WARDROP, A. B., (1949) What is reaction wood? Australian Forestry 13: 22—33.
— and —, (1955) The structure and properties of tension wood. Holzforschung 9: 97—104.
GRAF, O., (1938) Tragfähigkeit der Bauhölzer und der Holzverbindungen. Mitt. Fachaussch. Holzfragen No. 20, Berlin, VDI-Verlag.
HARTMANN, F., (1932) Untersuchungen über Ursachen und Gesetzmäßigkeit excentrischen Dickenwachstums bei Nadel- und Laubbäumen. Forstwiss. Centralbl. 54: 497—517, 547—566, 581—590, 622—634.
KEMP, A. E., (1959) Collapse in aspen during kiln drying. For. Prod. J. IX: 124—130.

## Literature Cited

KLAUDITZ, W. and STOLLEY, I., (1955) Über die biologisch-mechanischen und technischen Eigenschaften des Zugholzes. Holzforschung **9**: 5—10.

KNUCHEL, H., (1947) Holzfehler. Werner Classen Verlag, Zürich. 119 pp.

KOEHLER, A., (1929) Raised grain — its causes and prevention. Southern Lumberman **137**: p. 210 M.

—, (1940) More about loosened grain. Southern Lumberman **161**: 171—173.

LIMBACH, J. P., (1945) Compression failures in wood detected by the application of carbon tetrachloride to the surface. U. S. For. Prod. Lab. Report No. 1591.

LORENZ, R. C., (1944) Discolorations and decay resulting from increment borings in hardwoods. J. Forestry **42**: 37—43.

MALCOLM, F. B., (1956) (Reprinted 1963) A simplified procedure for developing grade lumber from hardwood logs. U. S. For. Prod. Lab. Report No. 2056.

NOSKOWIAK, A. F., (1963) Spiral grain in trees — a review. For. Prod. J. **XIII**: 266—275.

PILLOW, M. Y., (1950) Presence of tension wood in mahogany in relation to longitudinal shrinkage. U. S. For. Prod. Lab. Report No. D 1763.

— and LUXFORD, R. F., (1937) Structure, occurrence and properties of compression wood. Tech. Bull. No. 546, U. S. Dept. Agric., Washington, D. C.

RASMUSSEN, E. F., (1961) Dry Kiln Operator's Manual. Agric. Handbook No. 188. U. S. For. Prod. Lab. U. S. Govt. Printing Off. Washington, D. C.

ROTH, E. R., (1950) Discoloration in living yellow-poplar trees. J. Forestry **48**: 184—185.

SIIMES, F. E., (1944) Mitteilungen über die Untersuchung über die Festigkeitseigenschaften der finnischen Schnittwaren. Silvae Orbis **15**: 60.

SINNOTT, E. W., (1952) Reaction wood and the regulation of tree form. Am. J. Bot. **37**: 69—78.

TIEMANN, H. D., (1942) Wood Technology. Pitman Publ. Corp., New York. 316 pp.

TRENDELENBURG, R., (1939) Das Holz als Rohstoff. J. F. LEHMANNS Verlag, München and Berlin. 435 pp.

# 4. BIOLOGICAL DETERIORATION OF WOOD

## 4.0 Introduction

The biological causes of wood deterioration, during its processing or when it is in service, are of great concern to wood technologists. The economic importance of wood losses due to attack by decay fungi, wood-boring insects and marine borers demands that consideration be given to methods of conserving wood through a better understanding of these biological agencies.

To make the discussion more relevant to the main topic of wood technology, it is convenient to disregard the immeasurable losses of wood destroyed or degraded by diseases and insects in standing timber in the vast forests of the world. Even when these great volumes are discounted, there remains an inestimable amount of loss of wood in the form of logs, lumber, or wood in service. It would be virtually impossible to place an accurate monetary value on the world losses for a single year. Various attempts have been made, however, to provide estimates of these losses for certain countries. LINDGREN (1953) estimated that losses through decay approximate $ 300 million per year in the United States. ZABEL and St. GEORGE (1962) estimated the losses in wood products due to wood-boring insects to be $ 150 million in a single year. These figures must be increased considerably if the labor costs for repair and replacement of wood in structures are included.

LINDGREN pointed out that much of this loss through decay could be avoided if relatively simple precautions such as proper seasoning and simple wood preservation methods were employed. Chapter 5 of this volume deals with the subject of wood preservation in some detail and Chapter 8 covers the fundamentals of wood seasoning.

This chapter develops the basic points that must be considered in utilizing wood most effectively to reduce losses due to insect attack, decay, or marine-borer infestation. These measures center around a knowledge of the biological requirements of the various organisms that cause the deterioration and the means of identifying the source of damage.

## 4.1 Fungi causing Wood Deterioration

The deterioration of wood by fungi can take many forms. In the extreme case, it can mean the total disintegration of wood substance. At the other extreme, wood deterioration may be but a mild discoloration of the wood by sapstain fungi. Both represent economic loss to a degree dependent on the nature of the product attacked.

The pattern or character of attack is often the only evidence of the identity of the organism involved. In fact, wood-destroying fungi can be classified according to the gross structural changes caused by the organisms. Three types of wood decay are generally recognized: brown rot, white rot and soft rot. Brown rot and

98                     4. Biological Deterioration of Wood                [Ref. p. 133

white rot are caused by higher fungi of the class Basidiomycetes. The brown-rot fungi derive their nourishment from the wood holocellulose while the white-rot fungi metabolize both the holocellulose and lignin. The soft-rot fungi are members of the Ascomycetes and Fungi Imperfecti. They apparently limit their attack to cellulosic components of the cell wall.

In addition to these wood-destroying fungi, there are other fungi that inhabit and cause discoloration or staining of the sapwood. Mycelia of Ascomycetes and Fungi Imperfecti concentrate in the parenchymatous elements and contribute the familiar blue and gray coloration. Pigments of brighter colors can also diffuse from the hyphae of these sap-stain fungi and color the surrounding tissue.

### 4.1.1 Characteristics of Wood-destroying Fungi

The role of wood-rotting fungi in nature is the reduction of woody material to simpler organic form. Though this is in many ways a beneficial activity of fungi, it becomes an undesirable one when wood, either as a commercial raw material or in service, is attacked. The deterioration of wood by fungi can be recognized by certain characteristic alterations of its physical and chemical properties. These changes are of variable significance to the wood user depending upon the desired utilization of the wood.

**4.1.1.1 Comparison of Brown Rot and White Rot.** COWLING (1961) reviewed the characteristic differences between the two major types of wood decay, the brown rots and the white rots, in connection with a comparative biochemical study of these decays. Among the features of decayed wood that he compared are the following: color, dimensional stability, pattern of decay within the cell wall, strength properties, yield and quality of pulp products, chemical composition, alkali solubility and host-wood distribution of the causal organisms. Differences in each of these properties reflect differences in the enzymatic mechanism involved in the two types of decay. Results of his work on this topic are referred to later in this discussion.

A comparison of the major differences between the white rots and the brown rots is essential to an understanding of the nature of wood decay. Therefore, many of the features mentioned above are considered in the following and summarized in Table 4.1.

Table 4.1. *Comparison of the Characteristics of Brown-rotted and White-rotted Wood*

| Characteristic | White-rotted Wood | Brown-rotted Wood |
|---|---|---|
| Color | White, bleached appearance | Reddish-brown |
| Constituents removed | Holocellulose and lignin | Holocellulose |
| Shrinkage | Nearly normal | Abnormally high, especially longitudinal |
| Static strength | Reduced only to some extent | Reduced greatly |
| Toughness | Rapidly reduced, even in the early stages | Rapidly reduced, even in the early stages |
| Effect on D. P. | Gradual decrease | Rapid decrease |
| Yield in pulping (on weight basis) | Approximately the same as from sound wood | Low |
| Quality of fiber | Comparable to that of sound wood | Poor |
| Solubility in 1% aqueous sodium hydroxide | Slightly more than normal | High |
| Preferred host-woods | Hardwoods | Softwoods |

# 4.1 Fungi causing Wood Deterioration

Wood decayed by a white-rot fungus has a bleached appearance while brown-rotted wood has a reddish-brown color. These colors have been attributed to the selective removal of the carbohydrates or of the lignin from the wood. In the case of the brown rots, this is evidently true, and the residual brown color is a

Fig. 4.1. Comparison of the dimensional stability of wood blocks decayed by brown-rot and white-rot fungi. The five light-colored blocks were decayed by various white-rot fungi. All five samples are sapwood of sweetgum, *Liquidambar styraciflua* L. The seven dark-colored blocks are sapwood of southern yellow pine, *Pinus* sp., decayed by various brown-rot fungi. All blocks were cubes of the same dimension before attack. (Photograph courtesy of E. B. Cowling)

property of the lignin residue left after removal of the carbohydrates. Contrary to common understanding, however, the bleached appearance of the white rots is not related to preferential utilization of lignin. The white-rot fungi are able to metabolize both the major wood constituents.

Fig. 4.2. Typical cubical pattern of cross-checks in brown-rotted wood

White-rotted wood exhibits nearly normal shrinkage while brown-rotted wood shrinks abnormally when dried. This difference can be observed in Fig. 4.1. A cubical pattern of cross-checks on brown-rotted wood (Fig. 4.2) is related to its exceptionally high longitudinal shrinkage and its evidence of the similarity of this chemical attack to that of wood hydrolysis or combustion. Wood in advanced

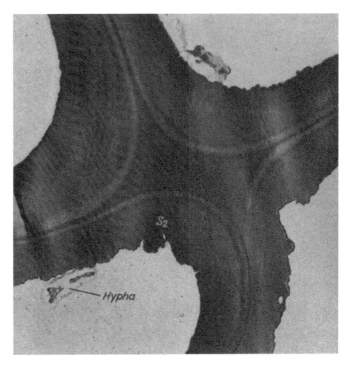

Fig. 4.3. Electron micrograph of an ultra-thin cross-section through portions of four tracheids of *Picea sitchensis* (Bong.) Carr. decayed by the brown rot, *Poria monticola*. Weight loss through decay, 65.7%. The general erosion of the surfaces of the cell walls brought about by brown rot is typical. (7,200×)

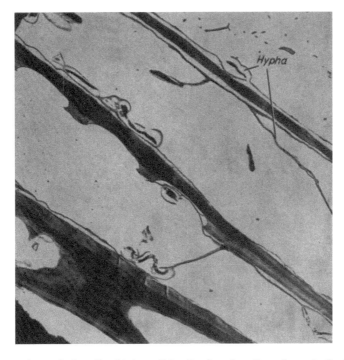

Fig. 4.4. Electron micrograph of an ultra-thin tangential section through portions of several cells of *Liquidambar styraciflua* L. decayed by the white rot, *Polyporus versicolor*. Weight loss through decay, 86.4%. The cell wall deterioration achieved by this white-rot fungus is of a localized nature. (2,800×)

stages of either of these types of decay is soft and weak. In earlier stages, however, there is a marked difference between the two types of attack at comparable weight loss. The brown-rotted wood becomes brittle and friable while the white-rotted wood retains much of its strength even at fairly advanced stages of decomposition. A rapid reduction in toughness occurs in the early stages of both types of decay.

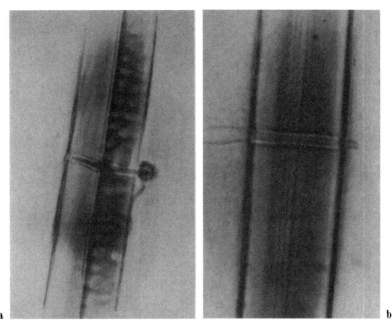

Fig. 4.5a. Ultra-violet photomicrograph of hypha nearing the point of emergence in bore hole formation through the tracheid walls of *Tsuga heterophylla* (Raf.) Sarg. The fungus is *Trametes serialis*. (1,600×)
(Photo by P. Proctor, Jr.)

Fig. 4.5b. Hypha of same species as Fig. 4.5a. with tip just emerging after bore hole formation through tracheid walls. Note constriction of hypha at point of entrance and swelling of its tip. (1,600×)
(Photo by P. Proctor, Jr.)

When decayed wood is used in pulping, brown-rotted material gives lower yields and poorer quality of fiber than when white-rotted wood is utilized. On a dry weight basis the yields from white-rotted wood do not differ greatly from those of sound wood and the quality is also nearly comparable.

According to CAMPBELL (1952), the two types of wood decay can be reliably distinguished on the basis of solubility of rotted wood in 1% aqueous sodium hydroxide. White-rotted wood is only slightly more soluble in this reagent than is normal, sound wood, but brown-rotted wood is much more soluble.

There appears to be only one way to differentiate between the two types of decay by microscopic examination. In a wood which has been decayed by a brown-rot fungus, there is no appreciable thinning of the cell walls until the late stages of decay are reached (Fig. 4.3). White-rotted wood, on the other hand, exhibits gradual cell wall thinning from the lumen toward the middle lamella, beginning in the earlier stages (Fig. 4.4).

Both types of fungi penetrate the cell wall via bore holes. P. PROCTOR (1940) showed that bore hole formation is accomplished without mechanical force. Ultra violet (365 mµ spectral line) photomicrographs revealed that enzymes secreted in advance of the hyphae produce local dissolution of the cell wall (Figs. 4.5a and 4.5b). Typical constriction of the hypha at the entrance of the bore hole can be seen in Fig. 4.5a. In Fig. 4.6 the hypha is constricted where it enters the bordered pit opening and is swollen to normal size at the exit. The hypha presumably moved through the pit membrane in this instance as there is no bore hole visible in the torus.

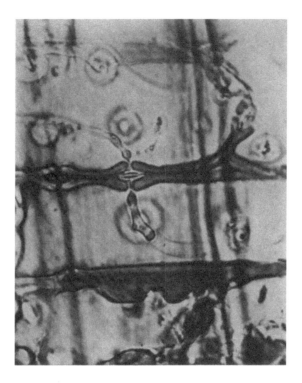

Fig. 4.6. The constriction phenomenon is illustrated in this ultra-violet photomicrograph of the hypha of *Fomes annosus* in *Pinus strobus* L. wood. (1,600×). (Photo by P. PROCTOR, Jr.)

(Figures 4.5a., 4.5b., and 4.6 reproduced by permission of P. PROCTOR, Jr., from PH. D. Dissertation, Yale Univ. School of Forestry, June 1940, "Penetration of the walls of wood cells by the hyphae of wood-destroying fungi")

Electron micrographs of bore holes in *Picea sitchensis* (Bong.) Carr. reveal no evidence of mechanical action in their formation. Enzymes of the white-rot fungus left smooth openings in the cell wall (Fig. 4.7). Deterioration of the cell wall by enzymes from a fungal hypha in a hardwood vessel segment is illlustrated in Fig. 4.8.

At the gross level, white-rotted wood retains its original shape and outward structure to a rather advanced stage of decay, probably because of the residual carbohydrate skeleton. The removal of cellulose through attack by brown-rot fungi results in disintegration of the wood at a much lower weight loss than in the case of white rots (MEIER, 1955).

An apparent host-wood preference has been found to exist for many of the brown- and white-rot fungi. COWLING (1961) states that white-rot fungi are associated most frequently with decay of hardwoods and brown-rot fungi with decay of softwoods. There seems to be no complete specificity involved, but a general statement of this relationship is valid.

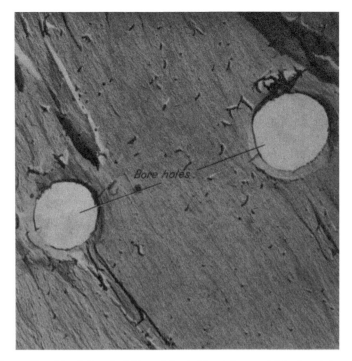

Fig. 4.7. Surface replica of tracheid wall of *Picea sitchensis* (Bong.) Carr. containing bore holes caused by the white-rot fungus *Polyporus versicolor*. Electron micrograph. (15,200×)

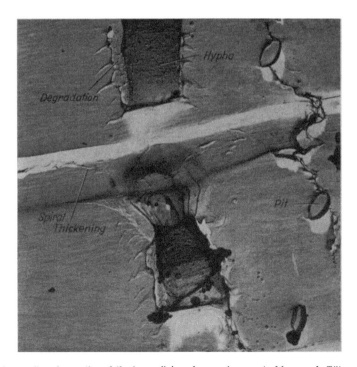

Fig. 4.8. Surface replica of a portion of the lumen lining of a vessel segment of basswood, *Tilia americana* L. showing deterioration of the cell wall in the vicinity of the fungal hypha. Note the collapse of the spiral thickening under which the hypha has passed. Electron micrograph. (6,200×)

A major point of contrast between these two types of decay is the chemical composition of the residual material. The brown-rot fungi utilize the carbohydrate fraction of the wood and metabolize lignin only to a slight extent. Though the lignin is not utilized, it is substantially altered and degraded. It has been shown quite clearly that many white-rot fungi utilize both the carbohydrates and the lignin though they derive their nourishment primarily from the carbo-

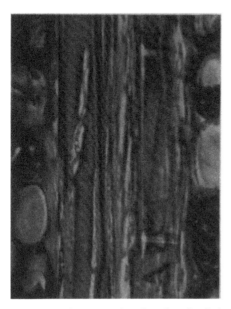
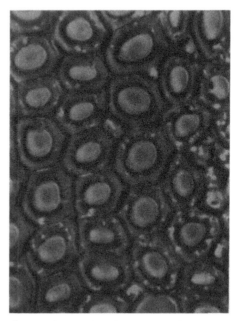

Fig. 4.9. Longitudinal section through soft-rotted wood of *Mimusops* sp. showing hyphae within spindle-shaped cavities which they formed in the secondary wall

Fig. 4.10. Cross section through soft-rotted wood of *Mimusops* sp. showing hyphae and bore holes concentrated in the secondary wall of the cells

(Figs. 4.9 & 4.10, Photomicrographs by N. KUTSCHA, by permission, from C. G. DUNCAN, "Wood-attacking capacities and physiology of soft-rot fungi." Report No. 2173, Jan. 1960, U.S. Forest Products Laboratory)

hydrates. COWLING (1961) reported the constant rate of this utilization using *Polyporus versicolor* (L.) Fr. with sapwood of *Liquidambar styraciflua* L. as the host-wood and several different chemical and physical procedures to determine the uniformity of the decay process.

**4.1.1.2 Soft Rot.** A third type of wood decay is the so-called soft rot. This type was not widely recognized until the past decade although the unusual pattern of attack, only within the secondary walls of the wood cells, has been known for a considerably longer time (BAILEY and VESTAL, 1937). Wood that has been attacked by this kind of decay organism is generally soft on the surface because of prolonged wetting as well as chemical breakdown. Firm, sound wood can usually be found when the soft-rotted exterior is scraped away.

FINDLAY and SAVORY (1950) were evidently responsible for generating the interest that currently prevails in the study of this group of decay organisms. They were seeking the cause of deterioration of wood slats in cooling towers and found fungi that limit their attack to the central portion of the secondary wall of the cells. Investigation by SAVORY (1954) and DUNCAN (1960b) showed that most of the soft rots are caused by fungi of the group called Fungi Imperfecti, although a few Ascomycetes have also been isolated from soft-rotted wood.

Soft rot can be distinguished from brown rot and white rot in several ways. As mentioned above, attack is generally made on the outer surfaces of wood which is exposed to very wet conditions, submerged in water or in contact with wet soil. The soft, decayed wood can be readily removed from the sound core when wet. When dried, cross checks appear on the surface much like those in brown-rotted wood.

Hyphae of soft-rot fungi move through the cell wall in a pattern that is considerably different from that of the common Basidiomycetes. The hyphae of this group are found characteristically within the secondary wall and oriented longitudinally, parallel to the microfibrils of the $S_2$ layer. Their enzymes break down the cellulosic components leaving diamond- or cylindrical-shaped cavities with pointed ends, as seen in longitudinal section using polarizing microscopy (Fig. 4.9). In cross-section, these cavities appear as bore holes which are concentrated in the $S_2$ layer of the cell wall (Fig. 4.10). They do not normally penetrate the cell walls with transverse bore holes as the brown rots and white rots usually do. Chemical analysis of soft-rotted wood, according to SAVORY and PINION (1958), shows carbohydrate depletion with little lignin attack, a situation that is characteristic of brown-rotted wood. The gradual increase in alkali solubility of soft-rotted wood is more analogous to white rot, however. From the evidence collected to date, softwoods do not seem to be as readily attacked by soft-rot fungi as are hardwoods. Deterioration is more gradual in softwoods attacked by this group of fungi than by the usual type of decay.

### 4.1.2 Characteristics of Wood-staining Fungi

The deterioration of wood by staining fungi is one of the important sources of financial loss in lumber processing. It is particularly critical for products that are to be given clear or natural finishes. Although many stains of wood are of biological origin, some are due to oxidation of wood constituents or caused by contact with chemically reactive materials such as iron. CARTWRIGHT and FINDLAY (1950) devote an entire chapter of their book on decay of timber to the various aspects of stains in wood. They consider the non-biological causes mentioned above, but concentrate on the discolorations brought about by incipient decay, surface molds and sap-staining fungi. The non-biological causes of discoloration can be of considerable economic significance in certain cases, but the fungal causes are of greater general concern to the wood processor.

Since the wood-staining fungi derive their nourishment only from the contents of parenchyma cells rather than from the polysaccharides and lignin, and since they bore only minute holes through the cell wall, their effects on the mechanical strength of wood are not as detrimental as is attack by wood-rotting fungi (Fig. 4.11). Consequently, the determination of the type of causal organisms responsible for staining can be very critical. In structural applications, a blue-stained member may be used without great concern unless shock loading is likely to occur. Incipient decay, however, may have already reduced the overall strength of the wood by a substantial amount. Yet both may appear, superficially, similar. Stained wood should always be suspected of containing decay organisms because the conditions that are favorable for the growth of sap-stain fungi are also favorable for the development of decay fungi.

A few comparative characteristics are useful in distinguishing between discolorations typical of decay-causing organisms and those due to the more innocuous staining fungi. Sap-stains, as implied by the name, are confined to sapwood. Stains associated with incipient wood decay are not always restricted to sapwood, but are likely to be more highly concentrated there. Decay is generally found in patches or streaks whereas overall discoloration is characteristic of staining fungi. Sap-stain can also appear as a wedge-shaped area on a log cross-section. This is due to the concentration of hyphae in the ray parenchyma cells.

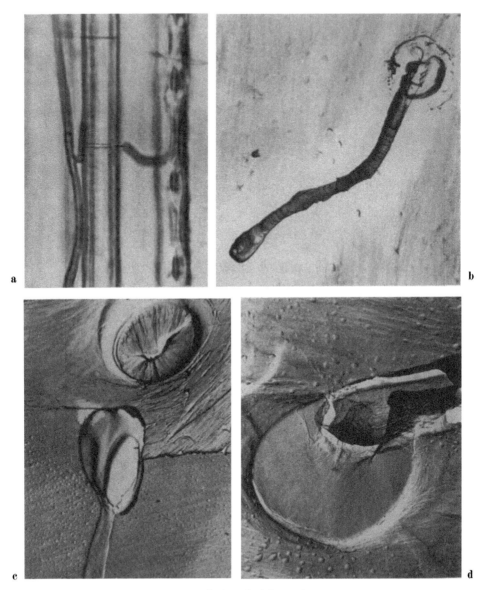

Fig. 4.11a. Tangential section through blue-stained wood of *Pinus silvestris* L. showing appressorium against tracheid wall and fine bore hole through the cell wall. (810×)

Fig. 4.11b. Hypha of stain fungus, *Ophiostoma coeruleum*, emerging from bore hole in tracheid wall of *Pinus silvestris* L. (13,500×). Electron micrograph, replica technique

Fig. 4.11c. Appressorium of stain fungus, *Ophiostoma piceae*, on warty lumen lining of a tracheid of *Pinus silvestris* L. Part of the warty layer was removed mechanically. (4,500×). Electron micrograph, replica technique

Fig. 4.11d. Hypha of stain fungus, *Ophiostoma coeruleum*, grown through the torus of a bordered pit in a *Pinus silvestris* L. tracheid. (6,300×). Electron micrograph, replica technique. From LIESE and SCHMID (1961) by permission, Holz als Roh- u. Werkstoff

Color differences, as well as pattern, provide further evidence for identification. Bluish-gray stain in sapwood is a good indication that sap-stain fungi are present. Colors ranging through green, pink, yellow, orange or blackish are also found. Discolorations by decay fungi in softwood are more typically reddish-brown while in hardwoods they may be white or dark-brown spots or streaks. Zone lines may be present with incipient decay, but these well-defined, dark outlines are not present with stain fungi.

A test for weakened fibers is suggested by CARTWRIGHT and FINDLAY (1950) as another method of determining whether a stain or a decay fungus has caused discoloration. A knife point inserted into the stained region will break fibers weakened by incipient decay, but prying of this sort reveals little weakening by staining fungi.

### 4.1.3 Physiological Requirements of Wood-destroying and Wood-inhabiting Fungi

The wood user has some very effective decay prevention methods at his disposal if he will give consideration to the physiological requirements of fungi that attack wood. Wood in service can generally be maintained at conditions that are unsuitable for fungal attack and growth. A good example of this control is the use of design and construction methods that do not utilize wood where it will be subject to soaking in water. Keeping the moisture content of the wood below the fiber saturation point is basic to preventive maintenance. It is true that wood is a material having properties desirable for use even in very wet locations. In situations such as these, it is advisable to select a species with decay resistant heartwood, or to use wood treated with non-leachable chemical preservatives.

It must be recognized that there is considerable variation in the ecology of wood-deteriorating fungi, and that general statements about all fungi with respect to all requirements cannot be made. Consequently, where evidence does exist, such differences will be noted.

**4.1.3.1 Temperature.** For most wood-rotting fungi the temperature for optimum growth lies in the range of 25° to 30 °C. Growth can proceed at lower temperatures, perhaps down to 0 °C. in some cases, but higher temperatures within this range promote more rapid growth (Fig. 4.12). Most wood rotters cannot tolerate temperatures in excess of 40 °C. Blue-stain fungi (*Ceratocystis* spp.) do not grow at temperatures above 35 °C. However, *Paeciliomyces varioti*, which causes yellow stain in oak, can make rapid growth at 40 °C. and continue to grow even at 45 °C. (CARTWRIGHT and FINDLAY, 1950). Evidently stain fungi can withstand high temperatures for brief periods, but prolonged exposure to elevated temperatures is fatal. On the other hand, some are very resistant to low temperatures and will survive exposure to frost.

DUNCAN (1960a) found that the soft-rot fungi as a group appear to grow well at temperatures which are higher than most wood-destroying Basidiomycetes prefer. Nearly half of the test group she investigated exhibited optimum growth at temperatures of 34 °C, or higher.

**4.1.3.2 Oxygen.** Since respiration is involved in the metabolism of fungal organisms, oxygen is essential for growth. The end products of respiration are carbon dioxide and water. Where there is insufficient oxygen, organic products such as alcohol and oxalic acid are formed. Experiments have shown that too high concentrations of carbon dioxide will retard growth, most fungi requiring free access to atmospheric oxygen for optimum growth. In the absence of oxygen, no fungus will grow.

Wood attacked by soft-rot fungi is often attacked under very wet conditions, such as in cooling towers. In this type of situation, the oxygen requirements of the organism can be satisfied by the oxygen dissolved in the water with which it is in contact.

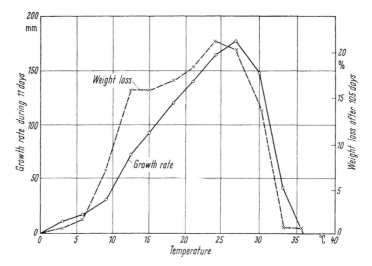

Fig. 4.12. The effects of temperature on the growth of *Polyporus* vaporarius (from E. GÄUMANN, 1939)

CARTWRIGHT and FINDLAY (1950) state that oxygen requirements of both staining and decay fungi are closely interrelated with moisture needs. If air and free water are available in the cell lumens, blue-stain fungi can grow over a wide range of moisture contents. However, if the supply of oxygen is inadequate, growth is restricted at the higher moisture levels. It is known that the storage of logs under water or under a continuous spray will protect them from blue stain.

**4.1.3.3 Moisture.** The determination of moisture requirements of fungi under laboratory conditions is difficult because water is produced during respiration. Therefore, the moisture content of the wood can be changed by the fungus. On the average, moisture contents from 35 to 50% seem to be most favorable for fungal growth. This condition finds the cell walls saturated with water and a layer of free water lining the cell lumens. The remainder of the cell cavity is then available for gas exchange and the liquid layer facilitates the diffusion of enzymes from the hyphae into the cell wall.

There have been reports of attack by fungi at moisture levels below the fiber saturation point, when there is no free water in the cell lumens. In the absence of moisture no fungi will grow, but at moisture contents as low as 15% some fungi have been reported to have become established. Generally, however, it is safe to assume that wood will not be subject to attack by the common fungi at moisture contents below the fiber saturation point. It should be borne in mind that moisture content figures are average levels for an entire board or sample and moisture gradients are present. It is possible, in cases where the circulation of air is uneven or restricted, that there will be pockets where moisture levels are much higher than the average value. This condition exists where wood is exposed to rain or is in contact with the soil and fungi can get a start under such conditions. In addition, it should be noted that the fiber saturation point for different woods ranges rather widely, 20 to 35% moisture content, with a few instances below or above these levels.

Another consideration is that the spread of decay and the continued growth of fungi probably require different conditions. Spores likely require a level of moisture content to germinate that is considerably different from that required for mycelial growth.

It is recommended that wood in service be maintained at a moisture content level 2 to 3 % below the known fiber saturation point to provide a desirable safety factor.

**4.1.3.4 Nutrients.** Carbon forms more than half of the dry weight of fungus cells, and compounds containing carbon are important in the nutrition of fungi. These organisms derive their energy by the oxidation of organic compounds (LILLY and BARNETT, 1951).

The basic nutritional requirements of wood-rotting and wood-staining fungi are satisfied by the carbohydrates and lignin present in wood. The wood rotters can use polysaccharides in a wide range of forms while the stain fungi evidently require simpler forms such as the more readily available soluble carbohydrates, proteins and other substances present in the parenchyma cells of sapwood. Stain fungi are not capable of utilizing the major cell wall constituents. This fact explains why the strength properties of blue-stained wood are relatively unaffected.

Cellulose in the form found in wood, in a complex with lignin, is more resistant to decay than chemically separated cellulose. Lignification of the cell wall reduces the number of organisms capable of degrading the tissue. Lignin-free cotton fabrics are more susceptible to decay than lignified wood exposed to similar conditions (SIU, 1951). The chemically isolated product known as lignin is not sufficient for nutrition. Although certain enzymes produced by fungi are capable of degrading lignin in wood, this degradation does not occur in the absence of the metabolism of wood carbohydrates.

It has been observed that the presence of nitrogen is necessary for the growth of fungi in wood. The requirements are evidently not very great since wood contains only about 0.01 to 0.03% nitrogen. However, the addition of organic or inorganic nitrogen accelerates the decomposition of wood by decay fungi. Studies on growth and nutrition of 42 different wood-rotting Basidiomycetes in synthetic media (JENNISON and co-workers, 1952) reveal that none require nitrogen in organic form, but growth is usually greater in organic nitrogen than in ammonium salts. Potassium nitrate and potassium nitrite do not support continued growth under these conditions.

The above statements on the need for nitrogen must be qualified. This study by Jennison, and other studies, shows that the nitrogen supply generally cannot be utilized in the absence of vitamin $B_1$, thiamine. This is the only vitamin essential for the growth of most fungi, although the addition of certain other vitamins may stimulate growth. Jennison states that biotin can be successfully substituted for thiamine in only a few instances.

In addition to the growth factors such as thiamine, certain micronutrients of an inorganic nature are needed. COCHRANE (1958) lists iron, zinc, copper, manganese and molybdenum in this category. He also includes phosphorus, potassium, sulfur and magnesium, but these are required in larger amounts.

**4.1.3.5 Hydrogen Ion Concentration.** Another condition that can be considered to be a physiological requirement is the hydrogen ion concentration of the wood medium. Wood-rotting fungi show a decided preference for a pH on the acid side, between 4.5 and 5.5. Optimum growth of both the white- and the brown-rot fungi takes place at this level. At the same time, it is interesting to note that both of these groups will render more acidic the wood in which they grow, simply as a result of their metabolism. Brown-rot fungi increase the acidity more than the white-rot fungi do.

**4.1.3.6 Natural Durability.** The heartwood of many tree species is naturally resistant to attack by decay organisms. Studies by a number of investigators have shown that this natural durability can be attributed to the presence of extractives toxic to fungi. These toxic materials protect the wood by actually killing the decay fungi as they attempt to grow within the wood cells. The effect of toxic extractives on prolonging the life of wood in service has been demonstrated unequivocally. The removal of extractives from wood and its subsequent rapid decay as compared with a natural, unextracted sample is convincing evidence of this.

Another example of the effect of extractives on the durability of heartwood in some species is the early decomposition of the extractives-free sapwood from the same piece of lumber. Actually, in some instances, the lower resistance of sapwood may be due to its greater permeability. Also this zone contains stored carbohydrates which can be a factor in promoting attack by fungi.

**4.1.3.7 Relationship of Wood Preservation to Physiological Requirements.** The principle of artificial decay resistance brought about by impregnating wood with preservative chemicals is based largely on rendering the medium toxic to fungi. Some of the chemicals used are oily and relatively water repellent, but the commonly employed commercial processes do not rely on this property for their effectiveness. Practically speaking, all of the physiological requirements of wood-deteriorating fungi could be satisfied in treated wood, but the organism must be living to take advantage of them. When desirable toxicity levels are provided, the fungi cannot become established.

### 4.1.4 Mechanism of Wood Decay

The contention of early pathologists that cellulose is attacked by brown-rot fungi remains basically accurate in the light of modern research on this subject. However, their theory on white rots was erroneous, as was indicated in section 4.1. White rots do not attack lignin exclusively as was once believed. There is ample evidence to support the concept of carbohydrate and lignin decomposition at approximately the proportions at which they occur in wood. CAMPBELL (1952) states that there are some white-rot fungi that do not attack the major components of wood simultaneously, but the majority do.

The decay of wood under attack by brown-rot fungi has been compared to acid hydrolysis in which cellulose and hemicellulose are the main wood constituents decomposed. The increase in solubility of the wood is a reflection of the degradation of carbohydrates into water-soluble molecular fragments. Solubility determinations also reveal changes in lignin caused by the brown-rot fungi since the rate of depolymerization of the major wood constituents exceeds the rate of utilization through respiration.

Brown-rot fungi secrete hydrolyzing exoenzymes which can diffuse through the film of moisture coating the cell lumens (at moisture contents above the fiber saturation point) and catalyze the depolymerization of the cellulosic components in the cell wall. These soluble, partial-degradation products can then be assimilated by the fungus and metabolized by oxidative means.

In reviewing the role of enzymes in decomposing wood, COWLING (1961) qualified his use of the term "depolymerization" because the exact mechanism of enzymatic cleavage has not been completely established. Cleavage of the polymeric chains could be occurring between monomeric units or within them. However, the question of end-wise versus random depolymerization of cellulose chains by cellulolytic enzymes is closer to realization following COWLING's research with the brown-rot fungus, *Poria monticola* Murr. The evidence clearly supports a random rather than end-wise mechanism.

The visual evidence of attack by a brown-rot fungus is the presence of bore holes in the cell wall. While this is not sufficient to acount for the high weight losses in the advanced stages of decay, even electron microscopic evidence cannot explain these losses at this time.

As mentioned earlier, white-rot fungi exhibit a general thinning of the cell wall from the lumen toward the middle lamella. This is consistent with the chemical evidence assembled by COWLING with the white-rot fungus, *Polyporus versicolor* L. ex Fries. This fungus utilizes each of the major constituents in all stages of decay and at an essentially constant rate. They are depolymerized at a rate equal to their conversion into volatile products of respiration. CAMPBELL (1952) states that oxidizing exoenzymes are necessary for the breakdown of lignin, so they can be presumed to be present in the secretions of white-rot fungi. Because of the uniform rate of utilization of all wood constituents by the white-rot fungus, it can be assumed that the lignin-rich middle lamella was reached by the enzyme early in the decay process. This indicates that the high molecular weight enzymes are not too large to diffuse through the cell wall, at least when the wood is in the swollen state.

### 4.1.5 Influence of Decay on Mechanical Properties

The weakening and embrittling effect of decay on wood is well known. In the more advanced stages of decay, wood is soft and crumbly. Under these conditions there is little danger that any degree of structural strength would be expected from the wood in service. However, in the less obvious state of decay, failure can occur suddenly and without warning, particularly when the wood is subjected to impact loading.

Toughness is the mechanical property most sensitive to decay. A decrease in toughness can be found even before weight loss is detected. Other strength values drop also, though not as rapidly as toughness figures are reduced. CARTWRIGHT and FINDLAY (1950) list bending strength, compression, hardness and elasticity as the approximate order in which mechanical strengths decrease during decay.

Brown-rot fungi cause a more rapid drop in strength properties than the white-rot fungi do. White-rotted wood shows an early reaction to attack only in toughness values. Analysis of decaying wood reveals that attack by brown-rot fungi brings about a rapid decrease in DP (degree of polymerization) in the cellulose, while white-rot fungi cause a more gradual drop in the average length of chains. This evidence supports the theory that DP is an indicator of the tensile strength of wood, although there is no experimental evidence for this in the literature.

Considering the different nature of attack by stain fungi, it is interesting to note that CROSSLEY (1956) found that this group of organisms affects the toughness values of wood. The five species of sapstain fungi used in this study vary considerably in their effects on toughness values. In the *Pinus strobus* L. sapwood used for the tests, the reduction in toughness could not always be directly correlated with intensity of staining. It was found that the wood stained gray or blue-gray was invariably lower in toughness strength than unstained material, but that brown-stained wood, presumably attacked by *Cytospora pini*, retained substantially all of its toughness strength. One species of fungus producing a gray sapstain, *Alternaria tenuis*, reduced the toughness of the test specimens by more than 43%.

ARMSTRONG and SAVORY (1959) found that soft rot also causes reductions in the impact strength of wood. These losses are more typical of white than of brown rot.

## 4.2 Wood-boring Insects

Insects are second only to decay fungi in the economic loss they cause to converted lumber and wood in service. Although periodic estimates have been prepared for certain countries, the true world-wide losses in wood destroyed and labor expended in replacement cannot be determined with any degree of accuracy. It is sufficient to state that the losses are extremely great and measures taken by wood users to reduce such damage are a sound investment.

The objectives of this brief review of wood-boring insects are: 1) to provide the wood technologist with some basic information about the insects which cause the greatest amount of damage to wood in service, and 2) to outline the features of insect physiology, nutritional requirements and habitat patterns which form the basis for control measures.

For convenience, the wood-boring insects which are responsible for the greatest economic losses can be separated into three categories: termites, insects causing powder-post-type of damage, and carpenter ants. They are listed in approximate order of economic importance. It should be noted that although some of the insects included in this discussion are known to attack living trees, emphasis is placed on those that are primarily wood products pests.

### 4.2.1 Termites

These important wood-boring insects, often erroneously called "white ants", are found in virtually all parts of the world with the exception of the arctic and antarctic regions. It is estimated that there may be as many as 5000 species of termites in the five families of the order, Isoptera (SNYDER, 1948).

The subterranean termites (family Rhinotermitidae) are easily of greatest significance in the United States since they cause about 95% of the termite damage; dry-wood termites (family Kalotermitidae) are very destructive, but more limited in distribution. In the tropical regions, where the preponderance of termites is found, members of the Termitidae are very common inhabitants of decaying wood.

Termites of the genus *Reticulitermes* (Fig. 4.13) are most widely distributed in the United States with representation in every state. *R. flavipes* Kollar is the subterranean species commonly found in the east and south while *R. hesperus* Banks occurs on the Pacific Coast. Dry-wood termites of the genus *Kalotermes* (Fig. 4.14) are found primarily in coastal regions of the southeast and southwest. Damp-wood termites of the genus *Zootermopsis* are restricted to the Pacific Coast of North America (SNYDER, 1949).

**4.2.1.1 Characteristics.** Termites are soft-bodied, social insects which feed on cellulose although the gastric juices of most species do not contain cellulases. The presence of cellulose-digesting protozoa in their hind gut enables them to utilize wood as food. Because these one-celled animals play such an important role, termites are indirectly vulnerable to high temperatures and oxygen pressures which would kill protozoa. Only in the higher forms of termites, the Termitidae, are protozoa unessential. This group can digest cellulose, but not lignin, a component of wood that none of the wood-boring insects can utilize (WIGGLESWORTH, 1947).

There is considerable evidence of the interrelationship between termites and fungi in wood. The presence of fungi in the wood may be the termites' main source of nitrogen. There is also much evidence that members of the Rhinotermitidae and the Kalotermitidae will eat sound, uninfected wood (PENCE, 1957). It is very obvious that the spread of decay is aided by termites who distribute spores, but doubtful that fungi are dependent upon these social insects.

Fig. 4.13. Termite castes of *Reticulitermes lucifugus*: a) secondary reproductive, b) worker, c) soldier (Photograph by G. BECKER)

Fig. 4.14. Termite castes of *Kalotermes flavicollis* Fabr.: a) soldier, b), c) & d) nymphal stages, e) swarming adult (Photograph by G. BECKER)

Subterranean termites nest in the ground or in wood in contact with the ground. They construct earthen tubes to provide covered passageways to the wood under attack. These tubes, which often span considerable distances over building foundations or other inedible materials, provide a connection to soil moisture, an important need of this type of termite.

Under the caste system in the termite society, only the workers are involved in chewing tunnels and chambers in the wood. In most species the soldiers and the reproductives have their specialized duties, as implied in their names. The alates or winged reproductives are the ones seen on swarming flights as they leave the home nest to begin new colonies. The other termites avoid light, as a rule, and do not emerge from their dark and moist habitat.

The worker termites are creamy-white, soft-bodied, maggot-like, wingless, sterile, blind and usually eyeless insects (SNYDER, 1948). Soldiers have very large,

Fig. 4.15a. Termite work in oak flooring    Fig. 4.15b. Termite work in softwood lumber

highly sclerotized heads equipped with large mandibles designed for defense against frontal attack. The soft body is just as vulnerable to attack by ants or other insects as is the worker's. Dry-wood termites are more primitive forms and the specialized worker class is lacking.

Although termites destroy all types of wood products and household items such as books and clothing, the damage done to lumber in wood structures is of primary concern (Fig. 4.15). Homes built without consideration to the habits of termites are particularly susceptible to attack in many regions of the world. In general, conditions leading to decay by wood-destroying fungi also invite termite attack. These two types of biological degradation are often found together in wood that has failed in service.

**4.2.1.2 Control.** The best approach against termite infestations is their prevention. This is more feasible in new construction, but can be employed in existing structures as well. Practical design and construction practices for building construction to protect against subterranean termite attack are outlined by ST. GEORGE et al. (1960). Since these insects get their start in wood in the soil or the ground, the removal of stumps, wood débris and other cellulosic material before construction is essential. Any termites found during excavation for the foundation should be killed by poisoning.

All wood used during layout work, in the construction of forms, and scraps should be removed from the site before filling around the foundation. Such materials left under porches or steps invite termite attack.

The foundation itself should be termite proof. It should have no cracks more than 1/32-inch ( — 1 mm.) in width or the small subterranean termite can gain an avenue of access to the wood. Hollow block construction should have a solid concrete cap with all joints filled with mortar. Provision should be made for good drainage and ventilation beneath the building. Where only a crawl space is allowed under a building, the soil should be poisoned and then covered with heavy tar paper. This reduces the condensation of soil moisture on the floor joists and subflooring.

Where possible, the wood sills of the building should be treated with preservatives. At least six inches of clearance should be provided between the ground and the lowest woodwork. Termite shields are recommended to prevent the construction of tubes from ground to wood, but these must be carefully constructed and installed in order to be effective. Small gaps between shields, poor joints or inadequate overhang will still permit termites to permeate or penetrate these barriers with their passageways.

Buildings supported by piers with adequate clearance from the ground to provide ready access for inspection need not be shielded. Routine inspections by personnel trained to recognize tubes and other symptoms of termite attack, fungal decay, or infestation by other pests should be considered as necessary preventive maintenance. When infestations are found, pest control specialists can recommend the procedure to follow in eliminating the problem. A detailed discussion on the control of termites is given by MALLIS (1960). Useful information is also available in the U. S. D. A. bulletin by St. GEORGE et al. (1960).

### 4.2.2 Powder-post Beetles

Although the true powder-post beetles are members of the family Lyctidae, the term powder-post is often used in describing the damage done by beetles of other families. Among them are Anobiidae, Cerambycidae and Bostrichidae. The name is derived from the type of damage produced, a fine powder-like dust in the case of *Lyctus* to a coarser frass or small pellets in the case of the other bettles. Insects from this broad group are widely distributed throughout the world and are the source of great damage to articles and structures made of wood (SNYDER, 1927).

**4.2.2.1 Lyctidae.** Adult *Lyctus* beetles are reddish to brownish in color and from 2 to 7 mm. in length, depending on species (Fig. 4.16). Most species have a rather similar life history whether found in North America, Europe or Australia. The wood attacked by *Lyctus* beetles is generally restricted to the sapwood of seasoned or partly seasoned hardwoods. Coniferous woods are evidently not susceptible to attack (most likely because tracheid diameters are too small). The larvae (Fig. 4.17), which are responsible for the damage caused, cannot digest cellulose, but require sugar or starch as a major constituent of their diet. Boring is therefore limited to the sapwood which is richer in starch than is heartwood.

Undoubtedly the extent of damage brought about by any single larva is a function of the starch content of the wood. WIGGLESWORTH (1947) points out that when cellulose cannot be digested by an insect, much greater quantities of wood must be ingested to obtain sufficient carbohydrate. The larvae are believed to require a certain amount of sugar and protein as well as starch. In addition,

8*

the moisture content of the wood under attack is a factor in the development of the larvae. FISHER (1928) states that *Lyctus* beetles can invade wood having a moisture content as high as 40%. Eggs and larvae develop at moisture contents ranging from 10 to 28%. They do not develop at a moisture level below 8%, however.

Fig. 4.16. Adult *Lyctus* beetle, (11×) (Photograph by J. B. SIMEONE)

Among the *Lyctus* beetles commonly found in households in the United states are *Lyctus brunneus* Steph., *L. linearis* Goeze, *L. parallelopipedus* Melsh., *L. planicollis* Lec., and *L. cavicollis* Lec. At least three of these species are cosmopolitan in distribution, the first two having been introduced from Europe and *L. planicollis* now found in England although originally a pest in the United States. *L. brunneus* is believed to be the only species of *Lyctus* in the Union of South Africa (TOOKE, 1949).

TOOKE's review of the European *Lyctus* beetle situation as it exists in South Africa, with a summary of its life history, characteristics, and food habits provides a representative case. The experience there has been that this powder-post beetle has increased and spread, due in large measure to the increased importation of highly susceptible hardwoods, such as limba (*Terminalia superba* Engl. & Diels) from central Africa. Timbers that have an unusually large percentage of sapwood or those that do not exhibit a clear color difference between sapwood and heartwood are more likely to be involved.

Fig. 4.17. Larvae of *Lyctus* beetle, (11×). (Photograph by J. B. SIMEONE)

The factor which is most significant in affecting the susceptibility of sawn lumber is the diameter of the vessels as viewed in cross-section. The openings must be large enough to permit the adult female's ovipositor to be inserted for egg laying. Timbers having only small-diameter vessels are not susceptible to attack. In the case of *L. brunneus*, the pores need to be 90 µm or more in diameter.

Once the eggs have hatched in the vessels, the larvae tunnel in the wood, following the grain direction in the early stages, criss-crossing at random later.

Fig. 4.18 a–e. a) Exit holes of *Lyctus* powderpost beetles in hardwood tool handle; b) *Lyctus* powderpost work in same piece as (a), longitudinal view; c) Cross-sectional view of same piece as (a) and (b); d) *Lyctus* powderpost work in hardwood lumber, cross-sectional view; e) Frass piles from *Lyctus* work in stacked hardwood lumber

This stage continues for several months and then the larva bores its way to within a few millimeters of the surface where it pupates. The adult can chew its way to the surface, leaving small round exit or flight holes. The cycle can then be repeated, the adults mating and the female depositing eggs in another piece of wood or in the same one.

It is interesting to note that finished furniture or other woodwork in houses is rarely attacked. Evidently the waxed, polished or painted surfaces leave no openings for oviposition. Flight holes found in furniture can usually be traced to egg laying when the lumber was in the raw stage. The holes packed with finely-ground and powdery wood, typical of powder-post work (Fig. 4.18), may be the product of many months to several years' activity, depending on environmental conditions such as starch content of the wood, moisture content and temperature.

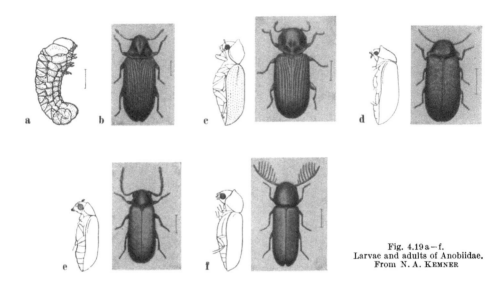

Fig. 4.19 a–f.
Larvae and adults of Anobiidae.
From N. A. KEMNER

**4.2.2.2 Wood-feeding Anobiidae.** Representatives of the family Anobiidae can be found in most parts of the world. They are important pests of wood products and structures, causing damage not greatly different from *Lyctus*. In Europe, attacks by the death-watch beetle, *Xestobium rufovillosum* DeG., (Fig. 4.19) and the furniture beetle, *Anobium punctatum* DeG., are widespread and serious as are those by *Ernobius mollis* L. and *Ptilinus pectinicornis* Geof. *Anobium punctatum*, perhaps because of its wide distribution in Africa, New Zealand and Europe, has been considered important in the United States. However, a survey of the distribution of Anobiidae in northeastern United States and adjacent Canada by SIMEONE (1960) places this species very low in frequency of occurrence in this region. In New York State and neighboring areas, *Hadrobrogmus carinatus* (Say) is the most common wood-feeding anobiid species encountered with *Ptilinus ruficornis* (Say) second in frequency. *Ernobius mollis* L., one of the common European species, also proved to be of more common occurrence than the death-watch beetle and the furniture beetle.

*Anobium punctatum*, typical of the anobiid wood-feeders, differs from the *Lyctus* beetle in its food preference. It has been found in both, softwoods and hardwoods, heartwood and sapwood. TOOKE (1949) reports that it is found only in pine in the Union of South Africa. It prefers old seasoned timber such as furniture, flooring, and similar woodwork. Typical damage caused by this insect is shown in Fig. 4.20.

The less specialized feeding habits of this group are related to the insect's digestive abilities. WIGGLESWORTH (1947) states that anobiids produce a true cellulase in the mid-gut without the aid of protozoa and can therefore digest cellulose and hemicellulose, a distinct advantage over the *Lyctus* beetle larvae.

The life cycle of anobiids is similar to that of the *Lyctus* beetles in many respects. There are four distinct stages: egg, larva, pupa, and adult. Eggs are

Fig. 4.20. Typical *Anobium punctatum* damage and frass in softwood
(Photograph by G. BECKER)

oviposited in cracks or crevices in wood rather than being limited to vessel openings. The larva does not bore along the grain as the powder-post larvae do initially. The frass is a mixture of small oval pellets and granular powder somewhat coarser than that of *Lyctus* (Fig. 4.21). The duration of a life cycle is from one to several years for the anobiids and a few months for *Lyctus* beetles, depending upon temperature.

**4.2.2.3 Cerambycidae. Long-horned Beetles or Round-headed Borers.** This very large family of beetles (approximately 13,000 species) consists primarily of forest insects. Only a few species are injurious to seasoned lumber and fewer still are found in wooden buildings. However, *Hylotrupes bajulus* L., the European house borer, is the most important destroyer of softwood in northern Europe. It has been particularly damaging to roof beams and other wood structures in attics of houses and public buildings. This beetle can be found in Russia, South Africa, Australia, South America and eastern North America.

Because of the economic importance of the old-house borer, considerable research data on its life history is available. It is known that oviposition and hatching can occur over a wide range of temperatures and relative humidities, but that the incubation period and percentage of eggs hatching are very sensitive to variations

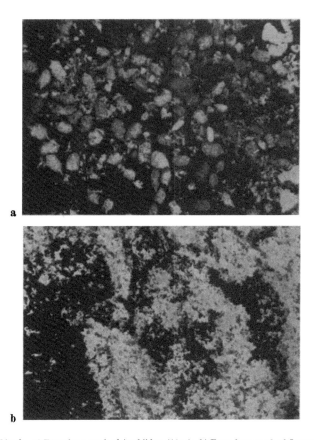

Fig. 4.21a, b. a) Frass from work of Anobiidae, (11×), b) Frass from work of *Lyctus*, (11×)
(Photographs by J. B. SIMEONE)

in these conditions. Incubation can be as short as six days or as long as forty-eight days, the longer period being at lower temperatures and relative humidities. Optimum growth takes place at moderate temperatures and high relative humidities.

The large larvae up to 35 mm long, bore in the sapwood of coniferous species for periods ranging from two to seventeen years before pupation. TOOKE (1949) suggests that the larval period is related to quality and quantity of food in the wood, temperature, and moisture content of the inhabited wood. Although the larvae can digest cellulose, their boring is limited to sapwood, evidently due to the need for some substance which is concentrated in this area. BECKER (1942) found that the albumin present in dead wood is needed for larval development. When albumin concentrations are reduced too greatly, the larvae die.

After a pupal period of about three weeks, the rather large, black adults (Fig. 4.22) chew their way to the surface and emerge from the oval-shaped holes to repeat the cycle. The larva, eggs, frass and work of this insect are illustrated in Figures 4.23, 4.24, 4.25 and 4.26, respectively.

**4.2.2.4 Bostrichidae. Auger or Shot-hole Borers.** This group of beetles is represented by relatively few wood-eaters in the north temperate zone, but are

Fig. 4.22a, b. a) European old-house borer, *Hylotrupes bajulus* L. Adult male, (2.8×). Note also the frass in borings, b) Adult female, (2.8×)

found in larger numbers in warm regions. Though less important than the beetles discussed earlier, they can be serious pests and are interestingly different. Both the adults and the larvae have the ability to bore through wood. Eggs are not oviposited in pores, cracks or checks, but in galleries bored by the adults under the surface of sea soned hardwoods. Like the powder-post beetles, starch is essential to the larvae.

One species of this group bears mentioning, *Dinoderus minutus* Fab., the bamboo borer. Although once restricted to tropical and sub-tropical regions, the increase in the trade of bamboo products from Japan, India and China has aided its wider distribution. Other Bostrichid beetles are of varying importance in the lumber trade, but are primarily of local interest.

**4.2.2.5 Control Measures.** The first control measures that should be considered by the wood technologist in reducing losses to wood-boring insects are those that suggest themselves by the characteristics and habits of these wood feeders. For example, in handling hardwoods known to be susceptible to *Lyctus* attack, sapwood should be eliminated to the extent that is economically feasible. Scraps cannot be allowed to accumulate in wood storage areas, but should be burned. Where the moisture content of the wood is known to be a critical element for a particular insect's development, kiln drying to a safe level should be done as early as possible, followed by immediate utilization or a closing of the vessel openings with a sealing finish (SNYDER, 1926).

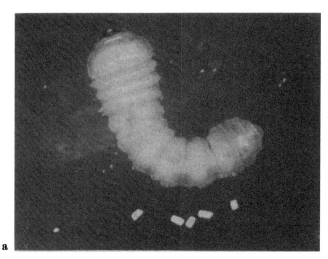

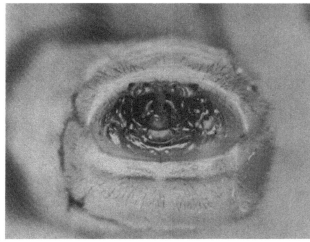

Fig. 4.23a, b
European old-house borer, *Hylotrupes bajulus* L.
a) Larva, (2.8×), b) Anterior end of larva, (13 ×)

The preservative treatment of wood to be used structurally, as in buildings, is an obvious preventive measure. The use of accepted commercial preservatives according to standards specified by various laboratories is recommended for protection against decay as well as wood-boring insects.

In cases where infestation has become established, the nature of the wood product will, of course, limit the procedure that can be used to eradicate the pests. Subjecting wood to elevated temperatures and high relative humidities (up to 180°F. (82°C). and 100% relative humidity) for periods up to six hours, depending upon thickness of stock, will kill insects at all stages, from egg to adult. Fine furniture would be damaged with such treatment, but sterilization can be applied to many other forms of wood products.

Fig. 4.24. Eggs, (6.5×)
European old-house borer. *Hylotrupes bajulus* L.

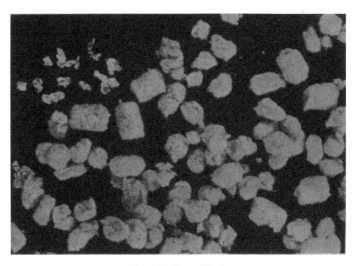

Fig. 4.25. Frass, (13×)
European old-house borer. *Hylotrupes bajulus* L.
(Photographs 4.22, 4.23, 4.24, 4.25 by J. B. SIMEONE)

The application of wood preservative to wood in service is possible, but often impractical. In cases where it can be used, liberal amounts of 5% pentachlorophenol in mineral spirits can be sprayed, brushed or otherwise injected into the wood. Since water-borne chemicals do not penetrate very effectively under these conditions, oily solvents are recommended. Insecticides such as DDT, chlordane, and lindane solutions are effective. MALLIS (1960) also suggests that paradichlorobenzene and orthochlorobenzene be used to kill wood-borers. Dip treatments should be employed in preference to less effective surface methods to provide longer periods of protection, whenever the nature of the wood products permits.

Fig. 4.26 a, b. Work of the European old-house borer in softwood
(Photographs by G. BECKER)

### 4.2.3 Carpenter Ants
*(Camponotus* spp.)

A second group of social insects responsible for considerable damage to wood in service is the carpenter ant, of the family Formicidae (Fig. 4.27). In contrast to the colonizing termite, this insect is not a wood-feeder, but inflicts its damage on poles, house porches, steps and other wood structures while excavating galleries for living quarters. The workers remove the softer zones of the wood and deposit the fragments outside of the nest (Fig. 4.28). The galleries are clean and smooth compared to those of termites.

SIMEONE (1954) points out that carpenter ants prefer moist wood for their nests under conditions in the northeastern United States. This is apparently a requirement for development of the immature forms. Since moisture contents greater than 15% are needed for establishment of the ants in wood, this suggests a vulnerable point which can be exploited in their prevention and control. SIMEONE

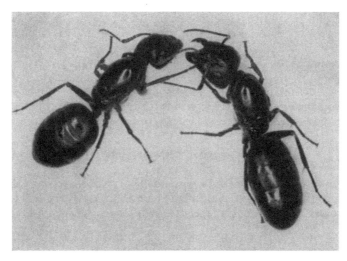

Fig. 4.27. Black carpenter ants (*Camponotus pennsylvanicus* (DeG.)) adult females, (4×)
(Photograph by J. B. SIMEONE)

Fig. 4.28 a, b. a) Carpenter ant nest in center of power pole (split open), b) Creosote-treated power pole, split along a radial check to reveal nest (light-colored area) where wood fragments were removed by carpenter ants
(Photographs by J. B. SIMEONE)

emphasizes that the intelligent use of wood is a potent tool for prevention of carpenter ant attack in wood structures. Care in design, construction and timely repairs to ensure that the wood will be kept as dry as possible is the basis for this approach. Wood members should be kept out of contact with the ground from which they will absorb moisture. If contact is necessary, wood adequately protected with chemical preservatives should be employed. Construction methods that allow free drainage from porch floors, steps, trim and other areas exposed to rain or melting snow are helpful in keeping the moisture content at a safe level.

When a colony of carpenter ants has become established, chemical poisons such as chlordane, DDT, or ground derris powder may be used to kill them. The ants carry the poison into the nest and distribute it through their grooming habits. After eradication of a colony the structural characteristics that brought about the wet conditions should be corrected in order to prevent future infestations.

### 4.2.4 Carpenter Bees. Order Hymenoptera

As in the case of carpenter ants, carpenter bees (*Xylocopa* spp.) use wood only for nesting, not for food. The female adults bore holes, about one-half inch (12 mm.) in diameter and an average of six inches (15 cm.) in length (Fig. 4.29). The eggs are

Fig. 4.29a–c. a) Carpenter bee (*Xylocopa virginica* (Drury)), adult female, (1.4×), b) Entrance hole to carpenter bee nest in power pole, c) Diagram of carpenter bee nest in power pole, entrance at lower right. Dimension A, 15–23 cm. (6–9 in.); dimension B, 5–10 cm. (2–4 in.) in typical nests
Figures 4.27, 4.28 and 4.29, by permission, J. B. SIMEONE, from: "Pole deterioration by wood-destroying insects" in Proceedings, Eastern Wood Pole Conference, State Univ. College of Forestry at Syracuse University, Sept. 12–14, 1961

laid in these galleries, are supplied with a mixture of pollen for use as food during the larval stage, and remain through the pupal stage. The danger of weakening structural timber in use is accentuated by their habit of forming their galleries in clusters around the outer shell, such as has been found in telephone and power poles (SIMEONE, 1961). Preservative treatment according to commercial standards will generally be adequate to discourage carpenter bees from excavating such wood. The injection of DDT, chlordane or dieldrin into nests is an effective control measure to kill eggs, larvae and pupae of this insect.

### 4.2.5 Horntails. (Siricidae)

The Siricidae (Order Hymenoptera) are primarily insect pests of forest trees, but under certain conditions they may damage wood products as well. The usual hosts are dead, dying or recently felled trees and logs. In some areas suppressed and burnt trees are heavily attacked.

Horntails are widely distributed through most parts of the world, having been introduced through timber shipments in many cases. For example, *Sirex juvencus* (L.) was introduced into New Zealand about the turn of the century (TILLYARD,

Fig. 4.30 a, b. Horntail adults a) *Sirex gigas* L. b) *Xeris spectrum* (L.)

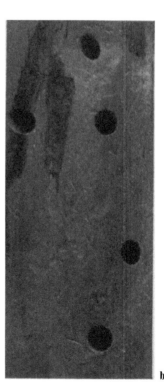

Fig. 4.31 a, b. Borings, frass and flight (exit) holes of horntails, *Sirex gigas* L.

1926). Some sixty years later *Sirex noctilio* Fab. was discovered in Australia where it threatens extensive *Pinus radiata* D. Don plantations (HOWICK, 1962).

Horntails are medium to large-sized insects, 1.5 to 5 cm. in length (Fig. 4.30). They have a cylindrical head, thorax and abdomen of nearly equal diameter. Most

128    4. Biological Deterioration of Wood    [Ref. p. 133

species are dark in color, blue or black with combinations of red, yellow or white. The adult females have a characteristic horn-like ovipositor extending beyond the abdomen and giving the insect its name (CRAIGHEAD, 1950). Males are considerably smaller than the females, but resemble them in having two pairs of long, slender wings.

Oviposition is carried out at suitable sites on the selected host after small holes are bored into the bark or wood. The young larvae work in the solid wood, boring circular holes as they feed and packing the boring dust in the tunnels behind them (Fig. 4.31). They move toward the heartwood at first and after periods of one to two years, the full grown larvae, up to 5 cm. long, work their way toward the outside and pupate just under the surface. Adult horntails emerge in early summer. If the infested tree is felled and sawn into lumber in the intervening period, adults can emerge from dressed lumber even after it is installed.

However, if the lumber is kiln dried according to normal schedules, all larvae are killed and subsequent emergence is prevented. Heat sterilization can also be used to kill them. Under other circumstances, lumber, wood products or buildings must be fumigated with methyl bromide to eradicate the horntail larvae.

Logs can be protected from attack by placing them in mill ponds and rolling them frequently. Prompt utilization of material helps to prevent attack.

## 4.3 Marine Borers

Deterioration of wood in service by marine borers is a third major biological cause of loss. In the coastal waters of the United States alone, damage caused by marine borers is estimated to be 500 million dollars per year (P. D. C. NEWSLETTER, Jan. 1958). In other parts of the world, especially in tropical waters where the number of marine borer species and individuals is much greater, the losses cannot even be estimated.

In spite of its susceptibility to attack by marine boring organisms, wood has properties which make its use advantageous for marine piling and other coastal construction. However, the wood technologist is faced with one of the most challenging problems of wood preservation in trying to prolong the service life of wood in situations where it is subject to attack by marine borers. Because of the medium in which they live and the lack of a weak link in their life history, shipworms, an important representative group of marine borers, appear to be virtually invulnerable (TURNER, 1959). Research on the biology of marine borers remains as a key to the future development of more effective wood preservation methods. Finding the reasons for the natural durability of certain woods would also contribute to this end (RAY, 1958).

Marine borers burrow in wood for two reasons, food and shelter. However, not all of them bore for both reasons. Members of some genera produce cellulase which enables them to digest wood, but some of the others cannot utilize wood for nutrition. Wood is an essential part of the diet of *Teredo, Bankia,* and *Nausitora* (Teredinidae), and *Limnoria* (Isopoda). The nitrogen requirements of some marine borers are known to be supplied by plankton, and fungi are believed to have a role in the life of certain others (BECKER, 1958).

The many species of marine animals responsible for the damage to wood in coastal structures can be separated into two categories, molluscan borers and crustacean borers. The family Teredinidae contains three important molluscan genera and the family Pholadidae is represented by one genus of economic importance. Two genera of crustacean marine borers belong to the order Isopoda and another is classed in the order Amphipoda (BECKER, 1958).

## 4.3.1 Molluscan Borers

The three Teredinien genera considered here are *Teredo, Bankia* and *Nausitora*. *Martesia*, family Pholadidae, is a fourth genus of economic importance among the molluscan borers. Species of *Teredo* and *Bankia* are found in the waters of the temperate zone as well as in tropical seas while the other two genera contain primarily warm water species.

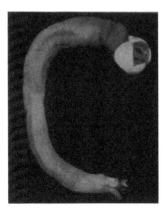

Fig. 4.32. Shipworm (*Teredo navalis* L.)

The teredinids are called "shipworms" because of their long, worm-like appearance (Fig. 4.32). They are equipped with a pair of shells at their anterior ends and the rasping or chewing to excavate in wood is accomplished with these. Under optimum conditions shipworms can grow to a meter in length, but they are often no more than a few centimeters long and a few millimeters in diameter in heavily infested wood because of the competition for space. They may also be smaller in size in wood of high natural durability or, their small size may be attributable to variations between species. Shipworms live on the products of digested wood, which they can produce with their own cellulase, and on plankton which is available to them in sea water (BECKER, 1958).

Once established, a shipworm never leaves the wood. It is free-swimming only as a young larva. After lodging on the surface of submerged wood, it eventually bores an entrance hole, burrows into the structure and develops into an adult with no opportunity to leave the initial site of attack. The entrance holes on the surface of the wood are rather small (Fig. 4.33), but with the development of the

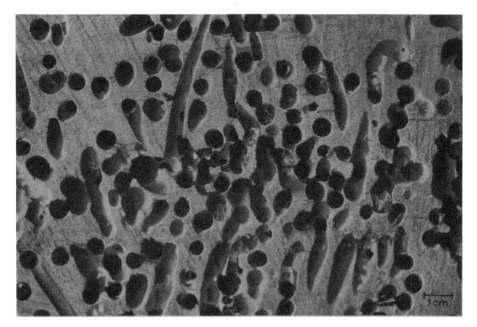

Fig. 4.33. Shipworm damage in wood (Photo by G. BECKER)

shipworm the holes inside the wood increase in diameter as burrowing proceeds. Shipworms bore for as long as they live and can find wood to move into.

Molluscan borers of the genus *Martesia* are clam-like creatures only three to four centimeters in length. The whole body is enclosed in the shells. They burrow in wood, but only to a depth somewhat greater than their length (MENZIES and TURNER, 1956). Upon reaching adult size, they cease boring. These marine animals are not wood feeders and no cellulase has been found in their gut. They ingest plankton and other microorganisms for nutrition.

### 4.3.2 Crustacean Borers

The genus *Limnoria* contains the principal Isopod marine wood borers. Members of this genus, of which there are more than twenty known species, are distributed throughout the world. Most species, the subgenus *Limnoria*, are xylophagous, having mandibles with a rasp and file arrangement for chewing wood.

Fig. 4.34a, b. *Limnoria* sp. female, a) dorsal side, b) ventral side, with eggs
(Photos by H. KÜHNE, courtesy of G. BECKER)

Members of the subgenus *Phycolimnoria* are kelp borers and they lack these specialized mandibles (MENZIES, 1959). Another crustacean genus, *Sphaeroma*, found chiefly in tropical and subtropical waters, can be very destructive. *Chelura*, generally considered to be less destructive than *Limnoria*, does not occur in tropical waters.

Crustaceans are not limited to burrowing in the piece of wood in which they first develop. The adults can move about and spread their attack from the original site by swimming or crawling, and by floating in infested driftwood. Attacks by these animals are generally concentrated along the length of the timber between the low tide and half tide levels. The gribbles do not burrow deeply, approximately one centimeter, but the borings parallel the surface of the timber. The holes are of small diameter since the animals themselves are less than one millimeter broad by three to five millimeters in length (Fig. 4.34). Nevertheless, damage is great because the wood is attacked by large numbers of borers (Fig. 4.35) and wave action aids the destruction. MENZIES and TURNER (1956) estimate that an infested, moderate-sized piling contains 200,000 adult animals, each of which can eat approximately 0.54 mg of untreated wood in ten days. Fresh attacks on the same timber (by the gribbles) are made possible by the removal of the shell of attacked wood by the crushing action of floating débris.

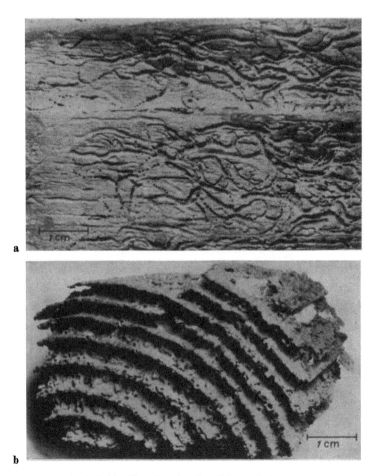

Fig. 4.35a, b. Wood attacked by *Limnoria*, a) surface of damaged pine wood with breathing holes, b) spring- wood zones heavily attacked (Photos by M. GERSONDE and G. BECKER)

It is known that *Limnoria* produce their own cellulase and that they consume nearly all of the polyuronide hemicellulose and noncellulosic carbohydrates as well as about half of the celluloses in their digestion of wood (RAY, 1959a). There is a lack of agreement on the other nutritional requirements of *Limnoria*. Marine borers of this genus can make no direct use of planktonic animals for food in the way that *Teredo* does, but they must rely on wood for all of their nutritional needs. BECKER et al. (1957) claim that the gribbles eat wood to provide themselves with fungal proteins which are always present in exposed wood.

BECKER (1959) suggests that marine fungi in submerged wood are essential sources of protein and vitamins for the nutrition of *Limnoria* in addition to aiding carbohydrate metabolism for these animals. RAY and STUNTZ (1959) and

RAY (1959b) are of the opinion that marine fungi are not necessary for *Limnoria* attack on submerged wood, but they point out that their experiments do not permit them to conclude that *Limnoria* requires nothing but wood for its nutrition.

Another report (KOHLMEYER *et al.*, 1959) indicates that *Limnoria lignorum* and *L. tripunctata* will, in fact, feed on sterile wood without conditioning by fungi. However, the life span of these animals is increased when marine Ascomycetes and fungi imperfecti and terrestrial Basidiomycetes are present in the wood. Also, it is noted that no eggs or young *Limnoria* are observed in the sterile wood.

Fig. 4.36a, b. *Chelura terebrans* Philippi, a) male, lateral view, b) female, dorsal view
(Photos by H. KÜHNE, courtesy of G. BECKER)

The isopods of the genus *Sphaeroma* are also small marine wood borers, measuring up to about twelve millimeters in length. Unlike *Limnoria*, they cannot digest the wood in which they burrow, but they use the timber for dwelling purposes and feed on plankton and other small marine organisms (BECKER, 1958).

The amphipod, *Chelura* (Fig. 4.36), is somewhat larger in size than *Limnoria*, but is not as important as a wood borer. It is usually associated with this genus and is active in moderately cool and subtropical waters.

### 4.3.3 Protection against Marine Wood Borers

The obvious choice for marine or coastal construction where marine borers are a problem is a timber species having natural resistance to attack. Tropical woods of high density and extractives content do have longer service life in waters of the temperate zone than species native to these regions. However, the natural durability of these same species, when used in tropical waters, seems to offer little advantage.

There seems to be no question but that the wood itself must be given mechanical or chemical protection against marine borer attack. Eradication of these marine animals by mechanical or electric shock, sonic and ultrasonic vibrations, chlorination through electrolysis of the sea water and poisoning with chemicals have been tried with varying degrees of success. However, all of these methods have certain objectionable features which make their general use impractical or unfeasible (BECKER, 1962).

In spite of its recognized limitations, the most successful wood preservative treatment for the protection of marine structures consists of coal-tar creosote or creosote and coal tar solutions applied with vacuum and pressure in a full-cell process. Recommended retentions of 14 to 20 lbs. per cu. ft. (225 to 320 kg/m³), or treatment to refusal, provide good protection (HUNT and GARRATT, 1953). Timber treated in this manner has a service life twenty to thirty years longer than untreated wood, in waters of the temperate zone. However, this same piling can fail in equatorial waters in less than ten years (VIND et al., 1956).

Trials on the use of solutions of creosote and coal tar have proven this combination to be very effective against *Limnoria*. The coal tar proportion of these mixtures ranges from 20 to 40% of the treating solutions (BECKER, 1962). Although many other preservatives and treatment methods have been investigated, creosote applied in full-cell processes will apparently remain as the standard for some time to come (RAY, 1958).

In estimating the probable service life of creosoted timber, an important factor to consider is the type of marine borer known to be active in the particular location. For example, *Limnoria tripunctata* is known to tolerate creosote better than any of the other borers, at least under the conditions where a certain amount of leaching takes place (VIND et al., 1956). The increased effectiveness of creosote against this same species through the addition of the synthetic contact insecticide, lindane ($\gamma$-hexachlor-cyclohexane) (BECKER, 1955) may, hopefully, offset this tolerance. Also, continued research on the use of heavy metal compounds, particularly copper (ROE et al., 1957) as additives to creosote, suggests that the future holds some promise for better protective measures against marine borer attack.

## Literature Cited

ARMSTRONG, F. H. and SAVORY, J. G., (1959) The influence of fungal decay on the properties of timber. Effect of progressive decay by the soft-rot fungus, *Chaetomium globosum* on the strength of beech. Holzforschung **13**: 84—89.

BAILEY, I. W. and VESTAL, M. R., (1937) The significance of certain wood-destroying fungi in the study of enzymatic hydrolysis of cellulose. J. Arnold Arboretum **18**: 196—205.

BECKER, G., (1942) Untersuchungen über die Ernährungsphysiologie der Hausbockkäfer-Larven. Z. Vergleich. Physiol. **29**: 315—388.

—, (1955) Über die Giftwirkung von anorganischen Salzen, o-Chlornaphthalin und Kontaktinsektiziden auf die Holzbohrassel Limnoria. Holz Roh- Werkstoff **13**: 457—461.

—, (1958) Holzzerstörende Tiere und Holzschutz im Meerwasser. Holz Roh- Werkstoff **16**: 204—215.

—, (1959) Biological investigations on marine borers in Berlin-Dahlem. In: Marine Boring and Fouling Organisms. Dixy Lee Ray, Ed. Univ. of Wash. Press. Seattle.

—, (1962) Status of biology and control of marine borers. Proceedings Fifth World Forestry Congress, Sept., 1960, Seattle, Wash. pp. 1522—1530.

—, KAMPF, W. D. and KOHLMEYER, J., (1957) Zur Ernährung der Holzbohrasseln der Gattung Limnoria. Naturwiss. **44**: 473—474.

134 Literature Cited

CAMPBELL, W. G., (1952) The biological decomposition of wood. In: Wood Chemistry. L. E. WISE and E. C. JAHN, Eds. 2nd. Ed., Vol. 2: 1061—1116. Rheinhold Publ. Corp. New York.

CARTWRIGHT, K. St. G. and FINDLAY, W. P. K., (1950) Decay of timber and its prevention. Chemical Publ. Co., Inc. Brooklyn, N. Y. 294 pp.

COCHRANE, V. W., (1958) Physiology of fungi. John Wiley & Sons, Inc. New York. 524 pp.

COWLING, E. B., (1961) Comparative biochemistry of the decay of sweetgum sapwood by white-rot and brown-rot fungi. Tech. Bull. No. 1258, U.S.D.A.

CRAIGHEAD, F. C., (1950) Insect enemies of eastern forests. U.S.D.A. Misc. Publ. No. 657. U.S. Govt. Printing Off. Washington, D.C. 679 pp.

CROSSLEY, R. D., (1956) The effects of five sapstain fungi on the toughness of eastern white pine. Unpublished M.S. thesis, State Univ. College of Forestry at Syracuse Univ., Syracuse, N.Y.

DUNCAN, C. G. (1960a) Wood-attacking capacities and physiology of soft-rot fungi. U.S. For. Prod. Lab. Report No. 2173.

—, (1960b) Soft-rot in wood, and toxicity studies on causal fungi. Proceedings A.W.P.A., 1960.

FINDLAY, W. P. K. and SAVORY, J. G., (1950) Breakdown of timber in water-cooling towers. Proceedings Intl. Bot. Congress 7: 315—316.

FISHER, R. C., (1928) Timbers and their condition in relation to *Lyctus* attack. Forestry (London) 2: 40—46.

GÄUMANN, E., (1939) Über die Wachstums- und Zerstörungsintensität von *Polyporus vaporarius* und von *Schizophyllum commune* bei verschiedenen Temperaturen Angew. Bot. 21: 59—69.

HOWICK, C. D., (1962) Sirex — A threat to the timber industry? C.S.I.R.O., Australia, Forest Products Newsletter, No. 292.

HUNT, G. M. and GARRATT, G. A., (1953) Wood Preservation. 2nd Ed. McGraw-Hill Book Co. New York. 417 pp.

JENNISON, M. W., (1952) Physiology of the wood-rotting fungi. Final report, ONR, Microbiology Branch. ONR Research Contract NR 132—159. Syracuse Univ., Syracuse, N.Y.

KEMNER, N. A., De ekonomiskt viktiga vedgnagande anobierna, Medd. 108, Centralanstalten för försöksväsendet, Ent. avd., Stockholm 1915.

KOHLMEYER, J., BECKER, G. and KAMPF, W. D., (1959) Versuche zur Kenntnis der Ernährung der Holzbohrassel *Limnoria tripunctata* und ihre Beziehung zu holzzerstörenden Pilzen. Z. Angew. Zool. 46: 457—489.

LIESE, W. and SCHMID, R., (1961) Licht- und elektronenmikroskopische Untersuchungen über das Wachstum von Bläuepilzen in Kiefern- und Fichtenholz. Holz Roh- Werkstoff 19: 329—337.

LILLY, V. G. and BARNETT, H. L., (1951) Physiology of the fungi. McGraw-Hill Book Co., Inc. New York, 464 pp.

LINDGREN, R. M., (1953) An overall look at wood deterioration. U.S. For. Prod. Lab. Report No. 1966.

MALLIS, A., (1960) Handbook of pest control. 3rd. Ed. MacNair-Dorland Co., New York. 1132 pp.

MEIER, H., (1955) Über den Zellwandabbau durch Holzvermorschungspilze und die submikroskopische Struktur von Fichtentracheiden und Birkenholzfasern. Holz Roh- Werkstoff 13: 323—338.

MENZIES, R. J., (1959) The identification and distribution of the species of *Limnoria*. In: Marine Boring and Fouling Organisms. Dixy Lee Ray, Ed. Univ. of Wash. Press. Seattle.

— and TURNER, R., (1956) The distribution and importance of marine wood borers in the United States. In: ASTM Spec. Techn. Publ. No. 200, Symposium on Wood for Marine Use and its Protection from Marine Organisms.

PENCE, R. J., (1957) The prolonged maintenance of the western subterranean termite in the laboratory with moisture gradient tubes. J. Econ. Entomol. 50: 238—240.

P. D. C. NEWSLETTER, (1958) Prevention of Deterioration Center, National Academy Science —National Research Council. Washington, D.C. Jan. 1958.

PROCTOR, P., JR., (1940) Penetration of the walls of wood cells by the hyphae of wood-destroying fungi. Unpublished Ph.D. dissertation, Yale Univ. School of Forestry, New Haven, Conn.

RAY, D. L., (1958) Recent research on the biology of marine wood borers. Proceedings A.W.P.A., 1958.

—, (1959a) Nutritional physiology of *Limnoria*. In: Marine Boring and Fouling Organisms. Dixy Lee Ray, Ed. Univ. of Wash. Press. Seattle.

—, (1959b) Marine fungi and wood borer attack. Proceedings A.W.P.A., 1959.

— and STUNTZ, D. E., (1959) Possible relation between marine fungi and *Limnoria* attack on submerged wood. Science 129: 93—94.

## Literature Cited

ROE, T., HOCHMAN, H. and HOLDEN, E. R., (1957) Performance tests of heavy metal compounds as marine borer inhibitors. ASTM Spec. Tech. Publ. No. 200: 29–32.

ST. GEORGE, R. A., JOHNSTON, H. R. and KOWAL, R. J., (1960) Subterranean termites, their prevention and control in buildings. U.S.D.A. Home and Garden Bull. No. 64. 30 pp.

SAVORY, J. G., (1954) Damage to wood caused by microorganisms. Proceedings Symposium on Microbial Spoilage in Industrial Materials, Paper III. J. Applied Bacteriol. **17**: 213–218.

— and PINION, L. C., (1958) Chemical aspects of decay in beech wood by *Chaetomium globosum*. Holzforschung **12**: 99–103.

SIMEONE, J. B., (1954) Carpenter ants and their control. Bull. No. **34**, State Univ. College of Forestry at Syracuse Univ. 19 pp.

—, (1960) Survey of wood-feeding Anobiidae in northeastern United States, including a study of temperature and humidity effects on egg development of *Hadrobregmus carinatus* (Say.). Proceedings XI. Internationaler Kongress für Entomologie, Wien, pp. 326–335.

—, (1961) Pole deterioration by wood destroying insects. Proceedings Eeastern Wood Pole Conference, State Univ. College of Forestry at Syracuse University, Sept. 12–14, 1961. pp. 15–22.

SIU, R. G. H., (1951) Microbial decomposition of cellulose. Rheinhold Publ. Corp. New York. 531 pp.

SNYDER, T. E., (1926) Preventing damage by *Lyctus* powder-post beetles. U.S.D.A. Farmers Bull. No. 1477. 13 pp.

—, (1927) Defects in timber caused by insects. U.S.D.A. Dept. Bull. No. 1490. 47 pp.

—, (1948) Our Enemy the Termite. Comstock Publ. Co., Ithaca, N.Y. Rev. Ed. 257 pp.

—, (1949) Catalog of the termites (Isoptera) of the world. Publ. No. 3953, Smithsonian Misc. Collections. Vol. 112 (Whole Volume).

TILLYARD, R. J., (1926) The insects of Australia and New Zealand. Angus & Robertson, Ltd. Publ. Sydney. 560 pp.

TOOKE, F. G. C., (1949) Beetles injurious to timber in South Africa. Science Bull. No. 293, Dept. of Agric., Entomology Series No. 28. Pretoria. 95 pp.

TURNER, R. D., (1959) The status of systematic work in the Teredinidae. In: Marine Boring and Fouling Organisms. Dixy Lee Ray, Ed. Univ. of Wash. Press. Seattle.

VIND, H., HOCHMAN, H., MURAOKA, J. and CASEY, J., (1956) Relationship between *Limnoria* species and service life of creosoted piling. In: ASTM Spec. Techn. Publ. No. 200, Symposium on Wood for Marine Use and its Protection from Marine Organisms.

WIGGLESWORTH, V. B., (1947) The principles of insect physiology. 3rd. Ed. Methuen & Co., Ltd. London.

ZABEL, R. A. and ST. GEORGE, R. A., (1962) Wood protection from fungi and insects during storage and use. Proceedings Fifth World Forestry Congress, Sept. 1960, Seattle, Wash pp. 1530–1540.

# 5. WOOD PRESERVATION

## 5.0 Introduction

In Chapter 4 the biological aspects of wood deterioration were discussed, These considerations of the susceptibility of wood to attack by wood-destroying fungi, wood-boring insects and marine borers emphasize the disadvantages of using untreated wood in applications where it is exposed to such agencies. In addition to the serious losses caused by these natural organisms, there is also much wood deterioration brought about by physical, mechanical and chemical means. VAN GROENOU *et al.* (1951) classify fire, heat and moisture as physical agents of wood deterioration; fracture, wear or permanent deformation as mechanical influences; and the action of acid, caustic and salt as chemical attack.

Though the natural durability of wood offers a certain degree of resistance to these agencies, there is evidently no ideal wood which can be used in any application advantageously and without eventual break-down. As wages continue to rise, the labor cost of replacing deteriorated wood is frequently greater than the value of the material replaced. Because of such economic considerations, the field of wood preservation is an important part of wood technology. Wood preservation includes every process of chemical or physical treatment which is undertaken to extend the life of wood in service by increasing its resistance to biological attack, fire, shrinkage and swelling due to changes in moisture content and, in fact, to any of the agencies mentioned above.

A wide range of methods for preservative treatment has been developed during the long history of wood preservation practices. An ever increasing list of wood preservatives, each having certain characteristic properties, is available to the field. A survey of the processes or application methods will be followed by a discussion of preservative materials and their properties. Fire retardants, their characteristics and their application are discussed as a separate topic, as are the treatments for improving the dimensional stability of wood.

## 5.1 General Considerations

If the wood preservative is to provide protection to all parts of the wood susceptible to attack or deterioration, it must be distributed in sufficient concentration throughout the vulnerable areas. Many methods of preservative application have been developed, some of them inefficient, some of them inadequate, but all intended to extend the service life of wood. The real measure of their effectiveness cannot be ascertained immediately after treatment, although various types of accelerated tests are designed to help predict the probable service life. From experience it is known that the factors of preservative retention and penetration, as well as distribution, can be used as indicators of the efficiency of treatment. This assumes, of course, that the preservative material is known to have some degree of effectiveness in resisting attack by the various wood deteriorating agencies.

Net retention of preservative by the treated wood is not an adequate index of the efficacy of treatment. As HUNT and GARRATT (1953) point out, the preservative can be concentrated in certain areas of the wood, leaving wide variations in depth of penetration. Neither is penetration alone a sufficient index because the concentration of toxicant in the treated zone is important. Even distribution is also desirable or there may be areas less well protected than others. In practice, attempts are made to measure all three factors to correctly evaluate the efficiency of the treatment, but not all are readily measurable.

*Retention* can be measured by determining the quantity of solution absorbed by the wood. In large operations net retentions are determined on a volume basis from tank gauge readings and then converted to weight of preservative per unit volume of wood. Weighing the wood before and after treatment is a direct means of determining retention. When treating with creosote, net retention is the actual average weight of effective preservative retained per unit volume of wood. Specifications for retentions of water-borne preservatives are given on the basis of weight of dry salt per unit volume. The concentration of the solution must be known in order to convert liquid retention to dry salt retention.

*Penetration*, the depth to which the preservative enters the wood, is measurable by borings. A single boring does not provide a reliable value for average penetration, but sampling can be used when deemed necessary. This system is also an aid in determining *distribution* of the preservative although it can never be more than an indication. The determination of the depth of penetration of colorless materials requires staining techniques.

These factors of impregnation are influenced by the anatomical structure of the wood being treated, pre-treatments of the wood, such as seasoning and incising, and the method used to introduce the preservative into the wood. A large degree of control can be exercised over treatment methods and preparation of the wood to be treated. However, there is no control possible over the anatomy of the wood except the freedom to select species.

### 5.1.1 Effect of Structure on Treatment

Since the structure of wood is obviously a very important factor in its impregnation with preservative, it would seem that its relationship to wood preservation would have been investigated very intensively. However, this is not the case. The literature contains many more references to specifications, practice and technique than to the critical factor of wood anatomy.

In the reports on the effect of wood structure, more deal with the bordered pit pairs, which serve as communication channels between adjoining coniferous tracheids, than with any other single topic. This is perhaps to be expected since softwoods, as a class, are far more important for structural applications and are treated with preservative in much greater volume than hardwoods, and these pits are the most critical, single feature influencing liquid movement in the relatively simple organization of tissues in coniferous woods.

The structure of the membrane in coniferous bordered pits is now well known and the new evidence supports the early work of BAILEY (1913). The pit membrane is definitely porous in softwoods and permits the flow of liquids between tracheids if the pit is not aspirated. LIESE (1954), FREY-WYSSLING *et al.* (1956), HARADA (1956), CÔTÉ (1958) and others all report confirming evidence of this porosity using electron microscopy. A detailed discussion of pit structure is given in Chapter 1.

The question of pit aspiration and its effect on the penetrability of softwood is still somewhat of a controversy. CÔTÉ and KRAHMER (1962) demonstrated the efficiency of pit aspiration in soft pine through the use of India ink carbon particles impregnated into the wood. Electron micrographs of ultra-thin sections through the treated wood (Fig. 5.1) show that the particles cannot move through an aspirated pit, but there is evidence that diffusion through the torus can continue. Further research on aspiration in fresh, green wood is handicapped by technique limitations, and conclusive evidence is not easily developed.

Unfortunately, pit membrane structure does not appear to be the only factor affecting the permeability of wood. Mountain-type Douglas-fir resists penetration even though many pits are supposedly unaspirated. In most species the sapwood is more readily penetrated than is the heartwood. This is generally explained on the basis of changes which occur in the wood as sapwood is transformed into heartwood. The accumulation of extractives, gums, tannins, resins and other organic materials usually accompanies the formation of heartwood. While these accumulations do not materially alter the structure of wood as seen with the light microscope, cell lumens become occluded in some cases. Hardwood vessels in some species are filled with tyloses by the time heartwood has developed. Where the cell cavities remain clear, no change in penetrability should be expected. Nevertheless, there is variation and at least some of it can be explained on the basis of pit membrane encrustation.

At the high resolution and magnification possible with the electron microscope, a sharp structural difference between sapwood and heartwood pit membranes can be observed in some species. The cellulosic strands of the membrane in coniferous sapwood are free of encrusting materials (see Fig. 1.31 and 1.32), while in the heartwood they are often encrusted (see Fig. 1.33). In some cases of encrustation the pores in the membrane are virtually closed while in others they are only partially occluded. The effect that this condition can have on the movement of liquids is obvious. However, this encrustation is not found in all refractory heartwood (KRAHMER and CÔTÉ, 1963). It should also be noted that encrustation is not limited to heartwood, but can be found in the transition zone between sapwood and heartwood in some cases.

In hardwoods, where pit membranes have the structure of primary wall, there are no membrane openings visible even at extremely high magnifications (see Fig. 1.37b). Diffusion through these membranes must be considered to be the principal method of movement from the vessels into surrounding fibers, fiber-tracheids, ray parenchyma cells or other contiguous elements. Encrustation of the membrane surface occurs in the heartwood of hardwood species so that diffusion rates are presumably reduced (see Figs. 1.37a and 1.38).

Resin canals are known to be channels for preservative movement in some coniferous woods, while in others they appear to be so badly occluded with tylosoids or hardened materials that flow must be limited. Some softwoods have no normal resin canals. When resin canals are present, they may be small or large, sparse or numerous, localized or well distributed. In addition, communication from the canals to the neighboring cells is variable so that, as a general rule, resin canals cannot be considered to be dependable avenues for the movement of preservatives.

The significance of rays in the penetration of wood by preservatives has received very little attention. Nevertheless, the general concensus is that rays are not of great importance, except in woods which have ray tracheids. However, in a study by SARGENT (1959), it was demonstrated that rays are indeed penetrable by creosote under mild pressure and cannot be dismissed as insignificant. The rays in heartwood are less frequently penetrated than those in sapwood of the same species. It was also noted that the characteristic ranges of ray penetrability correspond with the reputed ease of penetration of the species.

## 5.1 General Considerations

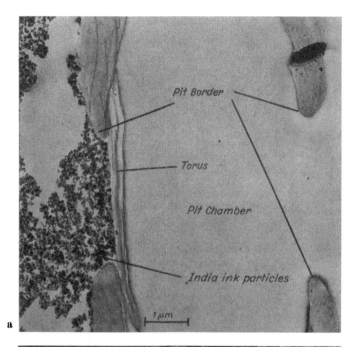

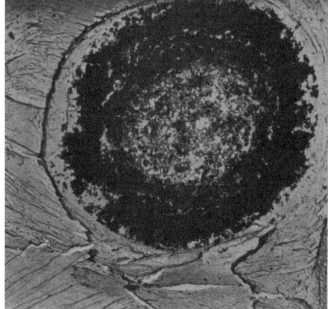

Fig. 5.1.a The torus in this aspirated bordered pit in white pine wood effectively seals the pit opening and prevent the movement of India ink (carbon) particles through the pit and into the adjoining tracheid. Electron micrograph of ultrathin tangential section of impregnated wood. (11,600×)

Fig. 5.1b. Face view of a bordered pit membrane in white pine wood impregnated with India ink. The accumulation of carbon particles in the membrane with relatively little concentration on the central torus indicates that some diffusion takes place even though the pit is in aspirated condition. Electron micrograph of pseudo-replica. (3,800×)
(Both photographs from CÔTÉ and KRAHMER (1962), by permission TAPPI)

The important structural factor of capillary size and its effect on penetration is complex and one which is extremely variable. Capillary size varies between earlywood and latewood of one growth ring. Since it is a function of cell type, it also varies markedly between species. The significance of capillary size in non-pressure treatment processes is great since penetration by liquid movement is largely dependent upon capillarity. In pressure treatment, on the other hand, larger capillaries are desirable as they offer less resistance to penetration.

### 5.1.2 Timber Preparation

Satisfactory penetrations cannot be obtained in wood which has not been adequately prepared for treatment. Round timbers must be debarked and seasoned to sufficiently low moisture content before most preservative treatment. Bark is

Fig. 5.2. A piece of Douglas-fir wood incised before pressure treatment with wood preservative of water-borne type

left on round timbers in the Boucherie process, however. For processes where treatment depends on diffusion of water solutions of salts, seasoning below the fiber saturation point is not critical. In some methods, the green wood can be conditioned to reduce the moisture content by vapor drying or boiling under vacuum as an integral part of the treatment process. Mechanical means of improving penetration without reducing strength can be used advantageously. For example, the incising of timber to increase penetration depth in heartwood has been found to be effective in increasing retention as well (Fig. 5.2).

## 5.2 Wood Preservation Processes

All of the wood preservation processes in current use can be placed under one of the following categories: non-pressure, diffusion, sap replacement, or pressure and vacuum impregnation. Each of these general methods will be taken up in some detail. The major factors influencing the choice of a particular method of application will be considered. Cost is often the major consideration. Simplicity of equipment is another. The simplest processing method generally requires the least expensive equipment, but at the same time, it is often the least effective for long term protection of the material. However, for some requirements it is not necessary to extend the service life of a pole, post or timber more than two or three years beyond the normal life of untreated material. So the choice of wood preservative and application method depends upon many factors, each of which should be considered from the technological and economic viewpoint.

### 5.2.1 Non-pressure Processes

Non-pressure processes include any method where no external pressure is applied to force the wood preservative into the timber. These include brushing, spraying, dipping, steeping, cold soaking and hot-and-cold bath. Diffusion methods can also be considered in this category.

**5.2.1.1 Brushing or Spraying.** The application of wood preservatives by brushing, painting or spraying is the simplest treatment available. It requires a minimum investment in equipment and can be employed for applying oil-borne and water-borne preservative chemicals, coal-tar creosote or other low viscosity oils. This method permits treatment to be carried out at the construction site or on wood parts already in service.

Even when used in treating well seasoned material, the effect of brush or spray treatments is superficial. They cannot be recommended except as temporary expedients. The moderate penetration that results is seldom more than a few millimeters in depth. If any physical damage should rupture the thin protective shell, the piece is subject to attack through the open area. Water-borne chemicals are readily leached out of such timbers if they are in the open, but oily materials should provide protection for somewhat longer periods. Under optimum conditions the normal service life of the wood can be extended one to three years, assuming that surface cracks and checks are thoroughly filled and that generous quantities of preservative are applied over the entire surface of the timber.

**5.2.1.2 Dipping.** Another non-pressure process for applying preservative materials is dipping. This method involves the immersion of the wood in a treating solution for a period of a few seconds to a few minutes. It provides little more effectiveness than brushing or spraying except that end penetration is frequently better in easily treated species. Complete immersion probably provides greater uniformity of coverage than brushing and gives more assurance that all checks are filled.

BROWN et al. (1956) found that dip treatment is not effective if the wood is at a moisture content above the fiber saturation point. Their studies also show that over half of the total absorption occurs in the first fifteen seconds of a ten-minute immersion time and that then the rate of absorption decreases with time. It should be emphasized that these tests were made on specimens having an end-to-side surface ratio of 1:20 and the results may not apply directly to dipping treatments of long length materials. Transverse penetration is limited so that retention is largely dependent upon end penetration.

Dipping has been found to be particularly well suited for the treatment of millwork such as window sash at the factory. Paintable preservative in a non-swelling carrier (NSP), of which pentachlorophenol is a good example, is widely used. Dipping times are approximately three minutes. Dip treatment extends service life two to four years when the wood is not subjected to physical damage.

VERRALL (1961) stressed that surface treatments must be supplemented by adequate structural design to provide good protection to wood in service.

**5.2.1.3 Steeping and Cold Soaking.** Steeping and cold soaking are merely prolonged immersions of wood in preservative solutions. The term cold soaking is generally applied to the use of oil solutions while steeping refers to the soaking of wood in water solutions of preservative. Kyanizing is a steeping process in which mercuric chloride in water is the solution used.

Cold soaking has been shown to be a rather effective method of treating seasoned material for farm use because of its simplicity. Fuel oil solutions of pentachlorophenol are commonly employed. The more viscous oils are not as satisfactory unless heated to reduce viscosity. Soaking times are not critical and may be extended for long periods although two to several days is usually sufficient. Actually, a large proportion of the absorption takes place during the first day of treatment, but prolonged soaking does increase depth of penetration and amount of retention. As would be expected from anatomical considerations, the sapwood of certain softwoods is easily treated by this method. Hardwoods having tylosisfree vessels show good end penetration, but generally poor transverse penetration.

In the steeping process it is possible to use green as well as seasoned timber because salt from the treating

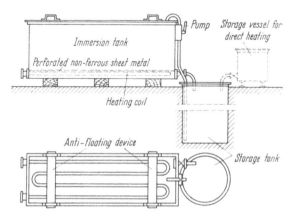

Fig. 5.3. Scheme of a small plant for Kyanizing (From A. NOWAK)

solution can move into wet wood by diffusion. Seasoned timber absorbs both water and salt so that lower concentrations of salt solutions can be used. When treating green material, stronger solutions should be used to offset the dilution and to speed up the rate of diffusion.

Though not so widely used in the Unites States, steeping, particularly with mercuric chloride in the Kyanizing process, has been used for many years in Europe. According to HUNT and GARRATT (1953), it is being replaced by other processes there while in the United States there is no longer any commercial operation using Kyanizing. Mercuric chloride is a highly toxic and effective wood preservative, but it is also very corrosive and deadly poisonous. These disadvantages probably account for its decreasing use. Only non-metallic materials can be employed in treating equipment, seriously limiting the design of pressure equipment for mercuric chloride application. Steeping with mercuric chloride has been done in concrete or wooden tanks because of this limitation (Fig. 5.3).

**5.2.1.4 Hot-and-Cold Bath.** The hot-and-cold bath process is undoubtedly the most effective of the so-called non-pressure treatments. The effectiveness of the method can actually be attributed to the mild vacuum which is produced by the process, though not by mechanical equipment. The poles, posts or other timbers are first heated in preservative solution, or in a dry kiln in some cases. This causes the air in the cells of the outer layers of the wood to expand. The heated material is then transferred to the cold preservative solution. The warm air in the wood cells contracts upon cooling and creates a partial vacuum in the outer portions of the wood. As atmospheric pressure tends to satisfy this mild vacuum, penetration of preservative into the wood is aided.

The mechanics of the operation are very flexible and can be adjusted to the conditions at hand. Heating can be accomplished in a kiln, in a tank of preservative or in water, depending on the choice of treating chemical. The material being treated need not be removed from the hot tank if the solution can be pumped out and quickly replaced by cold preservative. Though possibly not quite as effective, the hot liquid can simply be allowed to cool.

Though coal-tar creosote and other preservative oils are generally used in this treatment process, water-soluble salts can also be applied very effectively by this method. Care must be exercised to limit the temperature of the hot bath to a level that is safe for the particular solution being employed while the cold bath must be warm enough to ensure liquid flow. If too high temperatures are reached, the oily preservatives are likely to evaporate. Water solutions are subject to this danger as well as to the possibility of precipitating part of the salts out of solution.

Using Standard C10 of the A. W. P. A. Manual of Recommended Practice (1963) as an example, temperature ranges specified for the thermal (hot and cold) process treatment of full length lodgepole pine poles are as follows. The hot bath should be maintained at a temperature between 190 °F. (88 °C.) and 235 °F. (113 °C.) for a minimum treatment period of six hours. The temperature range of 90 °F. (32 °C.) to 150 °F. (65°C.) is indicated for a cold bath treatment of two hours. These temperatures are specified for creosote as well as for pentachloro-phenol-petroleum solution treatments.

A review of the effectiveness of the double-diffusion method by BAECHLER and ROTH (1964) reveals that this system provides deeper penetrations and greater retention than the single-diffusion method. In double-diffusion treatment, green wood is soaked first in one chemical solution and then in another. The water soluble chemicals diffuse into the wood where they react with each other to form compounds of relative insolubility. Careful selection of treating chemicals produces compounds that are non-leachable as well as toxic to wood deteriorating agents.

The same standard requires a penetration depth of 3/4 inch (2 cm) and 85% of the sapwood. Retentions of 20 lbs. per cu. ft. (32 kg/m³) of creosote or 1.00 lb. per cu. ft. (16 kg/m³) of dry penta in the outer 1/2 inch (12.5 mm) are also stipulated. The treatment times indicated above must be extended if minimum retention and penetration standards are not met.

**5.2.1.5 Diffusion Method.** In the diffusion method of treatment, green timber is gradually penetrated by a water-soluble salt which is generally applied in concentrated form. The best known example is the Osmose process in which the toxic chemicals in paste form are coated over the surface of green, peeled timber. Over a period of weeks the preservative diffuses into the green wood, provided the timber is stacked and carefully covered to prevent moisture loss. Variations of this method include the use of preservative bandages which are wrapped around individual poles, either after an application of chemicals, or with a layer of preservative lining the bandage itself. The purpose of the bandage is to prevent the loss of moisture from the unseasoned timber and to keep a supply of chemicals in contact with the wood. The steps in the treatment of poles by the Osmose process are illustrated in Fig. 5.4.

As with other non-pressure treatment methods, there is no great degree of control over depth of penetration except through duration of stacking. Duration of treatment can be adjusted from about 30 days for small material to 90 days for timber requiring greater penetration. The effect of stacking time on depth of penetration for pine and spruce is shown in Fig. 5.5.

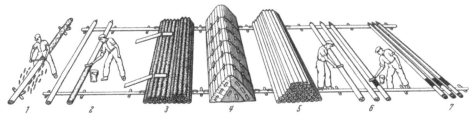

Fig. 5.4. Osmose process shown diagrammatically

Diffusion treatment is also employed in the protection of the groundline zone on standing poles. Holes are bored into the poles near the ground line, preservative is introduced into them and then the holes are plugged. The toxic chemical is believed to diffuse through this critical zone over a period of time.

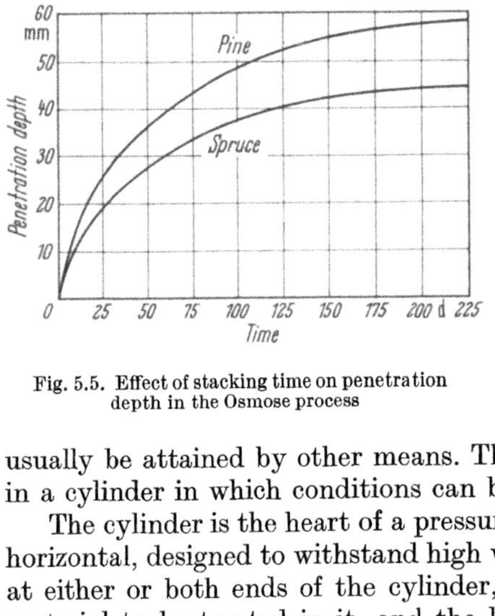

### 5.2.2 Pressure Processes

The preservative treatment of wood by pressure methods is the preferred commercial approach because of its greater efficiency and effectiveness. Its efficiency stems from the much closer control over treating conditions than is possible with the non-pressure processes. Its effectiveness is due to the more uniform, deeper penetration and greater absorption of preservative than can usually be attained by other means. The fact that the timber is totally enclosed in a cylinder in which conditions can be varied widely offers a great advantage.

Fig. 5.5. Effect of stacking time on penetration depth in the Osmose process

The cylinder is the heart of a pressure treating plant. It is a steel tank, usually horizontal, designed to withstand high working pressures. Doors may be installed at either or both ends of the cylinder, depending on its size, the nature of the material to be treated in it, and the loading system used. In the treatment of poles, piling, and other large timbers, the charge can be rolled into the cylinder on standard or narrow gauge rail trams and rolled out of the cylinder to the yards. Hand loading or crane systems are used for smaller material and small treating cylinders.

Accessory equipment must be provided for heating and storing preservative, for transferring it in and out of the treating cylinder and for measuring the amount of preservative consumed in treating a charge. In addition, compressors and pumps are required for vacuum and pressure phases of the treating schedule and gauges must be installed to indicate these conditions.

Steam is generally used for heating the preservative since it can also be employed for special treatments of the charge in the cylinder. The size and complexity of storage facilities for preservative at a treating plant depends on the size and scope of its operations.

In a small plant it may be possible to limit pumping equipment to a vacuum pump and a hydraulic pressure pump while in larger operations a larger number and variety of pumps are necessary. When a compressed air system is used, only a vacuum pump may be needed to supplement the air compressor.

Pre-treatment with steam, vacuum or air pressure permits a greater range of control over the final treatment with preservative. Each of the well-known treating methods using pressure is based on a variation in treating schedule of one or more of the above factors. The desired method is selected because of its characteristic preservative retention which in turn can be related directly to the cost of treatment chemicals.

The terms "empty-cell process" and "full-cell process" are frequently applied to treatments by pressure methods. Though these terms may not be strictly accurate, they can be applied in a relative way in describing the effect of a particular treatment schedule. Cell lumens in the penetrated portions of the wood treated by the full-cell process are supposedly full of preservative. In the empty-cell process the lumen walls are left with only a coating of chemical. Diffusion from the lumen into the cell wall must take place in both cases, but it would be expected that greater concentrations of preservative would ultimately be found in the walls of wood treated by the full-cell process.

**5.2.2.1 Full-cell Processes.** The primary objective of a full-cell treatment is to attain maximum retention of preservative in the treated portion of the timber. The factor which distinguishes it from empty-cell treatment is the preliminary vacuum which is designed to remove as much air from the cells as possible, thereby removing the air cushion which resists preservative penetration. A further advantage is that there is a minimizing of preservative release ("kick-back") caused by the expansion of trapped air when pressure is removed from the cylinder. The phases of a full-cell process are illustrated graphically in Fig. 5.6.

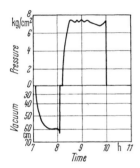

Fig. 5.6. Full-cell process

The Bethell process is a full-cell process that is employed in treating with oils. A preliminary vacuum is applied to the charge and held for a period up to one hour. Then, without releasing the vacuum, the cylinder is filled with preservative and a pressure of 125 to 200 lb. per sq. in. (8.75 to 14 kg. per sq. cm.) is applied and maintained until the desired absorption is reached. During the treatment, the preservative temperature is held between 180°F. and 210°F. (82°C. to 99°C.), depending upon the particular oil being used. After the preservative has been drained from the cylinder, it is customary to apply a mild final vacuum to reduce preservative dripping from the treated timber (MACLEAN, 1960).

Since seasoned timber is necessary for effective treatment with oily preservatives, treatment in an enclosed cylinder permits the pre-treatment of green material to reduce moisture content. Steaming-and-vacuum, vapor drying, or the Boulton boiling-under-vacuum process can precede the actual preservative treatment with oils. When water-borne preservatives are to be used in a full-cell process, seasoning below the fiber saturation point before treatment is less important, but air seasoning or kiln drying after treatment is, of course, necessary.

The Burnett process was developed for impregnating with zinc chloride solutions using full-cell treatment, but zinc chloride is no longer much used alone. Chromated zinc chloride and copperized chromated zinc chloride are used in place of zinc chloride because of their lower leachability and less corrosiveness. When these and other water-borne preservatives are used with the full-cell process, essentially the same schedule is followed as with oils. The concentration of the treating solution is varied to provide a retention of 1/3 to 1 1/2 lbs. of dry preservative per cu. ft. of wood (5.3 to 24 kg/cu m).

The high net retentions attainable with the full-cell process can result in rather high preservative costs. Its use can be justified for marine applications where maximum retentions of creosote are necessary for effective protection. In certain tropical situations the high costs are also justifiable. However, in many other uses, the service life obtained from timber treated by the empty-cell process is adequate and the treatment costs are lower.

HENRIKSSON (1954) has devised a method of rapid oscillating vacuum-pressure treatment as a means of increasing penetration in full-cell treatment of green timber with water-borne preservatives. The aqueous solution is applied at a pressure of 115 lbs. per sq. in. (8 kg. per sq. cm.), quickly released and then a vacuum of about 28 inches (70 cm) is drawn. A complete cycle can be as short as 1 minute or as long as 6 minutes and the cycling is controlled automatically so that it can be left overnight. The claim was made that treatment of unseasoned sapwood of some resistant species results in better penetration by this process than by conventional full-cell treatments. However, the evaluation reports do not indicate very great improvement (BLEW et al., 1961).

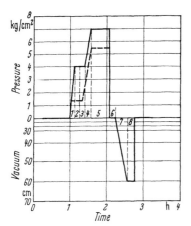

Fig. 5.7. Empty-cell process

**5.2.2.2 Empty-cell Processes.** The empty-cell process differs from the full-cell process in that some means of recovering much of the preservative is used, leaving no liquid preservative in the cell lumens of the treated portions of the wood. In the original Rüping process, this is accomplished by applying compressed air to the timber before forcing the preservative into it. The preservative can be admitted into the treating cylinder from an equalizing tank where the air in the cylinder can interchange with the preservative. This procedure traps air in the cells and when the pressure is released after treatment, the trapped air expands and forces the preservative out. A final vacuum serves to remove even more of the solution (Fig. 5.7).

In the Lowry process there is no preliminary air pressure applied, but the schedule is otherwise the same as for the Rüping process. This eliminates the need for an air compressor.

### 5.2.3 Miscellaneous Processes

The Boucherie process is the best known of the sap replacement treatment methods. It has been in use for more than a century and variations of the original system are still employed. Designed originally for the treatment of standing trees with copper sulfate solution, the system has evolved into one in which unpeeled poles are placed on horizontal supports with the butt ends slightly raised. A watertight cap is placed around the butt end of each pole and a hose connection from the preservative reservoir is attached to the cap. The reservoir is elevated

35 feet (10 m) or more, providing hydrostatic pressure to force the preservative through the sapwood and push the sap out of the green timber (Fig. 5.8).

GEWECKE (1957) reported on an adaptation of the Boucherie process which approaches vacuum and pressure treatment, but without the use of a pressure cylinder. Suitable connections are made to both butt end and top end of green spruce or fir poles. Vacuum is drawn at the top end of the pole while the solution of treating chemical is applied under pressure to the butt end. This method is designed to overcome the otherwise slow treatment afforded by pure hydrostatic pressure and to reduce the loss of preservative from the small end of the pole.

Another adaptation of the Boucherie technique was described by WIRKA (1945) for the treatment of green fence posts. In this case, an automobile tire tube is cut, placed over the end of the post, tightly fastened, and then the free section of the tube is used as a reservoir to hold the solution. This is an acceptable means of treating posts in a small operation such as a farm.

Wolman salts and other water-borne preservatives have largely replaced copper sulfate for sap replacement treatment in Europe. Copper sulfate proved to be too corrosive and was also very readily leached. It is now used in mixture with other salts.

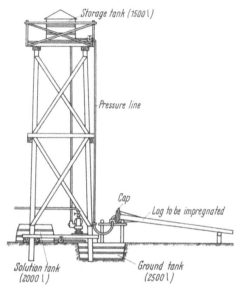

Fig. 5.8. Boucherie process

A new concept in the application of wood preservatives is reported by GOODWIN and HUG (1961). The method, patented as the "Cellon Process", consists of using liquefied gas as a carrier for pentachlorophenol and water-repellent chemicals. The solutions are forced into the wood and when pressure is released, they evaporate leaving the preservatives in the wood. The vapors are recovered and liquefied so that the wood is left free of carrier.

## 5.3 Wood Preservatives

In attempting to find the most efficient method of protecting wood from attack or deterioration, it would seem logical to consider the nature and requirements of the causal agency or organism. In the case of wood-destroying fungi, for example, the approach that can be taken is to eliminate one or more of the basic needs of the fungus: oxygen, moisture, raw material (food) and favorable temperature. These factors, and those for wood-boring insects and marine borers, are discussed in detail in Chapter 4. Wood preservation is concerned with the elimination or control of these essentials, but for the most part preservative treatments are limited to the introduction of toxic chemicals into wood.

### 5.3.1 Characteristics of Preservatives

The selection of chemicals poisonous or repellent to fungi and to wood-boring insects developed primarily on an empirical basis in the early history of wood preservation efforts. At the present time the effectiveness of a preservative is predictable to a certain extent and can be tested in the laboratory by one of several methods before being placed into field use. Of the hundreds of materials that have been evaluated over the years, only a relatively small number have proven to be practical.

The principal criteria used in evaluating a wood preservative are: 1. Degree of toxicity toward fungi, insects and marine borers. The concentration of chemical required to be effective against the more tolerant organisms must be known. While toxicity is the most important factor, it cannot be considered independently of many others. 2. Permanence. The preservative must continue to be effective for many years. If it is readily leached from the wood, as are many of the water-borne salts, the timber is left with reduced protection. 3. Penetrating ability. The chemicals must be well distributed deep inside the wood as well as at the surface. If the chemical is too viscous, or if it will not penetrate because of some other characteristic, high toxicity and permanence are of little real value. 4. Non-corrosiveness to metals. In cases where metal fasteners or fittings must be employed with treated timbers, the corrosive effects of certain preservatives can seriously weaken the structure. 5. Non-damaging to wood. Similarly, if the required concentration of chemical attacks the carbohydrates or lignin of the wood, degradation may weaken the timber. 6. Safety in handling. Preservation chemicals must be safe to handle in the treating process and should offer no serious health hazard when the treated timber is placed in service. Only a limited number of chemicals are safe enough to be used in the treatment of food containers, but this degree of safety is unnecessary for most applications. 7. Cost. In the final analysis, a chemical that fulfills all of the above requirements may be unfeasible to use because of excessive cost. It must be available in sufficient quantity and at a price that will permit its commercial use. 8. Fire resistance. When more than one benefit can be derived by treating with a particular preservative, it has a decided advantage in the market. Certain salts are adequately toxic and offer the added feature of retarding burning. In any case, preservatives should not increase the fire hazard.

The importance of any one of these criteria may deserve more consideration than others, but the preservatives that are used in greatest volume today meet most of these requirements.

### 5.3.2 Preservative Materials Toxic to Insects, Fungi and Marine Borers

There are three general classes of wood preservatives: oils, water-soluble chemicals and chemicals soluble in volatile oils or organic solvents. Each type has a particular suitability for a given application. Listing every known preservative in each of the categories would be of little value as well as repetitious since VAN GROENOU et al. (1951) devote an entire book to a survey of wood preservatives used throughout the world up to 1950. They also include coverage of trade names and proprietary materials, referencing all of their data with extensive bibliographic lists. It is interesting to note, however, that only a very small percentage of this long list of chemicals actually enjoys wide commercial use.

For example, the statistics for 1961 published in the 1962 Proceedings of the American Wood Preservers' Association (MERRICK, 1962) reveal that creosote, in one form or another, was used in treating 68% of the wood volume processed in commercial wood preservation plants in the United States. In the same year, pentachlorophenol was used for 19% of the wood treated. Solutions of creosote-pentachlorophenol accounted for 4% of the volume of wood treated while the remaining 9% of treated material was processed using one of several other preservatives.

Among this latter group of preservatives would be those listed in the A. W. P. A. "Manual of Recommended Practice" (1963): acid copper chromate, ammoniacal copper arsenate, chromated zinc arsenate and copperized chromated zinc arsenate, chromated zinc chloride and copperized chromated zinc chloride, and fluor chrome arsenate phenol, types A and B. This group is all of the water-borne type. Copper naphthenate, used as an oil-borne preservative, is also included.

## 5.4 Fire Retardant Treatment

### 5.4.1 General Remarks about the Combustibility of Wood

Wood and wood base materials consist of organic compounds which are composed mainly of carbon and hydrogen. For this reason they are combustible and it is impossible to make them incombustible. But incombustibility of building

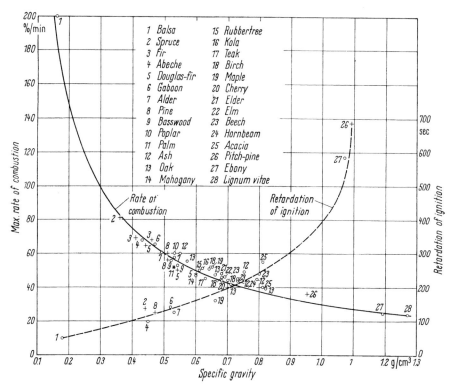

Fig. 5.9. Rate of combustion and retardation of ignition of various wood species and their relationship to the specific weight (From L. METZ)

construction is required only in rare, special cases. Most building components are suitable if spread and penetration of flame are low. In this connection it may be pointed out that the property of "incombustibility" of building and construction material alone does not secure fire protection in its highest possible sense. An incombustible material which melts at relatively low temperatures or develops dense smoke or even toxic vapors under the influence of heat should not be used. On the other hand incombustibility depends not only on the chemical composition but also on the size of the material. Building parts of normal cross section made of iron or steel are incombustible, but they have the unfavorable property of very high heat conductivity and plastic deformation at temperatures of a few hundred degrees centigrade. Steel wool on the other hand, due to its very large surface area with respect to its solid volume is combustible; a mixture of iron powder, with its still greater internal surface, and air is explosive.

For wood and wood base materials, form and dimensions also play an important part in fire behavior. A thin strip of wood, e.g. a match, is easily ignited, burns quickly and induces the inflammation of other fuels. Also a thin and dry veneer, if held freely in the air, may easily be ignited, but it loses this susceptibility if glued on solid wood, plywood or particle board. Apparently the ratio of surface to volume of combustible bodies is critical for their inflammability. The greater is this ratio, the easier can wooden parts be ignited and the quicker is the spread of flame. Many sharp edges or rough fibrous surfaces increase this ratio. The extreme case is the suspension of wood flour in air. Such a mixture is highly explosive.

Inflammability also depends on wood species. Speed of combustion and retardation of ignition, both tested in the fire tube according to TRUAX and HARRISON (1929), depend on density, as shown in Fig. 5.9. One can see that the maximum speed of combustion decreases hyperbolically with density while retardation of ignition increases parabolically with density. From a practical point of view, the speed of combustion for the very light balsa wood is 3 times as high as for pine wood, 5 times as high as for oak wood and nearly 10 times as high as for *Lignum vitae*. The maximum temperatures increase, to a limited extent, with the density (Fig. 5.10).

### 5.4.2 Developed Heat and Strength

Combustibility and heat or calorific value of wood are related to its moisture content. The heat value of oven-dry wood, practically without variation, averages 4500 kcal/kg. Fuels which contain hydrogen and water produce water vapor together with the combustion gases. The heat value is then diminished by the heat of evaporation. The so-called lower heat or calorific value, based on water vapor, is of technical interest. Fig. 5.11 shows the effect of the moisture content, based on oven-dry weight; the lower calorific value amounts to only about 3800 kcal/kg and for 30% moisture content only about 3300 kcal/kg. This rapid decrease of heat value with increasing moisture content is unfavorable for wood as a fuel material, but favorable for the fire protection of wooden building parts.

Any heating of wet or air dry wood results in some drying. In the hygroscopic range, that is, below 30% moisture content, stiffness and static strength values become higher with decreasing moisture content. If, for example, coniferous wood is dried from 20 to 10% moisture content, its crushing strength increases by about 100%, its ultimate bending strength by about 50%. Fire fighting units report unanimously that wooden building parts in fires show remarkable strength properties (cf. Anonymous, THOMPSON, 1958, FLEISCHER, 1960; SEDZIAK, 1961).

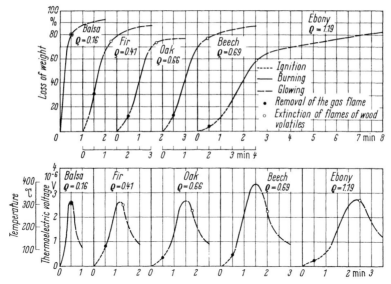

Fig. 5.10. Curves for the loss of weight and the temperature at the combustion of various wood species (From L. METZ)

### 5.4.3 Course of Temperature and Chemical Phenomena in Combustion of Wood

Temperatures below 100 °C, but above room temperature heat the wood and, as already stated, lead to drying. Chemical reactions are insignificant. Between 100 and 150 °C the chemical reactions of wood and wood base materials are, in general, still negligible. In reality, they may become rather important, but their reaction speed is still very slow. Between 150 and 200 °C gases are formed which consist, on the average, of 70% incombustible carbon dioxide and 30% combustible carbon monoxide. The wood becomes brown. Up to 175 °C the development of gases is still slow. Their composition remains about as stated and the heat value of the gases is only about 1200 kcal/m³.

Above 275 °C gases are rapidly developed, the amount of $CO_2$ goes down quickly, as does the amount of CO, and rather large amounts of readily combustible carbohydrates are produced. The calorific value of the gases rises to about 4000 kcal/m³ and higher. The reaction becomes exothermic, the color of the wood dark brown (AMY, 1961; BROWNE, 1958). The point at which the gases formed in the heating of wood may be ignited by any flame is important. This flame point may lie between 225 and 260 °C. The burning point follows between 260 and 290 °C. The ignited gases form a sustained flame. The course of all thermal reactions investigated in a charring stove by JUON (1907)

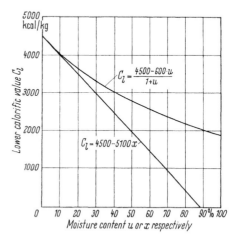

Fig. 5.11. Dependency of the lower calorific value on the wood moisture content, related to oven-dry and moist weight (From F. KOLLMANN)

is shown in Fig. 5.12. According to Japanese investigations on oak wood by HUDO and YOSHIDA (1957) a temperature of 150 to 180 °C pyrolizes the hemicelluloses

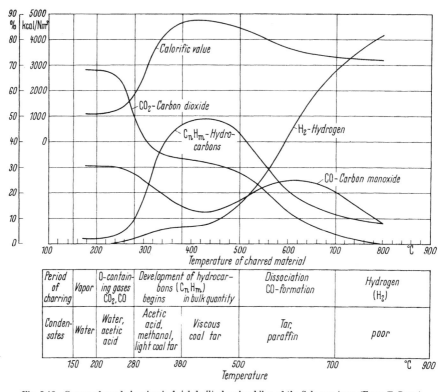

Fig. 5.12. Course of wood charring in brick-built charring kilns of the Schwarz-type (From E. JUON)

and in the range of 280° to 350°C, the cellulose. The pyrolisis of the lignin also starts at 280°C, reaches its maximum between 350 and 400°C and is completed between 450 and 500°C. In larger fires, temperatures between 800°C and 1300°C may be reached.

In the literature one can find repeatedly remarks that wood may be ignited by extended exposure to temperatures between 100 and 150°C. Fig. 5.13 shows how the time until ignition depends on the temperature of wood. Upon extrapolation one can see that the curve shown approaches the value of 150°C asymptotically. KOLLMANN (1960) has found that the exothermic reactions start in the wood of broadleaved species at a lower temperature than in coniferous species probably due to their higher content of pentosans.

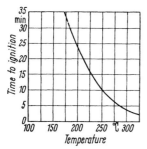

Fig. 5.13. Dependency of time until ignition of wood on the temperature (From National Fire Protection Association, USA, cf. L. METZ)

### 5.4.4 Effects and Properties of Fire Retardants

Fire retardants may be divided into the following classes according to METZ (1942):

### 1. Fire retardants acting mechanically

Coatings or deck layers produced by brushing or spraying exclude, to some extent, the oxygen of the air from the wood surface. Such coatings must be highly elastic, otherwise they will be delaminated quickly in a fire.

### 2. Melt-forming chemicals

These fire retardants remove the heat of melting from their surroundings, producing adhesive melt layers and decreasing the production of charcoal.

### 3. Foam-forming chemicals

In heating, these preservatives form porous charcoal-like foam layers which are highly thermal insulating and protect the wood from the effect of the heat by this characteristic.

### 4. Slake-gas-developing chemicals

The slake gases lower the concentration of the combustible gases and reduce their ignitibility.

### 5. Wood-charring preservatives

As METZ (1936) proved, all known fire retardants increase the charring of wood at relatively low temperatures, thus improving thermal insulation.

HARTMANN (1960), in a more condensed way, lists four definite functions of fire retardants which control the combustion of processed lumber. In practice effective fire retardants often combine some of the effects mentioned under numbers 1 to 5. They should have the following properties:

a) The combustibility of the wood should be decreased and its fire resistance increased. After the ignition of the wood the combustion should be slowed down.

b) The adhesion of the fire retardants to the wood should be high, long lasting, and not disturbed by the fire.

c) The properties of the wood should not be badly affected by the treatment; in particular, hygroscopicity and strength should not be changed.

d) The fire retardants must not be toxic for human beings and should not form toxic products under the influence of heat.

e) The treatment should not encourage attack by wood destroying fungi.

f) Water soluble fire retardants must have a high solubility (at least 20 to 25% at room temperature).

g) If the chemicals are not delivered ready for use, their preparation must be easy.

h) The fire retardants should be absorbed by the wood in suitable amounts at prescribed concentrations.

i) The application of the fire retardants should be economical; that is, the costs should be moderate.

There are very many fire retardants available on the market. Just a few of the most important may be characterized as follows:

**5.4.4.1 Water Soluble Salts.** Fire retardant salts consist mainly of ammonium phosphate and ammonium sulphate. The concentration of the solution may be between 10 and 40%; the effect of this type of fire retardant consists either of the development of inert gases e. g. carbon dioxide, ammonia, sulfuric acid, etc. or of the formation of melts which coat the wood (phosphates, borates, acetates etc.). In the USA (VAN KLEECK, 1948), the salt solutions are mixed with colloidal extracts such as alginates. Large amounts of this more viscous solution may be spread on the wood surfaces.

A few salts may be mentioned:

Diammonium phosphate $((NH_4)_2HPO_4)$ is the most effective ammonium salt and ensures fire protection for many years. It does not cause corrosion of iron;

Ammonium sulphate $((NH_4)_2SO_4)$ is cheap and very soluble, but it causes corrosion of metals;

Ammonium bromide $(NH_4Br)$ is very effective, but rather expensive, and leads to corrosion of iron;

Ammonium tetraborate $((NH_4)_2B_4O_7 \cdot 4H_2O)$ is effective, but rather expensive, and its solubility in water is poor;

Sodium tetraborate or Borax $(Na_2B_4O_7 \cdot 10H_2O)$ is effective, but its solubility is very low. It is suitable therefore only as an additive to other fire retardants.

Sodium acetate $(NaC_2H_3O_2 \cdot 3H_2O)$ is a fire retardant preservative expecially recommended by SCHWALBE and BERLING (1932). Using the immersion procedure, or several coatings, it results in adequate fire protection according to METZ (1942). The salt is cheap, does not promote corrosion, but does further decay. Some metal compounds, such as aluminum chloride $(Al_2Cl_6 \cdot 12H_2O)$ and aluminum sulphate $(Al_2(SO_4)_3 \cdot 18H_2O)$ may be effective as fire retardants if the amount of absorption is large enough. The solutions have a low pH value. Though they are corrosive, they are also fungicidal in character.

Potassium sulphate $(KAl(SO_4)_2 \cdot 12H_2O)$ was used as a fire retardant even in ancient Egypt. According to VAN KLEECK (1948) the fire protection produced is very low.

Zinc borate $(3 ZnO \cdot 2B_2O_3)$ can be used in mixture with white lead, linseed oil, turpentine and a drying agent as a good fire retardant preservative (VAN KLEECK, 1948).

**5.4.4.2 Alkali Silicates.** Potassium and sodium alkali silicates (water glass in solutions of about 40°Bé) spread on the surface of wood, act as fire retardants by forming insulating foam layers in melting. The main disadvantage of the coatings is that they are not weather-proof and that fire protection is greatly reduced in outdoor locations. In the United Kingdom during World War II (Brit. Standards Air Raid Precaution Series 39, Feb. 1940) the following formula was recommended:

Sodium water glass (density 1.41 to 1.42, 112 parts by weight)
Kaolin 150 parts by weight
Water 100 parts by weight
Three to four coatings should be applied.

**5.4.4.3 Foam Forming Organic Compounds.** This type of fire retardant consists of mixtures of salts and chemicals which form foam-like charcoal bubbles under the influence of fire and protect the wood by a very high thermal insulation. As examples of typical formulas may be mentioned mixtures of diammonium phosphate with borax and urea formaldehyde condensates or emulsions of formaldehyde, and dicyan diamide.

It is essential that coatings of these organic fire retardants not be painted or coated afterwards because then the foam layer cannot be produced.

**5.4.4.4 Other Fire Retardants.** It is impossible to list exhaustively all types of proposed fire retardants. Additional ones to be mentioned are asbestos paints, mixtures of magnesium oxychloride cement, of potassium fluoride with cement, and magnesium chloride with cement. An important goal is to combine in a single preservative, if possible, fire protection as well as preservation against fungi and/or insects.

## 5.4.5 Testing of Fire Retardants

Since the entire combustion process is rather complex and since many internal and external factors interfere with it, it is impossible to include all of the factors in one single combustion test. At the Fourth FAO Wood Technology Conference at Madrid (EGNER et al., 1958) is was agreed to aim at small scale tests. In accordance with this resolution, systematic investigations of small apparatus and simple methods for fire tests were carried out by KOLLMANN (1960). Six rather simple tests (application varying from country to country) were selected:

1) Fire tube test, 2) weight loss test for board (according to SEEKAMP, 1954), 3) slab chimney test according to DIN 4102, 4) hot box test according to BRAUNS (1956), 5) electric radiating panel test, and 6) inclined panel test according to Brit. Standards Institution Specification No 476. The weight loss and one or two characteristic temperatures were measured as a function of test time. In this evaluation, the second phase of the process is especially significant. The weight loss here is proportional to the heating time, meaning that the combustion process is stationary. The energy balance can be used for calculation of a dimensionless combustion figure which is the ratio between the energy released in combustion and the total energy. The higher this number, the more intense was the combustion process.

The combustion figure for the slab chimney test with unprotected woods and wood base materials may have a value between 0.73 and 0.87 whereas preserved wood base materials, with a low spread of flame, exhibit combustion figures as low as between 0.39 and 0.44.

Additional reference should be made to the brief description of some fire test methods used for wood and wood base materials worked out by MARKWARDT and co-workers (1954). Also a report about flame spread resistance of fiber insulation boards by VAN KLEECK and MARTIN (1950) should be mentioned. A revised draft test for combustibility of materials has been worked out by ISO TC 92 (Secretariat Paper 28–30 E).

## 5.5 Dimensional Stabilization

In efforts to improve some of its undesirable properties, wood has been modified by several means. Some of the methods that have been developed are still in the experimental stage, but others are in commercial production.

Improvement of the dimensional stability of wood has been of primary interest. Undoubtedly, this is because the tendency to shrink and swell with changes in humidity is the most unfavorable property of wood. If wood were an isotropic material, shrinking and swelling equally in all directions, the problem might be less serious. However, its variable morphology, gross and sub-microscopic, is the basis for unequal dimensional changes in the tangential, radial and longitudinal direction. In the tangential direction, shrinkage from the green condition to 6% moisture content ranges from 4% to about 9% while the radial shrinkage for the same conditions is as low as 1.8% to somewhat over 6%, on the average. Volumetric shrinkage can be very substantial in some woods, but perhaps the more important factor is the differential shrinkage in the various directions that can lead to warping and distortion of all sorts, and the swelling that accompanies the reabsorption of water.

## 5.5.1 Theory

STAMM (1960) lists the successful procedures or approaches that can be used to reduce these volumetric changes. One method is to coat the external or internal surfaces of wood with a substance that will retard moisture gain or loss, a substance that is insoluble in water. Another way is to bulk the cell walls while they are in the swollen state, thus physically preventing normal shrinkage. A third is to change the wood chemically by replacing hydroxyl groups in the cellulose with others that are less hygroscopic. The formation of chemical cross bridges between structural units is another approach.

## 5.5.2 Methods

Coating methods are not very effective in maintaining dimensional stability. They do have application in situations where the mere retardation of moisture absorption is adequate. External coatings have been proven to be more satisfactory than the so-called water-repellent treatments or internal coatings. In any case, neither coating method prevents moisture absorption by the cell wall constituents over an extended period of time and are not true modifications of the wood.

One bulking method has recently become more practical due to the use of higher molecular weight polyethylene glycol, a water-miscible substance which can be introduced into unseasoned wood. STAMM (1959) has shown that the use of this material of 1000 molecular weight can reduce the surface checking of wood during seasoning. In concentrations of about 30% of the dry weight of the wood, very high degrees of dimensional stabilization are attained in the treated wood.

Polyethylene glycol treatments are now used commercially for processing gunstock and wood carving blanks where they have greatly reduced losses due to splitting and checking (MITCHELL and WAHLGREN, 1959). This bulking process has no serious adverse effect on strength, gluing or finishing. Furthermore, in the higher concentrations used for dimensional stabilization, where shrinkage is reduced 50% or more, good decay resistance is indicated, probably because of inadequate moisture to support decay.

Another method based on the bulking technique uses resin formation within the cell wall. The most effective system found for this method uses water-soluble phenol formaldehyde which chemically blocks the hydroxyl groups in addition to bulking the cell walls. In aqueous solution the molecules of this substance are small enough and their polarity is great enough to provide good penetration into the cell walls by selective adsorption. This is indicated by a swelling which is greater than the maximum with water alone. According to STAMM and BAECHLER (1960) this treatment provides a reduction in hygroscopicity, swelling and susceptibility to decay.

In the product "Impreg", the resin in the resin-impregnated wood is cured by heating without pressure. A similar product of higher density, "Compreg", having a specific gravity of 1.3 or more, is made by curing under high pressure. Both products show great improvement in dimensional stability. The percentage of improvement is dependent upon the amount of resin added to the wood. When 30 to 35% of resin, based on the oven-dry weight of the wood, is cured within Impreg, shrinkage is reduced to 25—35% of that in normal, untreated wood.

In the case of Compreg, both swelling and recovery from compression must be considered. Taken together, these do not exceed 12% increase in the thickness of a panel made according to recent specifications and tested by a method developed at the U. S. Forest Products Laboratory (STAMM, 1960).

Unfortunately, not all properties of wood are improved by phenol-formaldehyde treatment. Toughness and impact strength are reduced although compressive strength is higher in material of high resin content. This embrittling effect is lacking in acetylated wood made by a bulking process which differs from the other methods described.

Acetylation of wood in lumber thickness is now possible commercially (Koppers Co., 1961). Of all the processes developed for imparting high dimensional stability to wood, this method appears to offer the greatest advantages with the least change in the desirable properties of wood.

Acetylation involves the introduction of acetyl groups in place of the hydroxyl groups in wood that has been swollen chemically. Acetic anhydride is used in the acetylation, and pyridine is used as a catalyst. Permanent dimensional and density changes are brought about in the process with weight increases as high as 28%. STAMM (1960) attributes the high dimensional stability of acetylated wood more to bulking action than to reduction of hygroscopicity.

Acetylated wood swells the least of any of the modified woods. In certain species, swelling is reduced more than 80%. Koppers Company evaluations (1961) show swelling reductions in the tangential direction, in wood treated to exhaustive acetylation, range from 70 to 84% over untreated material. This process imparts excellent decay and stain resistance to wood. Tests indicate good resistance to termite attack as well to damage by marine borers. Certain strength values are improved by acetylation, and none are reduced. In addition there is no darkening of the wood upon treatment, a characteristic which is typical of resin impregnation.

A method which seems promising because of the high anti-shrink efficiency it develops with low weight increase is the formaldehyde treatment. The process consists of heating wood in the presence of formaldehyde vapor and a mineral-acid catalyst. It is presumed that acetal cross-linkages are produced between the hydroxyl groups on adjacent cellulose chains. This approach to dimensional stabilization, while theoretically sound from the chemical standpoint, has been found to be impractical so far. A serious embrittling effect is produced in the wood when the acid catalyst is used in the quantities required to give the desired degree of dimensional stability (TARKOW and STAMM, 1953).

Another stabilization technique is the thermal modification of wood. The chemical changes brought about by subjecting the wood to heat under controlled conditions does reduce its swelling and shrinkage. However, there is always some loss in mechanical strength. The product "Staybwood" is based on this thermal principle, but it is not being manufactured commercially.

A second thermal method is used to make "Staypak". This involves heat and pressure and is reported to cause lignin to flow within the wood. Though not as dimensionally stable as Compreg, this product does have mechanical properties which are superior to Compreg. It can be used advantageously where high impact strength is needed. Its resistance to decay and attack by termites is inferior to other products such as Compreg, though it is better than untreated wood of the same species (STAMM, 1960).

## Literature Cited

American Wood Preservers' Association, (1963) Manual of recommended practice, Washington, D.C.

AMY, A., (1961) Les bases physico-chimiques de la combustion de la cellulose et des matériaux ligneux. Cahiers du Centre Technique du Bois 45: 1—30

158 Literature Cited

Anonymous, (1958) Le comportement du bois au feu. Cahiers du Centre Technique du Bois **14**: 1—28, 2ᵉ Ed.

BAECHLER, R. H. and ROTH, H. G., (1964) The double-diffusion method of treating wood: A review of studies. For. Prod. J. XIV: 171—178

BAILEY, I. W., (1913) The preservative treatment of wood. II. The structure of the pit membranes in the tracheids of conifers and their relation to the penetration of gases, liquids and finely divided solids into green and seasoned wood. Forestry Quarterly **11**: 12—20.

BLEW, J. O., JR., HENRIKSSON, S. T. and HUDSON, M. S., (1961) Oscillating pressure treatment of 10 U.S. woods, For. Prod. J. XI: 275—282

BRAUNS, O., (1956) Ein Gerät zur Bestimmung des Flammwiderstandes von Wandverkleidungsplatten. Holz Roh-Werkstoff, **14**: 271—281.

British Standards Institution Specification, No 476.

BROWN, F. L., MOORE, R. A. and ZABEL, R. A., (1956) Absorption and penetration of oilsoluble wood preservatives applied by dip treatment. State Univ. College of Forestry at Syracuse Univ. Tech. Publ. 79.

BROWNE, F. L., (1958) Theories of the combustion of wood and its control US For. Prod. Lab. Report No 2136.

CÔTÉ, W. A., JR., (1958) Electron microscope studies of pit membrane structure. For. Prod. J. VIII: 296—301.

—, and KRAHMER, R. L., (1962) The permeability of coniferous pits demonstrated by electron microscopy. TAPPI **45**: 119—122.

EGNER, K., KLAUDITZ, W., KOLLMANN, F. and NOACK, D., (1958) Bericht über die Vierte FAO-Konferenz für Holztechnologie in Madrid vom 22. April bis 2. Mai (1958), Bonn, Juli 1958, S. 51 bis 91.

FLEISCHER, H. O., (1960) The performance of wood in fire. US For. Prod. Lab., Report No 2202.

FREY-WYSSLING, A., BOSSHARD, H. H. and MÜHLETHALER, K., (1956) Die submikroskopische Entwicklung der Hoftüpfel, Planta **47**: 115—126.

GEWECKE,.H., (1957) Baumsaft als Lösungsmittel für Imprägniersalze bei der Tränkung von Mastenhölzern nach dem Trogsaug-, Kesseldrucksaug- oder Stapelsaug-Verfahren. Holz Roh-Werkstoff **15**: 416—417.

GOODWIN, D. R. and HUG, R. E., (1961) A new wood preserving process. For. Prod. J. XI: 504—507.

HARADA, H., (1956) Electron microscopic study of wood tissue with special reference to wart-like and bordered pit structure. Transactions 65th Mtg. Japan For. Soc. pp. 1—8.

HARTMANN, C. F., (1960) The past and the future of fire retardant lumber. Proc. A. W. P. A. **56**: 107—113.

HENRIKSSON, S. T., (1954) Holztränkung nach der Wechseldruckmethode. Holz Roh-Werkstoff **12**: 233—241.

HUDO, K. and YOSHIDA, E., (1957) The decomposition process of wood constituents in the course of carbonization I. The decomposition of carbohydrate and lignin in Mizunara. J. Japan Wood Research Soc. 3 (4) 125—127.

HUNT, G. M., and GARRATT, G. W., (1953) Wood Preservation, 2nd Ed. McGraw Hill Book Co., New York, 417 pp.

JUON, E., (1907) Stahl u. Eisen **27**: 733—771.

VAN KLEECK, A., (1948) Fire-retarding coatings. U.S. For. Prod. Lab. Report No R 1280.

VAN KLEECK, A. and MARTIN, T. J., (1950) Evaluation of flame-spread resistance of fiber insulation boards. US For. Prod. Lab. Report No D 1756.

KOLLMANN, F., (1960) Vergleichende Prüfungen des Brandgeschehens bei Holz und Holzwerkstoffen im unbehandelten und imprägnierten Zustand mittels verschiedener Kleingeräte. Svensk Papperstidning **63**: 208—217.

Koppers Company Inc., (1961) Koppers Acetylated Wood (RWD–400) New Materials Technical Information. E–106, Pittsburgh, Pa.

KRAHMER, R. L. and Côté, W. A., JR., (1963) Changes in coniferous wood cells associated with heartwood formation. TAPPI **46**: 42—49.

LIESE, W., (1954) Der Feinbau der Hoftüpfel im Holz der Koniferen. Proc. Intl. Conf. on Electron Microscopy, London, pp. 550—555.

MACLEAN, J. D., (1960) Preservative treatment of wood by pressure methods. Agric. Handbook No 40, U.S.D.A. 160 pp.

MARKWARDT, L. J., BRUCE, H. D. and FREAS, A. D., (1954) Brief decription of some fire test methods used for wood and wood base materials, U.S. For. Prod. Lab. Report No 1976.

MERRICK, G. D., (1962) Wood preservation statistics, 1961. Proceedings A. W. P. A.

METZ, L., (1936) Fire protection of wood. Z. Ver. deut. Ing. **80**: 660—667.

—, L., (1942) Holzschutz gegen Feuer. VDI-Verlag, 2nd Ed., Berlin.

MITCHELL, H. L. and WAHLGREN, H. E., (1959) New chemical treatment curbs shrink and swell of walnut gunstocks. For. Prod. J. XI: 437—441.

## Literature Cited

SARGENT, J. W., (1959) The significance of rays in the penetration of certain softwoods by creosote—an anatomical study. Unpublished M.S. thesis, State Univ. College of Forestry at Syracuse Univ., Syracuse N.Y.

SCHWALBE, C. G. and BERLING, K., (1932) Neue billige Flammenschutz-Mittel für Holz, Chemiker-Ztg. **56**: 909—911.

SEDZIAK, H. P., (1961) Fire retardant treatment of wood. Timber of Canada, Feb. 1961.

SEEKAMP, H., (1954) Die Klassifizierung der Brennbarkeit holzhaltiger Platten. Holz Roh-Werkstoff **12**: 189—197.

STAMM, A. J., (1959) Effect of polyethylene glycol on the dimensional stability of wood. For. Prod. J. **IX**: 375—381.

—, (1960) Modified woods. U.S. For. Prod. Lab. Report No 2192.

—, and BAECHLER, R. H., (1960) Decay resistance and dimensional stability of five modified woods. For. Prod. J. **X**: 22—26.

TARKOW, H. and STAMM, A. J., (1953) Effect of formaldehyde treatments upon the dimensional stabilization of wood. J. For. Prod. Res. Soc. **III**: 33—37.

THOMPSON, H. E., (1958) Economy in timber construction plus fire safety. Timber Technology **66**: 175—176.

TRUAX, T. R. and HARRISON, C. A., (1929) A new test for measuring the fire resistance of wood. A.S.T.M. **29**: II.

VAN GROENOU, H. B., RISCHEN, H. W. L. and VAN DEN BERGE, J., (1951) Wood preservation during the last 50 years. A.W. SIJTHOFF's Uitgeversmaatschappij N.V. Leiden, Holland, 318 pp.

VERRALL, A. F., (1961) Brush, dip and soak treatments with water-repellent preservatives, For. Prod. J. **XI**: 23—26.

WIRKA, R. M., (1945) Tire-tube method of fence post treatment. U.S. For. Prod. Lab. Report No R 1158 (rev.).

# 6. PHYSICS OF WOOD

## 6.1 Density and Specific Gravity

### 6.1.1 Density, Porosity, Specific Gravity of Wood Substance and of Wood Constituents

Of all the physical properties of wood its density or its specific gravity, as the ratio of the density of a substance to that of water at the same temperature, was the first to be investigated. The classical study by CHEVANDIER and WERTHEIM, "Mémoire sur les propriétés mécaniques du bois" (1848), contains a collection of data from the 18th and the beginning of the 19th century. It has been generally assumed that the quality of wood as a building material depends mainly on its density. There is, in reality, a close correlation between mechanical properties, hardness, abrasion resistance, and heat value of wood on the one hand, and density on the other.

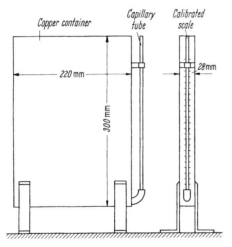

Fig. 6.1. Immersion vessel for determination of density. From NIETHAMMER (1931)

Since density is influenced to a large extent by the moisture content of the wood, comparisons of density figures can only be made at the same moisture content. Standard values are 0 and 12%. Density is defined as the mass per unit volume. The simplest method of determining the density of a piece of wood is to weigh it and then to find its volume. If the sample has regular dimensions and no cracks one can measure its length, width, and thickness and calculate the volume. For samples of irregular shape, an immersion method is more suitable. Fig. 6.1 shows an immersion vessel which is useful also for ovendry wood; according to NIETHAMMER (1931) the values thus obtained for the volume are, on the average, only 1.79% greater than the measured ones. The samples should have the approximate shape of a disc with a diameter of 12 cm ($4^3/_4''$) and a width of 2.5 cm ($1''$). The method is appropriate for determining the density of pulpwood. A method of very quickly estimating the density of woods with a density lower than 1.0 g/cm³ using a minimum of equipment has been developed by PAUL (1946). It consists of measuring the proportion of a test specimen (1 sq. in. in cross section and 10 in. long, marked into 10 equal divisions of 1 in.) that is submerged when the specimen is floated upright in a cylinder filled with water. For example, if five divisions of the piece are under water, the density is 0.5; intermediate levels may be rather closely estimated as 0.54, 0.57, and so on.

# 6.1 Density and Specific Gravity

An exact determination of density may be carried out by immersion in mercury. The apparatus according to Breuil (Fig. 6.2) consists of a cylindrical steel vessel $a$ with a screw-cap $b$, in the middle of which a glass capillary tube $c$ is fixed. Perpendicular to the vessel, a horizontal cylinder $e$ is mounted in which a ground piston can be moved back and forth by means of a micrometer-screw $g$. A wood sample of any shape is put into the vessel and pressed below the mercury level by means of the spring clamp $f$. Then the screw-cap is tightened and the piston has to be turned until the mercury in the capillary tube reaches the edge of the adjusting ring $d$. Now the position of the micrometer-screw is read, the screw-cap is opened and the wood sample is removed. After having tightened the screw-cap again, the micrometer screw must be turned until the mercury reaches the adjusting-ring once more. A second reading has to be made and the volume is obtained as the difference between the first and second readings.

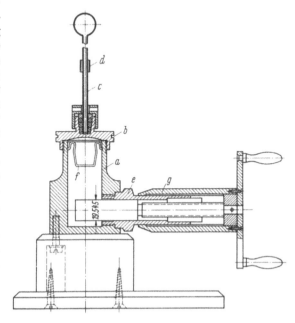

Fig. 6.2. Apparatus for determination of density by immersion in mercury. From Breuil

The use of radiation methods has in recent years become prominent for the measurement of material quality. Thus, the application of radiation methods (X-rays, beta rays, and gamma radiation) have gained in importance for the wood technologist for rapid, accurate, and non-destructive physical measurements of wood properties, such as density and moisture content. Loos (1961), investigating the absorption of gamma rays into wood, found that if specific gravity or moisture content of wood are known the other one can be rapidly and accurately determined by measuring gamma-ray attenuation through a given thickness. Because of their great penetrating power, gamma rays are used most effectively for measuring the density differences associated with the decay in large wood cross-sections, such as poles, timbers, and standing trees (Loos, 1965).

Wood is a porous material as has been shown in Chapter 1. A piece of oven-dry wood is composed of the solid wood substance of its cell walls and of the cell cavities containing air and very small amounts of sap constituents such as protein, minerals and other substances such as resins, gums, etc. The density of the solid wood substance has been found to be very similar in all timbers and is about 1.5 g/cm$^3$. This density is an ideal physical value for a lignified cellulosic cell wall which is completely non-porous. In reality the cell wall in the native state is characterized by minute cavities, capillaries, and irregularities. Jayme and Krause (1963) have described a method for determining the "packing-density". This is the ratio of the over-all density to the portion of the total volume occupied by cell walls (measured microscopically). They found values for the packing density ranging from 0.71 g/cm$^3$ for *Musanga Smithii* to 1.27 g/cm$^3$ for common beech (*Fagus sylvatica* L.). The density of different timbers (based on oven-dry weight and oven-dry volume) varies from about 0.04 for Balsawood (*Ochroma lagopus* Swartz), Fromage de Hollande (*Bombax malabaricum* D. C), *Erythrina sp.*, *Alstonia spp.* and *Leitneria spp.* to 1.4 for *Guajacum officinale* and *Brosimum aubletii* Poepp. This thirty-five-fold difference in density is mainly due to the differences in porosity.

The actual solid volume $m$ expressed as a ratio to the total volume can be calculated by the following equation:

$$m = 1 - c = 1 - \frac{\varrho_0}{\varrho_w} = 1 - \frac{\varrho_0}{1.50} = 0.667 \varrho_0, \tag{6.1}$$

where $c$ is the ratio of air spaces in dry wood to the total volume, $\varrho_0$ is the bulk density of the wood based on dry volume, and $\varrho_W$ is the density of wood substance. A graphical solution of Eq. (6.1) is given in Fig. 6.3 (TRENDELENBURG 1939); it shows that the void volume of the lightest woods may amount to 97%, whereas that of the heaviest may be as low as 7%. In Fig. 6.3 the influence of swelling

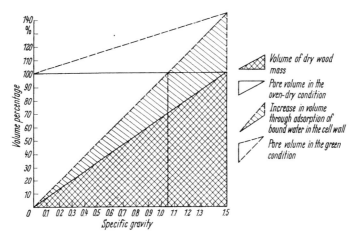

Fig. 6.3. Relationship between pore volume in the dry and in the wet condition, volume of dry wood mass, and specific gravity of wood. From TRENDELENBURG (1939)

on the ratios of void and solid volume to total volume is also shown under the assumption that the maximum external volumetric swelling $\alpha_v$ (or shrinking $\beta r$) is proportional to the density of oven-dry wood (based on oven-dry volume) $\varrho_0$ (or to the density, oven-dry weight, based on the volume, when green, $R_g$, respectively)

$$\alpha_v = u_f \cdot \varrho_0 \tag{6.2a}$$

or

$$\beta_v = u_f \cdot R_g. \tag{6.2b}$$

The proportionality constant $u_f$ is equal to the moisture content at the fiber saturation point expressed as a ratio to the oven-dry weight. On the average $u_f = 0.28$ may be used.

Whereas in physics density is generally based on both weight and volume at the same moisture content, in forest products research, especially in the United

States of America, it is often based on ovendry weight and volume at a specified moisture content. The following conditions are frequently used:

$$\varrho_0 = \frac{W_0}{V_0} = \frac{\text{oven-dry weight}}{\text{oven-dry volume}}, \qquad (6.3\,\text{a})$$

$$R_{12} = \frac{W_0}{V_{12}} = \frac{\text{oven-dry weight}}{\text{volume at 12\% m.c.}}, \qquad (6.3\,\text{b})$$

$$R_g = \frac{W_0}{V_g} = \frac{\text{oven-dry weight}}{\text{green volume}}. \qquad (6.3\,\text{c})$$

Thus we obtain for the solid wood substance at the fiber saturation point the highest possible value of

$$R_g = \frac{1.50}{1 + 0.28\,(1.50)} = 1.056.$$

The foregoing concepts are valid only if the volume of the cell cavities or lumina remains constant in the hygroscopical range. This is not true according to STAMM (1938). One has to take into consideration the compression of the adsorbed water on the internal surface of the wood and, above the fiber saturation point, the filling of cavities with liquid water. The following equation is valid for the void volume $c'$ of wood containing moisture:

$$c' = 1 - R\left(\frac{1}{\varrho_w} + \frac{u_h}{\varrho_s} + \frac{u_k}{\varrho}\right), \qquad (6.4)$$

where $R$ is the density of the wood based on the volume at any current moisture content. $\varrho_w$ is the density of wood substance, $u_h$ is the amount of hygroscopic moisture expressed as the weight of water per unit weight of dry wood, $u_k$ is the amount of free water, $\varrho$ is the normal density of water at the temperature of measurement, and $\varrho_s$ the density of the compressed adsorbed water. STAMM calculated at a temperature of 25°C or 77°F the following values: for $u_h = 0.30$, $\varrho_s = 1.113$; for $u_h = 0.20$, $\varrho_s = 1.145$; for $u_h = 0.10$, $\varrho_s = 1.20$; and for $u_h = 0$, $\varrho_s = 1.30$ (cf. Fig. 6.49). The variation of the pore or void volume $c'$ with moisture content is illustrated in Fig. 6.4. The true density of wood substance may not be determined by the water-displacement method; the values will become too high due to the compression of the adsorbed water. Benzene-displacement gives a better approximation, but the figures obtained are somewhat too low. Though benzene has a very small affinity for cellulosic materials, and therefore causes no swelling, it probably does not penetrate the void structure of the dry cell walls completely (STAMM and HANSEN 1937). HOWARD and HULETT (1924) have shown that at room temperature helium is practically not adsorbed on carbon, and DAVIDSON (1927) similarly found that

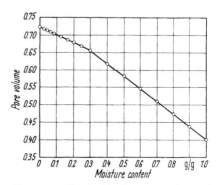

Fig. 6.4. Fractional void volume (pore volume) of a wood specimen with a specific gravity of 0.365 on a swollen volume basis and of 0.405 on a dry-volume basis for different moisture contents. From STAMM (1938)

164                               6. Physics of Wood                        [Ref. p. 285

helium is not adsorbed on cellulose. Furthermore, it completely penetrates the
void structure due to its low molecular weight.

Table 6.1. *Densities of Wood, Cellulose, and Lignin at 30°C*
(From STAMM and HANSEN, 1937)

| Substance | Displacing-Medium | | |
|---|---|---|---|
| | Helium | Water | Benzene |
| Extracted white spruce | 1.4603 | 1.5332 | 1.444 |
| Standard cotton linters | 1.585 | 1.6028 | 1.571 |
| Spruce sulfite pulp: | | | |
| Unbeaten | 1.570 | 1.590 | 1.555 |
| Beaten | 1.593 | 1.616 | 1.578 |
| Spruce lignin (modified sulfuric acid method) | 1.377 | 1.399 | 1.366 |
| Maple lignin (modified sulfuric acid method) | 1.406 | 1.422 | 1.388 |

The figures indicate that the densities of the main constituents of spruce are not
equal. The value of 1.58 represents the reliable specific gravity of wood cellulose,
while lignin values between 1.38 and 1.41 may be used. The density of wood is
certainly influenced by the proportion of cellulose to lignin. For simplicity, the
value $\varrho_w = 1.50$ for the density of wood substance is always assumed in this
book.

### 6.1.2 Effect of Moisture Content in Wood on its Density

The relationship between the density $\varrho_u$ at a given moisture content $u$ and the
density $\varrho_0$ of oven-dry wood is technically of the greatest importance. A series of
empirical formulae has been proposed, but it is not difficult to reach a theoretically correct solution (KOLLMANN 1932, 1933, 1934):
The moisture content $u$ is

$$u = \frac{W_u - W_0}{W_0} \qquad (6.5\,\mathrm{a})$$

where $W_u$ is the weight of the wood with the moisture content $u$ and $W_0$ is the
weight of the oven-dry wood. From Eq. (6.5a) it follows:

$$W_u = W_0(1 + u). \qquad (6.5\,\mathrm{b})$$

Below the fiber saturation point swelling accompanies the increase in weight
with growing moisture content. The volume $V_u$ at any given moisture content $u$
is

$$V_u = V_0(1 + \alpha_{vu}), \qquad (6.6)$$

where $V_0$ is the volume of the oven-dry sample and $\alpha_{vu}$ is the coefficient of volumetric swelling in the range between 0 and $u\%$ moisture content. We now can
combine both equations, (6.5b) and (6.6), and thus obtain for the density of wood
containing moisture $\varrho_u$

$$\varrho_u = \frac{W_u}{V_u} = \frac{W_0(1 + u)}{V_0(1 + \alpha_{vu})} = \varrho_0 \frac{1 + u}{1 + \alpha_{vu}}, \qquad (6.7\,\mathrm{a})$$

This equation is strictly valid, but the factor $\alpha_{vu}$ must be known at any moisture
content $u$. We know from many investigations that there is a linearity of swelling

between 0 and approximately 25% of moisture content (Fig. 6.5); at 25% moisture content, on the average, 75% of the maximum volumetric swelling $\alpha$ has been reached.

Therefore, and by substitution using Eq. (6.2), it follows that in the range between 0 and 25% of moisture

$$\varrho_u = \varrho_0 \frac{1+u}{1+0.84\,\varrho_c \cdot u}. \qquad (6.7\,\text{b})$$

Note should be made of the fact that there is a variation of the factor $m$ in the relationship $\alpha_{vu}/u = m \cdot \varrho_0$, but Fig. 6.6 indicates that $m = 0.84$ is the probable mean value.

The density at the fiber saturation point can be calculated by introducing the maximum volumetric swelling into Eq. (6.7a) and using $u = 0.28$. It is known from some observations that maximum swelling rarely occurs before $u = 0.40$. Above the maximum swelling point, increasing moisture content raises only the weight of the wood, but its volume is kept constant. In concluding the discussion on this problem, reference is made to the graphical Fig. 6.7 which has proved to be useful for practical application.

### 6.1.3 Density of Green Wood

As stated previously, for physical comparisons only the density (weight, oven-dry; volume, oven-dry) is advisable, but density (weight, oven-dry; volume, green) is generally used in the United States of America. This figure is not only easily determined, since the green volume is measurable by the displacement of water without any error, but it is informative from the point of view of forestry. One can obtain the

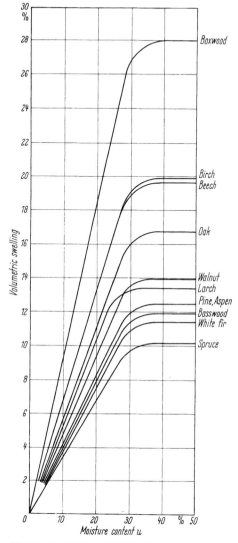

Fig. 6.5. Volumetric swelling as a function of moisture content for different wood species. From MÖRATH (1932)

dry-mass production of any forest by multiplying the yield in cubic meters with the density $R_g$ (cf. Eq. (6.3c) in kilograms per cubic meter).

Knowing the maximum volumetric shrinkage $\beta_v$, expressed as a decimal fraction, it follows that:

$$\varrho_0 = R_g \frac{1}{1-\beta_v} \quad \text{or} \quad R_g = \varrho_0(1-\beta_v). \qquad (6.8\,\text{a})$$

Proceeding from the maximum volumetric swelling $\alpha_v$ and making use of the formulae

$$\beta = \frac{\alpha}{1+\alpha} \quad \text{and} \quad \alpha = \frac{\beta}{1-\beta}. \qquad (6.9)$$

it can be written

$$\varrho_0 = R_g(1 + \alpha_v)$$

or

$$R_g = \varrho_0 \frac{1}{1 + \alpha_v}. \tag{6.8b}$$

The statistical relationship $\alpha_v = 0.28\varrho_0$ already referred to above may be introduced in Eq. (6.8b); similarly the corresponding relationship $\beta = 0.28 R_g$ into Eq. (6.8a).

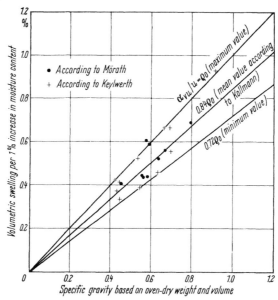

Fig. 6.6. Volumetric swelling per 1% increase in moisture content related to specific gravity

Table 6.2. *Shipping Weight of Commercial Timbers in Green Condition and after Storage in the Forest*
(From TRENDELENBURG, 1941)

| Species | Condition | Log diameter (measured without bark) cm | Shipping weight kg per cu. meter green | $u \approx 0.50$ |
|---|---|---|---|---|
| Spruce | debarked | >40<br>20···40<br><20 | 750<br>800<br>850 | 600···750 |
| Fir | debarked | >40<br>20···40<br><20 | 800<br>880<br>980 | 600···800 |
| Pine | debarked | >40<br>20···40<br><20 | 750<br>800<br>880 | 600···800 |
| Beech | with bark | >30<br><30 | 1080<br>1160 | 850···1100 |
| Beech | debarked | >30<br><30 | 1000<br>1080 | 800···1000 |
| Oak | with bark | >30<br><30 | 1180<br>1270 | 950···1200 |
| Oak | debarked | >30<br><30 | 1000<br>1000 | |

In practical application the density of green wood is important with respect to shipping weight. TRENDELENBURG (1941) has shown that the variation of the density of green wood is much smaller than that of oven-dry wood. During the storage of round logs in the forest, their moisture content decreases, influenced by different factors, such as season of felling, debarking, place and duration of storage and weather conditions.

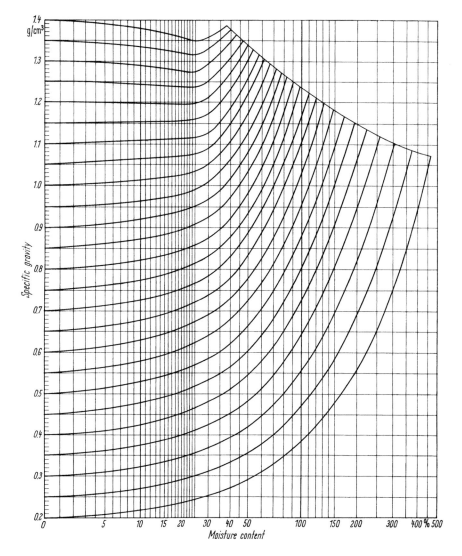

Fig. 6.7. Specific gravity-moisture content diagram. From KOLLMANN (1934)

Another important question which has to be dealt with is the density of tropical imported woods after the arrival in harbors of Europe or North America. During storage and shipping, the logs dry. Table 6.3 contains some figures, computed by MOMBÄCHER (1962).

168                    6. Physics of Wood                    [Ref. p. 285

Table 6.3. *Arrival Weights of Imported Tropical Timbers at Harbors in Europe
or North America*
(From MOMBÄCHER, 1962)

| Species | Arrival weight kg per cu. meter | Density | | Moisture content | |
|---|---|---|---|---|---|
| | | $\varrho_0$ (oven-dry volume) | $R_g$ (green volume) | kg per cu. meter | $u$ % |
| Balsa | 185 | 0.12 | 0.11 | 75 | 67 |
| Samba[1]) | 600 | 0.35 | 0.32 | 280 | 87 |
| Okoumé | 600 | 0.41 | 0.37 | 230 | 62 |
| Ilomba | 750 | 0.44 | 0.39 | 360 | 93 |
| Khaya | 750 | 0.45 | 0.405 | 345 | 85 |
| Limba | 800 | 0.52 | 0.47 | 330 | 70 |
| Makoré | 850 | 0.62 | 0.55 | 300 | 54 |
| Niangon | 850 | 0.64 | 0.56 | 290 | 51 |
| Abura | 900 | 0.52 | 0.46 | 440 | 69 |
| Sapelli | 900 | 0.59 | 0.53 | 370 | 70 |
| Teak | 1000 | 0.64 | 0.59 | 410 | 70 |
| Iroko | 1050 | 0.63 | 0.57 | 480 | 84 |
| Afzelia | 1150 | 0.70 | 0.65 | 500 | 77 |
| Bongossi | 1200 | 1.03 | 0.89 | 310 | 35 |
| *Lignum vitae* | 1400 | 1.20 | 1.04 | 360 | 34 |

[1]) The arrival weight of Abachi and Wawa is somewhat lower.

### 6.1.4 Variations in Density

Variations in the density of wood are due to differences in the structure and to the presence of extraneous constituents. The structure is characterized by the proportional amounts of different cell types such as fibers, tracheids, vessels, resin ducts, wood rays, and by their dimensions, especially the thickness of the cell walls. Hereditary tendencies, physiological and mechanical influences as well as factors of environment (soil, heat, precipitation, wind) affect the structure of wood and thus its density. The growth of a tree is affected not only by the site on which it grows but by its age. Finally the position in the tree-trunk has a considerable effect on the density of the wood.

The relationships are very complex and our knowledge is still incomplete. Nevertheless, some tendencies are evident (TRENDELENBURG and MAYER-WEGELIN 1953; BROWN, PANSHIN and FORSAITH, 1952). In Europe, the average density of coniferous species and of beechwood increases with decreasing elevation of the site and to some extent from the south to the north. These effects are more pronounced for pine and larch than for spruce. HALE and PRINCE (1940) concluded from their investigations on spruce and balsam fir in Eastern Canada that climatic conditions over broad regions may influence the density of wood, but local factors interrupt and even mask this general trend.

Fig. 6.8 shows some typical frequency curves for the specific gravity of spruce and pine from different European sites. In general, the density and strength of pine is higher than that of spruce, but Fig. 6.9 illustrates that both species have nearly the same specific gravity when grown on the plateau of Upper Bavaria. Swamp soils produce, according to SCHWAPPACH (1898), the lightest and poorest beechwood. An interesting example of a very distinct effect of an environmental factor on density is shown in Fig. 6.10 (PAUL and MARTS, 1934). The frequency curve has two peaks; it consists, therefore, of two normal frequency curves, one for the distribution of extremely light wood and one for wood of normal density. The wood with the very low specific gravity is formed in the swollen buttresses of ash trees (*Fraxinus pennsylvanica* var. *lanceolata* Sarg.) growing on the bottomlands of the lower Mississippi River valley where every year an inundation occurs. The wood of the swollen part has properties similar to those of roots while the wood above the butt swells normal; the tran-

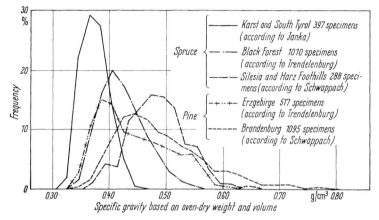

Fig. 6.8. Frequency curves for the specific gravity of spruce and pine from different European sites. From TRENDELENBURG (1935)

sition between these two kinds of wood is abrupt. Asymmetrical and uneven frequency curves are composed of symmetric normal frequency curves. Fig. 6.11 shows how the multinodal frequency curve for European pinewood is built up of three symmetrical curves, each one valid for parts of the stem at different heights (TRENDELENBURG, 1934, 1939).

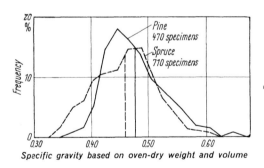
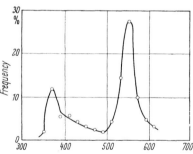

Fig. 6.9. Frequency curves for the specific gravity of spruce and pine grown on the plateau of Upper Bavaria. From TRENDELENBURG (1935)

Fig. 6.10. Frequency curve with two peaks for the specific gravity of as wood growing on the bottomlands of the lower Mississippi River valley. From PAUL and MARTS (1934)

The influence of position in the tree on density, and thereby on strength properties, has been studied by many scientists. The literature, as a general rule, states that the butt log contains wood of the greatest density and that the lowest density occurs in the upper portion. An exception already mentioned is the light wood found in trees with swollen bases growing under swampy conditions. Fig. 6.12 indicates that occasionally (e.g. spruce) there is little correlation between height in the tree and density. VOLKERT (1941) observed that the density in the base of trees with a cylindrical stem is greater than in one that is strongly tapered.

The variations in density throughout a particular cross-section of the stem are less pronounced than are those in height, and are very much affected by the width of the annual rings or the percentage of summerwood. However, there are some general rules (WANDT, 1937). In spruce trees, the wood of lowest specific gravity

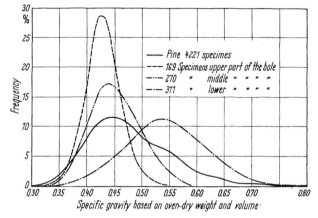

Fig. 6.11. Dissolution of the multinodal frequency curve for the specific gravity of pine wood in three symmetrical curves, each one valid for parts of the stem at different heights. From TRENDELENBURG (1934, 1939)

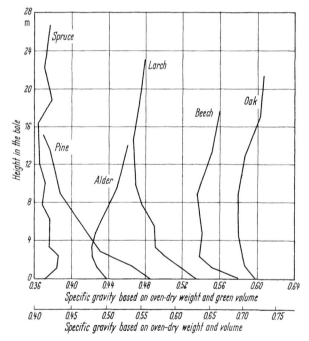

Fig. 6.12. Variation of specific gravity with height in the tree. From TRENDELENBURG (1939, p. 295)

is always produced near the pith of the tree where wide rings usually are formed. The highest density is produced in sapwood with narrow annual rings; a period of decrease may follow in very old trees. In pine and larch trees, the density increases outward from the center of the stem and reaches a maximum at a greater age, correlated with an optimum width of the growth rings; later, with the formation of less narrow rings, the density decreases. In a study of the properties of second-growth Douglas-fir by WANGAARD and ZUMWALT (1949), the density of wood near the pith as well as that at some distance from the pith was found to decrease with increasing height above ground. At any particular height, the wood adjacent to the pith was lighter than that nearer to the bark (Fig. 6.13). In broadleaved trees, as a rule, the highest density is produced near the center. HARTIG (1894) found that in an oak tree, 246 years old, the density 1.3 m above the soil amounted to 0.72 g/cm³ at the center and to 0.46 g/cm³ at the periphery. TRENDELENBURG (1939, p. 288) reported a similar trend for beech wood.

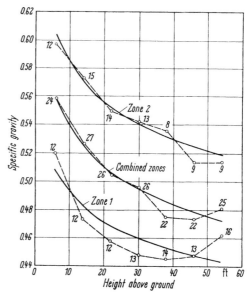

Fig. 6.13. Effect of position in the tree on the specific gravity of second-growth Douglas fir. From WANGAARD and ZUMWALT (1949)

In the United States of America, PAUL (1930) carried out the most comprehensive investigations on the effect of various conditions of growth on the quality of wood. A summary of his conclusions may be given: The growing space is the silvicultural tool of the forester in controlling the density of wood. In broadleaved species severe crowding in the stands resulted in a decrease in density, while relief from crowding is accompanied by an increase. Wood having the most uniform properties and the highest quality is produced in hardwoods when the trees are grown sufficiently close together while young. During the early years of coniferous stands, the size of the tree crown seems to be the main factor in determining the density of the wood. Where crowding in a southern pine stand had induced a decline in the density of the wood, a thinning either caused a subsequent increase in the rate of growth in conjunction with a remarkable increase in density when all of the growth conditions in the stand were especially favorable, or such thinning caused a decrease in density when the growth conditions were more favorable for springwood than for summerwood production. The closely crowded trees on the good sites continued to produce wood of high density. In such trees, narrow rings are correlated with a high percentage of summerwood. In order to produce timber of high quality, stands of coniferous species have to be thinned carefully.

Very informative are stem-growth diagrams showing the distribution of the density of wood over the plane determined by the stem axis and the north-south direction. In order to eliminate individual factors, growth diagrams for five or ten stems of a species are combined to represent the basic type of diagram. A few examples are given in Fig. 6.14.

There exists a definite influence of some anatomical or chemical properties on density. Compression wood is 20 to 40% heavier than normal wood. In general, the wood of the branches is heavier than that of the trunk. The density of heartwood is generally higher the darker the color. For this reason colored tropical

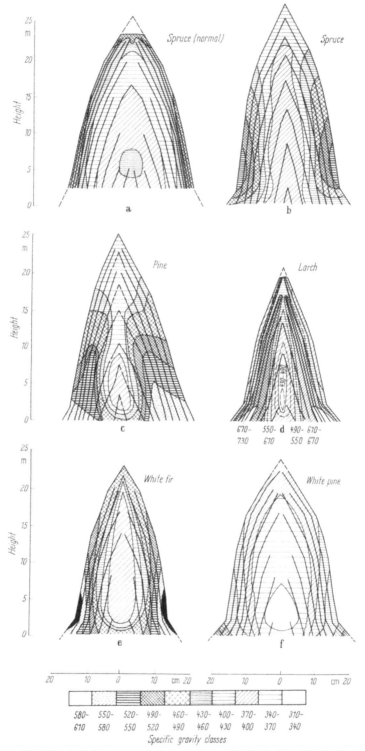

Fig. 6.14 a—f. Basic stem-growth diagrams showing the distribution of the density
a) Spruce, mountain site, b) Spruce, plaine site, c) Pine, d) Larch, e) Fir, f) Northern white pine. From VOLKERT (1941)

woods are rather heavy. According to PRÜTZ (1941) tropical woods are, on the average, 23% higher in density than species from the northern temperate and subtropical zone (Fig. 6.15). Wood containing a large amount of mineral constituents (e.g. calcium oxalate and silicates) are heavy and hard. The presence of tyloses in the vessels (e.g. in green-heart) increases the density as does a high resin

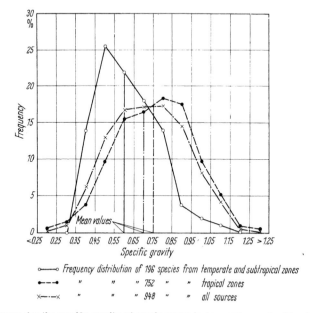

6.15. Frequency curves for the specific gravity of woods grown in temperature and subtropical zones and in tropical zones From PRÜTZ (1941)

content. Variability may be very marked even within the bole of a tree; for example, the sapwood of a longleaf pine (*Pinus palustris* Mill.) may contain about 2% resin, the average heartwood 7 to 10%, the heartwood in the butt log 15%, and the heartwood in the aged virgin stump 25% (KURTH and SHERRARD, 1931, 1932). The density of the resin of German coniferous trees is between 0.985 and 1.073 gm/cm³ according to MAYR (1894) and the resin of *Guajacum officinale* between 1.23 and 1.25 g/cm³, according to WIESNER (1927).

### 6.1.5 Density of Springwood and Summerwood, Correlation with Width of Annual Rings

It is apparent from the discussions in section 2 of Chapter 1 that the contrast in density between springwood and summerwood is more pronounced in conifers and ring-porous hardwoods than in diffuse-porous hardwoods. Some values are given in table 6.4, but the importance of these figures should not be overestimated. In some species the contrast in density between the two portions of the annual ring may be only slight, as for example in the white pines producing a wood "very uniform in texture and easy to work" (WANGAARD, 1950, p. 160). On the other hand, in the Southern pines and in Douglas-fir, the contrast is very marked.

Variability curves for the density $R_g$ of springwood and summerwood of four species of southern yellow pine are presented in Fig. 6.16. The curves show a considerable range in the density of both springwood and summerwood. Summer-

174 6. Physics of Wood [Ref. p. 285

Table 6.4. *Mean Values of Density of Springwood and Summerwood and Density Ratio for some Woods*

| Species | Density $r_0$ | | Density ratio | Reference |
|---|---|---|---|---|
| | Springwood | Summerwood | | |
| **Conifers** | | | | |
| *Abies pectinata* D. G. | 0.277 | 0.625 | 2.3 | WAHLBERG (1922) |
| *Larix europaea* D. C. | | | | |
| Sapwood | 0.35 | 0.88 | 2.8 | } TRENDELENBURG |
| Heartwood | 0.44 | 0.91 | 2.1 | (1939) |
| | 0.36 | 1.04 | 2.9 | VINTILA (1939) |
| Heartwood | 0.425 | 0.96 | 2.3 | MÜLLER-STOLL (1948) |
| *Larix leptolepis* Gord. | 0.35 | 0.77 | 2.2 | TRENDELENBURG (1939) |
| *Picea excelsa* Lk. | 0.307 | 0.601 | 1.96 | WAHLBERG (1922) |
| | 0.35 | 0.87 | 2.5 | |
| Light wood | 0.29 | 0.82 | 2.8 | TRENDELENBURG |
| Heavy wood | 0.38 | 0.91 | 2.4 | (1939) |
| Compression wood | 0.41 | 0.67 | 1.6 | |
| | 0.298 | 0.609 | 2.04 | JOHANSSON (1940) |
| *Pinus Banksiana* Lamb. | 0.33 | 0.69 | 2.1 | MÜLLER-STOLL (1940) |
| *Pinus caribaea* Morelet | 0.298 | 0.678 | 2.3 | } PAUL (1939) |
| *Pinus echinata* Mill. | 0.286 | 0.722 | 2.5 | |
| | 0.429 | 0.858 | 2.0 | JOHANSSON (1940) |
| *Pinus palustris* Mill. | 0.304 | 0.856 | 2.8 | PAUL (1939) |
| Heartwood | 0.45 | 1.02 | 2.3 | } MÜLLER-STOLL (1948) |
| *Pinus pinea* L. | 0.52 | 0.64 | 1.3 | |
| *Pinus sylvestris* L. | 0.30 | 0.92 | 3.1 | SIIMES (1938) |
| Sapwood | 0.36 | 0.90 | 2.5 | |
| Heartwood | 0.34 | 0.81 | 2.4 | } VINTILA (1939) |
| | 0.343 | 0.830 | 2.4 | |
| Sapwood | 0.327 | 0.877 | 2.7 | } MÜLLER-STOLL (1948) |
| Heartwood | 0.370 | 0.891 | 2.4 | |
| | 0.316 | 0.763 | 2.4 | JOHANSSON (1940) |
| *Pinus taeda* L. | 0.30 | 0.85 | 2.7 | PILLOW and LUXFORD (1937) |
| | 0.340 | 0.758 | 2.2 | PAUL (1939) |
| *Pseudotsuga taxifolia* Britt. | 0.29 | 0.82 | 2.8 | PILLOW and LUXFORD (1937) |
| Compression wood | 0.35 | 0.73 | 2.1 | |
| | 0.32 | 0.98 | 3.1 | SIIMES (1938) |
| | 0.30 | 0.79 | 2.5 | TRENDELENBURG (1939) |
| | 0.282 | 0.837 | 3.0 | VINTILA (1939) |
| | 0.344 | 0.958 | 2.8 | |
| **Ring-porous hardwoods** | | | | |
| *Fraxinus excelsior* | 0.385···0.506 | 0.721···0.803 | 1.55···2.03 | |
| *Quercus* sp. | 0.317···0.454 | 0.888···0.930 | 1.96···2.80 | } MÜLLER-STOLL (1948) |
| **Diffuse-porous hardwoods** | | | | |
| *Acer* sp. | 0.500···0.530 | 0.668···0.751 | 1.28···1.43 | |
| *Fagus sylvatica* L. | 0.502···0.536 | 0.748···0.883 | 1.34···1.76 | |
| *Tilia* sp. | 0.366···0.405 | 0.488···0.534 | 1.21···1.40 | |

wood was found to be more variable than springwood. PAUL (1939) also investigated the variation within the tree. The density of springwood either decreases slightly from pith to bark or keeps at about the same level. From butt to top it decreases. The density of summerwood increases from pith to bark and decreases from butt to top.

## 6.1 Density and Specific Gravity

As a rule, the great variability in density of coniferous species depends less on the variability in density of springwood and summerwood than on the variability of the percentage of summerwood. The significance of the percentage of summerwood to density in conifers is illustrated in Fig. 6.17. Theoretically, the following equation is valid according to YLINEN (1942):

$$\varrho_u = \varrho_{e_u} + (\varrho_{l_u} - \varrho_{e_u})s, \tag{6.10}$$

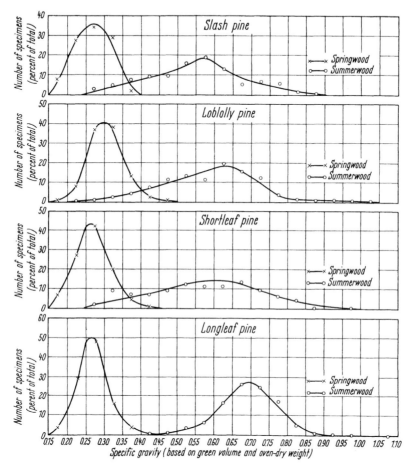

Fig. 6.16. Variability curves for the specific gravity of springwood and summerwood in four species of Southern pines. From PAUL (1939)

where $\varrho_u$ is the density of the wood at the moisture content $u$, $\varrho_{e_u}$ is the density of the springwood, $\varrho_{l_u}$ is the density of the summerwood (both at the moisture content $u$), and $s$ is the proportion of summerwood, expressed as a decimal fraction.

A graphical solution of Eq. (6.10) for Finnish pinewood is given in Fig. 6.18. A few experimental points are shown within the range of values usually found for proportion of summerwood. More complicated are the relationships existing between density and the formation of summerwood in hardwoods. Fig. 6.19 shows not only a considerable scattering of the test values, but also that the smoothed line drawn through the group means is curvilinear. It may be concluded that the percentage of summerwood becomes less of a factor as it reaches very high values.

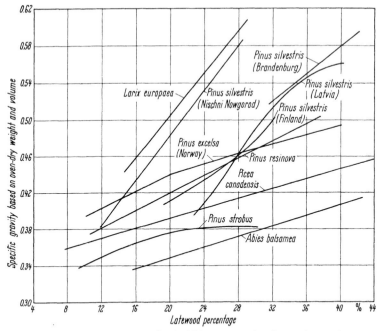

Fig. 6.17. Relationship between specific gravity (weight, oven-dry, based on volume, when oven-dry) and summerwood percentage for coniferous species. (Diagramm designed by KOLLMANN (1951, p. 345) using data published by JALAVA (1934), KLEM (1934), ROCHESTER (1933), SAVKOV (1930), TRENDELENBURG (1939, p. 283)

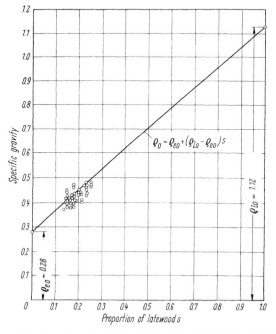

Fig. 6.18 Theoretical relationship between specific gravity and summerwood percentage, as compared with measured data for Finnish pine. From YLINEN (1942)

The rate of growth, expressed either as width of annual rings or as number of rings per inch, has frequently been investigated and discussed as a measure of

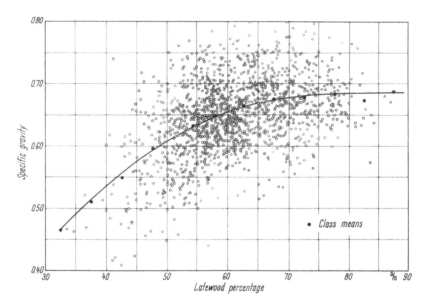

Fig. 6.19. Relationship between specific gravity and summerwood percentage for ash (*Fraxinus excilsior* L.). From KOLLMANN (1941)

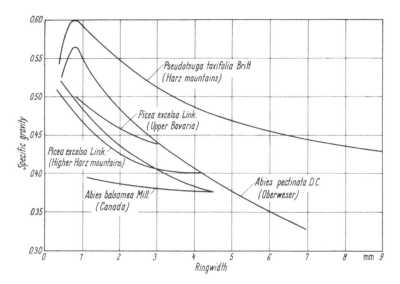

Fig. 6.20 Relationship between width of annual rings and specific gravity for Douglas fir, spruce, and fir. Diagram designed by KOLLMANN (1951, p. 348) using data published by ROCHESTER (1933) and TRENDELENBURG (1939, p. 283)

quality, i.e. the density of wood. Though there are some contradictory results and though the correlation between the percentage of summerwood and the density of the wood is closer than that between width of growth rings and density, some statements are in order:

1. Among the coniferous woods the density of spruce increases with decreasing width of annual rings (Fig. 6.20), whereas the density of pines and larches at first

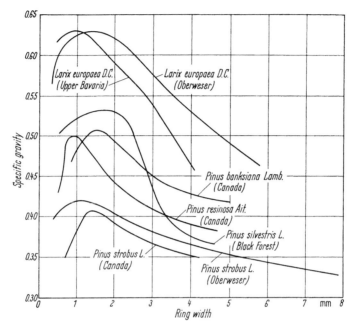

Fig. 6.21. Relationship between width of annual rings and specific gravity for different species of pine and larch. Diagram designed by KOLLMANN (1951, p. 349) using data published by ROCHESTER (1930) and TRENDELENBURG (1939, p. 283)

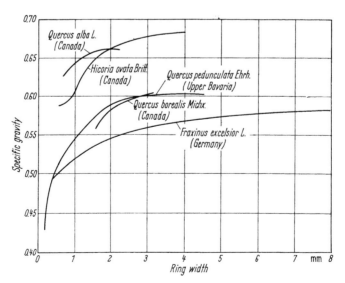

Fig. 6.22. Relationship between width of annual rings and specific gravity for ring-porous woods. Diagram designed by KOLLMANN (1951, p. 350) using own results and data published by ROCHESTER (1930) and TRENDELENBURG (1939, p. 285)

increases and then decreases again (Fig. 6.21). The pattern of fir and Douglas-fir apparently lies between the two. For pulpwood, the width of the annual rings is a usable measure of density. HALE and PRINCE (1940) found that the density of *Picea mariana B.S.P.* and of *Picea glauca Voss.*, estimated on the basis of the number of rings per inch, varied from the true value no more than 8%, whereas the probable error amounted to only 2.5%.

2. In ring-porous woods wide annual rings imply wood of high specific gravity (Fig. 6.22). There are, of course, exceptions mostly due to the amount and structure of summerwood.

3. In diffuse-porous woods, the ring width is not a significant criterion of density (Fig. 6.23).

### 6.1.6 Content of Solid Matter in Piles of Wood and Wood Residues

It is essential for forest industries to have reliable figures for the average weight per unit volume of different assortments of wood and wood residue. In the German forestry practice conversion factors for different assortments given in table 6.5 are used.

There is a great variability of wood residue due to the many different species, shapes, and sizes encountered. This means that there is also a great variability of the content of solid matter per unit volume of wood residue piles. A few of the more important figures are listed in table 6.7.

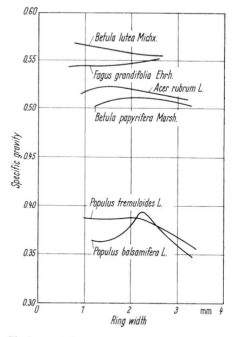

Fig. 6.23. Relationship between width of annual rings and specific gravity for diffuse-porous woods. Diagram designed by KOLLMANN (1951, p. 350) using data published by ROCHESTER (1930).

Table 6.5. *Conversion Factors for Different Assortments of Wood used in German Forestry Practice*

| Assortment | Content of solid matter in a pile of 1 cu.meter | | White peeled cu.meter |
|---|---|---|---|
| | with bark cu.meter | without bark cu.meter | |
| Round-wood or logs | 0.80 | 0.88 | — |
| Fuelwood | 0.70 | 0.77 | — |
| Brushwood (tops and branches) | 0.50 | — | — |
| Stump-wood | 0.50 | — | — |
| Bark | 0.30 | — | — |
| Mine timbers | | | |
| spruce, fir, beech | 0.80 | 0.88 | — |
| pine, larch, oak | 0.70 | 0.80 | |
| Pulp wood | | | |
| diameter >7 cm, length 1···2 m | 0.80 | 0.88 | 0.92 |
| diameter <7 cm, length 1 and 2 m, or diameter >4 cm length more than 2 m up to 4 m | 0.70 | 0.77 | 0.80 |

In table 6.6 a few experimentally determined figures are reported.

180 6. Physics of Wood [Ref. p. 285

Table 6.6. *Actual Content of Solid Matter in Pulpwood Piles*

| | | Average diameter of the pulpwood bolts in cm | | | | | | | Authority |
|---|---|---|---|---|---|---|---|---|---|
| | | 4 | 5 | 6 | 8···10 | 11···12 | 13···15 | 16···18 | |
| Length | 1 | — | — | — | 0.70 | 0.72 | 0.75 | 0.76 | } Aro (1931) |
| of the | 2 | — | — | — | 0.67 | 0.70 | 0.72 | 0.73 | |
| rolls | 3 | — | — | — | 0.64 | 0.67 | 0.70 | 0.72 | |
| in m | 4 | 0.50 | 0.59 | 0.65 | 0.72 | 0.74 | 0.75 | — | Schmidt (1940) |

Table 6.7. *Content of Solid Matter per Unit Volume of Wood Residues*
From Vorreiter (1943) and others as indicated)

| Kind of wood residue | Content of solid matter, cu.meter per piled cu.meter |
|---|---|
| 1. Residues from sawmilling | |
|   a) off-cuts (in random piles) | **0.37** |
|   b) slabs | 0.51···**0.58**···0.64 |
|   c) edgings and trimmings (length 1 m) medium | |
|     thick (according to Aro 1931) | 0.47···**0.56**···0.63 |
|     thin (according to Aro 1931) | 0.47···**0.52**···0.57 |
|     spruce and fir (according to Flatscher 1929) | 0.44···**0.47**···0.49 |
|   d) sawdust (according to Levon 1931) | 0.33 |
| 2. Residues from planer mills | |
|   shavings | 0.18···**0.20**···0.25 |
| 3. Residues from plywood factories | |
|   a) edgings from peeled veneers | **0.45** |
|   b) edgings from sliced veneers | **0.55** |
|   c) veneer cores (according to Jalava 1929 and Aro 1931) | 0.69···**0.70**···0.71 |
|     split into two halves (according to Aro 1931) | 0.69···**0.70**···0.71 |
| 4. Residues from woodworking | |
|   a) mixed residues (e.g. from the manufacture of vehicles) | 0.60···**0.65**···0.70 |
|   b) off-cuts from bobbins in layers | 0.20···0.40 |
|   c) sawdust | 0.32···0.38 |

## 6.2 Wood-Liquid Relations

### 6.2.1 Moisture Content, Definition

The moisture content, within a certain range, influences strength properties, stiffness, hardness, abrasion resistance, machinability, heat value, thermal conductivity, properties of gases produced in gasogens, yield and quality of pulp and resistance of wood against decay. Wood, like many other organic materials, shrinks as it loses moisture below the fiber saturation point and swells as it adsorbs moisture. Moisture content is important in drying, impregnation, finishing and bending of wood. Finally, costs of transportation and handling in lumber yards depend on the density of wood which is influenced by its moisture content. The determination of moisture content and the knowledge of wood-liquid relations are therefore of the utmost importance.

The moisture content $u$ is the weight of water contained in the wood, expressed as a ratio to the weight of the oven-dry wood and is computed as follows:

$$u = \frac{W_u - W_0}{W_0} \ [\text{kg/kg}] \tag{6.5a}$$

in which $W_u$ represents the weight of the wood with moisture content $u$ (original weight) and $W_0$ the weight of the oven-dry wood. It is customary to multiply this value by 100, thus obtaining the percentage of moisture based on the dry weight. In some instances, e.g. in the pulp industry, the moisture content is based on the original weight:

$$x = \frac{W_u - W_0}{W_u} \; [\text{kg/kg}]. \tag{6.11}$$

Between $u$ and $x$ exists the following relation:

$$u = \frac{x}{1-x} \quad \text{and} \quad x = \frac{u}{1+u} \tag{6.12a}$$

or when moisture content is expressed as a percentage

$$u = \frac{100\,x}{100-x} \quad \text{and} \quad x = \frac{100\,u}{100+u}. \tag{6.12b}$$

### 6.2.2 Determination of Moisture Content

There are five distinct methods of determining the moisture content of wood: oven-drying, distillation, titration, use of hygroscopic elements, and measurement of certain electrical properties.

**6.2.2.1 Oven-drying Method.** The oven-drying method as a rule is the most exact method, but it is slow and requires that samples be cut from the material. The samples should be about 15 to 22 mm ($^5/_8$ to $^7/_8''$) thick in the direction of the grain, they should be at least 30 cm (about 1 foot) from the end of the board (or other piece) to avoid the effect of rapid drying along the grain and they should be clear and free from defects. Each sample is immediately weighed and is then placed in an oven heated to 100 to 103 °C (212 to 217 °F) and kept there until constant weight is reached. Electrically heated ovens are used mostly, but only those with automatic temperature control are satisfactory. Good air circulation is essential, especially around the end grain of the samples. For weighing ordinary samples, balances having a capacity of about 200 g and a sensitivity to 0.1 g are useful. If many tests are to be carried out, much labor and time can be saved by using a semi-automatic balance which gives readings up to 5 g by a pointer moving across a scale.

The main disadvantage of the oven-drying method is its long duration. At a temperature of 100 to 103 °C, one-piece samples of 100 g require a drying time of 20 to 60 hrs (on the average about 30 hrs), depending on initial moisture content and density. The same sections chipped to match-size pieces by means of a chisel require 4 to 10 hrs. Samples of 20 g require 5 to 20 hrs as a single piece and 4 to 6 hrs chipped.

Oven-drying and high-vacuum drying yield correct results only with woods which, in addition to water, do not contain any volatile constituents like resins, oils, fats, or volatile preservatives like creosote. The application of both methods, therefore, has to be restricted to such wood species.

**6.2.2.2 Distillation Method.** When a sample of wood contains a significant amount of volatile constituents or preservatives, the distillation method is usually recommended. In this method a sample of wood, 20 to 50 grams, in the form of chips or sawdust is placed into the wide-necked flask (preferably of 500 to 1000-ml capacity) of a distillation apparatus (Fig. 6.24). A water-immiscible solvent (120 to 130 ml) is added and the flask is then heated by suitable means (usually an electrically heated sand-bath). The density of the solvent may be higher or lower than that of water. Some solvents used are given in Table 6.8.

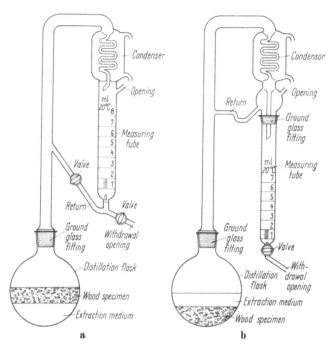

Fig. 6.24a, b. Distillation apparatus for the determination of the moisture content of wood. a) Density of the solvent higher than that of water. b) Density of the solvent lower than that of water. From KOLLMANN and HÖCKELE (1962)

Table 6.8. *Solvents for the Distillation Method* [1]

| Name | Formula | Density g/ml | Boiling point °C |
|---|---|---|---|
| Tetrachloroethane | $C_2H_2Cl_4$ | 1.600 | 146.3 |
| Tetrachlorethylene | $C_2Cl_4$ | 1.631 | 121.2 |
| Trichloroethylene | $C_2HCl_3$ | 1.456 | 87 |
| Chloroform | $CHCl_3$ | 1.499 | 61.3 |
| Xylene (mixture of isomers) | $C_8H_{10}$ | 0.869 | 139.4 |
| Toluene | $C_7H_8$ | 0.866 | 110.8 |
| Benzene | $C_6H_6$ | 0.879 | 80.1 |

[1] Data from the Handbook of Chemistry and Physics, 42nd. Ed., Chemical Rubber Publishing Co., Cleveland, Ohio, 1960/61.

The distillation apparatus is provided with a reflux condenser discharging into a calibrated trap connected to the flask. The condensed water is collected in the trap but the solvent returns to the flask. The distillation is continued until no more water is evaporated and condensed. This may last 4 to 24 h (Fig. 6.25). According to KOLLMANN and HÖCKELE (1962) the distillation method is not suitable for an exact determination of the water content due to its destructive influence on the wood tissues and to inaccuracies in reading. Only trichloroethylene, xylene and toluene are conditionally applicable as distillation liquids. Tetrachloroethane, chloroform, and benzene are not usable.

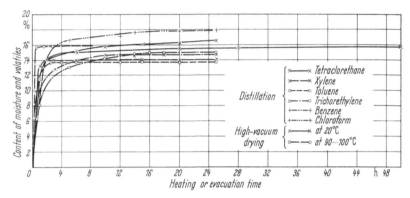

Fig. 6.25. Dependence of the content of moisture and volatiles, determined by means of the distillation method-upon type of solvent used and upon heating or evacuation time. From KOLLMANN and HÖCKELE (1962)

**6.2.2.3 Titration According to K. Fischer (1935); Eberius 1952, (1958).** The apparatus for the titration method is shown in Fig. 6.26. Since atmospheric humidity must be carefully excluded, all openings for the compensation of pressure must be secured by glass tubes filled with calcium chloride. The method is an iodometric titration in which elementary iodine reacts with sulfurdioxide and water to form hydrogen iodide and sulfuric acid:

$$SO_2 + I_2 + 2\,H_2O \rightarrow 2\,HI + H_2SO_4.$$

Probably the reaction is more complicated since the solvents pyridine and methanol take part. For a given amount of iodine used in the reaction the amount of water present can be computed with high accuracy according to the stoichiometric conditions. The titration method requires much time (6 hrs) and is rather expensive.

**6.2.2.4 Hygrometric Methods.** In a hole (6 mm diameter, 95 mm length) freshly drilled into a piece of wood the relative humidity corresponds to the moisture content of the surrounding wood. Thus it is possible to measure the moisture content of wood using a wood hygrometer. The method is much quicker (10 to 15 min.) than the oven-drying or the titration method but not as fast as the electric method. Measurements are restricted to

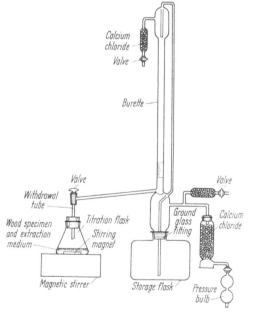

Fig. 6.26. Apparatus for the titration method according to FISCHER (1935)

the range between 3 and 25% of moisture content. Fig. 6.27 shows a wood hygrometer which is commercially available. The instrument consists simply of a perforated tube containing a string of hair which changes its length in response to changes in humidity. One end of this hygroscopic element is attached to the tube, the other to a light lever which moves a pointer across a dial scale.

Other pocket-size wood-hygrometers have been developed at the US Forest Products Laboratory in Madison, Wisc. One instrument consists of a piece of glass capillary tube containing mercury and closed at the bottom with a bulb of goldbeater's skin (DUNLAP, 1933). This thin membrane responds to changes in relative humidity, causing the bulb volume to vary and thereby forcing the mercury in the capillary to go up or down similar to a thermometer. For measurements the instrument base encasing the bulb is fitted closely to the wood surface for about 10 min. The relative humidity within the opening is indicated on a scale on the glass tube graduated in terms of moisture content. When not in use, the hygrometer is kept over anhydrous calcium chloride. The instrument has not been manufactured commercially. It is inexpensive but fragile; rough handling impaired calibration and it could not be shipped successfully since mercury is used. Therefore, another instrument has been designed based on the same principle (DUNLAP, 1936). A small piece of goldbeater's skin 0.25" (6.4 mm) wide and 0.30" (7.6 mm) long is mounted in a short perforated tube. One end of this moisture-sensitive element is fixed to the tube and the other to a target through a lever and a fine wire. The target is moved up or down in response to variations in relative humidity. The measurement is made by manually turning a screw into the tube until it touches the target as indicated by a small electric check light. Fastened to the screw is a pointer moving against a scale graduated in moisture content percentages.

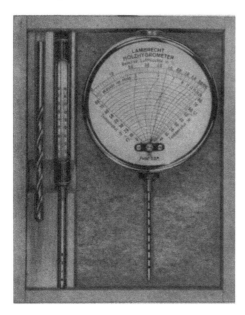

Fig. 6.27. Wood hygrometer, manufactured by Wilhelm Lambrecht A. G., Göttingen, W.-Germany

Another wood-hygrometer developed by ROTHER is an instrument useful for small factories. A mouthpiece is screwed into a freshly drilled hole. A glass-tube containing a plunger is attached to this mouthpiece and a paper-strip treated with cobalt-chloride is clamped to the lower end of the plunger. The test strips are kept in a tight glass tube over anhydrous calcium chloride. The "false air" in the hole is sucked out by means of a small plunger pump fastened with a rubber tube to the mouthpiece. Then the test strip is pushed into the hole. The color of the anhydrous cobalt chloride is pale blue. As it takes up moisture it changes successively to $CoCl_2 \cdot H_2O$, $CoCl_2 \cdot 2 H_2O$, $CoCl_2 \cdot 4 H_2O$ and finally to $CoCl_2 \cdot 6 H_2O$ accompanied by color changes to blue-violet, rose-violet, peach-blossom-red and finally to rose. Estimation of the moisture content in the range from 6 to 23% in steps of 2.5 (3% respectively) is possible by comparing the color of the test strip with a color-scale. Pale-rose characterizes wet, dark-blue very dry wood. The method is cheap, simple and quick (the whole test including drilling of the hole is done in 25 min.), but the results are influenced by variations in lighting conditions and the personal element in visual observations.

**6.2.2.5 Electrical Moisture Meters.** Electrical moisture meters first appeared on the American market about in 1930. Since then they have come into widespread use in woodworking industries (DUNLAP and BELL, 1951). The electrical methods make use of electrical properties of wood which depend considerably on moisture content, namely its electrical resistance, dielectric constant and radio-frequency power loss. Electrical moisture meters facilitate rapid moisture content determination for control purposes and are quite satisfactory in this regard.

*6.2.2.5.1 Resistance-type Moisture Meters.* The measurement is based on the fact that in the range of oven-dry condition to the fiber saturation point a nearly linear relationship exists between the logarithm of the electrical resistance and the moisture content. After drying from 30 to 0% of moisture content the electrical resistance is increased about a millionfold, but above the fiber saturation point an increase in moisture content up to complete filling of the coarser capillaries with liquid water the electrical resistance changes less than to a fifty fold. For this reason and as the electrical resistance becomes extremely high for very low moisture content values, resistance-type moisture meters are usually calibrated for measurements in the range from 7 to 25% of moisture content. A few instruments cover a wider range (up to 120%) but with lower accuracy. If moisture meters are maintained and used carefully, and if the necessary corrections for species and temperature are applied, an accuracy of $\pm 1\%$ may be expected in the range from 7 to 25% of moisture content. Moisture meters give the results instantaneously and allow testing of almost any board for wet spots that may be present. The necessary contact to the wood as a resistance element in the electrical circuit of the moisture meter is effected by electrodes. Usually needle or blade electrodes are driven into the wood. A hammer-extractor facilitates driving and removal of electrodes when testing hardwoods. Lumber which has just come from the kiln or which is drying in the lumber yard does not have a uniform moisture content. Fig. 6.28 shows the distribution of moisture at different times during the drying of a piece of hardwood, 2″ (51 mm) thick. Studies of moisture gradients in drying boards and planks have proven that, after the entire piece has reached a moisture content below the fiber-saturation point, "the moisture content in a plane located at one-fifth of the thickness of the material from its surface is usually very near the average of the piece" (DUNLAP and BELL, 1951). Instead of the regular electrodes two nails may be used for heavy structural timbers. Thin material, such as box board, veneer, and plywood, needs electrodes with very short needles or flat plates which are pressed to opposite faces of the sheet-like materials by a lever or a C-clamp (Fig. 6.29). A resistance type moisture meter is shown in Fig. 6.30. Such instruments are portable and generally operate on dry batteries, although some moisture meters can be plugged into ordinary electrical outlets.

*6.2.2.5.2 Capacity and Radio-frequency Power-loss Types of Moisture Meters.* The dielectric properties of wood change in proportion to its moisture content. Moisture meters of the capacity and radio-frequency power-loss type operate on the same principle. The radio-frequency current is supplied mainly by a circuit consisting of a battery and a vacuum-tube, and this current is applied to some type of condenser which is pressed as an electrode against the timber to be tested. The instruments are either calibrated directly in moisture content percentage or the scales are marked with arbitrary numbers and conversion tables for moisture content are provided. The range of capacity of power-loss types of moisture meters is from 0 to about 25% moisture content. Unfortunately readings depend on the density of the specimen. Since density varies considerably within a species, and the meters are calibrated for the average specific gravity of each species, an error may occur. This error is directly proportional to the difference between the density of the wood being tested and the value used for calibration. Density determinations of every sample (so that readings could be corrected) would be too

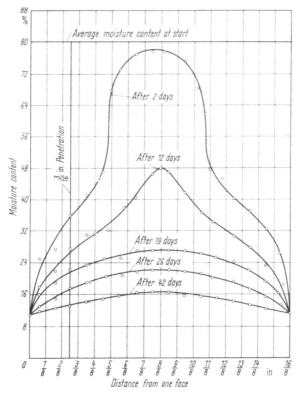

Fig. 6.28. Distribution of moisture content at different times during the drying of a piece of hardwood, 2″ (= 51 mm) thick. By courtesy of Delmhorst Instrument Co., Boonton, N. J.

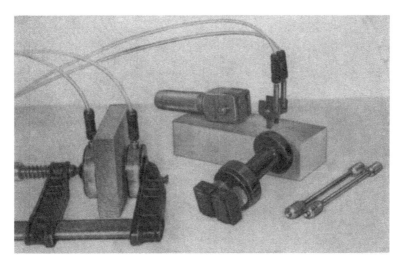

Fig. 6.29. Types of electrodes for resistance-type moisture meters; from right to left: Screw-type electrodes, stamp-electrodes for surface measurements on veneers, blade-type electrodes. C-clamp electrode for wood thicknesses between 20 and 100 mm. By courtesy of Siemens & Halske A. G., Munich, W.-Germany

cumbersome for general use (Division of Forest Products, Australia, 1954). Capacity-type moisture meters can be installed permanently on machines or conveyors where they can check automatically every piece of timber moving through. Any piece with a moisture content above a chosen limit is indicated by a flashing light or may be automatically marked or removed from the normal production line.

6.2.2.5.3 *Accuracy of Measurements.* The readings of both the resistance and capacity moisture meters depend upon the *species*. In the case of the resistance instruments this is probably due to the content of mineral substances acting as electrolytes and in the case of capacity and radio-frequency power-loss meters, due to density differences. Electrical moisture meters manufactured in the U.S.A. are calibrated for Douglas-fir and many instruments of European origin are calibrated for beech. This limitation may be overcome by means of correction figures, which the manufacturer of the meter should provide. For the same meter, between 8 and 24%, the corrections for two species were $+2$ to $-9\%$ and $+5$ to $+1\%$, respectively (Division of Forest Products, Australia, 1954).

These corrections, of course, are average values only, and differences of one or two percent on individual boards are frequent.

*Density* apparently has no influence on the accuracy of resistance-type moisture meters. The capacity method is principally dependent on density as mentioned earlier. Unequal moisture distribution may be another

Fig. 6.30. Resistance type moisture meter, manufactured by Siemens & Halske A.G., Munich, W.-Germany

source of severe errors. If the surface of the board is wet due to rain, dew, melted snow, dense fog or from recent reconditioning, either type of moisture meter will give readings which are too high. Similarly, if wet spots or water pockets exist within the wood the readings become too high. Surface moisture should be permitted to evaporate before the electrical meters are used. Water pockets can be detected by checking different parts of the board. If the moisture content of a board is higher in its core, readings will be lower than the average moisture content of the board. It has been noted previously how to overcome this trouble.

Resistance-type meters using needle or blade electrodes about $^3/_8''$ (9.5 mm) long give an average moisture content of a board $1''$ (25.4 mm) thick. For thicker material the moisture distribution should always be determined by driving two nail-electrodes to different depths, or the value obtained for the nails driven to one-fifth the thickness should be regarded as representative for the average moisture content of the wood. When using plate-type electrodes in testing veneer, a correction for thickness is necessary as Table 6.9 shows.

The electrodes of capacity-type moisture meters are pressed against one surface of the wood. There are various types of electrodes each intended for a definite board thickness. High moisture contents of the core cannot be determined as easily as with resistance-type instruments furnished with needle or blade electrodes. If the effective penetration is greater than the thickness of the timber tested the material should be laid on a suitable support (e.g. glass, but never on

Tabelle 6.9. *Corrected Moisture Contents for Douglas-fir Veneer of Various Thicknesses*
(Division of Forest Products, Australia, 1954)

| Meter reading % | | 10 | 11 | 12 | 13 | 14 | 15 | 16 | 17 | 18 | 19 | 20 | 21 | 22 | 23 | 24 |
|---|---|---|---|---|---|---|---|---|---|---|---|---|---|---|---|---|
| Veneer mm | Thickness in. | \multicolumn{15}{c}{Moisture content corrected for thickness} |
| 0.8 | 1/32 | 9 | 9 | 9 | 9 | 10 | 10 | 11 | 11 | 12 | 13 | 13 | 14 | 15 | 16 | 17 |
| 1.6 | 1/16 | 9 | 9 | 10 | 10 | 10 | 11 | 11 | 12 | 13 | 13 | 14 | 15 | 16 | 17 | 17 |
| 2.5 | 1/10 | 9 | 10 | 10 | 10 | 11 | 11 | 12 | 12 | 13 | 14 | 15 | 15 | 16 | 17 | 18 |
| 3.2 | 1/8 | 9 | 10 | 10 | 11 | 11 | 11 | 12 | 13 | 13 | 14 | 15 | 16 | 16 | 17 | 18 |
| 4.8 | 3/16 | 10 | 10 | 10 | 11 | 11 | 12 | 12 | 13 | 14 | 14 | 15 | 16 | 17 | 18 | 19 |

metal). Since the electrical resistance of wood decreases as the *temperature* increases, readings made with resistance-type moisture meters must be corrected when measuring hot or cold timber. Most resistance-type meters are calibrated at 21 °C (70 °F). For one moisture meter (Siemens & Halske design, Fig. 6.30), the following formula devised by KEYLWERTH and NOACK (1956) is valid:

$$\log\,[\log\,(R) - 4] = -(0.000132 u + 0.00335)\vartheta - 0.0296 u + 1.073, \qquad (6.13)$$

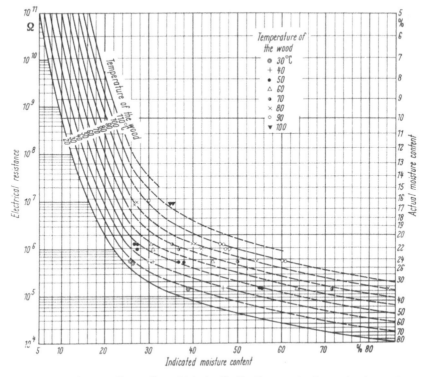

Fig. 6.31. Nomogram for correcting readings obtained with the Siemensmeter for varying temperatures and finding the electrical resistance. From KEYLWERTH and NOACK (1956)

in which $R$ represents the electrical resistance across the grain in [$\Omega$], $u$ the moisture content of the wood in [%], and $\vartheta$ the temperature (in the range between 20 and 110 °C) in [°C]. Fig. 6.31 makes it possible to correct readings obtained with the Siemens-meter and to find the electrical resistance. Example: measured moisture content, 20%; temperature of wood, 90 °C (193 °F); corrected moisture content, 9.8%, electrical resistance $7 \cdot 10^8$ $\Omega$. Readings of capacity-type moisture meters are not significantly affected by the temperature of the wood.

Firm contact is necessary for all electrical moisture meters since, if poor contact is made, readings with any type of electrodes will be too low.

The *grain direction* has an influence to some extent, since the resistance perpendicular to the grain is slightly higher than that measured along the grain. Indications of resistance-type moisture meters will generally be incorrect if the wood has been impregnated with preservatives or if it contains glue lines. The magnitude of the error will depend on the kind and amount of preservative or glue and on the moisture content. The moisture content of plywood as determined with moisture meters is significantly higher than that obtained by oven-drying (BELL and KRUEGER, 1949). The measurements of resistance-type moisture meters appear to be independent of the content of natural *resin*.

As much care as possible should be taken to carry out accurate measurements and, whenever possible, to use an average based on at least three individual readings.

### 6.2.3 Sorption and Equilibrium Moisture Content

Any piece of oven-dry wood will adsorb surrounding condensable vapors until an equilibrium content is reached. This phenomenon called sorption is typical

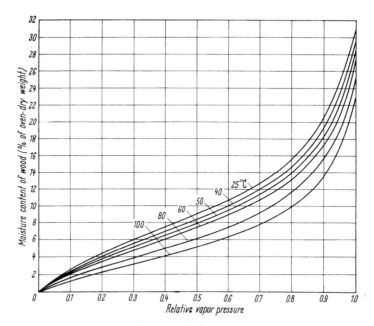

Fig. 6.32. Moisture content-relative humidity desorption-isotherms for Sitka spruce at different temperatures. From STAMM and HARRIS (1953)

for solids with a complex capillary structure. Sorption is influenced by a series of chemical and physical properties of the solid. Doubtless the affinity (e.g. of

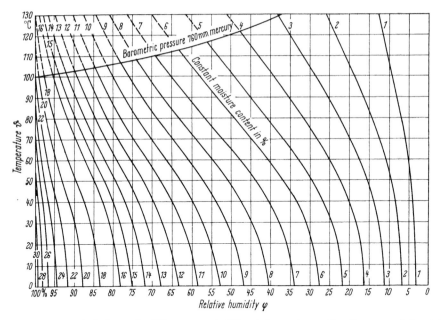

Fig. 6.33. Hygroscopic isotherms for Sitka spruce at different temperatures. The diagram worked out by LOUGHBOROUGH and published by HAWLEY, (1931), has been converted to centigrades, corrected and extrapolated for temperatures over 100 °C by KEYLWERTH (1949).

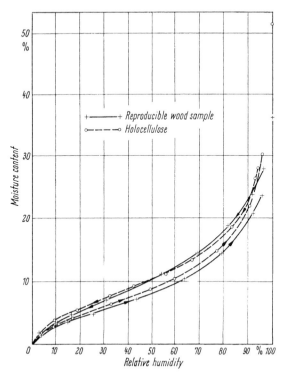

Fig. 6.34. Hygroscopic isotherms (water vapor 25 °C) for a reproducible sample for *Eucalyptus regnans* and for holocellulose, isolated from this wood. From CHRISTENSEN and KELSEY (1959)

cellulose in wood for water) and the size of the internal surface (e.g. for cellulose about $6 \cdot 10^6$ cm$^2$/cm$^3$) are contributory. Fig. 6.32 gives the moisture content-relative vapor pressure isotherms for Sitka spruce (in this case desorption curves, i.e. for a loss of moisture) at different temperatures. The $S$-shaped curves are quite similar for many sorbing substances (HERMANS, 1949; STAMM and HARRIS, 1953). They show that the sorptive power or hygroscopicity decreases with rising temperatures. For technical purposes Fig. 6.33, compiled and corrected by KEYLWERTH (1949) using Fig. 6.32, is valuable.

The isotherms for Sitka spruce are representative for most commercial timbers. The fiber saturation point corresponding to 100% relative humidity is generally 1 to 2% higher for hardwoods than for softwoods (STAMM and LOUGHBOROUGH, 1942), probably due to a higher hemicellulose content of the former (cf. Section 6.25). Among the main constituents of wood, the hemicelluloses have the highest sorptive capacity followed by cellulose and lignin (Fig. 6.34 to 6.36) (RUNKEL and LÜTHGENS, 1956; CHRISTENSEN and KELSEY, 1959). It has been estimated by the latter authors that cellulose contributes approximately 47%, the hemicelluloses 37%, and lignin 16% to the total sorption of wood.

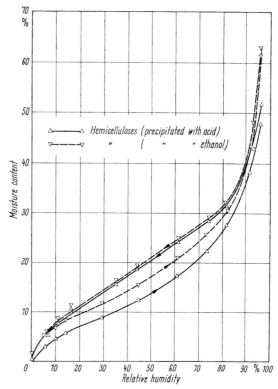

Fig. 6.35. Hygroscopic isotherms (water vapor 25 °C) for hemicelluloses, gained from *Eucalyptus regnans*. From CHRISTENSEN and KELSEY (1959)

The equilibrium moisture content of wood attained at any relative humidity depends on whether equilibrium has been reached from a higher or lower relative humidity or relative vapor pressure. The desorption and adsorption isotherms form a "loop" and the difference in moisture content is called hysteresis (Fig. 6.37). This phenomenon is well known since VAN BEMMELEN's hysteresis investigations (1896), which proved that many gel-like systems behave similarly. The sorption isotherms depend upon the pretreatment or the preliminary history of the "colloidal" body. If fresh cellulose fibers are dried from a highly swollen state, equilibria represented by curve *1* in the diagram of Fig. 6.38 are reached. If the same fibers, after having been dried at 0% relative humidity, again take up moisture curve *2* will result with a new equilibrium attained at 100% relative humidity. If the humidity is reduced again by steps, curve *3* results. Further repetitions of adsorption and desorption again reproduce curves *2* and *3*. The curves *2* and *3* of the hysteresis loop are demarcations for the entire range of sorption. Within this area any point can be reached by suitable operations as shown in Figs. 6.39 and 6.40 (HERMANS, 1949; KNIGHT and PRATT, 1935). The intermediate sorption curve in Fig. 6.37

represents the results of sorption experiments for comparatively large pieces of wood and not very accurate vapor pressure control so that both adsorption and desorption may have occurred at the same time ("oscillating desorption").

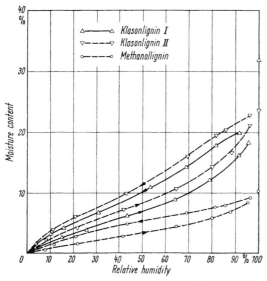

Fig. 6.36. Hygroscopic isotherms (water vapor 25 °C) for Klasonlignin and methanol-lignin, derived from *Eucalyptus regnans*. From CHRISTENSEN and KELSEY (1959)

The ratio of the adsorbed to the desorbed moisture content over the relative vapor pressure range between 0.10 and 0.95 is quite constant and amounts to about 0.85 (SEBORG, 1937; STAMM and WOODRUFF, 1941). Various explanations for the cause of hysteresis are found in literature. ZSIGMONDY (1911) attributed the phenomenon primarily to differences in wetting or contact angles of the water against the capillary walls in desorption and adsorption. This theory is not applicable at low relative vapor pressures where capillary condensation is impossible. Another more generally accepted explanation is that the hydroxyl groups of cellulose and lignin are responsible for hys-

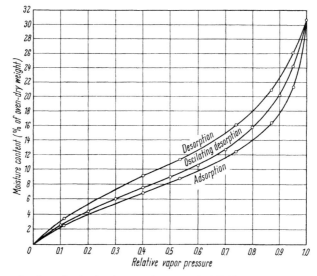

Fig. 6.37. Adsorption-desorption hysteresis curves for spruce at 25 °C. From STAMM and HARRIS (1954)

teresis. URQUHART and coworkers (1924, 1929) stated that the hydroxyl groups of cellulose and lignin in drying wood are drawn closely together and may satisfy each other instead of binding water molecules. When water is adsorbed some of the hydroxyl groups continue to satisfy each other and therefore are not available for taking up water.

There may exist still other causes of sorption hysteresis, but it should be mentioned that according to a more recent and interesting conclusion of BARKAS (1945, 1949) plasticity is the principal cause of sorption hysteresis. The hysteresis

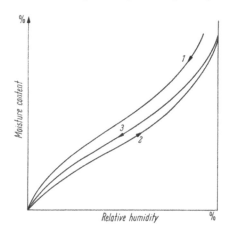

Fig. 6.38. Sorption isotherms of cellulose objects (diagrammatic). *1* Fresh fibers (desorption); *2* and *3*: adsorption and desorption after first drying. From HERMANS (1949)

Fig. 6.39. All points of condition within the domain of hysteresis can be attained by suitable operations. From HERMANS (1949)

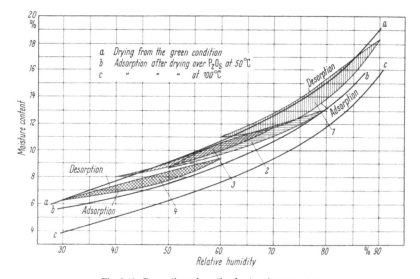

Fig. 6.40. Desorption-adsorption hysteresis curves for:
*1* Drying of green wood at relative humidities of 90% to 60%, re-wetting, *2* Drying at relative humidities of 80% to 50%, re-wetting, *3* Drying at relative humidities of 70% to 40%, re-wetting, *4* Drying at relative humidities of 60% to 30%, re-wetting. Plotted are average values for five species: oak, elm, munyana (*Khaya anthotheca* C. DC.), pine, and spruce. From KNIGHT and PRATT (1935)

cycles of both plasticity and sorption represent energy losses which are correlated and can be expressed in each case in cal per g.

Water is held within the structure of wood in four different phases: as water of constitution (by the formation of hydrates), as surface-bound water or water in an adsorbed monomolecular layer with a high apparent density of 1.3 g/cm³, as water adsorbed in multimolecular layers with a decreasing order of dipoles approaching the density of 1.0 g/cm³ for liquid water, and finally as capillary

13 Kollmann/Côté, Solid Wood

condensed water. The transition between the different phases is by no means sharp. Nevertheless, S-shaped isotherms seem to be made up of a combination of three components: at first a true adsorption curve according to LANGMUIR (1918) or FREUNDLICH (1926) which is parabolic and concave toward the relative humidity axis, followed by a curve for the "apparent" capillary condensation in the submicroscopic structure, then a curve for the real capillary condensation in the microscopic pores. Based on this assumption and statistical laws regulating the sizes both of submicroscopic and microscopic interstices, KOLLMANN (1963) developed an equation which mathematically describes the hygroscopic isotherms in the total range of relative humidity (Fig. 6.41).

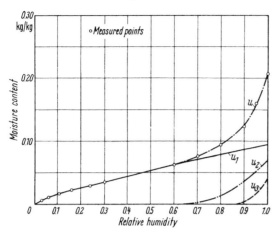

Fig. 6.41. Composition of the sorption isotherm $u$, at 100 °C, of three branches: $u_1$ for adsorption, $u_2$ for apparent capillary condensation of more or less polarized water in the submicroscopic capillaries, $u_3$ for capillary condensation of free water in the microscopic capillaries. From KOLLMANN (1963)

$$u = u_1 + u_2 + u_3 = a\varphi + c_1 e^{-\frac{1}{2}(b_1\psi - 1)^2} + c_2 e^{-\frac{1}{2}(b_2\psi - 1)^2} \qquad (6.14)$$

with the abbreviation $\psi = \varphi - 1$. The constants $a, n, c_1, b_1, c_2$ and $b_2$ must be determined experimentally for each sorbent. The equilibrium moisture content consists of three parts, where

$$u_1 = a^\varphi$$

describes the adsorption,

$$u_2 = c_1 e^{-\frac{1}{2}(b_1\psi - 1)^2}$$

the submicroscopic capillary condensation and

$$u_3 = c_2 e^{-\frac{1}{2}(b_2\psi - 1)^2}$$

the microscopic capillary condensation.

Additional theories of sorption are given by BRUNAUER, EMMETT and TELLER (1938) and more recently by MALMQUIST (1958, 1959). Two examples for the fact that the sorptive power depends upon the preliminary history of the object are shown in Fig. 6.42 and 6.43. The former proves that the application of high temperatures in artificial drying greatly reduces the sorption capacity (KOLLMANN and SCHNEIDER, 1963) and the latter elucidates the varying influence of irradiation with $\gamma$-rays, neutrons, or other powerful rays. While a small dose slightly reduces the hygroscopicity (Dept. Sci. Ind. Res., 1958, London, p. 46; WEICHERT, 1963), heavy doses considerably increase the uptake of moisture. The isotherms become quite similar to those obtained for hemicelluloses thus leading to the concept that the powerful rays split the chain molecules of cellulose into smaller units of about the same molecular weight as that of the hemicelluloses.

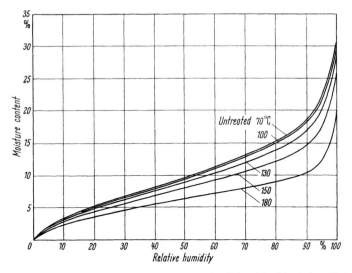

Fig. 6.42. Average adsorption-desorption isotherms for untreated and heat-treated wood at 20 °C. From KOLLMANN and SCHNEIDER (1963)

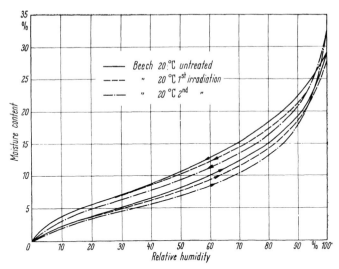

Fig. 6.43. Adsorption-desorption hysteresis curves for untreated and irradiation-treated beechwood at 20 °C From WEICHERT (1963)

### 6.2.4 Recommended Moisture Content for Wood in Service

The recommended moisture content for any piece of wood in service is intended to reduce changes in moisture content to a minimum, thereby minimizing dimensional movement due to swelling or shrinkage. The moisture-content requirements depend upon the use of lumber or wood products either in the interior of buildings or outdoors and also upon climate. In a subcontinent like the United States of America different moisture contents are recommended for various wood items in various regions (Table 6.10 and Fig. 6.44).

Table 6.10. *Recommended Moisture-content Values for Different Uses of Wood at Time of Installation in the USA*
(U.S. Dept. Agric. Forest Products Laboratory, 1955)

| Use of lumber | Moisture content %, based on oven-dry weight for ||||||
|---|---|---|---|---|---|---|
| | Dry South-Western States || Damp Southern Coastal States || Remainder of the United States ||
| | Average | Individual pieces | Average | Individual pieces | Average | Individual pieces |
| Interior-finish woodwork and flooring | 6 | 4···9 | 11 | 8···13 | 8 | 5···10 |
| Siding, exterior trim, sheating, and framing | 9 | 7···12 | 12 | 9···14 | 12 | 9···14 |

In general, the moisture content averages are less important than the range in moisture content permitted in individual pieces. It should be common commercial practice to dry wood for flooring, furniture, etc. to the lowest noted moisture content. During storage, manufacture, and use the moisture content will be increased and the distribution of moisture will be equalized. More detailed recommended moisture contents for various wood uses are shown in Table 6.11, compiled by HENDERSON (1951). The values should only be used as a guide.

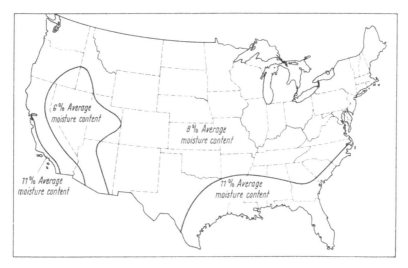

Fig. 6.44. Variation of recommended moisture content averages for interior-finishing woodwork for use in different parts of the USA. From U.S. Dept. Agric., For. Prod. Lab. (1955)

Some of the values listed in table 6.11 are not in accordance with the rules in central Europe. There the recommended values for staves are 18 to 20%; for wagons, and items exclusively used for exterior exposure 13 to 16%; for window frames, flush doors, wheels, wheel spokes, and parts of aircraft 12 to 15%; for furniture, parquetry and interior woodwork used in rooms heated by stoves, 10 to 12%; for the same items in buildings with central heating 8 to 12% and for plywood and laminated products 5 to 7%.

Investigations of the seasonal variation of the moisture content of wood in use have been carried out and published for some countries (PECK, 1932; MÖRATH, 1933; Commonwealth of Australia, Counc. Sci. Ind. Res., Div. For. Prod. 1934; REHMAN, 1942). More important than theoretical considerations are practical measurements of the equilibrium moisture content out of doors and in closed

Table 6.11. *Recommended Moisture Content Values for Various Wood Uses in the USA*

| Use of lumber | Moisture content %, based on ovendry weight | |
|---|---|---|
| | Average | Range |
| Baskets and fruit packages | 20 | 8···45 |
| Boxes and crates | 12 | 6···18 |
| Brush handles and backs | 5 | 2···10 |
| Car decking | 12 | 8···16 |
| Inside box car material | 9 | 7···12 |
| Caskets and coffins | 5.5 | 5··· 6 |
| Rough boxes | 10 | 6···14 |
| Chairs and chair stock | 6 | 5···12 |
| Cooperage: | | |
|    Tight staves and heading | 6 | 5··· 7 |
|    Slack staves and heading | 7 | 6···12 |
| Flooring | 6 | 6···10 |
| Furniture stock | 6 | 4···10 |
| Handles | 7 | 2···10 |
| Instruments, musical: | | |
|    Radios, piano cases, actions | 5 | 3··· 6 |
| Sash, doors, blinds, etc. | 6 | 4··· 8 |
| Ship and boat lumber | 12 | 12···15 |
| Shoe last blocks | 5 | 4··· 6 |
| Shoe heels | 6 | 3··· 9 |
| Shuttles and bobbins | 5 | 4··· 6 |
| Spools | 8 | 4···12 |
| Tanks and silos | 12 | 8···16 |
| Trunks and valises | 6 | 4··· 9 |
| Veneers: | | |
|    Face | 4 | 2··· 7 |
|    Crossband | 5 | 2···10 |
|    Cores | 5 | 4··· 6 |
|    Plywood | 6 | 2··· 9 |
| Shingles | | |
| Lumber: | 10 | 10···12 |
|    Rough construction | | |
|    Joists, studs, sub-flooring | | |
| Timbers, cross-ties, poles, etc. for treatment with creosote or other preservatives | 8 | 6···20 |
| | 25 | free water removed |

rooms over a sequence of several years. Fig. 6.45 shows the change in moisture content of yard-piled softwood lumber in Eastern Canada (FELLOWS, 1937). When air circulation is lacking, e.g. during ocean shipment of seasoned lumber, moisture content will remain almost constant (JENKINS and GUERNSEY, 1937). If lumber for construction purposes is insufficiently seasoned, a severe loss of moisture combined with shrinkage is unavoidable in new buildings after glazing and heating (Fig. 6.46).

### 6.2.5 Fiber Saturation Point, Maximum Moisture Content of Wood

The fiber saturation point of wood is the moisture content reached when all the fibers are completely swollen (saturated with colloidal water), but no liquid or free water exists in the coarser capillary structure. This point is often falsely

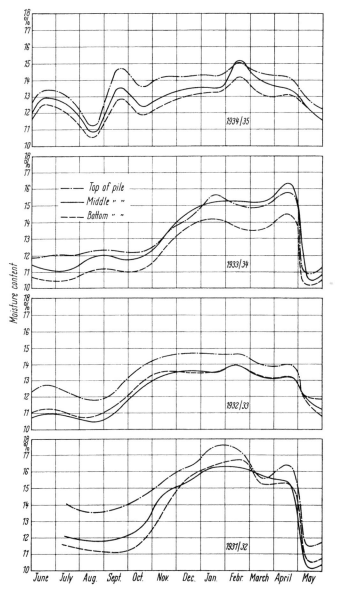

Fig. 6.45. Changes in moisture content of yard-piled softwood lumber in Eastern Canada. From FELLOWS (1937)

defined as the moisture content in equilibrium with 100% relative humidity. Since according to Thomson's law all capillaries depress the vapor pressure (although the effect may be minor), the real fiber saturation point can only be determined by extrapolation (STAMM, 1929; STAMM, 1959; KOLLMANN, 1959).

STAMM has pointed out that characteristic breaks or limiting values occur at the fiber saturation point in the relationships between moisture content and shrinkage, between moisture content and heat of wetting, moisture content and adsorption-compression, moisture content and electrical conductivity (Fig. 6.47), and moisture

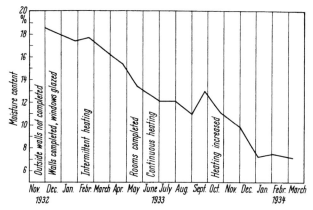

Fig. 6.46. Loss of moisture content of insufficiently seasoned lumber for construction purposes in a new building after glazing and heating. From Forest Products Research Laboratory, Princes Risborough

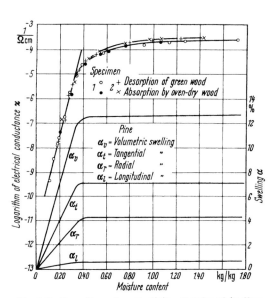

Fig. 6.47. Logarithm of conductivity of redwood in fiber direction (from STAMM 1927) and of swelling (from MÖRATH 1932); vs. moisture content

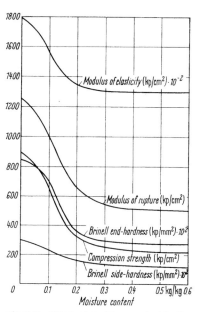

Fig. 6.48. Effect of moisture content in the hygroscopic range on mechanical properties for pine wood (*Pinus sylvestris* L.). From KOLLMANN (1944)

content and stiffness or strength (Fig. 6.48). The values obtained by these different methods are in fairly good agreement (e.g. for Sitka spruce at 25 °C between 24 and 30.5%). TRENDELENBURG (1939) showed that the fiber saturation point varies with structure and chemical composition of woods as follows:

1. Diffuse-porous broadleaved woods without distinct heartwood (lime-tree, willow, poplar, alder, birch, beech, and hornbeam) and the sapwood of ring-porous

and semi-ring-porous hardwoods with distinctly colored heartwood (black locust, chestnut, oak, ash, black walnut, and cherry) exhibit a very high fiber-saturation point from 32 to 35% (and even more).

2. Coniferous woods without significantly colored heartwood (fir and spruce) and the sapwood of coniferous species with distinctly colored heartwood, such as pine, white pine and larch, have a high fiber-saturation point from 30 to 34%.

3. Coniferous woods with distinctly colored heartwood:
a) with a moderate content of resin such as pine, larch, Douglas-fir have a fiber saturation point from 26 to 28%;
b) with high content of resin have a fiber saturation point from 22 to 24%.

4. Ring-porous and semi-ring-porous broadleaved woods with distinctly colored heartwood (black locust, chestnut, oak, ash, black walnut, and cherry) have a low fiber-saturation point from 22 to 24%.

Extremely light root-wood with very fine pores may reach a fiber-saturation point as high as 50%, but it is assumed that on the average the fiber saturation point is 28%. Due to the depression of vapor pressure in capillaries the moisture content in equilibrium with unit relative vapor pressure could far exceed the fiber-saturation point and could reach values even above 100% and more. This raises the question how great the maximum moisture content of wood may be.

For this computation we need the relative amount $c'$ of air space in the total volume. STAMM (1938) has derived the following formula

$$c' = 1 - R\left(\frac{1}{\varrho_w} + \frac{u_h}{\varrho_s} + \frac{u_k}{\varrho}\right), \quad (6.15\,\text{a})$$

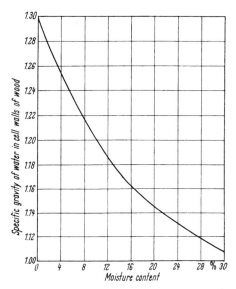

Fig. 6.49. Specific gravity of water adsorbed in cell walls of wood as a function of moisture content. From STAMM and HANSEN (1937)

where $R$ is the density based on dry weight and wet volume, $\varrho_W$ is the density of the oven-dry wood substance (assumed as 1.50), $u_h$ is the content on hygroscopic moisture, $\varrho_s$ is the density of the adsorbed and compressed water (Fig. 6.49), $u_k$ is the content of free water with the density $\varrho \approx 1.0\,\text{g/cm}^3$. In the case of maximum moisture content $u_{\max}$, $c' = 0$ and assuming $u_h = 0.28$ and approximating for the fiber-saturation point $\varrho_s \approx 1.0\,\text{g/cm}^3$, we obtain,

$$c' = 0 = 1 - R\left(\frac{1}{1.50} + u_h + \frac{u_{\max} - u_u}{1}\right) = 1 - \frac{R}{1.50} - R \cdot u_{\max}. \quad (6.15\,\text{b})$$

Consequently, the maximum moisture content of wood may be computed

$$u_{\max} = \frac{1.50 - R}{1.50\, R} \quad (6.15\,\text{c})$$

or based on the specific gravity on a dry volume-basis when (cf. formula 6.2, and 6.3c, and 6.5b)

$$R = \frac{G_0}{V_u} = \frac{G_0}{V_0(1 + 0.28\varrho_0)} = \frac{\varrho_0}{1 + 0.28\varrho_0} \,,$$

$$u_{\max} = \frac{1.50 - \dfrac{\varrho_0}{1 + 0.28\varrho_0}}{\dfrac{1.50 \cdot \varrho_0}{1 + 0.28\varrho_0}} = \frac{1.50(1 + 0.28\varrho_0) - \varrho_0}{1.50 \cdot \varrho_0} = \frac{1.50 - \varrho_0}{1.50 \cdot \varrho_0} + 0.28, \quad (6.15\,\mathrm{d})$$

where $\varrho_0$ is the density weight and volume ovendry.

### 6.2.6 Thermodynamics of Sorption

Water adsorption by wood is accompanied by the evolution of energy which appears in the form of considerable heat. This heat can be measured by adding water to dry wood in a calorimeter. The evolution of heat when 1 g of dry wood is moistened up to fiber saturation point is called the integral heat $H$ of wetting. Some figures are given in Tab. 6.12.

Table 6.12. *Integral Heat of Wetting of Wood and Other Cellulosic Materials*

| Material | Sorption of | Integral heat of wetting $H$ cal/g | Author |
|---|---|---|---|
| Spruce | Water | 17.0 | VOLBEHR (1896) |
| Wood | Water | 14.6···19.6 | DUNLAP (1913) |
| | Creosote | 4.6 | |
| | Turpentine | 2.3 | |
| Wood flour | | | |
| Sapwood | Water | 19.78 | |
| Heartwood | Water | 18.41 | |
| Sulphite pulpwood, bleached beaten: | | | |
| 0 min. | Water | 12.68 | ARGUE and MAASS (1935) |
| 20 min. | Water | 13.48 | |
| 40 min. | Water | 13.92 | |
| Cotton, pulp | Water | 10.54 | |
| Standard-cotton-cellulose | Water | 10.16 | |
| Cotton | Water | 20.8 | |
| Filter paper | Water | 9.6 | ROSENBOHM (1914) |
| Filter paper | Water | 10.7 | KATZ (1917) |

VOLBEHR (1896) had already shown how the integral heat of wetting $H$ depends upon moisture content $u$ and that a reduction of the volume of the adsorbed water also takes place (Fig. 6.50). According to KATZ (1917) the lower part of the curve for $H = f(u)$ can be represented by the empirical relation

$$-H = \frac{A \cdot u}{B + u} \,. \qquad (6.16)$$

KOLLMANN derived $A = 22$ and $B = 0.07$ from the data of VOLBEHR. Fig. 6.51 was designed with these values which is useful in determining heat consumption in kiln drying.

KELSEY and CLARKE (1956) developed for the curve of the integral heat of wetting the following equation

$$-H = \frac{u}{C + D \cdot u + E \cdot u^2} \tag{6.17}$$

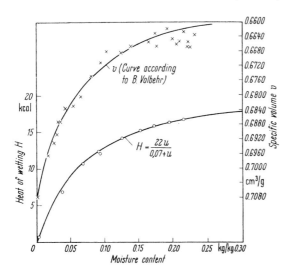

Fig. 6.50. Heat of wetting and contraction of volume for the swelling of spruce wood vs. moisture content. (Designed by KOLLMANN using data published by VOLBEHR, 1896)

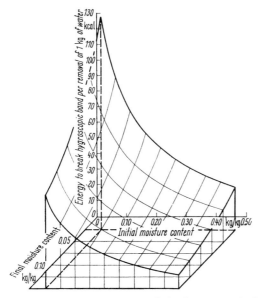

Fig. 6.51. Energy to break hygroscopic bond per removal of 1 kg of water in kiln drying as influenced by the difference of initial moisture content and final moisture content. From KOLLMANN (1935)

where $-H$ is the integral heat of wetting up to the moisture content $u$, and $C$, $D$, and $E$ are constants.

If a very small amount of water is added to wood or cellulosic fibers an intermediate heat of wetting can be measured. This differential heat of wetting $\Delta H$ can be deduced thermodynamically from the hygroscopic isotherms. The following equation is valid

$$\Delta H = RT^2 \frac{d \ln h}{dT}, \tag{6.18}$$

where $R$ is the gas constant $\left(R \times 1° = 8.314 \cdot 10^7 \text{ erg} = 1.986 \text{ cal} = 82.06 \text{ atmosphere} \cdot \text{cm}^3 = 0.8478 \frac{\text{mkp}}{\text{gmol}}\right)$, $T$ is the absolute temperature, and $h$ is the relative fugacity of the sorbed liquid at the moisture content for which a particular value of $\Delta H$ is calculated. The value of $h$ may be taken as equal to the relative humidity $\varphi$ with which the material is in equilibrium. The equation given above is related to the Clausius-Clapeyron equation for heat effects.

The differential heat of wetting reaches a maximum of about 260 cal per gram of water absorbed (950 BTU per pound of water) for oven-dry wood ($u = 0\%$) and it becomes zero at fiber saturation point. Generally speaking, the more moisture wood contains in the hygroscopic range, the more the heat effect is reduced and it approaches zero asymptotically. STAMM and LOUGHBOROUGH (1935) have computed that during the initial stages of adsorption the free energy $F$ is reduced less than the differential heat of wetting $\Delta H$. For constant pressure and temperature, we can write

$$\Delta F = \Delta H - T \Delta S \tag{6.19}$$

where $\Delta F$ denotes the change of free energy and $\Delta S$ the change of entropy.

Fig. 6.52 shows that the change of heat of wetting is much greater than the change of free energy. During the swelling process, initially a decrease of entropy occurs. This means that the adsorbed water occurs in a less probable condition, i.e. in a higher degree of orientation, a nearly crystalline association of the water molecules due to hydration of hydroxyl groups on the surfaces of micelles or crystallites. The water bound to the cellulose with some degree of orientation has, as has been mentioned previously, a density higher than 1.0 g/cm³. In this connection it must be pointed out that in the crystalline regions the hydroxyl groups are mutually satisfied. The hydroxyl groups available for water are statistically distributed in the amorphous regions, and therefore these regions make the predomi-

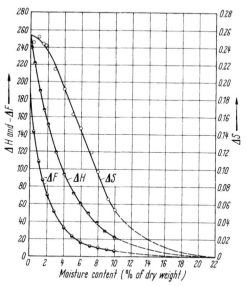

Fig. 6.52. Differential heat of wetting and free-energy and entropy changes in calories per gram of liquid adsorbed by an infinite amount of adsorbent vs. moisture content for soda boiled cotton, calculated by STAMM and LOUGHBOROUGH (1935) from the data of URQUHART and WILLIAMS

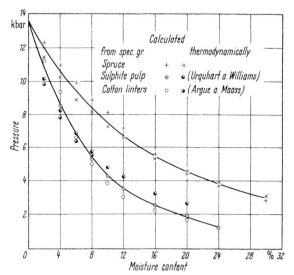

Fig. 6.53. External pressure which could produce the same compression of water which occurs as volume contraction in sorption by wood, pulp, and cotton. From STAMM and HANSEN (1937)

nant contribution to the sorption phenomena. STAMM and HANSEN (1937) calculated the external pressure which would produce the same compression of water which occurs as volume contraction in sorption by wood, pulp, and cotton (Fig. 6.53).

204                               6. Physics of Wood                        [Ref. p. 285

If sorption is performed in a reversible manner by the addition of a small amount
of water to a large amount of sorbent, the maximum work $\Delta A$ produced can
be computed from the sorption-isotherms. If we assume that this maximum work
could lift a piston and that we apply a pressure on its head for reaching an equilibrium, we come to a thermodynamically deduced formula for the swelling
pressure $P$:

$$P = - \frac{RT}{M} \ln \frac{h}{h_0}, \tag{6.20}$$

where $R$ is the gas constant, $T$ the absolute temperature, $M$ the molecular weight
of the liquid (for water $= 18$), and $h/h_0 = \varphi =$ the relative vapor pressure.
    For a temperature of $23\,°C$ ($73.4\,°F$) we obtain

$$P = -1395 \ln \varphi = - 2311 \log \varphi \ [\text{kp/cm}^2]. \tag{6.21}$$

Wood contains cell cavities which are very fine capillaries. In such capillaries
surface tension causes hemispherical menisci which are outwardly concave and a
capillary of radius $r$, according to THOMSON (1871), can retain water at a reduced
vapor pressure $h/h_0$ following the equation

$$r = - \frac{2\sigma M}{\varrho\,RT \ln \dfrac{h}{h_0}} \tag{6.22}$$

where $\varrho$ is the density and $\sigma$ the surface-tension of water.
    It is easily seen that the tension exerted on the walls of the capillaries, which
enforces shrinkage and collapse, is identical with the swelling pressure.

Table 6.13. *Interrelationships between Relative Vapor Pressure,
Radius of Capillaries and Swelling Pressure P*

| Relative vapor pressure $h/h_0$ % | Radius $r$ of capillaries cm | Swelling pressure $P$ | |
|---|---|---|---|
| | | kp/cm² | lb./sq.in. |
| 100.0 | | 0 | 0 |
| 99.9 | $1.06 \times 10^{-4}$ | 1.4 | 19.6 |
| 99.5 | $2.12 \times 10^{-5}$ | 6.9 | 98.5 |
| 99.0 | $1.06 \times 10^{-5}$ | 7.5 | 107 |
| 95.0 | $2.06 \times 10^{-6}$ | 71.5 | 1020 |
| 90 | $1.01 \times 10^{-6}$ | 146 | 2067 |
| 80 | $4.78 \times 10^{-7}$ | 311 | 4380 |
| 70 | $3.05 \times 10^{-7}$ | 497 | 6960 |
| 60 | $2.08 \times 10^{-7}$ | 712 | 10000 |
| 50 | $1.54 \times 10^{-7}$ | 999 | 13600 |
| 40 | $1.16 \times 10^{-7}$ | 1278 | 18000 |
| 30 | $8.85 \times 10^{-8}$ | 1677 | 23600 |

The theoretical formula is not applicable at low humidities.

### 6.2.7 Shrinkage and Swelling

**6.2.7.1 Maximum Volumetric Shrinkage and Swelling, Influence of Drying Tem-
perature.** If a gel like wood swells, the volume of the colloidal structure increases
approximately by the volume of water sorbed. The assumption is justified only
if the microscopic voids do not change in size during swelling or shrinking.

Apparently this is the case for many wood species while others behave differently. Very careful physical investigations have also shown that the adsorbed water is slightly compressed. PECK (1928) published Fig. 6.54 showing that the external volumetric shrinkage depends linearly upon moisture content between about 22 to 24% and the oven-dry condition. This relationship holds for different densities.

Similar measurements have been made for many woods in different countries. The curvature of the lines above 22% moisture content is due to the moisture content gradient in drying specimens. The outer layers fall below fiber saturation point when the core is still above this point. The result is an early shrinkage near the wood surface.

The higher the density of the wood, the greater is its volumetric shrinkage or swelling. Available data for a large number of different species show that, on the average, a linear equation holds

$$\alpha_v = u_f \cdot \varrho_0 \; [\%] \quad (6.2\,\text{a})$$

or

$$\beta_v = u_f \cdot R_g \; [\%] \quad (6.2\,\text{b})$$

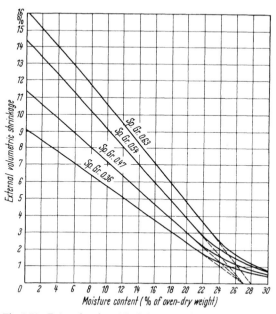

Fig. 6.54. External volumetric-shrinkage moisture-content relationship for 7/8 inch thick boards of Loblolly pine with different swollen volume specific gravities. From PECK (1928)

in which $\alpha_v$ is the volumetric swelling, $\beta_v$ is the volumetric shrinkage, $u_f$ is the moisture content at the fiber saturation point, $\varrho_0$ is the oven-dry density, and $R_g$ is the swollen volume density. NEWLIN (1919) published Fig. 6.55 which shows that the plotted points fit well to a straight line with a slope of $28 \cdot R_g$ which goes through the origin. The maximum deviations are identified. These deviations are caused either by a high content of water-soluble extractives (e.g. in Redwood and Eastern red cedar; STAMM and LOUGHBOROUGH, 1942; STAMM, 1962) or by internal stresses which may even lead to collapse and hence to abnormally high shrinkage. KEYLWERTH (1943) could confirm the average value of 28% for the fiber saturation point. GREENHILL (1940) computed a value of 27% for the mean ratio of the external volumetric shrinkage to the density based on green volume for 170 different Australian species before reconditioning. STAMM and LOUGHBOROUGH (1942) deduced $u_f = 26\%$ for 52 different softwood species and $u_f = 27\%$ for 106 different hardwood species, based on data by MARKWARDT and WILSON (1935).

STEVENS (1963) investigated the effects of temperature and rate of drying on the shrinkage of wood. He found that shrinkage tends to increase with increasing temperature and to decrease with increased rate of drying and becomes roughly proportional to moisture loss when the moisture content falls to about 30%. Below 25% of moisture content, however, the shrinkage values appear to be unaffected by either temperature or drying-rate.

**6.2.7.2 Anisotropy of Shrinkage and Swelling.** Shrinkage and swelling is not the same in the different grain directions. The greatest dimensional change occurs in a direction tangential to the annual rings. Shrinkage from the pith outwards, or radially, is considerably less than the tangential shrinkage, while shrinkage along the grain is so slight that it can nearly always be neglected. In the tangential direction the limits of $\beta_t$ for drying from the green to the oven-dry condition

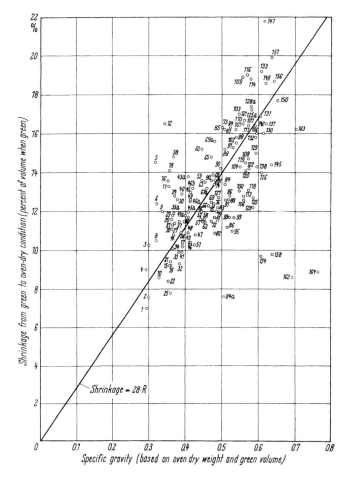

Fig. 6.55. Volumetric shrinkage of different wood species vs. specific gravity, based on oven dry weight and green volume. From NEWLIN (1919)

are 3.5 to 15.0% of the green dimension; the comparable figures for radial shrinkage $\beta_r$ are 2.4 to 11.0%. The longitudinal shrinkage of normal wood from the green to the oven-dry condition ranges between 0.1 and 0.9%. In general, tangential and radial shrinkage are the higher the heavier the woods are.

The volumetric shrinkage $\beta_v$ or the volumetric swelling $\alpha_v$ can be computed from the linear shrinkage or swelling values as follows:

$$\beta_v = 1 - (1 - \beta_t)(1 - \beta_r)(1 - \beta_l) \qquad (6.23\,\text{a})$$

and

$$\alpha_v = (1 + \alpha_t)(1 + \alpha_r)(1 + \alpha_l) - 1. \qquad (6.24\,\text{a})$$

If we neglect the rather small products of shrinkage and swelling coefficients, respectively, we obtain the following simplified equations:

$$\beta_v = \beta_t + \beta_r + \beta_l \approx \beta_t + \beta_r \tag{6.23b}$$

and

$$\alpha_v = \alpha_t + \alpha_r + \alpha_l \approx \alpha_t + \alpha_r. \tag{6.24b}$$

Typical swelling curves in the principal grain directions are shown in Fig. 6.56. One can see a straight-line relationship between moisture content and swelling from zero moisture content nearly up to fiber-saturation point. STEVENS (1938)

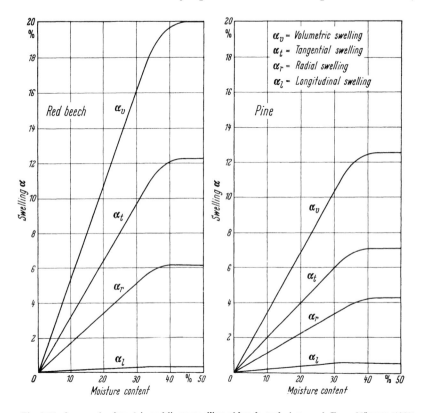

Fig. 6.56. Curves of volumetric and linear swelling of beech- and pine wood. From MÖRATH (1932)

found a linear relationship between shrinkage and moisture content only below 20% moisture content. The experiments of this author with beechwood are interesting. Slight shrinkage appeared just below 50% of moisture content; between 50 and 35% tangential shrinkage reached 2% (Fig. 6.57). STEVENS (1938) explained this phenomenon as the action of surface tension in the cell wall due to capillary forces (cf. Section 6.3.1) which may even cause collapse.

The curves for the weight increase due to swelling theoretically follow an asymptotic course; for the swelling rate, according to PASCHELES (1897), the differential equation for monomolecular reaction is valid:

$$\frac{du}{dt} = k(u_{h_{\max}} - u); \tag{6.25}$$

integrated and solved with respect to $u$

$$u = u_{h_{\max}} \frac{e^{kt} \pm 1}{e^{kt}} \qquad (6.26)$$

$u$ representing the moisture taken up during the time $t$, $u_h$ the moisture in the swelling maximum, and $k = \dfrac{1}{t} \ln \dfrac{u_h}{u_h - u}$ a constant.

The equation states that the swelling rate is higher, the farther the distance is from the equilibrium. A presupposition for the validity of this law is that the swelling gel layer is so thin that it swells uniformly. In thick gel bodies, of which wood is practically always representative, swelling of the outer layers takes place at a distinctly quicker rate than that of the inner ones. In this case, Eq. (6.26) is no longer applicable but the equation for the moisture uptake in capillaries

$$u^n = k \cdot t \qquad (6.27)$$

seems to provide useful results with the exponent $n = 2$, at least during the first half of the swelling procedure.

The difference between radial and tangential shrinkage, formerly exclusively explained by a restraining influence of the wood rays in the radial direction, is now additionally explained, at least so far as coniferous woods are concerned, by a different helical arrangement of fibrils in radial and tangential cell walls. In the radial cell walls of the tracheids of conifers there are 50 to 300 pits and in their vicinity the fibrils are deflected from their normal course. Doubtless, one cause of the anisotropic properties of wood is the orientation of micelles, fibrils and wood fibers.

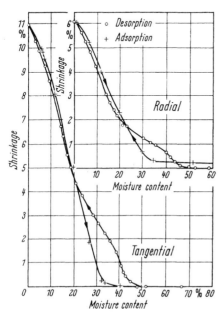

Fig. 6.57. Radial and tangential shrinkage respectively swelling curves of beechwood. From STEVENS (1938)

Summarizing, it may be pointed out that the differential transverse shrinkage of wood is related to the alternation of late and early wood increments within the annual ring (MÖRATH 1931), the influence of wood rays (McINTOSH, 1954, 1955, 1957), the features of the cell wall structure as are fibril angle modifications (FREY-WYSSLING, 1940) and pits (COCKWELL, 1946) and to the chemical composition of the middle lamella (FREY-WYSSLING, 1940). PENTONEY (1953) states clearly: "Due to the complexity and diversity of structure found in wood, it is believed that the mechanism of greatest influence depends upon the wood under consideration, and that this mechanism is modified by one or several of the other mechanisms."

KEYLWERTH (1948) stated that there are also purely physical causes for the swelling anisotropy. Under the assumption that the pore volume remains approximately constant during shrinkage, the polar arrangements of the cells in the stem leads to shifting conditions, the superposition of which causes anisotropic deformations.

Assuming that Eq. (6.2a) approximately describes the swelling process, according to KEYLWERTH (1962), in sufficient distance below the fiber saturation point — i.e. within the moisture variations frequently found in practice — a swelling gradient

$$q = \frac{\Delta \alpha}{\Delta u} = \text{const} \quad \text{for} \quad \varrho_0 = \text{const} \tag{6.28}$$

is reached which is of considerable practical importance. A survey on the volumetric and linear swelling gradients of different wood species is given in Tab. 6.14.

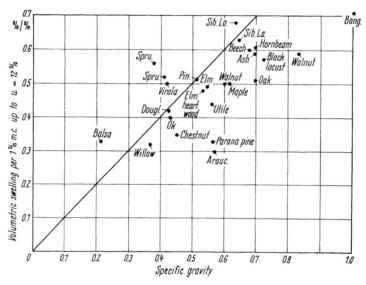

Fig. 6.58. Volumetric swelling per 1% change of moisture content vs. specific gravity. From KEYLWERTH (1932)

With the mean values of Tab. 6.14, Fig. 6.58 could be drawn. It can be seen that on the statistical average the volume variation with the swelling of wood occurs in effect in proportionality to the density and the relative water weight, and that, correspondingly, with swelling the pore volume remains nearly constant. Of course, there are wood species which swell less and others which swell more than complies with the water volume taken up. In the first case, this can be attributed partly to the sorptional contraction of water and material. Wood species which swell more strongly than corresponds to the water volume no longer behave like solid gels; their cavities increase with swelling.

KEYLWERTH (1962) showed that not only a volume contraction of the water occurs within chemisorption ($0 < u < 5\%$), but minute cavities in the cell wall close. As a consequence, the volume increase within a larger moisture range takes a linear course (Fig. 6.59).

This linearity absolutely necessitates a constant swelling gradient, i.e. a constant differential within the corresponding wood moisture range. The increase to be computed from the water volume $\beta'/\varrho_0 = 0.01 \text{ cm}^3/\%\text{g}$ is plotted in Fig. 6.59 in the form of a broken line. Most of the wood species group rather closely around this normal line, though balsawood swells double, Araucariawood, however, only half as strongly. The peculiar behavior of Balsawood is based on its anatomical structure; the cell walls apparently perform only a moderate crossband-effect, thus allowing a remarkable enlargement of the void volume. In contrast to this phenomenon, the crossband-effect within the cells of Araucariawood seems to be much more pronounced and larger internal swelling is effected.

Table 6.14. *Volumetric and Linear Swelling Gradients of Different Wood Species*

| Wood species | Density $\varrho_0$ g/cm³ | Border and mean values of swelling per 1% wood moisture increase within $0 < u < 12\%$ [%/%] | | | Number of samples [1] |
|---|---|---|---|---|---|
| | | $\alpha_{tang}$ | $\alpha_{rad}$ | $\alpha_{vol} \approx \alpha_{tang} + \alpha_{rad}$ | |
| Maple | 0.53···**0.615**···0.67 | 0.25···**0.30**···0.32 | 0.17···**0.20**···0.23 | **0.50** | 42 |
| Araucaria | **0.571** | **0.21** | **0.09** | **0.30** | 5 |
| Balsa | **0.214** | **0.22** | **0.11** | **0.33** | 5 |
| Bongossi | **1.038** | **0.40** | **0.31** | **0.71** | 5 |
| Brazilian Pine | 0.52···**0.555**···0.60 | 0.19···**0.23**···0.28 | 0.08···**0.10**···0.13 | **0.33** | 30 |
| Box | 0.75···**0.830**···0.98 | 0.36···**0.40**···0.44 | 0.15···**0.19**···0.23 | **0.59** | 17 |
| Oak | 0.60···**0.695**···0.77 | 0.25···**0.32**···0.38 | 0.15···**0.19**···0.23 | **0.51** | 36 |
| Ash | 0.64···**0.695**···0.79 | 0.34···**0.38**···0.39 | 0.18···**0.21**···0.22 | **0.59** | 30 |
| Fir | 0.37···**0.410**···0.48 | 0.24···**0.33**···0.41 | 0.12···**0.19**···0.24 | **0.52** | 35 |
| Pine | 0.44···**0.510**···0.58 | 0.28···**0.32**···0.35 | 0.17···**0.19**···0.22 | **0.51** | 30 |
| Siberian Larch | 0.61···**0.635**···0.67 | 0.41···**0.44**···0.46 | 0.22···**0.24**···0.26 | **0.68** | 34 |
| Siberian Larch (wide-ringed) | 0.60···**0.645**···0.76 | 0.24···**0.43**···0.49 | 0.10···**0.20**···0.25 | **0.63** | 28 |
| Walnut | 0.53···**0.600**···0.72 | 0.28···**0.30**···0.33 | 0.19···**0.20**···0.22 | **0.50** | 34 |
| Okoumé | 0.40···**0.430**···0.50 | 0.22···**0.24**···0.29 | 0.14···**0.16**···0.21 | **0.40** | 34 |
| Oregon Pine | 0.40···**0.425**···0.46 | 0.23···**0.27**···0.31 | 0.13···**0.15**···0.17 | **0.42** | 36 |
| Black Locust | 0.67···**0.720**···0.79 | 0.28···**0.33**···0.37 | 0.21···**0.24**···0.27 | **0.57** | 34 |
| Common Horsechestnut | 0.42···**0.450**···0.50 | 0.23···**0.25**···0.26 | 0.09···**0.10**···0.12 | **0.35** | 36 |
| Common Beech | 0.64···**0.675**···0.72 | 0.36···**0.38**···0.40 | 0.21···**0.22**···0.25 | **0.60** | 34 |
| Elm | 0.48···**0.540**···0.62 | 0.25···**0.29**···0.32 | 0.19···**0.20**···0.23 | **0.49** | 40 |
| Elm, heart | 0.48···**0.530**···0.58 | 0.25···**0.28**···0.30 | 0.19···**0.20**···0.21 | **0.48** | 30 |
| Spruce | 0.35···**0.375**···0.41 | 0.33···**0.37**···0.40 | 0.17···**0.19**···0.22 | **0.56** | 36 |
| Utile | 0.48···**0.560**···0.66 | 0.23···**0.25**···0.28 | 0.17···**0.19**···0.21 | **0.44** | 85 |
| Virola | 0.40···**0.415**···0.43 | 0.29···**0.32**···0.36 | 0.16···**0.18**···0.21 | **0.50** | 30 |
| Willow | 0.32···**0.365**···0.42 | 0.21···**0.23**···0.25 | 0.07···**0.09**···0.12 | **0.32** | 12 |
| Hornbeam | 0.65···**0.695**···0.73 | 0.33···**0.35**···0.37 | 0.24···**0.26**···0.28 | **0.61** | 28 |

[1] Square samples with length of 3 cm, thickness of 1 cm. Careful drying in vacuum. Air-conditioning and measurement at normal climate 20/65 DIN 50014.

[Ref. p. 285]

External and internal swelling are not only caused by the structure of the cell wall but also by the inhomogeneous structure of the coarser wood tissues, consisting of tracheids, wood fibers, vessels with varying diameters, of parenchymatous cells and wood rays, of earlywood and latewood. The differences in swelling of these elements produce a reciprocal restraint which leads to a super-position of stresses

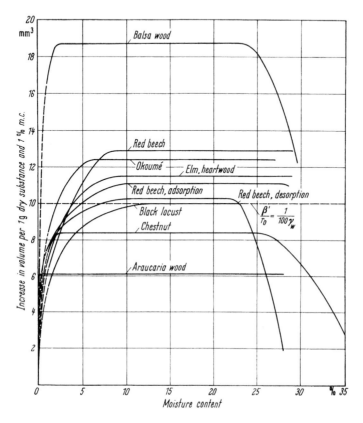

Fig. 6.59. Specific increase of volume of different wood species vs. moisture content. From KEYLWERTH (1962)

(PENTONEY, 1953; SCHNIEWIND, 1959). Additional stresses and deformations of wood cells are induced by the unavoidable gradients of moisture content in the samples. The specific course of differential swelling in the range of monomolecular chemosorption (from zero up to about 5 per cent moisture content) and in the case of true capillary condensation ($u > 22\%$) may be noted. Also the individual radial and tangential components show significant differences in the course of swelling. The anisotropy of tangential and radial swelling, expressed either as a ratio of the components or as their difference depends upon the moisture content (Fig. 6.60). KEYLWERTH (1962) emphasized that the course of swelling of different wood species is influenced by varying creep-recovery phenomena. For the ratio of tangential shrinkage to radial shrinkage, KEYLWERTH (1943) designed Fig. 6.61. A statistical analysis leads to equation

$$\beta_t = 1{,}65\,\beta_r. \tag{6.29}$$

In reality, $\beta_t/\beta_r$ is not independent from density as MÖRATH (1932) had already found. Tab. 6.15 gives evidence of this fact.

Table 6.15. *Ratio $\varepsilon$ of Tangential Shrinkage to Radial Shrinkage or Swelling*

| Density g/cm³ | $\varepsilon = \dfrac{\beta_t}{\beta_r}$ calculated by KOLLMANN using data given by MATHEWSON (1930) | $\varepsilon = \dfrac{\alpha_t}{\alpha_r}$ according to MÖRATH (1932) |
|---|---|---|
| 0.3 ···0.5 | 3.68···1.52 | 2.22···1,89 |
| 0.51···0.7 | 2.26···1,41 | 1.92···1.66 |
| 0.71···0.9 | 2.08···1.29 | 1.75···1.39 |
| 0.91···1.1 | 1.76···1.23 | 1.55···1.30 |
| 1.11···1.3 |  | 1.41···1.19 |

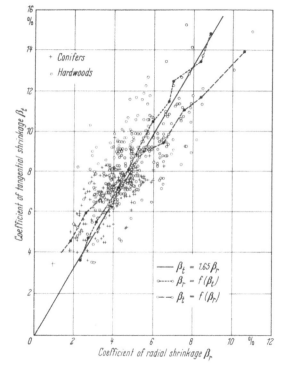

Fig. 6.60. Interrelationship between tangential and radial shrinkage. (Diagram designed by KEYLWERTH, 1943, using data published by MARKWARDT and WILSON, 1935, and by MÖRATH, 1932)

From Eq. (6.9) follows

$$\frac{\alpha_t}{\alpha_r} = \frac{\dfrac{\beta_t}{1-\beta_t}}{\dfrac{\beta_r}{1-\beta_r}} = \frac{\beta_t - \beta_t \cdot \beta_r}{\beta_r - \beta_t \cdot \beta_r} \approx \frac{\beta_t}{\beta_r}. \qquad (6.30)$$

The relatively very small longitudinal shrinkage $\beta_l$ may be estimated as follows (KEYLWERTH, 1943):

$$\beta_l = \beta_t/23. \qquad (6.31)$$

Longitudinal shrinkage of wood may ordinarily be neglected but there are some types of abnormal wood which shrink excessively. Fig. 6.62 gives longitudinal shrinkage values for 30 different wood species at 3 different percentages of moisture content; namely about 12%, about 7%, and at 0%. Shrinkage was assumed to

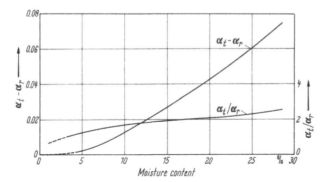

Fig. 6.61. Anisotropy of swelling of red beechwood vs. moisture content. From KEYLWERTH (1962)

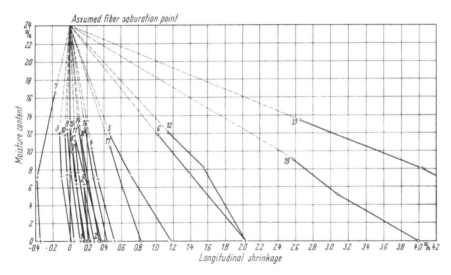

Fig. 6.62. Coefficient of longitudinal shrinkage of different woods as affected by moisture content. From KOEHLER (1931)

1 *Pinus palustris* Mill. (high specific gravity)
2 *Pinus palustris* Mill. (low specific gravity)
3 *Pinus echinata* Mill. (latewood)
4 *Pinus echinata* Mill. (earlywood)
5 *Pinus echinata* Mill. (latewood, abnormal)
6 *Pinus echinata* Mill. (earlywood, abnormal)
7 *Pinus echinata* Mill. (latewood, minimum)
8 *Pseudotsuga taxifolia* Britt. (latewood)
9 *Pseudotsuga taxifolia* Britt. (earlywood)
10 *Sequoia sempervirens* Endl. (latewood)
11 *Sequoia sempervirens* Endl. (earlywood)
12 *Cypress* (abnormal)
13 *Pinus ponderosa* Laws. (compressionwood, maximum)
14 *Fraxinus americana* L. (high specific gravity)
15 *Fraxinus americana* L. (medium specific gravity)
16 *Fraxinus americana* L. (low specific gravity).
17 *Quercus borealis* Michx. (low specific gravity)
18 *Ochroma lagopus* Sw. (Balsa wood)
19 *Sambucus callicarpa* Greene (heartwood)

have begun at a moisture content of 26%, which was taken as an average fiber saturation point (KOEHLER, 1931, 1946). For most uses a total uniform shrinkage potential of 0.3% is near the maximum permissible.

Compression wood formed on the lower side of branches and leaning trunks of coniferous trees has a greater tendency to shrink longitudinally than normal wood.

The maximum longitudinal shrinkage so far found at the U.S. Forest Products Laboratory was 5.78% in wood from the lower side of a limb of Western yellow pine. The maximum shrinkage determined for compression wood from a tree trunk is 5.42%. The longitudinal shrinkage of tension wood is also abnormally high and may range up to 1.7%. The longitudinal shrinkage was found to vary inversely to the density. It follows that species of wood which are abnormally light in weight shrink more along the grain than heavier wood of the same species. The early wood shrinks more along the grain than the latewood. The wood near the pith also exhibits a greater longitudinal shrinkage.

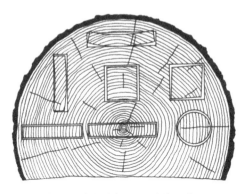

Fig. 6.63. Characteristic shrinkage and distortion of flats, squares, and rounds as affected by the direction of the annual rings. From US Forest Products Laboratory

In fast growing conifers the longitudinal shrinkage is usually excessive. Any piece of cross-grained wood has a high longitudinal shrinkage. Some woods may even show elongation in the first and second stages of drying, but oven-dry they are shorter than when wet.

**6.2.7.3 Super-position of the Components of Swelling, Restrained Swelling.** The joint effects of radial and tangential shrinkage cause a distortion of flats, squares, and rounds cut from a stem, in drying from the green condition (Fig. 6.63). In general the edges of a cross-section are bent in proportion to their deviation from the principal direction of extension (radial and tangential); for an intermediate direction (45° to the radius) the distortion reaches a maximum.

The linear coefficient of shrinkage $\beta_\varphi$ at any angle $\varphi$ to the radius can be computed from the shrinkage values $\beta_t$ and $\beta_r$ (KEYLWERTH, 1948):

$$\beta_\varphi = \beta_t \sin^2 \varphi + \beta_r \cdot \cos^2 \varphi. \quad (6.32)$$

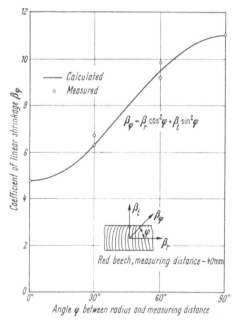

Fig. 6.64. Coefficient of linear shrinkage of beechwood at any given angle $\varphi$ to the radial direction. From KEYLWERTH (1948)

Eq. (6.32) neglects the squares of shrinkage values (which is not precisely allowable) but the correspondence of the calculated curve with measured points is sufficiently accurate (Fig. 6.64).

If swelling of dry wood is restrained by the influence of external forces, the anatomical structure of the wood will be changed. Hence it follows that subsequent re-drying to the original moisture content is accompanied by a reduction of the dimensions, that is, by a permanent shrinkage. This phenomenon is of practical importance for all kinds of wooden framework. KNIGHT and NEWALL

(1938) investigated swelling and shrinkage of temporarily restrained and of free wood, exposed to wetting and drying cycles. Fig. 6.65 shows the behavior of a clamped specimen and of a free control specimen, first put into a wet room and then both exposed to atmospheric conditions which enforced drying and shrinking. Repetition of the cycles (9 times) with restrained swelling caused a reduction of the dimensions by about 22%, while the free specimen underwent practically no change in volume.

PERKITNY (1937/38, 1960) et al. (1957) during the years published many valuable contributions to the different aspects of restrained swelling. The influence of grain orientation is interesting; mechanical restraint of the radial swelling of pine-sapwood increased its normal tangential swelling by about 5%, restraint of the tangential swelling increased its normal radial swelling by about 88%. The longitudinal swelling remained nearly constant in both cases. Repeated wetting

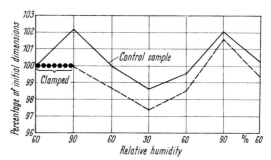

Fig. 6.65. Swelling and shrinkage of temporarily clamped and of free control samples on repeated wetting and re-drying cycles. From KNIGHT and NEWALL (1938)

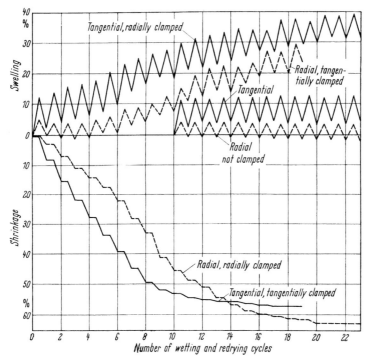

Fig. 6.66. Swelling and shrinkage of clamped and free wood samples in the course of repeated wetting and re-drying. From PERKITNY (1937/38)

and re-drying of wood with continuous unilaterally restrained swelling led to the deformations shown in Fig. 6.66. The summarized values of the shrinkage in the direction of restraint are extraordinarily high. After removal of the restraint and wetting a recovery comes about, but in the case of the tested pine sapwood a tangential shrinkage of 12.9% and a radial shrinkage of 24.1% remained. In practice restrained swelling (or shrinkage) occurs frequently and may be caused by internal forces, such as an inevitable moisture gradient, inside of drying or moistened boards or in glued or coated parts.

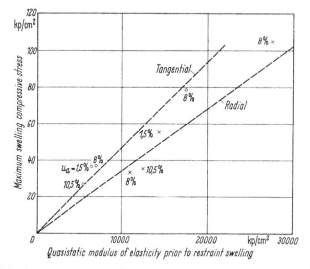

Fig. 6.67. Correlation between maximum swelling compressive stress and the modulus of elasticity perpendicular to the grain at the beginning of the restraint swelling. From KEYLWERTH (1962)

New investigations by KEYLWERTH (1962) proved that uniaxial, totally or partially restrained swelling corresponds to a pressure strain. The clamping conditions, the swelling velocity and the rheological behavior of the wood determine the deformations. Apparently a correlation exists between the maximum swelling compressive stress and the modulus of elasticity perpendicular to the grain at the beginning of the restrained swelling (Fig. 6.67). A substantial part of the deformation is, in the case of restrained swelling, transformed into plastic deformation. Even very long lasting steaming of green beechwood has only a small effect, i.e. an insignificant decrease, in the swelling pressure maximum.

**6.2.7.4 Swelling in Aqueous Solutions and Organic Liquids.** Wood and other cellulosic materials swell in most aqueous solutions in a similar manner as in water. For some chlorides and potassium salts the following order of swelling effects has been found (TOLMAN and STEARN, 1918):

$$K < NH_4 < Na < Ba < Mn < Mg < Ca < Li < Zn,$$

$$ClO_3 < SO_4 < NO_3 < Cl < Br < CrO_4 < J < CNS.$$

The increase in swelling runs parallel to an increase of solubility in water, with an increase of the fractional volume occupied by the salt in the solution, with an increase of the surface-tension of the solution, and with a decrease of the relative vapor pressure over the solution.

The high swelling of wood in strong alkalies is well known. Dilute alkalies also swell cellulosic materials beyond the swelling in water. Dilute mineral acids in the pH range between 2 and 6 on the other hand, show practically the same swelling as in water. Concentrated mineral acids (such as sulfuric acid) swell wood to a high degree before hydrolysis dissolves the material. Between volumetric swelling and equilibrium pH of water exists a well defined interrelationship (STAMM, 1934) as Fig. 6.68 illustrates. The mechanism of swelling in solutions is explained by different theories; none of them is generally applicable (cf. STAMM, 1964, pp. 251—256).

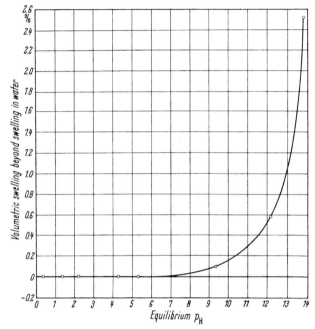

Fig. 6.68. External volumetric swelling of thin cross sections of white pine beyond swelling in distilled water vs. equilibrium pH of water. From STAMM (1934)

Wood and cellulose swell not only in electrolyte solutions but also in some organic liquids more than in water. The following examples may be listed: n-Butylamine, Piperidine, Morpholine, Formamide, Pyridine, Diethylamine. OBERG and HOSSFELD (1960) described the swelling behavior of wood in dioxane-water mixtures and found nearly linear trends in the relationships between vapor pressure and swelling, and dielectric polarization and swelling for systems containing either large amounts of water or of dioxane. According to KATZ (1933) the presence of halogens advance the swelling in the following order:

$$\text{Iodides} > \text{Bromides} > \text{Chlorides}.$$

The swelling of wood and other cellulosic materials in water-free organic liquids is normally less than in water (HASSELBLATT, 1926). There is a trend for increased swelling with increased polarity of the liquid, as indicated by the dielectric constant. There are, however, many exceptions. DE BRUYNE (1938, 1939) developed the following equations, based on theoretical considerations, for the coefficient of linear swelling $\alpha$

$$\alpha = \left(c\varepsilon^{\frac{1}{2}} - 1\right) \tag{6.33}$$

where $\varepsilon =$ dielectric constant and $c =$ constant. Fig. 6.69 shows that Eq. 6.33 with $c = 0.025$ fits tolerably the data measured by HASSELBLATT.

A series of alcohols exhibits decreasing swelling with increasing molecular weight (Fig. 6.70; STAMM, 1935). Similar observations were made for the sorption of vapors of alcohols (SHEPPARD and NEWSOME, 1932) and for swelling in fatty acids.

**6.2.7.5 Dimensional Stabilization of Wood.** The term "dimensional stabilization" refers to mechanical or chemical treatments or modifications that reduce the tendency of wood to shrink or swell with accompanying changes in its moisture content.

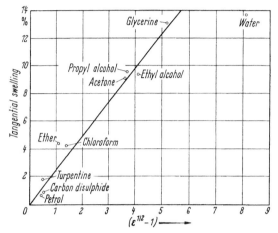

Fig. 6.69. Interrelationship between tangential swelling of birch wood and the dielectric constant of the swelling medium. Plotted points from HASSELBLATT (1926) regression line from DE BRUYNE (1939)

All methods for attaining improved dimensional stability of wood fall into one or more of five different types (STAMM, 1964):

1. Laminating of anisotropic sheets so as to restrain the dimensional changes of one sheet in the direction in which swelling is the greatest by cross-sheets that swell less in this direction, as in plywood.

2. Applying water-resistant surface and internal coatings to retard moisture adsorption or loss.

3. Reducing the hygroscopicity of the wood, thus reducing water adsorption and swelling.

4. Bulking the fibers, that is introducing a nonvolatile material into the cell walls thus reducing their capacity for water.

5. Cross-linking the cellulose chains of the fibers so that their separation by water adsorption is minimized. The most successful methods at the present time for permanently stabilizing the dimensions of wood that are suitable for commercial use are: resin impregnation, acetylation and treatment with polyethylene glycols. All of these involve bulking and are sufficiently expensive to limit their use to specialty items.

A reduction in equilibrium swelling and shrinking to about 70% may be accomplished quite simply by the introduction of water-soluble phenol-formaldehyde resinoids, followed by hardening the resins within the wood tissues (STAMM, 1955). Wood treated in this way has been given the name Impreg. Because of the difficulty of properly impregnating, drying and curing the resin in larger pieces of wood, the treatment has been confined mainly to veneers.

Treatment of wood with resins embrittles the products to such a degree that they cannot be used for some purpose where they might otherwise be suitable. The embrittlement with resins is due to the high rigidity developed within the fibers by the three-dimensional resin network that is probably chemically bonded to the available hydroxyl groups. It thus appeared that embrittlement could be avoided if single molecules with only one reactive group would replace the hydroxyl groups of the wood (STAMM, 1964, 31a). This aim is accomplished by the method of acetylating the wood which involves both a replacement of hygroscopic hydroxyl groups by less hygroscopic acetyl groupe and a bulking effect due to the introduction of the more bulky acetyl groups in place of the hydroxyls. This acetylation may be done in either the liquid or the vapor phase, the latter being preferred, using acetic anhydride, with pyridine as a combination of catalyst and swelling agent. Comparative studies have been made of acetylation, butyralation, crotonylation, crotylation, allylation, and phthaloylation (RISI and ARSENEAU, 1957). None of the other reactants showed any advantage in comparison to acetylation.

The third effective method for dimensional stabilization of wood outlined above involves bulking the fibers with polyethylene glycols (PEG). Polyethylene glycols are a series of polymers of glycol in which water is eliminated between two or more glycol groups giving ether linkages between ethyl groups. These polymers are highly water soluble liquids (up to a molecular weight of 600), plastic solids (molecular weight about 1000), or hard solids (molecular weight above 6000). The molecular weight of polyethylene glycol in the range of 200 to 1000 has only a limited effect upon the extent of bulking of wood under identical soaking and drying conditions. When the molecular weight exceeds 1000, however, the extent of bulking is considerably decreased, as the molecular weight increases, when the soaking and drying are carried out at room temperature. At somewhat elevated temperatures the extent of bulking with the higher molecular weight polyethylene glycols can be increased (STAMM, 1964).

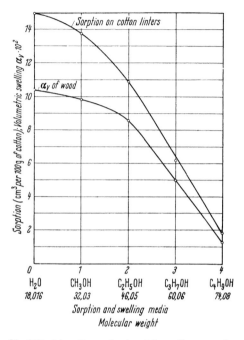

Fig. 6.70. Adsorption and volumetric swelling vs. molecular weight of alcohols. From STAMM (1935)

When optimum bulking is desired, it is important that the wood takes up at least 35% of polymer of the dry weight of the untreated wood. The soaking time or the number of surface applications for wood to attain this amount of take-up depends upon a number of factors, such as the thickness, grain direction, specific gravity of the specimen, concentration and temperature of the polymer solution, and the moisture content of the wood.

Measurements of the bending strength properties of polyethylene glycol-treated wood indicate a slight loss with increasing content of PEG (STAMM, 1959). The toughness of wood is virtually unaffected by treating it with amounts of PEG up to 45%, which corresponds to a reduction in swelling of 80%.

## 6.3 Capillary Movement and Diffusion in Wood

### 6.3.0 General Considerations on the Movement of Water in Wood Above and Below Fiber Saturation Point

Any movement of water (or other liquids and gases) in wood involves the permeability of its microscopic and submicroscopic structure:

1. Above fiber-saturation point the coarser capillaries contain free liquid. The molecules of water adjacent to the capillary walls are — as shown in section

6.2.3 — not free but bound by chemosorption. Fig. 6.71 shows that these sorbed molecules are to a certain extent parts of the capillary walls. The movement of liquid water above the fiber saturation point is caused by capillary forces, but in HAGEN-POISEUILLE's law for capillary movement a term must be introduced to consider the disturbance of the flow by the bound molecules at the capillary walls.

2. Below fiber saturation point bound water moves through the cell walls due to moisture gradients set up across the cell walls. This movement is a diffusion phenomenon.

3. The movement of water occurs not only as a liquid but also as a vapor. Within the microscopic capillary structure there is, according to THOMSON's law (Eq. 6.22), a small reduction in vapor pressure.

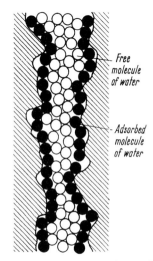

Fig. 6.71. Molecules of water in a small capillary. Scheme from KRÖLL (1951)

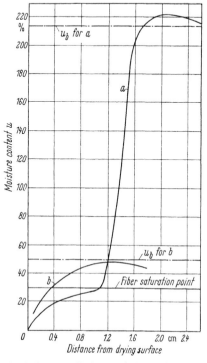

Fig. 6.72. Moisture gradient set up in tangential drying of Sitka spruce at 40 °C temperature and 0% relative humidity of the air
$a$ curve for water saturated wood; $b$ curve for green wood; $u_b$ = average moisture content before the beginning of the drying. From BATEMAN, HOHF and STAMM (1939)

Consequently, movement of water vapor is possible both above and below the fiber saturation point due to relative vapor pressure gradients in accordance with the diffusion law.

STAMM (1946) has pointed out that natural green wood is never completely filled with water so that practically all fiber cavities contain considerable air. BATEMAN, HOHF and STAMM (1939), in drying a water-saturated block of Sitka spruce, obtained a moisture distribution curve as shown in Fig. 6.72a. The moisture distribution curve $b$ in Fig. 6.72 is for a natural green specimen of the same wood. Remarkable is the sharp break in curve $a$ at the fiber saturation point due to the restricted motion of the (virtually) free water, whereas curve $b$ shows no such break. If all air bubbles could have been completely removed from the water-saturated block, the moisture-distribution curve above the fiber saturation point would have followed the vertical dotted line to the original moisture content.

"The fact that no break is obtained in the moisture-distribution curves for the drying of most green woods, which contain appreciable amounts of air in the fiber cavities, indicates that the moisture gradient below the fiber saturation point must control the rate of free water movement above the fiber saturation point. If the free water moved faster than the bound water and water vapor below the fiber saturation point, the moisture content near the wet line would tend to build up. This would require that the air bubbles in the fiber cavities near the wet line be under compression and decrease rather than increase in size. Under such conditions flow would tend to be reversed. The moisture movement above the fiber-saturation point, though not in itself a diffusion phenomenon, will thus be controlled by the diffusion below the fiber saturation point and will appear as if it were a diffusion phenomenon. It is for this reason that TUTTLE (1925), KOLLMANN (1936), SHERWOOD (1929, 1935) and LOUGHBOROUGH (1946, unpublished data) were able to consider the drying of wood a diffusion phenomenon over the total moisture content range." (STAMM, 1946, p. 48.) A series of experiments were carried out by CHOONG (1963) to investigate the mechanism and the rate of moisture movement through a typical softwood (Western fir) in the hygroscopic range of wood moisture content. The experimentally determined contributions of the bound-water and vapor movement mechanisms in the wood varied with moisture content, temperature and structural directions of movement. The fraction of total movement taking place in bound-water form increased with growing moisture content and temperature. The bound-water fraction was highest for radial movement and least for longitudinal movement.

### 6.3.1 Capillary Movement of Water in Wood

In filled capillary tubes movement of water is produced by differences in tension due to surface forces in the menisci within the capillaries. The tension force $T$ in the balanced meniscus within a tube with the internal radius $r$ [cm], neglecting the differences in air-pressure within the capillary, is:

$$T = H \cdot \varrho = \frac{2\beta}{\varrho \cdot r} \varrho = \frac{2\beta}{r} \left[\frac{\text{g}}{\text{cm}^3}\right] \qquad (6.34)$$

where $H$ [cm] = height of rise, $\varrho$ [g/cm³] = density of water, and $\beta$ = surface tension of the water [g/cm]. For the temperature $\vartheta$ [°C], according to WEINBERG (1892),

$$\beta = 0.0769(1 - 0.00225\vartheta) \text{ [g/cm]}. \qquad (6.35)$$

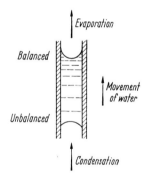

Fig. 6.73. Scheme of movement of water in capillaries. From KRISCHER (1938)

If a balanced and an unbalanced meniscus exist in a capillary tube, e.g. as shown in Fig. 6.73, the differences in tension cause movement. In a network of connected capillary tubes with different widths, different tension forces occur in the balanced menisci. Hence it follows that the narrow tubes draw liquid from the wide ones.

Considering the friction within the tube and introducing POISEUILLE's law we can compute the apparent height of capillary rise $H'$

$$H' = \frac{2\beta}{r\left(\varrho + \frac{8\eta V}{r^4 \pi}\right)} \text{ [cm]} \qquad (6.36)$$

where $\eta$ = coefficient of viscosity [gs/cm²] and $V$ = volume escaping per unit time [cm³/s] (KRISCHER, 1938).

According to THOMSON's equation the vapor pressure over spherical water-surfaces is lower than over plane surfaces. The vapor pressure gets lower as the diameter of the capillary tube decreases (cf. section 6.2.3). KRISCHER (1938) has pointed out that for capillary movement as for all equalizing processes the following characteristic equation holds:

$$W_m = k_m \cdot \frac{du}{dx} \qquad (6.37)$$

where $W_m$ = weight of moisture moved through a unit area per unit time [g/cm²s or in technical calculations kg/m²h], $k_m$ = moisture conductivity [g/cm s % or kg/m h %] and $\frac{du}{dx}$ = gradient of moisture over the cross section [%/cm or %/m]. He further showed that the moisture conductivity like the thermal conductivity and the diffusion coefficient is not constant even for a specific case but depends

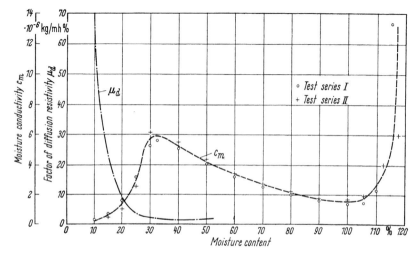

Fig. 6.74. Moisture conductivity and diffusion resistance figure as affected by moisture content. From VOIGT, KRISCHER and SCHAUSS (1940)

on the distribution of capillaries in the substance, the surface tension and viscosity of the liquid, and on the moisture content. The change of moisture conductivity with changes in moisture content is shown in Fig. 6.74. Approaching zero moisture content the moisture conductivity must also tend to become zero.

For the highest possible values of moisture content, when all pores are filled with water, moisture conductivity values approach infinity.

Due to the strong decrease of the coefficient of viscosity with rising temperature (the surface tension $\beta$ changes only to a limited extent) the moisture conductivity will increase considerably with temperature.

Generally the following equation holds:

$$k_{m1} = \frac{\beta_1}{\beta_2} \cdot \frac{\eta_2}{\eta_1} \cdot k_{m2} \; [\text{g/cm s \%}]. \qquad (6.38)$$

A numerical evaluation of the capillary movement of moisture in wood is nearly impossible, particularly since the varying dimensions of cell cavities and pit membrane pores would have to be considered. In this connection it may be recalled that the moisture movement below the fiber saturation point is regulated by diffusion and thus may apparently be treated as a diffusion problem itself.

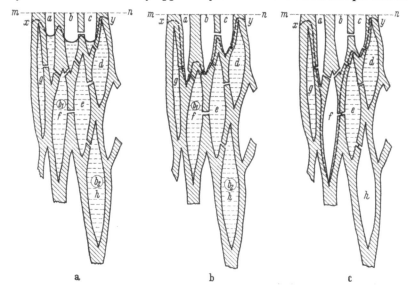

Fig.6.75a—c. Schematic representation of moisture movement in the capillaries of drying wood. a) Initial state; b) Second state; c) Third state. From HAWLEY (1931)

Nevertheless a general description of the capillary movement of moisture in wood is of interest. HAWLEY (1931) explained this process during drying. The simplified and schematic picture of wood structure in Fig. 6.75 may be discussed. One can clearly see the higher level of water in the narrower capillaries, the general decrease of the mean water level as drying proceeds, and the presence of air bubbles in individual cells still filled with water which expand and gradually fill the entire cell cavities.

STAMM (1946) investigated the collapse in wood, a well-known severe drying defect in some species. When completely water-saturated wood is dried in air, free evaporation occurs until menisci are formed in the pores of the pit-membrane of fibers near the surface. According to THOMSON's equation a reduction in vapor pressure is caused which will produce a hydrostatic tension. The elastic limit for the compressive strength perpendicular to the grain of green softwoods is about 34 kp/cm². This corresponds to a relative vapor-pressure reduction of 0.025 and capillary radius of $4.2 \cdot 10^{-6}$ cm. Capillary radii of about this size may occur here and there in the pit membranes of some woods.

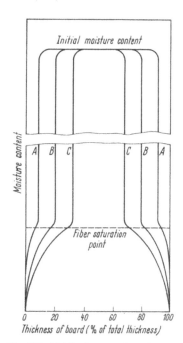

Fig. 6.76. Distribution of moisture in drying wood at three different times when no movement of free water exists. From HAWLEY (1931)

It is to be mentioned that in the case of some timbers the movement of the free water by capillary forces can take place in a very limited manner, only

224                           6. Physics of Wood                      [Ref. p. 285

e.g. within a zone of several fiber lengths. The free water in the interior of the board then finds itself completely at rest and the main evaporation occurs in the outer layers, where the moisture content already decreased below the fiber saturation point. If one plots the moisture distribution of such a timber over the thickness of the board as a function of time, then one obtains groups of curves according to Fig. 6.76. Such a discontinuous course of the moisture gradient has been observed in severe drying of water-saturated Sitka spruce (cf. Fig. 6.28 and Fig. 6.72). Woods which in drying behave according to these figures, tend to be susceptible to severe case-hardening.

### 6.3.2 Diffusion of Water in Wood

Drying of wood below the fiber saturation point has been explained by different scientists as a pure diffusion process. The phenomenon is nevertheless complicated because diffusion of water vapor through the void structure and bound water diffusion through the cell walls occur simultaneously. Experiments made by STAMM (1960) on Sitka spruce showed that in the pit system, bound-water diffusion through the pit membranes predominates in effect over the vapor diffusion through the permanent pit membrane pores and also over the vapor diffusion through the pit chambers. Bound water diffuses intermittently through pit membranes and cell walls and continuously through the rest of the wall. TUTTLE (1925) assumed, in the case of unidimensional movement of moisture, that the phenomenon of diffusion can be expressed by FICK's first law of diffusion as follows:

$$\frac{dm}{dt} = -D\frac{du}{dx} \tag{6.39}$$

where $\frac{dm}{dt}$ = the rate of transfer of mass (moisture) per unit time in kg/m²h, $D$ = diffusion coefficient in kg/m h%, and $\frac{du}{dx}$ = the rate of change of concentration in the $x$ direction (moisture gradient in wood) in %/m. MARTLEY (1926), STILLWELL (1926), KOLLMANN (1933, 1935), EGNER (1934) and other authors worked with Eq. (6.39). When the physical centimeter-gram-second system is applied $D$ is in cm²/s, and the rate of change of concentration in the $x$ direction is in mol/cm³ per cm. A method for measuring bound-water movement through the fibers independent of liquid and vapor movement was conceived by STAMM (1960).

If moisture passes through wood under the influence of a relative vapor pressure gradient, each imaginary layer of which the drying slab consists is swollen to a varying degree. In this case the diffusion coefficient is not a constant, but increases greatly with increasing moisture content. The moisture gradient set up in the wood is no more a linear gradient, as for a constant $D$, but a parabolic gradient.

Diffusion going on under unsteady-state conditions, but with continuously changing boundary conditions, may be mathematically described by FICK's second law of diffusion:

$$\frac{du'}{dt} = h^2 \frac{\partial^2 u'}{\partial x^2}. \tag{6.40}$$

New investigations by HART (1964) showed that the concept of Fickian diffusion in wood is complicated by the apparent presence of a second rate-controlling phenomenon, namely, a time-dependent deformation of the cellulose-lignin matrix. It seems probable that future unsteady-state models of moisture

[Ref. p. 285]

# 6.3 Capillary Movement and Diffusion in Wood

movement may require that both Fickian diffusion and deformation rate be combined, but the exact relationship remains to be worked out.

### 6.3.3 Drying of Wood as a Diffusion Problem

**6.3.3.1 Analogy to Fourier's Analysis for Heat Conduction.** TUTTLE (1925) has shown that Eq. (6.40) may be deduced for the unidimensional drying of wood as an analogy to FOURIER's theory for heat conduction.

In Eq. (6.40) $u'$ means the moisture content surmounting the hygroscopic equilibrium moisture content $u_g$ measured in the middle of the drying slab in the direction of the thickness $s$ of the slab (that is in a distance $\left(\frac{s}{2} - x\right)$ from the surface of the slab), $t$ the drying time, and $h^2$ the moisture conductivity in the wood. If we assume uniform distribution of the moisture content $u_b$ within the slab at the beginning of drying, the differential equation (6.40) leads to the solution (TUTTLE, 1925):

$$u' = \frac{4}{\pi} u_b \left[ e^{-\left(\frac{h\pi}{s}\right)^2 t} \sin\left(\frac{\pi x}{s}\right) + \frac{1}{3} e^{-9\left(\frac{h\pi}{s}\right)^2 t} \cdot \sin\left(\frac{3\pi x}{s}\right) + \cdots \right]. \tag{6.41}$$

TUTTLE could prove that Eq. (6.41) fits experimental data well for the drying of spruce boards in a limited range of moisture content. From the practical point of view, it is more interesting and important to deal with the average moisture content $u'_m$ at the end of the drying process. We obtain the average moisture content $u'_m = u_m - u_g$ (where $u_m$ = actual average moisture content, $u_g$ = hygroscopic equilibrium moisture content corresponding to the climate of the surrounding air, $u'_m$ = average moisture content based on $u_g$-level) at drying time $t$ by integration of Eq. (6.40) into the boundaries from $x = 0$ to $x = \frac{s}{2}$ and dividing the result into $s/2$. If the drying process was carried out carefully and led therefore to a very small, or even completely absent moisture gradient, then $\lim u'_m = u_g = u_e$ is valid. Fig. 6.77 shows how this phenomenon is being approached step by step.

$$u_m = \frac{8}{\pi^2} u_b \left[ e^{-\left(\frac{h\pi}{s}\right)^2 t} + \frac{1}{9} e^{-9\left(\frac{h\pi}{s}\right)^2 t} + \frac{1}{25} e^{-25\left(\frac{h\pi}{s}\right)^2 t} + \cdots \right]. \tag{6.42}$$

The expansion equation (6.42) is converging so rapidly that it is permissible to consider only the first term and, introducing a small safety factor of about 20% for the drying time, we take the simplified form $\frac{8}{\pi^2} \approx 1$, then we obtain as a rough approximation of Eq. (6.42)

$$u_m = u_e = u_b \cdot e^{-\alpha t}. \tag{6.43}$$

From this Eq. (6.43) we can deduce a simple solution for the drying time $t$ as follows

$$t = \frac{1}{\alpha} (\ln u_b - \ln u_e) = \frac{1}{\alpha} \ln \frac{u_b}{u_o}. \tag{6.44}$$

KOLLMANN (1936) could show that Eq. (6.44) is well applicable to practical drying cases. Using conventional drying methods (temperature 65 °C) he calculated the

15 Kollmann/Côté, Solid Wood

following statistical figures for, on the average, 1" thick softwood boards, $\alpha = 0.0477$ and for 1" thick hardwood boards, $\alpha = 0.0265$.

Fig. 6.78 and 6.79 show diagrams for the drying times of 1" thick softwood boards and 1" thick hardwood boards, respectively.

### 6.3.3.2 Approximated Calculation of the Drying Time.

If we assume that the drying time $t_2$ in a drying case 2 can be calculated from the drying time $t_1$ in a drying case 1 by multiplication with factors of correction $f_a, f_b, f_c, \ldots, f_m, f_n$, considering the changed conditions (e.g. species, density, thickness, temperature, velocity of the drying medium, etc.) we can write

$$t_2 = t_1 \cdot f_a \cdot f_b \cdot f_c \cdots f_m \cdot f_n. \quad (6.45)$$

This procedure is of course not at all exact but the results give usable approximations for practical application under the assumption that the factors of correction are based on proper physical investigations on the influence on the rate of drying or drying time (KOLLMANN, 1935).

#### 6.3.3.2.1 Influence of Specific Gravity and of Structure.

KOLLMANN (1933) found that the rate of diffusion is approximately inversely proportional to the square root of the specific gravity $\varrho_0$. If we consider that the amount of water to be removed by drying is proportional to the specific gravity $\varrho_0$, we can write:

$$t_2 = t_1 \frac{\varrho_{02} \sqrt{\varrho_{02}}}{\varrho_{01} \sqrt{\varrho_{01}}} = t_1 \left(\frac{\varrho_{02}}{\varrho_{01}}\right)^{1.5}. \quad (6.46)$$

The results of such calculations correspond well with practical observations (MOLL, 1930; SONNLEITHNER, 1933). TUOMOLA (1943), investigating the kiln drying of Finnish pine wood, came to the following formula

$$t_2 = t_1 \left(\frac{R_2}{R_1}\right)^n \quad (6.47)$$

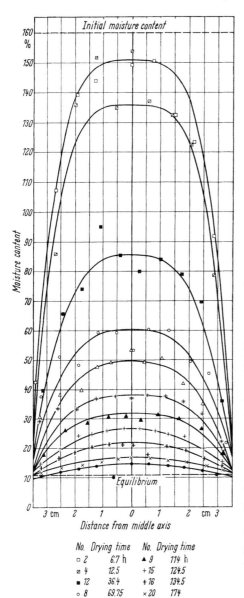

| No. | Drying time | No. | Drying time |
|---|---|---|---|
| □ 2 | 6.7 h | ▲ 9 | 114 h |
| ⌀ 4 | 12.5 | + 15 | 124.5 |
| ■ 12 | 36.4 | + 16 | 134.5 |
| ○ 8 | 69.75 | × 20 | 174 |
| △ 11 | 93.5 | ● 25 | 213.5 |
| + 13 | 107.5 | | |

Fig. 6.77. Distribution of moisture in a drying spruce board (radial-diffusion) at different drying times. From SONNLEITHNER (1933)

where $R$ = specific gravity, oven-dry weight based on volume when wet. He found for tangential movement of moisture $n = 1.8$; for radial movement $n = 2.0$. SCHLÜTER and FESSEL (1940) ascertained for kiln drying processes with especially high quality requirements that instead of 1.5 the higher exponent 2.4 should be introduced into Eq 6.46.

The transmission of moisture in wood is strongly affected by its structure. The radial rate of diffusion in woods of the temperate zones is for usual drying

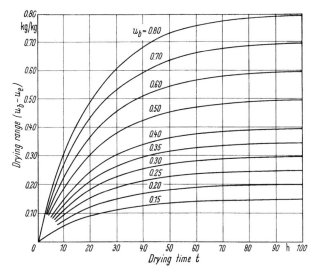

Fig. 6.78. Diagram for the determination of the drying time of 1" softwood boards (temperature 65 °C) as affected by the initial moisture content and by the difference initial moisture content minus final moisture content. From KOLLMANN (1936)

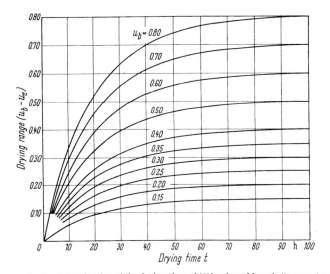

Fig. 6.79. Diagram for the determination of the drying time of 1" hardwood boards (temperature 65 °C) as affected by the initial moisture content and by the difference in initial moisture content minus final moisture content. From KOLLMANN (1936)

temperatures approximately 20 to 50% greater than the tangential rate due to the effects of wood rays. The effect of the drying temperature on the ratio of tangential rate to radial rate (Fig. 6.80; TUOMOLA, 1943) is interesting.

In some hardwoods, having a larger proportion of wood rays, the radial diffusion may be even more than double of the diffusion in tangential direction

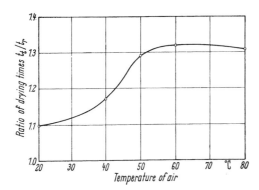

Fig. 6.80. Ratio of the drying times for tangential and radial diffusion vs. temperature of the air (pine wood, $u_b = 0{,}20$, $u_e = 0{,}15$). From TUOMOLA (1943)

(BURR and STAMM, 1947). For all dense woods with a very uniform structure the differences may nearly disappear.

Along the grain the resistance against diffusion is low in comparison to the resistance across the grain. The rate of diffusion in longitudinal direction in woods with densities between $\varrho_0 = 0.4$ and $\varrho_0 = 0.6$ g/cm³ (in the hygroscopic range, drying temperatures above 50 °C, and usual gradients of vapor pressure) may be 5 to 8 times higher than the rate perpendicular to the grain. For a low moisture content the ratio may be increased to 16. TUOMOLA (1943) obtained the curves plotted in Fig. 6.81. He explained the very severe influence of very low relative humidities of the drying air on the formation of cracks in the

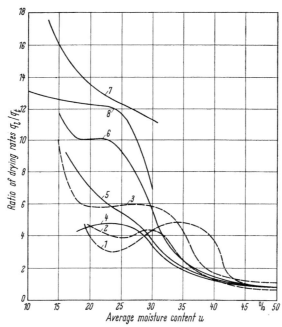

Fig. 6.81. Ratio of the longitudinal and the radial drying rate vs. average moisture content

| | | | | | | |
|---|---|---|---|---|---|---|
| Curve 1: | $r = 0{,}39$ g/cm³ | $\vartheta = 21°$ | $\varphi = 64{,}5\%$ | Curve 5: | $r = 0{,}45$ g/cm³ | $\vartheta = 50{,}3°$ $\varphi = 69{,}5\%$ |
| Curve 2: | $r = 0{,}46$ g/cm³ | $\vartheta = 21°$ | $\varphi = 64{,}5\%$ | Curve 6: | $r = 0{,}45$ g/cm³ | $\vartheta = 50{,}3°$ $\varphi = 56\%$ |
| Curve 3: | $r = 0{,}40$ g/cm³ | $\vartheta = 20{,}2°$ | $\varphi = 27\%$ | Curve 7: | $r = 0{,}45$ g/cm³ | $\vartheta = 40°$ $\varphi = 10\%$ |
| Curve 4: | $r = 0{,}43$ g/cm³ | $\vartheta = 50{,}1°$ | $\varphi = 78\%$ | Curve 8: | $r = 0{,}45$ g/cm³ | $\vartheta = 80°$ $\varphi = 2\%$ |

Finnish pine wood, annual ring width 1.1 to 1.7 mm, thickness of the board 52 mm, air speed 1.33 m/s. From TUOMOLA (1943)

wood. NARAYANAMURTI (1936) measured average ratios of the longitudinal rate of diffusion to the mean rate across the grain for three Indian woods (2.17···6.42) and of the radial to the tangential rate (1.24 ··· 1.94) which agree well with the figures mentioned above. The width of annual rings seems to be of no marked importance.

**6.3.3.2.2 Influence of Board Thickness.** The rate of diffusion is proportional to the gradient of vapor pressure, and since this gradient is inversely proportional to the thickness of the drying board, the rate of diffusion must also be inversely proportional to the board thickness $s$ or the drying time proportional to the square of the board thickness. Hence it follows that

$$t_2 = t_1 \frac{(s_2)^2}{s_1} = t_1 \left(\frac{s_2}{s_1}\right)^n. \quad (6.48)$$

One should expect that theoretically $n = 2$, but especially for thinner boards the exponent decreases, MOLL (1930) proposed $n = 1.5$, EGNER (1934) $n = 1.7$, and SCHLÜTER and FESSEL (1940) even found $n = 1.25$ (Fig. 6.82). For veneers the value $n = 1$ is recommended.

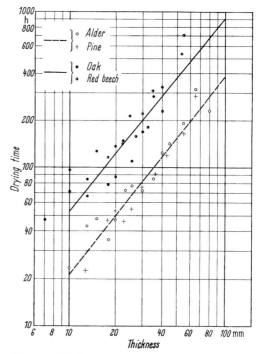

Fig. 6.82. Relationship between drying time and board thickness. From SCHLÜTER and FESSEL (1940)

**6.3.3.2.3 Influence of Moisture Content.** For the influence of mois-

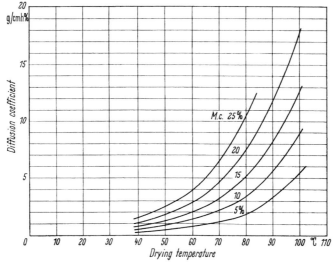

Fig. 6.83. Diffusion coefficients of spruce wood ($\varrho_0 = 0.404$ g/cm³) for radial moisture movement vs. drying temperature. From EGNER (1934)

ture content on the drying rate or drying time empirical formulae have been derived by different authors (KOLLMANN, 1935; TUOMOLA, 1943; SCHLÜTER and FESSEL, 1939). All these formulae are based on simplified assumptions and the results can only be rough approximations. Nevertheless, even such rough approximations may be useful tools for the engineer who has to control kiln drying processes.

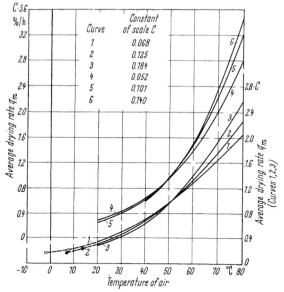

Fig. 6.84. Relationship between the average drying rate of freshly felled pine sapwood and the air temperature, when the average moisture content decreases from 25 to 20% (*curves 1 to 3*) or from 20 to 15% (*curves 4 to 6*). From TUOMOLA (1943)

Curve 1: $u_g = 16\%$  Curve 3: $u_g = 7\%$  Curve 5: $u_g = 7\%$
Curve 2: $u_g = 12\%$  Curve 4: $u_g = 12\%$  Curve 6: $u_g = 2\%$

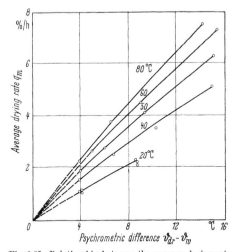

Fig. 6.85. Relationship between the average drying rate of freshly precipitated pine sapwood and the psychrometric difference of the dry air at different air temperatures. $\varrho_0 = 0.42$ g/cm³, $u_b = 140\%$, $u_e = 40\%$, air speed 80 m/min. From TUOMOLA (1943)

If we take Eq. (6.44) as generally valid for equal drying conditions we can write coefficient $\alpha = $ const in Eq. (6.44) for two drying cases:

$$t_2 = t_1 \frac{\ln \frac{u_{b2}}{u_{e2}}}{\ln \frac{u_{b1}}{u_{e1}}}. \qquad (6.49)$$

**6.3.3.2.4 Influence of Temperature and Psychrometric Difference.** EGNER (1934) has shown that the rate of diffusion in drying sprucewood is strongly increased by rising moisture contents and temperatures (Fig. 6.83). The curves are quite similar to that for water vapor pressure. This observation is another argument for the fact that at least below about 25% of moisture content, the diffusion of water vapor practically controls the whole drying process. TUOMOLA (1943) found similar relationships (Fig. 6.84) and developed the empirical formula

$$t_2 = t_1 \left(\frac{\vartheta_1}{\vartheta_2}\right)^n. \qquad (6.50\mathrm{a})$$

He determined exponents $n$ between 1.5 and 2.5. KOLLMANN (1935) found $n = 1$ more suitable in many cases. SCHLÜTER and FESSEL (1939) came to about the same result. Fig. 6.85 shows the influence of the psychrometric difference on the average drying rate according to TUOMOLA (1943). KRISCHER (1942) studied the

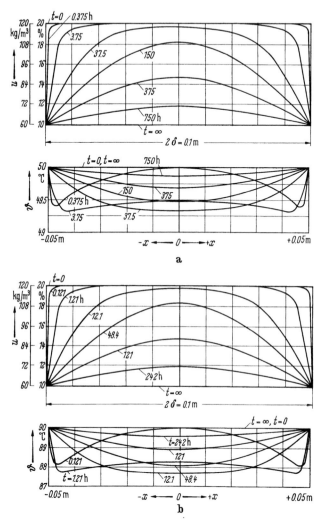

Fig. 6.86a, b. Field of moisture content curves ($u$) and field of temperature curves ($\vartheta$) in beechwood after sudden change of the boundary moisture from 20 to 10%. a) for medium temperature. b) for high temperature. From KRISCHER (1942)

movement of heat and moisture during the drying of wood and treated it mathematically. He found that the field of temperature curves is approximately a reflected image of the field of moisture content curves (Fig. 6.86). For comparison KRISCHER recommended the half-value times $t_h$, i.e. the times required to remove half of the total moisture. He found (Fig. 6.87)

$$t_{h_2} = t_{h_1} \left(\frac{\vartheta_1}{\vartheta_2}\right)^{1,86}. \tag{6.50b}$$

*6.3.3.2.5 Influence of the Speed of the Drying Medium.* LEWIS (1922) suggested a close interrelationship between the figure of evaporation $k'_2$ and the coefficient of heat transmission $\alpha$. According to JÜRGES (1924) the coefficient of heat transmission $\alpha$ may be estimated for air streaming with velocities $v \leq 5$ m/s along plane walls with rough surfaces as follows:

$$\alpha = 5 + 3.4\,v \quad [\text{kcal/m}^2\text{h}\,°]. \tag{6.51}$$

So we may write

$$\frac{k'_2}{k'_1} = \frac{\alpha_2}{\alpha_1} = \frac{5 + 3.4\,v_2}{5 + 3.4\,v_1} \quad (\text{for } v \leq 5 \text{ m/s}). \tag{6.52}$$

KOLLMANN (1935) and SCHLÜTER and FESSEL (1939) obtained a fairly good agreement between values calculated with Eq. (6.52) and experimental data.

The technical development in kiln drying tends towards higher air speeds. In the forties, air speeds between 1.5 and 2.0 m/s seemed to be the optimum, but in modern kilns values up to 4 (and even more) m/s are reached. LÖFGREN (1947) showed that power consumption is approximately increased with the third power of the air speed, but KOLLMANN and SCHNEIDER (1960)

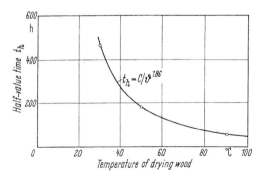

Fig. 6.87. Relationship between the half-value time when drying beechwood ($\varrho_0 = 0.60$ g/cm³, $u_b = 20\%$, $u_e = 10\%$, wood thickness $= 10$ cm) and the initial temperature. From KRISCHER (1942)

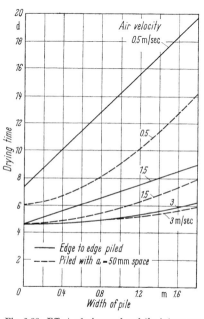

Fig. 6.88. Effect of air speed and the interspace between the single boards on the drying time for red oak when drying from 80 to 40% moisture content at 64 °C and relative humidity of 80%. From TORGESON (1941)

proved that a rather high ratio of the mechanical losses through air friction is regained as heat supporting the drying. High air speed also favors turbulence which equalizes the temperatures within a dry kiln and which reduces the drying times. TORGESON (1941) published the curves given in Fig. 6.88. One can see the influence of varying air speeds, of spaces between the boards to be dried, and of the width of the piles.

**6.3.3.3 Stamm's Theoretical Drying Diffusion Coefficients.** The rate of commercial drying of wood can be estimated and approximated as is shown in the previous section. A more scientific approach must be considered for calculating the diffusion coefficients since they affect the drying rate and control the moisture movement, as already mentioned, even above the fiber saturation point. STAMM (1946) first analyzed, in an excellent study, the diffusion of solutes into solvent-

Ref. p. 285]        6.3 Capillary Movement and Diffusion in Wood        233

saturated wood and showed that diffusion of an electrolyte through the complex network of water-filled capillaries of wood is analogous to electrical conduction through the same structure. Diffusion through cell cavities in parallel is additive, according to KIRCHHOFF's law for conduction. The reciprocals of diffusion through paths in series are additive. STAMM (1946) developed the following equation for the combined conductivity or diffusion through wood in fiber direction:

$$
C_l = \left( \cfrac{1}{\cfrac{1}{A} + \cfrac{n_l}{\cfrac{1}{\cfrac{L_\nu}{q_l + R(q_p - q_t) S_t} + \cfrac{L - L_p}{R q_p}} + \cfrac{(1 - q_i - q_p) R S_t}{L}} + (l - A) S_l} \right) \tag{6.53}
$$

where $C_l$ = conduction or diffusion in the fiber direction, $A$ = fractional cross-section of wood made up of voids, $S_l$ = fraction of the cell wall cross-section effective for conduction in the fiber direction (= 0.0154), $S_t$ = fraction of the cell wall cross-section effective for conduction in transverse directions (= 0.0078), $l$ = average fiber length of a softwood (= 0.38 cm), $L$ = double cell wall thickness (= 0.00072 cm for wood with a swollen volume specific gravity of 0.4), $L_p$ = average thickness of pit membranes (= 0.0001 cm), $q_p$ = fraction of cell wall area covered with pit openings (= 0.014), $q_l$ = fractional cross-sectional area of permanent pit membrane pores effective for transverse passage from electroosmose measurements (= 0.0038), $q_t$ = fractional cross-section area of permanent pit membrane pores effective for transverse passage from electro-osmotic measurements (= 0.00052), $R$ = ratio of the effectiveness of pits to diffusion in the fiber direction to that in the across-fiber direction (= $q_l/q_t$ = 7.3), $n_l$ = average number of communicating structures traversed per cm in fiber direction (= 2.5).

All of the structural dimensions in Eq. (6.53), except for $A$ and $L$, are considered constants. $A$ varies with the swollen-volume specific gravity of the wood, $R_u$ at current moisture content $u$ on a fractional basis, hence

$$
A = V = 1 - R_u \left( v_0 + \frac{u}{\varrho_s} \right) \tag{6.54}
$$

where $V$ = fractional void volume, $v_0$ = dry-specific volume of wood substance determined in helium (= 0.685 according to STAMM; in this book the value 0.667 is used), $\varrho_s$ = average specific gravity of the adsorbed water (for $u = 0.30$, $\varrho_s = 1.113$ cf. p. 163.). For a softwood with a swollen-volume specific gravity of 0.4, $A$ is equal to 0.618 according to STAMM[1]. $L$ also varies with the specific gravity, based on volume when green, as follows

$$
L = \frac{1 - \sqrt{A}}{n_t} \tag{6.55}
$$

where $n_t$ = average number of fibers per cm in the transverse direction of wood. For a softwood with a swollen-volume specific gravity of 0.4, $L = 7 \cdot 2 \cdot 10^{-4}$ cm. STAMM (1962) points out that when the indicated structural dimensions are substituted in Eq. (6.53) a value of 0.616 is obtained for $C_l$. This is almost the same as $A$, the fractional void cross-section. Eq. (6.53) can be transformed to

---

[1] The values of STAMM are based on recent measurements of the specific gravity of the wood substance in helium and in toluene (average 1.44 to 1.46). When we use the specific gravity of wood as determined in water 1.50 (to 1.54) the term $\varrho_s$ becomes 1.0 and $A = 0.614$.

calculate the conduction or diffusion through softwoods perpendicular to the grain in tangential direction. The following equation in which the proper structural dimensions are to be inserted is valid:

$$C_t = \left( \cfrac{1}{1 + \cfrac{1}{\cfrac{1}{\cfrac{L_p}{q_t + (q_p - q_t) S_t} + \cfrac{L - L_p}{q_p}} + \cfrac{(\sqrt{A} - q_t - q_p) S_t}{L}}} + (1 - \sqrt{A}) S_t \right) \tag{6.56}$$

When the structural dimensions for a softwood with a swollen-volume specific gravity of 0.4 are substituted in Eq. (6.56), a value for $C_t$ of 0.0445 is obtained. This is of the same order of magnitude as the experimental value 0.03 (BURR and STAMM, 1947). The analogy between electrical conduction in wood and diffusion in wood has also been applied by STAMM (1946), and STAMM and NELSON (1961) to the drying of wood. In this case the calculations become complicated since two different types of diffusion, namely vapor diffusion through the voids and bound water diffusion through the wood substance, occur simultaneously and must be taken into account. The free water vapor diffusion coefficients $D_v$, normally expressed in grams of water vapor per unit of space, can be obtained from the experimentally developed equation (STAMM, 1946)

$$D_v = 0.220 \left( \frac{T}{273} \right)^{1.75} \cdot \frac{760}{P} \tag{6.57}$$

where $T$ = absolute temperature, and $P$ = atmospheric pressure in mm of mercury. "These values were put on the same moisture gradient basis as the bound water diffusion coefficients by multiplying them by the number of grams of water vapor in a cubic centimeter of space at the temperature under consideration and dividing them by the quotient of the grams of water per gram of dry wood and the volume of swollen wood substance per gram of dry wood. In order to convert the combined diffusion to the wood volume basis it must be multiplied by $v/v_0$, the average specific volume of the wood per unit specific volume of the wood substance plus water." (STAMM, 1964, p. 437).

For tangential diffusion of water through wood Eq. (6.56) becomes:

$$D_t = \frac{v}{v_0} \left[ \cfrac{1}{D_{v_1}} + \cfrac{1}{\cfrac{1}{\cfrac{L_p}{q_t D_{v_2} + (q_p - q_t) D_b} + \cfrac{L - L_p}{q_p D_{v_1}}} + (\sqrt{A} - q_t - q_p) D_b} + (1 - \sqrt{A}) D_b \right] \tag{6.58}$$

where $D_{v_1}$ = free diffusion coefficient for water vapor [from Eq. (6.57)], $D_{v_2}$ = hindered water vapor diffusion coefficient (a value $D_{v_2} = 0.025 D_{v_1}$ has been used), $D_b$ = bound water diffusion coefficient (experimentally determined from diffusion measurements on metal-impregnated wood). The symbols have the same significance as those given in Eq. (6.56).

Calculated theoretical drying diffusion coefficients in the tangential and in the radial direction are plotted in Fig. 6.89 against the swollen volume specific gravity of the wood at five different temperatures between 50 °C and 120 °C (STAMM and NELSON, 1961, STAMM 1962).

Experimental data for tangential drying all fall below the theoretical lines. The agreement for radial drying is somewhat better. Nevertheless, in general the agreement between theoretical and experimental values is quite good and indicates that the electrical conductivity analogy is applicable to the drying of wood under circulating and heating condition in that manner such a these factors do not control the rate. Thus STAMM's theory is an encouraging example of modern principles of wood science: the choice of suitable physico-chemical analogies and the introduction of statistically secured values for structural dimensions.

Solving Eq. (6.58) stepwise, STAMM (1962) showed that the combined vapor diffusion through fiber cavities in series with bound water diffusion through the discontinuous portion of the cell walls is in general, especially for low and average specific gravities, the predominant portion of diffusion. The pit membrane permeability has a minor effect upon the combined diffusion coefficients.

## 6.4 Physical Aspects of Wood Impregnation

### 6.4.1 Nonpressure Processes

A detailed description on the nonpressure processes used today is given in chapter 5.2.

In the present chapter only the physico-chemical aspect of the osmotic procedure is referred to. As was

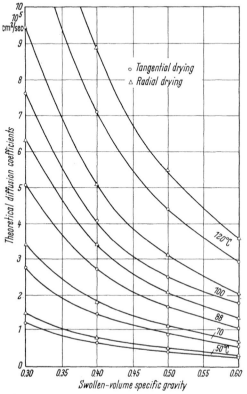

Fig. 6.89. Average theoretical drying diffusion coefficients for softwoods dried at five different temperatures vs. the swollen volume specific gravity of the wood. From STAMM (1962)

mentioned above, during this treatment the preservative, coated in a paste form over the peeled timber penetrates due to osmosis into the wood, in the course of a correspondingly long storage time. After this period the paste-layer has disappeared, the diffusion in the interior of the wood continues until equalization of concentrations of some depth is reached. The depth of penetration depends on wood species, grain orientation, moisture content, variation in density within the log, kind and concentration of the preservative, etc. The concentration $S_x$ at the end of the treatment in the depth $x$ under the wood surface may be calculated as follows (LUDWIG, 1947):

$$S_x = S_0 \cdot e^{-kx} \qquad (6.59)$$

where $S_0$ = concentration at the log surface, and $k$ = constant.

It is impossible to calculate theoretically the constant $k$ in advance, but its value can be determined easily by experiment. Fig. 6.90 shows the good agreement of a calculated curve with the analytically determined curve. The influences of grain orientation and initial moisture content on the penetration as a function of time is illustrated in Fig. 6.91. It may be concluded that the diffusion penetration depth along the grain is about sixfold the penetration depth perpendicular to the grain.

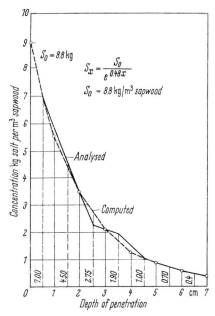

Fig. 6.90. Distribution of NaF by osmosis in green pine sapwood. Computed by Ludwig (1947)

### 6.4.2 Pressure Processes

**6.4.2.0 General Considerations.** Wood may be rapidly impregnated by the application of pressure. In the simplest "full-cell" process the wood is put, loaded on special tram cars, into a treating cylinder and the liquid preservatives are forced into the wood by externally applied pressures. There exist many different processes, already described in chapter 5.

The penetration may be increased by incising. Incising consists of cutting slit-like holes (6 to 19 mm long, about 3 mm wide, 13 to 19 mm deep) into the surfaces of the wood with special machines. The slits expose end-grain wood which offers a much higher permeability than transverse wood. The slits should be spaced in such a way that the strength of the wood is not affected.

### 6.4.2.1 Theory of Pressure Treatment of Wood.

*6.4.2.1.1 Calculation of the Pressure.* The impregnation of wood by liquids under pressure consists practically of the movement of free liquid through the cell cavities only. Vapor movement and bound-liquid movement are negligible. For these reasons the flow of the preservative follows Poiseuille's law, governing the flow of viscous liquids through capillaries:

$$P = \frac{8Q\eta L}{t \cdot \pi \cdot r^4} \tag{6.60}$$

where $P$ = pressure in dyn/cm², $Q$ = volume of liquid that has flowed in cm³, $t$ = time of flow in s, $\eta$ = viscosity of the liquid in dyns/cm², $L$ = length of the tube in cm, $r = d/2$ = radius of the tube in cm.

The continuity equation is then introduced:

$$Q = v \cdot t \cdot A \tag{6.61}$$

with the new symbols $v$ = velocity of flow in cm/s, and $A$ = cross-section of the capillary in cm².

Substituting Eq. (6.61) in Eq. (6.60) we obtain

$$P = \frac{8 \cdot v \cdot \eta L}{r^2} = \frac{32 v \eta \cdot L}{d^2}. \tag{6.62}$$

The equation is not directly applicable to the pressure treatment of wood since it is valid only in the case where all capillaries are already filled with liquid.

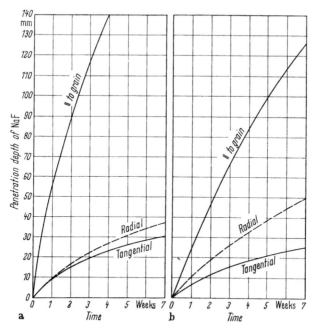

Fig. 6.91a, b. Relationship between grain orientation and penetration depth of NaF as a function of time. a) For green pine sapwood, b) For air-dried pine sapwood. From LIESE and SCHUBERT (1941)

The impregnation of wood starts with the flow of the preservative into empty capillaries (HAWLEY, 1931, p. 16). Nevertheless, Poiseuille's equation permits some conclusions. The pressure to be applied in impregnation depends mainly on the diameter $d$ of the capillaries in the pit-membranes. The velocity of flow $v_c$ in these capillaries is unknown, but one can deduce it by comparing the flow through the cross-section $A_t$ of a tracheid with the flow through a pore with the cross-section $A_c$ in a pit-membrane. According to Fig. 6.92 one can write:

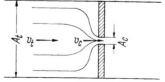

Fig. 6.92. Schematic representation of fluid motion through a cross-sectional reduction

$$Q = v_t \cdot A_t \cdot t = v_c \cdot A_c \cdot t \tag{6.63}$$

or

$$v_c = v_t \frac{A_t}{A_c} \tag{6.64}$$

Substituting Eq. (6.64) into Eq. (6.62) one obtains

$$P = \frac{32 \cdot v_t A_t \eta L_p}{A_c \cdot d^2}. \tag{6.65}$$

This formula describes only the flow through one opening in the pit-membrane. Pressures great enough to overcome the effect of surface tension in the liquid-air menisci that tend to form in the fine pit openings are required to displace liquid by entrapped air and permit treating liquids to enter. These pressures may range from about 1 to 50 kp/cm² depending on the size of the pit openings. Because of this, it is much easier to treat wood with non-condensable gases or vapors, at temperatures sufficiently high so as to avoid condensation (STAMM, 1963). The total pressure for impregnation can be expressed by the Equation (JOHNSTON and MAASS, 1929):

$$P_t = \frac{P \cdot m}{n \cdot o} \qquad (6.66)$$

where $m = s/2 \cdot l$, $s =$ thickness of the wood in cm, $l =$ average length of fibers in cm, $n =$ average number of pits per fiber, and $o =$ average number of openings per pit-membrane.

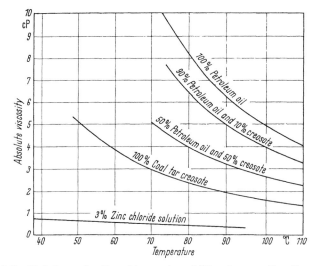

Fig. 6.93. Relationship between viscosity and temperature of different preservatives. From MACLEAN (1926)

Combining Eq. (6.66) with Eq. (6.65) and introducing usual units, one obtains, considering that $A_c = \dfrac{\pi d^2}{4}$

$$P_t = 4.16 \frac{v_t A_t L_p \eta m}{d^4 \cdot n \cdot o} \cdot 10^{-5} \; [\text{kp/cm}^2] \qquad (6.67)$$

where $v_t =$ velocity of flow along the grain in μm/s (e.g. $= 15$ μm/s), $A_t =$ lumen cross-section of one tracheid in μm² (e.g. 900 μm²), $L_p =$ length of the capillary $=$ thickness of pit-membrane in μm (e.g. $= 1$ μm), $\eta =$ viscosity of the preservative in dyn·s/cm² or poise, $d =$ diameter of one pit-membrane pore in μm (e.g. $= 4 \cdot 10^{-2}$ μm). Meanings and dimensions of $m$, $n$ and $o$ are mentioned above. A usable value for $n$ may be 30, for $o$ about 150.

Though Eq. (6.67) leads to reasonable values for the pressure, direct calculations of it are not justified. For comparison Poiseuille's equation is instructive. It follows from it that under otherwise constant conditions the ratio of pressures is proportional to the ratio of viscosities. Thus one has the possibility of estimating necessary changes of pressure if kind, concentration, and temperature of a preservative are changed. How the viscosity of different preservatives depend on temperature is given by Fig. 6.93 (MACLEAN, 1926).

### 6.4.2.1.2 Kinetics of Flow.
HAWLEY (1931) considers the flow of the preservative during impregnation "as a series of discontinuous flows, the conditions remaining constant while one set of cell cavities is being filled and the resistance increasing stepwise by a practically constant amount as each set of cell cavities is filled and the liquid begins to flow through the next set of orifices. Accordingly, if it takes one unit of time to fill the first set of cavities through one set of orifices, it will take two units of time to fill the second set of cavities, since for the second filling the liquid must flow through two similar sets of orifices in series; the third set of cavities will require three units of time, and so on."

If $y$ is the depth of penetration, $l$ the dimension of cell cavity in the direction of flow, $t$ the time to penetrate the depth $y$, and $a$ the time required to fill the first cell, then

$$t = ay\frac{y+l}{2l^2} \quad \text{or} \quad \frac{t}{a} = \frac{1}{2}\left[\left(\frac{y}{l}\right)^2 + \frac{y}{l}\right]. \tag{6.68}$$

Fig. 6.94 represents a general graphical solution of Eq. (6.68).

Air bubbles trapped in the cell cavities may complicate the very complex hydraulic phenomena in impregnation. Therefore Eq. (6.68) can give only some qualitative hints but MACLEAN (1924) found curves showing the relationship between time and penetration from the type shown in Fig. 6.94.

If we assume, in accordance with the reality, that $y \gg l$, than the penetration $y$ in Eq. (6.68) may be considered to vary inversely as the square root of $a$ $(\sqrt{a})$. According to Eq. (6.60) the time $a$, required to fill the first cell, varies inversely with the pressure $P$. Therefore, other conditions remaining equal, the penetration $y$ should vary proportional to $\sqrt{P}$.

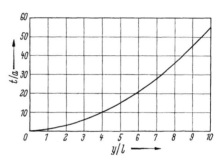

Fig. 6.94. Dependence of the penetration depth upon the time of impregnation. From HAWLEY (1931)

Measurements by MACLEAN (1924) showed that in general the penetration varies more nearly proportional to $P$ than to $\sqrt{P}$. The causes for this discrepancy may lay in a stretching of the pit-membranes by the pressure, thus enlarging the pores to some extent. This effect may be increased by the higher plasticity of the wood tissues due to steaming prior to impregnation. According to Eq. (6.68) the penetration $y$ should vary inversely as the square root of the time required to fill the first cell $\left(y = \frac{\text{const}}{\sqrt{a}}\right)$ and according to Eq. (6.60) the time $a$ should vary directly as the viscosity $\eta$ $(a = \text{const } \eta)$. Hence il follows that:

$$y = \frac{\text{const}}{\sqrt{\eta}} \tag{6.69}$$

MACLEAN (1924, 1926, 1927), using various mixtures of coal tar creosote with petroleum oil, found relationships between viscosity and penetration well in agreement with HAWLEY's (1931) theoretical deductions (Fig. 6.95). For waterborne preservatives the law is not applicable.

### 6.4.2.1.3 Influence of Moisture Content, Grain Orientation, and Structure.
JOHNSTON and MAASS (1929) found that the velocity of penetration increases directly in accordance with the theory, in green wood and in sapwood, with the

pressure applied. A curvilinear (progressive) increase was observed with dry woods, especially for higher pressures. The permeability in dry wood, at first higher than in wet wood, soon drops to a fraction of its original value. The flow in sapwood is approximately one hundred times faster than in heartwood; of about the same order is the ratio of longitudinal to transverse penetration velocity. These figures may not be generalized. ERICKSON, SCHMITZ, and GORTNER (1938) determined by pressure-impregnating small cubes of dry wood that the permeability

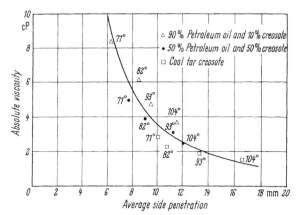

Fig. 6.95. Relationship between absolute viscosity of different preservatives and penetration depth into air-dried Eastern hemlock thresholds. From MACLEAN (1927)

along the grain in sapwood is one-millionfold and in heartwood fifty-thousand-fold higher than across the grain. There is no correlation between density of wood and depth of penetration. Some woods with high density, such as beech- and oakwood, are very permeable while many particularly light woods are refractory. Differences between heartwood and sapwood, the nature of vessels and tracheids, the presence of resin ducts and so on, have much more influence on the permeability than does the density. The maximum amount of absorption, of course, is a function of density or void volume.

## 6.5 Thermal Properties of Wood

### 6.5.1 Thermal Expansion

Since an increase in the temperature of solid bodies is accompanied by an increase in the oscillations of their molecules, the distances between the molecules become greater at higher temperatures. This means linear and volumetric expansion. The elongation $dl$ of a rod with the original length $l_1$ by a change in temperature $d\vartheta$ may be computed as follows:

$$dl = \alpha_{th} \cdot l_1 \cdot d\vartheta \tag{6.70a}$$

where $\alpha_{th}$ is the coefficient of thermal expansion.

From Eq. (6.70a) follows

$$\alpha_{th} = \frac{1}{l_1} \cdot \frac{dl}{d\vartheta}. \tag{6.70b}$$

Since $dl = l_2 - l_1$ and $d\vartheta = \vartheta_2 - \vartheta_1$ in which the indices 2 and 1 express the final and initial conditions, respectively, one obtains from Eq. (6.70a) by substitution

$$l_2 = l_1[1 + \alpha_{th}(\vartheta_2 - \vartheta_1)]. \tag{6.71}$$

Table 6.16. *Coefficients of Linear Thermal Expansion of Various Woods per Degree Centigrade*

| | Wood species | Thermal expansion | | $\dfrac{\alpha_{W\perp}}{\alpha_{W\parallel}}$ | Reference |
|---|---|---|---|---|---|
| | | $\alpha_{W\parallel} \times 10^6$ | $\alpha_{W\perp} \times 10^6$ | | |
| *Porous* | Maple | 6.38 | 48.4 | 7.6 | 1 |
| *woods* | Sugar maple | | | | |
| | (*Acer saccharum* Marsh.) | 2.16 | — | — | 2 |
| | (*Acer saccharum, r* = 0.62) | 4.22 | 60.2 | 14.2 | 3 |
| | Yellow birch (*Betula lutea* Michx.) | 1.98 | rad. 26.3 | — | 2 |
| | Boxwood (*Buxus sempervirens* L.) | 2.57 | 61.4 | 23.9 | 1 |
| | Ebony | 9.70 | — | — | 4 |
| | Oak | 4.92 | 54.4 | 11.1 | 1 |
| | Northern red oak | 3.43 | rad. 28.3 | — | 2 |
| | (*Quercus borealis* Michx.) | | tg. 42.3 | 12.3 | |
| | Ash | 9.51 | — | — | 4 |
| | White ash (*Fraxinus americana* L.) | 11.00 | 45.8 | 4.2 | 3 |
| | $r = 0.64$ | | | | |
| | Hornbeam | 6.04 | — | — | 4 |
| | Chestnut | 6.49 | 32.5 | 5.0 | 1 |
| | Basswood (*Tilia glabra* Vent.) | 5.46 | 44.4 | 8.1 | 3 |
| | $r = 0.38$ | | | | |
| | Cucumber Magnolia | 5.95 | 42.9 | 7.2 | 3 |
| | (*Magnolia acuminata* L.) $r = 0.41$ | | | | |
| | Mahogany | 3.61 | 40.4 | 11.2 | 1 |
| | Mahogany | 7.84 | — | — | 4 |
| | Walnut | 6.55 | 48.4 | 7.4 | 1 |
| | Rosewood | 6.08 | — | — | 4 |
| | Poplar | 3.85 | 36.5 | 9.5 | 1 |
| | Elm | 5.65 | 44.3 | 7.8 | 1 |
| *Coni-* | Spruce | 5.41 | 34.1 | 6.3 | 1 |
| *ferous* | Spruce | 6.08 | — | — | 4 |
| *woods* | Fir | 3.71 | 58.4 | 15.8 | 1 |
| | Fir | 3.55 | — | — | 5 |
| | White pine (*Pinus strobus* L.) | | | | |
| | $u = 0$ | 3.65 | 63.6 | 17.4 | } 3 |
| | $r = 0.39, u = 4\%$ | 4.00 | 72.7 | 18.2 | |

1 VILLARI, 1868; 2 MENZEL, 1935; 3 HENDERSHOT, 1924; 4 GLATZEL, 1877; 5 STRUVE, 1850.

Eq. (6.71) is exactly valid only if the thermal coefficient of expansion does not depend on temperature. This is not the case since the coefficient increases with rising temperature, but this phenomenon may be neglected for the small range of temperatures which are of interest for wood.

As has been shown in Chapter 2, wood contains about 50% of cellulose. The long-chain molecules of cellulose form rod-like crystalline parts with a length to width ratio of about 10. In consequence the amplitudes of the vibrations of the molecules must be approximately tenfold greater perpendicular to the chain-direction than along this direction. Due to the microscopic structure the coefficient of thermal expansion in the tangential direction is, in general, slightly greater than in the radial direction.

16  Kollmann/Côté, Solid Wood

The dimensional changes of wood caused by differences in temperature are small in comparison to the dimensional changes due to swelling or shrinking. In most cases, therefore, the thermal expansion or contraction of wood may be neglected. This fact and the poor heat conduction of wood are of advantage in fire. Wooden members support their load for a remarkably long period without failing; under the same conditions they may be superior to steel girders.

In frozen wood stresses occur due to unequal contraction at different depths. The outer layers of the stem contract as the temperature falls, before the inner layers are affected in the same sense. Frost cracks as radial splits in the bark and wood below are frequent in hardwoods growing in cold climates. (See chapter 3.1.5, p. 86).

Some thermal expansion data for natural wood published by different authors are listed in Table 6.16.

Very exact measurements were carried out by WEATHERWAX and STAMM (1946) at the U.S. Forest Products Laboratory. The coefficients of linear thermal expansion (hereafter called "$\alpha_{th}$" for brevity) were determined in the three structural directions. Nine different species of solid wood with specimens of low, medium, and high density for each species were tested over a temperature range from $-50°$ to $+50°C$ by means of a quartz dilatometer. The experimental data for yellow

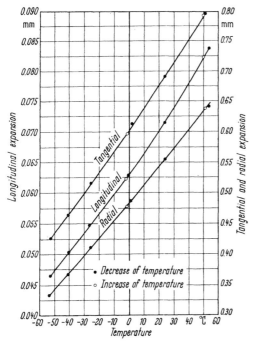

Fig. 6.96. Expansion of yellow birch ($\varrho_0 = 0.593$ g/cm³) between $-55°C$ and $+55°C$ in the three structural directions. From WEATHERWAX and STAMM (1946)

birch are presented graphically in Fig. 6.96. The thermal expansion is slightly greater at high than at low temperatures. The $\alpha_{th}$-values for nine wood species measured between $0°C$ and $+50°C$ are plotted in Fig. 6.97. The values of $\alpha_{th}$ in the radial and tangential directions increased in all cases with the specific gravity and straight lines through the origin represent the data satisfactorily. The low $\alpha_{th}$-values parallel to the grain are apparently independent of specific gravity and approximately the same for all species. A comparison of the lines for $\alpha_{thr}$ and $\alpha_{tht}$ shows that the species fall into three groups: Sitka spruce, white fir, Douglas fir, redwood, cottonwood, and yellow-poplar all have nearly the same $\alpha$-density relationship; birch and maple belong to another group with relatively lower values of $\alpha$, while balsa has considerably higher values of $\alpha$ than would be expected on the basis of density. The redwood specimens showed low values of $\alpha_{th\perp}$ and very high values of $\alpha_{th\parallel}$. This may be explained by a high content of compression wood. In swelling, compression wood also exhibits low values of extension perpendicular to the grain and high values along the grain (cf. Section 6.2.7.2).

Table 6.17 gives the data for the coefficients of linear thermal expansion in the three structural directions determined between $-50°C$ and $+50°C$ and between $0°C$ and $+50°C$ for each of the nine species.

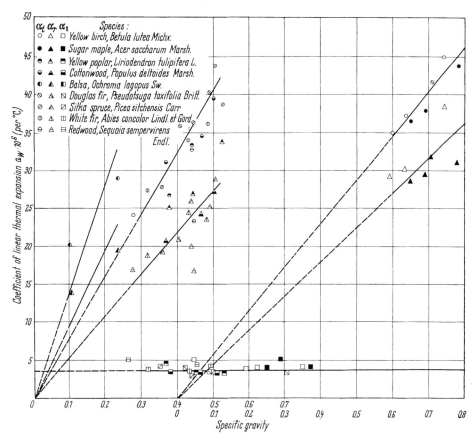

Fig. 6.97. Dependence of the coefficients of linear thermal expansion in the three structural directions for nine wood species between 0 °C and +50 °C on specific gravity. From WEATHERWAX and STAMM (1946)

Table 6.17. *Average Coefficients of Linear Thermal Expansion per °C of nine Species of Solid Wood*
(From R. C. WEATHERWAX and A. J. STAMM, 1946)

|  | Wood species | Density $\varrho_0$ g/cm³ | $\alpha_t \cdot 10^6$ −50 to +50° | $\alpha_t \cdot 10^6$ 0 to +50° | $\alpha_r \cdot 10^6$ −50 to +50° | $\alpha_r \cdot 10^6$ 0 to +50° | $\alpha_\| \cdot 10^6$ −50 to +50° | $\alpha_\| \cdot 10^6$ 0 to +50° |
|---|---|---|---|---|---|---|---|---|
| Porous woods | Sugar maple (*Acer saccharum* Marsh.) | 0.68 | 35.3 | 37.6 | 26.8 | 28.4 | 3.82 | 4.16 |
|  | Balsa (*Ochroma lagopus* Sw.) | 0.17 | — | 24.1 | — | 16.3 | — | — |
|  | Yellow birch (*Betula lutea* Michx.) | 0.66 | 38.3 | 39.4 | 30.7 | 32.3 | 3.36 | 3.57 |
|  | Cottonwood (*Populus deltoides* Marsh.) | 0.43 | 32.6 | 33.9 | 23.2 | 23.3 | 2.89 | 3.17 |
|  | Yellow poplar (*Liriodendron tulipifera* L.) | 0.43 | 29.7 | 31.4 | 27.8 | 27.2 | 3.17 | 3.55 |
| Coniferous woods | Douglas fir (*Pseudotsuga taxifolia* Britt.) | 0.51 | 42.7 | 45.0 | 27.9 | 27.1 | 3.16 | 3.52 |
|  | Sitka spruce (*Picea sitchensis* Carr.) | 0.42 | 32.3 | 34.6 | 23.8 | 23.9 | 3.15 | 3.50 |
|  | Redwood (*Sequoia sempervirens* Endl.) | 0.42 | 35.1 | 35.8 | 23.6 | 23.9 | 4.28 | 4.59 |
|  | White fir (*Abies concolor* Lindl. et Gordon) | 0.40 | 32.6 | 31.6 | 21.8 | 21.7 | 3.34 | 3.90 |

Pieces of wood are seldom cut so that their faces correspond exactly to the principal structural directions. Fig. 6.98 shows the three simple cases of a rectangular block of wood with the grain sloping only in one plane with respect to

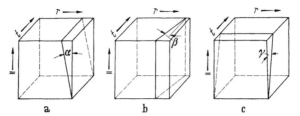

Fig. 6.98a—c. Scheme of a rectangular block of wood with the grain sloping only in one plane with respect to the faces. a) $\alpha$ = grain angle to the tangent plane, b) $\beta$ = angle resulting from the slope of the annual rings with respect to the tangent plane, c) $\gamma$ = grain angle to the radial plane

the faces of the block. The following equations (in analogy to those derived from stress-strain relations in wood by MARCH, 1944) hold:

For Fig. 6.98a
$$\alpha_\| = \frac{\alpha_{\|0} \cdot \cos^2 \alpha - \alpha_{r0} \cdot \sin^2 \alpha}{\cos 2\alpha}, \tag{6.72a}$$
$$\alpha_r = \frac{\alpha_{r0} \cdot \cos^2 \alpha - \alpha_{\|0} \cdot \sin^2 \alpha}{\cos 2\alpha}; \tag{6.73a}$$

For Fig. 6.98b
$$\alpha_t = \frac{\alpha_{t0} \cdot \cos^2 \beta - \alpha_{r0} \cdot \sin^2 \beta}{\cos 2\beta}, \tag{6.74a}$$
$$\alpha_r = \frac{\alpha_{r0} \cdot \cos^2 \beta - \alpha_{t0} \cdot \sin^2 \beta}{\cos 2\beta}; \tag{6.73b}$$

For Fig. 6.98c
$$\alpha_\| = \frac{\alpha_{\|0} \cdot \cos^2 \gamma - \alpha_{t0} \cdot \sin^2 \gamma}{\cos 2\gamma}, \tag{6.72b}$$
$$\alpha_t = \frac{\alpha_{t0} \cdot \cos^2 \gamma - \alpha_{\|0} \cdot \sin^2 \gamma}{\cos 2\gamma}. \tag{6.74b}$$

In these equations the subscript "0" indicates true coefficients of linear expansion. The equations are approximations. The exact solutions of the problem are elliptical equations whose numerical evaluation is extremely difficult. The expressions above are obtained by algebraic and trigonometrical simplification and by elimination of all second, third and fourth degree terms. The results are reliable within the limits of the experimental error. The case becomes far more complicated when the grain is inclined to each of the planes of the rectangular specimen. The reader interested in the derivation of the corresponding equations may refer to the papers of MARCH (1944) and WEATHERWAX and STAMM (1946). The percentage change in $\alpha_{tht}$ and $\alpha_{thr}$ in all cases is small, but under some circumstances the percentage change in $\alpha_{thl}$ is large.

WEATHERWAX and STAMM (1946) also developed theoretical equations from which the $\alpha_{th}$-values for laminates and birch plywood, either uncompressed or compressed, may be calculated. Additional $\alpha_{th}$-data were obtained for Impreg, Compreg, and molded hydrolyzed-wood plastics. Reference is made to Volume II of this book dealing with wood products.

Experiments by MACLEAN showed that when green wood is heated in steam, water, or in preservative oils it expands tangentially and shrinks radially.

Ref. p. 285]                    6.5 Thermal Properties of Wood                         245

This often leads to the development of burst checks or shakes that occur nearer the surface between earlywood and latewood or to the development of star checks radiating from the pith. Table 6.18 shows the maximum percentage of tangential expansion in samples of different species and the related percentage of radial shrinkage.

Table 6.18. *Maximum Tangential Swelling and Corresponding Radial Shrinkage obtained in Softwoods and Hardwoods under Different Heating Conditions*
(From MacLean, 1952)

| Species | Maximum tangential swelling % | Corresponding radial shrinkage % | Heating conditions |
|---|---|---|---|
| *Softwoods* | | | |
| Southern pine | +1.29 | −0.39 | Heated in steam at 149 °C for 10 minutes |
| Western larch | +1.90 | −0.13 | Heated in water at 107 °C for 1 hour |
| California red fir | +1.86 | −0.19 | Heated in steam at 121 °C for $^3/_4$ hour |
| Sitka spruce | +1.66 | −0.15 | Heated in water at 107 °C for 1 hour |
| Coast Douglas fir | +1.51 | −0.17 | Heated in steam at 121 °C for 1 hour |
| *Hardwoods* | | | |
| Red oak | +1.84 | −0.40 | Heated in steam at 149 °C for 10 minutes |
| Hickory | +1.44 | −0.29 | Heated in steam at 121 °C for $1^1/_2$ hours |
| Hard maple | +1.36 | −0.23 | Heated in steam at 121 °C for $1^1/_2$ hours |
| Yellow poplar | +1.32 | −0.01 | Heated in water at 121 °C for 20 minutes |
| Beech | +1.18 | −0.08 | Heated in water at 99 °C for $1^1/_2$ hours |

### 6.5.2 Specific Heat of Wood

The specific heat of a substance is the ratio of its thermal capacity to that of water at 15 °C. If a quantity of heat of $Q$ calories is necessary to raise the temperature of $m$ grams of a substance from $\vartheta_1$ to $\vartheta_2$ °C, the specific heat $c$ is defined as:

$$c = \frac{Q}{m(\vartheta_2 - \vartheta_1)} . \tag{6.75}$$

The specific heat of wood is low, which is of importance for many technical processes, such as seasoning, impregnation, destructive distillation, wood hydrolysis, etc.

The true specific heat $c$ of wood at the temperature $\vartheta$ is indicated by the following equation (Dunlap, 1912):

$$c = 0.266 + 0.001116\vartheta \left[\frac{\text{cal}}{\text{g °C}}\right] . \tag{6.76}$$

In technical calculations for solid and liquid substances mainly the average specific heat $c_m$ is of significance. Between 0° and $\vartheta$°, $c_m$ is expressed by the equation

$$c_m = \frac{1}{\vartheta} \cdot \int_0^\vartheta c\, d\vartheta \tag{6.77}$$

or applied to Eq. (6.76) and the range of temperatures between 0° and 100 °C:

$$c_m = \frac{1}{100} \cdot \int_0^{100} (0.266 + 0.001116\vartheta)\, d\vartheta = 0.324 \left[\frac{\text{cal}}{\text{g °C}}\right] . \tag{6.78}$$

246                           6. Physics of Wood                    [Ref. p. 285

The average specific heat determined experimentally by DUNLAP (1912) for twenty
species between $0\,°C$ and $106\,°C$ is 0.324, the minimum and maximum values being
0.317 and 0.337. The average specific heat is independent of both wood species
and specific gravity. The specific heat of cellulose is 0.37 and that of charcoal 0.16.

The average specific heat of wood varies considerably with moisture content.
Assuming a simple additive effect of the dry wood and of the water, one can write

$$c_x = x \cdot c_w + (1 - x)c_0 \tag{6.79}$$

where $x$ is the moisture content, based on wet weight, and $c_0$ is the average specific
heat of oven-dry wood. Since in wood science and technology the moisture content $u$ is based on oven-dry weight, the relationship $x = \dfrac{1}{1 + u}$ must be substituted in Eq. (6.79):

$$c_u = \frac{u}{1 + u} \cdot c_w + \left(1 - \frac{u}{1 + u}\right) c_0 = \frac{u \cdot c_w + c_0}{1 + u} \tag{6.80a}$$

or with $c_w = 1$ for water and $c_0 = 0.324$ for dry wood

$$c_u = \frac{u + 0.324}{1 + u}. \tag{6.80b}$$

### 6.5.3 Thermal Conductivity of Wood

**6.5.3.0 General Considerations.** Wood and other cellulosic materials are poor
conductors of heat due to the paucity of free electrons which are responsible for
an easy transmission of energy (such as in metals) and due to their porosity.
Wood and wood-based materials are therefore used to a large extent as heat insulating materials in building construction, refrigerator cars, beer barrels, etc.

The thermal conductivity varies with the direction of heat flow with respect
to the grain, with density, with kind and quantity of extractives, with defects,
and especially with moisture content. The thermal conductivity $\lambda$ is the thermal
energy $Q$ per unit time $t$ which flows through a thickness $s$ of a substance with
a surface area $A$ under a steady-state temperature difference between faces of
$(\vartheta_2 - \vartheta_1)$. Hence,

$$\lambda = \frac{Q \cdot s}{A \cdot t(\vartheta_2 - \vartheta_1)} \left[\frac{cal}{cm\,s\,°C}\right]. \tag{6.81}$$

To convert physical c.g.s.-units [cal/cms $°C$] into English units (BTU per square
foot per second for a temperature gradient of $1\,°F$ per inch), multiply by 0.80620.
To convert English units, as above, into technical c.g.s.-units [kcal/m h $°C$]
multiply by 0.12404.

**6.5.3.1 Influences of Structure and Density, Moisture Content and Temperature
on the Thermal Conductivity of Wood.** The thermal conductivity of wood in the
radial direction is about 5 to 10% greater than in the tangential direction (GRIF-
FITHS and KAYE, 1923; WANGAARD, 1940). Conductivity in the longitudinal
direction is (in the moisture content range from about 6 to 15%) about 2.25
to 2.75 times the conductivity across the grain. Some authors (MACLEAN, 1941;
NARAYANAMURTI and RANGANATHAN, 1941; THUNELL and LUNDQUIST, 1945;
KOLLMANN, 1951) stated that the conductivity $\lambda$ of wood is proportional to its
density $\varrho$ according to the equation

$$\lambda = A\varrho + B. \tag{6.82a}$$

MACLEAN (1941) calculated for the constants in this equation — English units assumed — for oven-dry wood $A = 1.39$ and $B = 0.165$. When $\varrho$ is zero, $\lambda$ is equal to 0.165, which is the thermal conductivity of dry air at room temperature. This relationship holds with great accuracy for all species tested ranging in specific gravity from 0.11 to 0.76. The results of the tests are shown in Fig. 6.99. The figures opposite the respective points give the number of runs in each of the test groups.

Moisture in wood increases its capacity to conduct heat. MACLEAN (1941) has modified equation (6.82a) as follows:

$$\lambda = (1{,}39 + 0{,}028 u_1)\varrho + 0.165 \quad (6.83)$$

where $u_1$ is the moisture content below about 40% and

$$\lambda = (1.39 + 0.038 \cdot u_2)\varrho + 0.165 \quad (6.84)$$

where $u_2$ is the moisture content above about 40%. The conductivity $\lambda$ in Eqs. (6.83) and (6.84) is taken as the BTU that flow in 1 hour through 1 in. thickness of material 1 sq.ft. in area, when the temperature difference between two surfaces is $1°F$.

For $u_1 = 12\%$ and after conversion into c.g.s.-units we obtain

$$\lambda = 0.177 \varrho_{12} + 0.0205. \quad (6.82\text{b})$$

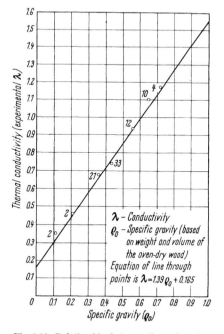

Fig. 6.99. Relationship between thermal conductivity and specific gravity of wood species in the range of 0.11 to 0.76 g/cm³. From MACLEAN (1941). The figures opposite the respective points are the numbers of runs in each of the test groups

This is in fair accordance with the equation statistically derived by KOLLMANN (1951) for the density range $0.2 < \varrho < 0.8$:

$$\lambda = 0.168 \varrho_{12} + 0.022. \quad (6.82\text{c})$$

MAKU (1954) has published an equation of the second degree for $\lambda$ in the density range $0.3 < \varrho < 1.50$

$$\lambda_\perp = 0.02 + 0.0724 \varrho_0 + 0.0931 \varrho_0^2. \quad (6.85)$$

Based on statistical evaluations of data presented in literature, MAKU calculated for the thermal conductivity of wood substance ($\varrho_w \approx 1.50$) in longitudinal direction $\lambda_{w\|}$

$$\lambda_{w\|} = 5.62 \ [\text{kcal/m h °C}]$$

and perpendicular to the grain $\lambda_{w\perp}$

$$\lambda_{w\perp} = 0.362 \ [\text{kcal/m h °C}].$$

KOLLMANN and MALMQUIST (1956) discussed the effect of the fiber orientation on the heat transmission in wood and wood-based materials. There are two boundary-cases: parallel arrangement of the fibers in the direction of heat flow according to Fig. 6.100a which creates maximum "heat bridges" ($\lambda_{max}$) and parallel arrangement of the fibers perpendicular to the direction of the heat flow according to Fig. 6.100b which creates minimum heat bridges ($\lambda_{min}$).

For a fibrous body with a mixed arrangement of layers one can write

$$\lambda = \xi \lambda_{max} + (1 - \xi) \cdot \lambda_{min} \quad (6.86)$$

where

$$0 \leq \xi \leq 1$$

and

$$\lambda_{min} \leq \lambda \leq \lambda_{max}.$$

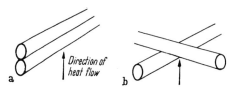

Fig. 6.100a, b. Scheme of arrangement of the fibers in wood. a) Parallel arrangement. b) Cross-shaped arrangement. From KOLLMANN and MALMQUIST (1956)

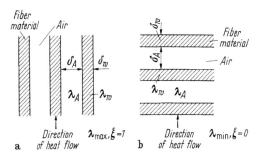

Fig. 6.101a, b. Idealistic image of the boundary cases of layer-shaped sequence of fiber material and air. a) Parallel arrangement of layers, parallel to direction of heat flow (maximum heat bridges). b) Arrangement of layers transverse to direction of heat flow (minimum heat bridges). From KOLLMANN and MALMQUIST (1956)

For minimum heat bridges the factor $\xi$ is zero, for maximum heat bridges the factor $\xi$ becomes one. $\xi$ is called "bridge-factor".

For both boundary cases (Fig. 6.101) one obtains the following equations for the thermal conductivity

$$\lambda_{min} = \frac{\lambda_A \cdot \lambda_w (\delta_A + \delta_w)}{\lambda_A \cdot \delta_w + \lambda_w \cdot \delta_A} \quad (6.87\text{a})$$

and

$$\lambda_{max} = \frac{\lambda_A \cdot \delta_A + \lambda_w \cdot \delta_w}{\delta_A + \delta_w} \quad (6.88\text{a})$$

where $\lambda_A$ = conductivity of the air, $\delta_A$ = thickness of the air layer, $\lambda_w$ = conductivity of the poreless wood substance, and $\delta_w$ = thickness of the wood substance layer. One can further define the density $\varrho$ of the wood, as follows

$$\varrho = \frac{\delta_w}{\delta_A + \delta_w} \varrho_w. \quad (6.89)$$

If one introduces Eq. (6.89) into Eqs. (6.87a) and (6.88a) respectively, the following terms result

$$\lambda_{min} = \frac{\lambda_A \cdot \lambda_w}{\lambda_w - \frac{\varrho}{\varrho_w}(\lambda_w - \lambda_A)} \quad (6.87\text{b})$$

and

$$\lambda_{max} = \lambda_A + \frac{\varrho}{\varrho_w}(\lambda_w - \lambda_A). \quad (6.88\text{b})$$

The effective thermal conductivity can be computed by combining the equations (6.87b) and (6.88b) with Eq. (6.86)

$$\lambda = \xi \left[\lambda_A + \frac{\varrho}{\varrho_w}(\lambda_w - \lambda_A)\right] + \frac{(1-\xi)\lambda_A \cdot \lambda_w}{\lambda_w - \frac{\varrho}{\varrho_w}(\lambda_w - \lambda_A)}. \quad (6.90)$$

The hitherto existing observations justify the assumption that fibrous materials of the same kind, but with different density, have a nearly constant bridge-factor.

This means that Eq. (6.90), under the supposition that $\xi$ itself does not essentially depend upon density, represents the relationship between thermal conductivity and density. The bridge-factor $\xi$ characterizes the network of heat-bridges within the structure. For heat conduction along the grain in solid wood practically $\xi = 1$,

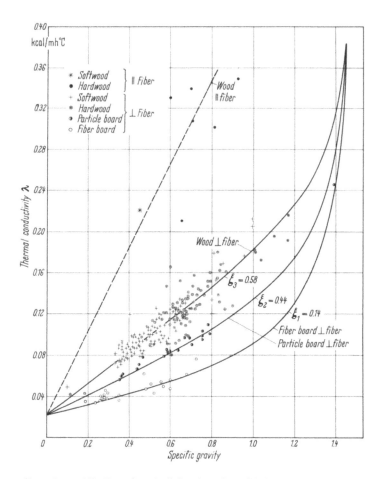

Fig. 6.102. Dependence of the thermal conductivity of wood, particle boards and fiberboards upon the specific gravity. From KOLLMANN and MALMQUIST (1956)

perpendicular to the grain $\xi = 0.58$ was calculated. In fiberboards where fibers and bundles of fibers are felted in criss-crossing patterns as shown in Fig. 6.100b, a low bridge-factor ($\xi = 0.14$) is given. The structure of particleboards consisting of flakes may be placed between that of solid wood and fiberboards. Fig. 6.102 shows data for the thermal conductivity from many sources and curves computed with Eq. (6.90).

KOLLMANN and MALMQUIST (1956) also derived the following approximation equation for the thermal conductivity $\lambda_{wu}$ of wet wood

$$\lambda_{wu} \approx 1.75 \left[ \frac{1 + 2.14 u}{1 + 1.5 u} + \frac{1 + 1.5 u}{1 + 1.05 u} \right]. \tag{6.91}$$

The density of the wet cell wall substance $\varrho_{wu}$ is

$$\varrho_{wu} = \frac{1.5 + u}{1 + u}. \tag{6.92}$$

By substitution of the values calculated for various moisture contents, using equations (6.91) and (6.92), into Eq (6.90) and taking the figures for $\varrho_u$ out of the graphical representation given in Fig. 6.7, the family of curves in Fig. 6.103 could be designed. It may be mentioned that the extrapolation above the fiber saturation point is to some extent uncertain.

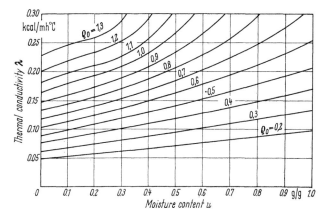

Fig. 6.103. Dependence of the thermal conductivity on the moisture content of wood. From KOLLMANN and MALMQUIST (1956)

In the range of moisture content $5 < u < 35\%$ one can assume

$$\lambda_2 = \lambda_1 [1 - 0.0125 (u_1 - u_2)]. \tag{6.93}$$

The thermal conductivity of wood increases with rising temperatures; the following empirical formula derived by KOLLMANN (1951) may be applied within the range of temperature $-50° < \vartheta < +100\,°C$

$$\lambda_2 = \lambda_1 \left[ 1 - (1.1 - 0.98 \varrho_0) \frac{\vartheta_1 - \vartheta_2}{100} \right] \tag{6.94}$$

where $\varrho_0$ is the oven-dry density of the wood.

### 6.5.4 Diffusivity of Wood, Change of Temperature in Heated Wood

Knowledge of the heat conduction in wood is of interest in wood-drying and in wood-preserving. An analysis of the time-temperature relationship first needs a calculation of the diffusivity, or thermal conductivity constant.

The diffusivity factor $a$ (or in American papers $h^2$) is as follows:

$$a = \frac{\lambda}{c \cdot \varrho} \left[ \frac{m^2}{h} \right] \tag{6.95}$$

where $\lambda$ = thermal conductivity in kcal/mh°C, $c$ = specific heat in kcal/kg°C, and $\varrho$ = density in kg/m³.

Table 6.19 shows the influence of density and moisture content on the diffusivity. The possible changes of $a$ are small. The factor decreases slightly with increasing density and with increasing moisture content.

Table 6.19. *Diffusivity Perpendicular to the Grain as Influenced by Density and Moisture Content*

| Density $\varrho_0$ | Moisture content $u$ | Density $\varrho_u$ according to Fig. 6.7 | Thermal conductivity $\lambda$ according to Eq (6.82c), (6.93), and (6.94) | Specific heat $c$ according to Eq (6.80) | Diffusivity $a$ according to Eq (6.95) |
|---|---|---|---|---|---|
| g/cm³ | % | g/cm³ | kcal/mh °C | kcal/kg °C | m²/h |
| 0.20 | 10 | 0.217 | 0.057 | 0.385 | 0.00068 |
|  | 20 | 0.233 | 0.064 | 0.436 | 0.00063 |
|  | 30 | 0.252 | 0.071 | 0.480 | 0.00059 |
|  | 50 | 0.287 | 0.085 | 0.549 | 0.00054 |
|  | 100 | 0.380 | 0.122 | 0.662 | 0.00052 |
| 0.40 | 10 | 0.427 | 0.092 | 0.385 | 0.00056 |
|  | 20 | 0.451 | 0.104 | 0.436 | 0.00053 |
|  | 30 | 0.480 | 0.115 | 0.480 | 0.00050 |
|  | 50 | 0.540 | 0.138 | 0.549 | 0.00047 |
|  | 100 | 0.720 | 0.195 | 0.662 | 0.00041 |
| 0.60 | 10 | 0.627 | 0.124 | 0.385 | 0.00051 |
|  | 20 | 0.657 | 0.140 | 0.436 | 0.00049 |
|  | 30 | 0.690 | 0.155 | 0.480 | 0.00047 |
|  | 50 | 0.776 | 0.186 | 0.549 | 0.00044 |
|  | 100 | 1.030 | 0.264 | 0.662 | 0.00038 |
| 0.80 | 10 | 0.825 | 0.156 | 0.385 | 0.00049 |
|  | 20 | 0.846 | 0.176 | 0.436 | 0.00048 |
|  | 30 | 0.880 | 0.195 | 0.480 | 0.00046 |
|  | 50 | 0.985 | 0.234 | 0.549 | 0.00043 |

MacLEAN (1930) has experimentally determined diffusivity factors for different timbers of various diameters. He found as average $h^2 = 0.000271$ sq. in. sec. If we convert $1\ \dfrac{\text{sq.in.}}{\text{s}} = 2.32\ \dfrac{\text{m}^2}{\text{h}}$, then $h^2 = 0.000271$ corresponds to $a = 0.000629$. MacLEAN (1930, 1932) calculated the diffusivity factors from typical temperature-time curves actually obtained in steaming experiments. For wood the diffusivity factor depends on the direction of the grain. "In long timbers, heating from the ends does not extend sufficient distance to be of importance. Most of the wood in round timbers must then be heated radially while in sawed material heating may be from both the radial and tangential directions or largely in one of these directions depending upon the cross-sectional dimensions and the direction of the annual growth ring." (MacLEAN, 1930, p. 307.)

The change of temperature in a homogeneous isotropic body is expressed according to FOURIER by the following partial differential equation:

$$\frac{\partial \vartheta}{\partial t} = a \left( \frac{\partial^2 \vartheta}{\partial x^2} + \frac{\partial^2 \vartheta}{\partial y^2} + \frac{\partial^2 \vartheta}{\partial z^2} \right) \tag{6.96}$$

where $\vartheta =$ temperature of a point with the coordinates $x, y, z$ at time $t$. Assuming an infinitely long beam in which temperature changes in a transverse plane through

252                          6. Physics of Wood              [Ref. p. 285

the point at which the temperature was taken are therefore independent of heat conduction in the longitudinal direction, Eq. (6.96) becomes simplified

$$\frac{\partial \vartheta}{\partial t} = a \left( \frac{\partial^2 \vartheta}{\partial x^2} + \frac{\partial^2 \vartheta}{\partial y^2} \right). \tag{6.97a}$$

Since wood is anisotropic the diffusivity factor $a_r$ in radial direction and $a_t$ in tangential direction must be introduced, thus modifying Eq. (6.97a) to the form

$$\frac{\partial \vartheta}{\partial t} = a_r \frac{\partial^2 \vartheta}{\partial x^2} + a_t \frac{\partial^2 \vartheta}{\partial y^2}. \tag{6.97b}$$

Eq. (6.97b) can be solved subject to the following boundary conditions (in agreement with MacLean, 1930):

$$\Theta = (\vartheta - \vartheta_1),$$
$$\vartheta = \vartheta_0 \quad \text{for} \quad t = 0,$$
$$\Theta = (\vartheta_0 - \vartheta_1) \quad \text{for} \quad t = 0,$$
$$\Theta = 0 \quad \text{for} \quad x = 0, \, y = 0 \atop x = w, \, y = h \quad \Big\} \quad \text{and} \quad t > 0.$$

In the foregoing, $\vartheta$ is the temperature at any point $x$ and $y$ in cross-section of the timber, $\vartheta_0$ is the initial temperature of the wood prior to heating, $\vartheta_1$ is the surface temperature of the wood while heat is being applied (in this case, the steam temperature), $x$ and $y$ are the coordinates of the point, $w$ and $h$ are the dimensions of the timber in the $x$ and $y$ directions, respectively, $t$ is the time in seconds.

Substituting $\Theta$ in Eq. (6.97b), renders

$$\frac{\partial \Theta}{\partial t} = a_r \frac{\partial^2 \Theta}{\partial x^2} + a_t \frac{\partial^2 \Theta}{\partial y^2}. \tag{6.97c}$$

Considering the boundary conditions one obtains the following solution of Eq. (6.97c):

$$\Theta = \sum_{m=1}^{m=\infty} \sum_{n=1}^{n=\infty} A_{m,n} e^{-\pi^2 t \left( \frac{a_r m^2}{w^2} + \frac{a_t n^2}{h^2} \right)} \sin \frac{m \pi x}{w} \sin \frac{n \pi y}{h} \tag{6.98}$$

where $A_{m,n}$ is the indefinite integral

$$A_{m,n} = \frac{4}{wh} \int_0^w \int_0^h f(\lambda, \mu) \sin \frac{m \pi \lambda}{w} \sin \frac{n \pi \mu}{h} d\lambda \, d\mu.$$

In this case $f(\lambda, \mu)$ is a constant and equal to $(\vartheta_0 - \vartheta_1)$ and therefore

$$A_{m,n} = \frac{4(\vartheta_0 - \vartheta_1)}{wh} \int_0^w \int_0^h \sin \frac{m \pi \lambda}{w} \sin \frac{n \pi \mu}{h} d\lambda \, d\mu$$

$$= \frac{4(\vartheta_0 - \vartheta_1)}{mn\pi^2} (1 - \cos m\pi) (1 - \cos n\pi) = \frac{16(\vartheta_0 - \vartheta_1)}{mn\pi^2}$$

when $m$ and $n$ are both odd, and $A_{m,n} = 0$ when either $m$ or $n$ is even. The general expression for temperature $\vartheta$ at any point $x$ and $y$ for a given time $t$ then yields:

$$\vartheta = \vartheta_1 + (\vartheta_0 - \vartheta_1)\frac{16}{\pi^2}\left[e^{-\pi^2 t\left(\frac{a_r}{w^2}+\frac{a_t}{h^2}\right)}\sin\frac{\pi x}{w}\sin\frac{\pi y}{h} + \right.$$

$$+\frac{1}{3}e^{-\pi^2 t\left(\frac{9a_r}{w^2}+\frac{a_t}{h^2}\right)}\sin\frac{3\pi x}{w}\sin\frac{\pi y}{h} + \frac{1}{3}e^{-\pi^2 t\left(\frac{a_r}{w^2}+\frac{9a_t}{h^2}\right)}\sin\frac{\pi x}{w}\sin\frac{3\pi y}{h} +$$

$$+\frac{1}{5}e^{-\pi^2 t\left(\frac{25a_r}{w^2}+\frac{a_t}{h^2}\right)}\sin\frac{5\pi x}{w}\sin\frac{\pi y}{h} + \frac{1}{5}e^{-\pi^2 t\left(\frac{a_r}{w^2}+\frac{25a_t}{h^2}\right)}\sin\frac{\pi x}{w}\sin\frac{5\pi y}{h} +$$

$$\left.+\frac{1}{7}e^{-\pi^2 t\left(\frac{49a_r}{w^2}+\frac{a_t}{h^2}\right)}\sin\frac{7\pi x}{w}\sin\frac{\pi y}{h} + \frac{1}{7}e^{-\pi^2 t\left(\frac{a_r}{w^2}+\frac{49a_t}{h^2}\right)}\sin\frac{\pi x}{w}\sin\frac{7\pi y}{h} + \cdots\right]$$

(6.99)

Eq. (6.99) is rapidly converging, so that in most cases the first term gives a good approximation. A few examples for computed temperatures at the centers of green round and green sawn Southern pine, when steamed at 127 °C, for temperature distribution and for isothermal contours in squares show Figs. 6.104 to 6.107.

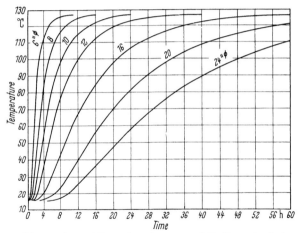

Fig. 6.104. Computed temperatures at the centers of green round Southern pine timbers of different sizes when steamed at 127 °C. From MACLEAN (1930)

Table 6.20. *Temperatures in the Interior of Green Pine Lumber after Various Steaming Periods* (From MACLEAN, 1932)

| Cross section in mm × mm | Initial temperature of the lumber = 15.5 °C   Temperature of the steam = 127 °C Average initial moisture content of the lumber $u_b$ = 0.41 Temperature in the interior after a steaming period of h ||||||||||||| 127 °C reached after h |
|---|---|---|---|---|---|---|---|---|---|---|---|---|---|
| | 0.5 | 1 | 2 | 3 | 4 | 6 | 8 | 10 | 15 | 24 | 36 | 48 | 60 | |
| 51 · 102 | 88 | 118 | 126 | — | — | — | — | — | — | — | — | — | — | 2.5 |
| 51 · 203 | 84 | 114 | 126 | — | — | — | — | — | — | — | — | — | — | 2.5 |
| 102 · 102 | — | 74 | 111 | 122 | 126 | 127 | — | — | — | — | — | — | — | 6 |
| 102 · 254 | — | 50 | 86 | 106 | 116 | 124 | 126 | 127 | — | — | — | — | — | 10 |
| 203 · 203 | — | 18 | 34 | 56 | 74 | 98 | 111 | 118 | 125 | — | — | — | — | 22 |
| 203 · 406 | — | 18 | 26 | 38 | 51 | 72 | 88 | 100 | 116 | 125 | — | — | — | 35 |
| 305 · 305 | — | — | 18 | 25 | 29 | 48 | 66 | 80 | 103 | 119 | 125 | 127 | — | 48 |
| 356 · 356 | — | — | 18 | 20 | 21 | 34 | 47 | 61 | 86 | 110 | 122 | 125 | 126 | 66 |
| 406 · 406 | — | — | 17 | 18 | 18 | 25 | 34 | 45 | 69 | 98 | 115 | 122 | 125 | 86 |

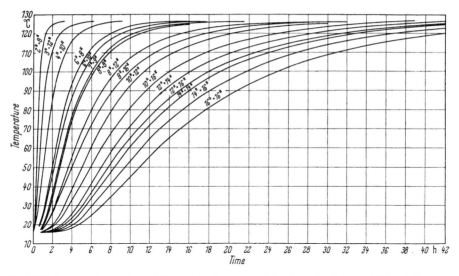

Fig. 6.105. Computed temperatures at the centers of green sawed Southern pine timbers of different dimensions when steamed at 127 °C. From MacLean (1932)

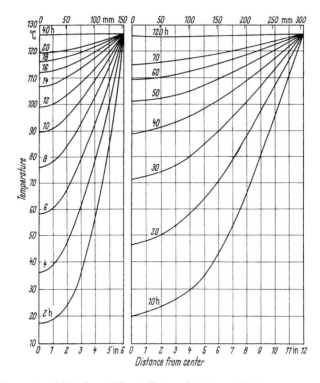

Fig. 6.106. Temperature distribution at different distances in timbers with *12* in. and *24* in. diameter at *2* and *10* hour periods respectively. From MacLean (1930)

Table 6.20 contains values of the calculated temperatures in the interior of green pine lumber as a function of various steaming periods. The values of MacLean (1930) are partly interpolated, converted to metric units, and rounded off.

If only the dimensions of lumber are varying for otherwise equal conditions, the following equation may be used

$$\frac{t_1}{t_2} = \frac{\dfrac{1}{w_2^2} + \dfrac{1}{h_2^2}}{\dfrac{1}{w_1^2} + \dfrac{1}{h_1^2}} \qquad (6.100\,\mathrm{a})$$

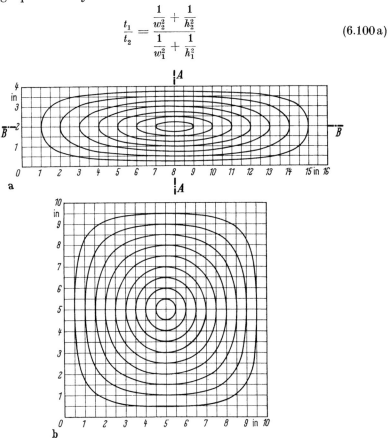

Fig. 6.107a, b. a) Isothermal contours in a timber 4 in. × 16 in. (10 cm × 40 cm) in cross-section. Contours shown for one inch intervals on axis B–B; b) Isothermal contours in a square timber 10 in. × 10 in. (25 cm × 25 cm) in cross-section. Contours shown for half an inch. From MacLean (1932)

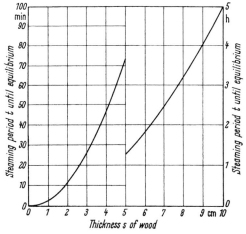

Fig. 6.108. Relationship between steaming period until equilibrium and thickness of wood for a square timber at 100°C. From Kollmann (1951)

where $t_1$ and $t_2$ are the steaming periods, $w_1$ and $h_1$ the widths and the height of the piece of lumber 1 and $w_2$ and $h_2$ the corresponding dimensions of the piece of lumber 2. For a quadratic cross-section ($w = h = s$) is valid

$$\frac{t_1}{t_2} = \left(\frac{s_1}{s_2}\right)^2. \tag{6.100b}$$

Fig. 6.108 shows the interrelationship between steaming period (until equilibrium is reached) and thickness of wood squares for a steam temperature of 100 °C.

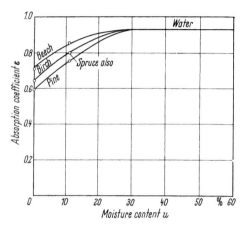

Fig. 6.109. Dependence of the absorption coefficient for beech, birch and pine wood on the moisture content. From KOLLMANN and MALMQUIST (1955)

### 6.5.5 Radiation of Heat with Respect to Wood

Like any other body wood does not emit radiant energy at absolute zero temperature. The rate of emission $E$ of radiant energy per unit area and unit time is determined by STEFAN's and BOLTZMANN's law as follows:

$$E = \varepsilon \cdot C_r \cdot T^4 \tag{6.101}$$

where $\varepsilon$ = emissivity of the surface (this value is zero for a perfect reflector and 1.0 for a "black body", i.e. a body which absorbs, for all values of the wave length of the incident radiant energy, all of the energy which falls upon it), $C_r$ = constant for the black body the numerical value of which depends upon the units of measurement used, e.g. = 4.96 × $\times 10^{-8}$ kcal/m²h (°K)⁴, and $T = 273.18 + \vartheta$ = absolute temperature.

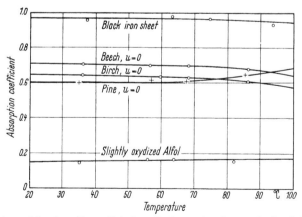

Fig. 6.110. Dependence of the absorption coefficients on the temperature for oven-dry beech, birch and pine wood compared with a black iron sheet and slightly oxydized aluminium foil. From KOLLMANN and MALMQUIST (1955)

The law which can be deduced thermodynamically shows that radiation is entirely a surface phenomenon. The darker and rougher a wood surface is, the greater is its absorption or emissive power with respect to radiation. There is no correlation between the amount of radiation and the density of the wood.

SCHMIDT (1927) published the first data for the emissivity of wood at room temperature ($\vartheta = 21\,°C$). He found that for oak wood with planed surface $\varepsilon = 0.895$. Later (1950) he quoted (for unspecified species of wood) $\varepsilon = 0.935$ at $70\,°C$ of temperature. KOLLMANN and MALMQUIST (1955) investigated the dependence of the emissivity from the moisture content for different wood species. The drier the wood is, the lower are the $\varepsilon$-values (Fig. 6.109). It is assumed that above the fiber saturation point the emissivity of wood is approximately the same as for water ($\varepsilon = 0.93$). The influence of the temperature in the range between 20 and $100\,°C$ on the emissivity of oven-dry wood is small (Fig. 6.110).

## 6.6 Electrical Properties of Wood

### 6.6.1 Direct-current Properties: Electrical Resistance and Electrical Conductivity

Very dry wood is an excellent insulator. Assuming that the wood can be kept in very dry or, better, oven-dry condition it approaches the most effective

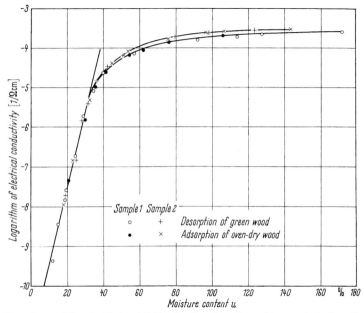

Fig. 6.111. Dependence of the electrical resistivity of redwood on the moisture content. From STAMM (1929)

insulators, such as phenol-formaldehyde resins. Unfortunately the electrical resistance of wood is lowered by increasing moisture content. Especially below the fiber saturation point the direct-current electrical resistance of wood decreases rapidly as the moisture content increases. Even traces of water increase the conductivity considerably. The mechanism of electrical conduction depends upon the pressence of ions in the wood. A model for ionic conduction has been proposed by LIN (1965) to explain the electrical conduction through the cell wall of wood. He pointed out that the number of charges-carriers in wood is the major factor affecting the conduction mechanism over the moisture content range from 0 to 20%. At higher moisture contents, the degree of dissociation of absorbed ions is sufficiently high so that the mobility of ions may become the major factor in determining the electrical conductivity. Therefore, any change in ion concentration, distribution, or both, will also change the electrical conductivity of the wood.

CLARK and WILLIAMS (1933) determined the specific resistivity (resistance in ohms per cm of length per cm² of cross-section, or ohm · cm) of various samples of wood, dried at 105 °C for several weeks, and extrapolated the results to room temperature. By this method, values ranging from $3 \cdot 10^{17}$ to $3 \cdot 10^{18}$ $\Omega$cm were obtained.

Measurements of direct-current electrical conductivity of wood and other cellulosic materials have been made by several scientists using different methods. STAMM (1964, p. 359/360) has pointed out how important it is, in determining the

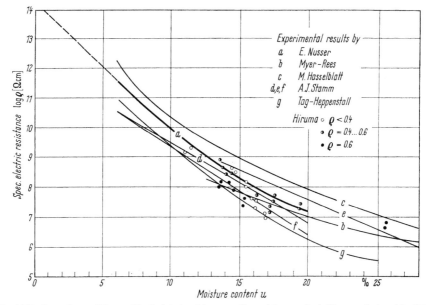

Fig. 6.112. Dependence of the specific electrical resistance on the moisture content. Diagram designed by NUSSER (1938) using data published by different authors

electrical conductivity of cellulosic materials, to obtain good electrical contact. The voltage range in which the measurements are made also affects the electrical resistance readings. TAKECHI and INOSE (1953) found that the readings for wood seem to be quite constant above a potential of 30 volts.

The effect of moisture content of wood on its electrical resistivity has been investigated by HIRUMA (1915), HASSELBLATT (1926), MEYER and REES (1926), STAMM (1927, 1929, 1930, 1932, 1960), NUSSER (1938), WEATHERWAX and STAMM (1945), and LIN (1965).

STAMM (1929) published Fig. 6.111 which gives a plot of the logarithm of the electrical resistivity of redwood plotted against the moisture content from above 8% to 170%. Below fiber saturation point (about 30%), the curves show a linear relationship between the logarithm of the electrical conductivity (reciprocal of resistivity) and the moisture content of the wood. The following equations express this relationship:

$$\log \varkappa = \log \frac{1}{\varrho} = a \cdot u - C \qquad (6.102\,\text{a})$$

or

$$\log \varrho = -a \cdot u + C \qquad (6.102\,\text{b})$$

or

$$\varrho = 10^{C-a \cdot u} = D \cdot e^{-b \cdot u} \qquad (6.102\,\text{c})$$

where $\varkappa$ = conductivity in $1/\Omega\,\text{cm}$, $\varrho$ = resistivity in $\Omega\,\text{cm}$, $u$ = moisture content in %, $a, b, C$ and $D$ are constants. For the curves in Fig. 6.107, $a = 0.2$ and $C = 11.5$.

NUSSER (1938) came to the same general relationship and determined $a = 0.32$, $b = 0.736$, $C = 13.25$, and $D = 1.78 \cdot 10^{13}$.

Eqs. (6.102b) and (6.102c) with these values for oak, beech, maple, cherry, chestnut, mahogany, pine, spruce, and fir fit a temperature of 15 °C within a range of moisture content from 8% to 18% with current flowing across the grain.

Fig. 6.112 allows a comparison of the results of different measurements. Most curves are slightly bent but one can approximate a linear relationship between logarithm of specific resistivity and moisture content for average hygroscopic moisture contents. In a wider range of moisture content (up to about 28%) according to SUITS and DUNLAP (1931), a double logarithmic relationship is more suitable:

$$\log(\log \varkappa) = c \cdot u. \qquad (6.103)$$

Fig. 6.113 illustrates how well Eq. (6.103) fits experimental data obtained by SUITS and DUNLAP (1931) and by NARAYANAMURTI (1936). In any case above the fiber saturation point the relationship of the logarithm of the conductivity to the moisture content is nonlinear. Furthermore, the influence of moisture content on conductivity becomes less and less pronounced. STAMM (1929) noted, for a change of moisture content from 0 to about 30%, a millionfold increase of conductivity but less than a fiftyfold

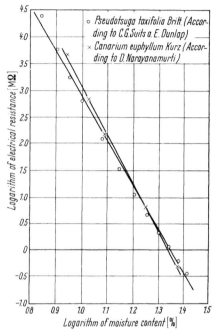

Fig. 6.113. Relationship between electrical resistance and moisture content for Douglas-fir, from SUITS and DUNLAP (1931) and Indian canarium, from NARAYANAMURTI (1936)

increase in conductivity from this point to complete water saturation. BROWN et al. (1963) and LIN (1965) applied HEARLE's theory (1953) that the electrical conduction in textile materials is ionic rather than electronic to wood and compared with experimental data. The relationship between $\log \varrho$ and moisture content $u$ of wood at room temperature is shown in the following equation:

$$\log \varrho = G + \frac{L}{\varepsilon_w} \qquad (6.104)$$

where the dielectric constant of wood substance $\varepsilon_w = 3.93 \cdot 10^{0.0242\,u}$, according to results of BROWN et. al. (1963). The constants $G$ and $L$ were evaluated by substituting the resistivity at two different moisture contents. The results (Fig. 6.114) show that HEARLE's theory explains fairly well the dependence of resistivity on moisture content between the oven-dry condition and about 15%. Comparative resistance measurements were obtained by KATZ and MILLER (1963) on red pine and Western red cedar at moisture content values ranging from 15 to 200% and following short-term and long-term sea-and fresh-water submersions. Sodium chloride concentrations indicated that the increase in conductivity was roughly proportional to the square of the salt content. Fresh-water soaking increased the resistance of wood by at least a factor of 2; a time dependence of this resistance increase was clearly noted. The resistance increase lessens exponentially with the time elapsed since the end of the soaking, and it becomes asymptotic at a level still above the

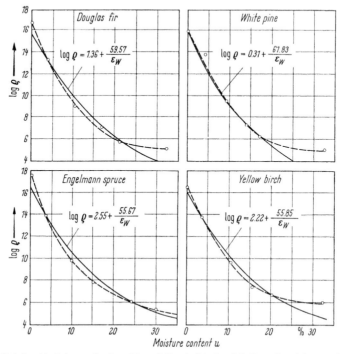

Fig. 6.114. Relationship between the logarithm of the electrical resistivity and moisture content for different wood species. From LIN (1965)

resistance of unsoaked wood. The net long-term increase in resistance appeared to be a function of the amounts and types of water-soluble extractives present in the wood.

HIRUMA (1915) has found that wood shows less resistance to electric current passing along the grain than to current directed across the grain. STAMM (1927) confirmed these observations as shows Table 6.21.

Table 6.21. *Electrical Resistivity of Woods as Affected by Grain Orientation*
(According to STAMM, 1927)

| Species | Moisture content % | Resistivity longitudinal MΩ | across the grain radial MΩ | tangential MΩ |
|---|---|---|---|---|
| Western red cedar (*Thuja plicata* D. Don) | 14.0 | 9 | 22 | 24 |
| Sitka spruce (*Picea sitchensis* Carr.) | 15.7 | 10 | 18 | 20 |
| Alaska yellow cedar (*Chamaecyparis nootkatensis* Sudw.) | 15.6 | 18 | 27 | 27 |
| Douglas-fir (*Pseudotsuga taxifolia* Britt.) | 15.3 | 11 | 21 | 23 |

One can see that the resistivity along the grain is roughly half that perpendicular to the grain, and that the resistivity in the radial direction is about equal or up to 10% less than that in the tangential direction. Of minor importance are species and specific gravity with respect to electrical resistance.

A rise in temperature causes a decrease in the direct-current resistivity of oven-dry wood. CLARK and WILLIAMS (1933) found a straight-line relationship between the logarithm of resistivity of oven-dry birch wood in tangential direction and the reciprocal of absolute temperature.

The curve for oven-dry birch wood can be expressed mathematically by the equation

$$\frac{1}{\varkappa} = \varrho = 10^{\left(0.8 + \frac{5000}{T}\right)} \quad (6.105)$$

where $\varkappa$ = conductivity, $\varrho$ = resistivity in $\Omega$ cm, and $T$ is the absolute temperature in degrees Kelvin.

KEYLWERTH and NOACK (1956) found that a linear relationship exists between moisture content, as shown by the electrical moisture meter in the hygroscopic range, and temperature (Fig. 6.115).

Experimental data by LIN (1965) showing the influence of temperature and moisture content on the electrical resistivity of wood are represented in Fig. 6.116, where log $\varrho$ is plotted against the reciprocal $(1/T)$ of the absolute temperature for various moisture contents. The curves are practically linear at low moisture con-

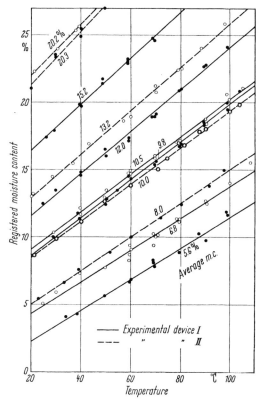

Fig. 6.115. Relationship between the moisture content registered by a Siemens & Halske moisture meter and the temperature of wood. From KEYLWERTH and NOACK (1956)

tents; curvilinearity was observed at high moisture contents, however. The curves for the highest moisture contents (near the fiber saturation point) showed clear discontinuities at temperatures between 0 °C and −10 °C. This phenomenon may be caused by water freezing out of the cell walls to ice crystals in the cell cavities ("coldness shrinkage", KÜBLER, 1962) in moist wood.

In Section 6.2.2.5.3, Eq. (6.13) gives the relationship between the electrical resistance across the grain in $\Omega$, the moisture content of the wood in %, and the temperature in °C in the range between 20 and 110 °C. This formula and Fig. 6.31 make it possible to correct readings obtained with the Siemens moisture meter (cf. p. 185). When wood contains water-soluble electrolytes its electrical conductivity is significantly increased. Untreated wood normally contains on the average only 0.2 to 0.6% of mineral substances, though much higher values may be found, e.g. 1.17% in beech wood (SCHWALBE and BECKER, 1919) or 2.12% in balsa wood (RITTER and FLECK, 1922) and a large part of it may be deposited in water-insoluble condition. Thus the normal ash content should hardly influence the electrical conductivity of most wood species. Treatment of wood with water-soluble preservatives against decay or with fire-retardant salts has, in contrast, a very significant influence on the electrical conductivity. Treatment with nonhygroscopic materials of low conductivity, e.g. phenolic resins, reduces the electrical conductivity of wood at a given humidity; the effect is especially pronounced if the surrounding relative humidity of the air is high.

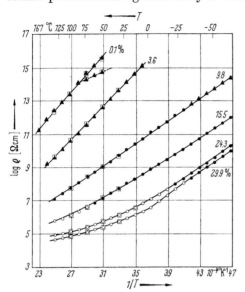

Fig. 6.116. Logarithm of resistivity versus reciprocal value of absolute temperature for various moisture contents of Engelmann spruce. From LIN (1965)

### 6.6.2 Alternating-Current Properties of Wood

**6.6.2.1 Resistivity.** The high-frequency resistivity of wood is very low in comparison to the direct-current resistivity. For example, SKAAR (1948) determined the transverse high-frequency resistivity for oven-dry wood of $18 \cdot 10^6$ $\Omega$cm at 2 megacycles per second; this is less than a billionth of the direct-current resistivity. The effect of moisture content and grain orientation, at the frequencies indicated, on the specific high-frequency resistivity of wood is shown in Fig. 6.117. It is worthwhile to note that the effect of moisture content on high-frequency resistivity is very small compared to the direct-current resistivity. Going from 0 to about 30% of

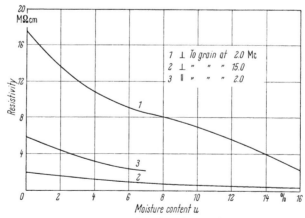

Fig. 6.117. Effect of moisture content and grain orientation at different frequencies on the specific high-frequency resistivity of wood. From SKAAR (1948)

moisture content, the former decreases to a value of about $10^{-13}$, the latter only to about $10^{-1}$. A few values for the specific high-frequency conductivity and the dielectric constant of different wood species under varying conditions are tabulated in Table 6.22 and 6.23.

**6.6.2.2 Dielectric Constant.** If a piece of wood as a poor electrical conductor is placed between two parallel plates (electrodes) in an alternating-current circuit,

Table 6.22. *Specific Electrical Conductivity* [$\Omega^{-1}$cm$^{-1}$] *and Dielectric Constant* $\varepsilon$
*of Various Woods at Different Moisture Contents and Frequencies*
(From HEARMON, 1944)

$u$ = moisture content %; $f$ = frequency kcps; $\varepsilon$ = dielectric constant;
$\varkappa$ = specific electrical conductivity $\Omega^{-1}$cm$^{-1}$;  temperature 18 °C

| Species | $u$ | | | | | | | |
|---|---|---|---|---|---|---|---|---|
| *Fagus sylvatica* | 60 | $f$ | 801 | 6,530 | 20,600 | 32,100 | | |
| Annual rings | | $\varepsilon$ | 61.6 | 32.9 | 26.9 | 26.3 | | |
| parallel to lines | | $\varkappa \cdot 10^6$ | 26.4 | 63.2 | 127 | 171 | | |
| of force | 23 | $f$ | 9 | 100 | 1,040 | 10,000 | 31,400 | 46,000 |
| | | $\varepsilon$ | 46.3 | 25.0 | 16.6 | 13.9 | 13.5 | 13.4 |
| | | $\varkappa \cdot 10^6$ | 0.55 | 0.87 | 2.40 | 10.9 | 34,3 | 52.5 |
| | 15 | $f$ | 1 | 1,240 | 11,000 | 37,000 | 55,500 | |
| | | $\varepsilon$ | 18.0 | 10.0 | 9.7 | 9.6 | 9.5 | |
| | | $\varkappa \cdot 10^6$ | 0.016 | 0.46 | 3.93 | 16.3 | 32.3 | |
| *Fagus sylvatica* | 54 | $f$ | 1 | 9 | 600 | 6,700 | 21,000 | 32,600 |
| Annual rings | | $\varepsilon$ | 5,360 | 940 | 70 | 34.6 | 30.4 | 30.3 |
| across lines | | $\varkappa \cdot 10^6$ | 28.6 | 35.2 | 48.4 | 111 | 200 | 209 |
| of force | 23 | $f$ | 90 | 1,000 | 8,200 | 30,000 | 46,000 | |
| | | $\varepsilon$ | 34.2 | 22.4 | 14.5 | 12.0 | 12.0 | |
| | | $\varkappa \cdot 10^6$ | 0.76 | 4.26 | 18.2 | 42.7 | 64.2 | |
| | 16 | $f$ | 1 | 90 | 1,000 | 8,200 | 30,000 | 46,000 |
| | | $\varepsilon$ | 15.8 | 10.6 | 9.4 | 9.1 | 8.5 | 8.5 |
| | | $\varkappa \cdot 10^6$ | 0.011 | 0.055 | 0.31 | 2.40 | 10.7 | 17.6 |
| *Pinus sylvestris* L. | 45 | $f$ | 1 | 9 | 70 | 600 | 6,700 | 21,000 | 3,260 |
| | | $\varepsilon$ | 264 | 61.6 | 36.4 | 21.3 | 15.4 | 14.4 | 14.2 |
| | | $\varkappa \cdot 10^6$ | 3.44 | 3.65 | 4.26 | 5.30 | 15.6 | 30.0 | 41.0 |
| | 22 | $f$ | 90 | 1,000 | 8,200 | 30,000 | 46,000 | |
| | | $\varepsilon$ | 14.7 | 11.3 | 10.4 | 10.1 | 9.9 | |
| | | $\varkappa \cdot 10^6$ | 0.42 | 1.09 | 4.32 | 14.0 | 25.7 | |
| | 15 | $f$ | 1 | 90 | 1,000 | 8,200 | 30,000 | 46,000 |
| | | $\varepsilon$ | 11.2 | 9.1 | 8.1 | 8.1 | 7.7 | 7.3 |
| | | $\varkappa \cdot 10^6$ | 0.011 | 0.045 | 0.168 | 2.12 | 9.1 | 16.7 |
| *Quercus pedun-* | 62 | $f$ | 70 | 600 | 6,700 | 21,000 | 32,600 | |
| *culata* Ehrh. | | $\varepsilon$ | 75 | 47 | 31 | 27 | 27 | |
| | | $\varkappa \cdot 10^6$ | 6.2 | 15.0 | 49.8 | 111 | 168 | |

Very remarkable are the extraordinarily high dielectric constants observed for specimens at high moisture content and low frequencies. The highest value is 5360 for the flat-sawn beech, a figure from 20 times of dielectric constants of water itself, but as HEARMON stated — it is not possible to explain the effect.

the system of electrodes and wood is equivalent to a condenser with the capacitance $C$, in farads, in parallel with a resistance $R$, in ohms. The current $I$ in the system consists of two components: $I_C$ through the condenser (capacitance) and $I_R$ through the resistance. In the vector-diagram (Fig. 6.118) $\varphi$ is the phase angle and $\delta = 90° - \varphi$ is the power loss angle.

Under the influence of the alternating voltage the molecules in the wood undergo cyclic motions which cause friction within the individual molecules and between different molecules. The friction induces heat which is dissipated. The magnitude of the power $P$ in the circuit is

$$P = E \cdot I_R = E I \cos \varphi = \frac{E^2}{R} ; \qquad (6.106)$$

$\cos \varphi$ is called "power factor" since it expresses the ratio of the power dissipated to the total power led into the circuit.

## Table 6.23. *Specific Electrical Conductivity and Dielectric Constant for Various Woods, Impregnated and not Impregnated*
(From Hearmon, 1944)

| Species | Moisture content % | Preservative | Treating method | Hardened (h) not hardened (n) | Specific electrical conductivity $\varkappa \cdot 10^6$ | | | | Dielectric constant $\varepsilon$ | | | |
|---|---|---|---|---|---|---|---|---|---|---|---|---|
| | | | | | 1 mcps | 10 mcps | 50 mcps | 100 mcps | 1 mcps | 10 mcps | 50 mcps | 100 mcps |
| *Betula* sp. Birch | 0 | none | none | — | | 0.5 | 2 | 3 | 2.0 | 2.0 | 1.9 | 1.8 |
| | 4.65 | none | none | — | | 1.0 | 4 | 6.5 | 2.3 | 2.25 | 2.2 | 2.15 |
| | 12.8 | none | none | — | | 1.5 | 8 | 17.5 | 4.4 | 4.2 | 3.9 | 3.8 |
| | | B | 1 | n | | 2 | 8.5 | 12.5 | 3.7 | 3.5 | 3.0 | 2.8 |
| | | E | 1 | n | | 2 | 8 | 13 | 3.5 | 3.3 | 2.9 | 2.6 |
| *Betula* sp. Birch, sap-wood | 0 | none | none | — | | 1 | 2 | 3 | 2.2 | 2.2 | 2.2 | 2.1 |
| | 5.2 | none | none | — | | 1.2 | 5 | 8 | 2.5 | 2.4 | 2.3 | 2.3 |
| | 7.5 | none | none | — | | 1.5 | 6 | 12 | 2.7 | 2.6 | 2.5 | 2.5 |
| | 10.7 | none | none | — | | 1.7 | 9 | 19 | 4.2 | 4.1 | 3.8 | 3.8 |
| | | A | 1 | n | | 4 | 13 | 22 | 5.5 | 4.7 | 4.0 | 3.7 |
| | | A | 2 | n | | 2 | 8.5 | 18 | 3.8 | 3.6 | 3.2 | 3.0 |
| | | A | 3 | n | | 1.5 | 6 | 13 | 3.5 | 3.4 | 3.0 | 3.0 |
| | | A | 1 | h | | 8 | 28 | 42 | 9 | 7.5 | 6.5 | 6.5 |
| | | A | 2 | h | | 7 | 25 | 35 | 8.2 | 7.0 | 6.1 | 6.0 |
| | | A | 3 | h | | 11 | 36 | 55 | 12 | 9.5 | 8.0 | 7.5 |
| *Fagus sylvatica* | 15 | none | none | — | 0.4 | 3.6 | 27 | | 10.2 | 9.8 | 9.5 | |
| | 23 | none | none | — | 2.4 | 11.0 | 58 | | 16.6 | 13.9 | 13.3 | |
| | 60 | none | none | — | 29 | 85 | 220 | | 57 | 30 | 26 | |
| *Fagus sylvatica* | 16 | none | none | — | 0.31 | 3.0 | 20 | | 9.4 | 9.0 | 8.5 | |
| | 23 | none | none | — | 4.3 | 21 | 75 | | 22 | 14 | 12 | |
| | 54 | none | none | — | 59 | 130 | 270 | | 59 | 33 | 30 | |
| *Khaya* sp. | | A | 1 | n | | 2.5 | 9.7 | 17 | 3.5 | 3.3 | 3.1 | 3.0 |
| | | C | 1 | n | | 1 | 6 | 11 | 3.0 | 2.8 | 2.7 | 2.6 |
| | | D | 1 | n | | 1.5 | 6.6 | 12 | 3.2 | 3.1 | 2.8 | 2.7 |
| *Liriodendron tulipifera* | 0 | none | none | — | | 0.3 | 1.5 | 2.5 | 2.0 | 1.9 | 1.8 | 1.8 |
| | 5.5 | none | none | — | | 0.8 | 4 | 7.5 | 2.4 | 2.3 | 2.2 | 2.1 |
| | 6 | none | none | — | | 0.7 | 3.7 | 7.2 | 2.6 | 2.5 | 2.3 | 2.3 |
| | 11 | none | none | — | | 1 | 6 | 13.5 | 2.75 | 2.7 | 2.6 | 2.6 |
| | | A | 1 | n | | 2 | 8 | 12 | 3.4 | 3.1 | 2.7 | 2.8 |
| | | D | 1 | n | | 1 | 5.5 | 8.5 | 2.7 | 2.6 | 2.3 | 2.3 |
| Mahogany (species not determined) | 0 | none | none | — | | 0.15 | 1.25 | 2.5 | 1.7 | 1.7 | 1.8 | 1.8 |
| | 2.6 | none | none | — | | 0.3 | 2.0 | 4.0 | 2.0 | 2.0 | 2.0 | 1.95 |
| | 7 | none | none | — | | 0.5 | 3.5 | 6.0 | 2.3 | 2.3 | 2.2 | 2.2 |
| | 12.6 | none | none | — | | 0.7 | 5.5 | 10 | 2.5 | 2.4 | 2.3 | 2.3 |
| *Pinus sylvestris* | 15 | none | none | — | 0.17 | 2.8 | 19 | | 8.1 | 8.1 | 7.3 | |
| | 22 | none | none | — | 1.1 | 5 | 30 | | 11.3 | 10.3 | 9.9 | |
| | 45 | none | none | — | 7 | 20 | 58 | | 20 | 15 | 14 | |
| *Quercus* sp. | 62 | | | — | 18 | 64 | 260 | | 44 | 30 | 27 | |
| *Swietenia* sp. | | A | 1 | n | | 2.5 | 10 | 15.5 | 3.7 | 3.5 | 3.1 | 3.0 |
| | | C | 1 | n | | 1.3 | 7.5 | 11 | 3.2 | 3.1 | 2.8 | 2.8 |
| | | D | 1 | n | | 1.5 | 8.5 | 13 | 3.6 | 3.4 | 3.0 | 3.0 |

Resin A, B, C  middle polymer $\left.\right\}$ phenol formaldehyde resins
Resin D, E      low polymer

*Treating method 1:* Veneers dried, subsequently dipped into preservative and finally seasoned to desired moisture content
*Treating method 2:* Veneers 24 h soaked in water, subsequently 12 h dipped into preservative
*Treating method 3:* Veneers 24 h soaked in water, subsequently 24 h dipped into preservative

The reactance $X_C$ of the condenser can be expressed in terms of its capacitance $C$ and angular frequency $\omega$ or frequency $f$ in cycles per second, respectively,

$$X_C = \frac{1}{\omega C} = \frac{1}{2\pi f C} \ [\Omega]. \tag{6.107}$$

The capacitance $C$ of the condenser can be calculated as follows

$$C = \frac{\varepsilon A}{36\pi d} \cdot 10^{-11} \ [\text{farad}] \tag{6.108}$$

where $A$ = area in cm², $d$ = thickness in cm, and $\varepsilon$ = dielectric constant of the dielectric material (wood) between the plates.

Substituting Eq. (6.108) in Eq. (6.107) one obtains

$$X_C = \frac{1.8 \cdot 10^{12} \cdot d}{f \cdot \varepsilon \cdot A}. \tag{6.109}$$

Now we introduce

$$\frac{X_C}{R} = \frac{I_R}{I_C} = \mathrm{tg}\,\delta = (90° - \varphi) \tag{6.110}$$

where $R$ = resistance and we assume for small angles

$$\mathrm{tg}\,(90° - \varphi) = \sin(90° - \varphi) = \mathrm{tg}\,\delta = \cos\varphi.$$

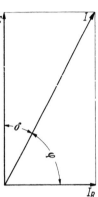

Fig. 6.118. Vector-diagram of the current in a system of electrodes and wood; $I_C$ capacitance, $I_R$ current through condenser, $\varphi$ power loss angle

Thus

$$\frac{1}{R} = \frac{\mathrm{tg}\,\delta}{X_C} = \frac{f \cdot \varepsilon \cdot A \cdot \mathrm{tg}\,\delta}{1.8 \cdot 10^{12} d} \ [\Omega^{-1}]. \tag{6.111}$$

The specific alternating-current conductivity $\varkappa$ is defined as follows

$$\varkappa = \frac{1}{R} \cdot \frac{d}{A} = \frac{f \cdot \varepsilon \cdot \mathrm{tg}\,\delta}{1.8 \cdot 10^{12}} \ [\Omega^{-1}\mathrm{cm}^{-1}]. \tag{6.112}$$

The dielectric behavior of wood is characterized by its dielectric constant $\varepsilon$, the specific conductivity $\varkappa$, and the power loss factor $\mathrm{tg}\,\delta$. Each of these units can be calculated from the other two by means of Eq. (6.112). The specific conductivity is especially important since it determines the power losses and thus the rise of temperature in the wood:

$$P = E^2 \varkappa \frac{A}{d} \tag{6.113}$$

where $E$ is the effective voltage in $V \cdot \mathrm{cm}^{-1}$.

From Eqs. (6.112) and (6.113) follows:

$$P = E^2 \frac{f \cdot \varepsilon \cdot \mathrm{tg}\,\delta \cdot A}{1.8 \cdot 10^{12} d} = 0.556 \cdot E^2 \cdot f \cdot \varepsilon \frac{A}{d} \cdot \mathrm{tg}\,\delta \cdot 10^{-12} \tag{6.114a}$$

or, if one relates the power loss to 1 cm³ material

$$P = 0.556 \cdot E^2 \cdot f \cdot \varepsilon\, \mathrm{tg}\,\delta \cdot 10^{-12} \ [\mathrm{W} \cdot \mathrm{cm}^3] \tag{6.114b}$$

in English units, respectively

$$P = 1.415 \cdot E'^2 \cdot f \cdot \varepsilon \cdot \cos\varphi \cdot 10^{-12} \ [\mathrm{W/cu} \cdot \mathrm{inch}] \tag{6.114c}$$

where $E'$ is the effective voltage in $V \cdot \mathrm{inch}^{-1}$.

The dielectric constants of commercial woods in the average range of specific gravity and in oven-dry condition vary between about 1.8 and 3.0 for flat-sawn specimens in which the direction of the field is radial to the annual rings. When

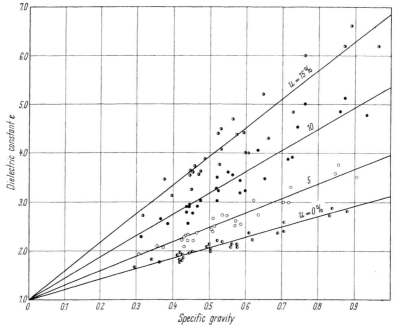

Fig. 6.119. Relationship between dielectric constant and specific gravity of wood for different moisture contents
From SKAAR (1948)

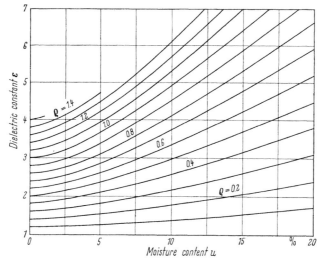

Fig. 6.120. Combined influence of specific gravity and moisture content of wood on its dielectric constant
From SKAAR (1948)

the field is directed along the grain the values of $\varepsilon$ become 30 to 60% greater than when it is perpendicular to the grain. The dielectric constant $\varepsilon$ is directly proportional to the specific gravity $\varrho$ (Fig. 6.119). SKAAR (1948) studied the combined influence of specific gravity and moisture content and the results are shown in Fig. 6.120. It is evident that the moisture content of wood can be measured fairly well by capacitance and power-factor meters only when the specific gravity is taken into consideration. Thorough calibration to correct variations of specific gravity of the wood tested is indispensable. The dielectric constant increases at high frequencies in the hygroscopic range in a curvilinear manner with increasing moisture content, and above the fiber saturation point a straight-line relationship holds (Fig. 6.121). The limiting value of the dielectric constant for lightest woods in the water-saturated condition may approach the dielectric constant for water ($\varepsilon_W = 81.0$). KRÖNER (1944) found for lower moisture contents ($u < 12\%$) and frequencies in the range between $10^6$ and $10^9$ cycles per second, a nearly linear relationship between dielectric constant and moisture content (Fig. 6.122). He assumed a logarithmic blending rule:

$$\log \varepsilon = (1 - x) \log \varepsilon_W + x \log \varepsilon_{H_2O} = \frac{1}{1+u} \log \varepsilon_W + \frac{u}{1+u} \log \varepsilon_{H_2O} \quad (6.115)$$

where $\varepsilon$ = dielectric constant of the theoretical non-swelling mixture of wood and water, $\varepsilon_W$ = dielectric constant of oven-dry wood, $\varepsilon_{H_2O}$ = 81.0 = dielectric constant of water, $x$ = moisture content, based on the wet weight, $u$ = moisture content, based on the oven-dry weight.

The influence of frequency on the dielectric constant, on power factor, and on electrical resistivity is exhibited in Fig. 6.123. Additional information is given in Figs. 6.124, 6.125 and 6.126. It can be stated that the dielectric constant generally decreases with increasing frequency.

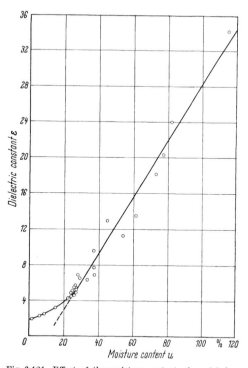

Fig. 6.121. Effect of the moisture content of wood below and above fiber saturation point on the dielectric constant at high frequencies. From SKAAR (1948)

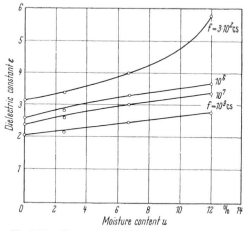

Fig. 6.122. Change of dielectric constant when varying the moisture content of cross cut beech wood at different frequencies. From KRÖNER (1944)

**6.6.2.3. Power Factor.** The power factor of a dielectric material, such as wood, "is the ratio of the electrical energy dissipated per cycle of oscillation in a condenser, using the material as the dielectric, to the total electrical energy stored in the condenser during the cycle" (BROWN, PANSHIN, FORSAITH, 1952, Vol. II, p. 169). If in the condenser a vacuum dielectric is employed no molecules are available which, under the influence of the alternating voltage, could cause

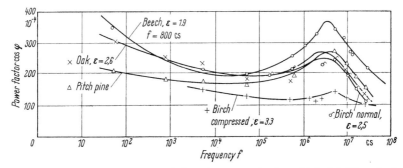

Fig. 6.123. Dependence of the power factor of different wood species (oven-dried at 20 °C) on the frequency. From BRAKE and SCHÜTZE (1935)

molecular friction dissipated in the form of heat. The power factor in this special case is therefore zero. On the other hand, the power factor cannot exceed a value of unity for the phase angle $\pi/2$ radians. As has been shown in the foregoing section, a wood-dielectric condenser can be represented electrically by a circuit consisting of a parallel combination of a capacitance $C$ and a resistance $R$. The alternating voltage $E$ causes a total current flow $I$ in the system. This total effective current $I$ consists of two components, namely, the component $I_C$ flowing through the capacitance $C$, and $I_R$ coursing through the resistance $R$. A phase relationship exists between these current components shown schematically in Fig. 6.118. The total current $I$ leads the voltage $E$ by the phase angle $\varphi$. The resistance current component $I_R$ is in phase with the voltage, while the reactance component $I_C$ leads the voltage by a phase angle of $\pi/2$ radians.

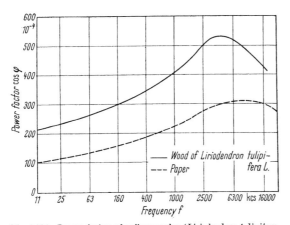

Fig. 6.124. Power factor of yellow poplar (*Liriodendron tulipifera* L.) and paper made of it. From SHARPE and O'KANE (1935)

The apparent power $P_T$ existing in the circuit is determined by the product of the voltage $E$ and the total current $I$:

$$P_T = EI. \tag{6.116}$$

A portion of the apparent power $P_T$ is dissipated as heat. The magnitude of this power loss $P$ is directly proportional to the amount of current $I_R$ flowing through the resistance $R$

$$P = EI_R. \tag{6.106}$$

It is evident from Fig. 6.118 that
$$I_R = I \cos \varphi$$
and therefore Eq. (6.106) may also be written in the following form

or
$$\left. \begin{array}{l} P = E \cdot I \cdot \cos \varphi \\ \cos \varphi = \dfrac{P}{E \cdot I} = \dfrac{P}{P_T} \cdot \end{array} \right\} \quad (6.106)$$

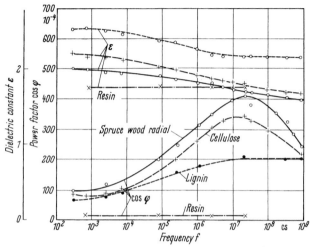

Fig. 6.125. Power factor and dielectric constant of spruce wood, cellulose, lignin and resin as a function of frequency. From KRÖNER (1944)

The trigonometric function $\cos \varphi$ represents the "power factor" as the ratio of the power dissipated $P$ to the total power $P_T$ existing in the circuit. The power factor can be expressed using the vector diagram in Fig. 6.118, in terms of the current $I_R$ through the resistance and the total effective current $I$:

and since
$$\left. \begin{array}{l} \cos \varphi = \dfrac{I_R}{I} \\ I = \sqrt{I_R^2 + I_C^2} \, . \\ \cos \varphi = \dfrac{I_R}{\sqrt{I_C^2 + I_R^2}} \end{array} \right\} \quad (6.117)$$

In most woods with a moisture content below about 12% the leakage current $I_R$ is very small and the squared value $I_R^2$ may be neglected. In this case Eq. (6.117) can be simplified to
$$\cos \varphi = \frac{I_R}{I_C} \cdot \quad (6.118)$$

Applying Ohm's law ($I_R = E/R$ and $I_C = E/X_C$) Eq. (6.118) can be written as follows:
$$\cos \varphi = \frac{X_C}{R} \cdot \quad (6.119)$$

If we substitute the term for the reactance $X_C$ according to Eq. (6.107) in Eq. (6.119) we obtain
$$\cos \varphi = \frac{1}{2\pi f C R} \cdot \quad (6.120)$$

270 6. Physics of Wood [Ref. p. 285

Between the electrical resistance $R$ in $\Omega$ and the electrical resistivity $\varrho$ of a cubic centimeter, in $\Omega \cdot$ cm, holds the relationship

$$R = \frac{\varrho \cdot d}{A} \tag{6.121}$$

where $d = $ length of resistor in cm, $A = $ cross-sectional area of resistor in cm². Assuming for small angles tg $\delta = \cos \varphi$ and using Eqs. 6.112 and 6.121 we can rearrange Eq. 6.120

$$\text{tg } \delta = \cos \varphi = \frac{1.8 \cdot 10^{12}}{\varrho \cdot f \cdot \varepsilon} . \tag{6.122}$$

This equation is instructive insofar as, solved for $\varrho$, we can see that for a constant frequency the resistivity $\varrho$ varies inversely with the product of the power factor ($\cos \varphi$) and the dielectric constant $\varepsilon$. The power factor of wood depends on its density only with a weak positive correlation, irrespective of kind and chemical nature. The parallel-to-grain power factor for a piece of wood (from oven-dry condition up to about 7% of moisture content) is roughly twice the corresponding across-the-grain value.

The power factor of a wood increases, at a given frequency, with its moisture content. From his experiments SKAAR (1948) concluded that the power factor of wood is greater at higher frequencies (15 megacycles instead of 2.0 megacycles) if the moisture content is below 15 or 16%. Above this moisture content the reverse condition is valid.

For various oven-dry wood species BRAKE and SCHÜTZE (1935) found the curves shown in Fig. 6.123 for the dependence of the power factor on the frequency. There are distinct maxima for frequencies between about 4 and 5 megacycles. SHARPE and O'KANE (1935) observed a similar trend for the wood of yellow poplar and for paper manufactured from this wood (Fig. 6.124). STOOPS (1934) obtained similar curves for cellophane, SHARPE and O'KANE (1935) for oven-dry vulcanized fiber. The uniformity of all these results and the fact that cellulose is the main constituent of the various materials mentioned led to the conclusion that the cellulose determines mainly the dielectric properties of these organics. KRÖNER (1944) gave full evidence for this theory investigating the power factor as a function of frequency separately for spruce wood and its constituents: cellulose, lignin, and resin (Fig. 6.125). One can see that the curves for cellulose and wood are quite similar whereas the curve for the lignin has another character and the frequency-independent values for the resin are so low that they do not influence the dielectric behavior of the wood.

According to DEBYE's theory of polar molecules, power factor and dispersion are based on the asymmetrical electrical structure of matter. The chain-molecule of cellulose is a complicated polar structure. One can assume that free rotation of the three OH groups of each glucose residue is possible and on this assumption, based on DEBYE's theory and considering the coupling of the dipoles, one can estimate the value for the dielectric constant $\varepsilon$. One obtains $\varepsilon = 7.7$ for water-free pure cellulose in this way. The measured value for oven-dry standard cellulose is $\varepsilon = 6.4$ (DE LUCA, CAMPBELL and MAASS, 1938).

### 6.6.3 Magnetic Properties of Wood and Wood Constituents

The magnetic susceptibility $\varkappa$, based on unit volume, is measured by the ratio of the intensity of magnetization $I$ produced in a substance to the magnetizing force or intensity of field $H$ to which it is subjected

$$\varkappa = I/H . \tag{6.123}$$

The relation of volume susceptibility $\chi$ is shown by the equation

$$\chi = \varkappa/\varrho \qquad (6.124)$$

where $\varrho$ is the density of the substance. The values are positive for para-magnetic bodies, negative for diamagnetic ones to which wood belongs (Table 6.24).

NILAKANTAN (1938) has shown that the magnetic susceptibility of cellulose (of teakwood) is higher ($-0.51 \cdot 10^{-6}$) than that of lignin ($-0.42 \cdot 10^{-6}$) and that the diamagnetic susceptibility along the grain reaches a maximum. Wood ashes have ferromagnetic properties (REISCHEL and WEDEKIND, 1932, 1933). The insignificant magnetization of wood stimulated, especially in Germany in the Thirties, the erection of wooden antenna masts and later on the construction of a large series of nonmagnetic wooden mine sweepers.

Table 6.24. *Magnetic Susceptibility of Some Substances*

| Substance | Magnetic susceptibility $10^6 \cdot \varkappa_v$ * |
|---|---|
| Wood | $-0.2 \cdots -0.4$ |
| Shellac | $-0.3 \cdots -0.39$ |
| Paraffin | $-0.6$ |
| Ebonite | $+0.6 \cdots 1.1$ |

\* Values condensed from LANDOLT-BÖRNSTEIN (1955) and Handbook of Chemistry and Physics (1948).

### 6.6.4 Piezoelectric Properties of Wood

The piezoelectric effect in wood was discovered by SHUBNIKOV (1940) in the USSR, and studied by BAZHENOV (1950, 1953, 1959, 1960, 1961) in the USSR, by FUKADA (1955, 1956, 1957, 1964) in Japan, and by GALLIGAN and coworkers (1961, 1963) in the USA. Piezoelectricity occurs not only in monocrystalline materials (e.g. quartz, Rochelle salt, barium titanate, and lithium sulphate), but in the so-called piezoelectric textures. The piezoelectric effect appears either directly as an electric polarization by application of mechanical stress or inversely as a mechanical strain by application of an electric field. In the earliest publications on the piezoelectricity of wood this property was attributed to the cellulose as the crystalline constituent in wood. But the dissimilarity between wood (with its fiber structure and with its cellular mixture of crystalline, paracrystalline, and amorphous regions in the submicroscopic ranges) and typical inorganic crystalline materials is evident. Nevertheless wood is a typical anisotropic material and by analogy crystalline materials are always anisotropic. In crystals the an isotropy is caused by the structure of the crystal lattice, in wood, as already mentioned, by its complex structure, "macroscopic and microscopic structure, structure of the cell wall, structure of the basic constituent of wood-cellulose—and so on" (BAZHENOV, 1961). By means of an electrometer BAZHENOV measured the electrical charges developed in specimens during bending and compression tests. The samples were small and had well defined grain orientation. He adopted vector analysis to describe the polarization and stress-strain relationships. He examined also the inverse piezoelectric effect in wood, in other cellulosic materials and in

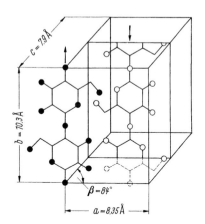

Fig. 6.126. Unit cell of crystal of cellulose. From FUKADA (1965)

272                                6. Physics of Wood                          [Ref. p. 285

some protein substances, such as wool and bone. BAZHENOV and FUKADA both pointed out that the piezoelectric effects are dependent upon the presence of some molecular order and came to the conclusion that the cellulose offers such a "piezoelectric texture". Both investigators demonstrated that the mechanical action required to cause a piezoelectric effect is shear strain between the fibers.

The general relation between electrical polarization $P$ and applied stress $\sigma$ is represented by[1]

$$\left.\begin{array}{l} P_x = d_{11}\sigma_x + d_{12}\sigma_y + d_{13}\sigma_z + d_{14}\tau_{yz} + d_{15}\tau_{zx} + d_{16}\tau_{xy}, \\ P_y = d_{21}\sigma_x + d_{22}\sigma_y + d_{23}\sigma_z + d_{24}\tau_{yz} + d_{25}\tau_{zx} + d_{26}\tau_{xy}, \\ P_z = d_{31}\sigma_x + d_{32}\sigma_y + d_{33}\sigma_z + d_{34}\tau_{yz} + d_{35}\tau_{zx} + d_{36}\tau_{xy} \end{array}\right\} \qquad (6.125)$$

where $P_x$, $P_y$, $P_z$ are the $x, y, z$ components of polarization $P$, $\sigma_x, \sigma_y, \sigma_z$ are the tension or pressure stresses in $x, y, z$ directions, and $\tau_{yz}, \tau_{zx}, \tau_{xy}$, are the shear stresses in $yz, zx, xy$ planes. The coefficient $d_{mn}$ which connects the components of $P$ and $\sigma$ or $\tau$ is called piezoelectric modulus. The most general piezoelectric tensor is, therefore, expressed as

$$\begin{pmatrix} d_{11} & d_{12} & d_{13} & d_{14} & d_{15} & d_{16} \\ d_{21} & d_{22} & d_{23} & d_{24} & d_{25} & d_{26} \\ d_{31} & d_{32} & d_{33} & d_{34} & d_{35} & d_{36} \end{pmatrix} \qquad (6.126)$$

In a crystal with a certain crystallographic symmetry, some of the $d_{mn}'s$ become zero due to the symmetry. As mentioned above, the piezoelectric effect in wood is caused by the crystalline structure of the native cellulose fibrils. The crystals are characterized by the monoclinic symmetry. In this case the possible piezoelectric tensor is

$$\begin{pmatrix} 0 & 0 & 0 & d_{14} & d_{15} & 0 \\ 0 & 0 & 0 & d_{24} & d_{25} & 0 \\ d_{31} & d_{32} & d_{33} & 0 & 0 & d_{26} \end{pmatrix} \qquad (6.127)$$

where $x, y$ and $z$ axes correspond to $a, c$ and $b$ directions in the unit cell of cellulose crystal (Fig. 6.126).

The intensity of piezoelectricity in wood depends on the degree of crystallinity and the degree of orientation of crystallites.

GALLIGAN and coworkers (1961, 1963) verified the piezoelectricity in wood. They used long specimens of wood and generated piezoelectric signals by the impact of an air hammer system (6.127). The signals were received from point electrodes and photographed by means of an oscilloscope. Fig. 6.128 shows three piezoelectric voltage profiles at different cross-sections of a specimen. The polarity is marked by $+$- and —-signs.

As pointed out already by BAZHENOV, there is a distinct influence of wood variables on piezoelectric properties. GALLIGAN et al. (1963) introduced as an average piezoelectric parameter $\Delta A_{\max}$ the arithmetic average of the average maximum positive signal throughout the length of the specimen and the average

---

[1] The following explanations are based on a paper "Piezoelectric Effect in Wood and other Crystalline Polymers" by E. FUKADA (1965).

maximum negative signal without regard to sign. The value $\Delta A_{max}$ is an integrated measure of the ability of the specimen to generate a piezoelectric voltage. The average $\Delta A_{max}$ parameter is directly proportional to density (Fig. 6.129), but no correlation was found between width of annual rings and average $\Delta A_{max}$. There is a linear increase of average $\Delta A_{max}$ with the static modulus of elasticity $E_{st}$ (Fig. 6.130). BAZHENOV also has demonstrated an increase in the piezoelectric stress constant with increase in modulus of elasticity. In summary: piezoelectricity may become a tool for nondestructive testing of local strain in wood and of the internal influences, such as density, knots and other features, on a dynamic system (GALLIGAN and BERTHOLF 1963).

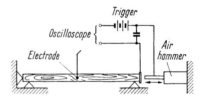

Fig. 6.127. Schematic diagram of the air hammer impact system used for generating piezoelectric signals. From GALLIGAN and BERTHOLF (1963)

The piezoelectric behavior of old timbers is remarkable. KOHARA (1958) investigated the change of compressive strength with age in old timbers of Hinoki or

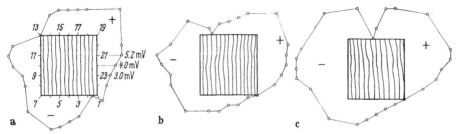

Fig. 6.128a—c. Three piezoelectric voltage profiles at different cross sections of a specimen. Magnitude of voltage (mV) is represented by displacement of data points from wood surface. a) At cross section 7 in. from impact end; b) at cross section 23 in. from impact end; c) at cross section 25 in. from impact end. From GALLIGAN and BERTHOLF (1963)

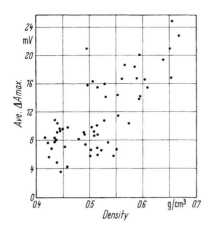
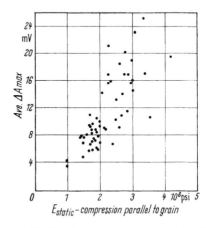

Fig. 6.129. Piezoelectric voltage generation versus density for 60 Douglas-fir specimens. From GALLIGAN and BERTHOLF (1963)

Fig. 6.130. Relationship between piezoelectric voltage generation, average $\Delta A_{max}$, and static modulus of elasticity. From GALLIGAN and BERTHOLF (1963)

Japanese cypress (*Chamaecyparis obtusa* Endl.). This wood has been used extensively as a building material since old times in Japan. Each point of the curve for $\sigma_{cb}$ in

18 Kollmann/Côté, Solid Wood

Fig. 6.131 is the average of 20 specimens, reduced to 13% of moisture content. It can be seen that the strength increases first to a maximum at the end of about 350 years and then decreases approximately in a linear manner with time. FUKADA and coworkers (1957) measured on Kohara's specimens the dynamic Young's modulus $E_d$ and the piezoelectric modulus $d_{25}$.

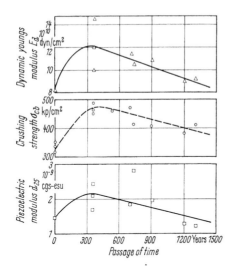

Fig. 6.131. Change of crushing strength, dynamic Young's modulus, and piezoelectric modulus with age in old timbers of Japanese cypress (*Chamaecyparis obtusa* Endl). From FUKADA and coworkers (1957)

Maxima were observed again at the end of 360 years. This phenomenon may be explained as follows: The degree of crystallinity of cellulose increases very slowly for a period of about 300 to 350 years at room temperature. As time passes, the decomposition of cellulose begins to reduce the crystallinity. X-ray diffraction studies have proved this fact. Sharper diffraction images were observed for specimens aged 350 years than for those 8 years of age. The pattern for wood 1300 years in use is very diffuse, clearly indicating the decrease of crystallinity.

## 6.7 Acoustical Properties of Wood

### 6.7.0 General Considerations

Wood is a unique material for musical instruments and acoustic purposes. The highest possible increase in value in woodworking has without doubt been reached in the violins manufactured by such Italian masters as were Amati, Guarneri, and Stradivarius. Beside violins, other string instruments, sounding boards in pianos and grand pianos, organ pipes, and xylophones make use of the sound-radiating properties of vibrating wood. It is possible to contribute to noise reduction in buildings with wood and wood-base materials.

All phenomena of vibrations and waves generated in elastic media with a frequency between about 16 and 20000 Hz (cycles per second, abbreviation cps) which are subjectively perceived are called sound. If a piece of wood is struck, e.g. by a hammer, the wood is set into vibration at its resonant or natural frequency. This vibration emits sound waves into the surrounding air. Sound as a wave phenomenon may be characterized by the following physical properties: wave length, amplitude, frequency, and velocity of sound. Since a sound wave moves at a definite speed $v$, there is obviously the following relationship to frequency $f$ and wave length $\lambda$

$$v = \lambda f \quad \text{or} \quad \lambda = \frac{v}{f}. \tag{6.128}$$

A piece of wood vibrates if periodic forces act upon it. The amplitude of vibration depends on the frequency and the driving force. There are certain frequencies to which the particular wood object responds with maximum amplitude of vibration. These frequencies are the resonant or natural frequencies of the piece. As mentioned before, the same characteristic frequencies occur if the piece of wood is hit by a hammer blow.

In a solid body such as wood three kinds of vibratory resonant motion are possible: a) Longitudinal resonant vibrations which can be taken as the dynamic analogue of axial stresses acting in a short column. The longitudinal resonance methods can be applied for determining the elastic constants, the sound velocity, and the logarithmic decrement (cf. Sections 7.1.5.2 and 7.1.6); b) Transverse resonant vibrations or flexural vibrations, corresponding to static bending in a wooden beam, are the most frequent dynamic stresses in structural parts, such as trusses, joists and spars. Flexural vibrations are common also in poles, diving springboards and as sound-radiating phenomena in the sounding boards of pianos. Wooden beams or plates may be supported at two points or they may be clamped as cantilevers. The kind of supporting or fastening determines, together with the dimensions of the object and the sound velocity, the lowest resonant frequency; c) Torsional resonant vibrations as the dynamic counterpart to static torsion. Decisive for the internal stress and the fundamental resonant frequency is the modulus of rigidity $G$ of the wood. HEARMON (1948) used a modified torsional vibration technique for measuring the modulus of rigidity. As shown in Fig. 6.132 a wood rod of circular cross-section (radius $r$, length $L$) is clamped at one end. To the other end of the specimen a weight of known mass is attached. The mass is initially twisted and then released. This results in an oscillation in accordance with the laws of simple harmonic motion. The attached mass has a high moment of inertia $I$ (in comparison to which the moment of inertia of the wood is negligible). Since the attached mass lowers the frequency of vibration $f_r$, a stop watch instead of a complicated electronic device can be used to measure the frequency. The method is a typical "non-destructive testing method".

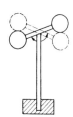

Fig. 6.132. Torsional vibrations in a wood member induced by the moment of Inertia I of the attached mass. From BROWN, PANSHIN and FORSAITH (1952)

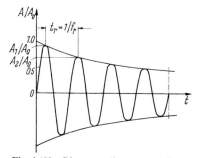

Fig. 6.133. Diagrammatic representation of the decay of free vibrations in a wood member. The sinusoidal curve represents a plot, as a function of time $t$, of the position of a point on a vibrating wood member which has a natural frequency $f_r$ and a period $t_r$. The logarithmic curves show the decay of the successive peaks of vibration due to damping. The ratio $A/A_0$, when $A$ is the peak amplitude of a particular vibration of the time $t$ and $A_0$ is the peak amplitude at the time $t = 0$, is plotted as a function of $t$. From BROWN, PANSHIN and FORSAITH (1952)

$$G = 8\pi I L \left(\frac{f_r}{r^2}\right)^2. \tag{6.129}$$

If no external periodic forces are acting on a solid vibrating body it returns to the static condition. The successive amplitudes become lower and lower (Fig. 6.133) since the original inherent energy is dissipated partly by radiation of sound, partly by internal friction, which produces heat and which is called "damping capacity" because it decays the free vibrations in the wooden member. In the case of free vibrations the decrease in amplitudes of two successive cycles of vibration

276 6. Physics of Wood [Ref. p. 285

follows a logarithmic law. Therefore the logarithmic decrement $\varLambda$ may be computed as follows

$$\varLambda = \ln \frac{A_1}{A_2} \tag{6.130}$$

where $A_1$ and $A_2$ are the amplitudes of two succeeding cycles.

The damping capacity of wood is higher than it is for most other structural materials. More details on this important property are given in Sections 6.7.1.3 and 7.1.6.3.

### 6.7.1 Sound Transmission in Wood

**6.7.1.1 Sound Velocity.** In long solid rods, the thickness of which may be neglected compared with the wave length in the case of the propagation of longitudinal waves along the axis of the rod, the sound velocity $v$ is

$$v = \sqrt{\frac{E}{\varrho}} \tag{6.131}$$

where $E$ = modulus of elasticity (Young's modulus) in kp/cm² and $\varrho$ = density in kps²m⁻⁴.

The propagation velocity of elastic torsional waves depends on the modulus of rigidity $G$.

Hence

$$v = \sqrt{\frac{G}{\varrho}} \, . \tag{6.132}$$

The density $\varrho$ for a particular piece of wood is constant. Therefore the ratio of the sound velocity along the grain $v_{\parallel}$ to that across the grain $v_{\perp}$ can be deduced as follows:

$$\frac{v_{\parallel}}{v_{\perp}} = \sqrt{\frac{E_{\parallel}}{E_{\perp}}} \, . \tag{6.133}$$

Table 6.25. *Sound Velocity in Woods and other Materials*
(From Hesehus (1908) and Lübcke (1927))

| Wood species or material | Average specific gravity | Sound velocity parallel to grain $v_{\parallel}$ m/s | $\dfrac{v_{\parallel}}{v_{\perp}}$ |
|---|---|---|---|
| Beech | 0.70 | 3412 | 1.34 |
| Cedar | 0.55 | 4400 | — |
| Oak | 0.65 | 3381···4310 | 1.36 |
| Ash | 0.65 | 3900 | — |
| Cherry | 0.60 | 4400 | — |
| Walnut | 0.55 | 4700 | — |
| Fir | 0.40 | 5256 | 2.2 |
| Fir, compression wood | 0.45 | 4180 | 1.5 |
| Iron | 7.85 | 5000 | — |
| Copper | 8.90 | 3900 | — |
| Lead | 11.34 | 1320 | — |
| Glass | 2.5 | 5100···6000 | — |
| Ebonite | 1.15 | 1570 | — |
| Cork | 0.25 | 430···530 | — |

Values for the sound velocity in different wood species and in some other solid materials are listed in Table 6.25. It may be mentioned that in the literature for the ratio $v_{\parallel}/v_{\perp}$ mostly very low figures are reported which are probably due to incorrect grain orientation in the samples tested.

More reliable data calculated on the basis of elastic constants using Eqs. (6.131) and (6.133) are available from Table 6.26.

Table 6.26. *Average Sound Velocity in Woods* (calculated from the elastic constants)

| Wood species | Average specific gravity g/cm³ | Average density kp s² m⁻⁴ · 10⁶ | Average modulus of elasticity parallel to grain kp/cm² | Average modulus of elasticity perpend. to grain kp/cm² | Average sound velocity $v$ parallel to grain m/s | Average sound velocity $v$ perpend. to grain m/s | $\dfrac{v_\parallel}{v_\perp}$ |
|---|---|---|---|---|---|---|---|
| Spruce | 0.47 | 0.479 | 110,000 | 5,500 | 4,790 | 1,072 | 4.47 |
| Pine | 0.52 | 0.530 | 120,000 | 4,600 | 4,760 | 932 | 5.11 |
| Fir | 0.45 | 0.459 | 110,000 | 4,900 | 4,890 | 1,033 | 4.73 |
| Maple | 0.63 | 0.642 | 94,000 | 9,150 | 3,826 | 1,194 | 3.21 |
| Beech | 0.73 | 0.744 | 160,000 | 15,000 | 4,638 | 1,420 | 3.27 |
| Oak | 0.69 | 0.703 | 130,000 | 10,000 | 4,304 | 1,193 | 3.61 |
| Lime | 0.53 | 0.541 | 74,000 | (2,500) | 3,700 | (680) | (5.44) |

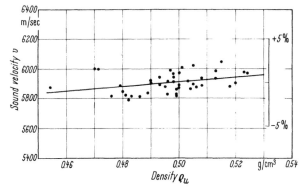

Fig. 6.134. Correlation between sound velocity parallel to fibers and density of spruce wood. From KOLLMANN and KRECH (1960)

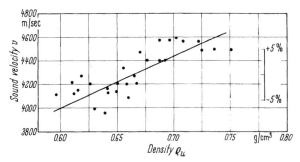

Fig. 6.135. Correlation between sound velocity parallel to the fibers and density of oak wood. From KOLLMANN and KRECH (1960)

Since the modulus of elasticity $E$ is directly proportional to the density (cf. Section 7.1.6.2) the sound velocity $v$ should be independent from the density $\varrho$ according to Eq. (6.132). KOLLMANN and KRECH (1960) found that for spruce wood the correlation between $v$ and $\varrho$ is very weak (0.13) as Fig. 6.134 shows. For oak wood a stronger correlation (0.64) has been established (Fig. 6.135). The logarithmic decrement $\varLambda$ is not influenced by the density $\varrho$. The wide scattering of the points of measurement in Figs. 6.136 and 6.137 is remarkable. The relationship between velocity of sound $v$ and the moisture content $u$ of wood is represented

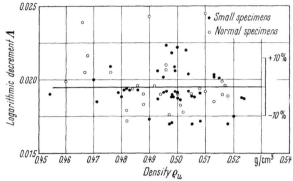

Fig. 6.136. Relationship between logarithmic decrement and density of spruce wood.
From KOLLMANN and KRECH (1960)

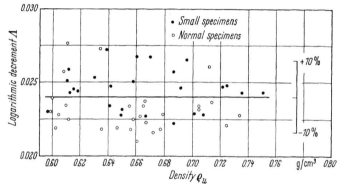

Fig. 6.137. Relationship between logarithmic decrement and density of oak wood.
From KOLLMANN and KRECH (1960)

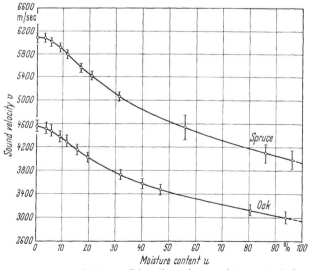

Fig. 6.138. Dependence of sound velocity parallel to the grain on moisture content of spruce and oak wood. The vertical lines show the statistical confidence limits of 99% for any point of measurement.
From KOLLMANN and KRECH (1960)

Ref. p. 285] 6.7 Acoustical Properties of Wood 279

by Fig; 6.138, the vertical lines show the statistical 99% confidence limits for any point of measurement.

**6.7.1.2 Sound Wave Resistance, Damping of Sound Radiation and Internal Friction.** Sound velocity in wood, parallel to the fiber, has about the same magnitude as sound velocity in metals except lead. The density of wood amounts to only 1/20 to 1/10 of the density of technically used metals. Consequently the sound wave resistance $w$, which is decisive for the propagation of sound and especially for the reflection of sound at the boundary between two media is quite different for woods and metals since

$$w = \varrho \cdot v = \varrho \sqrt{\frac{E}{\varrho}} = \sqrt{\varrho E}. \qquad (6.134)$$

BRILLIÉ (1919) has published some observed values (Table 6.27).

The decay of the free vibrations of any plates is caused, as mentioned in the foregoing section, partly by damping due to internal friction and partly by

Table 6.27. *Sound Wave Resistance in Different Media* (From BRILLIÉ, 1919)

| Medium | Sound wave resistance $w$ dyn.s/cm |
|---|---|
| Steel | $395 \cdot 10^4$ |
| Cast iron | $258 \cdot 10^4$ |
| Brass | $234 \cdot 10^4$ |
| Bronze | $168 \cdot 10^4$ |
| Lead | $82.5 \cdot 10^4$ |
| Teak | $37 \cdot 10^4$ |
| Fir | $20 \cdot 10^4$ |
| Beech | $22 \cdot 10^4$ |
| Water | $14 \cdot 10^4$ |
| Rubber | $1 \cdot 10^4$ |
| Air | $0.004 \cdot 10^4$ |

damping due to sound radiation. Damping due to sound radiation depends mainly on the ratio of sound velocity to density for a particular material. Comparing woods with metals the acoustical superiority of the former becomes evident. HÖRIG (1929) has designed Fig. 6.139. In musical instruments, e.g. sounding boards of pianos or violins, low damping due to internal friction and high damping due to sound radiation is desirable.

Careful selection of the wood according to its density and elasticity is advisable (HANSEN, 1934). A sounding board assembled in a proper way, clamped at the edge and set in its center into vibrations by means of alternating current (430 cps) exhibits clear symmetrical sound figures (Fig. 6.140), whereas a less uniform plate shows for the corresponding resonance (460 Cps) very irregular knot lines and a knot point (Fig. 6.141).

Losses of radiating energy in the interior of solid vibrating bodies due to molecular friction are inevitable. For the mathematical treatment of the internal friction there exist two principally different theories. VOIGT (1892) and subsequently HONDA and KONNO (1921) proceed from an analogy to the internal friction in liquids whereas BOLTZMANN (1876) deduced the internal friction from the elastic after-effect. According to the first theory the coefficient of internal friction is proportional to the period of the oscillation applied in the test; according to BOLTZMANN's theory the coefficient of internal friction is independent on the frequency. Already VOIGT has pointed out that his theory is valid for some materials whereas BOLTZMANN's theory applies to other materials. Apparently there are transitions between both theories. Experimental data obtained by JAMES (1961) on Douglas-fir suggest that the internal friction in wood may result from two basic mechanisms; one is the characteristic of completely dry wood, the other one appears in the presence of hygroscopically bound water only. The first mechanism is reduced in effect at temperatures between $-18\,°C$ ($0\,°F$) and $95\,°C$ ($200\,°F$) by adding water, whereas the second one increases with moisture content and temperature. The relationships between these mechanisms and their

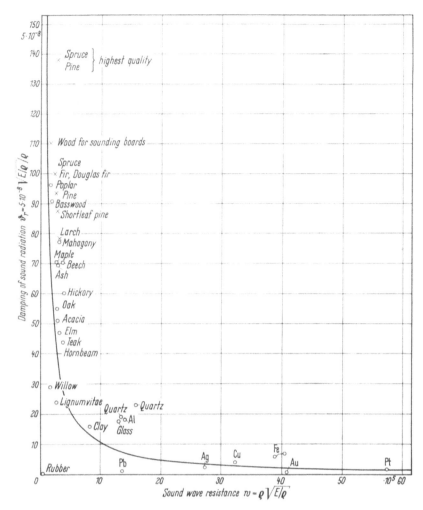

Fig. 6.139. Dependence of the damping of sound radiation on the sound wave resistance for different wood species and other materials. From HÖRIG (1929)

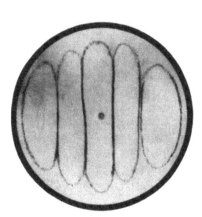

Fig. 6.140. Clear symmetrical sound figures on a clamped sounding board assembled in a proper way and caused to vibrate by means of alternating currents. (430 cps). (Phot. Grotrian-Steinweg, Braunschweig)

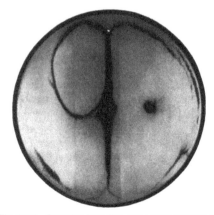

Fig. 6.141. Irregular knot line and a knot point on a clamped sounding board assembled in an improper way and caused to vibrate by means of alternating currents. (460 cps). (Phot. Grotrian-Steinweg, Braunschweig)

relative dependence on temperature and moisture content result in internal friction minima in the temperature range between $-18\,°C$ $(0\,°F)$ and $95\,°C$ $(200\,°F)$ and in moisture contents from about $2\ldots28\%$. At room temperature, an internal friction minimum appears at about 7% of moisture content. Other data show a strength maximum at about the same moisture content. Both the strength maximum and the internal friction minimum may result from the same basic process.

### 6.7.2 Acoustics of Buildings

**6.7.2.1 Sound Energy.** Audible sound in air is energy in motion. The threshold of human hearing is limited to an energy of $10^{-16}$ w/cm² (equivalent to zero de-

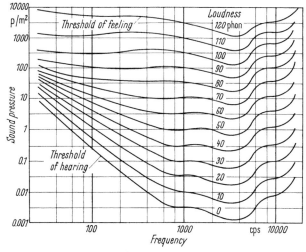

Fig. 6.142. Threshold of hearing and feeling and curves of equal loudness as function of frequency. From FLETCHER and MUNSON (1933)

Table 6.28. *Intensity of Sound*

| Source of sound | Intensity of sound | |
|---|---|---|
| | Arbitrary units | Decibels |
| Threshold of pain | 1,000,000,000,000 | 120 |
| Airplane engine, nearby Riveter at 1.5 m | 100,000,000,000 | 110 |
| Airplane cabin, normal flight | 10,000,000,000 | 100 |
| Heavy traffic, pneumatic | 1,000,000,000 | 90 |
| Loud radio music in room | 100,000,000 | 80 |
| Ordinary traffic, noisy office | 10,000,000 | 70 |
| Department store | 1,000,000 | 60 |
| Average office | 100,000 | 50 |
| Ordinary conversation | 10,000 | 40 |
| College classroom, quiet house | 1,000 | 30 |
| Whisper (audible at 1.2 m) | 100 | 20 |
| Rustle of leaves | 10 | 10 |
| Threshold of hearing | 1 | 0 |

cibels, db), the threshold of upper limit, that is of pain, is $10^{-4}$ w/cm² (120 decibels). The range of sound that must be tolerated by the human ear is very great as Table 6.28 shows.

The number of decibels between two intensity levels is obtained by computing ten times the logarithm of the ratio of the intensities. The threshold of hearing and feeling and the curves of equal loudness as a function of frequency are shown in Fig. 6.142, based on the fundamental investigations of FLETCHER and MUNSON (1933). Satisfactory acoustics of buildings or architectural acoustics are very important and must be well planned in advance. The noise in buildings must be controlled by: a) Separating noise sources from quiet spaces, b) Reducing the transmission loss (in decibels) through the construction elements, c) Absorbing sound within a space.

**6.7.2.2 Sound Transmission Loss for Various Type of Constructions.** The isolation of a kind of construction against transmission of sound is characterized by an aver-

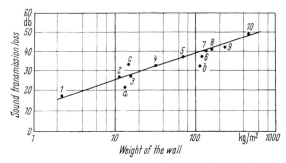

Fig. 6.143. Dependence of the sound transmission loss through single walls on the weight of the wall. *1* Plywood 5 mm; *2* plate glass 3 to 4 mm; *3* plate glass 7 to 8 mm; *4* metal lath wall 2,5 cm.; *5* pressed and plastered straw 9 cm.; *6* plastered porous bricks 12 cm.; *7* plastered pumice concrete with cork filler 11 cm.; *8* 1/4 brick wall, plastered 9 cm.; *9* 1/2 brick wall, plastered 15 cm.; *10* 1/1 brick wall, plastered 27 cm.; *a.* pressed straw, non-plastered; *b.* 1/4 brick wall, non-plastered; *c.* sheet iron 2 mm. From MEYER (1931)

Table 6.29. *Conversion Factors from Phones into Decibels* (Valid up to 90 phones with ±8% error)

| Frequency Cps | Average Conversion factor $\frac{\text{phone}}{\text{db}}$ |
|---|---|
| 50 | 2.8 |
| 70 | 2.28 |
| 100 | 1.85 |
| 150 | 1.54 |
| 200 | 1.36 |
| 300 | 1.18 |
| 500 | 1.04 |
| 1,000 | 1.0 |
| 2,000 | 1.03 |
| 4,000 | 1.07 |
| 10,000 | 1.04 |

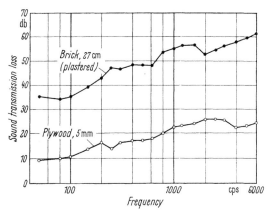

Fig. 6.144. Relationship between sound transmission loss and frequency of a plywood and a brick wall. From MEYER (1931)

age transmission loss figure $D$, measured in decibels. Only for the frequency of 1000 cps and at a distance greater than 1 m from the ear the loudness in phons is identical with the transmission loss figure in decibels. Table 6.29 allows the transition from phones to decibels.

The transmission loss figure $D$ can be calculated from the sound pressures $p_1$ and $p_2$ in the space before and behind the wall or construction element as follows:

$$D = 20 \cdot \log \frac{p_1}{p_2} - 10 \log \frac{A}{F}. \tag{6.135}$$

The second term in Eq. (6.135) takes into consideration the sound absorption. $A$ = total sound absorption area of the room to be isolated, $F$ = area of the wall, transmitting the sound.

The sound transmission loss through single walls is larger for larger pressure drops across the wall. This means that the walls should be heavy and rigid. Due to the law of inertia the wall can follow the variations of the sound pressure

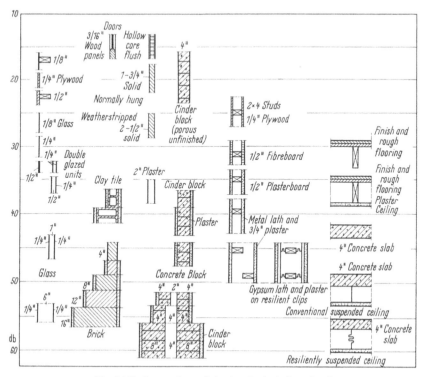

Fig. 6.145. Average transmission loss for typical constructions

less the quicker the impulses follow one another, i.e. the higher the frequency is. The sound isolation is therefore better for higher sounds than for lower ones. Relatively high noise isolation can be achieved by separation of construction elements (multi-skin walls).

The sound transmission loss $D$ for single walls is governed by their weight and by the frequency $f$ of the sound:

$$D \approx 20 \log (0.004\, W \cdot f) \text{ in db} \quad (6.136)$$

where $W$ = weight of the wall per unit area (kg/m²) and $f$ = frequency in cps. According to German standard specifications (DIN 4110) airtight single walls must have a minimum transition loss of 53 db in the frequency range between 100 and 3,000 cps. Examples for transmission losses are given in Figs. 6.143, 6.144 and 6.145.

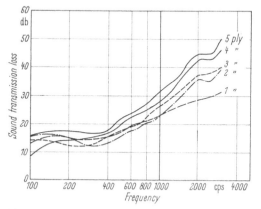

Fig. 6.146. Sound transmission loss for 1-, 2-, 3-, 4- and 5-ply Plywood walls (weight of each wall 2 kg/m²; distance 3 cm.); cutoff (limiting) frequency 480 cps. From MEYER (1931)

Walls consisting of two or more elements which are airtight and not in direct contact with each other provide efficient sound-isolation. The transition loss of such constructions can not be calculated on the basis of the transition loss of their components; it must be determined experimentally. However, the transmission of sound is governed essentially by the weight of the components. The transition loss of plywood walls consisting of 1, 2, 3, 4 and 5 layers is illustrated by Fig. 6.146.

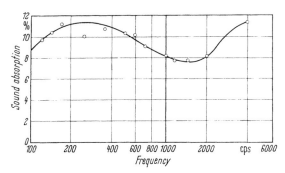

Fig. 6.147. Sound absorption for fir, 20 mm. From SABINE (1927)

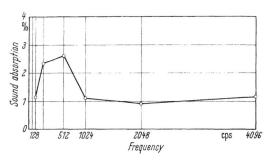

Fig. 6.148. Relationship between sound absorption and frequency for plywood 3 mm., plate size 1,21 m × 1,52 m, distance from wall 2,5 cm. From MEYER (1931)

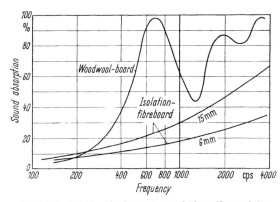

Fig. 6.149. Relationship between sound absorption and frequency for insulation fiber boards and woodwool-boards. From KOLLMANN (1951)

Careful attention must be paid to proper joints, inserted elements, connections, pipe lines, etc. Any small leaks or cracks can destroy the noise isolation.

Impact sounds from spaces overhead — e.g. from footfall noises, pianos or machinery — are transmitted directly through the floor and ceiling construction. Carpets or resilient flooring can help in overcoming this trouble.

**6.7.2.3 Sound Absorption.** The problem of sound absorption is quite different from that of sound isolation. Sound isolation demands heavy, airtight materials, whereas sound absorption requires soft, porous materials, such as carpets, heavy fabrics, hair felt, isolation fiberboard, woodwool board, and acoustical tiles. The molecules of air move in and out freely. The friction between the air molecules and the walls of the pores produces heat the amount of which is very small. The coefficient of sound absorption for an open window is 1.0 or 100%. The sound absorption depends on the frequency (cf. Table 6.30 and Figs. 6.147, 6.148 and 6.149).

Practically every material in a room absorbs some sound, but for an effective reduction of reverberation, as mentioned above, soft porous materials should be used. As Fig. 6.147 shows, solid pinewood has, on the average, an absorption of only 10%. This means that about 90% of the incident energy will be reflected back into the room. The conditions for plywood in the range of higher frequencies are quite similar (Fig. 6.148). Isolation fiberboard absorbs more sound at higher frequencies. The absorption also will be increased with increasing board thickness, but only up to 20 mm. Wood wool board, due to its great pore volume and the

Table 6.30. *Sound Absorption of Various Building Materials*

| Material | Thickness cm | Sound absorption in percent of incident energy | | | | | | Tested by |
|---|---|---|---|---|---|---|---|---|
| | | 128 | 256 | 512 | 1024 | 2048 | 4096 | |
| Stone wall, painted | 45 | 1.2 | 1.3 | 1.7 | 2.0 | 2.3 | 2.5 | W. C. Sabine |
| Masonite | 1.1 | 10 | 21 | 29 | 30 | 29 | — | V. O. Knudsen |
| Fiberboard | 3 · 1.5 = 4.5 | 39 | 51 | 52 | 53 | 59 | 53 | Heinr. Herz Institute |
| Wood Wool Board | 2.5 | 13 | 17 | 36 | 80 | 70 | 54 | Bureau of Standards |

porosity of its surface, has a high sound absorption capacity; the sensitivity to the frequency of the sound waves is remarkable. In rooms with a large auditorium (e.g. churches, theaters, cinemas, etc.) the acoustical conditions may be improved by the application of acoustical tiles. Isolation-fiberboard with perforations or slits made artificially are effective and economical in this respect.

## Literature Cited

Anonymous, (1954) Testing timber for moisture content. Division of For. Prod., Trade Circ. No. 50, CSIRO, Melbourne, Australia.
Aro, P., (1931) Über den Festgehalt der wichtigsten finnischen Schnittholzsortimente. Comm. ex Inst. quaest. Finl., 14th ed., Issue 15, Helsinki.
Argue, G. H., and Maass, O., (1935), Can. J. Res. **12**: 54.
Baechler, R. H., (1941) Proc. A. W. P. A. 37: 23.
—, (1953) J. Forest Prod. Res. Soc., **3** (5): 170.
—, (1958) How to treat fence posts by double diffusion. U. S. Dep. Agr. For. Prod. Lab. Rep. No 1955.
Barkas, W. W., (1938) Recent work on the moisture in wood in relation to strength and shrinkage. Dep. Sci. Ind. Res., For. Prod. Res., Spec. Rep. No 4, London, H. Maj. Stat. Off.
—, (1945) Swelling Stresses in Gels. For. Prod. Res. Spec. Rep. No 4, London, H. Maj. Stat. Off.
—, (1949) The swelling of wood under stress. London. H. Maj. Stat. Off.
Bateman, E., Hohf, J. P. and Stamm, A. J. (1939) Ind. Eng. Chem. **31**: 1150.
Bazhenov, V. A., (1953) Piezoelectric effect in wood (in Russian) Tr. Inst. Les. **9**.
—, (1959) Piezoelectric properties of timber (in Russian) M. Publ. A. N. SSR.
—, (1960) Piezoelectric effect of wood. Paper prepared for delivery 5th World Forestry Congress.
—, (1961) Piezoelectric properties of wood. Consultants Bureau, New York.
Bazhenov, V. A. and Konstantinova, V. P., (1950) Piezoelectric properties of wood. Dokl. Acad. Nauk SSR **71** (From Chem. Abstr. 45, 2747, 1951).
Bell, E. R. and Krueger, N. T., (1949) Effect of Plywood Glue Lines on the Accuracy of Moisture-Meter Indications, Proc. of the National Ann. Meet. For. Prod. Res. Soc. **3**: 85—93.
van Bemmelen, J. M., (1896) Z. Anorg. Allgem. Chem. **23**: 233.
Boltzmann, L., (1876) Pogg. Ann., Erg. **7**: 624.
Brake, E. and Schütze, H., (1935) Elektr. Nachr. Techn. **12**: 120.
Brillié, H., (1919) Génie Civ. **75**: 171, 194 and 218.
Brown, J. H., Davidson, R. W. and Skaar, C., (1963) For. Prod. J. **13**: 455.
Brown, H. P., Panshin, A. J. and Forsaith, C. C. (1952) Textbook of Wood Technology, Vol. II, 1st Ed., McGraw-Hill Book Company Inc., New York, Toronto, London.
Brunauer, S. J., Emmett, P. H. and Teller, E., (1938) J. Am. Chem. Soc. **60**: 309.
Burr, H. K. and Stamm, A. J., (1947) J. Phys. and Colloid. Chem. **51**: 240.

286 Literature Cited

CHOONG, C. T., (1963) Movement of moisture through a softwood in the hygroscopic range. For. Prod. J. **13**: 489–498.

CHRISTENSEN, G. N. and KELSEY, K. E., (1959) Die Sorption von Wasserdampf durch die chemischen Bestandteile des Holzes. Holz als Roh- und Werkstoff **17**: 189–203.

CLARK, J. D. and WILLIAMS, J. W., (1933) J. Phys. Chem. **37**: 117.

COCKWELL, A., (1946) Influence of fibril angle on longitudinal shrinkage of ponderosa pine wood. J. Forestry **44**: 876–878.

Commonwealth of Australia, C. Sci. Ind. Res., Div. For. Prod. (1934) The working of wood. Trade Circ. No 24, Melbourne.

DAVIDSON, G. F., (1927) J. Textile Inst. **18** T 175.

DE BRUYNE, N. A., (1938) Nature **142**: 570.

—, (1939) Aircraft Eng. **2**: 44.

DE LUCA, H. A., CAMPBELL, W. B. and MAASS, O., (1938) Can. J. Res. **16**: Part B, 273.

Dept. Sci. Ind. Res., For. Prod. Res. (1957), 46, London 1958, H. Maj. Stat. Off.

DUNLAP, F., (1912) The specific heat of wood. U. S. Dep. Agr. Bull. No 110, Washington, D. C.

—, (1913) Heat of absorption of water by wood. U. S. For. Prod. Lab. File Rep. Proj. No 169.

DUNLAP, M. E., (1933) A new moisture indicator for wood. Instruments.

—, and BELL, E. R., (1951) Electrical moisture meters for wood. U. S. Dep. Agr., For. Prod. Lab. Rep. No R 1660, Madison, Wisc.

EBERIUS, E., (1952) Wasserbestimmung nach Karl Fischer in der Sprengstoffchemie. Angew. Chem. **64**: 195.

—, (1958) Wasserbestimmung mit Karl Fischer-Lösung, 2nd. Ed., Weinheim/Bergstr., Verlag Chemie G. m. b. H.

EGNER, K., (1934) Beiträge zur Kenntnis der Feuchtigkeitsbewegung in Hölzern, vor allem in Fichtenholz, während der Trocknung unterhalb des Fasersättigungspunktes. Forsch.-Ber. Holz H. 2, Berlin.

ERICKSON, H. D., SCHMITZ, H. and GORTNER, R. A., (1938) J. Agr. Res. **56**: 711.

FELLOWS, E. S., (1937) The change in moisture content of yard-piled softwood lumber in Eastern Canada. Dep. Mines and Res. Can., Dom. For. Circ. No 52, Ottowa.

FISCHER, K., (1935) Neues Verfahren zur maßanalytischen Bestimmung des Wassergehalts von Flüssigkeiten und festen Körpern. Angew. Chem. **48**: 394.

FLATSCHER, J. H., (1929) Handbuch des Sägebetriebs, Berlin.

FLETCHER, H., and MUNSON, W. A., (1933) J. Acoust. Soc. Am. **5**: 82.

FREUNDLICH, H., (1926) Colloid and capillary chemistry. Translat. by H. S. Hatfield from the 3rd German Ed., New York., N. Y., Dutton.

FREY-WYSSLING, A., (1940) Die Anisotropie des Schwindmaßes auf dem Holzquerschnitt. Holz als Roh- und Werkstoff. **3**: 43–45.

FUKADA, E., (1955) Piezoelectricity of wood. J. Phys. Soc. of Japan, **10**: 149–154.

—, (1956) J. Phys. Soc. Japan **11**: 1301.

—, (1965) Piezoelectric effect in wood and other crystalline polymers Proc. of 2nd Symposium on Nondestructive Testing, Washington State University, Spokane, April 1965.

—, and coworkers, (1957) The dynamics Youngs modulus and the piezoelectric constant of old timbers. J. Appl. Phys. Japan **26**: 25.

—, and YASUDA, I., (1957) J. Phys. Soc. Japan **12**: 1158.

—, and YASUDA, I., (1964) J. Appl. Phys. Japan **3**: 117.

GALLIGAN, W. L., and JAYNE, B. A., (1961) Development of a nondestructive test for the adhesive bond in wood scarf joints. Final Report for the Department of the Navy, Bureau of Ships, Contract NO bs-78 267.

—, and BERTHOLF, L. D., (1963) Piezoelectric effect in wood. For. Prod. J. **13**: 517–524.

GLATZEL, P., (1877) Poggendorfs Annalen der Physik **160**: 497–514.

GREENHILL, W. L., (1940) The shrinkage of Australian timbers — Part III, Australian. C. Sc. Ind. Res., Div. of For. Prod., Techn. Paper 35, Pamphlet 97, Melbourne.

GRIFFITHS, E. and KAYE, G. W. C., (1923) Proc. Roy. Soc. Lond. Ser. A **104**: 71.

HALE, J. D. and PRINCE, J. B., (1940) Density and rate of growth in the spruces and balsam fir of Eastern Canada. Can. Dept. Mines and Resources, For. Serv. Bull. 94.

HANSEN, CH., (1934) Umschau, **38**: 843.

HART, C. A., (1964) Principles of moisture movement in wood. For. Prod. J. **14**: 207–214.

HARTIG, R., (1894) Forstl.-Naturw. Z. **3**: 1, 49, 172, and 193.

HASSELBLATT, M., (1926) Z. Anorg. Allg. Chem. **154**: 375.

HAWLEY, L. F., (1931) Wood-liquid relations. U. S. Dep. Agr., Tech. Bull. No 248, Washington, D. C.

HEARMON, R. F. S., (1944) The high frequency electrical properties of wood and wood resin combinations. Rep. of For. Prod. Res. Lab.

—, (1948) For. Prod. Res. Spec. Rep. No 7, H. Maj. Stat. Off., London.

HENDERSHOT, O. P., (1924) Thermal expansion of wood. Science **60**: 456–467.

# Literature Cited 287

HENDERSON, H. L., (1951) The air seasoning and kiln drying of wood. 5th Ed., Albany, New York.

HERMANS, P. H., (1949) Physics and chemistry of cellulose fibers. New York-Amsterdam-London-Brussels, Elsevier Publishing Co., Inc., 180—196.

HESEHUS, N., (1908) J. Tech. Phys. USSR **40**: 112.

HIRUMA, J., (1915) Extracts Bull. Forestry, Exp. Stat., Merguro, Tokyo, **59**.

HODGMAN, C. D., (1948) Handbook of Chemistry and Physics, rev. 30th Ed., Chemical Rubber Publishing Co., Cleveland, Ohio, 1788.

HONDA, K. and KONNO, S., (1921) Phil. Mag. **42**: 115.

HÖRIG, H., (1929) Schalltechn., p. 70, also DRP 534516.

HOWARD, H. C. and HULETT, G. A., (1924) J. Phys. Chem. **28**: 1082.

JALAVA, M., (1929) Comm. ex Inst. quaest. Forest. Finl., Ed. 13, Issue 8, Helsinki.

—, (1934) Investigation into the influence of the position of a tree in the stand upon the properties of wood. Acta Forestalia Fennica 40. Helsinki (Finnish with English summary).

JAMES, W. L., (1961) Internal friction and speed of sound in Douglas fir. For. Prod. J. **11**: 383—390.

JAYME, G. and KRAUSE, TH., (1963) On the packing-density of the cell walls in deciduous woods. Holz als Roh- und Werkstoff **21**: 14—19.

JENKINS, J. H., and GUERNSEY, F. W., (1937) Ocean shipment of seasoned lumber, Dep. Mines Res., Dom For. Sen. Circ. No 49, Ottawa.

JOHANSSON, D., (1940) Über Früh- und Spätholz in schwedischer Fichte und Kiefer und über ihren Einfluß auf die Eigenschaften von Sulfit- und Sulfatzellstoff. Holz als Roh- und Werkstoff **3**: 73/78.

JOHNSTON, H. W., and MAASS, O., (1929) Penetration studies. For. Prod. Lab. Can. Res. Notes Vol. 2.

JÜRGES, W., (1924) Der Wärmeübergang an einer ebenen Wand. Beih. zu Gesundh.-Ing., Reihe 1, Nr. 19, Oldenburg.

KATZ, J. R., (1917) Kolloidchem. Beih. **9**: 64.

—, (1933) Faraday Soc. Trans. **29**: 279.

—, and MILLER, D. G., (1963) Effect of water storage on electrical resistance of wood. For. Prod. J. **13**: 255—259.

KELSEY, K. E. and CLARKE, L. N., (1956) The heat of sorption of water by wood. Australian J. Appl. Sci. **7**: 160—175.

KEYLWERTH, R., (1943) Das Schwinden und seine Beziehungen zu Rohwichte und Aufbau des Holzes. Diss. T. H. Berlin.

—, (1948) Beitrag zur Mechanik der Holzschwindung. Reinbek.

—, (1949) Holz-Zentr. **75**: No 75.

—, (1962) Untersuchungen über freie und behinderte Quellung von Holz — Erste Mitteilung: Freie Quellung. Holz als Roh- und Werkstoff **20**: 252—259.

—, and NOACK, D., (1956) Über den Einfluß höherer Temperaturen auf die elektrische Holzfeuchtigkeitsmessung nach dem Widerstandsprinzip. Holz als Roh- und Werkstoff **14**: 162—172.

—, and —, (1956) Über den Einfluß höherer Temperaturen auf die elektrische Holzfeuchtigkeitsmessung nach dem Widerstandsprinzip. Holz als Roh- und Werkstoff, **14**: 162.

KLEM, G. G., (1934) Untersuchungen über die Qualität des Fichtenholzes. Meddelelser fra det Norske Skogsforsoksvesen **5** (2 No 17): 197, 333 (Norwegian with German summary).

KNIGHT, R. A. G. and NEWALL, R. J., (1938) Forestry (Great Britain) **12**: 125.

—, and PRATT, G. H., (1935) The humidity-moisture content relations of wood. Eng. May, 3.

KOEHLER, A., (1931) The longitudinal shrinkage of wood. Amer. Soc. Mech. Eng. Trans., Wood Ind. Div. **53**, No 5: 17—20.

—, (1946) Longitudinal shrinkage of wood. U. S. Dep. Agr. For. Serv., For. Prod. Lab., Rep. No R 1093, Madison, Wisc.

KOHARA, I., (1958) Research Reports of Faculty of Eng., Chiba University, IX, No 15/16.

KOLLMANN, F., (1932) Die Abhängigkeit des Raumgewichts der Hölzer vom Feuchtigkeitsgehalt. Zentr. ges. Forstwes. **58**: 276.

—, (1933) Neue Erkenntnisse über einige physikalische und technologische Eigenschaften der Hölzer. Forstwiss. Zentr. **55**: 149.

—, (1934) Holzgewicht und Feuchtigkeit. Z. VDI **78**: 1399.

—, (1936) Wege und Ergebnisse der mechanisch-technologischen Holzforschung. Forstarch. **12**: 1.

—, (1933) Die Abhängigkeit der Trocknungsgeschwindigkeit des Holzes vom Raumgewicht und der Struktur. Forsch.-Ber. Holz H. 1, Berlin.

—, (1934) Über die wärmetechnischen Eigenschaften der Hölzer. Gesundheits-Ingenieur **57**: 224.

288 Literature Cited

KOLLMANN, F., (1935) Der Wärmeverbrauch bei der künstlichen Holztrocknung. Arch. Wärmewirtsch. **16**: 329.

—, (1935) Rechnerische Verfolgung der künstlichen Holztrocknung. Forsch. Gebiete Arb. Ingenieurw. **6**: 169.

—, (1936) Ein neues Verfahren zur Berechnung der Trockenzeit am Beispiel der Holztrocknung. Forsch. Gebiete Ingenieurw. **7**: 113.

—, (1941) Die Esche und ihr Holz. Springer-Verlag, Berlin.

—, (1951) Technologie des Holzes und der Holzwerkstoffe 2nd Ed., Vol. 1, Springer-Verlag, Berlin-Göttingen-Heidelberg.

—, (1959) Über die Sorption von Holz und ihre exakte Bestimmung. Holz als Roh- und Werkstoff **17**: 165—171.

—, (1963) Zur Theorie der Sorption. Forsch. Gebiete Ingenieurw. **29**: 33—41.

—, and HÖCKELE, G., (1962) Kritischer Vergleich einiger Bestimmungsverfahren der Holzfeuchtigkeit. Holz als Roh- und Werkstoff **20**: 461—493.

—, and KRECH, H., (1960) Dynamische Messung der elastischen Holzeigenschaften und der Dämpfung. Holz als Roh- und Werkstoff **18**: 41—51.

—, and MALMQUIST, L., (1955) Untersuchungen über das Strahlungsverhalten trocknender Hölzer. Holz als Roh- und Werkstoff **13**: 249.

—, and —, (1956) Über die Wärmeleitzahl von Holz und Holzwerkstoffen. Holz als Roh- und Werkstoff **14**: 201—204.

—, and SCHNEIDER, A., (1960) Einfluß der Belüftungsgeschwindigkeit auf die Trocknung von Schnittholz mit Heißluft-Dampf-Gemischen. Holz als Roh- und Werkstoff **18**: 81—94.

—, and SCHNEIDER, A., (1963) Über das Sorptionsverhalten wärmebehandelter Hölzer. Holz als Roh- und Werkstoff **21**: 77—85.

KRISCHER, O., (1938) Z. VDI **82**: 373.

—, (1942) Der Wärme- und Stoffaustausch im Trocknungsgut. VDI-Forschungsheft 415. Berlin.

KRÖLL, K., (1951) Die Bewegung der Feuchtigkeit in Nadelholz während der Trocknung bei Temperaturen um 100°. Holz als Roh- und Werkstoff **9**: 176—181.

KRÖNER, K., (1944) Über dielektrische Untersuchungen an Naturhölzern und deren mechanischen und chemischen Abbaustoffen im großen Frequenzgebiet. Diss. T. H. Braunschweig.

KÜBLER, H., (1962) Schwinden und Quellen des Holzes durch Kälte. Holz als Roh- und Werkstoff **20**: 364—368.

KURTH, E. F. and SHERRARD, E. C., (1931) Ind. Eng. Chem. **23**: 1156.

—, and —, (1932) Ind. Eng. Chem. **24**: 1179.

LANDOLT-BÖRNSTEIN, (1955) Zahlenwerte aus Physik, Chemie, Abnormier, Geophysik, Technik. 6. Aufl. Berlin/Göttingen/Heidelberg: Springer.

LANGMUIR, I., (1918) J. Am. Chem. Soc. **40**: 1361.

LEVON, M., (1931) Comm. ex Inst. quaest. Forest. Finl., Ed. 16, Helsinki.

LEWIS, W. K., (1922) Mech. Eng. **44**: 445.

LIESE, J. and SCHUBERT, R., (1941) Beiträge zum Osmose-Holzschutz-Verfahren. Holz als Roh- und Werkstoff **4**: 93—101.

LIN, R. T., (1965) A study on the electrical conduction in wood. For. Prod. J. **15**: 506—514.

LÖFGREN, B., (1947) Moderne Sågverkstorkar. Fläkten, No 1.

LOOS, W. E., (1961) The relationship between Gamma ray absorption and wood moisture content and density. For. Prod. J. **11**: 145—149.

—, (1965) Determining moisture content and density of wood by nuclear radiation techniques. For. Prod. J. **15**: 102—106.

LÜBCKE, E. ,(1927) in: Handbuch der Physik **8**: 247, Berlin.

LUDWIG, W., (1947) Holz-Zentr. **73**, No 35.

McINTOSH, D. C., (1954) Some aspects of the influence of rays on the shrinkage of wood. J. For. Prod. Res. Soc. IV: 39—42.

—, (1955) The effect of the rays on the radial shrinkage of beech. For. Prod. J. **5**: 67—71.

—, (1955) Further information on the shrinkage of red oak and beech. For. Prod. J. **5**: 355 to 359.

—, (1957) Transverse shrinkage of red oak and beech. For. Prod. J., VII: 114—120.

MacLEAN, J. D., (1924) Relation of temperature and pressure to the absorption and penetration of zinc chloride solution into wood. A. W. P. A. Proc. **20**: 44—73.

—, (1926) Effect of temperature and viscosity of wood preservative oils on penetration and absorption. Proc. A. W. P. A. **20**: 147—167.

MacLEAN, J. D., (1927) Relation of treating variables to the penetration and absorption of preservatives into wood. Proc. A. W. P. A. **23**: 52—70.

—, (1930) Studies of heat conduction in wood. Results of steaming green round southern pine timbers. Proc. A. W. P. A. p. 197.

## Literature Cited

MacLean, J. D., (1932) Studies of heat conduction in wood. Results of steaming green round southern pine timbers. Proc. A. W. P. A. p. 303.

—, (1941) Heating, piping and air conditioning. **13**: 380.

—, (1942) Effect of temperature on the dimensions of green wood. Proc. A. W. P. A.

—, (1952) Effect of temperature on the dimensions of green wood. Proc. A. W. P. A. **48**: 136—157.

Maku, T., (1954) Studies on the heat conduction in wood. Wood Res. Bull. No 13, Wood Res. Inst. Kyoto University, Kyoto, Japan.

Malmquist, L., (1958) Sorption as deformation of space. Kylteknisk Tidskr. Issue 4, p. 1—11.

—, (1959) Die Sorption von Wasserdampf vom Standpunkt einer neuen Sorptionstheorie. Holz als Roh- und Werkstoff **17**: 171—178.

March, H. W., (1944) Stress-strain relations in wood and plywood considered as orthotropic materials. U. S. For. Prod. Lab. Rep. No. 1503, Madsion, Wisc.

Markwardt, L. J. and Wilson, T. R. C., (1935) Strenght and related properties of woods grown in the United States. U. S. Dep. Agr. Tech. Bull. No. 479, Washington, D. C.

Martley, J. F. (1926) Moisture movement through wood, the steady state. For. Prod. Res. Tech. Paper 2, London.

Mathewson, J. S., (1930) The air seasoning of wood. U. S. Dep. Agr., Tech. Bull. No. 174, Washington D. C.

Mayr, H., (1894) Das Harz der Nadelhölzer. Berlin, p. 5.

Menzel, C. A., (1935) in: U. S. Dep. Agr. Wood Handbook, table 3, Washington D. C., p. 43.

Meyer, E. (1931) Grundlegende Messungen zur Schallisolation von Einfach-Wänden. Sitz.-Ber. Preuss. Akad. Wiss.

—, and Rees, L. W., (1926) N.Y. S. Coll. For. Syracuse Univ. Tech. Bull. No. 19.

Mörath, E., (1931) Kolloidchem. Beihefte **33**: 131.

—, (1932) Studien über die hygroskopischen Eigenschaften und die Härte der Hölzer, Darmstadt.

—, (1933) Holzind. **13**: 73.

Moll, F., (1930) Künstliche Holztrocknung. Berlin.

Mombächer, H., (1962) Holzgewichte der Praxis. Holz-Zentr. **88**: 1579.

Müller-Stoll, W. R., (1948) Photometrische Holzstrukturuntersuchungen. Part II: Über die Beziehungen der Lichtdurchlässigkeit von Holzschnitten zu Rohwichte und Wichtekontrast. Forstwiss. Centr. **68**: 21—63.

Narayanamurti, D., (1936) Current Sci. **5**: 79.

—, (1936) Versuche über die Feuchtigkeitsbewegung in Holz und anderen Körpern beim Trocknen und über das Wärmeleitvermögen feuchten Holzes. Beitr. Verfahrenstechnik 2, Z. VDI.

—, and Ranganathan, V., (1941) Proc. Ind. Acad. Sci. **13**, No. 4, Sec. A.

Newlin, J. A. (1919) The relation of the shrinkage and strength properties of wood to its specific gravity. U. S. Dep. Agr. Bull. No. 676, Washington, D. C.

Niethammer, H. (1931) Paper-Fabrikant **29**: 557.

Nilikantan, P., (1938) Proc. Indian Acad. Sci., Sect. A 7, No. 1, p. 38.

Nusser, E., (1938) Die Bestimmung der Holzfeuchtigkeit durch Messung des elektrischen Widerstandes. Forschungsber. Holz, H. 5, Berlin.

Oberg, J. C. and Hossfeld, R., (1960) Hydrogen bonding and swelling of wood. For. Prod. J. **10**: 369—372.

Pascheles, W., (1897) Pflügers Archiv **67**: 219.

Paul, B. H., (1930) The application of silviculture in controlling the specific gravity of wood. U. S. Dep. Agr. Tech. Bull. 168, Washington, D. C.

—, (1939) Variation in the specific gravity of the springwood and summerwood of four species of southern pines. J. For. **37**: 478—482.

Paul, B. H., (1946) The flotation method of determining the specific gravity of wood. U. S. For. Prod. Lab., Madison/Wisc., Rep. No. 1398.

—, and Marts, R. O., (1934) J. Forestry (San Francisco) **32**: 861.

Peck, E. C., (1928) Am. Lumberman, July 14, 52.

—, (1932) Moisture content of wood in dwellings. U. S. Dep. Agr. Circ. No 239, Washington D. C.

Pentoney, R. E., (1953) Mechanisms affecting tangential vs. radial shrinkage. For. Prod. J. **3**: 27—32, 86.

Perkitny, T., (1937/38) Holz als Roh- und Werkstoff **1**: 449.

—, (1960) Druckschwankungen in eingeklammerten Holzkörpern. Holz als Roh- und Werkstoff **18**: 200—210.

—, Stefaniak, J. and Chudzinski, Z., (1947) Ber. d. Poln. Inst. f. Holztechnologie. **3** (4): 34—51.

Pillow, M. Y., and Luxford, R. F. (1937) Structure, occurence, properties of compression wood. U. S. Dep. Agr., Tech. Bull., No. 546, Washington, D. C.

PRÜTZ, G., (1941) Kolonialforstl. Mitt. **4**: 298.

REHMANN, M. A., (1942) A survey of the seasonal variation of the moisture content of Indian woods. Ind. For. Res. Utilization (New series), **2**, No. 10, Delhi.

REISCHEL, and Wedekind, E., (1932) Silva **20**: 377.

—, and —, (1933) Naturwiss. **21**: 24.

RISI, J. and ARSENEAU, D. F., (1957) Dimensional stabilization of wood. For. Prod. J. **7**: 210—213; 245—246; 261—265; 293—295.

RITTER, G. J. and FLECK, L. C., (1922) Ind. Eng. Chem. **14**: 1950.

ROCHESTER, G. H., (1933) The mechanical properties of Canadian woods, together with their related physical properties, Dep. Int. For. Serv. Bull. 82, Ottawa.

ROSENBOHM, E., (1914) Kolloidchem. Beih. **6**: 177.

RUNKEL, R. O. H. and LÜTHGENS, M., (1956) Untersuchungen über die Heterogenität der Wassersorption der chemischen und morphologischen Komponenten verholzter Zellwände. Holz als Roh- und Werkstoff **11**: 424—441.

SAVKOV, E. S., (1930) Die Erforschung der physikalisch-mechanischen Eigenschaften des Kiefernholzes für den Flugzeugbau. Arbeiten des Aerohydrodynamischen Instituts Moscow, Issue 62 (Russian with German Summary).

SABINE, W. C., (1927) Collected Papers on Acoustics. p. 93, Cambridge.

SCHLÜTER, R. and FESSEL, F., (1939) Holz als Roh- und Werkstoff **2**: 169.

—, and —, (1940) Holz als Roh- und Werkstoff **3**: 109.

SCHNIEWIND, A. P., (1959) Transverse anisotropy of wood. For. Prod. J. **9**: 350.

SCHMIDT, E., (1927) Gesundh. Ing. Beih. 20.

SCHMIDT, E., (1940) Holz als Roh- und Werkstoff **3**: 228.

SCHWALBE, C. G. and BECKER, E., (1919) Z. Angew. Chem. **32**: 229.

SCHWAPPACH, A., (1898) Untersuchungen über Raumgewicht und Druckfestigkeit des Holzes wichtiger Waldbäume. II. Fichte, Weißtanne, Weymouths-Kiefer und Rotbuche, Berlin.

SEBORG, C., (1937) Ind. Eng. Chem. **29**, 169.

SHARPE, B. A. and O'KANE, B. J., (1935) Eng. **140**: 403.

SHEPPARD, S. E. and NEWSOME, P. T., (1932) J. Phys. Chem. **36**: 3306.

SHERWOOD, T. K., (1929) Ind. Eng. Chem. **21**: 12.

—, and E. W. COMINGS, (1935) Vsessayanznovo Teplotechn. Inst. No. 8.

SHUBNIKOV, A. V., (1940) On the tensor piezoelektric moduli of noncrystalline anisotropic media. (in Russian). Report in the Division of Physical-Mathematical Sciences of the Academie of Sciences of the USSR (from Bazhenov, V. A., Piezoelectric Properties of Wood, Consultant Bureau, New York, 1961).

SIIMES, F. E., (1938) On the structural and physical properties of Finnish plywood, especially the phenomenon of shrinking and swelling affected by changing the moisture content of wood, Helsinki.

SKAAR, C., (1948) N.Y. S. Coll. For. Syracuse Univ. Tech. Pub. No. 69.

SONNLEITHNER, E., (1933) Verlauf der Feuchtigkeit innerhalb des Holzes während der Trocknung. Forsch.-Ber. Holz, H. 1, VDI-Verlag, Berlin.

STAMM, A. J., (1927) Ind. Eng. Chem., Anal. Ed. **19**: 1021—1025.

—, (1929) Ind. Eng. Chem., Anal. Ed. **1**: 94—97.

—, (1930) Ind. Eng. Chem., Anal. Ed. **2**: 240—244.

—, (1932) J. Phys. Chem. **236**: 312.

—, (1934) J. Am. Chem. Soc. **56**: 1195.

—, (1935) Ind. Eng. Chem. **27**: 401.

—, (1938) Calculations of the void volume in wood. Ind. Chem. **30**: 1281.

—, (1946) Passage of liquids, vapors and dissolved materials through softwoods, U. S. Dep. Agr. Tech. Bull. 929, Washington, D. C.

—, (1959) Verfahren zur Abschätzung der Wasserdampfsorption am Fasersättigungspunkt von Holz und Papier. Holz als Roh- und Werkstoff **17**: 203—209.

—, (1959) Effect of polyethylene glycol on the dimensional stability of wood. For. Prod. J. **9**: 375—381.

—, (1960) Bound-water diffusion into wood in across-the-fiber directions. For. Prod. J. **10**: 524—528.

—, (1960) Combined bound-water and water-vapor diffusion into Sitka spruce. For. Prod. J. **10**: 644—648.

—, (1962) Wood and cellulose-liquid relationships. North Carolina Agricultural Experiment Station. Tech. Bull. No. 150.

—, (1963) Permeability of wood to fluids. For. Prod. J. **13**: 503—507.

—, (1964) Wood and cellulose science. The Ronald Press Company, New York.

—, (1964) Factors affecting the bulking and dimensional stabilization of wood with PEG. For. Prod. J. **14**: 403—408.

# Literature Cited

STAMM, A. J., and HANSEN, L. A., (1937) The bonding force of cellulosic materials for water (from specific volume and thermal date) J. Phys. Chem. **41**: 1007—1016.
—, and HARRIS, E. E., (1953) Chemical processing of wood. New York, N.Y., Chem. Publ. Co. Inc. p. 113—138.
—, and LOUGHBOROUGH, W. K., (1935) J. Physic. Chem. **39**: 121—132.
—, and LOUGHBOROUGH, (1942) Variation in shrinking and swelling of wood. Trans. Am. Soc. Mech. Eng. **64**: 379—386.
—, and NELSON, R. M., JR., (1961) Comparison between measured and theoretical drying diffusion coefficients for southern pine. For. Prod. J. **11**: 536—543.
—, and SEBORG, R. M., (1955) U. S. Dep. Agr. Forest. Prod. Lab. Mimeo No 1380.
—, and WOODRUFF, S. A., (1941) Ind. Eng. Chem. Anal. Ed. **13**: 336.
STEVENS, W. C., (1938) Forestry (Great Britain) **12**: 38.
—, (1963) The transverse shrinkage of wood. For. Prod. J. **13**: 386—389.
STILLWELL, S. T. C., (1926) The movement of moisture with reference to timber seasoning. For. Prod. Res. Tech. Paper No 1, London.
STOOPS, W. N., (1934) Proc. J. Am. Chem. Soc. **56**: 1481.
STRUVE, W., (1850) Fortschritte der Physik **6**: 48—52.
SUITS, C. G. and DUNLAP, E., (1931) Gen. El. Rev. **1931**: 706.
THOMSON, W., (1871) Phil. Mag. J. Sci. **42**: 448.
THUNELL, B. and LUNDQUIST, H., (1945) Trätorkning I, Svenska Träforskningsinstitutet, Trätekniska Avdel., Medd. 4, Stockholm.
TOLMANN, R. C. and STEARN, A. E., (1918) J. Am. Chem. Soc. **40**: 264.
TORGESON, O. W., (1941) Timberman **42**: 42—44.
TRENDELENBURG, R., (1934) Untersuchungen über das Raumgewicht der Nadelhölzer. I. Grundlagen und vergleichende Auswertung bisheriger Forschungen. Tharandter Forstliches Jahrbuch **85**, 649—747.
—, (1935) Schwankungen des Raumgewichts wichtiger Nadelhälzer nach Wuchsgebiet, Standort und Einzelstamm. Z. VDI **79**: 85—89.
—, (1939) Das Holz als Rohstoff. 1st ed, München, J. F. Lehmann Verlag.
—, (1941) Verladegewicht von Rohholz. Dtsch. Holzanzeiger Nos 100 and 101.
—, and MAYER-WEGELIN, H., (1953) Das Holz als Rohstoff. 2nd ed. München, C. Hanser-Verlag.
TUOMOLA, T., (1943) Über die Holztrocknung. Diss. T. H., Helsinki.
TUTTLE, F., (1925) J. Franklin Inst. **200**: 609.
URQUHART, A. R., (1929) J. Textile Inst. **20** (1929) T 125.
—, and WILLIAMS, A. M. (1924) J. Textile Inst. **15**: 559—572; (1929) **20**: 125—132.
VILLARI, F. E., (1868) Ann. Chim. Physique **14**: 503.
—, (1868) Poggendorfs Annalen der Physik **133**: 400.
VINTILA, E., (1939) Untersuchungen über Raumgewicht und Schwindmaß von Früh- und Spätholz bei Nadelhölzern. Holz als Roh- und Werkstoff **2**: 345—357.
VOIGT, H., KRISCHER, O. and SCHAUSS, H., (1940) Holz als Roh- und Werkstoff **3**: 305—321.
VOIGT, W., (1892) Ann. d. Physik **283**: 671.
VOLBEHR, B., (1896) Untersuchungen über die Quellung der Holzfaser. Inaugural Dissertation, University of Kiel (Germany).
VOLKERT, E., (1941) Untersuchungen über Größe und Verteilung des Raumgewichts in Nadelholzstämmen. Mitt. Akad. dtsch. Forstwiss., Frankfurt, **2**: 1—133.
VORREITER, L., (1943) Handbuch für Holzabfallwirtschaft, 2nd Ed., Neudamm.
WAHLBERG, H. E., (1922) Svensk. Papp. Tidn. **35**: 25.
WANDT, R., (1937) Mitt. Fortwiss. u. Forstwirtsch. **8**: 343.
WANGAARD, F. F., (1940) Heat. Pip. Air Condit. **12**: 459.
—, (1950) The mechanical properties of wood. New York, London. p. 160.
—, and ZUMWALT, E. V., (1949) Some strength properties of second-growth Douglas fir. J. For. Washington D. C., **47**: 18—124.
WEATHERWAX, R. C. and STAMM, A. J., (1945) Elec. Eng. **64**: 833.
—, and —, (1946) The coefficients of thermal expansion of wood and wood products. U. S. For. Prod. Lab. Rep. No 1487 Madison, Wisc.
WEICHRET, L., (1963) Untersuchungen über das Sorptions- und Quellungsverhalten von Eiche, Buche und Buchen-Preßvollholz bei Temperaturen zwischen 20°C und 100°C. Holz als Roh- und Werkstoff **21**: 290—300.
WEINBERG, B., (1892) Z. physik. Chem. **10**: 34.
WIESNER, J. v. (1927) Die Rohstoffe des Pflanzenreiches, 4th ed. Leipzig **1**: 1051.
Wood Handbook (1955) U. S. Dep. Agr., For. Prod. Lab., Madison, Wisc., p. 329.
YLINEN, A., (1942) Acta Forestalia Fennica **50**.
ZSIGMONDY, R., (1911) Z. Anorg. Allg. Chem. **71**: 356.

# 7. MECHANICS AND RHEOLOGY OF WOOD

## 7.1 Elasticity, Plasticity, and Creep

### 7.1.1 Hooke's Law, Modulus of Elasticity

Elasticity means that deformations produced in a solid body by low stresses are completely recovered after unloading. The elastic properties are characteristic for solid bodies below a certain limit of stress; above this limit plastic deformations or failure will occur. Apparently the elastic limit is a rather arbitrary conception. In wood the influence of hygroscopic moisture is also important. Small elastic deformations imposed for a period of time may turn into plastic deformations. If deformations increase, structural wooden members may fail rapidly since there is no yielding of stresses as in steel. The relations between elastic deformations or strains and stresses within certain limits of stresses are therefore of great importance.

According to BACH and BAUMANN (1923) the following exponential law is valid for all building materials with the exception of marble and rubber:

$$\varepsilon = \alpha \sigma^n \qquad (7.1)$$

where

$\varepsilon = \dfrac{\text{elongation}}{\text{original length}} = \dfrac{\Delta l}{l}$ = relative elongation (strain),

$\sigma$ = stress in kp/cm², and

$\alpha$ and $n$ are material constants.

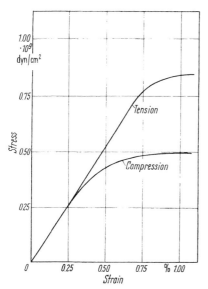

Fig. 7.1. Stress-strain curves for wood. From MEREDITH (1953)

Experiments proved that $n > 1$ for cast iron, copper, cast zinc, granite, sandstone, and concrete; and $n < 1$ for leather and hemp ropes. For other materials, e.g. steel, aluminium, and wood, $n = 1$ with high accuracy. In this case Hooke's law states that the strain $\varepsilon$ is proportional to the stress $\sigma$:

$$\varepsilon = \alpha \sigma. \qquad (7.2)$$

where $\alpha = \varepsilon/\sigma$ is the compliance, i.e., the strain per unit stress. In the technical literature normally the reciprocal value $1/\alpha = E$ is used. $E$ is called the modulus of elasticity or Young's modulus. It expresses that hypothetical stress [kp/cm²] by which a rod would be extended to double its initial length. The weakness of this definition is evident since the tensile strength for most materials is much lower than the modulus of elasticity. The moduli of elasticity in tension, compression and bending of wood are approximately equal, but the elastic limit is considerably lower for compression than for tension (Fig. 7.1). The calculation of the average, $E_m$, of two moduli of elasticity ($E_1$ and $E_2$) must be based on the compliances ($\alpha_1$ and $\alpha_2$)

$$\alpha_m = \frac{\alpha_1 + \alpha_2}{2} \tag{7.3}$$

$$E_m = \frac{1}{\alpha_m} = \frac{2}{\alpha_1 + \alpha_2} = \frac{2 E_1 E_2}{E_1 + E_2}. \tag{7.4}$$

### 7.1.2 Rhombic Symmetry of Wood. Systems of Elastic Constants

As has been shown in Chapter 1, wood is an anisotropic material but the trunk of a tree consists more or less of concentric cylindrical shells thus imparting a cylindrical symmetry to the wood. This symmetry is reflected in most physical properties of the wood as in the elastic properties, the strength values, the thermal and electrical conductivity values.

If a small rectangular sample is cut out from a trunk, at some distance from its pith, with a pair of faces tangential to the annual rings, as in Fig. 7.2, this sample has three axes of symmetry. These are parallel to the longitudinal ($y$, $L$, or, $l$)[1], radial ($z$, $R$, or $r$), and tangential ($x$, $T$, or $t$). directions, and are approximately perpendicular to one another. In reality the radial axes in wood

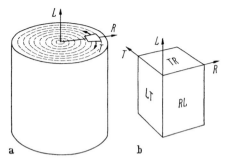

Fig. 7.2 a, b. Principal axes and planes in wood. From MEREDITH (1953)

are not parallel but diverging and $yx$- or $LT$-surface is not plane, but almost cylindrical. Nevertheless it is customary, since the investigations of SAVART (1829), v. HELMHOLTZ (1902) and CARRINGTON (1921, 1922, 1923), to assume three mutually perpendicular axes of elastic symmetry thus allowing the treatment of the system mathematically as a rhombic one.

A material with rhombic symmetry possesses 9 independent elastic constants and Hooke's law for this case states:

$$\left.\begin{aligned}\varepsilon_x &= s_{11}\sigma_x + s_{12}\sigma_y + s_{13}\sigma_z,\\ \varepsilon_y &= s_{21}\sigma_x + s_{22}\sigma_y + s_{23}\sigma_z,\\ \varepsilon_z &= s_{31}\sigma_x + s_{32}\sigma_y + s_{33}\sigma_z,\\ \gamma_{yz} &= s_{44}\tau_{yz},\\ \gamma_{zx} &= s_{55}\tau_{zx},\\ \gamma_{xy} &= s_{66}\tau_{xy}\end{aligned}\right\} (7.5)$$

where $s_{ik} = s_{ki}$;

$$\left.\begin{aligned}\sigma_x &= c_{11}\varepsilon_x + c_{12}\varepsilon_y + c_{13}\varepsilon_z,\\ \sigma_y &= c_{21}\varepsilon_x + c_{22}\varepsilon_y + c_{23}\varepsilon_z,\\ \sigma_z &= c_{31}\varepsilon_x + c_{32}\varepsilon_y + c_{33}\varepsilon_z;\\ \tau_{yz} &= c_{44}\gamma_{yz},\\ \tau_{zx} &= c_{55}\gamma_{zx},\\ \tau_{xy} &= c_{66}\gamma_{xy},\end{aligned}\right\} (7.6)$$

where $c_{ik} = c_{ki}$.

The 6 stress components $\sigma_x$, $\sigma_y$, $\sigma_z$, $\tau_{yz}$, $\tau_{zx}$, $\tau_{xy}$ are linear variables of the 6 strain components $\varepsilon_x$, $\varepsilon_y$, $\varepsilon_z$, $\gamma_{yz}$, $\gamma_{zx}$ and $\gamma_{xy}$. The $s_{ik}$- and $c_{ik}$-values are "elastic constants", from which the $s_{ik}$-figures have the dimensions of compliances [cm²/kp], and the $c_{ik}$-figures the dimensions of stresses [kp/cm²].

---

[1] The indices for the grain orientation vary in the literature partly as capital letters, partly as miniscule letters.

Table 7.1. *Systems of Elastic Constants for Different Species* (condensed from R. KEYLWERTH, 1951)

| Species | Density $\varrho_u$ | Moisture content $u$ | $s_{11}$ $s_{22}$ $s_{33}$ | $s_{44}$ $s_{55}$ $s_{66}$ | $-s_{12}$ $-s_{13}$ $-s_{23}$ | $E_x = \frac{1}{s_{11}}$ $E_y = \frac{1}{s_{22}}$ $E_z = \frac{1}{s_{33}}$ | $G_{yz} = \frac{1}{s_{44}}$ $G_{zx} = \frac{1}{s_{55}}$ $G_{xy} = \frac{1}{s_{66}}$ | $\frac{E_y}{G_{yz}}$ | $\frac{E_y}{G_{xy}}$ | Author |
|---|---|---|---|---|---|---|---|---|---|---|
| | g/cm³ | % | cm²/kp | cm²/kp | cm²/kp | kp/cm² | kp/cm² | — | — | |
| | | | | *Coniferous Species* | | | | | | |
| Douglas fir | 0.59 | 9.5 | 0.000109 0.00000599 0.0000755 | 0.0000835 0.00123 0.000108 | 0.0000023 0.000045 0.0000022 | 9,200 166,900 13,200 | 12,000 800 9,300 | 13.9 | 17.9 | HÖRIG, 1935 STAMER, 1935 STAMER and SIEGLERSCHMIDT, 1933 |
| Douglas fir | 0.45···0.51 | 11···13 | 0.000126 0.00000625 0.0000994 | 0.000111 0.00111 0.000111 | 0.00000279 0.0000427 0.00000190 | 8,000 160,000 10,100 | 9,000 900 9,000 | 17.8 | 17.8 | DOYLE, DROW, MCBURNEY, 1945/46 |
| Spruce | 0.44 | 9.8 | 0.000250 0.00000616 0.000143 | 0.000159 0.00270 0.000129 | 0.0000033 0.000060 0.0000027 | 4,000 162,300 7,000 | 6,300 400 7,800 | 25.8 | 20.8 | HÖRIG, 1935 STAMER, 1935 |
| Spruce | 0.43 | 12 | 0.000204 0.00000727 0.000110 | 0.000136 0.00307 0.000196 | 0.00000390 0.0000615 0.00000329 | 4,900 137,600 9,100 | 7,300 300 5,100 | 18.8 | 27.0 | HEARMON, 1948 |
| Spruce | 0.50 | 12 | 0.000155 0.00000587 0.000121 | 0.000157 0.00279 0.000115 | 0.0000033 0.0000517 0.0000022 | 6,500 170,400 8,300 | 6,400 400 8,700 | 26.6 | 19.6 | CARRINGTON, 1923 |
| Sitka spruce | 0,39 | 11···13 | 0.000196 0.00000846 0.000109 | 0.000131 0.00252 0.000136 | 0.00000395 0.0000480 0.00000315 | 5,100 118,300 9,200 | 7,600 400 7,300 | 15.6 | 16.2 | DOYLE, DROW, MCBURNEY, 1945/46 |
| Pine | 0.54 | 9.7 | 0.000172 0.00000602 0.0000890 | 0.0000563 0.00148 0.000146 | 0.0000027 0.000054 0.0000028 | 5,800 166,100 11,200 | 17,800 700 6,800 | 9.3 | 24.4 | HÖRIG, 1935 STAMER, 1935 |

*Broad-leaved Species*

| Species | Density | | | | | | | | | | Reference |
|---|---|---|---|---|---|---|---|---|---|---|---|
| Balsa wood | 0.10 | 9 | 0.00258 | 0.000791 | 0.0000215 | 400 | 1,300 | 19.2 | 27.7 | | DOYLE, DROW, McBURNEY, 1945/46 |
| | | | 0.0000402 | 0.00704 | 0.000594 | 24,900 | 100 | | | | |
| | | | 0.000860 | 0.00115 | 0.0000124 | 1,200 | 900 | | | | |
| Balsa wood | 0.20 | 9 | 0.000926 | 0.000314 | 0.00000798 | 1,100 | 3,200 | 20.1 | 30.6 | | DOYLE, DROW, McBURNEY, 1945/46 |
| | | | 0.0000156 | 0.00298 | 0.000219 | 64,200 | 300 | | | | |
| | | | 0.000327 | 0.000483 | 0.00000474 | 3,100 | 2,100 | | | | |
| Alpine maple | 0.58 | 9.6 | 0.000112 | 0.0000805 | 0.0000046 | 8,900 | 12,400 | 8.2 | 9.1 | | HÖRIG, 1935; STAMER, 1935 |
| | | | 0.0000098 | 0.000336 | 0.000049 | 102,000 | 3,000 | | | | |
| | | | 0.0000645 | 0.0000894 | 0.0000048 | 15,500 | 11,200 | | | | |
| Birch | 0.62 | 8.8 | 0.000159 | 0.0000836 | 0.0000026 | 6,300 | 12,000 | 13.9 | 17.9 | | HÖRIG, 1935; STAMER, 1935 |
| | | | 0.0000060 | 0.000527 | 0.000064 | 166,700 | 1,900 | | | | |
| | | | 0.0000888 | 0.000108 | 0.0000029 | 11,300 | 9,300 | | | | |
| Yellow birch | 0.64 | 13 | 0.000136 | 0.0000925 | 0.00000318 | 7,300 | 10,800 | 13.5 | 14.7 | | DOYLE, DROW, McBURNEY, 1945/46 |
| | | | 0.00000686 | 0.000409 | 0.0000600 | 145,800 | 2,400 | | | | |
| | | | 0.0000876 | 0.000101 | 0.00000336 | 11,400 | 9,900 | | | | |
| Red beech | 0.74 | 10.5 | 0.0000862 | 0.0000610 | 0.0000037 | 11,600 | 16,400 | 8.5 | 13.0 | | HÖRIG, 1935; STAMER, 1935; STAMER und SIEGLERSCHMIDT, 1933 |
| | | | 0.00000714 | 0.000215 | 0.000031 | 140,100 | 4,700 | | | | |
| | | | 0.0000438 | 0.0000929 | 0.0000032 | 22,800 | 10,800 | | | | |
| Oak | 0.67 | 11.6 | 0.0001015 | 0.0000760 | 0.0000087 | 9,900 | 13,200 | (4.4)[1] | (7.4)[1] | | HÖRIG, 1935; STAMER, 1935; STAMER und SIEGLERSCHMIDT, 1933 |
| | | | (0.0000172)[1] | 0.000250 | 0.000030 | (58,100)[1] | 4,000 | | | | |
| | | | 0.0000457 | 0.000128 | 0.0000055 | 21,900 | 7,800 | | | | |
| Ash | 0.68 | 9.2 | 0.000122 | 0.0000731 | 0.0000032 | 8,200 | 13,700 | 11.8 | 17.7 | | HÖRIG, 1935; STAMER, 1935 |
| | | | 0.00000621 | 0.000363 | 0.000045 | 161,000 | 2,800 | | | | |
| | | | 0.0000651 | 0.000110 | 0.0000029 | 15,400 | 9,100 | | | | |
| Ash | 0.80 | 14 | 0.000101 | 0.000114 | 0.00000425 | 9,900 | 8,800 | 17.4 | 24.7 | | HEARMON, 1948 |
| | | | 0.00000654 | 0.000392 | 0.0000395 | 152,900 | 2,500 | | | | |
| | | | 0.0000598 | 0.000161 | 0.00000347 | 16,700 | 6,200 | | | | |
| Khaya | 0.44 | 11 | 0.000192 | 0.000109 | 0.00000616 | 5,200 | 9,200 | 11.3 | 17.0 | | DOYLE, DROW, McBURNEY, 1945/46 |
| | | | 0.00000962 | 0.000467 | 0.0000511 | 104,000 | 2,100 | | | | |
| | | | 0.0000868 | 0.000164 | 0.00000288 | 11,500 | 6,100 | | | | |

[1] The figures in parentheses are taken from the original publication, but the primary value for $E_y$ ($= 58,100$ kp/cm²) is extraordinarily low. For oak with average density at the same moisture content $E_y$ should amount to about 130,000 kp/cm².

Table 7.1 (*Continued*)

| Species | Density $\varrho_u$ | Moisture content $u$ | $s_{11}$ $s_{22}$ $s_{33}$ | $s_{44}$ $s_{55}$ $s_{66}$ | $-s_{12}$ $-s_{13}$ $-s_{23}$ | $E_x = \frac{1}{s_{11}}$ $E_y = \frac{1}{s_{22}}$ $E_z = \frac{1}{s_{33}}$ | $G_{yz} = \frac{1}{s_{44}}$ $G_{zz} = \frac{1}{s_{55}}$ $G_{zy} = \frac{1}{s_{66}}$ | $\frac{E_y}{G_{yz}}$ | $\frac{E_y}{G_{zy}}$ | Author |
|---|---|---|---|---|---|---|---|---|---|---|
| | g/cm³ | % | cm²/kp | cm²/kp | cm²/kp | kp/cm² | kp/cm² | | | |
| *Broad-leaved Species* | | | | | | | | | | |
| Mahogany | 0.53 | 13 | 0.000204 0.00000791 0.000101 | 0.000161 0.000654 0.000209 | 0.00000443 0.0000844 0.00000244 | 4,900 126,400 9,900 | 6,200 1,500 4,800 | 20.4 | 26.3 | HEARMON, 1948 |
| Mahogany | 0.50 | 12 | 0.000134 0.00000861 0.0000804 | 0.0001001 0.000307 0.000131 | 0.00000457 0.0000463 0.00000266 | 7,400 116,200 12,400 | 10,000 3,300 7,600 | 11.6 | 15.3 | DOYLE, DROW, McBURNEY, 1945/46 |
| Quipo[1] | 0.10 | 10 | 0.00240 0.0000934 0.000419 | 0.000869 0.00252 0.00166 | 0.0000685 0.000253 0.00000197 | 400 10,700 2,400 | 1,200 400 600 | 8.9 | 17.8 | DOYLE, DROW, McBURNEY, 1945/46 |
| Quipo | 0.20 | 10 | 0.000677 0.0000297 0.000250 | 0.000456 0.00136 0.000701 | 0.0000205 0.0001016 0.00000902 | 1,500 33,600 4,000 | 2,200 700 1,400 | 15.3 | 24.0 | DOYLE, DROW, McBURNEY, 1945/46 |
| Sweet Germ | 0.54 | 10···11 | 0.000166 0.00000838 0.0000727 | 0.0000846 0.000363 0.000124 | 0.00000359 0.0000505 0.00000299 | 6,000 119,300 13,800 | 11,800 2,800 8,100 | 10.1 | 14.7 | DOYLE, DROW, McBURNEY, 1945/46 |
| Walnut | 0.59 | 11 | 0.000156 0.00000876 0.0000824 | 0.000102 0.000427 0.000140 | 0.00000557 0.0000585 0.00000429 | 6,400 114,200 12,100 | 9,800 2,300 7,100 | 11.7 | 16.1 | HEARMON, 1948 |
| Whitewood | 0.38 | 11 | 0.000239 0.0000101 0.000110 | 0.000136 0.000892 0.000146 | 0.00000425 0.0000781 0.00000328 | 4,200 98,900 9,100 | 7,300 1,100 6,800 | 13.5 | 14.5 | DOYLE, DROW, McBURNEY, 1945/46 |

[1] *Cavanillesia platanifolia* H. B. K.

Since about 1920 many theoretical investigations on the anisotropic elasticity of wood were published by JENKIN (1920), CARRINGTON (1921, 1922, 1923), PRICE (1928), HÖRIG (1931, 1933, 1935, 1936, 1937, 1943, 1944), SCHLÜTER (1931), STAMER and SIEGLERSCHMIDT (1933), STAMER (1935), HEARMON (1948), and KEYLWERTH (1951). The application of the elastic constants in connection with the system of equations (7.5) is simple and of practical value. Table 7.1 shows complete systems of elastic constants for wood condensed after KEYLWERTH (1951).

Table 7.1 shows that across the grain ($z$ or $R = r$ and $x$ or $T = t$ directions) the moduli of elasticity are much lower than along the grain ($y$ or $L = l$ direction). The order is $E_l \gg E_r > E_t$. The ratio $E_l/E_r$ varies for the coniferous species between 40.6 and 182 and for the broad-leaved species between 12.1 and 62.

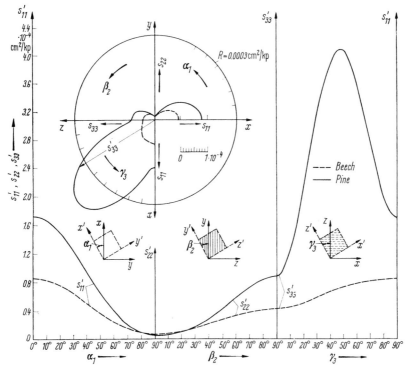

Fig. 7.3. Sections through the deformation planes in drilling beech and pine wood. From KEYLWERTH (1951)

As HEARMON (1953) pointed out, these high figures emphasize, the strongly anisotropic nature of wood; among inorganic rhombic crystals the ratio of greatest to least modulus of elasticity is never larger than 2.

The anisotropy of coniferous woods is much more pronounced than that of broad-leaved woods; this is especially valid for the quadrant $z-x$, as Fig. 7.3 shows.

### 7.1.3 Poisson's Ratios

An elastic longitudinal strain $\varepsilon_l$ is accompanied by contraction in a perpendicular direction or a lateral strain. The contraction $\varepsilon_q$ is proportional to the elastic elongation $\varepsilon_l$:

$$\frac{\varepsilon_l}{\varepsilon_q} = m \quad \text{or} \quad \frac{\varepsilon_q}{\varepsilon_l} = \mu = \frac{1}{m}. \tag{7.7}$$

The ratio of lateral strain to longitudinal strain is called Poisson's ratio. For wood in the case of a rhombic system 6 Poisson's ratios can be calculated using the following scheme given by HÖRIG (1933):

| Axis of the rod | Directions of lateral strain | | | Compressive stress parallel to the axis of the rod |
|---|---|---|---|---|
| | $y$ longitudinal | $z$ radial | $x$ tangential | |
| Longitudinal | — | $-s_{32}$ | $-s_{12}$ | $\sigma_y$ |
| Radial | $-s_{23}$ | — | $-s_{13}$ | $\sigma_z$ |
| Tangential | $-s_{21}$ | $-s_{31}$ | — | $\sigma_x$ |

HÖRIG (1935) has calculated Poisson's ratios for some species using data given by CARRINGTON, STAMER and SIEGLERSCHMIDT (Table 7.2).

Table 7.2. *Poisson's Ratios for Wood*

| $\mu$ | CARRINGTON (1922, 1923) Spruce | STAMER und SIEGLERSCHMIDT (1933) | | | | | | | |
|---|---|---|---|---|---|---|---|---|---|
| | | Douglas-fir | Spruce | Pine | Maple | Birch | Oak | Ash | Beech |
| $\dfrac{s_{21}}{s_{11}}$ | 0,021 | 0.021 | 0.013 | 0.015 | 0.041 | 0.017 | 0.085 | 0.026 | 0.043 |
| $\dfrac{s_{13}}{s_{33}}$ | 0.43 | 0.60 | 0.42 | 0.61 | 0.76 | 0.72 | 0.66 | 0.70 | 0.71 |
| $\dfrac{s_{32}}{s_{22}}$ | 0.37 | 0.37 | 0.43 | 0.46 | 0.49 | 0.49 | 0.32 | 0.30 | 0.45 |
| $\dfrac{s_{12}}{s_{22}}$ | 0.57 | 0.38 | 0.53 | 0.44 | 0.47 | 0.44 | 0.50 | 0.52 | 0.52 |
| $\dfrac{s_{31}}{s_{11}}$ | 0.33 | 0.42 | 0.24 | 0.31 | 0.44 | 0.40 | 0.30 | 0.37 | 0.36 |
| $\dfrac{s_{23}}{s_{33}}$ | 0.018 | 0.029 | 0.019 | 0.031 | 0.074 | 0.033 | 0.12 | 0.045 | 0.073 |

The theoretically postulated condition

$$s_{12} = s_{21}; \quad s_{23} = s_{32}; \quad s_{31} = s_{13}$$

is rather well fulfilled. Elastic constants with the suffixes $l$ = longitudinal, $r$ = radial, and $t$ = tangential are often used instead of the $s_{ik}$-values known from crystal physics according to VOIGT (1928). In this case the first suffix denotes the direction of the applied stress, the second the direction of the strain measured. Thus the following Eqs. may be written:

$$\left. \begin{aligned} \frac{\mu_{23}}{\mu_{32}} &= \frac{s_{32} \cdot s_{33}}{s_{22} \cdot s_{23}} = \frac{s_{33}}{s_{22}} = \frac{\mu_{l,r}}{\mu_{r,l}} = \frac{\alpha_r}{\alpha_l} = \frac{E_l}{E_r}, \\ \frac{\mu_{21}}{\mu_{12}} &= \frac{s_{12} \cdot s_{11}}{s_{22} \cdot s_{21}} = \frac{s_{11}}{s_{22}} = \frac{\mu_{l,t}}{\mu_{t,l}} = \frac{\alpha_t}{\alpha_l} = \frac{E_l}{E_t}, \\ \frac{\mu_{31}}{\mu_{13}} &= \frac{s_{12} \cdot s_{11}}{s_{33} \cdot s_{31}} = \frac{s_{11}}{s_{33}} = \frac{\mu_{r,t}}{\mu_{t,r}} = \frac{\alpha_t}{\alpha_r} = \frac{E_r}{E_t}. \end{aligned} \right\} \quad (7.8)$$

In the English literature—e.g. HEARMON and MEREDITH (1953)—the symbol for extensional strain is $e$, for stress is $d$, for Poisson's ratio is $\sigma$; the suffixes for the co-ordinates for the principal axes and planes in wood are $L, R, T$.

HEARMON gives the following equations expressing strains in terms of stresses for wood:

$$\left.\begin{array}{l} e_L = -\dfrac{1}{E_L}(dL - \sigma_{LR}\,dR - \sigma_{LT}\,dT),\\[4pt] e_T = -\dfrac{1}{E_T}(-\sigma_{TL}\,dL + dT - \sigma_{TR}\,dR),\\[4pt] e_R = -\dfrac{1}{E_R}(-\sigma_{RL}\,dL - \sigma_{RT}\,dT + dR). \end{array}\right\} \quad (7.9)$$

If we take compressive stresses and expansive strains as negative—in contrast to HEARMON who has taken them as positive by a special consideration—and replace the symbols by those used here, Eq. (7.9) is identical with the terms for $\varepsilon_x, \varepsilon_y, \varepsilon_z$ in Eq. (7.5).

### 7.1.4 Compressibility (Bulk Modulus)

The isothermic, cubic compressibility $\varkappa$ (of a body under the influence of hydrostatic pressure $p$ [kp/cm²]) is defined as

$$\varkappa = \frac{\Delta V}{V} \cdot \frac{1}{p} \qquad (7.10)$$

where $\dfrac{\Delta V}{V}$ is the relative change of volume.

For the anisotropic wood the compressibility $\varkappa$ can be calculated as follows:

$$\varkappa = s_{11} + s_{22} + s_{33} + 2(s_{23} + s_{31} + s_{21}) > 0. \qquad (7.11)$$

In table 7.1 the elastic constants for different species are given. One can calculate from these bulk moduli ranging from 0.0000915 to 0.0002672 cm²/kp for coniferous species and from 0.0000613 to 0.0022655 cm²/kp for broad-leaved species.

From Eq. (7.11) it follows that:

$$\mu_k = -\frac{s_{31} + s_{12} + s_{23}}{s_{11} + s_{22} + s_{33}} < 0.5 \qquad (7.12)$$

where $\mu_k$ might be called a bulk Poisson's ratio.

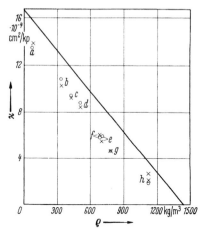

Fig. 7.4. Relationship between apparent compressibility $\varkappa$ and density $\varrho$. From KOLLMANN (1963). *a* balsa, *b* poplar, *c* lime, *d* Brazilian pine, *e* beech, *f* maple, *g* boxwood, *h* mora. Theoretical straight line (——), measured points after two orthogonal pressures (o o) and after the application of a hydrostatic pressure (× ×) respectively

HÖRIG (1935) computed from STAMER's (1935) figures $\mu_k$-values between 0.17 and 0.31.

As KOLLMANN (1963) has shown, compressibility is an essential mechanical property, not only below but also far above the elastic limit. With increasing hydrostatic pressure—and increasing temperatures up to a critical value — the moduli tend to reach maxima which are a function of density (Fig. 7.4).

### 7.1.5 Determination of Elastic Constants

**7.1.5.1 Determination by Static Tests.** For longitudinal tension and compression in a range of stresses $\Delta\sigma = \sigma_2 - \sigma_1 = \dfrac{P_2 - P_1}{A_0}$ (where $P_2$ and $P_1$ are the external forces, producing the stresses, and $A_0$ is the area of a cross section of the specimen prior to the test) a relative deformation $\dfrac{\Delta l}{l}$ (where $\Delta l$ is the elongation of the original gage length $l$ of the extensometer) is produced. According to Hooke's law the following is valid for tension as well as for compression tests

$$E = \frac{\sigma_2 - \sigma_1}{\varepsilon} = \frac{\Delta\sigma}{\dfrac{\Delta l}{l}} = \frac{(P_2 - P_1)\, l_1}{A_0 \Delta l}. \tag{7.13}$$

For compression tests the samples are clear wooden prisms, 2.5 cm × 2.5 cm × 7.5 cm up to 5 cm × 5 cm × 20 cm, carefully cut and surfaced. Deformation and load readings are taken and recorded for reasonable load increments until the

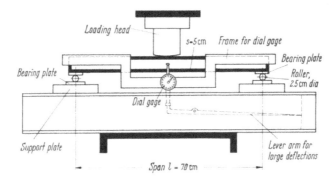

Fig. 7.5. Set-up of a beam for static bending by a central load. From SCHLYTER and WINBERG (1929)

maximum load is reached. The values are then plotted and a line is drawn through the points. The longitudinal Young's modulus $E$ can be derived from the slope of the straight part of the curve going through the origin. Some testing machines are equipped with automatic stress-strain recorders.

For tension tests shouldered samples—cf. Section 7.2.2—are used and the modulus of elasticity $E$ can be obtained in an analogous manner.

Predominant in bending is the determination of the modulus of elasticity $E_b$. Fig. 7.5 shows the set-up of a beam for static bending by a central load. The dimensions of the clear beam are 2 cm × 2 cm × 36 cm up to 5 cm × 5 cm × 76 cm. The ratio of span to depth of the beam should be at least 14 : 1. The modulus of elasticity can be calculated from the straight part of the load-deflection curve. Selecting corresponding load-deflection data (load $P$ and deflection $y$); in the interest of the reader, dealing with international literature it may be mentioned that the symbols for deflection are either $y$ or $f$, or $d$, equally in the same mechanical sense. They can be substituted in the following equation.

$$E_b = \frac{Pl^3}{4ywh^3} \tag{7.14}$$

where $l$ is the span [cm], $w$ is the width [cm], and $h$ is the depth (thickness) [cm] of the beam. The value for $E_b$ obtained as above is too low since the deflection is not only caused by axial strains but also by longitudinal shear. The influence of shear stresses may be neglected only if $l/h \geq 20$. BAUMANN (1922) has shown how the compliance $\alpha$ depends on the ratio of span to depth $l/h$. The ratio of Young's modulus $E$ to the modulus of rigidity $G$ is the deciding factor. BAUMANN has taken the value of 17 as an average. The large variations which are possible can be seen in table 7.1. When $E/G = 17$, the moduli of elasticity obtained by equations (7.14) for $l/h$ values of 14, 15, and 20 are too low by about 9.5, 8.8, and 4.8%, respectively.

The influence of shear stresses can be considered approximately by the following equation:

$$E_b = \frac{Pl^3}{4ywh^3}\left(1 + 1.2\frac{E}{G}\frac{h^2}{l^2}\right). \tag{7.15}$$

Another approximation is given by a formula developed by v. KÁRMÁN (1927) and SEEWALD (1927):

$$E_b = \frac{Pl^3}{4ywh^3}\left[1 + \frac{h^2}{l^2}\left(\frac{1.5E}{G} - \frac{1.5}{m} - 0.6\right)\right]. \tag{7.16}$$

While the measurement of longitudinal strains and lateral strains (and thus of the elastic constants $s_{11}$, $s_{22}$ and $s_{33}$ as well as $s_{12}$, $s_{23}$ and $s_{31}$) can be done easily, the determination of the shear strains (and thus of the elastic constants $s_{44}$, $s_{55}$ and $s_{66}$) is very difficult for anisotropic materials.

With special precautionary measures moduli of rigidity can be obtained from torsion tests using the method of evaluation given by HÖRIG (1923, 1935, 1936). Apparatus for the direct measurement of the moduli of rigidity are described by SCHLÜTER (1932), HÖRIG (1943) and YLINEN (1942).

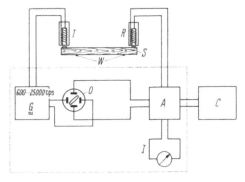

Fig. 7.6. Scheme of the electronic instrument for measurement of natural frequency "Elastomat". From KOLLMANN and KRECH (1960)

**7.1.5.2 Determination by Dynamic Tests.** In addition to the standard static methods of determining the elastic constants of wood, there are dynamic methods based on the measurement of natural frequency. Studies of wood in vibration were published by KRÜGER and ROHLOFF (1938), THUNELL (1941), POULIGNIER (1942), WUOLIJOKI (1947), HEARMON (1948), KOLLMANN and KRECH (1960), and SCHNEIDER (1965).

The scheme of a suitable electronic instrument according to FÖRSTER is shown in Fig. 7.6. The specimen $S$ rests at the nodes on two thin, tightly stretched wires $W$. Immediately above the steel plates glued to the ends of the specimen are the magnetic transmitter $T$ and the receiver $R$, respectively. The transmitter is connected to the audiogenerator $G$. The amplitude and frequency (600 to 25,000 cps) of its sinusoidal output voltage can be varied continuously and is also fed to the oscilloscope $O$. The voltage produced in the receiver during resonance is fed through the amplifier $A$ to the oscilloscope, the indicator $I$ and the counter $C$. The frequency is varied until the indicator shows a maximum and the counter determines the number of oscillations of the specimen during a period of 1, 10 or 100 sec. The resonance frequency can thereby be determined to an accuracy of 1, 0.1, or 0.01 cps.

The vibration test allows the determination of the modulus of elasticity, the modulus of rigidity, the velocity of sound and the logarithmic decrement as a measure of the damping of the vibrations (cf. section 6.7.1). The methods of calculation, partly based on considerations published by GOENS (1931), are discussed by HEARMON (1948), KOLLMANN and KRECH (1960), and SCHNEIDER (1966). The accuracy of the determination of the elastic constants of wood by vibration tests is doubtless higher than that of static tests. Another advantage of dynamic measurements is their short duration of only about 1 min for each sample. Though generally the moduli of elasticity obtained by vibration tests are somewhat higher than those from static experiments (cf. table 7.3), the differences are small and mostly negligible.

Table 7.3. *Comparison of Moduli of Elasticity along the Grain in Spruce and Oak Obtained by Vibration and Static Tests* (From KOLLMANN and KRECH, 1960)

| Property | Dimension | Spruce Average | Spruce Coefficient of variation % | Oak Average | Oak Coefficient of variation % |
|---|---|---|---|---|---|
| Density | g/cm³ | 0.501 | 2.3 | 0.653 | |
| $E_t$ from transverse vibrations | kp/cm² | 161,800 | 3.6 | 115,800 | 11.7 |
| $E_{tc}$ after correction of shear influence | kp/cm² | 178,800 | 3.6 | 123,000 | 11.7 |
| $E_l$ from longitudinal vibrations | kp/cm² | 180,500 | 3.9 | 125,500 | 12.1 |
| $E_b$ from static bending | kp/cm² | 151,400 | 4.0 | 110,500 | 10.4 |
| $E_{bc}$ after correction of shear influence | kp/cm² | 171,000 | 4.0 | 118,400 | 10.4 |

No significant influence of frequency on the dynamic measurements has been found.

### 7.1.6 Influences Affecting the Elastic Properties of Wood

**7.1.6.1 Grain Angle.** In an isotropic material the elastic constants and strength properties are the same in all directions, but in anisotropic materials like crystals

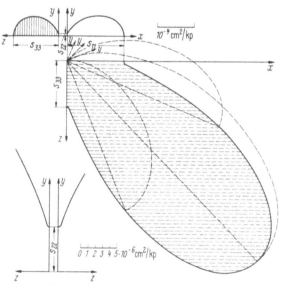

Fig. 7.7. Lines of intersection of the stress-strain plane for spruce. Polar-diagram from HÖRIG (1931) using measurements of CARRINGTON (1922)

and wood the elastic constants and other mechanical properties vary with the direction relative to the grain. The complete equations are given by VOIGT (1928),

HÖRIG (1931, 1933, 1935, 1936, 1937, 1943, 1944), HEARMON (1948) and KEYL-WERTH (1951); they will not be discussed here in detail. If the axis of a specimen has any arbitrary position $y'$ to the principal axes $x$, $y$, $z$ and if we represent the elastic constant $s'_{22}$ (for longitudinal extension, $\dfrac{1}{s'_{22}}$ = Young's modulus in the direction $y'$) by the vector $y'$, then we obtain surfaces of fourth order which—symmetrical in its octants—represent the elastic properties of the wood. Fig. 7.7 shows sections throug these surfaces in selected planes for spruce wood, designed by HÖRIG (1951) based on measurements of CARRINGTON (1922). Similar graphs were designed by THUNELL (1941) for swedish pine and by KEYLWERTH (1951) for pine and beech.

As an example we may take a test sample from a quarter-sawn plank by rotating around the fixed $x$-axis (Fig. 7.8). In the case of rhombic symmetry we find (KEYLWERTH, 1951):

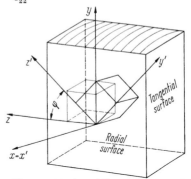

Fig. 7.8. Rotation of the system of coordinated around the fixed $x$-axis. From KEYLWERTH (1951)

$$\left.\begin{aligned}
s'_{11} &= s_{11}, \\
s'_{22} &= s_{22} \cos^4 \varphi + (2s_{23} + s_{44}) \sin^2 \varphi \cos^2 \varphi + s_{33} \sin^4 \varphi, \\
s'_{33} &= s_{22} \sin^4 \varphi + (2s_{23} + s_{44}) \sin^2 \varphi \cos^2 \varphi + s_{33} \cos^4 \varphi;
\end{aligned}\right\} \quad (7.17)$$

$$\left.\begin{aligned}
s'_{21} &= s_{21} \cos^2 \varphi + s_{31} \sin^2 \varphi, \\
s'_{31} &= s_{21} \sin^2 \varphi + s_{31} \cos^2 \varphi, \\
s'_{23} &= (s_{22} + s_{33}) \sin^2 \varphi \cos^2 \varphi + s_{23}(\sin^4 \varphi + \cos^4 \varphi) - \\
&\qquad - s_{44} \sin^2 \varphi \cos^2 \varphi;
\end{aligned}\right\} \quad (7.18)$$

$$\left.\begin{aligned}
s'_{44} &= 4(s_{22} + s_{33} - 2s_{23}) \sin^2 \varphi \cos^2 \varphi + s_{44} (\cos^2 \varphi - \sin^2 \varphi)^2, \\
s'_{55} &= s_{55} \cos^2 \varphi + s_{66} \sin^2 \varphi, \\
s'_{66} &= s_{55} \sin^2 \varphi + s_{66} \cos^2 \varphi.
\end{aligned}\right\} \quad (7.19)$$

For the calculation of $s_{22}$, important for practical purposes, we need a shear compliance (here $s_{44}$). If such values are missing, tension or compression tests on diagonal samples ($\varphi = 45°$) are sufficient according to KEYLWERTH (1951). The equation for $s'_{22}$ (7.17) will then be transformed into

$$s'_{22} = (s_{22} \cos^2 \varphi - s_{33} \sin^2 \varphi) \cos 2\varphi + \sin^2 2\varphi \, s^*_{22} \qquad (7.20)$$

where $s^*_{22}$ is the experimentally determined compliance at $\varphi = 45°$, or written in a technical manner

$$E_\varphi = \dfrac{1}{\left(\dfrac{\cos^2 \varphi}{E_0} - \dfrac{\sin^2 \varphi}{E_{90}}\right) \cos 2\varphi + \dfrac{\sin^2 2\varphi}{E_{45}}}$$

Fig. 7.9 proves how well measured points for beech fit the calculated curve. According to KOLLMANN (1934), a rough approximation is possible by use of the following formula:

$$s'_{22} \approx s_{22} \cos {}^n\varphi + s_{33} \sin {}^n\varphi. \qquad (7.21)$$

The power $n$ must be determined by experiments; KOLLMANN (1934) derived from data published by BAUMANN (1922) for basswood and white fir $n \approx 3$. Fig. 7.10 shows fairly good agreement of experimental points (widely scattered due to improper orientation of sample axes to the grain) and calculated curves.

Similar aspects as for tension exist for torsion if we cut specimens at any arbitrary direction $y'$ to the principal axes $x, y, z$ and measure the torsion. The sums

$$a_0 = \frac{1}{2}(s'_{55} + s'_{66}), \quad b_0 = \frac{1}{2}(s'_{66} + s'_{44})$$

and $c_0 = \frac{1}{2}(s'_{44} + s'_{55})$ represent the moduli of twist for a right cylinder. If we twist circular cylinders around the $x'$, $y'$ or $z'$ axes of a crystal, we obtain the tortion of two cross-sections spaced apart by a distance $l$.

$$\left.\begin{aligned}\varphi_x &= \vartheta_x l = \frac{2M_x l}{\pi r^4}\frac{s'_{55}+s'_{66}}{2}, \\ \varphi_y &= \vartheta_y l = \frac{2M_y l}{\pi r_4}\frac{s'_{66}+s'_{44}}{2}, \\ \varphi_z &= \vartheta_z l = \frac{2M_z l}{\pi r^4}\frac{s'_{44}+s'_{55}}{2}.\end{aligned}\right\} \quad (7.22)$$

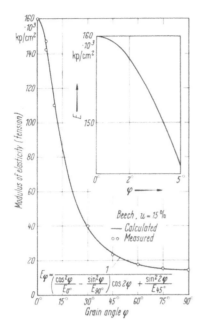

Fig. 7.9. Modulus of elasticity (tension) versus grain angle. From KEYLWERTH (1951)

The surface of torsion, calculated by KEYLWERTH (1951) for data measured by STAMER (1935) is shown in Fig. 7.11. The elastic constants for shear $s_{44}$, $s_{66}$ and $s_{55}$ can be calculated from the moduli of twist as follows:

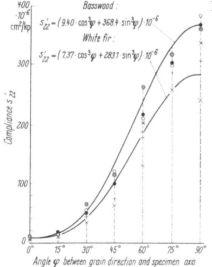
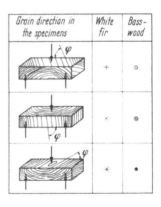

Fig. 7.10. Effect of grain orientation to the compliance for deflection tests. From BAUMANN (1922)

$$s_{44} = b_0 + c_0 - a_0, \qquad s_{66} = c_0 + a_0 - b_0, \qquad s_{55} = a_0 + b_0 - c_0. \quad (7.23)$$

**7.1.6.2 Density.** BAUMANN (1922), testing small samples, observed the relationship between compliance (in bending) and density $\varrho$ as shown in Fig. 7.12.

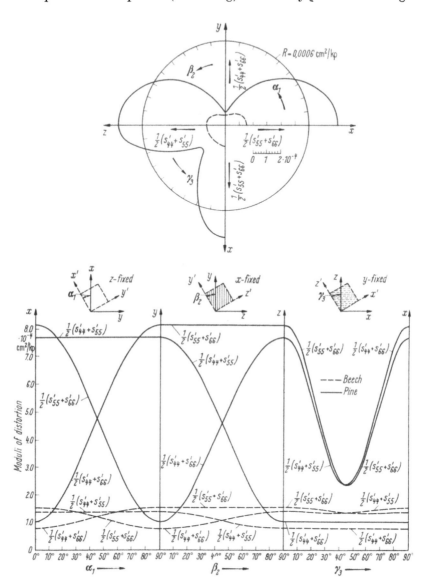

Fig. 7.11. Sections through the surfaces of torsion for beech and pine in polar diagram. From KEYLWERTH (1951)

The curves may be described by a hyperbolic equation

$$\alpha = \frac{a}{\varrho - b} \quad \text{or} \quad E = \frac{1}{\alpha} = \frac{\varrho - b}{a} = c(\varrho - b) = cb\left(\frac{1}{b}\varrho - 1\right) \quad (7.24)$$

where $a$, $b$ and $c$ are constants.

SCHLYTER and WINBERG (1929) found for Swedish pine:

| Moisture content $u\%$ | Young's modulus $E$ $kp/cm^2$ |
|---|---|
| 23···30 | $E = 460{,}000\,(R - 0.20) = 92{,}000\,(5R - 1)$, |
| 15···19 | $E = 475{,}000\,(R - 0.20) = 95{,}000\,(5R - 1)$, |
| 6···10 | $E = 490{,}000\,(R - 0.20) = 98{,}000\,(5R - 1)$ |

(7.25)

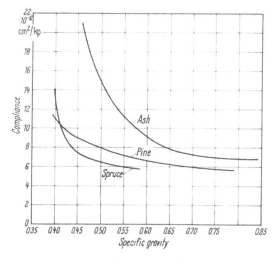

Fig. 7.12. Dependence of the compliance upon specific gravity. From BAUMANN (1922)

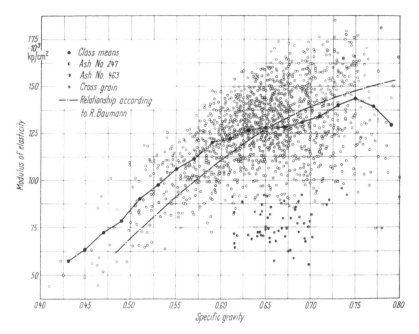

Fig. 7.13. Effect of density on the modulus of elasticity of ash (*Fraxinus excelsior* L.). From KOLLMANN (1941)

where $R$ is density, based on oven-dry weight and volume when green. The variability in these relationships apparently is more pronounced for hardwoods than for softwoods. KOLLMANN's (1941) results for the effect of density on the modulus of elasticity of ash are plotted in Fig. 7.13. Relationships between Young's modulus and density, which are evidently linear, were obtained for spruce and oak from vibration tests carried out by KOLLMANN and KRECH (1960) as illustrated in Fig. 7.14 and 7.15. According to the same authors the relationships for the

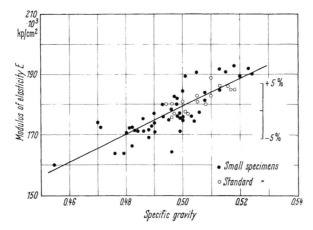

Fig. 7.14. Effect of density on the modulus of elasticity parallel to the grain of spruce. From KOLLMANN and KRECH (1960)

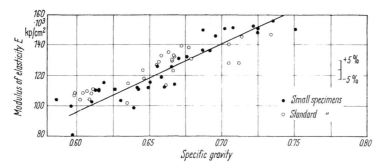

Fig. 7.15. Effect of density on the modulus of elasticity parallel to the grain of oak. From KOLLMANN and KRECH (1960)

moduli of rigidity or shear moduli are less clear, especially for spruce (Fig. 7.16), but for oak the rigidity moduli increase with density. The equations in Table 7.4 were calculated.

Table 7.4. *Elastic Constants of Spruce and Oak as a Function of Density* (From KOLLMANN and KRECH, 1960)

| Property | Dimension | Spruce | Oak |
|---|---|---|---|
| Density $\varrho_u$ | g/cm³ | 0.497 | 0.663 |
| Moisture content $u$ | % | 14.3 | 15.1 |
| Modulus of elasticity $E_y$ | kp/cm² | $E_y = 459{,}000\varrho_u - 49{,}990$ | $E_y = 444{,}000\varrho_u - 169{,}900$ |
| Modulus of rigidity $G_{xy}$ | kp/cm² | 6,350 | $G_{xy} = 10{,}300\varrho_u + 137$ |
| Modulus of rigidity $G_{zy}$ | kp/cm² | 5,960 | $G_{zy} = 15{,}800\varrho_u + 111$ |

For extremely light balsa wood DRAFFIN and MÜHLENBRUCH (1937) also found a linear proportion between Young's modulus and density (Fig. 7.17). The relation of the modulus of elasticity $E_{yc}$ (along the grain in compression) to the density $\varrho_u$ holds as follows:

or
$$E_{yc} \text{ [lb/sq. in.]} = 346{,}000\,(560\,\varrho_u - 1) \quad \text{when } \varrho_u \text{ in [lb/cu. in.]}$$
$$E_{yc} \text{ [kp/cm}^2\text{]} = 24{,}300\,(20.2\,\varrho_u - 1) \quad \text{when } \varrho_u \text{ in [g/cm}^3\text{]}.$$
(7.26)

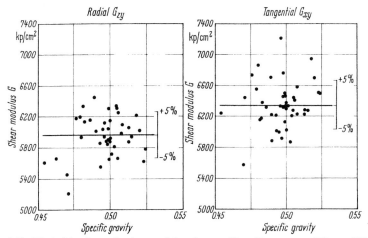

Fig. 7.16. Effect of density on the shear modulus of spruce. From KOLLMANN and KRECH (1960)

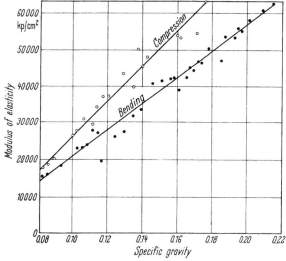

Fig. 7.17. Relationship between the moduli of elasticity (in compression and bending) and density of balsa (Ochroma lagopus Sw.). From DRAFFIN and MÜHLENBRUCH (1937)

YLINEN (1942) has shown that Eq. (7.24) can be established also by theoretical considerations. He deduced the following equation:

$$\alpha_u = \frac{1}{E_u} = \frac{\dfrac{\varrho_{lo} - \varrho_{eo}}{E_{lu} - E_{eu}}}{\varrho_0 + \dfrac{E_{eu}\varrho_{lo} - E_{lu}\varrho_{eo}}{E_{lu} - E_{eu}}} \tag{7.27}$$

where $\varrho_{l0}$ = density of oven-dry latewood, $\varrho_{e0}$ = density of oven-dry earlywood, $E_{lu}$ = Young's modulus of latewood at moisture content $u$, $E_{eu}$ = Young's modulus of earlywood at moisture content $u$ and $\varrho_0$ = density of clear oven-dry wood. For Finnish pine YLINEN (1942) determined the following mean values:

$$\varrho_{l0} = 1{,}12 \text{ g/cm}^3, \qquad \varrho_{e0} = 0{,}28 \text{ g/cm}^3,$$
$$E_{l0} = 385{,}000 \text{ kp/cm}^2, \quad E_{e0} = 68{,}000 \text{ kp/cm}^2.$$

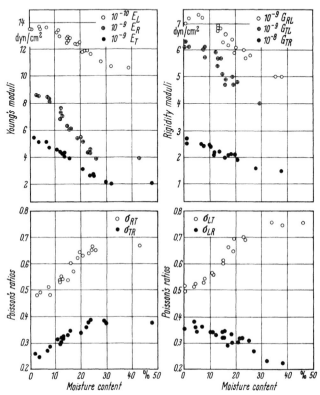

Fig. 7.18. Effect of moisture content on elastic constants of Sitka spruce. From CARRINGTON (1922)

Substituting these numerical values into Eq. (7.27) we obtain in conformity with Eq. (7.24)

$$\alpha_u \approx \frac{2{.}65}{\varrho_0 - 0{.}1} \cdot 10^{-6} \text{ [cm}^2\text{/kp]}.$$

**7.1.6.3 Moisture Content.** Above the fiber saturation point free liquid water filling the coarser capillaries in vessels, tracheids and other elements of the wooden tissues does not affect strength and elastic properties. Below the fiber saturation point shrinkage or swelling occurs thus increasing or reducing cohesion and stiffness. CARRINGTON's results (1922b) for the effect of moisture content on the elastic constants of spruce are plotted in Fig. 7.18 according to HEARMON (1953). Similar relationships between Young's modulus and moisture content were found by WILSON (1932), THUNELL (1941), and more recently, based on vibration tests, by KOLLMANN and KRECH (1960), Fig. 7.19 and 7.20. The considerable influence of moisture on the elasticity of wood makes it necessary to reduce all values to a standard moisture content normally of 12%. This can be done in the following manner. In the moisture-content range between approximately 8 and 22% the curves may be replaced by straight lines which are extended to the abscissa. The point of intersection will have the abscissa value $b$. Then we may write:

$$E_2 = E_1 \frac{b - u_2}{b - u_1} \text{ [kp/cm}^2\text{]}. \tag{7.28}$$

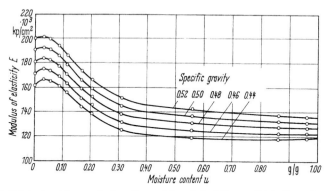

Fig. 7.19. Effect of moisture content on the modulus of elasticity parallel to the grain of spruce. From KOLLMANN and KRECH (1960)

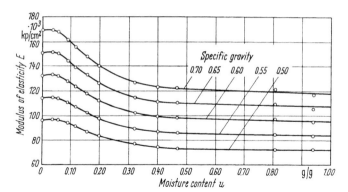

Fig. 7.20. Effect of moisture content on the modulus of elasticity parallel to the grain of oak. From KOLLMANN and KRECH (1960)

Example. From Fig. 7.19 the intersection point extrapolated graphically for $E = 0$ has the value $b = 82$. For spruce with an oven-dry density of $\varrho_0 = 0.44 \text{ g/cm}^3$ the modulus of elasticity for the moisture content $u_1 = 20\%$ was determined to be $E_1 \approx 140{,}000 \text{ kp/cm}^2$. The reduction to the standard moisture content $u_2 = 12\%$ leads to

$$E_2 = 140{,}000 \cdot \frac{82 - 12}{82 - 20} = 158{,}000 \text{ [kp/cm}^2\text{]}.$$

This value fits well into the diagram.

Of the elastic constants, Young's modulus along the grain $\left(E_y = E_l = \dfrac{1}{s_{22}}\right)$ is apparently the least sensitive to moisture content. The ratios in Table 7.5 are calculated from data in a number of publications and compared with mean values obtained at the U. S. Forest Products Laboratory (WILSON, 1932).

Ref. p. 414]                    7.1 Elasticity, Plasticity, and Creep                    311

Table 7.5. *Variations of Elastic Constants with Moisture Content*

| Elastic constant | Ratio to the value at $u \geq 30\%$ | | Reference |
|---|---|---|---|
| | $u = 0\%$ | $u = 12\%$ | |
| YOUNG's moduli | | | |
| $E_l = \dfrac{1}{s_{22}}$ (spruce) | 1.28 | 1.24 | CARRINGTON, 1922b |
| $E_l$ average of 54 softwood species | — | 1.28 | WILSON, 1928 |
| $E_l$ average of 113 hardwood species | — | 1.31 | |
| $E_l$ spruce $\varrho_0 = 0.52$ g/cm³ | 1.29 | 1.21 | |
| ,,      $\varrho_0 = 0.44$   ,, | 1.32 | 1.29 | KOLLMANN and |
| oak    $\varrho_0 = 0.70$   ,, | 1.42 | 1.29 | KRECH, 1960 |
| ,,      $\varrho_0 = 0.50$   ,, | 1.35 | 1.28 | |
| $E_r = \dfrac{1}{s_{33}}$ spruce | 2.17 | 1.83 | |
| $E_t = \dfrac{1}{s_{11}}$ spruce | 2.66 | 2.11 | |
| Rigidity moduli | | | |
| $G_{lr} = G_{yz} = \dfrac{1}{s_{44}}$ | 1.29 | 1.36 | CARRINGTON, 1922b |
| $G_{tr} = G_{xz} = \dfrac{1}{s_{55}}$ | 1.85 | 1.54 | |
| $G_{tl} = G_{xy} = \dfrac{1}{s_{66}}$ | 1.56 | 1.44 | |

The POISSON's ratios $\mu_{rt}$, $\mu_{tr}$ and $\mu_{lt}$ increase with increasing moisture content, the POISSON's ratio $\mu_{lr}$, however, decreases with increasing moisture content (cf. Fig. 7.18).

Remarkable is the dependence of the logarithmic decrement on moisture content. KOLLMANN and KRECH (1960) came, in the lower range of moisture content ($u \leq 16\%$), to corresponding results with those of PENTONEY (1955).

The following empirical parabolic equations may be used for the calculation of the logarithmic decrement $\varLambda$:

$$\left.\begin{array}{ll} \text{Spruce} & \varLambda = 0.0225 - 0.116\,u + 0.747\,u^2 \\ \text{Oak} & \varLambda = 0.0314 - 0.171\,u + 0.989\,u^2 \end{array}\right\} \quad 0 < u < 0.16 \text{ g/g}. \quad (7.29)$$

**7.1.6.4 Temperature.** The influence of temperature on strength, elasticity and plasticity of wood has been examined to some extent during the last two decades. We know that for wood, as for any other solid body (below the melting point or below the thermal decomposition), strength and stiffness decrease with increasing temperature due to thermal expansion of the crystal lattice of the cellulose and due to the increased intensity of the thermal molecular oscillations.

THUNELL (1941) has published a curve (Fig. 7.21) which shows a continuous decrease of the modulus of elasticity along the grain in bending with increasing temperature in the range from $-20\,°$C to $+50\,°$C. The most extensive investigations on the change of the elastic properties with temperature were carried out by SULZBERGER (1943, 1948, 1953). His data obtained from static bending tests (under loading) with Australian wood species of different moisture contents show a general decrease in Young's modulus $E$ with increasing temperature, the effect increasing with moisture content. At zero moisture content all species — except

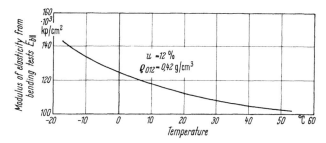

Fig. 7.21. Relationship between modulus of elasticity from bending tests of swedish pine and temperature From THUNELL (1941)

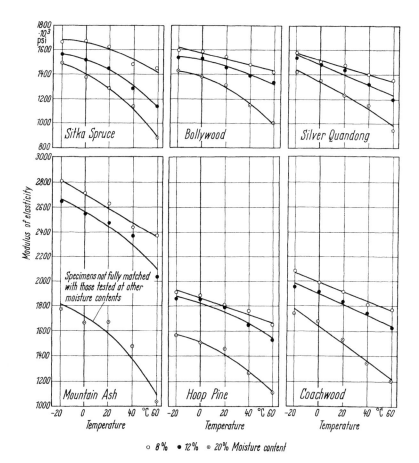

Fig. 7.22. Effect of temperature on the modulus of elasticity of various wood species. From SULZBERGER (1953)

Sitka spruce — exhibited a linear regression between modulus of elasticity and temperature. The effect of temperature is then expressed by the equation

$$E_2 = E_1[1 - \alpha(\vartheta_2 - \vartheta_1)] \tag{7.30}$$

where $\vartheta_2$ and $\vartheta_1$ are temperatures in °C ($\vartheta_2 > \vartheta_1$), $\alpha$ is the coefficient of the influence of temperature per °C. Coachwood (*Ceratopetalum apetalum* D. Don) showed linearity also for higher moisture contents but other species showed a general trend to curvilinearity, which became more pronounced at higher moisture content values (Fig. 7.22).

SULZBERGER (1953) plotted the average percentage moduli of elasticity of six species for each moisture content against temperature, so that modulus of elasticity at 20°C is equal to 100% (Fig. 7.23).

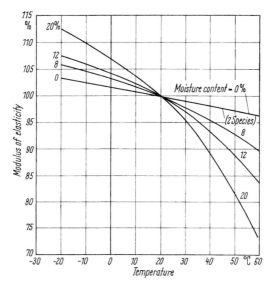

Fig. 7.23. Average percentage of the modulus of elasticity of six species plotted against temperature. From SULZBERGER (1953)

HEARMON (1953) pointed out that under practical conditions higher temperatures tend to dry out the wood, "so that the increase in modulus due to drying may well outweigh the reduction due to temperature alone". SALAMON (1963) reported about the effect of high-temperature drying on strength properties of Douglas-fir. When a high-temperature air-steam mixture was applied, strength properties of the wood under equalized air conditions were reduced as follows: fiber stress at proportional limit 7%; modulus of rupture 13%; modulus of elasticity 8%; and maximum crushing strength 2.5%. Superheated steam (temperatures up to 106°C dry bulb) affected strength properties more adversely, when time was the same. The decrease in strength properties was less when material of a higher density was used.

**7.1.6.5 Knots and Notches.** The elastic properties of wood are influenced considerably by knots, as they are by cross grain or curled fibers. BAUMANN (1922) determined the following modulus of elasticity values: for clear air-dry pinewood ($\varrho_u = 0.60$ g/cm³) $E_l = 161{,}800$ kp/cm², but for the same wood with one knot on the compression-side $E_l = 150{,}200$ kp/cm²; for clear ashwood ($\varrho_u = 0.76$ g/cm³) $E_l = 113{,}600$ kp/cm², but for the same wood with a knot on the tension-side $E_l = 107{,}600$ kp/cm², and with a knot on the compression-side $E_l = 100{,}700$ kp/cm². In teakwood ($\varrho_u = 0{,}73$ g/cm³) the modulus of elasticity decreased from $135{,}000$ kp/cm² for clear wood to $83{,}300$ kp/cm² for wood with one knot. THUNELL (1943) deduced a formula for the modulus of elasticity of Swedish pinewood as a function of a "knot-quotients sum".

The stress distribution in the vicinity of a knot is very complicated. YLINEN (1942) recorded that the elastic compliance in a knot of pinewood is about seven

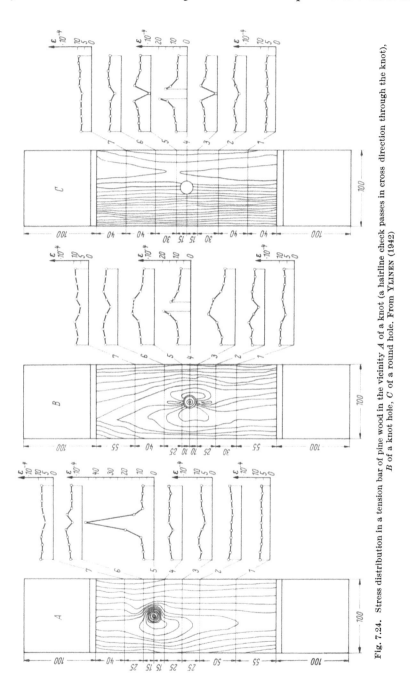

Fig. 7.24. Stress distribution in a tension bar of pine wood in the vicinity A of a knot (a hairline check passes in cross direction through the knot), B of a knot hole, C of a round hole. From YLINEN (1942)

times as high as in the surrounding wood. The influence of a round hole is much less prononuced (Fig. 7.24).

## 7.1.7 Plasticity and Creep

**7.1.7.1 Stress-strain Behavior.** Hooke's law (cf. Sect. 7.11) cannot be expected to be valid in a wide range for such comparatively complicated materials as wood and other natural high polymers. The stress-strain diagram is therefore

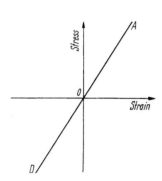

Fig. 7.25. Stress-strain diagram for an ideal elastic body

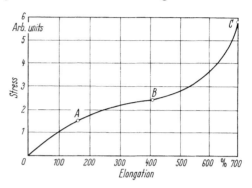

Fig. 7.26. Stress-strain curve of soft vulcanized Hevea rubber (0.5···1.0% sulfur). From SCHMIDT and MARLIES (1948)

not the same as for an ideal elastic body (Fig. 7.25). For wood, mainly for compressive stresses, the stress-strain function is rarely linear or only in a small region (cf. Fig. 7.1). It must be mentioned that the stress-strain curve for an elastic material which obeys Hooke's law needs not be a straight line, but it must be reversible. If a soft vulcanized Hevea rubber is stretched at 20 °C to a 700% elongation, the stress-strain curve may be of the type shown in Fig. 7.26, but the entire deformation is within the elastic limit (SCHMIDT and MARLIES, 1948). If ideal elastic behavior is attributed to rubber, then a loading-unloading cycle can be carried out without energy loss, and time is not a factor. For any stage of loading the same change of shape, which vanishes immediately on removal of the load, can always be measured independent of the duration of loading. For wood and other elastic-plastic materials a stress-strain cycle as in Fig. 7.27 takes place. Initial loading gives the curve $OA'$, and unloading, the curve $A'B'$, which shows that a permanent set $OB'$ occurred.

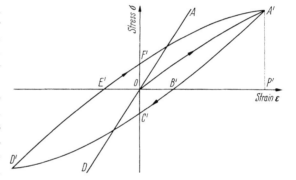

Fig. 7.27. Stress-strain curves for wood and other elastic-plastic materials. From BARKAS (1949). $AD$-ideal elastic body (Hooke's straight line)

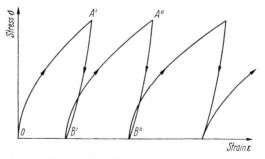

Fig. 7.28. Stress-strain cycles for repeated loading and unloading with increasing plasticity. From BARKAS (1949)

For recovering this permanent set, i.e. for again obtaining the original form of the body, one must apply a stress $OC'$ of opposite sign as the stress which results in curve $B'C'$. Further stressing to the (absolute) maximum $A'P'$ leads to curve $C'D'$. Removal of the load leads to the line $D'E'$, where now the stress-free

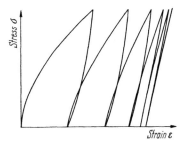

Fig. 7.29. Stress-strain cycles for repeated loading and unloading with increasing approach to the ideal elastic behavior. From BARKAS (1949)

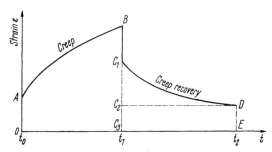

Fig. 7.30. Deformation of an elastic-plastic body as a function of time. Loading during time $t_0$ to $t_1$, followed by unloading (time $t_1$ to $t_2$). From KOLLMANN (1962)

state is characterized by a negative permanent set $E'O$. Through reversed stress $OF'$ the original form is restored. By increased loading the curve closes with $F'A'$ to a loop. Thermodynamically the area of the loop $C'D'F'A'$ represents the energy loss during the entire cycle.

If the stress always acts only in one direction and if the stress-strain cycles are repeated, then the permanent set may be increased as can be seen in Fig. 7.28. Thereby the energy loss also increases with each cycle; which means that more

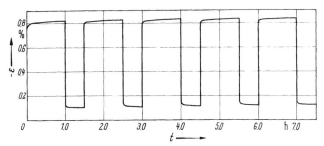

Fig. 7.31. Compressive strain $\varepsilon$ of oven-dry red beech in compression test parallel to the grain in dependence on time $t$ under loading with stress $\sigma = 700$ kp/cm² for 60 min and following unloading (loading/unloading cycles). From KOLLMANN (1962)

work is done in deformation, and the plasticity of the material is being increased step by step. Another possibility, as BARKAS (1949) has explained, is that the energy loss from cycle to cycle decreases more and more and finally the loading and unloading curves are identical (Fig. 7.29). This would mean that the previously plastic material has been transformed to an elastic one. In reality materials with the trend to increasing internal order with a loss of entropy in the course of repeated stress-strain cycles do not exactly reach the ideal state. KOLLMANN (1957) has shown that wood under certain circumstances may behave as in Fig. 7.28, while under other conditions transformation to a more elastic state can be observed. The stress level at which the set first begins to occur appears to vary considerably between species. KING (1961) found that the average maximum stress level at which no set occurs is approximately 42% of the ultimate tensile strength, the individual specimen values ranging from 21 to 80%.

**7.1.7.2 Creep and Creep Recovery.** Wood, according to the present experimental results and the results of the investigation of its fine structure possesses both elastic as well as plastic properties. If we apply a stress at zero-time there is an instantaneous elastic deformation $OA$ (Fig. 7.30). This is followed by a

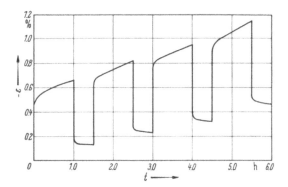

Fig. 7.32. Compressive strain $\varepsilon$ of red beech the (moisture content 12%) in compression test parallel to grain in dependence on time $t$ under loading with stress 500 kp/cm² for 60 min and following unloading (loading/unloading cycles). From KOLLMANN (1962)

retarded deformation (creep) $AB$ under constant stress. On removal of the stress at time $t_1$ an instantaneous elastic recovery $BC_1 \,(= \overline{OA})$ takes place, followed by a retarded partial creep recovery from $C_1$ to $D$ at time $t_2$. After that time $t_2$, further recovery is so insignificant that it can be neglected. Thus $DE$ represents the permanent deformation left at the end of the loading-unloading cycle. The

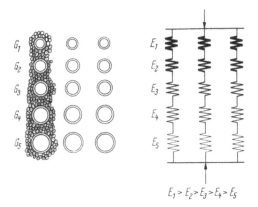

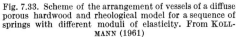

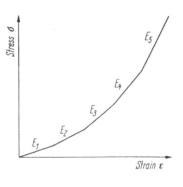

Fig. 7.33. Scheme of the arrangement of vessels of a diffuse porous hardwood and rheological model for a sequence of springs with different moduli of elasticity. From KOLLMANN (1961)

Fig. 7.34. Stress-strain diagram for an elastic model as given in Fig. 7.34. From KOLLMANN (1961)

mechanical behavior of a body according to Fig. 7.30 is classified as "viscoelastic". Wood and many high-polymer solids belong to this category. The large differences between individual wood specimens in structure and resultant mechanical properties are reflected in very different forms of creep and creep recovery curves (Figs. 7.31 and 7.32).

The creep $AB$ consists of two components which can be separated during creep recovery. The elastic, although retarded component $C_1C_2$ is called primary creep and the nonrecoverable component $C_2C_3$ ($=DE$) is termed secondary creep.

Based on the results of short-term tests made on 11 species of wood, KING (1961) could prove that creep is proportional to the logarithm of time at all load levels for the first 30…40 min. of the sustained-load period. At longer periods, creep tends to grow more than proportionally with the logarithm of time.

Investigations carried out by KELLOG (1950) indicate that ultimate tensile strain of wood including accumulated creep increases after repeated stressing in tension parallel to the grain. There is an indication that this increase in ultimate strain is a result of the increased strain due to the creep which occurred during the cycling period.

**7.1.7.3 Rheological Models and Mathematical Considerations.** In rheology, i.e. the theory of stress and strain, elastic behavior is compared to that of a helical steel-spring and plastic behavior to that of a dash pot. There are many possible combinations of these hypothetical rheological models or units. In wood elastic regions and plastic regions are combined, and are represented by different anatomical elements and chemical constituents of the cell wall. Fig. 7.33 shows the rheological model for a sequence of vessels within an annual ring of a diffuse-porous hardwood. Such a model leads to a stress-strain diagram as given in Fig. 7.34. Eearlywood and latewood, and small and broad wood rays and their combined mechanical action may be represented by the models of Fig. 7.35. The success of rheological models in elucidating the stress-strain behavior of wood depends on the proper choice and arrangement of the elements of the models. There are of course many basic difficulties. Highly crystalline cellulose regions in an isolated state would have the properties of a spring. Amorphous lignin would have the properties of an isolated dashpot. Naturally there are no isolated elements in wood tissue. Furthermore there is certainly a very small number of elastic and plastic elements clearly acting partly in series and partly in parallel. Actually large

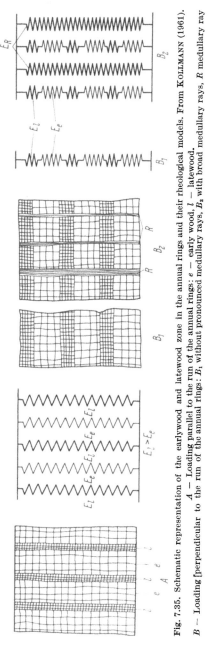

Fig. 7.35. Schematic representation of the earlywood and latewood zone in the annual rings and their rheological models. From KOLLMANN (1961). $A$ — Loading parallel to the run of the annual rings: $e$ — early wood, $l$ — latewood. $B$ — Loading [perpendicular to the run of the annual rings: $B_1$ without pronounced medullary rays, $B_2$ with broad medullary rays, $R$ medullary ray

distributions of both mechanical elements exist without a clear-cut differentiation of rheological behavior.

Nevertheless the rather simple mechanical model of Fig. 7.36 is quite satisfactory for wood and some high-polymer solids. The model consists of elastic and viscous elements acting in series and in parallel and it depicts the stress-strain behavior according to Fig. 7.30. If the model of Fig. 7.36 is loaded then a total strain $\varepsilon$ is caused which is made up of the following components

$$\varepsilon = BC_1 + C_1C_2 + C_2C_3 = \varepsilon_{el} + \varepsilon_{ea} + \varepsilon_{pl} \quad (7.31)$$

where $\varepsilon_{el}$ = pure elastic deformation, $\varepsilon_{ea}$ = elastic after-effect (primary creep), and $\varepsilon_{pl}$ = plastic deformation (= secondary creep). For the elastic deformation Hooke's law holds

$$\frac{\sigma}{E} = \varepsilon_{el} \quad (7.2)$$

Now we consider constant stress-flow curves of an elastic and a viscous element in series and alternatively in parallel. For an ideal elastic body under constant stress the deformation does not change with time and one can thus from Eq. (7.2) derive the flow curve $E = f(t) = $ const:

$$\frac{d\sigma}{dt} = E \cdot \frac{d\varepsilon}{dt} \quad \text{or} \quad \frac{d\varepsilon}{dt} = \frac{1}{E} \cdot \frac{d\sigma}{dt}. \quad (7.32)$$

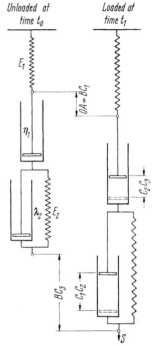

Fig. 7.36. Elastic and viscous elements acting in series and in parallel. From SCHMIDT and MARLIES (1948)

If the body on the other hand is viscous and behaves like Newtonian liquid, the deformation is proportional to the time, thus

$$\frac{d\varepsilon}{dt} = \frac{\sigma}{\eta} \quad (7.33)$$

where $\eta$ is the viscosity.

When a material has both an elastic as well as a plastic element in series (Fig. 7.37) the flow velocity can be calculated as follows:

$$\frac{d\varepsilon}{dt} = \frac{1}{E} \cdot \frac{d\sigma}{dt} + \frac{\sigma}{\eta}$$

or

$$\frac{d\sigma}{dt} = E\frac{d\varepsilon}{dt} - \frac{E}{\eta} \cdot \sigma \quad (7.34)$$

and

$$\frac{d\sigma}{dt} = E \cdot \frac{d\varepsilon}{dt} - \frac{\sigma}{\lambda}$$

where $\lambda = \eta/E$ is a constant, $t$ the relaxation time. This yields, on integration,

$$\varepsilon = \frac{\sigma}{E} + \frac{\sigma}{E} \cdot \frac{t}{\lambda}. \quad (7.35)$$

If the elastic and viscous elements are connected in parallel then the piston starts with a velocity $\frac{\sigma}{\eta}\eta$ which, however, asymptotically approaches zero the more the

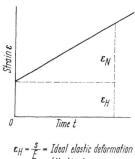

Fig. 7.37. Constant-stress flow curve of an elastic and a viscous element acting in series. From SCHMIDT and MARLIES (1948). For stress SCHMIDT and MARLIES (1948) use the symbol "s". We refer to the fact, that in this book stresses are designated with the symbol "$\sigma$"

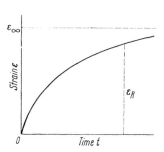

Fig. 7.38. Constant-stress flow curve of an elastic and a viscous element acting in parallel. From SCHMIDT and MARLIES (1948)

stress is gradually transmitted to the spring. The stress, which acts in the viscous element is thus equal to the applied external stress, less the stress of the elastic element. The flow curve takes a course according to Fig. 7.38 and the following equation holds:

$$\frac{d\varepsilon_R}{dt} = \frac{\sigma}{\eta} - \frac{\varepsilon_R \cdot E}{\eta} = \frac{\sigma - \varepsilon_R E}{\eta} \tag{7.36}$$

or after integration for the retarded elastic deformation $\varepsilon_R$ holds:

$$\varepsilon_R = \frac{\sigma}{E}\left(1 - e^{-\left(\frac{E}{\eta}\right)t}\right) = \varepsilon_\infty\left(1 - e^{-\frac{t}{\lambda}}\right) \tag{7.37}$$

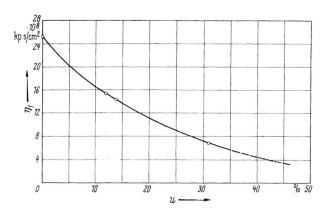

Fig. 7.39. Decrease of internal viscosity of red beech with increased moisture content in the hygroscopic range for compression tests parallel to the grain. From KOLLMANN (1962)

where $\varepsilon_\infty = \dfrac{\sigma}{E}$ is the elongation after infinite time. We obtain a mathematical description of the stress-strain behavior of a system as shown in Fig. 7.36 by summing up Eqs. (7.35) and (7.37):

$$\varepsilon = \frac{\sigma}{E_1} + \frac{\sigma}{E_2}\left(1 - e^{-\frac{t}{\lambda_2}}\right) + \frac{\sigma}{E_1} \cdot \frac{t}{\lambda_1} \qquad (7.38)$$

KOLLMANN (1962) calculated, from experimental strain-time curves, values for the viscosity $\eta_1$ (as a measure of internal friction) in dependence on moisture content (Fig. 7.39).

## 7.2 Tensile Strength

### 7.2.1 Tensile Strength of Cellulose Molecules, of Single Wood Fibers, and Breaking Length

MEYER and MARK (1930) have carried out interesting calculations on the tensile strength of a hypothetical rod consisting solely of endless cellulose molecules. Based on the work required for splitting the primary valence bonds a tensile strength of about 80,000 kp/cm$^2$ was determined. They concluded that a body built up exclusively by primary valence chains would exceed in strength all known materials. More recently MARK (1952) estimated that a sample consisting of uninterrupted parallel glucosidic chains would exhibit tenacities of 100,000 to 400,000 kp/cm$^2$. In practice pure cellulose fibers do not consist of endless molecules and the crystalline regions have only a length of about 500 Å. The breaking does not occur by tearing all the single chains, but through slipping. "If we consider the substance as having mixed crystalline and amorphous areas and submit it to a stress, some of the chains which are in a very unfavorable position may have to take the largest part of the tension, and hence may break or slip along one another in an early stage of deformation. Others, which are curled or wound up, may disentangle and gradually become parallel. This leads to an additional crystallization which increases the internal viscosity and makes it possible to accumulate more and more potential energy in the substance. As soon as the stress becomes still higher, the parallelized chains start to slip along one another and produce a flow, in consequence of which the cross-section of the material gradually decreases. This leads to an increase of the stress and finally to the break. Obviously it will not be easy to predict the exact moment of the break and the stress necessary to produce it. Calculations which have been made lead to a value of 12,500 to 15,000 kp/cm$^2$ as a limit for the case of completely parallelized overlapping chains having an average length (e.g. degree of polymerization) of 500 glucose units" (MARK, 1952). This value is also considerably higher than the highest tensile strengths of wood parts. This fact is not surprising since organic materials are built up by biological processes which obey statistical laws. Thus strong and weak spots are distributed statistically in organic tissues and we have to consider that the chain always breaks at its weakest link.

Nevertheless the tensile strength of wood parallel to the grain is extremely high and may reach, for some species in the air-dry condition ($u \approx 12\%$), a maximum of 3,000 kp/cm$^2$. The tensile strength of separated wood fibers is even higher and may vary between about 2,000 and 13,000 kp/cm$^2$. KLAUDITZ and co-workers (1947) determined for spruce fibers — based on the cell wall without lumen

Table 7.6. *Stress-Strain Properties of Wood Fibers*
(From JAYNE, 1959, 1960)

| | Young's Modulus kp/cm² · 10⁵ | Tensile Strength kp/cm² | | Young's Modulus kp/cm² · 10⁵ | Tensile Strength kp/cm² |
|---|---|---|---|---|---|
| Redwood Earlywood Fibers | | | White Fir Earlywood Fibers | | |
| Average | $1.75 \cdot 10^5$ | 4.850 | Average | 2.09 | 5.130 |
| Range | 1.41···2.38 | 3.730···6.750 | Range | 1.59···2.56 | 3.160···6.120 |
| Redwood Latewood Fibers | | | White Fir Latewood Fibers | | |
| Average | 2.99 | 9.140 | Average | 2.88 | 7.310 |
| Range | 1.95···4.29 | 6.890···12.020 | Range | 2.09···4.04 | 5.340···11.810 |
| Sitka Spruce Earlywood Fibers | | | Western Redcedar Earlywood Fibers | | |
| Average | 2.99 | 8.230 | Average | 1.30 | 3.440 |
| Range | 2.30···4.33 | 5.410···12.300 | Range | 0.780···2.33 | 1.900···4.850 |
| Sitka Spruce Latewood Fibers | | | Western Redcedar Latewood Fibers | | |
| Average | 3.61 | 9.070 | Average | 2.43 | 4.780 |
| Range | 2.40···5.24 | 5.830···11.600 | Range | 1.46···4.69 | 3.940···6.050 |
| White Spruce Earlywood Fibers | | | Cypress Earlywood Fibers | | |
| Average | 2.65 | 5.410 | Average | 1.82 | 5.690 |
| Range | 1.49···3.49 | 4.640···6.330 | Range | 1.15···2.66 | 4.640···7.730 |
| White Spruce Latewood Fibers | | | Cypress Latewood Fibers | | |
| Average | 3.75 | 5.980 | Average | 3.75 | 10.400 |
| Range | 2.12···6.13 | 5.410···6.890 | Range | 2.00···6.07 | 8.370···13.150 |
| Slash Pine Earlywood Fibers | | | Western White Pine Earlywood Fibers | | |
| Average | 1.16 | 3.300 | Average | 1.860 | 4.220 |
| Range | 0.55···1.88 | 2.460···5.480 | Range | 1.40···3.07 | 2.950···5.130 |
| Slash Pine Latewood Fibers | | | Western White Pine Latewood Fibers | | |
| Average | 2.20 | 6.470 | Average | 1.67 | 4.640 |
| Range | 1.03···3.78 | 4.290···13.780 | Range | 0.72···2.91 | 3.520···5.340 |
| Douglas-fir Earlywood Fibers | | | Western Hemlock Earlywood Fibers | | |
| Average | 1.86 | 3.590 | Average | 2.19 | 5.910 |
| Range | 1.18···3.35 | 2.950···4.080 | Range | 0.72···3.40 | 4.290···7.450 |
| Douglas-fir Latewood Fibers | | | Western Hemlock Latewood Fibers | | |
| Average | 4.46 | 9.980 | Average | 3.13 | 9.210 |
| Range | 2.90···6.73 | 8.010···13.010 | Range | 1.83···4.37 | 6.610···12.440 |

Ref. p. 414]                    7.2 Tensile Strength                         323

(cell cavity) —a tensile strength of 4,630 kp/cm², and for fibers from *Pinus mercusii* Jungh of 5,780 and of 6,920 kp/cm². JAYNE (1959, 1960), using an electrical-mechanical system for determining force and deformation, accurately measured Young's modulus and tensile strength of single fibers of ten coniferous species. The data presented in Table 7.6 demonstrate not only vast differences between fibers of different species but also regularly between earlywood and latewood fibers of a single species. Ultimate stresses are quite close to the theoretically computed values. Random depolymerization of the cellulose fraction in Douglas-fir wood by means of various doses of gamma irradiation was used by IFJU (1964) to study the influence of cellulose chain length on the mechanical behavior of

Table 7.7. *Tensile Strength and Breaking Length of some Materials*

| Material | Tensile strength kp/cm² | Breaking length km |
|---|---|---|
| Steel wire, maximum | 32,000 | 41 |
| Iron wire, hard drawn | 5,500··· 8,400 | 7  ···10.8 |
| Steel, building | 5,200··· 6,200 | 6.7···  8 |
| Copper wire, hard drawn | 4,200··· 4,900 | 4.7···  5.4 |
| Flax, Irish | 6,000···11,000 | 40···75 |
| Rayon, Acetate | 10,000 | 75 |
| Silk | 3,500 | 25 |
| Cotton | 2,800··· 8,000 | 18···53 |
| Hemp | 8,000··· 9,000 | 52···58 |
| Coniferous woods | 500··· 1,500 | 11···30 |
| Broadleaved woods | 200··· 2,600 | 7···30 |
| Bamboo | 1,000··· 2,300 | 10···35 |

wood in tension parallel to the grain. His experimental results indicate that tensile-strength properties of latewood are not only distinctly higher than those of earlywood, but also the response of the two growth zones to changes in the degree of polymerization of cellulose, moisture content, and temperature is different. Strength of wood with a low degree of polymerization of cellulose is more sensitive to moisture changes than that of wood having cellulose of long chain structure. Therefore, agencies which depolymerize cellulose will not only weaken wood but will also make it more sensitive to variations in atmospheric humidity.

On a density basis, expressed as breaking length (the ratio of tensile strength to density meaning the length of a strip of the material which will break under its own weight) as introduced by F. REULEAUX in 1861, most woods used for construction are relatively stronger than many metals (Table 7.7).

Unfortunately the high tensile strength of wooden parts cannot be utilized in construction for several reasons. First of all, the shearing strength along the grain is extremely low (only about 6 to 10%) in comparison to the tensile strength along the grain. Therefore the wood tends toward shear failures or cleavage at the fastenings or joints. Secondly the tensile strength is considerably reduced by knots, cross grain and/or any other irregularity in growth. Finally, tensile tests parallel to the grain are difficult: the manufacture of the test specimens needs much skilled manual labor, and clamping the samples in the machine implies the possibility of torque or of compressive stresses perpendicular to the grain which are too high. The failure in the sample is often not entirely tensile. Tensile tests along the grain are therefore scarce in comparison to compression or bending tests. Figures on tensile strength along the grain in the literature are limited in number and reliability.

21*

## 7.2.2 Determination of Tensile Strength along the Grain

Since, as mentioned above, the tensile strength of wood along the grain is much greater than the compressive strength perpendicular to the grain and the shearing strength, it is difficult to carry out satisfactory tests in tension parallel to the grain. The ends of the test samples, which are stressed both by compression and shear must be stronger than the smaller central cross-section subjected to pure tensile stress. A few recommended specimens are shown in Fig. 7.40. KUFNER

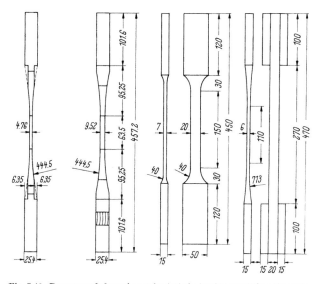

Fig. 7.40. Recommended specimens for tests in tension parallel to the grain

(1963) has proved by photoelastic tests that dangerous stress concentrations in tensile test specimens are caused not so much by their geometrical shape as by the sequence of earlywood and latewood. Fig. 7.41 shows a very dangerous stress peak on a latewood lamella cut in producing the reduced cross-section.

All specimens should be clear and straight grained. The annual rings should be so oriented that they are parallel to the small face of the cross-section.

A universal testing machine equipped with self-aligning and self-tightening grips and an extensometer are used for the tensile test. If only the ultimate stress $\sigma_{tb}$ in kp/cm² is desired, it can be computed by dividing the ultimate load $P$ in kp by the cross-section area $A_0$ of the slenderest position of the samples prior to the test in cm²:

$$\sigma_{tb} = \frac{P}{A_0}. \tag{7.39}$$

If the modulus of elasticity in tension is required, load-elongation readings must be taken and the modulus can be obtained by substitution of respective values for stress and strain at or below the proportional limit in Eq. (7.13).

ASTM Designation: D 143-52 recommends the following procedure in determining the tension-parallel-to-grain strength of wood. Size and shape of the specimens is shown on the left side of Fig. 7.40. The annual rings shall be perpendicular to the greater cross-sectional dimension. The specimen shall be held

Ref. p. 414]  7.2 Tensile Strength  325

in special grips. The load shall be applied continuously throughout the test at a rate of motion of the movable crosshead of 1 mm per min. The failure shall be sketched on the data sheet.

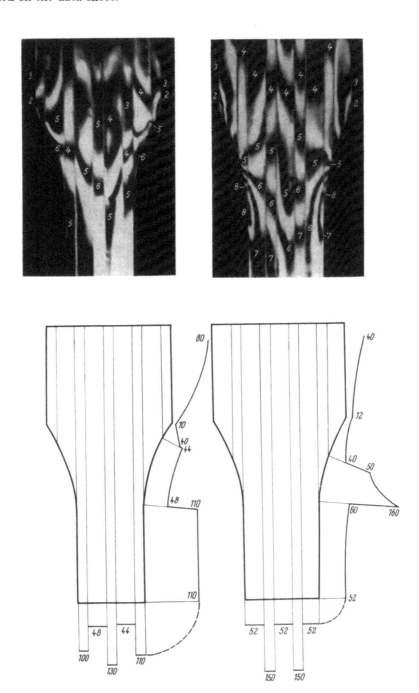

Fig. 7.41. Photoelastic stress pattern (above) and the derived boundary stresses (below, kp/cm²) on laminated models. The numbers above indicate the order of the isochromatic fringes. From KUFNER (1963)

### 7.2.3 Factors Affecting the Tensile Strength along the Grain

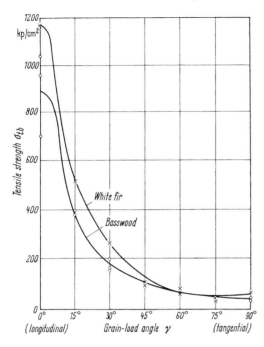

Fig. 7.42. Dependence of the tensile strength upon the angle between the tension direction and the direction of the grain for white fir and for basswood. From BAUMANN (1922)

**7.2.3.1 Grain Angle.** The following formula — known as HANKINSON's formula — is suitable for computing the tensile strength of wood $\sigma_\gamma$ in which the direction of the grain is inclined at an angle $\gamma$ to the direction of the load:

$$\sigma_\gamma = \frac{\sigma_\parallel \cdot \sigma_\perp}{\sigma_\parallel \cdot \sin^n\gamma + \sigma_\perp \cdot \cos^n\gamma} \quad (7.40)$$

in which $\sigma_\parallel$ represents the tensile strength parallel to the grain ($\gamma = 0$), $\sigma_\perp$ the tensile strength perpendicular to the grain ($\gamma = 90°$) and $n$ is a constant.

KOLLMANN (1934) has shown that values of $n$ between 1.5 and 2 are satisfactory. BAUMANN (1922) has published the curves in Fig. 7.42 and 7.43; from the latter it can be seen that tensile strength is much more reduced by cross grain than bending or compressive strength.

**7.2.3.2 Density.** There is, at least for coniferous species, a linear tensile strength-density relationship, as shown in Fig. 7.44. YLINEN (1942) deduced the following equation

$$\sigma_{tb} = \sigma_{tbe} + (\sigma_{tbl} - \sigma_{tbe}) \cdot \frac{s}{100} \quad (7.41)$$

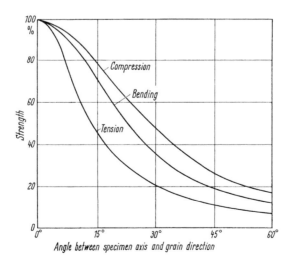

Fig. 7.43. Dependence of the relative strength properties upon the angle between specimen axis and grain direction. From BAUMANN (1922)

where $\sigma_{tb}$ represents the tensile strength along the grain of a piece of wood which contains earlywood and latewood, $s$ = percentage of latewood, $\sigma_{tbe}$ = tensile strength of the earlywood, and $\sigma_{tbl}$ = tensile strength of the latewood.

For Finnish pinewood (moisture content $u = 8 \cdots 10\%$) the following values were determined

$$\sigma_{tbe} = 650 \text{ kp/cm}^2$$
and  $\sigma_{tbl} = 3{,}940 \text{ kp/cm}^2$.

Fig. 7.45 illustrates how well the theoretical linear function fits the experimental data plotted in the diagram. It may be mentioned that the tensile strength along the grain of compression wood is only 50 to 60% that of normal wood due to the large fibril angle in the cell wall (TRENDELENBURG, 1932).

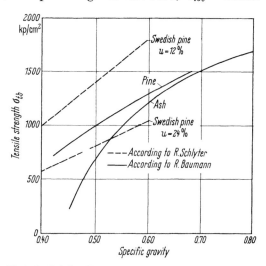

Fig. 7.44. Relationship between tensile strength parallel to the grain and specific gravity. From SCHLYTER (1929) and BAUMANN (1922)

### 7.2.3.3 Moisture Content.

Tensile strength along the grain increases as wood dries below the fiber saturation point. Since tensile strength is very sensitive to internal factors, such as growth features and defects, wide scattering of test results is unavoidable. In addition, only few tests have been made. SCHLYTER and WINBERG (1929) reported that from 10% moisture content up to the fiber saturation point there is a linear decrease of tensile strength (Fig. 7.46). According to the U.S. Forest Products Laboratory a 1% increase of moisture content lowers the tensile strength along the grain about 3%. Investigations of GABER (1937), GRAF (1937/38) and KÜCH (1943) brought evidence that there is a peak of tensile strength between 8 and 10% moisture content. The physical reasons for this maximum are still not clear.

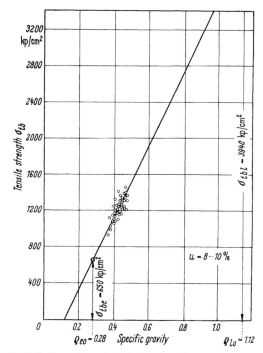

Fig. 7.45. Dependence of the tensile strength upon the specific gravity for Finnish pine. From YLINEN (1942)

### 7.2.3.4 Temperature.
The thermal phenomena described in section 7.1.6.4, influencing strength and stiffness of solid bodies, affect the tensile strength along the grain less than other mechanical properties (Fig. 7.47).

**7.2.3.5 Knots and Notches.** Knots reduce strength for the following reasons: (1) the grain of the wood is distorted in passing around them, (2) the fibers of the knots are nearly at a right angle to the fibers of the wood (3). Checks are fre-

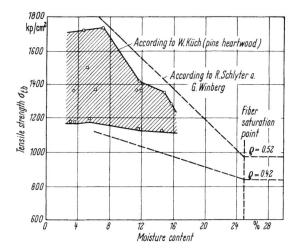

Fig. 7.46. Dependence of the tensile strength parallel to the grain upon the moisture content for pine. From KÜCH, SCHLYTER and WINBERG (1929)

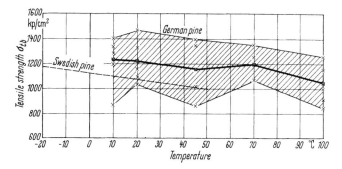

Fig. 7.47. Dependence of the tensile strength parallel to the grain upon the temperature for pine. From KOLLMANN and SCHULZ (1944)

Table 7.8. *Influence of Knots on Tensile and Compressive Strength* (From GRAF, 1929, 1938)

| Pinewood | Density g/cm³ | Tensile strength kp/cm² | Reduction % | Compressive strength kp/cm² | Reduction % |
|---|---|---|---|---|---|
| Clear | 0.50 | 780 | | 403 | |
| Few, small. knots | 0.53 | 384 | 51 | 361 | 10 |
| Many, larger knots | 0.57 | 119 | 85 | 314 | 22 |

quent in dry knots. The weakening effect of the knots is especially great on the tensile strength. Fig. 7.48 shows some results obtained by GRAF (1929, 1938). The wood shown under a) is selected structural material (grade I according to German standard specification DIN 4074), the wood in the center under b) belongs to grade II, while the wood at the left under c) is not admitted for structural purposes. It is extremely interesting to learn from the figures given in Table 7.8 that the tensile strength of the selected wood containing only a very small knot

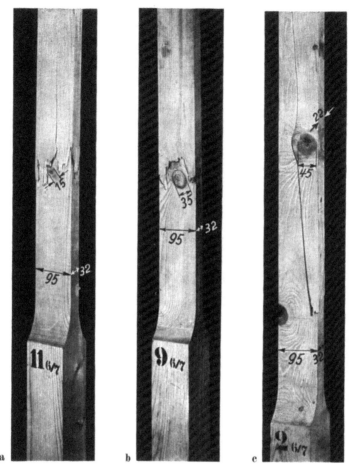

Fig. 7.48. Results from tension tests of specimens with knots of different sizes. From GRAF (1929, 1938). a) Spruce grade I, tensile strength 384 kg/cm², b) pine grade II, tensile strength 258 kp/cm², c) spruce not admitted for structural purposes, tensile strength 119 kp/cm²

reached only about 50% of the tensile strength of clear wood without any knot. It is further to be seen that the compressive strength is much less affected by knots.

Notches induce an irregular distribution of stresses with rather high peaks over the cross-section. Due to these peaks the tensile strength is lowered. REIN (1942) has investigated the sensitivity of wood, laminated woods, and Compreg to notches. There are two main types of notches: a) angular or round notches at the edges and b) notches (holes) in the inner parts. The ultimate stress $\sigma_N$ of a notched tension member may be calculated as follows

$$\sigma_N = \frac{P}{A_N} = \alpha_N \cdot \sigma_{tb} \qquad (7.42)$$

where $P$ = maximum load [kp], $A_N$ = area of the cross-section without notches, $\alpha_N$ = notch factor and $\sigma_{tb}$ = tensile strength of the intact cross-section without notches.

Generally edge-notches reduce the tensile strength more than internal notches. Laminated woods are more affected than solid wood. The shape of the notches apparently has little or no influence. Fig. 7.49 shows how the notch-factor is diminished with increasing ratio of the internal notch area to the unweakened cross-section. A few ranges of values for notch-factors are listed in Table 7.9.

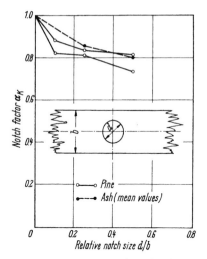

Fig. 7.49. Dependence of the notch factor upon the relative notch size. From REIN (1943)

Table 7.9. *Range of Notch-Factors*
(From REIN, 1942, and WINTER, 1944)

| Material | Notch-factor $\alpha_N$ | |
|---|---|---|
| | Internal notches | Edge notches |
| Pine | 1···0.73 | 1···0.77 |
| Ash | 1···0.80 | — |
| Laminated beech | 1···0.58 | 1···0.51 |
| Compreg | 1···0.73 | 1···0.47 |

### 7.2.4 Determination of Tensile Strength perpendicular to the Grain, Cleavage

In any structural parts tensile stresses perpendicular to the grain are carefully avoided since not only is the tensile stress across the grain very low but checks

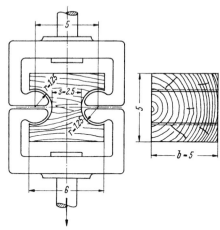

Fig. 7.50. Specimen with testing grips for tensile tests perpendicular to the grain according to ASTM D 143–52

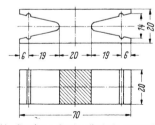

Fig. 7.51. Specimen for tensile tests perpendicular to the grain according to the French standard (specifications AFNOR)

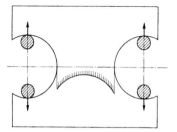

Fig. 7.52. Stress distribution over the medium cross-section of a double-cleavage specimen according to American and British standards. From COKER and COLEMAN (1930)

or shakes due to shrinkage can completely destroy the tensile strength across the grain. In Germany tensile tests perpendicular to the grain were rarely carried out though they are quite common (as optional tests) in USA, England, France, and other countries. The size and shape of the specimens according to ASTM D 143-52 is shown in Fig. 7.50. The testing grips used are presented in Fig. 7.50. The French

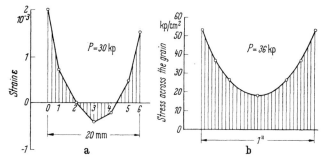

Fig. 7.53 a, b. Stress distribution in double-cleavage specimens a) from strain measurements on specimens according to the French standards. From KEYLWERTH (1944/45). b) From photoelastic tests on specimens according to the American standards. From COKER and COLEMAN (1930)

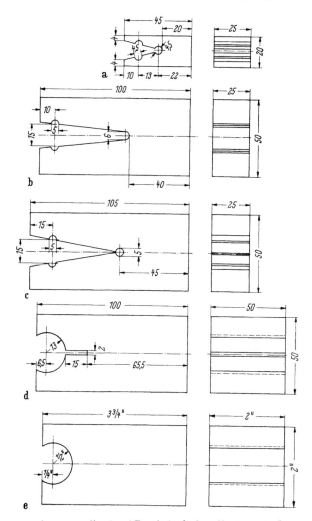

Fig. 7.54 a-e. Cleavage specimens according to: a) French standard specifications; b) NÖRDLINGER; c) SCHWANKL; d) UGRENOVIČ; e) American and English standard specifications

standard specimen (similar to that earlier used in Germany) is illustrated in Fig. 7.51. Tests with specimens as shown in Fig. 7.50 and 7.51 are not exact tensile tests perpendicular to the grain but double-cleavage tests. JENKIN, as early as 1921, criticized such tests. The stress is not evenly distributed over the minimum cross-section. Photoelastic investigations by COKER and COLEMAN (1930) clarified this. The stress distribution in Fig. 7.52 is valid for an isotropic material, but KEYLWERTH (1944/45) has proved by strain measurements that similar conditions are given for the anisotropic wood (Fig. 7.53). Doubtless the values obtained with specimens as shown in Fig. 7.50 and 7.51 are not ultimate stresses for tension across the grain, but technological figures closely related to cleavage stresses. Cleavage is the resistance against splitting determined with notched blocks of wood as indicated in Fig. 7.54. The stresses are not uniformly distributed over the surface of failure (Fig. 7.55). A comparison of Fig. 7.55 with Fig. 7.53 shows that the stress curve in the latter is approximately produced by adding a reflected image to the former. Under these circumstances a linear relationship between cleavage resistance of single and double notched specimens is self-evident (Fig. 7.56) and it is not necessary to work with both methods. Retaining one of them, the simple notched block should be preferred since it is

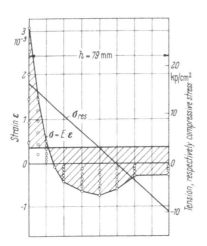

Fig. 7.55. Analytically obtained and actual strain distribution in the cross-section of cleavage specimens. According to the American standards. From KEYLWERTH (1944/45)

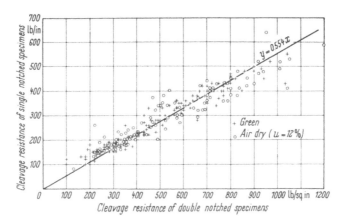

Fig. 7.56. Cleavage resistance of single notched specimens related to the cleavage resistance of double notched specimens. Diagram designed by KEYLWERTH (1944/45), using data published by the US. Forest Products Laboratory

easier to manufacture, since the test itself is performed more easily, and since the results may not be misunderstood. Perhaps cleavage tests are generally useless since their results are proportional to tensile strength, bending strength, and shock resistance values across the grain (Fig. 7.57).

The Forest Products Research Laboratory, Princes Risborough, recommends turned specimens shaped like bobbins (Fig. 7.58) for getting true figures for the

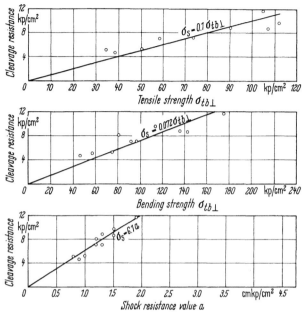

Fig. 7.57. Cleavage resistance as a function of tensile strength, bending strength, and shock resistance value perpendicular to the grain of different wood species. From KEYLWERTH (1944/45)

Table 7.10. *Mechanical Properties of some Commercial Woods Across the Grain* (From KEYLWERTH 1944/45)

| Species | Tensile strength $\sigma_{tb}$ kp/cm² | Bending strength $\sigma_{bb}$ kp/cm² | Shock resistance $a$ kpcm/cm³ | Cleavage $\sigma_{cl} = \dfrac{P_{max}}{F}$ kp/cm² Cleavage specimen | Double cleavage specimen | Density $\varrho_u$ g/cm³ | Moisture content $u$ % |
|---|---|---|---|---|---|---|---|
| Oak | 83···97 / 90 | 128···141 / 135 | 1.2···1.4 / 1.3 | 8.4···9.5 / 8.8 | 20.2···31.8 / 26.3 | 0.64···0.68 / 0.66 | 12.1···12.6 / 12.4 |
| Beech | 96···118 / 107 | 135···146 / 142 | 1.4···1.6 / 1.5 | 8.3···8.8 / 8.6 | 24.6···38.1 / 31.1 | 0.67···0.73 / 0.69 | 11.5···11.8 / 11.7 |
| Hornbeam | — / — | 58···91 / 80 | 0.9···1.3 / 1.2 | 7.1···9.7 / 8.1 | 27.1···33.3 / 30.2 | 0.76···0.78 / 0.77 | 10.8···11.3 / 11.2 |
| Ash | 101···127 / 112 | 62···167 / 113 | 1.3···1.8 / 1.5 | 8.5···10.0 / 9.6 | 25.4···42.3 / 30.6 | 0.67···0.79 / 0.76 | 8.6···9.1 / 0,8 |
| Walnut | 98···114 / 105 | 169···183 / 175 | 1.6···2.2 / 1.9 | 8.8···12.7 / 11.7 | 21.1···32.6 / 29.7 | 0.58···0.66 / 0.60 | 10.9···11.4 / 11.2 |
| Lime | 50···60 / 58 | 90···92 / 91 | 1.0···1.7 / 1.3 | 6.6···7.3 / 7.0 | 20.7···25.4 / 22.8 | 0.57···0.59 / 0.58 | 9.8···10.1 / 10.0 |
| Alder | 69···79 / 73 | 92···99 / 96 | 0.9···1.4 / 1.2 | 6.4···7.4 / 7.0 | 24.5···26.5 / 25.9 | 0.53···0.56 / 0.55 | 10.4···10.8 / 10.7 |
| Spruce | 33···40 / 38 | 42···49 / 46 | 0.7···1.0 / 0.9 | 4.1···5.1 / 4.6 | 15.0···20.4 / 17.3 | 0.48···0.69 / 0.54 | 11.5···11.6 / 11.7 |
| Larch | 48···52 / 50 | 69···77 / 75 | 0.9···1.1 / 1.0 | 4.8···5.5 / 5.2 | 11.3···27.9 / 20.1 | 0.66···0.72 / 0.68 | 11.6···12.3 / 12.1 |
| Pine | 32···37 / 34 | 53···64 / 57 | 0.7···1.0 / 0.8 | 4.7···5.4 / 5.1 | 15.1···19.6 / 17.4 | 0.51···0.58 / 0.55 | 10.9···13.7 / 12.0 |

tensile strength across the grain. The values thus obtained are about 50% greater for the same wood than those obtained with double-cleavage tests. KEYLWERTH (1944/45), using prismatic specimens, has found even higher breaking stresses (Table 7.10).

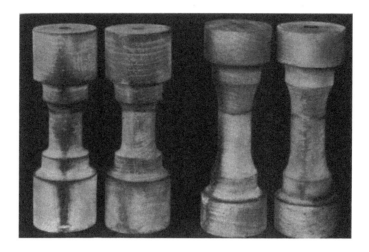

Fig. 7.58. Turned specimens for tensile strength tests across the grain. From the Forest Products Research Laboratory, Princes Risborough

## 7.2.5 Fatigue in Tension Parallel to the Grain

The fatigue resistance of wood parts is a very important factor in design. Most failures in practice are caused by repeated, reversed, or oscillating stresses or by impact loads. The resistance to fatigue may be remarkably reduced by defects such as knots and slope of grain or by notches (e.g. by abrupt changes in cross-section). If the load is not reversed but constant static loading acts over a long period ("dead load") the phenomena of creep and creep recovery (described in Section 7.1.7.2) determine the behavior of the wood. Fig. 7.59 shows that pinewood will bear a dead load in the order of 60% of the ultimate short time tensile strength.

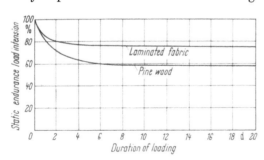

Fig. 7.59. Static endurance load in tension of pine and laminated fabric. From RIECHERS (1938)

Apparently laminated fabric has more favorable properties in this respect (RIECHERS, 1938). Doubtless the resistance of wood to dead loads, causing tensile stresses, depends on the beginning of plastic deformations. IVANOV (1938) has analyzed the situation. ROTH (1935) found in testing oak wood and spruce in compression and bending tests that dead loads produce greater and more enduring effects than corresponding alternating loads. From this point of view more attention should be paid to dead load tests.

It is possible to test specimens as shown in Fig. 7.60 by means of special machines either with tensile stresses alternating for each cycle between zero and a maximum value, between a minimum and maximum value, or with repetitive

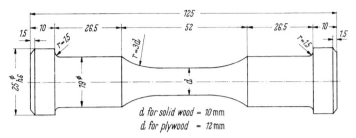

Fig. 7.60. Specimen for tension and compression tests. From KOLLMANN and DOSOUDIL (1944, unpublished)

tensile and compressive stresses. EGNER and ROTHMUND (1944) using the first method, determined an endurance stress for clear sprucewood with a high density ($\varrho = 0.54 \, \text{g/cm}^3$) of about 60% of the static tensile strength and for clear sprucewood with a low density ($\varrho = 0.42 \, \text{g/cm}^3$) of about 52% of the static tensile strength. According to the Wood Handbook (1955, p. 82) "tests in fatigue in tension parallel to the grain indicate an endurance load for 30 million cycles of stress of about 40% of the strength of air-dry wood as determined by a standard static test".

## 7.3 Maximum Crushing Strength and Stresses in Wood Columns

### 7.3.0 General Considerations

The maximum crushing strength plays an important role in the utilization of wood as a building and construction material. The tests of short wood columns ("blocks") or even of cubes, are easily carried out. The results elucidate a characteristic mechanical property and permit conclusions on the biological quality of the wood, and to some extent on its strength in general. The maximum crushing strength parallel to the grain reaches, on the average, for air-dry woods only

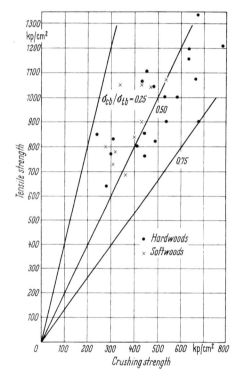

Fig. 7.61. Relationship between the average tensile strength and the average crushing strength of various wood species. From KOLLMANN (1951)

about 50% of the tensile strength along the grain. However, this ratio is extremely variable (Fig. 7.61) and is influenced by the moisture content (KOLLMANN, 1964, Fig. 7.62). The different behavior of wood in compression and tension may be explained by its fibrous structure. Its tightly wedged and cemented fibers sustain, as discussed in Section 7.2.1, very high tensile stresses; in compression probably an early buckling of individual fibers occurs, starting the failure.

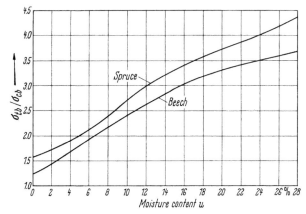

Fig. 7.62. Dependence of the ratio crushing strength to tensile strength upon the moisture content for beech and spruce wood. From KOLLMANN (1964)

### 7.3.1 Testing in Compression Parallel to Grain

A universal static testing machine and a compressometer are required. At least one plate of the testing machine should be equipped with a spherical bearing to obtain uniform stress distribution over the cross-section of the specimen. According to ASTM D 143-52 the compression-parallel-to-grain tests shall be made on 2 in. × 2 in. × 2.8 in. (5 cm × 5 cm × 20 cm) specimens. According to the German standard specification DIN 52 185 2 cm × 2 cm are allowed as minimum cross-section. The actual cross-section dimensions and the length shall be measured. If the cross-section is increased over 5 cm × 5 cm, the crushing strength decreases as a rule due to unavoidable irregularities and defects. The end grain surfaces of the blocks must be parallel to each other and at right angles to the longitudinal axis. The maximum crushing strength depends on the ratio of length $l$ to least dimension $a$. BAUMANN (1922) obtained the following results for coniferous air-dry species:

| Ratio $l/a$ | 0.5 | 1 | 3···6 |
|---|---|---|---|
| relative crushing % strength | 103 | 100 | 0.93 |

Similar values are given by RYSKA (1932). These phenomena are caused by the friction between the end surfaces of the blocks and the plates which impedes the lateral strain. A recommendable value for $l/a$ is 4.

In the USA the load shall be applied continuously throughout the test at a rate of motion of the movable crosshead of 0.003 in. per in. (cm per cm) of specimen length per min. The German standard specification DIN 52185 prescribes as speed of loading 200 to 300 kp/cm² per min. In France the duration of a compression-parallel-to-grain test shall be at least 2 min. YLINEN (1944) states that it is better to prescribe the speed of strain than the speed of stress. KOLLMANN (1944) showed that in the range between 100 and 600 kp/cm² per min there was no distinct influence of the speed of testing on the values of the crushing strength of pine wood. PEREM (1950) carried out experiments in a wide range of testing speeds. As Fig. 7.63 shows, the lowest speed of testing amounted to 2 kp/cm² per min, and the highest speed to 25,000 kp/cm² per min.

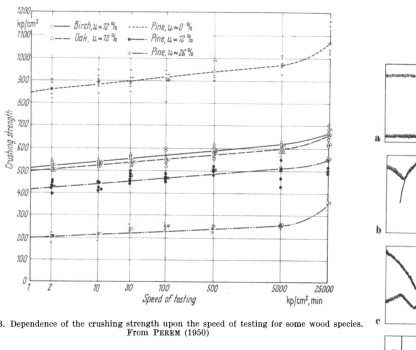

Fig. 7.63. Dependence of the crushing strength upon the speed of testing for some wood species. From PEREM (1950)

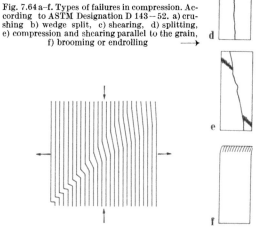

Fig. 7.64 a–f. Types of failures in compression. According to ASTM Designation D 143–52. a) crushing b) wedge split, c) shearing, d) splitting, e) compression and shearing parallel to the grain, f) brooming or endrolling ⟶

Fig. 7.65. Failure in compression test. From KOLLMANN (unpublished)

Fig. 7.66. Scheme of the disappearance of a plane of shear through transition in elastic deformation. From KUNTZE (1933, 1934, 1935)

One can see that crushing strength increases slowly and approximately in a linear manner with increasing speed of testing up to about 5,000 kp/cm² per min. From then on it increases progressively.

Crushing strength $\sigma_{cb}$ is determined as $\quad \sigma_{cb} = \dfrac{P_{max}}{F_0} \quad$ (7.43)

where $P_{max}$ = maximum load in kp and $F_0$ = initial cross-section of the specimen in cm².

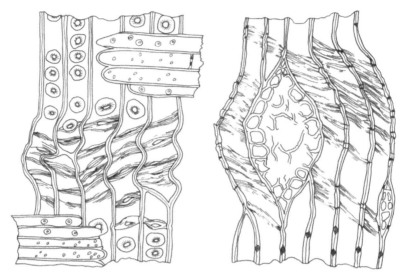

Fig. 7.67. Microscopic structural fractures in dry sprucewood (Picea rubra Link) under compressive stress parallel to the grain. From TIEMANN (1944)

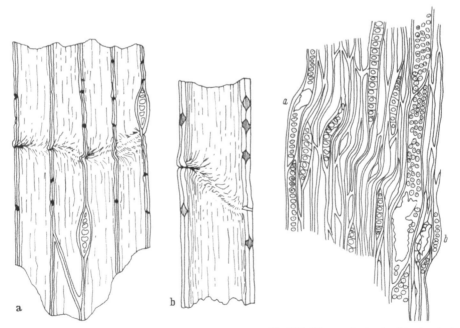

Fig. 7.68 a,b. Microscopic structural fracture in dry sprucewood (Picea rubra Link) under compressive stress parallel to the grain. a) failure start by buckling of the cell wall, ca. 300:1, b) failure of a single tracheid, ca. 600:1. From TIEMANN (1944)

Fig. 7.69. Microscopic structural fracture in dry chestnutwood (Castanea dentata Borkh.) under compressive stress parallel to the grain. The buckling areas of the fibers run together in a plane of shear into direction of the greatest shear stress a−b, ca. 150:1. From TIEMANN (1944)

[Ref. p. 414] 7.3 Maximum Crushing Strength and Stresses in Wood Columns

In order to obtain exact and reliable results, it is necessary that the failure be made to develop in the body of the specimen. ASTM Designation D 143-52 recommends that this result can best be obtained when the ends are at a slightly lower moisture content than the body. Compression failures shall be described and sketched on the data sheet according to the appearance of the fractured surface (Figs. 7.64 and 7.65). It is possible for a plane of shearing rupture in wood to disappear without forming splits (Fig. 7.66, KUNTZE, 1933, 1934, 1935).

The physical-mechanical treatment of the problem is not sufficient. A clear understanding of how compression failure occurs needs discussion of the phenomena of fracture in anatomical elements and tissues of wood. TIEMANN (1944, 1951) gave evi-

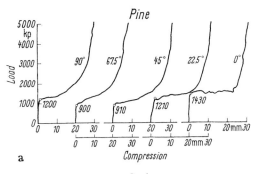
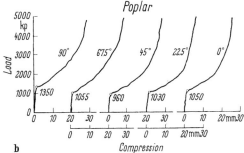
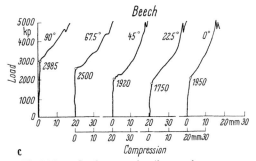

Fig. 7.70 a—c. Load-compression diagrams for compression perpendicular to the grain. From GABER (1940)

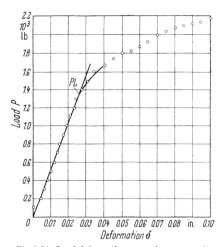

Fig. 7.71. Load deformation curve for compression perpendicular to the grain. From BROWN, PANSHIN and FORSAITH (1952). $PL$ — Proportional limit

dence that under compressive stresses a crumpling of the cell walls in the tracheids of coniferous species is caused (Fig. 7.67 and 7.68). The bordered pits are weak spots in the cell wall and crumble first under the stress. In woods of broad-leaved species the failure starts by a buckling of the cells themselves (Fig. 7.69).

### 7.3.2 Testing in Compression Perpendicular to Grain

The resistance of wood against perpendicular-to-grain compression is practically important for much building constructions and for railway ties. But one has to keep in mind that normally a crushing strength across the grain does not exist. The wood only will be densified under the influence of compressive force acting perpendicular to the grain. Fig. 7.70 shows typical load-deformation curves for perpendicular-to-grain compression of three wood species according to GABER (1940). Thus only the fiber stress at the proportional limit (Fig. 7.71) or the stress which causes 1% deformation are reliable. Other data should be omitted. One has to distinguish two different causes of loading: either on the whole surface of the specimen (it may be a cube or a short wood column) or on a part of its surface. In the last case the load may act on a stamp or like a rail on a sleeper. The wood behaves quite different under these circumstances.

Tests in compression across the grain are made with the same machines as those used in testing in compression parallel to the grain. The relationship between load and deformation should be read by means of a deflectometer. According to ASTM D 143—52 the samples undergoing test should be clear and should measure

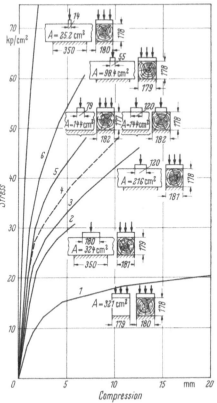

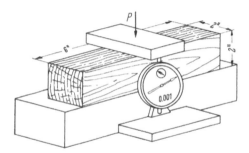

Fig. 7.72. Compression perpendicular to the grain, indicating bearing block and deflectometer. From BROWN, PANSHIN and FORSAITH (1952)

Fig. 7.73. Stress-compression diagrams for compression tests on wood columns loaded perpendicular to the grain compared with the results of cube-tests. From GRAF (1921)

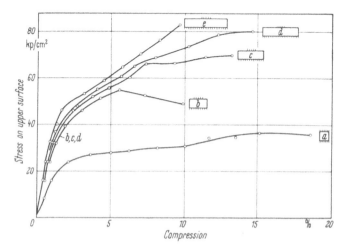

Fig. 7.74. Increase of compression perpendicular to the grain with increase of stress for wood specimens as shown. Specimens length: $a = 15$ cm, $b = 30$ cm, $c = 45$ cm, $d = 60$ cm, $e = 75$ cm. From SUENSON (1937/38)

2 in. × 2 in. × 6 in. The force is applied to the tangential face through a centrally located rectangular steel plate as shown in Fig. 7.72. According to German standard specifications (DIN 52185) the test is carried out either on cubes (length of one edge at the minimum 3 cm) or on short wood columns with a quadratic cross-section (length of one edge at the minimum $a = 5$ cm) and a length $l = 3 \cdot a$. The speed of testing in Germany is 10 kp/cm² per min. In the USA and in the United Kingdom the rate of descent of the machine head should be 0.024 inch per minute corresponding to a constant rate of deformation of $0.012 \pm 20\%$.

GRAF (1921) has investigated how the resistance of wooden columns loaded perpendicular to grain depends on the relative area of compression. The values are plotted in Fig. 7.73 and compared with the results of a cube-test. SUENSON (1937/38) came to similar results as illustrated in Fig. 7.74.

### 7.3.3 Influences Affecting the Crushing Strength

**7.3.3.1 Grain Angle.** For the design and statical computation of engineering structures in wood the knowledge of the dependence of crushing strength on grain orientation is often very important. The results of some experiments are plotted in Figs. 7.75 and 7.76. For very dry or oven-dry wood even rather small angles

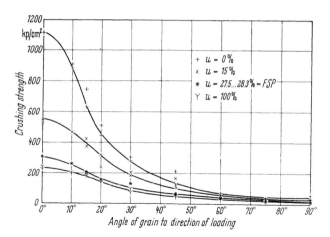

Fig. 7.75. Dependence of the crushing strength upon the angle of grain to direction of loading for pine wood at different moisture contents. From KOLLMANN (1951)

between fiber direction and direction of the force applied reduce the strength remarkably. The reason is the increased brashness due to drying and a more pronounced difference between Young's moduli in very dry earlywood and latewood. Again Hankinson's formula Eq. (7.40) (Section 7.2.3.1) can be used for calculating the compressive strength of wood $\sigma_\gamma$ at an angle $\gamma$ to the direction of the load. The exponent $n$ in Eq. (7.40) may be chosen as 2.5.

The differences between the compressive strength along the grain $\sigma_\parallel$ ($\gamma = 0$) and the compressive strength perpendicular to the grain $\sigma_\perp$ ($\gamma = 90°$) become smaller the denser and the more homogeneous are the woods. FÖPPL (1904) determined as ratio $\sigma_\perp/\sigma_\parallel$ for oakwood 0.294 and BAUMANN (1922), for *Lignum vitae* 0.895. A similar trend characterizes compressed wood with increasing specific gravity.

**7.3.3.2 Density.** The compressive strength of wood along the grain increases with density not only for single species but also for the total range of densities of all species. Fig. 7.77 shows the interrelationship between compressive strength and density for selected pine wood ($u = 12\%$), used in aircraft construction.

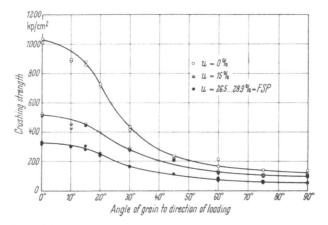

Fig. 7.76. Dependence of the crushing strength upon the angle of grain to direction of loading for beech wood at different moisture contents. From KOLLMANN (1951)

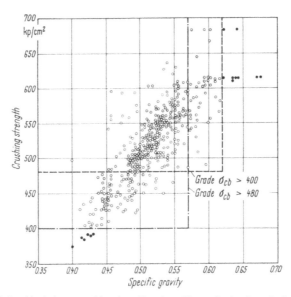

Fig. 7.77. Interrelationship between crushing strength and specific gravity for air-craft pinewood. From HAUBER. (1942)

A similar graph for oven-dry German ashwood is shown in Fig. 7.78. The field of scattering points of measurement may be equalized by the following linear function

$$\sigma_{cb} = I' \cdot \varrho + b. \tag{7.44}$$

BAUSCHINGER (1883), JANKA (1900) and BAUMANN (1922) had stated the validity of this simple equation for air-dry softwood. For ashwood the equation for the regression line in Fig. 7.78 is

$$\sigma_{cb} = 1{,}739 \cdot \varrho_0 - 126.15. \tag{7.44a}$$

It may be assumed, if simplified, that the crushing strength $\sigma_{cb}$ is directly proportional to density:
$$\sigma_{cb} = I \cdot \varrho. \tag{7.45}$$

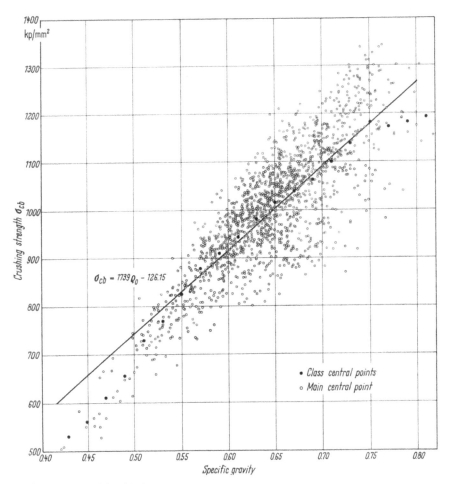

Fig. 7.78. Interrelationship between crushing strength and specific gravity for oven-dry ashwood. From KOLLMANN (1941)

This law was first discussed comprehensively by MONNIN (1919, 1932) in France. Later on SCHLYTER (1927) found the following values for Swedish pinewood

| Moisture content | Quality factor |
|---|---|
| $u\%$ | $I$ |
| 6⋯10 | 1250 |
| 10⋯19 | 850 |
| 23⋯30 | 550 |

MAYER-WEGELIN and BRUNN (1932) determined for Pitch pine $I = 820$, for Douglas fir $I = 830$, and for the heartwood of German pines $I = 730\cdots900$. KOLLMANN (1934) obtained for normal German pinewood ($u = 15\%$) $I = 900$, but the value $I$ dropped to 650 for samples having a high resin content. NEWLIN and WILSON (1919) published the following relationship of specific gravity $R$ (weight oven-dry, volume at the moisture condition indicated) and crushing strength $\sigma_{cb}$, which is based on average results of strength tests of more than 160 species:

$$\sigma'_{cb} = \quad\quad \text{Green wood} \quad\quad \text{Air-dry } (u = 12\%)$$
$$\quad\quad\quad\quad 6{,}730\,R \quad\quad\quad 12{,}200\,R \quad \text{p.s.i}$$

or converted to metric units

$$\sigma_{cb} = \quad\quad 470\,R \quad\quad\quad 860\,R \quad\quad \text{kp/cm}^2$$

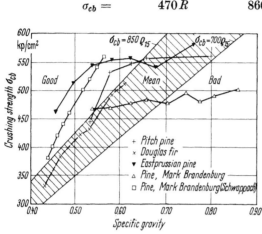

Fig. 7.79. Quality zones of air-dry pines Diagram designed by KOLLMANN (1934) using measurements by MAYER-WEGELIN, BRUNN and SCHWAPPACH (1932)

In his fundamental investigations MONNIN pointed out that the ratio of crushing strength to specific gravity is well suited to the comparison of qualities. He advised the treatment of softwoods and hardwoods separately. The quality figure or "crushing length"

$$I = \frac{\sigma_{cb}}{100\,\varrho_u} \text{ [km]} \quad (7.45\text{a})$$

where $\sigma_{cb}$ in kp/cm$^2$, $\varrho_u$ in g/cm$^3$, may be understood by analogy to the "breaking length" (cf. Section 7.2.1) as the length of a rod of the material which will crush under its own weight. Table 7.11 shows the classification system for woods by MONNIN according to their quality figures.

Table 7.11. *Quality Figures and Properties of Wood*
(From MONNIN, 1919, 1932)

| Wood species | $I = \dfrac{\sigma_{dB}}{100 \cdot \varrho_{15}}$ [km] | | | | Properties |
|---|---|---|---|---|---|
| | weak | medium hard | hard | very hard | |
| Coniferous species | <8 | <7 | 6< | — | poor |
| | 8...9.5 | 7...8.5 | 6...7.5 | — | fair |
| | >9.5 | >8.5 | >7.5 | — | good |
| Broadleaved species | <7 | <6 | <6 | <7 | poor |
| | 7...8 | 6...7 | 6...7 | 7...8 | fair |
| | >8 | >7 | >7 | >8 | good |

The application of Table 7.11 to the results of strength tests for air-dry pine woods is demonstrated by Fig 7.79. It is very essential that only the results for woods with the same moisture content or values reduced to the same moisture content be compared. Fig. 7.80 shows the relationship between crushing strength and specific gravity for oven-dry, air-dry and water-saturated (green) wood species in the natural range of density.

# 7.3 Maximum Crushing Strength and Stresses in Wood Columns

The suitability of wood and wood base materials for light weight construction is illustrated by Fig. 7.81 which shows their quality figures in the main axes of

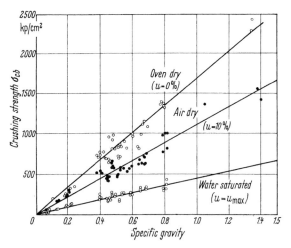

Fig. 7.80. Relationship between crushing strength and specific gravity for oven-dry, air-dry and water-saturated wood species in the natural range of density, temperature 20 °C. From KOLLMANN (1940)

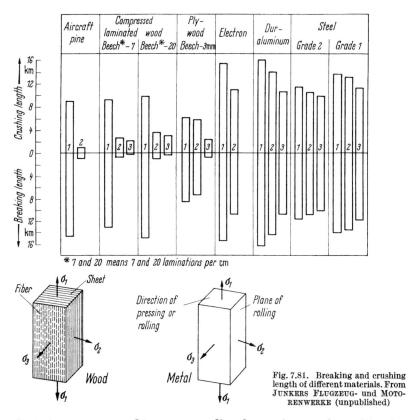

\* 7 and 20 means 7 and 20 laminations per cm

Fig. 7.81. Breaking and crushing length of different materials. From JUNKERS FLUGZEUG- und MOTORENWERKE (unpublished)

fiber orientation as compared to corresponding figures for metals used in aircraft construction.

The proportion of latewood and the width of annual rings has a clearly defined effect on the crushing strength. Fig. 7.82 gives an example for Finnish pinewood. Assuming equal moisture content, Eq. (7.41) may be used introducing the following values:

Crushing strength ($u = 8 \cdots 10\%$) of latewood $\sigma_{cbl} = 1{,}660$ kp/cm$^2$

earlywood $\sigma_{cbe} = 325$ kp/cm$^2$

According to CLARKE et al. (1933) for a given density the crushing strength of English ashwood is the lower the higher is the proportion of late wood (Fig. 7.83).

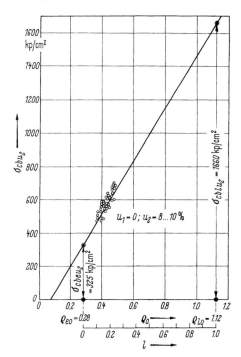

Fig. 7.82. Relationship between crushing strength and specific gravity for Finnish pine wood. From YLINEN (1942). $l$ — proportion of latewood

Since the crushing strength is proportional to density and the density of very many species varies at the same moisture content with a normal distribution curve according to Gauss (cf. Section 6.1.4), similar relationships are to be found for the frequency distribution of the crushing strength (Fig. 7.84). Table 7.12 lists a few figures on the effect of position in the tree and the moisture content on the crushing strength of ashwood.

**7.3.3.3. Moisture Content.** Below the fiber-saturation point the strength of wood increases as it dries. Above the fiber-saturation point the effect of moisture content on static strength is negligible. The decrease of the crushing strength of wood in swelling is easily understood if we consider that water molecules are deposited between the micelles thus increasing the distances between them. This causes a reduction of the intermicellar attractive forces and therefore of the cohesion.

In Fig. 7.85 the results of compressive tests on four different species are plotted. The measured points do not scatter to a remarkable extent. The curves are smooth and may be substituted in the moisture content range $8\% \leq u \leq 18\%$

Table 7.12. *Effect of Position in the Tree and of Moisture Content on the Crushing Strength of Ashwood in kp/cm$^2$*

| Tested by | CLARKE, CHAPLIN and ARMSTRONG (1933) | | KALNINŠ (1941) | KOLLMANN (1941) | | |
|---|---|---|---|---|---|---|
| | Site 13 and 21 | Site 97 | | Stem I | Stem II | Stem III |
| Moisture content | Green wood | Green wood | 13⋯15% | 0% | 0% | 0% |
| Butt | 195⋯214 | 242⋯313 | 463⋯560 | 714⋯1,102 | 834⋯1,242 | 750⋯1,080 |
| Middle of stem | 242⋯248 | 281⋯328 | 494⋯612 | 824⋯1,208 | 852⋯ 979 | 890⋯1,126 |
| Top | 249⋯274 | 288⋯348 | 476⋯599 | 955⋯1,345 | 736⋯1,112 | 1,004⋯1,149 |

## 7.3 Maximum Crushing Strength and Stresses in Wood Columns

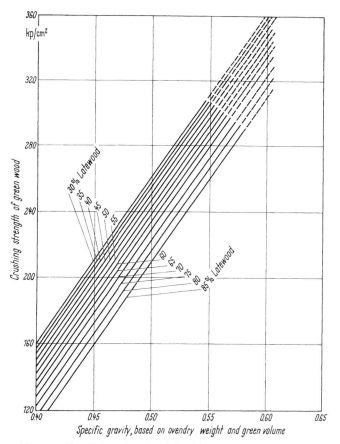

Fig. 7.83. Effect of the proportion of latewood on the relationship between crushing strength and specific gravity. From CLARKE et al. (1933)

by straight lines. Similar conclusions were drawn by JANKA and HADEK (1900), SCHLYTER and WINBERG (1929), KÜCH and TELSCHOW (1942). For similar species, e.g. for woods of conifers, the point of intersection of this straight line with the abcissa is approximately $u = 32\%$. Then the following formula—similar Eq. (7.28)—holds

$$\frac{\sigma_2}{\sigma_1} = \frac{b - u_2}{b - u_1} = \frac{32 - u_2}{32 - u_1} \quad (7.46)$$

where $\sigma_1$ = crushing strength at the moisture content $u_1$, $\sigma_2$ = = crushing strength at the moisture content $u_2$. A large series of calculations proved the validity of

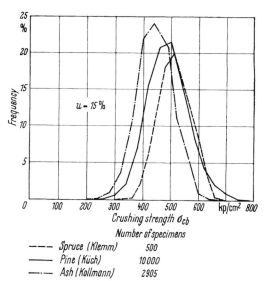

Fig. 7.84. Frequency distribution of the crushing strength of spruce, pine and ash

Eq (7.46) which gives rather good results. This formula is therefore incorporated into the German standard specification DIN 52185: Testing of crushing strength.

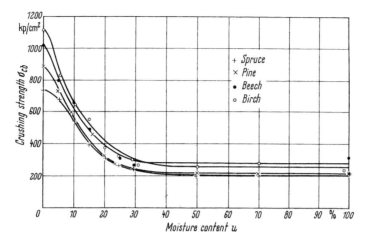

Fig. 7.85. Effect of moisture content on the crushing strength of some wood species. From KOLLMANN (1951)

In the USA, for rough adjustments of values for maximum crushing strength parallel to grain, an average increase (or decrease) of 6% for a 1%-decrease (or increase) in moisture content is assumed (Wood Handbook, 1955, p. 85). A more accurate adjustment can be made by using the exponential formula, devised by the US Forest Products Laboratory:

$$\log \sigma_3 = \log \sigma_1 + \frac{u_1 - u_3}{u_1 - u_2} \log \frac{\sigma_2}{\sigma_1} \qquad (7.47)$$

where $\sigma_1$ and $u_1$ are one pair of corresponding strength and moisture content values as found from test, $\sigma_2$ and $u_2$ are another pair, and $\sigma_3$ is the strength value adjusted to the moisture content $u_3$.

If one strength value is for green wood the moisture content value that must be used is somewhat lower than the fiber saturation point. This value, designated as $u_p$ is listed in Tabl. 7.13.

Table 7.13. $u_p$-Values for Green Species, to be Used in Eq (7.47)

|  | $u_p$ % |  | $u_p$ % |
| --- | --- | --- | --- |
| Ash, white | 24 | Pine, longleaf | 21 |
| Birch, yellow | 27 | Pine, red | 24 |
| Chestnut, American | 24 | Redwood | 21 |
| Douglas fir | 24 | Spruce, red | 27 |
| Hemlock, western | 28 | Spruce, Sitka | 27 |
| Larch, western | 28 | Tamarack | 24 |
| Pine, loblolly | 21 | Other species | 25 |

The exponential formula is not applicable if the moisture content varies to a large extent within the test specimen. The "crushing length" or quality figure I

—cf. Eq. (7.45a)—decreases even above the fiber saturation point, since the specific gravity increases with increasing moisture content (Fig. 7.86). The resistance of wood against compression perpendicular to grain is less influenced by moisture content than the crushing strength along the grain (STAMER, 1920). KOLLMANN (1951) published a diagram (Fig. 7.87) which allows the application of Eq. (7.46) in which as the abcissa of the intersection point instead of $u = 32$ the value $u = 40\%$ should be introduced.

### 7.3.3.4 Temperature.
A series of experiments carried out by KOLLMANN (1940) in the range of temperatures between $-191\,°C$ and $+200\,°C$ has shown that the crushing strength $\sigma$ of oven-dry wood, tested at different temperatures which are below the level where thermal decomposition starts, decreases as the temperature is increased according to a straight-line relationship (Fig. 7.88)

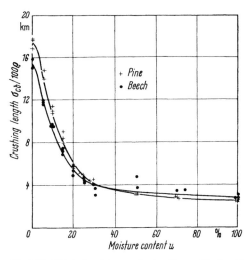

Fig. 7.86. Effect of moisture content on the crushing length (quality figure) of pine and beech. From KOLLMANN (1951)

$$\sigma_2 = \sigma_1 - n(\vartheta_2 - \vartheta_1) \quad (7.48)$$

where $\sigma_2$ and $\sigma_1$ represent crushing-strength values at temperatures $\vartheta_2$ and $\vartheta_1$, respectively. The experiments also showed that the slope $n$ of the strength-temperature curves varies directly with specific gravity $\varrho_0$ according to equation

$$n = 4.76 \varrho_0 \quad (7.49)$$

With respect to $0\,°C$ Eq (7.48) becomes

$$\sigma_\vartheta = \sigma_0 - 4.76 \varrho_0 \cdot \vartheta \quad (7.50\text{a})$$

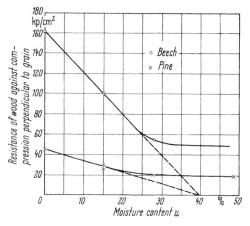

Fig. 7.87. Effect of moisture content on the resistance of wood against compression perpendicular to the grain. From KOLLMANN (1951)

where $\sigma_\vartheta$ is the crushing strength along the grain in kp/cm² at the temperature $\vartheta$. It may be convenient to express the equation in English units:

$$\sigma'_\vartheta = \sigma'_{32} - 37.61 \varrho_0 (\vartheta - 32) \quad (7.50\text{b})$$

where the strength values are given in psi.

Investigations by SULZBERGER (1948) completely confirmed and extended KOLLMANN's findings. Table 7.14 gives some relations statistically derived from the experimental points of both researchers between maximum crushing strength and density at various moisture contents and at 20 °C temperature. It can be seen that the proportionality factors at the same moisture contents are practically identical.

Table 7.14. *Relation Between Crushing Strength and Density of Wood at 20°C*

| Moisture content of wood % | According to SULZBERGER (1948) Crushing strength kp/cm² (density ϱ in g/cm³) | According to KOLLMANN (1940) Crushing strength kp/cm² (density ϱ in g/cm³) |
|---|---|---|
| 0 | 1,654.7 · $\varrho_0$ | 1,640 · $\varrho_0$ |
| 8 | 1,188.9 · $\varrho_8$ | — |
| 10 | — | 1,076 · $\varrho_{10}$ |
| 12 | 959.3 · $\varrho_{12}$ | — |
| 20 | 616.4 · $\varrho_{20}$ | — |
| 24 | 504.9 · $\varrho_{24}$ | — |
| Saturation | 431.1 · $\varrho_{u\max}$ | 442 · $\varrho_{u\max}$ |

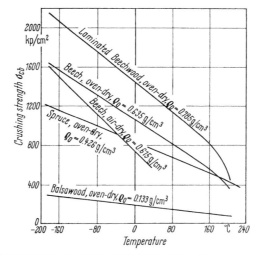

Fig. 7.88. Effect of temperature on the crushing strength of solid and laminated wood. From KOLLMANN (1940)

SULZBERGER (1948) also calculated similar temperature coefficients $n$ depending on the density of wood at different moisture contents and found the values assembled together in Table 7.15.

Table 7.15. *Temperature Coefficients n for Crushing Strength of Wood*

| Moisture content of wood % | $n$ kp/cm² °C | Moisture content of wood % | $n$ kp/cm², °C |
|---|---|---|---|
| 0 | 5.44 · $\varrho_0$ | 20 | 6.37 · $\varrho_{20}$ |
| 8 | 7.73 · $\varrho_8$ | 24 | 4.53 · $\varrho_{24}$ |
| 12 | 7.90 · $\varrho_{12}$ | Saturation | 3.15 · $\varrho_u$ max |

Generally if one puts $n = C\varrho_u$, then in the region $0 \leq u \leq 24$ the constant $C$ can be very well represented by a quadratic function. From the equation set up by SULZBERGER it can be deduced that $C$ is maximum at about 11%, thus at this moisture content the temperature influence is highest. More important is the fact that the experimental results vary so little that actually for any wood the temperature coefficient for the change of crushing strength can be calculated only from its density. The curves in Fig. 7.89 which show the isotherms of the crushing strength as influenced by moisture in the hygroscopic region and at saturation for one beech wood are interesting. In these figures it is conspicuous that the crushing strength of wood frozen at $-20\,°C$ in the entire region of hygroscopic moisture was considerably higher than the crushing strength at temperatures above $20\,°C$ and that the difference is especially large with water saturated wood.

For wood containing moisture the increase in maximum crushing strength, as influenced by a decrease in temperature, does not follow a straight line. The phenomena are complicated due to the very different effect of the freezing or melting ice lattice included in the minute interstices of the fibers. The ice itself strengthens the cell wall, but may be melted under the influence of compressive strains.

Fig. 7.89 allows the comparison of two series of compression tests on beech wood at various moisture contents between the oven-dry state and about 115%

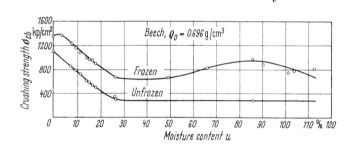

Fig. 7.89. Dependence of the crushing strength of beech upon moisture content and temperature. From KOLLMANN (1940)

moisture. The curve below is for unfrozen wood tested at room temperature ($\vartheta = 20\,°C$); the upper curve is for a corresponding test series of frozen samples ($\vartheta = -42\,°C$). The curve for unfrozen wood depicts the well-known decrease of the crushing strength of wood at a uniform rate down to the fiber saturation point; above the fiber saturation point the crushing strength is independent of the amount of free water. The upper curve shows an increasing crushing strength of the frozen wood up to 5% moisture level. It is assumed that the water molecules, bound in the fibrils by chemosorption, form a continuous film of minimum thickness and maximum strength, when the wood freezes. At this point the entire inner surface is covered with a continuous ice film which stiffens the structure of the wood. Therefore a maximum strength is reached at approximately the 5% moisture level. Beyond this value the additional water in the cell walls is in the state of capillary condensation between structural wall elements. This water entering between the fibrils and micelles pushes them apart and causes swelling. Further accumulation of ice will neither enhance the stiffness nor the strength, because ice itself has a small inherent cohesion. Beyond the 5% moisture content level up to the fiber saturation point frozen wood behaves in a manner similar to unfrozen wood. Therefore, in the range of moisture content between 5 and 25% both curves run in parallel, on account of the increased swelling. The difference between the two curves is caused by the increase in strength with cooling of the wood.

Beyond the fiber saturation point the cell cavities fill progressively with free water. As was stated above, there is no more change in compressive strength for unfrozen wood. A region of no change also occurs up to about 40% for frozen wood. As moisture increases beyond 40%, more and more cells are filled with ice as a strength-giving solid. Testing the frozen wood one has to overcome the combined resistance of wood and ice. As indicated in the upper curve of Fig. 7.89 B, at about 85% moisture content a maximum occurs. By calculation for red beech it can be proved that at 85% moisture a continuous ice-lattice first exists. All the vessels are filled with ice and the ice-lattice extends from plate to plate on the compressing testing machine.

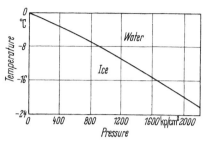

Fig. 7.90. Effect of moisture content on the crushing strength parallel to the grain for frozen and unfrozen beech. From KOLLMANN (1942)

Since ice has a higher Young's modulus than wood along the grain, the stiff rigid ice-lattice carries a greater proportion of the load than the wood and increasing pressure is produced within the ice-skeleton. The pressure lowers the melting temperature of ice as shown in Fig. 7.90. Therefore, as the

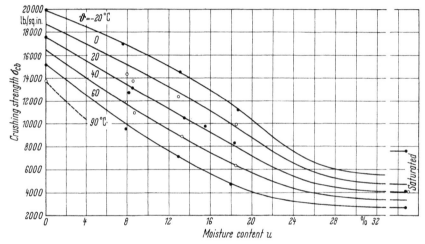

Fig. 7.91. Effect of pressure on the melting temperature of ice. From SULZBERGER (1948)

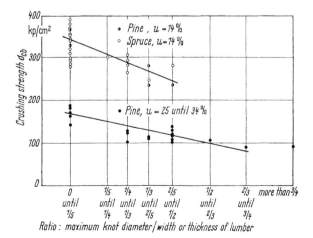

Fig. 7.92. Effect of knot size on the crushing strength for large specimens. From GRAF (1938)

load is increased the ice melts due to the pressure and due to the heat engendered by the pressure. The strength will fall off when the most highly stressed ice melts and the more plastic wood then shares a greater proportion of the load. The wood

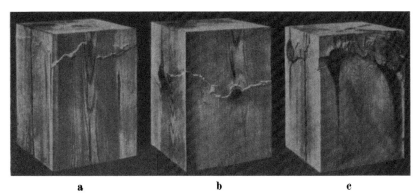

Fig. 7.93 a—c. Influence of knots on the crushing strength. From MPA Stuttgart. a) clear specimen $\sigma_{cb} = 403$ kp/cm²; $\varrho = 0.51$ g/cm³; b) specimen with small knots $\sigma_{cb} = 361$ kp/cm²; $\varrho = 0.53$ g/cm³; c) specimen with large knots $\sigma_{cb} = 314$ kp/cm²; $\varrho = 0.57$ g/cm³

will be compressed until a new ice-framework assumes the load and melts under the pressure. This process is repeated with a continuous loss of strength.

The specimens used for the experiments described above were small blocks, 2 cm × 2 cm × 3 cm in dimension. Compressive tests on wood-props showed that the crushing strength of completely frozen pine props was increased about 100%, of beech props about 50% as compared with unfrozen props. The influence of the temperature $\vartheta$ and the moisture content $u$ on the crushing strength $\sigma_{cb}$ of coachwood is shown in Fig. 7.91 according to SULZBERGER (1948).

**7.3.3.5 Knots and Notches.** The effect of knots on the crushing strength is not nearly as great as on the tensile strength; nevertheless this problem should not be underestimated. Fig. 7.92 shows the results of experiments carried out by GRAF (1938) with large samples. The values are markedly scattered since the densities of the woods were not uniform and since cracks and cross grain were present. Fig. 7.93 also illustrates the influence of knots on the crushing strength. It is significant that in green wood knots reduce the low strength considerably. Table 7.16 contains some test values for structural lumber with different moisture content.

Table 7.16. *Effect of Knots on the Crushing Strength of Structural Lumber with Varying Moisture Content* (From GRAF, 1938)

| Species | Specific gravity $\varrho_0$ | Moisture content % | Dimensions of the specimen $a$ cm | $b$ cm | $h$ cm | Spiral grain (divergence of fibers cm per 1 m length) | Maximum diameter of knots on one surface cm | Crushing strength kp/cm² |
|---|---|---|---|---|---|---|---|---|
| Pine | 0.48 | >25 | 6.54 | 10.09 | 34.91 | 9 | without knots | 185 |
| Pine | 0.48 | >25 | 6.32 | 10.54 | 28.80 | 6 | 2.2 | 113 |
| Pine | 0.56 | >25 | 6.29 | 10.06 | 49.72 | 10 | 3.3 | 105 |
| Pine | 0.60 | >25 | 6.36 | 10.38 | 49.30 | 20 | 5.0 | 95 |
| Pine | 0.41 | ~14 | 6.42 | 7.54 | 39.48 | 8 | without knots | 329 |
| Pine | 0.48 | ~14 | 7.53 | 11.44 | 34.90 | 11 | 2.8 | 279 |
| Pine | 0.46 | ~14 | 6.39 | 7.51 | 49.57 | 8 | 3.0 | 235 |
| Spruce | 0.43 | ~14 | 5.92 | 11.51 | 29.60 | 0 | without knots | 354 |
| Spruce | 0.40 | ~14 | 7.00 | 12.10 | 50.16 | 8 | 1.4 | 292 |
| Spruce | 0.42 | ~14 | 7.97 | 10.95 | 49.80 | 9 | 2.9 | 263 |

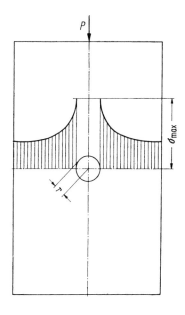

Fig. 7.94. Stress distribution over the cross-section of a notched compression specimen of isotropic material. From KOLLMANN (1951)

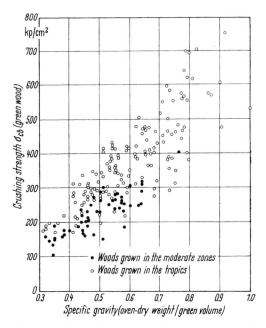

Fig. 7.95. Relationship between crushing strength and specific gravity for green woods grown in the moderate zones and in the tropics. From CLARKE (1939)

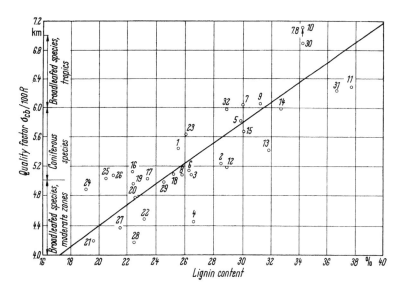

Fig. 7.96. Crushing length (quality factor) in relation to lignin content for broad-leafed and coniferous wood grown in the moderate zones and in the tropics. From TRENDELENBURG (1939)

*1* Pinus caribea   *2* P. echinata   *3* P. monticola   *4* P. ponderosa   *5* P. sylvestris   *6* P. strobus   *7* Picea excelsa   *8* Picea sitchensis   *9* Pseudotsuga taxifolia   *10* Sequoia sempervirens   *11* Libocedrus decurrens   *12* Chamaecyparis nootkatensis   *13* Cham. thyoides   *14* Taxodium dystichum   *15* Tsuga canadensis   *16* Hicoria glabra   *17* H. ovata   *18* Fraxinus americana   *19* Fr. excelsior   *20* Fagus sylvatica   *21* Betula verrucosa   *22* Liriodendron tulipifera   *23* Alnus rubra   *24* Tilia glabra   *25* Populus tremuloides   *26* Acer saccharum   *27* Catalpa speciosa   *28* Quercus alba   *29* Q. pedunculata   *30* Tectona grandis   *31* Chlorophora excelsa   *32* Khaya spec.

Fig. 7.94 shows a notched compression specimen of isotropic material; the stress distribution over the cross section is drawn. For the peak stress $\sigma_{max}$ approximately the following relation holds:

$$\sigma_{max} = 3\sigma_0 \qquad (7.51)$$

where $\sigma_0$ is the average normal stress calculated by neglecting the hole. With anisotropic and heterogeneous wood the same distribution cannot be expected; however, even here at the notch apex the greatest stress peaks will occur. Generally every notch causes a local stress concentration and the stress peaks increase with decreasing diameters of curvature of the notches. The result is a break at a lower load than with unnotched specimens. Unfortunately the available experimental material in the case of wood is scarce. Doubtless the notch sensitivity of any material increases with its brittleness.

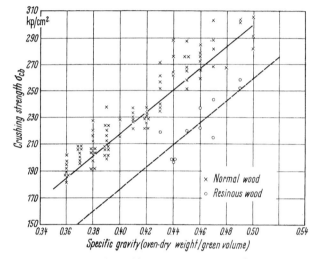

Fig. 7.97. Effect of specific gravity on the crushing strength of normal and resinous green pine. From TIEBE (1940)

**7.3.3.6 Chemical Constituents.** CLARKE (1939) proved that tropical woods with high lignin content have a higher crushing strength (but a lower shock resistance) than woods grown in the temperate zones (Fig. 7.95). TRENDELENBURG (1939), evaluating the figures of many scientists, has shown that the quality factor $\sigma_{cb}/100R$ increases, though widely scattering, in a linear manner with increasing lignin content (Fig. 7.96). Resin deposited in resin ducts or resin pockets does not improve the crushing strength of wood, but because resin increases the specific gravity of the wood its quality factor decreases. Results in this sense obtained by KOLLMANN (1934) for pine wood are stated in Section 7.3.2.2. TIEBE (1940) came to similar observations which are plotted in Fig. 7.97.

### 7.3.4 Fatigue in Compression Parallel to the Grain

GRAF (1928, 1929, 1930) tested clear air-dry specimens of fir and oak in the dimensions 10 cm × 10 cm × 20 cm with oscillating loading progressively up to the failure. The crushing strength was reduced to 84% and 81% respectively. ROTH (1935) found similar results but stated that constant duration of load is more effective than corresponding oscillating load. For long duration load the fatigue resistance in compression parallel to the grain is at the minimum 60% of the crushing strength determined in short time tests.

## 7.3.5 Stresses in Solid Wood Columns

Solid wood columns are classified according to length as follows:

1. Short columns in which the ratio $L/d$ ($L$ is the unsupported length and $d$ is the shortest dimension) is less than 11. Stresses in such columns may be calculated using Eq. (7.43); failure is by crushing.

2. Intermediate columns in which the ratio $L/d$ is between 11 and $K$. For such columns the following formula has been developed by NEWLIN and GAHAGAN (1930)

$$\sigma = \sigma_{cb}\left[1 - \frac{1}{3}\left(\frac{L}{K \cdot d}\right)^4\right] \qquad (7.52)$$

in which $\sigma$ is the stress in the column, $\sigma_{cb}$ is the crushing strength parallel to the grain of the wood composing the column, $L$ and $d$ have the same meaning as before and $K$ is a constant depending upon $\sigma_{cb}$ and Young's modulus $E$.

Eq (7.52) is known as the Forest Products Laboratory fourth-power formula. Fig. 7.98 shows the $FPL$-curve versus the ratio $\frac{L}{i} = \lambda$, where $i$ is the radius of gyration. The parabolic $FPL$-curve is so drawn that the stress reaches the crushing stress when $\lambda$ is zero. The parabola is tangent to the Euler curve at the proportional limit of the wood. Under this assumption the value for $K$ is given by the following expression:

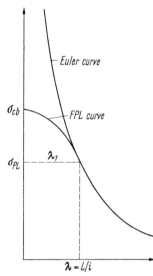

Fig. 7.98. Diagram showing interrelation of Euler and FPL curves. From BROWN, PANSHIN and FORSAITH (1952)

$$K = \frac{\pi}{2}\sqrt{\frac{E}{2\sigma_{cb}}} \qquad (7.53\,\mathrm{a})$$

where $E$ is the modulus of elasticity. Taking $E = 100{,}000$ kp/cm² as an average value for structural softwoods Eq (7.53a) yields

$$K = \frac{351.25}{\sqrt{\sigma_{cb}}}. \qquad (7.53\,\mathrm{b})$$

3. Long columns (Euler columns) have $L/d$ ratios equal to $K$ or greater. The well known Euler formula

$$\sigma = \frac{\pi^2 E}{\lambda^2} \qquad (7.54\,\mathrm{a})$$

has to be applied with $\pi^2 = 9.87$ and

$$\lambda = \frac{L}{i} = \frac{L}{\sqrt{\frac{I}{A}}} \qquad (7.55)$$

where $i$ is the radius of gyration, $I$ is the least moment of intertia and $A$ represents the cross-sectional area of the column.

## 7.3 Maximum Crushing Strength and Stresses in Wood Columns

For squat columns in the range $\lambda < 100$ TETMAJER (1884, 1903, 1905) has formulated an empirical law which is still used

$$\sigma = \sigma_{cb}(1 - a\lambda + b\lambda^2). \tag{7.56a}$$

For structural lumber the constant $b$ is zero and one obtains the equation of the so called "Tetmajer straight line":

$$\sigma = 293(1 - 0.00662\lambda) = 293 - 1.94\lambda. \tag{7.56b}$$

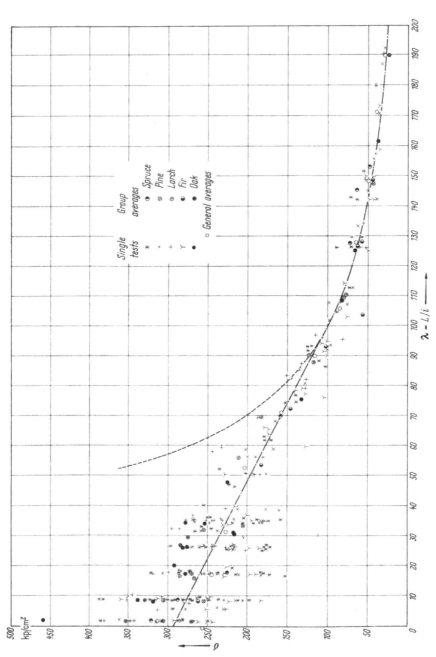

Fig. 7.99. Results of buckling tests with beams carried out by TETMAJER (1903, 1905)

Fig. 7.99 contains the plotted values of the tests carried out by TETMAJER with 305 beams of square cross section in the moisture content range between 11 and 25%. As shown, the Tetmajer line is tangent to Euler's curve when $\lambda = 100$. For long columns ($\lambda > 100$), taking again $E = 100{,}000$ kp/cm² Eq. (7.54a) becomes

$$\sigma = \frac{987{,}000}{\lambda^2} = 987{,}000 \frac{I}{AL^2}. \tag{7.54b}$$

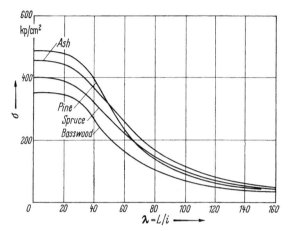

Fig. 7.100. Stress curves for various wood species using the equation of NATALIS. Designed by WINTER (1944)

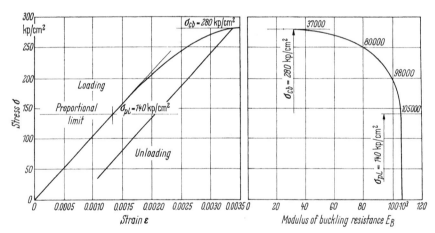

Fig. 7.101. Graphically determined modulus of buckling resistance from the stress-strain diagram. From ROŠ and BRUNNER (1932)

The desire to calculate the stresses in columns of all length classes by one uniform method has led to the development of various formulas. That published by NATALIS (1918, 1919) gives stress values which for wood are in a good agreement with the results of tests. The equation of NATALIS is

$$\sigma = \sigma_{cb} \frac{1+A}{1+A+A^2} \tag{7.57}$$

where $A = \dfrac{\sigma_{cb}}{\pi^2 \cdot E} \cdot \lambda^2$. Using Eq. (7.57) and basic values for $\sigma_{cb}$ and $E$ of wood species important for aircraft construction, WINTER (1944) has designed the curves shown in Fig. 7.100.

As early as 1889 ENGESSER stated that Euler's formula may also be applied in the inelastic range. In this case the modulus of elasticity $E$ has to be substituted by a "modulus of buckling resistance $E_B$", obtained as tangent $d\sigma/d\varepsilon$ to the stress-strain diagram in compression above the elastic limit $\sigma_{pL}$ (cf. Fig. 7.101). The theory of ENGESSER, at first ardently attacked, became generally recognized after the careful experiments made by v. KÀRMÀN (1910). Therefore for $\sigma > \sigma_{pL}$ the following holds

$$\sigma = \frac{\pi^2 E_B}{\lambda^2}. \qquad (7.58)$$

YLINEN's (1938) contribution to the problem of calculating approximate values for $d\sigma/d\varepsilon$ may be mentioned but not discussed in detail.

Eccentric loading has a severe influence on the stresses set up in columns. Reference may be made to the general formulas developed by NEWLIN (1950, quoted in the "Wood Handbook, 1955, p. 219). Roš and BRUNNER (1932) investigated not only the effect of eccentricity but also of varying moisture content on the stresses in columns. Fig. 7.102 clarifies that higher moisture only reduces the stresses in short and intermediate columns. In the Euler range the

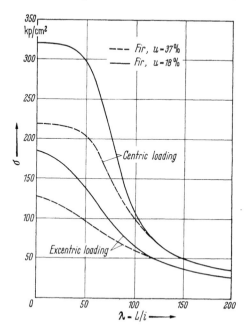

Fig. 7.102. Effect of moisture content and eccentric loading on the stresses in pine wood columns. From ROŠ and BRUNNER (1932)

strength of wood columns is practically independent of the moisture content. Similar to some extent is the effect of knots; it becomes more or less negligible for $\lambda \geq 150$ if the knots are distributed, sound and not extraordinarily numerous.

As the title of this section shows, only the conditions for solid wood columns are dealt with. This limitation is justified in a textbook devoted to the principles of wood science and technology of solid wood. Fabricated, built up, and spaced columns are of great importance for construction purposes, but the methods of calculating working stresses for them — though based on modifications of the formulae developed for solid columns — are of interest only to the building engineer.

## 7.4 Bending Strength (Modulus of Rupture)

### 7.4.0 General Considerations

The difference between the tensile strength and the crushing strength of wood determines the characteristic behavior of a wood beam in bending. Fig. 7.103 shows that on the average the ratio of bending strength $\sigma_{bb}$ to crushing strength $\sigma_{cb}$ amounts to 1.75 for common European pinewood in the air-dry condition. This quotient is not a ratio of exact breaking stresses because, following the usual standardized tests, the so-called bending strength is computed using Navier's formula, which is based on the assumption that the stresses are distributed linearly and symmetrically over the cross section of the bent beam. This assumption is justified for isotropic and homogeneous beams but for the anisotropic

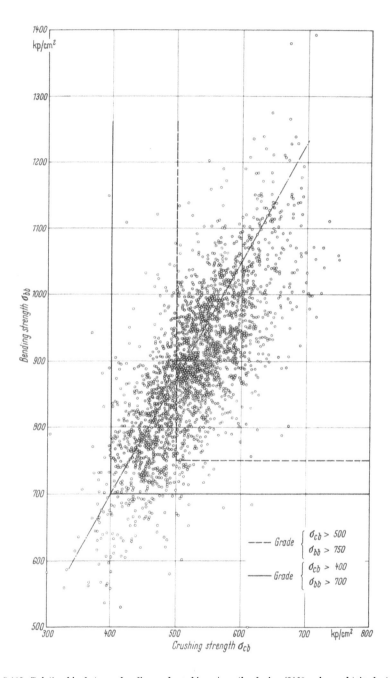

Fig. 7.103. Relationship between bending and crushing strength of pine (2160 values, obtained at the Fieseler aircraft factory)

## 7.4 Bending Strength (Modulus of Rupture)

wood beams only up to the proportional limit. This is the reason why in the USA the term "modulus of rupture" is accepted as a criterion of strength, although it is not a true stress as explained above.

For higher loads, especially as far as compressive forces are concerned, the strains are not proportional to the stresses and occasionally the moduli of elasticity for tension and compression are not quite equal. BAUMANN (1922) has solved the problem graphically. He used the stress-strain diagrams (Fig. 7.104) and plotted stresses

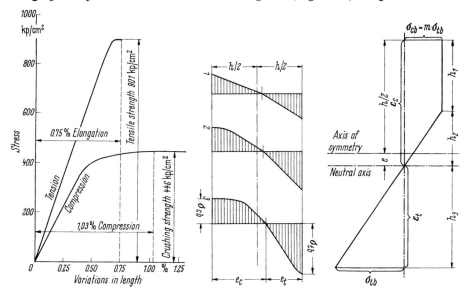

Fig. 7.104. Stress-strain diagram for tensile and compressive tests. From BAUMANN (1924)

Fig. 7.105. Stress distribution in a bent wood beam at increased stresses. From BAUMANN (1924)

Fig. 7.106. Stress diagram for trapezoid-like distribution of the stresses over the cross-section of a wood beam. From THUNELL (1939, 1940)

belonging to the different moments of force (Fig. 7.105). The hatched areas must be equally large. The sum of the internal moments (each one must have the same sign) must be equal to the moment of the external faces. The steep decrease of the compressive stresses towards the edge under the influence of higher loads is caused by the fact that Hooke's law is valid in compression only in a very limited range.

The difference between the real distribution of stress in a bent wood beam and the linear distribution according to NAVIER is evident. In the literature there exist some proposals for at least more realistic approximations. A trapezoid-like distribution of the stresses has been assumed by some researchers, e.g. Roš (1936) and THUNELL (1939, 1940). With the symbols used in Fig. 7.106 equal areas of stress are obtained with the following statement

$$m \cdot \sigma_{tb} \cdot \varepsilon_c - \frac{m \cdot \sigma_{tb} \cdot \dfrac{m \cdot \sigma_{tb}}{E}}{2} = \frac{\varepsilon_t \cdot \sigma_{tb}}{2} \tag{7.59a}$$

furthermore $m = \sigma_{cb}/\sigma_{tb}$ and $\varepsilon_t = \sigma_{tb}/E$; hence holds

$$\varepsilon_c = \frac{\sigma_{tb}}{E} \cdot \frac{1+m^2}{2m} \tag{7.59b}$$

or

$$\frac{\varepsilon_c}{\varepsilon_t} = \frac{1+m^2}{2m}. \tag{7.59c}$$

The position of the neutral axis dependent on the ratio $\sigma_{bb}/\sigma_{cb}$ is as follows

$$\frac{x}{h} = \frac{1 + m^2}{(1 + m)^2}. \qquad (7.59\,\text{d})$$

Using Eq. (7.59d) the following values were computed:

| $\sigma_{tb}/\sigma_{cb}$ | 1.5 | 2 | 2.5 | 3 |
|---|---|---|---|---|
| $m$ | 0.67 | 0.5 | 0.40 | 0.33 |
| $x/h$ | 0.52 | 0.56 | 0.60 | 0.62 |

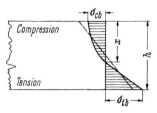

Fig. 7.107. Stress diagram for the cross-section of a bent beam ($\sigma_{tb} = 2 \cdot \sigma_{cb}$). From SUENSON (1941)

It can be assumed

$$\sigma_{bb} = k \cdot \sigma_{tb} = \frac{k}{m} \cdot \sigma_{cb}. \qquad (7.60)$$

THUNELL (1939, 1940), testing 400 specimens of pinewood, has found the following average values for the breaking stresses

$$\sigma_{cb} = 470\,\text{kp/cm}^2, \quad \sigma_{bb} = 1040\,\text{kp/cm}^2,$$

or from theses figures $m$ becomes 0.45 and $k$ becomes 0.80 and therefore

$$\sigma_{bb} = \frac{0.80}{0.45} \cdot \sigma_{cb} \approx 1.78\,\sigma_{cb} \;(\text{cf. Fig. 7.103}).$$

SUENSON (1941) assumed a parabola as a curve for the compressive stresses and a linear distribution of tensile stresses, as illustrated in Fig. 7.107. It follows from equilibrium conditions

$$\frac{2}{3} \cdot \sigma_{cb} \cdot x = \frac{1}{2} \cdot \sigma_{tb}(h - x). \qquad (7.61\,\text{a})$$

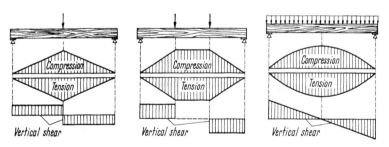

Fig. 7.108. Distribution of tensile, compressive and shear stresses along the axis of a bent beam for different cases of loading. From the U.S. Forest Products Laboratory

Introducing $\sigma_{cb}/\sigma_{tb} = m$ in Eq. (7.61a) one obtains

$$\frac{x}{h} = \frac{3}{4m + 3} \qquad (7.61\,\text{b})$$

and one can deduce

| $\sigma_{tb}/\sigma_{cb}$ | 1.5 | 2 | 2.5 | 3 |
|---|---|---|---|---|
| $x/h$ | 0.53 | 0.60 | 0.65 | 0.69 |

## 7.4 Bending Strength (Modulus of Rupture)

For a critical examination of the phenomena in bending a knowledge of the distribution of tensile, compressive, and shear stresses along the axis of the beam between the supports is also essential. Fig. 7.108 shows schematically this distribution for three important cases of loading.

### 7.4.1 Testing of Small Wood Beams under Static Center Loading

We will later (Section 7.4.2.5) see that the modulus of rupture for small clear wood beams with cross sections from 2 cm by 2 cm up to about 6 cm by 6 cm is practically equal. Taking into consideration this fact, the German standard

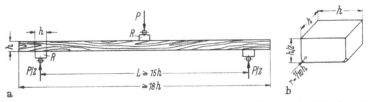

Fig. 7.109 a, b. a) Set-up of a beam for static bending test according to DIN 52186. b) Rider R for the test.

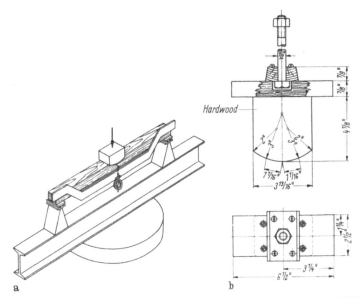

Fig. 7.110 a, b. a) Set-up of a beam for static bending test according to A.S.T.M., b) Rider

specification DIN 52186 permits any cross section from 2 cm by 2 cm up to 4 cm by 4 cm. The length of the test sample shall be $L = 18h$, when $h$ = height of the sample. The symbol for the width of the test sample is $w$. As ratio of span to height the value 15 is prescribed. The load $P$ should be applied tangential to the annual rings. The speed of testing shall be 400 to 500 kp/cm² per minute. Similar rates are used in France and in the CSSR.

In the USA (ASTM D 143-52) the dimensions of the test sample are 2 inches by 2 inches by 30 inches, providing a 28-inch span. The load is applied to the tangential face. The machine head causing the deflection should descend at a constant rate of 0.1 inch per minute.

The German and the American set-up of beams for static bending showing the necessary details and dimensions of the assemblies are illustrated in Figs. 7.109 and 7.110. The figures are self explanatory.

During a complete bending test deflection readings should be carefully recorded — modern testing machines are equipped with an automatic recording device — and subsequently plotted on graph paper. The stress-strain curve will be straight up to the proportional limit. As stated above, as a rule Navier's theory is applied in evaluating the bending test data. Thus the numerical value for the bending strength, or better, modulus of rupture $\sigma_{bb}$, can be obtained as follows:

$$\sigma_{bb} = \frac{3PL}{2w \cdot h^2}. \qquad (7.61)$$

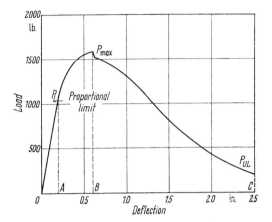

Fig. 7.111. Load-deflection diagram for a complete bending test. From WANGAARD (1950)

How to obtain the modulus of elasticity is explained in Section 7.15. For a complete bending test the work to the proportional limit $W_{PL}$ and the work to maximum load $W_{P\max}$ can be computed using and integrating (with a planimeter) the load-deflection diagrams as shown in Fig. 7.111. In the USA the work to the proportional limit $W_{PL}$ (area $OP_LA$), the work to maximum load $W_{P\max}$ (area $OP_{\max}B$), and the work to ultimate load $W_{PUL}$ (in

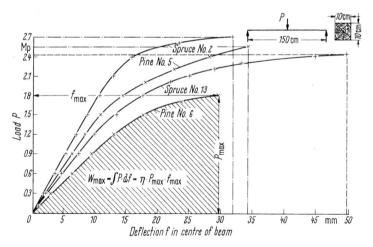

Fig. 7.112. Load-deflection diagram for bending tests on spruce and pine beams 200 cm · 10 cm · 10 cm. From KOLLMANN (1951)

Fig. 7.111 for a test load about 200 pounds) (area $OP_{UL}C$) are computed by finding the respective areas under the load-deflection curve and dividing the results into the volume of the span portion of the beam.

In Germany and France the work to maximum load ($W_{P\max} = \int P \cdot df$) usually is divided into the product of maximum load ($P_{\max}$) × maximum deflection ($f_{\max}$)

(Fig. 7.112). The resulting value is called the factor of completeness $\eta$ (Völligkeitsgrad):

$$\eta = \frac{W_{max}}{P_{max} \cdot f_{max}}. \tag{7.62}$$

For structural lumber of average quality $\eta$ is about 0.7, but its value can decrease due to defects, knots, cracks, etc. down to 0.5 (Roš, 1925).

### 7.4.2 Influences Affecting the Bending Strength (Modulus of Rupture)

**7.4.2.1 Grain Angle.** BAUMANN (1922) showed that the angle between load and a plane tangential to the annual rings is practically without any importance for the modulus of rupture on which the effect of angle between load and fiber direction is about as great as it is for tensile strength. Hankinsons's Formula (Eq 7.40) may be used again with the exponent $n \approx 2$.

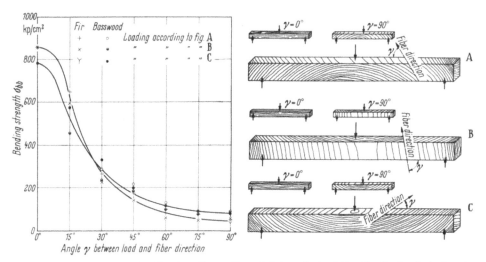

Fig. 7.113. Dependence of the bending strength upon the angle between load and fiber direction for air-dry fir and basswood. From BAUMANN (1922)

A few results of tests with fir and basswood are illustrated in Fig. 7.113 according to BAUMANN (1922). Fig. 7.114 shows corresponding curves for high grade pine and ash wood, computed by WINTER (1944). He applied Eq. (7.40) and he used the strength values parallel and perpendicular to grain (as listed for pine or shown for ash) and he introduced the exponents $n$ as listed in Fig. 7.114a. One can see that the weakening of slope of grain is much more pronounced for tensile and bending strength than for crushing strength. WANGAARD (1950) states that the decrease in tensile strength does not become appreciable until a slope of 1 : 25 is reached; in compression, the slope may approach 1 : 10 before the reduction of crushing strength is evident. In shear the weakening effect of cross grain may be neglected.

WILSON (1921) conducted a series of tests at the Forest Products Laboratory to determine the effect of spiral grain on the strength of wood. Table 7.17 shows his results, expressed as average percentage deficiency with respect to straight-grained wood with 7% moisture content.

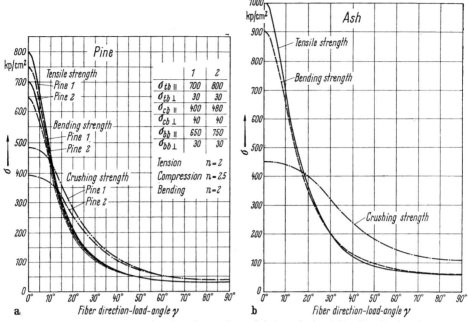

Fig. 7.114. Dependence of the bending strength upon the angle between load and fiber direction for selected pine and ash. From WINTER (1944)

Table 7.17. *Average Percentage Deficiency in Strength Properties of Spiral- and Diagonal-Grained Lumber of Various Slopes with Respect to Straight-Grained Material* (From WILSON, 1921)

| Species and slope of grain | Modulus of elasticity | Static bending — Modulus of rupture | Static bending — Work to maximum load | Impact bending maximum drop | Crushing strength grain |
|---|---|---|---|---|---|
| White ash |  |  |  |  |  |
| 1:25 | 2 | 4 | 9 | 6 | 0 |
| 1:20 | 3 | 6 | 17 | 12 | 0 |
| 1:15 | 4 | 11 | 27 | 22 | 0 |
| 1:10 | 7 | 18 | 43 | 37 | 1 |
| 1:5 | 22 | 36 | 61 | 59 | 7 |
| Sitka spruce |  |  |  |  |  |
| 1:25 | 2 | 2 | 14 | 8 |  |
| 1:20 | 4 | 4 | 21 | 13 |  |
| 1:15 | 7 | 8 | 33 | 22 |  |
| 1:10 | 13 | 17 | 55 | 45 |  |
| 1:5 | 36 | 44 | 76 | 69 |  |
| Douglas-fir |  |  |  |  |  |
| 1:25 | 4 | 7 | 17 | 1 |  |
| 1:20 | 6 | 10 | 24 | 4 |  |
| 1:15 | 8 | 15 | 34 | 13 |  |
| 1:10 | 14 | 25 | 46 | 31 |  |
| 1:5 | 40 | 54 | 68 | 65 |  |
| Average for three species |  |  |  |  |  |
| 1:25 | 3 | 4 | 13 | 5 |  |
| 1:20 | 4 | 7 | 21 | 10 |  |
| 1:15 | 6 | 11 | 31 | 19 |  |
| 1:10 | 11 | 19 | 48 | 38 |  |
| 1:5 | 33 | 45 | 68 | 64 |  |

**7.4.2.2 Density.** YLINEN (1942) taking into consideration that wood, at least such as it is grown in the temperate zones, is built up of earlywood and latewood, finally came, based on Bernoulli's formula for bending and St. Venant's theory on constant breaking strain, to the following equation

$$\sigma_{bb2} = \frac{1}{\varrho_{l1} - \varrho_{e1}} [(\sigma_{bbe2} \cdot \varrho_{l1} - \sigma_{bbl2} \cdot \varrho_{e1}) + (\sigma_{bbl1} - \sigma_{bbe2})\varrho_{1}], \qquad (7.63)$$

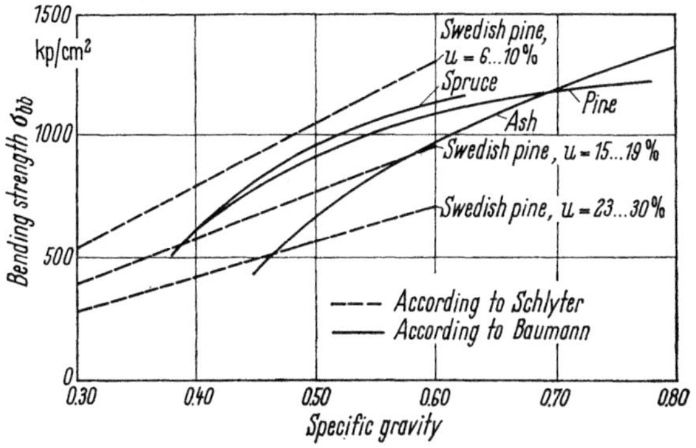

Fig. 7.115. Effect of specific gravity on the bending strength. From BAUMANN (1922) and SCHLYTER (1927)

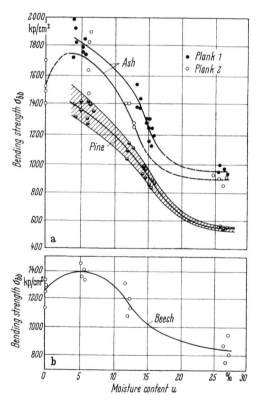

where $\sigma_{bb2}$ = modulus of rupture at moisture content $u_2$, $\sigma_{bbe2}$ = modulus of rupture of earlywood at moisture content $u_2$, $\sigma_{bbl2}$ = modulus of rupture of latewood at moisture content $u_2$, $\varrho_1$ = density at moisture content $u_1$, $\varrho_{e1}$ = density of earlywood at moisture content $u_1$ and $\varrho_{l1}$ = density of latewood at moisture content $u_1$. Once again the validity of a linear law between ultimate breaking strength and density is proved.

Fig. 7.115 shows that SCHLYTER (1927) in Sweden has found linear functions between bending strength and density at various moisture content ranges, whereas BAUMANN (1922) established curvilinear relationships between bending strength and density. The explanation of this contradiction is not difficult. A higher amount of resin in coniferous species increases the density but lowers the cohesion to some extent (cf. Fig. 7.97). YLINEN (1942) pointed out that Eq (7.63) remains a linear function only if the values for $\varrho$ and $\sigma_{bb}$, introduced in this equation, are true constants. If there is any functional dependence between them the function becomes curvilinear.

Fig. 7.116 a, b. Effect of moisture content on the bending strength of a) ash and pine, b) beech. From KÜCH (1943)

### 7.4.2.3 Moisture content.
KÜCH (1943) showed that the relation between modulus of rupture (of ash and beechwood) and moisture content is characterized

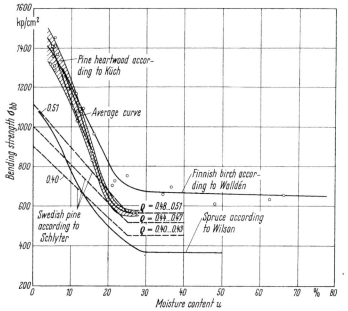

Fig. 7.117. Effect of moisture content on the bending strength of pine wood, from SCHLYTER and WINBERG (1929) and KÜCH (1943), and birch. From WALLDÉN (1939)

by a peak at about $u = 5\%$ (Fig. 7.116). A similar phenomenon is observed in testing the tensile strength (cf. Section 7.2.3.3). In the range between 8 and 15% moisture content a linear relationship is justified. SCHLYTER and WINBERG (1929) and WALLDÉN (1939) proved linear functions for Swedish pine wood and Finnish

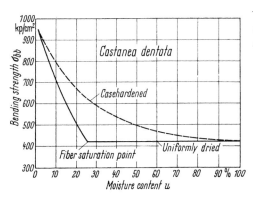

Fig. 7.118. Effect of casehardening on the bending strength. From BETTS (1919)

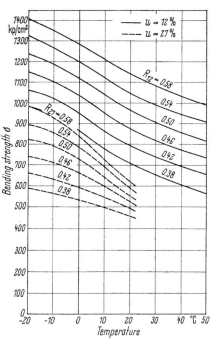

Fig. 7.119. Relationship between bending strength, temperature and density for pine at 12 and 27% moisture content. From THUNELL (1942)

birchwood respectively (Fig. 7.117). In this case Eq. (7.64) (cf. Eq. 7.46) may be applied.

$$\frac{\sigma_2}{\sigma_1} = \frac{b-u_2}{b-u_1}. \tag{7.64}$$

The following numerical values for the constant $b$ may be deduced:
1. from SCHLYTER's and WINBERG's measurement for pine $b = 46$;
2. from WALLDÉN's measurements for birch $b = 42$;
3. from KÜCH's measurements for pine and ash $b = 35$; for beech $b = 45$.

For rough approximations a uniform value $b = 42$ can be used.

The modulus of rupture of case-hardened wood is normally much higher than that of wood with a uniform distribution of moisture content (BETTS, 1919).

The reason is evident: due to the steep moisture gradient from the interior to the exterior the more highly stressed outer layers of the wood are drier and therefore stronger (Fig. 7.118).

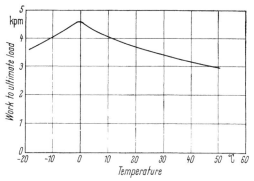

Fig. 7.120. Effect of temperature on the work to ultimate load. From THUNNELL (1942)

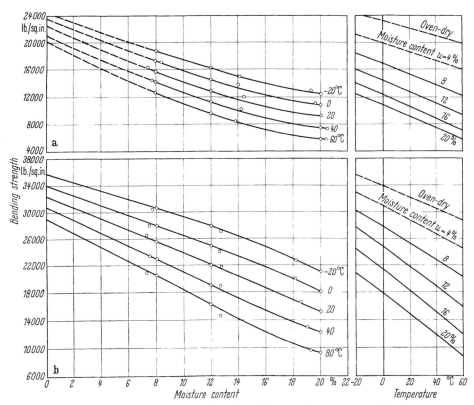

Fig. 7.121 a, b. Dependence of the bending strength upon moisture content and temperature, a) Sitka spruce, b) mountain ash. From SULZBERGER (1948)

**7.4.2.4 Temperature.** The effect of temperature on the modulus of rupture is well known. The general law that strength and stiffness decrease with increasing temperature (cf. Section 7.1.6.4) is valid. THUNELL (1942) published equalized curves for the modulus of rupture of pinewood with various densities and moisture contents as affected by temperature (Fig. 7.119). He summarized his results as follows:

1. The bending strength of wood decreases with increasing temperature and decreasing density;
2. The bending strength decreases with increasing moisture content above 0 °C;
3. The modulus of elasticity in bending increases with decreasing temperature, decreasing moisture content, and increasing density.
4. The largest deformation on failure occurs at 0 °C; for this reason the work to ultimate load also reaches a maximum at 0 °C (Fig. 7.120).

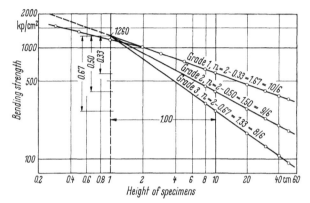

Fig. 7.122. Effect of the height (thickness) of specimens and wood quality on the bending strength according to presumptions of MONNIN; graphical evaluation by ČIŽEK (1932)

SULZBERGER (1948) arrived at similar curves for two woods (Fig. 7.121 and Fig. 7.122). As is evident for the region of moisture content between 6 and 20%, as also for the temperature region between $-20$ °C and $+60$ °C, a linear relationship with modulus of rupture can be assumed. This means that the bending strength for every temperature and every region of moisture can be calculated sufficiently accurately if one knows the bending strength at the same temperature but at another moisture content (within the region mentioned).

Another very informative property is the deflection in the bending test. According to SULZBERGER (1948) the deflection at fracture below 12% moisture content shows no great change with temperature while at higher moisture contents it increases considerably with temperature. Moreover, the effect varies very considerably with species.

**7.4.2.5 Shape and Size of Beams, Knots and Notches.** It was recognized very early that the application of Eq. (7.62) to thick beams containing defects leads to extraordinarily low calculated moduli of rupture. Considering this fact, in 1906 TANAKA[1] proposed the following variation of Eq. (7.61):

$$\sigma_{bb} = \frac{3PL}{2w \cdot h^n}. \tag{7.65}$$

---

[1] At the IVth Congress of the "Intern. Verband für die Materialprüfungen der Technik" in Brussels, Belgium.

According to TANAKA $n \leq 2$ depends on the wood species. Later on MONNIN (1919, 1932) in France investigated this problem thoroughly. He came to the conclusion that $n$ is independent of the species and is caused only by the technological properties of the wood. Based on numerous tests he found the following values for $n$

First grade: woods without defects (for aircraft and carriages)     11/6···10/6
Second grade: commercial woods (for structural purposes and
    masonry)     10/6··· 9/6
Third grade: waste woods with many knots     9/6··· 8/6

ČIŽEK (1932) published the instructive Fig. 7.123. For squat beams also vertical shearing reduces the bending strength. BAUMANN (1922) showed how the modulus of rupture decreases with decreasing ratio of span $L$ to depth $h$ of the specimens (Fig. 7.124). Only for values $L/h \geq 20$ the bending strength becomes approximately constant. If one knows the bending strength $\sigma_{bbl}$ of a relatively very long beam ($L/h \geq 40$) one can estimate the bending strength of short beams $\sigma_{bbs}$ (but only in the range $7.5 \leq \leq L/h \leq 40$)

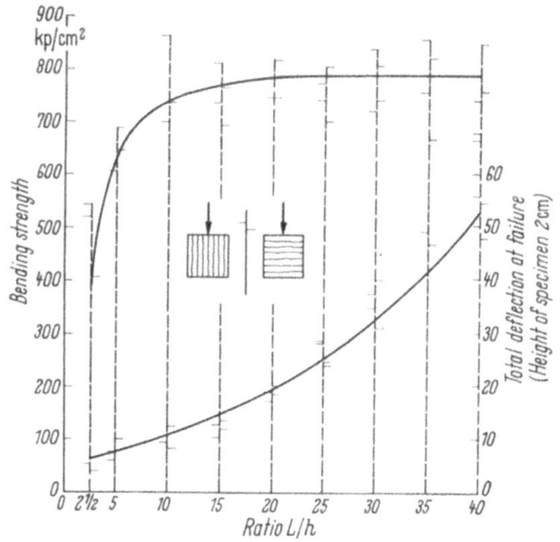

$$\sigma_{bbs} = \sigma_{bbl} \frac{L}{L+h} \quad (7.66)$$

Fig. 123. Dependence of the bending strength and the total deflection at failure upon the ratio of span $L$ to height $h$ of the specimens. From BAUMANN (1922)

The deflection is roughly proportional to the ratio $(L/h)^2$ (cf. Fig. 7.124).

The effect of size on the modulus of rupture is based on the heterogeneous structure of the wood. YLINEN (1943) showed how stresses are distributed by jumps in a bent beam due to the variations of Young's moduli in earlywood and latewood (Fig. 7.125). If the number of annual rings is large enough the beams may be treated as homogeneous bodies. The ratio of the bending strength $\sigma_{bbh}$ of an assumed homogeneous beam to the bending strength $\sigma_{bb}$ of the heterogeneous wood may be expressed, according to YLINEN, by the formula

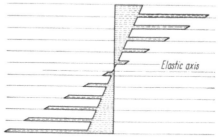

Fig. 7.124. Stress distribution over the cross section in a wood beam loaded with a constant bending moment. From YLINEN (1943)

$$\frac{\sigma_{bb}}{\sigma_{bbh}} = \frac{1 + c_1 h}{1 + c_2 h} \quad (7.66\text{a})$$

where $c_1$ and $c_2$ are constants, determined by experiments. YLINEN found, for clear Finnish pinewood for aircraft construction, that $c_1 = 0.16$ cm$^{-1}$ and

$c_2 = 0.2$ cm$^{-1}$, for pinewood with knots and checks, $c_1 = 0.03$ cm$^{-1}$ and $c_2 = 0.08$ cm$^{-1}$.

Fig. 7.126 permits the conclusion that curves calculated using Eq. (7.66) fit the experimental data well. The formula developed by NEWLIN and TRAYER (1924) gives results which are in fairly good agreement with the data obtained using Ylinen's theory, while a curve according to Eq. (7.65) with Monnin's factor $n = 10/6$ decreases much too steeply (cf. Fig. 7.124).

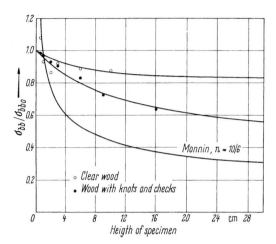

Fig. 7.125. Ratio of the bending strength of a heterogeneous beam $\sigma_{bb}$ to the bending strength of a homogeneous beam $\sigma_{bb_0}$ as function of the height of specimen. From YLINEN (1943)

Checks and splits in radial planes more or less influence the bending strength. Since it is very difficult to obtain beams of otherwise the same properties but without or with checks, the results of tests may be compared only with some reservations. This is valid, especially for the figures given in Table 7.18, where moduli of rupture for beams with checks and corresponding figures for small clear specimens cut from the undestroyed portions of the broken beams are listed.

Planks, having a long shake near their center may be split into two parts by exceeding the shear strength during bending. According to Roš (1925), the capacity of loading first reaches a minimum but increases again if the two separated parts work together due to friction.

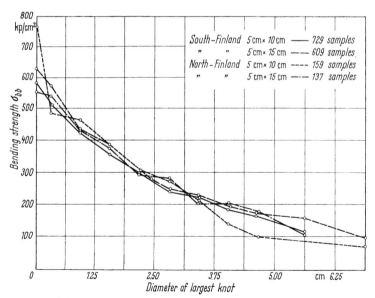

Fig. 7.126. Average effect of the diameter of the largest knot on the bending strength of Finnish pine beams. From SIIMES (1944)

Knots reduce the modulus of rupture of wood considerably if they are located in the tension zone near the critical cross section. SIIMES (1944), in Finland, proved that the largest knot in the middle part of the span characterizes the reduction in modulus of rupture better than the sum of knot diameters belonging

Table 7.18. *Bending Strength of Beams with Checks and of Clear, Small Specimens*
(From GRAF, 1940)

| Characteristics of Specimen | Whole beam with checks | | Small clear specimens | |
|---|---|---|---|---|
| | Cross section cm² | Bending strength kp/cm² | Cross section cm⁴ | Bending strength kp/cm² |
| Specimen with wane | 28 · 28 | 505 | 8 · 8 | 808 |
| Specimen of rectangular cross section | 21 · 21 | 534 | 8 · 8 | 787 |
| Specimen of rectangular cross section, with cross grain | 21 · 21 | 657 | 8 · 8 | 814 |

to the same group of knots (Fig. 7.127 and 7.128). Theoretical investigations on the effect of different types of knots on the bending strength of wood were carried out by GABER (1937) and THUNELL (1942). The results are of limited value because the moment of inertia is mostly reduced by several knots and since slope of grain around the knots and notch effects are unknown factors. In bent round woods, knots are of minor importance. GRAF (1939, 1940, 1944) showed that the allowable bending stresses for beams with rectangular cross section and for those with wane

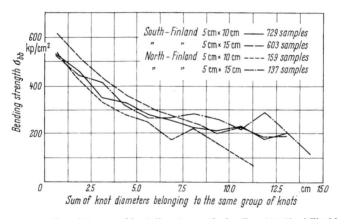

Fig. 7.127. Average effect of the sum of knot diameters on the bending strength of Finnish pine beams. From SIIMES (1944)

are approximately the same. Though the upper limits of modulus of rupture for green spruce decrease with increasing amounts of wane, the lower limits are independent of the relative portion of wane. The average values remain nearly constant. For air dry sprucewood ($u = 17 \cdots 21\%$) the lower limits of modulus of rupture even increased with higher amount of wane.

Notched beams are very frequently used in construction, mainly with notches at the ends over the supports. SCHOLTEN (1935), testing the strength of short, relatively deep beams notched at the lower side at the ends (Fig. 7.128) found that the breaking loads due to stress concentration at the notches were considerably reduced. If $V$ is the maximum vertical shear the following equation is recommended:

$$V = \frac{2}{3} \frac{b \cdot d_e^2}{d} \cdot \tau_s \tag{7.67}$$

where $b$ is the width of the beam, $d_e$ the actual end depth above the notch, $d$ the total depth of the beam, and $\tau_s$ the shearing strength in horizontal shear. In designing beams with square-cornered notches the desired bending load should

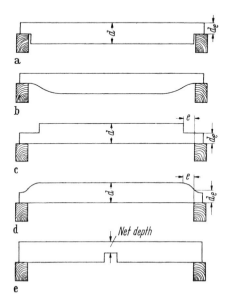

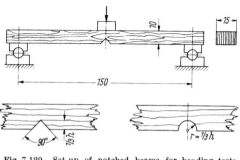

Fig. 7.128 a–e. Notched beams: a) notched at ends to one half depth of beam; b) beam with gradual change in cross-section at ends to eliminate concentration of stress; c) notched on the upper side; d) beveled on the upper side; e) notched to one half depth of beam at center. $d$ — depth of the beam, $d_e$ — depth below the notch, $e$ — distance that the notch extends inside the inner edge of the support. From US Forest Products Laboratory

Fig. 7.129. Set-up of notched beams for bending tests. From Küch (1937)

be checked against the load obtained by Eq. (7.67) and introducing instead of shearing strength $\tau_s$ the working stress $q$ in horizontal shear (cf. Wood Handbook 1955, p. 207/208).

The effect of notches with sharp corners is very pronounced. The breaking load for the ratio $d_e/d = 1/2$ is only half the breaking load for the same beam without notches. The breaking loads are raised very markedly if a gradual change is cut at the end notches (Fig. 7.129 b).

For beams notched or beveled on the upper side the shear stress—according ot Wood Handbook, 1955, p. 209—should be checked with the formula

$$V = \frac{2}{3} b \left[ d - \left(\frac{d - d_e}{d_e}\right) e \right] \tau_s. \tag{7.68}$$

The meaning of the symbols used in Eq. (7.68) is shown in Fig. 7.128c and 7.128d. If $e$ exceeds $d$, Eq. (7.68) is not applicable. The depth of a notch on the upper side should never exceed the value $0.4 d$.

When notches are at or near the middle of the length of a beam (Fig. 7.128e) breaking loads may be calculated using Eq (7.62) and introducing the net depth. The U.S. Forest Products Laboratory mentions (Wood Handbook, 1955, p. 209) that a notch on the top or the bottom of a beam and near the point of maximum moment may lower the proportional limit load and start "failure at lower loads than would be expected of an unnotched beam of a depth equal to the net depth of the notched beam". A few test data obtained by KÜCH (1937) and listed in Table 7.19 support this statement.

The shape of notch apparently had no effect. The ratio of modulus of rupture for unnotched beams varied (not systematically) between 0.72 and 0.93.

Table 7.19. *Modulus of Rupture for Unnotched and Notched Beams of Pine, u = 12%* (From KÜCH, 1937, details of tests are shown in Fig. 7.129)

| Part of stem | Direction of force | Modulus of rupture kp/cm² |||
| | | unnotched | notched with sharp notch | round notch |
|---|---|---|---|---|
| Heartwood | tangential | 871 | 630 | 680 |
|  | radial | 846 | 790 | 762 |
| Sapwood | tangential | 832 | 607 | 603 |
|  | radial | 681 | 583 | 603 |

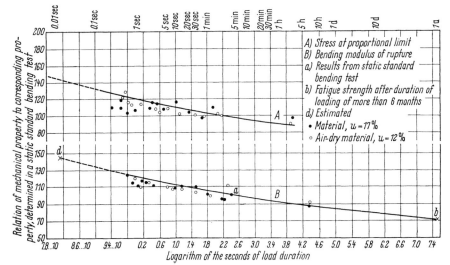

Fig. 7.130. Dependence of stress at the proportional limit (curve A) and modulus of rupture (curve B) upon logarithm of load duration for Sitka-spruce. From MARKWARDT (1930)

Table 7.20. *Reduction of Modulus of Rupture by the Effect of Square-Cornered Notches at the Ends of Beams, 4″ × 2″ Cross Section, 4′ Length, 3′ 6″ Span, Four-Point Loading* (From LANGLANDS, 1936)

| Depth of notch || Notch ratio (cf. Fig. 7.128) $d_e/d$ | Ratio $\frac{\text{breaking load, notched}}{\text{breaking load, unnotched}}$ ||
| inch | cm | | by experiment | according to Eq. (7.67) |
|---|---|---|---|---|
| 1/2 | 1.3 | 0.125 | 0.73 | 0.77 |
| 1 | 2.5 | 0.25 | 0.39 | 0.56 |
| 1 1/2 | 3.8 | 0.375 | 0.26 | 0.39 |
| 2 | 5 | 0.50 | 0.19 | 0.25 |
| 3 | 7.5 | 0.75 | 0.07 | 0.06 |

LANGLANDS (1936) in Australia carried out bending tests on beams with square-cornered notches at the ends. He used Red tulip oak (*Tarrietia argyrodendron* var. *peralata*), a hardwood with a very uniform structure. The dimensions of his beams and his results are given in Table 7.20. One can see that the use of Eq. (7.67) shows at least the trend of the geometrical notch effect.

**7.4.2.6 Fatigue in Bending.** It has been demonstrated in the chapters dealing with tensile and crushing strength that the endurance of load or stress has a marked effect on the mechanical behavior of wood. MARKWARDT (1930) pointed out that in structural parts of wooden aircraft during diving for a few seconds stresses may occur which would, lasting over a longer time, lead to certain failure. During impact bending, where the load is acting only for a few thousandths of a second,

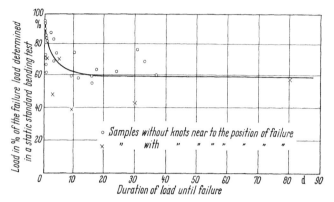

Fig. 7.131. Effect of duration of load until failure on the bending strength of wood beams. From GRAF (1938)

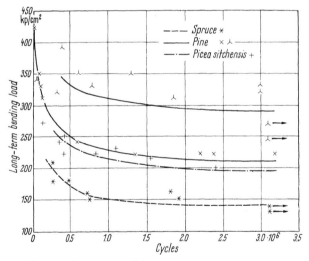

Fig. 7.132. Wöhler-curves from repeated flat bending of coniferous species. From SCHLYTER (1932)

about twice the ultimate stresses of those in normal static bending test are caused. On the other hand beams with a dead load over a period of years, e.g. in ceilings of stores and warehouses, may fail at 50 to 75% of the static modulus of rupture. Fig. 7.130 shows the relationship between stress at proportional limit and modulus of rupture on the one hand and endurance of load on the other. After one year of dead load the modulus of rupture was reduced to about 50%. GRAF (1938) arrived at similar results for structural beams containing knots near the point of failure (Fig. 7.131). ROTH (1935) determined, as the endurance dead load for fir, 65 to 70% of the short time static breaking load. Another problem is the ability of wood to sustain repeated, reversed or vibrational loads without failure. It is known from fatigue tests on metals that notches—e.g. abrupt changes in cross section—reduce the fatigue resistance. Wood as a fibrous material with a porous structure has an "inherent internal notch effect" and is therefore less sensitive to external notches and to repeated loads than more crystalline homogeneous solid bodies. Endurance limits in proportion to the ultimate strength of wood

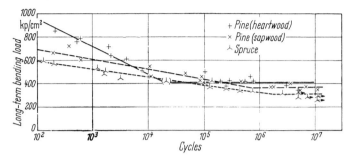

Fig. 7.133. Wöhler-curves from repeated long-term bending tests on small rotating coniferous specimens.
From KRAEMER (1930)

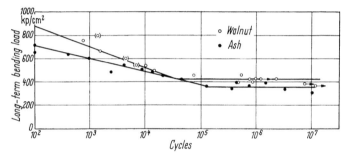

Fig. 7.134. Wöhler-curves from repeated long-term bending test on small rotating specimens of walnut and ash
From KRAEMER (1930)

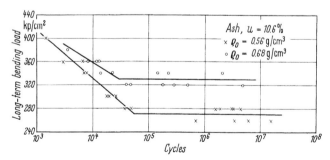

Fig. 7.135. Wöhler-curves from repeated long-term bending test on small rotating ash specimens.
From KOLLMANN (1941)

seem to be, as a rule, higher than for most metals. SCHLYTER (1932) published the results from repeated flat bending of small air-dry specimens (Fig. 7.132). The results of tests on small rotating beam specimens, carried out by KRAEMER (1930) and KOLLMANN (1941), are plotted in Figs. 7.133, 7.134, and 7.135. Fatigue resistance values for some species of wood are listed in Table 7.21.

It can be expected in analogy to static mechanical properties that the fatigue limit stress $\sigma_{fb}$ also increases in proportion to the specific gravity $\varrho$. R. SCHLYTER

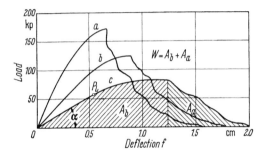

Fig. 7.136. Load deflection diagrams for static bending tests on wood at different moisture contents: *a* ovendry; *b* airdry; *c* green. From MONNIN (1932). W-Total work, $A_b$-work before rupture, $A_a$-work after rupture, $\alpha$-angle of the straight line in the elastic limit, $P_L$-proportional limit

(1932) could verify this fact and developed the formula

$$\sigma_{fb} = C \cdot R_u \quad (7.69\text{a})$$

where $C$ is a constant and $R_u$ is the ratio of ovendry weight to volume with the moisture content $u$ at test. SCHLYTER (1932) found for $R_{12}$ (at moisture content $u = 10 \cdots 12\%$ and with the dimension g/cm³) $C = 440$ for Swedish pine wood and $C = 480$ for spruce. The values for $\sigma_{fb}$ then will be obtained in kp/cm². Instead of $R_u$ we can use $\varrho_u$ (= weight at moisture content $u$ per volume at moisture content $u$). Values for the "quality factor" $C' = \dfrac{\sigma_{fb}}{100\varrho_u}$ are presented in Table 7.21 Using them Eq. (7.69a) must be rewritten as follows:

$$\sigma_{fb} = 100 \cdot C' \cdot \varrho_u. \quad (7.69\text{b})$$

Table 7.21. *Fatigue Resistance Values for Some Species of Wood*

| Species | Specific gravity $\varrho u$ | Moisture content $u$ % | Modulus of rupture $\sigma_{bb}$ kp/cm² | Fatigue limit stress $\sigma_{fb}$ kp/cm² | $\dfrac{\sigma_{fb}}{\sigma_{bb}}$ | Ratio $C' = \dfrac{\sigma_{fb}}{100\varrho_u}$ | Source |
|---|---|---|---|---|---|---|---|
| *Coniferous woods* | | | | | | | |
| Douglas fir [*Pseudotsuga taxifolia* (Poir.) Britt.] | — | 14.3 | 1,052 | 281 | 0.27 | — | US For. Prod. Lab. |
| Douglas fir [*Pseudotsuga taxifolia* (Poir.) Britt.] | — | 23.8 | 900 | 273 | 0.30 | — | |
| Spruce (*Picea excelsa* Link.) | 0.44 | 10···12 | 780 | 195 | 0.25 | 4.4 | R. SCHLYTER |
| Spruce (*Picea* sp.) | 0.48 | 10.8 | 1020 | 300 | 0.29 | 6.3 | O. KRAEMER |
| Sitka spruce (*Picea sitchensis* Carr.) | 0.44 | 10···12 | 780 | 185 | 0.24 | 4.2 | R. SCHLYTER |
| Sitka Spruce (*Picea sitchensis* Carr.) | — | 13.8 | 850 | 224 | 0.26 | — | US For. Prod. Lab. |
| Pine (*Pinus sylvestris* L.) | | | | | | | |
| Heartwood | 0.65 | 10.8 | 1160 | 420 | 0.36 | 6.5 | O. KRAEMER |
| Sapwood | 0.56 | 11.3 | 955 | 360 | 0.38 | 6.4 | O. KRAEMER |
| Pine, Swedish | 0.49 | 10···12 | 880 | 195 | 0.22 | 4.0 | R. SCHLYTER |
| Pine, Finnish | | | | | | | |
| Light sapwood | 0.36 | 13 | 522 | 160 | 0.31 | 4.4 | |
| Heartwood from the top | 0.46 | 13 | 763 | 240 | 0.31 | 5.2 | E. WEGELIUS |
| Heartwood from the butt | 0.47 | 13 | 817 | 285 | 0.35 | 6.1 | |
| Heartwood, resinous | 0.62 | 13 | 863 | 320 | 0.37 | 5.2 | |
| *Broadleaved species* | | | | | | | |
| Birch (*Betula verrucosa* Ehrhardt) | 0.67 | — | 1400 | 350 | 0.25 | 5.2 | O. KRAEMER |
| Birch (*Betula pubescens* Ehrhardt) | 0.63 | 12 | 1300 | 320 | 0.25 | 5.1 | E. WEGELIUS |
| Oak (*Quercus alba* L.) | — | saturated | 745 | 224 | 0.30 | — | US For. Prod. Lab. |
| Ash (*Fraxinus excelsior* L.) | 0.56 | 10.6 | — | 270 | — | 4.8 | F. KOLLMANN |
| | 0.68 | 10.6 | — | 330 | — | 4.9 | F. KOLLMANN |
| | 0.65 | 9.5 | 1220 | 360 | 0.30 | 5.5 | O. KRAEMER |
| Walnut (*Juglans regia* L.) | 0.60 | 8.1 | 1400 | 420 | 0.30 | 7.0 | O. KRAEMER |

It is not possible to predict the number of cycles after which the endurance load or fatigue limit stress (in the range between 22 and 38% and averaging 28 to 30% of the modulus of rupture) will be reached. However, experiments have shown, endurance limits between $3 \cdot 10^4$ and $5 \cdot 10^6$ cycles.

## 7.5 Shock Resistance or Toughness

### 7.5.0 General Considerations

Wood members of aircrafts, carriages, machines, sporting goods, ladders, tool handles as well as elements in building construction fail more frequently under the influence of impact stresses than under static overloading.

An impact stress acts for only a very short period of time, for example a few microseconds. The behavior of wood against impact stress is called shock resistance. High shock resistance of wood is equated with toughness, low shock resistance with brashness.

A beam loaded for a very short time, as when struck with a hammer or by the impact of a falling body, will offer much higher breaking strength than the same beam under static, short time bending test. Curve $B$, in Fig. 7.130, extrapolated in this sense, shows that for a very brief period a beam can sustain about 150% the load as under standardized bending tests. TIEMANN (1951), based on the observation that a beam in impact tests "may be deflected about twice as much before it fails as under ordinary tests", drew the conclusion that the breaking strength also is about doubled.

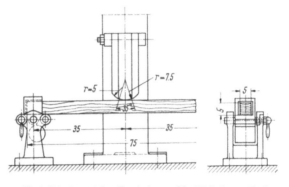

Fig. 7.137. Impact bending test assembly (Hatt-Turner-Test)

The theory of impact bending is very complicated. More recent contributions, applied to the impact loading of wood, have been published by KRECH (1960) and SCHNEIDER (1965). The amount of shock resistance of a solid body depends directly on its ability to absorb energy and to dissipate it by deformations. In this regard tough materials are superior to brittle or brash ones. MONNIN (1932) pointed out that the work to ultimate load is equal to the total energy absorbed in impact bending. Fig. 7.136 shows how this total work W (in kpm) is influenced by changes in moisture content. The numerical value of the area below the total load-deflection curve, determined by means of a planimeter, according to MONNIN depends on:

1. the slope tg $\alpha$ of the straight line in the elastic range, that is on the modulus of elasticity,
2. the position of the proportional limit $P_L$,
3. the modulus of rupture which itself depends on tensile strength and crushing strength,
4. the deflections at the proportional limit and at the moment of failure,
5. the resistance of the fibers after failure, which depends on the cohesion perpendicular to the grain and therefore on tensile strength perpendicular to the grain, on cleavage resistance, and on maximum stress in shear parallel to the grain.

380                    7. Mechanics and Rheology of Wood                [Ref. p. 414

Doubtless Monnin's considerations are interesting and prove the far reaching importance of impact tests. GHELMEZIU (1937/38), to be sure, objected that some details in static and in impact bending are very distinct.

### 7.5.1 Determination of Shock Resistance

**7.5.1.1 Single blow impact test.** The single blow impact test on small specimens (2 cm $\times$ 2 cm $\times$ 28 cm) is a valuable test. Several types of test machines are available. Central loading and a span length of 24 cm (9.47 in.) are standardized internationally. In Europe frequently a 10 kpm-pendulum hammer (e.g. Amsler type) is used. Occasionally 15 kpm —pendulum hammers are applied. The weight of the hammer and its velocity, when hitting the specimen, has very little influence on the test results (CASATI, 1932). Complete failure of the specimen must be obtained on one drop. According to German and French standard specifications the blow shall act tangential to the annual rings. The absorbed energy $W$ is measured in kpm. Usually it is related to the cross section $A = w \cdot h = 2\,\mathrm{cm} \times 2\,\mathrm{cm}$ of the specimen and designated as impact work a:

$$a = \frac{W}{A} = \frac{W}{2.2} = 0.25\ W. \tag{7.70}$$

It is possible to determine the reaction force $\frac{P}{2}$ under one support by means of suitable devices. A simple assembly is as follows: One support is rigid, the other can give way in the direction parallel to the direction of the blow. At the moment of the blow, a steel ball with a diameter of 1 cm is pressed against a bar made of soft aluminum having a defined Brinell-hardness $H_B$. The diameter $d$ of the ball impression is measured and using Brinell's formula one obtains

$$P = \pi\left(1 - \sqrt{1 - d^2}\right) H_B. \tag{7.71}$$

If the value for $P$ is substituted in the classical formula for the modulus of rupture —Eq. (7.61)— and if the standardized dimensions of the beams ($L = 24\ \mathrm{cm}$, $w = h = 2$ cm) are introduced, the dynamic modulus of rupture $\sigma_{bbd}$ is

$$\sigma_{bbd} = 4.5 P = 14.14\left(1 - \sqrt{1 - d^2}\right) H_B. \tag{7.72}$$

KOLLMANN (1940) first used piezoelectric indicators for the measurement of the reaction force. KRECH (1960) applied them too and compared the results with those obtained by the use of strain gauges.

The U.S. Forest Products Laboratory at Madison, Wisc., developed a toughness machine in which the kinetic energy of a pendulum weight acts through a flexible steel cable on an aluminum tup having a radius of $^3/_4$ in. (18 mm). ASTM Designation: D 143-52 (reapproved in 1961 without change) prescribes under paragraphs 74 and 75:

"The machine shall be adjusted before test so that the pendulum hangs truly vertical and shall be adjusted to compensate for friction. The cable shall be adjusted so that the load is applied to the specimen when the pendulum swings to 15 deg from the vertical so as to produce complete failure by the time the downward swing is completed. The weight position and initial angle (30, 45, or 60 deg) of the pendulum shall be chosen so that complete failure of the specimen is obtained on one drop. Most satisfactory results are obtained when the difference between the initial and final angle is at least 10 deg. The initial and final angle shall be read to the nearest 0.1 deg by means of the vernier attached to the machine. The toughness shall then be calculated as follows:

$$T = W_P \cdot L(\cos A_2 - \cos A_1) \tag{7.73}$$

where

$T$ = toughness (work per specimen), in inch-pounds or kpcm,
$W_P$ = weight of pendulum in pounds (kp),
$L$ = distance from center of the supporting axis to center of gravity of the pendulum, in inches (cm),
$A_1$ = initial angle, in degrees, and
$A_2$ = final angle the pendulum makes with the vertical after failure of the test specimen, in degrees.

Basic information on the toughness of the wood was obtained by JAMES (1962) using the toughness machine with slight modifications. Investigating the load deflection relations of 3 wood specimens (red oak, yellow birch, and sweetgum) under rapid loading rates, he found that for an increase in the rate of deflection by a factor of about 10,000, the modulus of rupture of air-dry wood increased by about 31% and that of green wood by about 44%. The moduli of elasticity increased in the order of 14%, the extreme fiber stress at the proportional limit averaged 85% higher for rapid deflection. A comparison of the total work under static and under rapid deflection is only approximate because of the difficulty in defining the proper termination of static bending tests.

### 7.5.1.2 The Hatt-Turner Test (Successive Blows Impact Test).

In the Hatt-Turner or similar impact machines the test is made by incremental drops of a 50-lb (22.5-kg) hammer or a

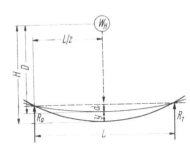

Fig. 7.138. Evaluation scheme for the Hatt-Turner test diagram. From WILSON (1922)

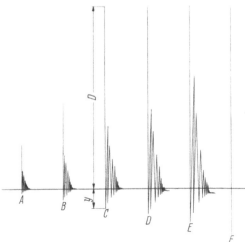

Fig. 7.139. Record of successive falls ($A$ to $F$) of hammer on center-loaded beam. $D$ = height of fall of the hammer; $y$ = added deflection caused by the impact. From BROWN, PANSHIN and FORSAITH (1952)

100 lb (45-kg) hammer respectively. The impact bending tests shall be made on 2 in. by 2 in. by 30 in. (5 cm × 5 cm × 76 cm) specimens using center loading and a span length of 28 in. (70 cm). The test assembly and the dimensions of the bearing block are shown in Fig. 7.137. The specimen is tested in impact bending by dropping the hammer from heights which are increased regularly. The first drop shall be 1 in. (2.5 cm), after which the drop shall be increased by 1-in. (2.5-cm) increments until a height of 10 in. (25 cm) is reached. A 2-in. (5-cm) increment shall than be used until complete failure occurs or until a 6-in. (15-cm) deflection is reached.

WILSON (1922) derived an analysis for the impact stresses in a wooden beam. Fig. 7.138 shows schematically which data are necessary for the evaluation. The following symbols may be used:

$W_H$ = weight of the hammer,
$L$ = span,
$d$ = deflection caused by the hammer when it acts as a static load,
$y$ = added deflection caused by the impact,
$\Delta$ = $d + y$ = total deflection,
$D$ = height of fall of the hammer to the position of the beam under static load $W_H$,
$H$ = $D + y$ = total fall of the hammer.

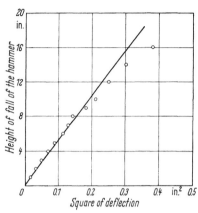

Fig. 7.140. Relationship between height of fall of the hammer and the square of deflection of the beam. According to ASTM

Under the two necessary assumptions, that 1.) up to the proportional limit the energy absorbed by the wood is recoverable on release of the stress, and 2.) the form of the elastic curve under impact is the same as that under static loading. WILSON finally came to the following equation:

$$D = \frac{y^2}{2d}. \qquad (7.74)$$

The Hatt-Turner machine is equipped with an automatic recorder for the $D$ and $y$ values. Fig. 7.139 shows such a sample drum record. The D values are plotted versus the $y^2$ values; the resulting curve is a straight line up to the proportional limit (Fig. 7.140). The total deflection $\Delta = d + y$ of the beam caused by the impact of the hammer could also be effected by some static load $P$. It can be written

$$P = \frac{2 \cdot W_H \cdot H}{\Delta} \qquad (7.75)$$

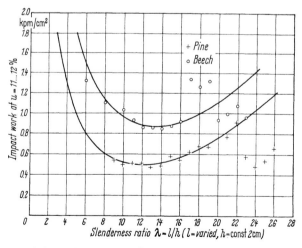

Fig. 7.141. Effect of slenderness ratio $l/h$ on the impact work. From KOLLMANN (1944)

Ref. p. 414]                7.5 Shock Resistance or Toughness                383

The equation for fiber stress $\sigma_{PL}$ at the proportional limit under impact for a beam rectangular in cross section and loaded at the center is therefore

$$\sigma_{PL} = \frac{3PL}{2w \cdot h^2} = \frac{3L}{2w \cdot h^2} \cdot \frac{2W_H \cdot H}{\varDelta} = \frac{3W_H \cdot HL}{w \cdot h^2 \cdot \varDelta} . \tag{7.76}$$

The modulus of elasticity under impact of such a beam can be obtained in a similar way, as follows

$$E = \frac{W_H \cdot H \cdot L^3}{2w \cdot h^2 \varDelta^2} . \tag{7.77}$$

### 7.5.2 Comparison of Impact Test Results

Due to the variety of methods and apparatus for impact tests their results may not be readily compared. Nevertheless PETTIFOR (1936), based on numerous investigations, could establish the following relationships between maximum drop $H$ (in.) of the 50-lb. hammer of the Hatt-Turner machine and the absorbed energy $W$ (in.-lb.) determined with the F.P.L. toughness-testing machine:

$$\left. \begin{aligned} H &= 0.146\,W + 9.3 \text{ for ash}, \\ H &= 0.128\,W + 12.7 \text{ for 37 other species, mainly hardwoods.} \end{aligned} \right\} \tag{7.78}$$

Between the maximum drop for 50-lb. and 100-lb. hammer the following equation was found:

$$H_{50} = 2.16\,H_{100} . \tag{7.79}$$

WILSON (1922) also plotted the results of toughness tests with the pendulum hammer versus maximum drops observed in Hatt-Turner tests. By this technique he came to the equation

$$W = 0.85 \cdot H^{5/4} . \tag{7.80}$$

Finally a contribution to the problem made by ELMENDORF (1916, 1922) may be mentioned. He derived the acceleration-time diagram by double differentiation from the deflection-time curves during one blow. Knowing the acceleration, it is possible to calculate, given weights of hammer and specimen, the acting force on the center of the beam which is equivalent to a static bending load. It develops that the breaking loads in impact tests are about 75% higher than in static tests.

### 7.5.3 Influences Affecting the Shock Resistance

**7.5.3.1 Shape and Size of Beams, Notches (Izod-test).** In impact bending and toughness tests mostly prismatic specimens with square cross section are used. The dimensions are variable. The volume of the specimens for the Hatt-Turner test between the two supports (volume of the beam span) is 1750 cm³. The corresponding volume for the toughness tests with the pendulum hammer is 96 cm³. Until 1949 in the USA, toughness test specimens $^5/_8$ in. (1.5 cm) square and 10 in. (25 cm) long, supported over a span of 8 in. (12 cm) were used, having as volume of the beam span only 27 cm³.

MONNIN (1919, 1932) stated that in toughness tests the absorbed energy reaches a minimum for the slenderness ratio $\lambda = \frac{L}{h} = 12$. KOLLMANN (1944) published the curves in Fig. 7.141 and showed that they obey the following equations:

$$W = W_0 \cdot e^{-\beta \frac{L}{h}} + \gamma \frac{L}{h} \tag{7.81}$$

where $W$ is the absorbed energy in kpm, $W_0$ is the potential energy of the pendulum hammer prior to the test, $\beta$ and $\gamma$ are constants.

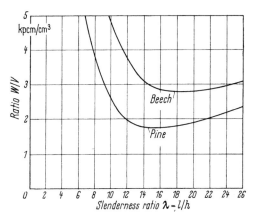

Fig. 7.142. Ratio of adsorbed energy $W$ to volume $V$ as a function of slenderness ratio $\lambda$ for beech and pine. From KOLLMANN (1944)

YLINEN (1944), based on theoretical considerations, pointed out that the absorbed energy should be proportional to the volume $V$ of the beam span. In Fig. 7.142 the ratio $W/V$ as a function of slenderness ratio $\lambda$ is plotted for beech and pine wood.

The wide scattering of the values for the impact work beyond $L/h = 16$ in Fig. 7.141 is remarkable. In this range of slenderness, beams, especially those cut from very tough woods, do not break into two parts, the friction on the supports and the braking of the pendulum hammer yielding much too high values of apparently absorbed energy. On the other hand a cleavage of the beams may occur inducing very low values. The whole problem of impact bending is complicated. Nevertheless a comparison between the results plotted in Figs. 7.141 and 7.142 shows that there

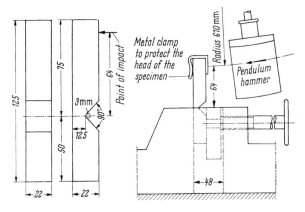

Fig. 7.143. Specimen and apparatus for Izod impact test

exists a similar tendency. Furthermore the following conclusions could be drawn: the potential energy of the hammer has only a limited effect on the test results; dimensions and span of the beam are critical.

In the United Kingdom notched specimens are tested by means of a pendulum hammer. Specimen and grip for this Izod-test are shown in Fig. 7.143.

PETTIFOR (1937) found the Izod-values for different species listed in Table 7.22.

Table 7.22. *Izod-values for Different Species with Varying Moisture Content* (From PETTIFOR, 1937)

| Different species | Izod-value ft. lb. |
|---|---|
| Green | 3.3···8.2···18.1 |
| Air-dry, $u = 12\%$ | 2.7···8.1···18.3 |

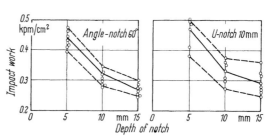

Fig. 7.144. Dependence of the impact work upon the depth of notch for pine. From REIN (1943)

PETTIFOR (1937) proved that there is a significant relationship between toughness and Izod impact values. Toughness tests on notched prismatic beams were carried out also in the U.S. Forest Products Laboratory and in Germany by REIN (1943). REIN used sticks with 2 cm by 2 cm cross section, with a span of 22 cm, and with a notch on the tension side. A few of his results are shown in Fig. 7.144 and later in Figs. 7.151 and 7.152.

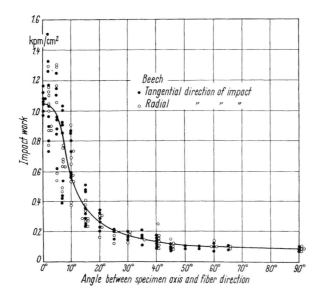

Fig. 7.145. Decrease in toughness with increase of the angle between specimen axis and fiber direction. From GHELMEZIU (1937/38)

**7.5.3.2 Grain Angle.** Generally the values for toughness and impact bending of coniferous species are higher for radial than for tangential blows (MARKWARDT and WILSON, 1935; SEEGER, 1937). No such clear differences could be found for broadleaved species. Slope of grain reduces the shock resistance of wood remarkably. An angle of only 5° causes a 10% decrease in toughness, an angle of 10° leaves only half the shock resistance (Fig. 7.145). Drawing an analogy to Han-

25 Kollmann/Côté, Solid Wood

kinson's formula for strength—cf. Eq. (7.40)—one can express the absorbed energy $W\gamma$ under the angle $\gamma$ of grain direction:

$$W\gamma = \frac{W_\| W_\perp}{W_\| \sin^n \gamma + W_\perp \cos^n \gamma} \qquad (7.82)$$

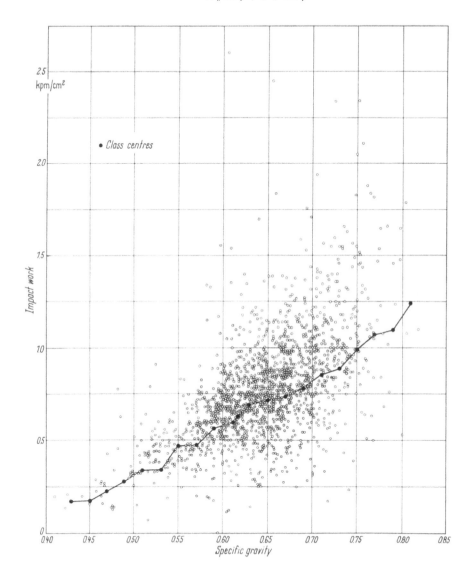

Fig. 7.146. Effect of the specific gravity on the impact work for air-dry ash. From KOLLMANN (1941)

where $W_\|$ is the toughness for $\gamma = 0°$, and $W_\perp$ is the toughness for $\gamma = 90°$. The exponent $n$ varies between 1.5 and 2.

**7.5.3.3 Density.** NEWLIN and WILSON (1919), analyzing the very numerous results of impact bending tests at the U.S. Forest Products Laboratory, found that the maximum height of drop in the Hatt-Turner test is proportional to the square of specific gravity, weight oven-dry based on green volume ($R_g^2$). MONNIN (1932) introduced a coefficient $k$ of shock resistance $\left(k = \dfrac{W}{w \cdot h^{10/6}} = 0.157 \text{ W kpm}\right)$ and stated that the ratio $k/\varrho_{15}^2$ suitably qualifies the mechanical properties of wood. He called the ratio $k/\varrho_{15}^2$ "dynamic quality factor" and remarked that this

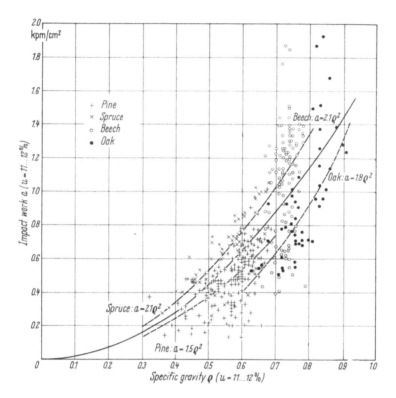

Fig. 7.147. Effect of the specific gravity on the impact work for pine, spruce, beech, and oak wood. From GHELMEZIU (1937/38)

figure for the toughest species reaches the value 2; for brash (e.g. partly rotted woods the value may be reduced to 0.2. KOEHLER (1933), investigating the causes of brashness in wood by means of the FPL-pendulum machine, also reported a parabolic function between toughness and specific gravity. SEEGER (1937) and THUNELL (1941) chose a linear relationship for pine and beech wood (similar to Eq. 7.44). As compared with the results of static tests, the wide scattering of the values is typical.

KOLLMANN (1941) obtained the points plotted in Fig. 7.146 for 1500 impact bending tests on ashwood. The mean values were calculated for density classes of 0.02 g/cm³. The curve through these mean values may be equalized through a straight line, but a cubic parabola ($a = 2.33 \cdot \varrho_0^3$) fits better. GHELMEZIU (1937/38) testing spruce, pine, beech, and oak having 11 to 12% moisture content with the Amsler pendulum hammer obtained the results shown in Fig. 7.147. He found good agreement with the observations of the U.S. Forest Products Laboratory:

$$a = C \cdot \varrho_u^n \tag{7.83}$$

where $\varrho_u$ = density at moisture content $u$. The constant $C$ varies between 1.5 and 2.1 (on the average 1.8, and for oven-dry condition 1.9) at 12% moisture content and for a tangential blow. The exponent $n$ as a rule is 2, except for oak and ash ($n \approx 3$).

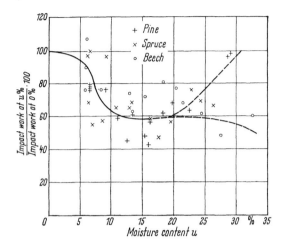

Fig. 7.148. Effect of moisture content on the impact work. From GHELMEZIU (1937/38)

### 7.5.3.4 Moisture Content.
In the preceding chapters we always found that wood increases in static strength as it dries below the fiber saturation point. This statement does not apply to properties that represent toughness, or shock resistance. Dry wood is not as flexible as green wood. This fact is illustrated in Fig. 7.136. The areas below the load-deflection curves up to complete failure, which are proportional to the total work, are hardly influenced by the moisture content. MARKWARDT and WILSON (1935) came to the conclusion that in evaluating the impact bending tests of air dry specimens the moisture constant may be neglected. In the "Wood Handbook" (1955, p. 85) the average increase in height of drop causing complete failure in impact bending per 1% decrease in moisture content is given by the value 0.5%. The same value is valid for work to maximum load in static bending.

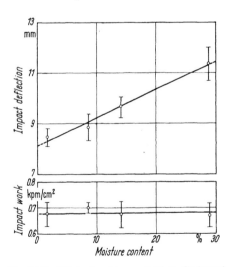

Fig. 7.149. Effect of moisture content on the impact deflection and impact work for beech. From KRECH (1960). The vertical lines show the statistical 99% confidence limits for any point of measurement

GHELMEZIU (1937/38), using his test data, designed the curve represented in Fig. 7.148. The graph shows that toughness is independent of moisture content in the range between 10 and 20%. More recent results by KRECH (1960) showed that for beechwood in the hygroscopic range the impact bending work was independent on the moisture content. Increasing moisture content means, of course, increasing deflection (Fig. 7.149).

REIN (1943) has investigated the toughness of notched specimens. His curves (Fig. 7.150) for pine are similar to those obtained by GHELMEZIU. For fir a rather steep decrease in toughness with decreasing moisture contents below about 5% is interesting. "Toughness is dependent upon both strength and pliability" (Wood Handbook, 1951, p. 85). Some woods become rather rigid, less pliable, and brittle as they approach the oven-dry state.

### 7.5.3.5 Temperature.
As was shown in the previous sections, the deformations increase with temperature, but on the other hand, the cohesion decreases. One must, therefore, expect more complicated relationships between impact work and temperature than those for the static strength properties. The curves of Fig. 7.151 were taken from a three-dimensional diagram by THUNELL (1941). One can deduce from them that the impact work of air-dry pinewood at low temperature (below $-20\,°C$) increases considerably. Very wet wood ($u \approx 70\%$) shows an opposite trend. The effect of temperature was essentially greater with denser than with lighter woods.

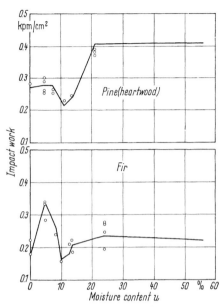

Fig. 7.150. Effect of moisture content on the impact work for notched specimens of pine and fir. From REIN (1943)

Fundamental investigations on the same question are credited to SULZBERGER (1948) in Australia. It became evident that the temperature effect varies with various species. Figs. 7.152 to 7.154 give some of the curves of SULZBERGER. Certain woods, e.g. *Eucalyptus regnans*, behave quite abnormally.

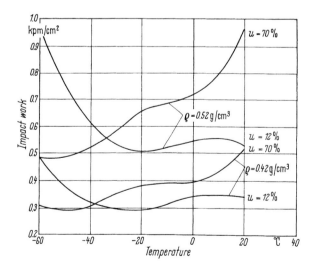

Fig. 7.151. Effect of temperature on the impact work for pine with different density in the air-dry and nearly green condition. From THUNELL (1941)

Results obtained by REIN (1943), testing notched specimens of pine heartwood and sapwood in the temperature range −40 °C and +70 °C, are shown in Fig. 7.155. The course of the curves is similar to those of unnotched specimens. Between −20 °C and +20 °C the influence of temperature is practically negligible.

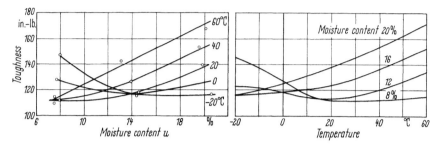

Fig. 7.152. Effect of moisture content and temperature on the toughness of spruce tangential to the annual rings. From SULZBERGER (1948)

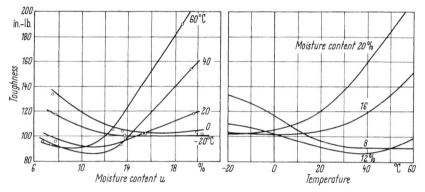

Fig. 7.153. Effect of moisture content and temperature on the toughness (tangential to the annual rings) of hoop pine wood. From SULZBERGER (1948)

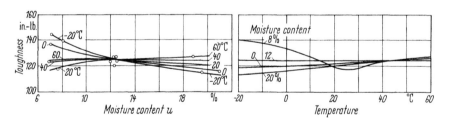

Fig. 7.154. Effect of moisture content and temperature on the toughness (tangential to the annual rings) of coach-wood. From SULZBERGER (1948)

**7.5.3.6 Anatomical Properties, Chemical Constituents, Decay.** The width of annual rings allows conclusions on static strength properties to some extent (cf. Section 7.3.3.2) but not on shock resistance. Fig. 7.156 gives evidence for this fact with the amendment that very wide annual rings in coniferous species are always associated with extremely low toughness. This may be easily understood since such very broad rings in coniferous species produce wood of extremely low densities (cf. Fig. 7.82 and 7.83). In the ring-porous woods such as ash and oak, broader rings are generally a feature of denser wood which is also tougher. As a rule, ashwood with very narrow rings (width less than 1 mm) is brash and brittle (KOLLMANN, 1941). In the diffuse-porous beechwood, GHELMEZIU (1937/38)

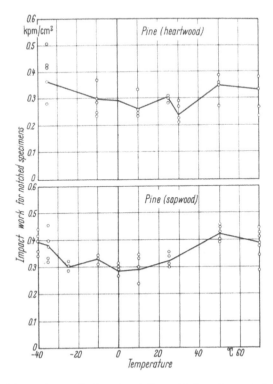

Fig. 7.155. Effect of temperature on the impact work for notched specimens of pine. From REIN (1943)

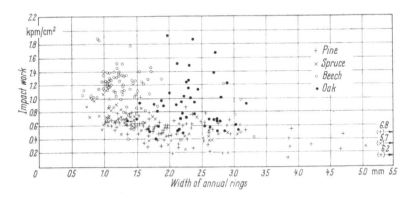

Fig. 7.156. Effect of the width of annual rings on the toughness of various wood species. From GHELMEZIU (1937/38)

found maximum values of toughness for ring widths somewhat below 1 mm. The wood anatomy could not completely explain the internal causes of brashness or toughness. Nevertheless, a few statements are possible: thin-walled tracheids in coniferous species and in broad-leaved species a large number of vessels and

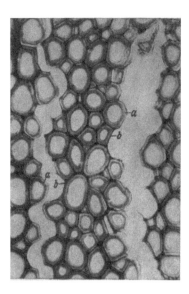

Fig. 7.157. Micro-section perpendicular to the fibers in a test sample of ash after an impact test. *a* Secondary cell walls, *b* rupture in the middle lamellae. From CLARKE (1935, 1936)

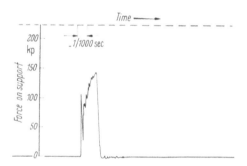

Fig. 7.159. Temporal course of the force on support in impact bending test on pine, $a = 0{,}55$ kpm/cm$^2$, 2 cm × 2 cm × 30 cm, span between support 24 cm, hammer energy 15 kpm, piezoelectrically recorded. From KOLLMANN (1940)

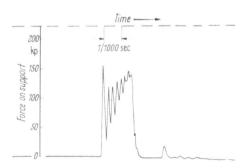

Fig. 7.160. Temporal course of the force on support in impact bending test on ash, $a = 0{,}54$ kpm/cm$^2$. Other test conditions see Fig. 1.159. From KOLLMANN (1940)

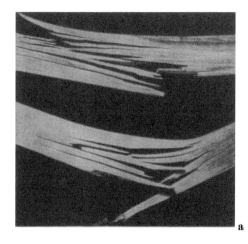

Fig. 7.158 a—c. Types of rupture in ash specimens after impact bending test. a) Wood with exceptionally high shock resistance ($a = 1.55$ resp. 1.71 kpm/cm$^2$), b) normal wood ($a = 0.80$ resp. 0.98 kpm/cm$^2$), c) brash wood ($a = 0.40$ resp. 0.31 kpm/cm$^2$). From KOLLMANN (1951)

parenchyma cells or libriform-fibers with thin walls cause brashness, but it is not possible to formulate quantitative relationships (KOEHLER, 1933). The length of fibers and the diameter of cell cavities have little or no effect on toughness, assuming equal density (GERRY, 1915; FORSAITH, 1921; KOEHLER, 1933). Apparently the principal source of brashness or toughness lies in the fine structure of the wood.

Shock resistance, just as tensile strength, is greater the more parallel to the fiber axis are the fibrils arranged. The condition of the middle-lamella which cements the single fibers is very important. It consists primarily of lignin which is brittle as compared with cellulose. CLARKE (1935, 1936) showed that after impact bending tests on ash wood, failure occurred almost exclusively in the middle lamellae whereas the secondary cell walls, which are destroyed in compression tests, remained intact (Fig. 7.157). Decay, and even incipient thermal decomposition markedly reduce toughness.

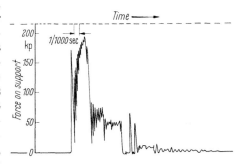

Fig. 7.161. Temporal course of the force on support in impact bending test on ash, $a = 1.51$ kpm/cm². Other test conditions see Fig. 7.159. From KOLLMANN (1940)

**7.5.3.7 Types and Phenomena of Failures in Impact Bending.** Types of failures in static tests will be described and classified (cf. ASTM Designation: D 143-52) but normally they do not allow conclusions on the ultimate stresses. In impact bending or toughness tests, types of rupture clearly characterize wood quality in the following way. Wood with exceptionally high shock resistance shows long splits and coarse splinters. Often, on the compressive side, a layer several fibers thick, remains intact (Fig. 7.158a). Wood with average properties as far as tough-

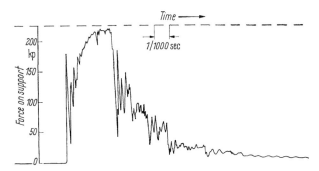

Fig. 7.162. Temporal course of the force on support in impact bending test on ash, $a = 1.84$ kpm/cm². Other test conditions see Fig. 7.159. From KOLLMANN (1940)

ness is concerned develops shorter, more "fibrous" splinters which are usually longer on the tensile than on the compression side (Fig. 7.158b). In brash woods the fractured surfaces are plane, rather smooth, occasionally wavy or stepped like a staircase (Fig. 7.158c).

The variety of types of rupture in impact bending shows that fracture in tough and brash wood occurs in a very different manner. Brash wood goes to failure with very small deformations and very quickly, whereas the fracture in tough wood obviously is accompanied by rather large deformations and vibrations, phenomena which require more time.

KOLLMANN (1940), using a piezoelectric indicator, found the diagrams shown in Figs. 7.159 to 7.162 for the temporal course of the reaction force under one support of the pendulum hammer machine. One can see the following: Failure occurs in brash pine and ashwood within a period of 2 to 3 msec. In tough woods the pendulum hammer cannot destroy the stick at once. Hammer and stick form

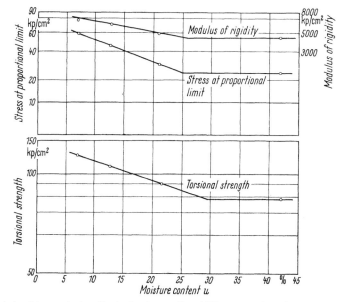

Fig. 7.163. Effect of moisture content on the torsional properties of Sitka spruce. From TRAYER and MARCH (1930)

an oscillating system. Complete rupture is effected in about 10 to 15 msec. Two phases must be distinguished: the event prior to ultimate load and the event after this threshold. For the first phase the ability of a material to absorb energy and to convert it into heat is critical; in the second phase the type of failure controls the further progress to complete failure. In this phase splinters and splits cause unpredictable friction resistance. KRECH (1960) has repeated and confirmed Kollmann's findings.

## 7.6 Torsional Properties and Shear Strength

### 7.6.0 General Considerations

It is clearly stated in the Wood Handbook (1955, p. 81), that "the torsional strength of wood is seldom needed in design, but data are available that permit calculation of torsional properties if they are required".

The moduli of rigidity in the longitudinal-radial ($LR$), longitudinal-tangential ($LT$), and radial-tangential ($RT$) plane, discussed in Section 7.1, determined the torsional deformation of wood. In twisting a wood member about an axis parallel to the grain only the moduli of rigidity $G_{LR}$ and $G_{LT}$ are involved. For many woods these moduli do not vary greatly, so that a "mean modulus of rigidity", taken according to BAUMANN as $G = \dfrac{E}{17}$ where $E$ is Young's modulus, may be used for computing deformation and strength of twisted wood members. Baumann's ratio is based on measurements on pine, hemlock, and ash. He found the greater value 35 for spruce, and for Okoumé the value of 10. THUNELL (1941) obtained $E/G = 20$ for Swedish pine.

### 7.6.1 Determination of Torsional Strength

The torsion problem for an isotropic material is so complicated that it cannot be discussed in this book. Especially for twisted sticks with square cross section the curvature of the cross section is remarkable. Testing of torsion is simple but the evaluation of the results is rather doubtful since St. Venant's formula for isotropic materials is used. In this case the torsional strength $\tau_{tb}$ for square cross section is calculated as follows:

$$\tau_{tb} = 4.80 \frac{M_t}{a^3} \tag{7.84}$$

where $M_t$ = torsional moment in cmkp, and $a$ = length of the edge of the cross section in cm. Table 7.23 contains the results of some measurements, including the deformation at failure.

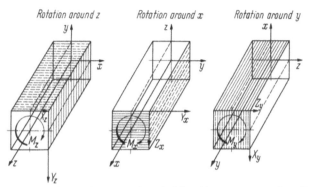

Fig. 7.164. Scheme of the possible torsion cases of wood sticks with square cross section. From HÖRIG (1944)

TRAYER and MARCH (1930) have shown that the modulus of rigidity, stress at proportional limit and torsional strength depend on moisture content in a manner similar to other static mechanical properties (Fig. 7.163). Strength values calculated using formula (7.84) have no physical sense. They are useful for the designer of course, since he is interested in the ratio of the moment at failure to the cubic value of "$a$".

HÖRIG (1944) has published a survey on the possible cases of wood sticks with square cross section (Fig. 7.164). The position of planes with elastic symmetry is as shown in Fig. 7.2. Especially interesting from the technical point of view is the torsion around the $y$-axis. HÖRIG (1944) developed the following formulas for the ultimate stresses at failure strains:

Case 1) torsion around $z$-axis:

$$(\tau_{zx})_{\max} = 3.52\, M_z/a^3, \qquad (\tau_{zy})_{\max} = 11.04\, M_z/a^3;$$

Case 2) torsion around $x$-axis:

$$(\tau_{xy})_{\max} = 12.41\, M_x/a^3, \qquad (\tau_{xz})_{\max} = 3.44\, M_x/a^3; \tag{7.85}$$

Case 3) torsion around $y$-axis:

$$(\tau_{yx})_{\max} = 4.60\, M_y/a^3, \qquad (\tau_{yz})_{\max} = 5.05\, M_y/a^3.$$

Table 7.23. *Torsional Strength of Woods*

| Species | Denisty $\varrho$ g/cm³ | Moisture content $u$ % | Torsional strength kp/cm² | | Deformation at failure °/cm | | Source |
|---|---|---|---|---|---|---|---|
| | | | Sticks with fibers parallel to the longitudinal axis | Sticks with fibers perpendicular to the longitudinal axis | Sticks with fibers parallel to the longitudinal axis | Sticks with fibers perpendicular to the longitudinal axis | |
| *Coniferous woods* | | | | | | | |
| Spruce | 0.35···0.42 | 11.9 | 87···101 | 32···36 | 0.96···1.33 | 1.8···2.0 | K. Huber |
| Spruce | 0.45 | 12 | $\{$ $\tau_1 = 186$ $\tau_2 = 181$ | 25 | — | — | H. Carrington |
| Spruce | 0.48 | 10.8 | 141···162 | 30···62 | 1.1 | 2.3 | O. Kraemer |
| Pine | 0.50···0.55 | 12.2 | 134···140 | 43···50 | 0.95···1.8 | 1.37···1.42 | K. Huber |
| Pine, Heartwood | 0.65 | 10.8 | 163···178 | 39···64 | 0.7 | 1.0 | O. Kraemer |
| Pine, Sapwood | 0.56 | 11.3 | 135···167 | — | 0.6···0.9 | — | O. Kraemer |
| *Broadleaved species* | | | | | | | |
| Birch | 0.67 | 12 | 200[1] | — | — | — | O. Kraemer |
| Beech | 0.66···0.69 | 11.3 | 246···250 | 151···156 | 2.6···3.0 | 2.0 | K. Huber |
| Oak | 0.67···0.71 | 11.7 | 190···220 | 112···114 | 1.2···2.0 | 1.0···1.5 | K. Huber |
| Ash | — | 9 | 158···**213**···250 | — | 1.5···2.5 | — | F. Kollmann |
| Ash | — | 12.0···16.3 | 140···**188**···238 | — | 1.6···3.0 | — | F. Kollmann |
| Ash | — | air dry | 210···245 | — | — | — | R. Baumann |
| Ash | 0.81···0.83 | 11.3 | 258···277 | 156···158 | 1.8···2.1 | 1.2···1.5 | K. Huber |
| Ash | 0.65 | 9.5 | 179···**262**···345 | 136···**167**···209 | 1.3 | 1.3···1.7 | O. Kraemer |
| Walnut | 0.60 | 8.1 | 275···**303**···326 | 135···**152**···162 | 1.0···1.2 | 1.4 | O. Kraemer |

[1] determined on round sticks.

Data of references: Baumann (1922), Carrington (1921), Huber (1928, 1929), Kollmann (1941), Kraemer (1933).

It is evident that the values calculated using Eq. (7.84) differ widely from the true failure stresses, especially for specimens in which the fibers are oriented perpendicular to the axis. Reference also may be made to the contribution by TRAYER and MARCH (1930) who used Prandtl's soap-bubble method in determining the torsion stresses in prisms consisting of anisotropic materials. Another valuable contribution dealing with the torsion properties of laminated woods has been published by REISSNER (1938).

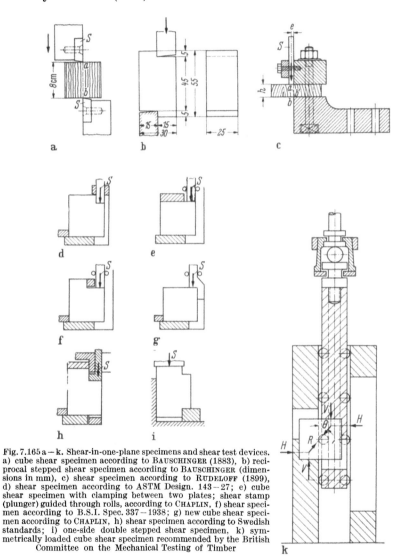

Fig. 7.165 a—k. Shear-in-one-plane specimens and shear test devices. a) cube shear specimen according to BAUSCHINGER (1883), b) reciprocal stepped shear specimen according to BAUSCHINGER (dimensions in mm), c) shear specimen according to RUDELOFF (1899), d) shear specimen according to ASTM Design. 143—27; e) cube shear specimen with clamping between two plates; shear stamp (plunger) guided through rolls, according to CHAPLIN, f) shear specimen according to B.S.I. Spec. 337—1938; g) new cube shear specimen according to CHAPLIN, h) shear specimen according to Swedish standards; i) one-side double stepped shear specimen, k) symmetrically loaded cube shear specimen recommended by the British Committee on the Mechanical Testing of Timber

### 7.6.2 Determination of Shearing Strength Parallel to Grain

The ultimate shearing strength parallel to grain is related to the torsional properties but the shear test is problematic due to superposed, mostly bending, stresses. Compressive stresses, stress concentrations and internal checks are others which mask a clear picture of the shear phenomenon. As a result, the ultimate shearing strength is always remarkably lower than the torsional strength. Considering this, the Wood Handbook (1955, p. 82) states: "For solid wood members, the allowable ultimate torsional shear stress may be taken as the shear stress

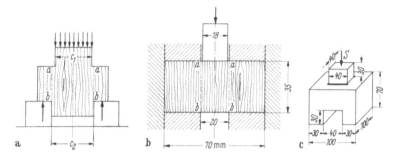

Fig. 7.166 a—c. Shear-in-two-planes specimens and shear test devices. a) Shear-cross, according to INOKUTY, TJADEN and DIN 52187, b) according to LANG, c) according to BROTERO

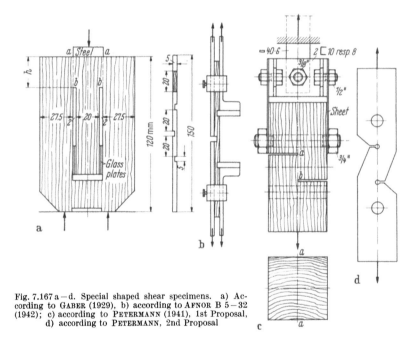

Fig. 7.167 a—d. Special shaped shear specimens. a) According to GABER (1929), b) according to AFNOR B 5—32 (1942); c) according to PETERMANN (1941), 1st Proposal, d) according to PETERMANN, 2nd Proposal

parallel to the grain and two-thirds of this value may be used as the allowable torsional shear stress at the proportional limit." The shearing stress perpendicular to the grain, a very rare case of stress in practice, is about 3 to 4 times higher than that parallel to grain.

The above mentioned unavoidable superposition of additional stresses to pure shear makes any shear test questionable and the results of different test methods may not be compared without special precautions.

Basically, one has to distinguish between shear in one plane (Fig. 7.165) and shear in two planes (Fig. 7.166). There are also specially shaped specimens (Fig. 7.167) and devices using bolts (Fig. 7.168) causing shear stresses. EHRMANN

(1937) has shown that the shape of the specimens and the kind of clamping has a great effect on the results of the tests. COKER and COLEMAN (1930) using photoelastic models, have shown the distribution of the shear stresses in a cube, loaded

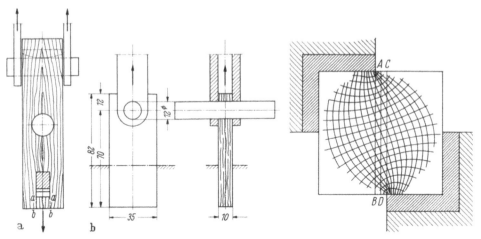

Fig. 7.168 a, b. Shear specimens using bolts. a) According to JOHNSON (USA), b) according to Junkers-Flugzeugbau

Fig. 7.169. Stress distribution in a symmetrically loaded cube. From COKER and COLEMAN (1930)

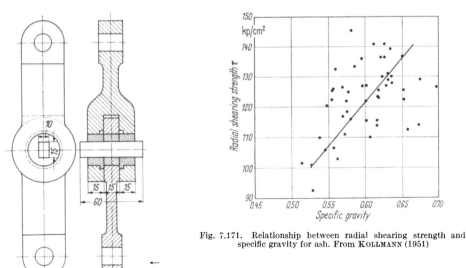

Fig. 7.171. Relationship between radial shearing strength and specific gravity for ash. From KOLLMANN (1951)

← Fig. 7.170. Device for shear-in-two-planes tests. From Junkers Flugzeugbau

symmetrically (Fig. 7.169). In planes $AB$ and $CD$ the vertical stresses are distributed rather uniformly but remarkable deviations occur near the grips. A device, enforcing under reproducible conditions shear in two planes, is illustrated in Fig. 7.170. Numerous tests carried out by KOLLMANN and KEYLWERTH (1944, unpublished) lead to the following relationships:

Ultimate stresses in shear tests according to:

| Fig. 7.170 | Fig. 7.166 a | Fig. 7.165 d | Fig. 7.168 | Fig. 7.167 d |
|---|---|---|---|---|
| 1.0 | 0.53···**0.68**···0.85 | 0.52···**0.66**···0.83 | 0.47···**0.63**···0.85 | 0.33···**0.43**···0.65 |

**400**          7. Mechanics and Rheology of Wood          [Ref. p. 414

The so-called "shear-cross" according to German standard specifications and the shear-parallel-to-grain test according to ASTM D 143-52, on the average as well as in the variation, deliver about the same values.

The shearing strength is proportional to density. SCHLYTER and WINBERG (1929) and later EHRMANN (1937) developed the following formula for the shearing strength $\tau_s$

$$\tau_s = a \cdot R - b$$

where the former found for Swedish pine:

| | Range of moisture content % | | |
|---|---|---|---|
| constants | 6···10 | 15···19 | 23···30 |
| $a$ | 225 | 225 | 160 |
| $b$ | 22.5 | 22.5 | 16 |

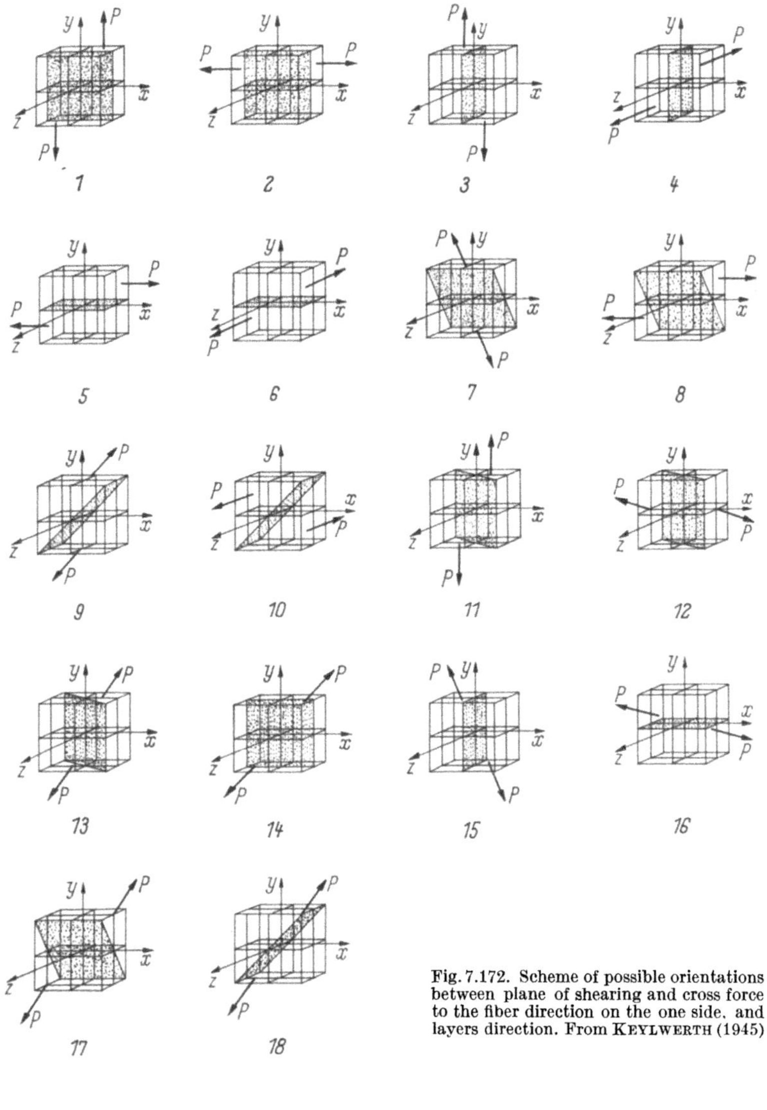

Fig. 7.172. Scheme of possible orientations between plane of shearing and cross force to the fiber direction on the one side, and layers direction. From KEYLWERTH (1945)

Ref. p. 414]     7.6 Torsional Properties and Shear Strength     401

Scattering is more pronounced if the failure occurs in a tangential plane than in a radial one. NEWLIN and WILSON (1919) found a parabolic function between shearing strength $\tau_s$ and density $R$ —weight oven-dry based on green volume

$$\tau_s = A \cdot R^{1/3}.\tag{7.87}$$

The average values for $A$ for all tested commercial timbers of the USA (tangential-parallel-to-grain-shearing) are as follows: green condition $A = 193$, air-dry condition $(u = 12\%)$ $A = 281$; (radial-parallel-to-grain-shearing) green condition $A = 179$, air-dry condition $(u = 12\%)$ $A = 255$. Using these constants one obtains $\tau_s$ in kp/cm$^2$.

Fig. 7.171 shows an example of the relationship between shearing strength and specific gravity. The influence of moisture content is less pronounced than that on tensile or crushing strength, but the orientation between plane of failure and cross force is critical, KEYLWERTH (1945) designed a scheme of possible orientations to grain in shearing (Fig. 7.172). It may be described as follows:

*case*

| | | |
|---|---|---|
| *1* | plane of shearing = tangential plane | cross force ∥ grain |
| *2* | plane of shearing = tangential plane | cross force ⊥ grain (tangential) |
| *3* | plane of shearing = radial plane | cross force ∥ grain |
| *4* | plane of shearing = radial plane | cross force ⊥ grain (radial) |
| *5* | plane of shearing = cross cut plane | cross force ⊥ grain (tangential or ∥ to the layers) |
| *6* | plane of shearing = cross cut plane | cross force ⊥ grain (radial or ⊥ to the layers) |
| *7* | plane of shearing diagonal to grain and perpendicular to radial plane | cross force diagonal to grain |
| *8* | plane of shearing diagonal to grain and perpendicular to radial plane | cross force ⊥ grain (tangential or parallel to the layers) |
| *9* | plane of shearing diagonal to grain and perpendicular to tangential plane | cross force diagonal to grain |
| *10* | plane of shearing diagonal to grain and diagonal to tangential plane | cross force ⊥ grain (radial or perpendicular to the layers) |
| *11* | plane of shearing perpendicular to grain and diagonal to the layers | cross force ⊥ grain |
| *12* | plane of shearing perpendicular to grain and diagonal to the layers | cross force ⊥ grain and diagonal to the layers |
| *13* | plane of shearing perpendicular to grain and diagonal to the layers | cross force diagonal to grain and diagonal to the layers |
| *14* | plane of shearing = tangential plane | cross force diagonal to grain |
| *15* | plane of shearing = radial plane | cross force diagonal to grain |
| *16* | plane of shearing = cross cut plane | cross force ⊥ grain, diagonal to the layers |
| *17* | plane of shearing diagonal to grain and perpendicular to radial plane | cross force diagonal to grain and to the layers |
| *18* | plane of shearing diagonal to grain and perpendicular to tangential plane | cross force diagonal to grain and to the layers |

Figs. 7.173 and 7.174 show the influence of moisture content on shearing strength for varying grain orientations. The average increase of shearing strength

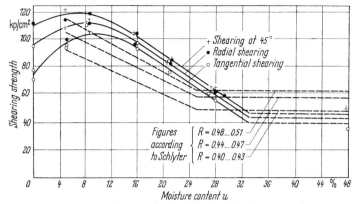

Fig. 7.173. Effect of moisture content on the shearing strength (shear-in-one-plane specimens) for pine. From EHRMANN (1937) and SCHLYTER and WINBERG (1929)

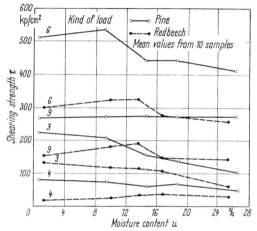

Fig. 7.174. Effect of moisture content on the shearing strength (shear-in-two-planes specimens) for pine and common beech. From KOLLMANN and KEYLWERTH (1944)

parallel to grain in the hygroscopic range with decreasing moisture content is lower than that of crushing strength or bending strength. For some grain orientations there is practically no effect of moisture content on shearing strength. Limit values for the shearing strength of three woods are listed in Table 7.24.

The influence of the angle between shearing plane and fiber direction can be calculated approximately using Hankinson's formula [Eq. (7.40)] or more precisely using a formula corresponding to Eq. (7.20).

Table 7.24. *Values of Shearing Strength for Woods* ($u = 8.8\cdots15.6\%$) *Determined in Using the "Shear-iron"* (Fig. 7.170) (From KEYLWERTH, 1945)

| Case of stress No | Shearing strength kp/cm² | | |
|---|---|---|---|
| | pine | beech | ash |
| 1 | 89$\cdots$112 | 147$\cdots$166 | 157$\cdots$185 |
| 2 | 25$\cdots$40 | 44$\cdots$58 | 55$\cdots$ 75 |
| 3 | 110$\cdots$119 | 159$\cdots$179 | 165$\cdots$198 |
| 4 | 18$\cdots$36 | 65$\cdots$73 | 76$\cdots$ 92 |
| 5 | 267$\cdots$324 | 416$\cdots$447 | 473$\cdots$516 |
| 6 | 276$\cdots$355 | 368$\cdots$429 | 397$\cdots$447 |
| 7 | 189$\cdots$290 | 297$\cdots$460 | 297$\cdots$369 |
| 8 | 158$\cdots$236 | 325$\cdots$438 | 269$\cdots$325 |
| 9 | 233$\cdots$257 | 222$\cdots$315 | 203$\cdots$314 |
| 10 | 250$\cdots$291 | 162$\cdots$291 | 199$\cdots$312 |
| 11 | 93$\cdots$113 | 161$\cdots$179 | 182$\cdots$205 |
| 12 | 18$\cdots$25 | 58$\cdots$89 | 43$\cdots$ 81 |
| 13 | 39$\cdots$48 | 78$\cdots$103 | 122$\cdots$169 |
| 14 | 35$\cdots$56 | 89$\cdots$130 | — |
| 15 | 28$\cdots$34 | 88$\cdots$111 | — |
| 16 | 278$\cdots$306 | 355$\cdots$428 | — |
| 17 | 143$\cdots$204 | 235$\cdots$358 | — |
| 18 | 155$\cdots$221 | 216$\cdots$298 | — |

## 7.7 Hardness and Abrasion Resistance

### 7.7.0 General Considerations

The testing of hardness of wood is a rather nebulous question. What is hardness? It is defined as the resistance of a solid body against the entering of another solid body by force. For metals, the impression of a steel-ball according to Brinell on plane, smooth surfaces, not too near the edges, delivers clear, reproducible figures. For the anisotropic, heterogeneous, hygroscopic wood, the hardness value is more than doubtful. NÖRDLINGER already (1881) has pointed out that the hardness of wood essential in woodworking depends on the type of tool employed. He came to the conclusion that no "absolute" but only a "relative" hardness value is reasonable. BÜSGEN (1904) tested the hardness of wood by impressing a steel-needle into it to a depth of 2 mm. The load producing this effect was called hardness. This procedure never found general application. The sources of error are evident: the tip of the needle is of the size of the wood fibers and therefore one cannot expect an average value of hardness. The results also are influenced by friction and cleavage. Nevertheless the hardness test as a non-destructive or a hardly destructive test, especially for finished wood parts, remained attractive.

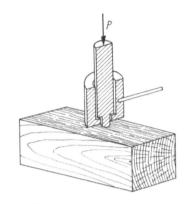

Fig. 7.175. Hardness tester and test block. From BROWN, PANSHIN and FORSAITH (1952)

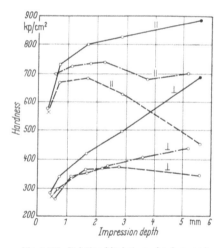

Fig. 7.176. Relationship between hardness and impression depth for common beech. From HUBER (1937/38)

### 7.7.1 Hardness Tests

JANKA (1906, 1908, 1915) proposed and developed a modified Brinell-hardness test for wood. The force required by static loading to embed a steel hemisphere 0.444 inch in diameter, corresponding to 1 cm² of hemisphere surface, completely into the wood is determined. The hardness tester and the test block, in the USA 6 inches in length and 2 by 2 inches in cross section, are shown in Fig. 7.175. There is no distinction between side hardness on a radial or a tangential face, but only between side and end hardness.

JANKA, evaluating the results of hardness tests on 280 wood species, found the following empirical relationships between hardness $H_J$ and crushing strength $\sigma_{cb}$:

$$H_J = 2\sigma_{cb} - 500 \ [\text{kp/cm}^2] \tag{7.88}$$

The formula is only suitable for rough estimations but it proves that the JANKA-hardness test is nothing else than a modified impression test influenced by effects such as friction, shearing and cleavage.

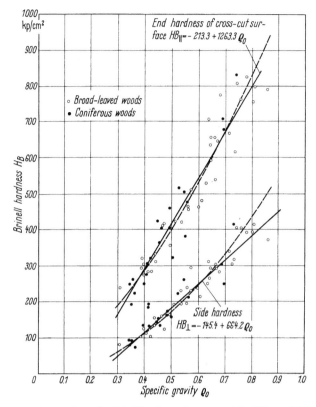

Fig. 7.177. Effect of specific gravity on the Brinell hardness. Measurements by MÖRATH (1932) on American wood species at 12 resp. 15% moisture content, equalized according to TRENDELENBURG and YLINEN

STAMER (1929) and later HUBER (1937/38) showed that hardness depends in an irregular manner on the depth of penetration (Fig. 7.176). The many objections against the JANKA-hardness test led MÖRATH (1932) to apply the well-known BRINELL-hardness test (1900) to wood. He took a steel ball of 10 mm diameter and a normal load of 50 kp — (for extremely hard species 100 kp and for very soft ones 10 kp) —. The maximum load $P$ should be reached within 15 sec, kept constant over a period of 30 sec, and than reduced to zero within another 15 sec. If $P$ is the maximum load, $D$ the diameter of the steel ball, and $d$ the diameter of the impression the value of the Brinell-hardness $H_B$ can be calculated as follows:

$$H_B = \frac{2P}{\pi D(D - \sqrt{D^2 - d^2})} \tag{7.89}$$

Table 7.25 facilitates the use of Eq. (7.87).

PALLAY (1937/38) has criticized the method for the following reasons: The three loads do not create a uniform basis for comparison. The calculation of hardness, based on load and diameter of impression, especially for the side-hardness is of doubtful validity. A steel-ball with only 10 mm diameter is too small with respect to the heterogeneous structure of the wood.

Table 7.25. *Brinell Hardness*

| Very soft woods $P = 10$ kp | | Normal woods $P = 50$ kp | | Very hard woods $P = 100$ kp | |
|---|---|---|---|---|---|
| $d$ mm | $H_B$ | $d$ mm | $H_B$ | $d$ mm | $H_B$ |
| 4.8 | 0.52 | 7.3 | 1.01 | 6.2 | 2.96 |
| 4.7 | 0.54 | 7.2 | 1.04 | 6.1 | 3.05 |
| 4.6 | 0.57 | 7.1 | 1.08 | 6.0 | 3.15 |
| 4.5 | 0.60 | 7.0 | 1.11 | 5.9 | 3.30 |
| 4.4 | 0.62 | 6.9 | 1.15 | 5.8 | 3.43 |
| 4.3 | 0.65 | 6.8 | 1.19 | 5.7 | 3.57 |
| 4.2 | 0.69 | 6.7 | 1.24 | 5.6 | 3.71 |
| 4.1 | 0.72 | 6.6 | 1.28 | 5.5 | 3.86 |
| 4.0 | 0.76 | 6.5 | 1.33 | 5.4 | 4.02 |
| 3.9 | 0.80 | 6.4 | 1.38 | 5.3 | 4.19 |
| 3.8 | 0.85 | 6.3 | 1.43 | 5.2 | 4.37 |
| 3.7 | 0.90 | 6.2 | 1.48 | 5.1 | 4.55 |
| 3.6 | 0.95 | 6.1 | 1.53 | 5.0 | 4.75 |
| 3.5 | 1.01 | 6.0 | 1.59 | 4.9 | 4.96 |
| 3.4 | 1.07 | 5.9 | 1.65 | 4.8 | 5.19 |
| 3.3 | 1.14 | 5.8 | 1.72 | 4.7 | 5.44 |
| 3.2 | 1.21 | 5.7 | 1.79 | 4.6 | 5.68 |
| 3.1 | 1,29 | 5.6 | 1.86 | 4.5 | 5.95 |
| 3.0 | 1.38 | 5.5 | 1.93 | 4.4 | 6.24 |
| 2.9 | 1.48 | 5.4 | 2.01 | 4.3 | 6.55 |
| 2.8 | 1.59 | 5.3 | 2.09 | 4.2 | 6.88 |
| 2.7 | 1.72 | 5.2 | 2.18 | 4.1 | 7.24 |
| 2.6 | 1.85 | 5.1 | 2.28 | 4.0 | 7.62 |
| 2.5 | 2.00 | 5.0 | 2.38 | 3.9 | 8.04 |
| 2.4 | 2.18 | 4.9 | 2.48 | 3.8 | 8.49 |
| 2.3 | 2.38 | 4.8 | 2.59 | 3.7 | 8.97 |
| 2.2 | 2.60 | 4.7 | 2.71 | 3.6 | 9.50 |
| 2.1 | 2.86 | 4.6 | 2.84 | 3.5 | 10.07 |
| 2.0 | 3.15 | 4.5 | 2.97 | 3.4 | 10.68 |
| 1.9 | 3.50 | 4.4 | 3.12 | 3.3 | 11.37 |
| 1.8 | 3.91 | 4.3 | 3.27 | 3.2 | 12.10 |
| 1.7 | 4.35 | 4.2 | 3.44 | 3.1 | 12.91 |
| | | 4.1 | 3.62 | 3.0 | 13.84 |
| | | 4.0 | 3.81 | 2.9 | 14.81 |
| | | 3.9 | 4.02 | 2.8 | 15.92 |
| | | 3.8 | 4.24 | 2.7 | 17.16 |
| | | 3.7 | 4.48 | 2.6 | 18.51 |
| | | 3.6 | 4.75 | 2.5 | 20.02 |
| | | 3.5 | 5.03 | 2.4 | 21.80 |
| | | 3.4 | 5.34 | 2.3 | 23.76 |
| | | 3.3 | 5.68 | 2.2 | 25.99 |
| | | 3.2 | 6.05 | | |
| | | 3.1 | 6.46 | | |
| | | 3.0 | 6.92 | | |
| | | 2.9 | 7.40 | | |
| | | 2.8 | 7.96 | | |

KRIPPEL (PALLAY 1937/38, 1939) proposed to use a steel-ball of a greater diameter (31.831 mm) and to impress it to an exact depth of accurately 2 mm producing a calotte-surface of exactly 2 cm². The method did not find a wider application. In France, the test according to CHALAIS-MEUDON is standardized. A

406                    7. Mechanics and Rheology of Wood                    [Ref. p. 414

steel-cylinder of 30 mm diameter is impressed with a maximum load of 200 kp on a radial section over a period of 5 sec. Since it is difficult to measure the depth of penetration $t$, it is deduced from the width $b$ of the impression:

$$t = 15 - 0.5 \sqrt{900 - b^2}. \tag{7.90}$$

The hardness figure $N$ is defined as

$$N = \frac{1}{t}. \tag{7.91}$$

This figure increases with decreasing hardness of the wood; the range from 0.2 to 2.0 is much more expanded than for the Janka- or Brinell-hardness and according to EGNER (1941), the method is very feasible in determining the side-hardness. BAUMANN (1922), SCHWARZ and BUES (1929), GABER (1935) and PEVZOFF (1935) developed and recommended dynamic hardness tests. The procedure of dropping a steel-ball on a wood surface is easy, but the sources of error are greater and there is also a strong correlation between dynamic end-hardness and crushing strength along the grain. Summarizing, it may be said that hardness tests cannot deliver valuable additional characteristics of the mechanical behavior of wood. However, since they are extensively applied the following section shall discuss the different factors affecting hardness.

### 7.7.2 Factors Influencing the Hardness of Wood

JANKA as early as 1906 and v. LORENZ (1909) stated that the hardness is approximately proportional to the density of the wood. NEWLIN and WILSON (1919), based on numerous measurements, determined for the Janka-hardness $H_J$, the following parabolic formula

$$H_J = A \cdot R^{5/4} \; [\text{kp/cm}^2], \tag{7.92}$$

where $A$ is a constant (in metric units for air-dry wood the end-hardness $A = 336$ and for the side-hardness $A = 260$; for green wood the end-hardness $A = 260$ and for green wood $A = 232$) and $R$ is the density, oven-dry weight based on volume at test.

TRENDELENBURG (1933) studying MÖRATH's (1932) hardness tests developed the equation

$$H_B = a \cdot \varrho_0^b, \tag{7.93}$$

where $H_B$ ist the Brinell-hardness, $\varrho_0$ is the density in the oven-dry condition and $a$ and $b$ are constants.

Table 7.26. *Values for the Constants a and b to be Used in Eq. (7.93)*

|  | End-hardness | | Side-hardness | |
|---|---|---|---|---|
|  | $a$ | $b$ | $a$ | $b$ |
| European woods | 1,180 | 1.62 | 670 | 2.14 |
| American woods | 1,200 | 1.53 | 680 | 2.0 |

(In TRENDELENBURG's publication, by mistake, all $a$-values are one tenth of the correct figures.)

The relationships between hardness and density using different tests methods are tabulated in Table 7.27.

Table 7.27. *Relationships Between Hardness and Density*

| Density $\varrho_0$ | NÖRDLINGER | | JANKA | | BRINELL-MÖRATH | | | CHALAIS-MEUDON | | |
|---|---|---|---|---|---|---|---|---|---|---|
| | Species | Category | $H_{J||}$ kp/cm² | Category | $H_B$ | $B_{B_2}$ kp/mm² | Category | Coniferous species | Broad-leaved species |
| 0.20···0.55 | Willow, Basswood, Aspen, Poplar, White Pine | very soft | < 350 | very soft | 1···4 | 0.5···2 | very soft | — | 0.2···1.5 |
| 0.35···0.65 | Fir, Spruce Pine, Alder Birch | soft | 350···500 | soft | 2···6 | 1···3 | soft | 1···2 | 1.5···3 |
| 0.50···0.70 | Walnut, Pear, Rowan tree Sweet chestnut, Oak, Beech | some-what hard | 500···650 | medium hard | 4···6.5 | 2···4 | medium hard | 2···4 | |
| 0.60···0.80 | Ash, Plane Plum, Elm, Robinia | pretty hard | | | | | | | 3.6 |
| 0.70···0.85 | Hornbeam, Yew | hard | 650···1000 | hard | 6···10 | 3···6 | hard | 4···9 | |
| 0.80···0.92 | Cornel Dogwood White Thorn | very hard | | | | | | | 6···9 |
| 0.90···1.05 | Boxwood Privet Lilac | as hard as bone | 100···1500 | very hard | 10···13 | 5···8 | very hard | — | 9···20 |
| 1.00···1.40 | Ebony Lignum vitae | as hard as stone | >1500 | as hard as bone | 12···20 | 7···14 | | | |

YLINEN (1943) also investigated the problem of the hardness of wood and in this connection, reviewing TRENDELENBURG's publication, showed that in practically the whole range of densities of commercial timbers a linear relationship between Brinell-hardness $H_B$ and (oven-dry) density $\varrho_0$ (quite similar to that between crushing strength $\sigma_{cb}$ and density $\varrho$, according to Eq. (7.44)] is applicable:

$$H_B = \beta \cdot \varrho_0 + \alpha. \tag{7.94}$$

Based on the results of MÖRATH and calculating with the method of least squares, YLINEN obtained the following figures

| End-hardness | | Side-hardness | |
|---|---|---|---|
| $\alpha$ | $\beta$ | $\alpha$ | $\beta$ |
| −213.3 | 1,263.3 | −145.4 | 664.2 |

Fig. 7.177 shows how well YLINEN's straight line fits MÖRATH's points. PALLAY (1937/38) checked the variation of hardness on one single cross section or one single side-board of one tree. Fig. 7.178 and 7.179 illustrate the results. He found that the variations of hardness for such a single arbitrary piece of wood are about of the same magnitude as for the whole species of wood concerned. This fact, the irregular course of the frequency curves, the correlation of hardness to crushing strength, and finally the variety of hardness tests not at all testing the resistance of the surface against penetration, should indicate the lack of validity of hardness tests. From this point of view, the similarity of the curves showing the dependence of hardness on the one hand, crushing strength on the other on the moisture content of the wood (Fig. 7.180) also supports the proposal to eliminate all hardness tests as they are now conceived.

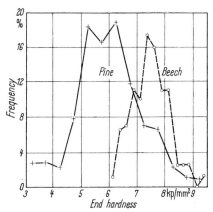

Fig. 7.178. Frequency distribution of the Brinell hardness for pine and common beech. From PALLAY (1937/38)

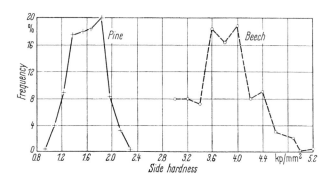

Fig. 7.179. Frequency distribution of the Brinell side hardness. From PALLAY (1937/38)

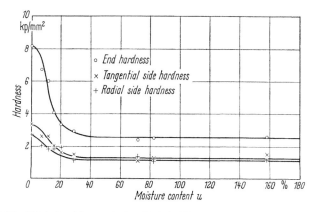

Fig. 7.180. Effect of moisture content on the Brinell hardness of pine. From KOLLMANN (1951)

### 7.7.3 Abrasion Resistance

Abrasion resistance is a very important mechanical property for many wood items, e.g. pavement, flooring, parts of machines, etc. Abrasion is caused by various factors: walking, carrying, friction, blows, oscillations, the influence of

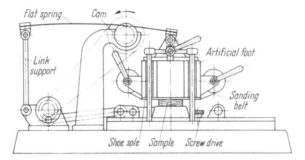

Fig. 7.181. Schematic design of the abrasion test machine according to KOLLMANN. Construction: Chemisches Laboratorium für Tonindustrie, Abt. Prüfmaschinenbau

sand, dirt and other extraneous bodies and by chemicals, moisture, and changing temperatures. Preservatives such as oil, lacquers and sealers reduce abrasion. The phenomenon of abrasion is so complex and so different that a uniform standardized test is hardly conceivable.

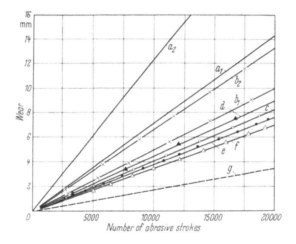

Fig. 7.182. Wear lines of various woods and wood-based materials, obtained with the Kollmann-abrader. $a_1$ heavy pine, $a_2$ light pine, $b_1$ heavy oak, $b$ light oak, $c$ common beech, $d$, $e$, $f$ fiberboards, $g$ special hardened fiberboard

It is only possible to simulate the continually repeated service loads on the parts mentioned above. Therefore tests for abrasion are of comparative value only (WANGAARD, 1950). They may be carried out by determining the wear (loss of weight or loss of thickness) of wood when scrubbed by abrasives such as are fine quartz sand propelled either by compressed air or by means of a superheated steam jet, sandpaper, grinding discs, hard metal scrapers, steel brushes or a combination of such tools. The conditions must be controlled. Some of abrasion test machines or abraders simulate the wear produced on floors in service. "Straight and twisting slippages as well as a stepping effect are produced. Pressure equal to that caused by the average person is applied to the sample by a leather surface which in conjunction with siliceous grit effects the wear" (GEISTER, 1926). The samples may be rather small (e.g. 50 mm × 50 mm × 25 mm for the Kollmann-abrader, with an annular abrasion surface of 3400 mm² for the Taber-abrader, with 1 inch square by 1/4 in thickness for the National Bureau of Standards abrader, with 2 cm square by 1 cm thickness for the du Pont de Nemour and Co. abrader) or something larger (e.g. 200 mm × 200 mm for the Stuttgart abrader according to DIN 51954) or they may consist of whole panels (e.g. 36 inches by 21 inches for the abrasion machine developed in the British Forest Products Laboratory or 1500 mm × 1800 mm used by THUNELL and PEREM in Sweden).

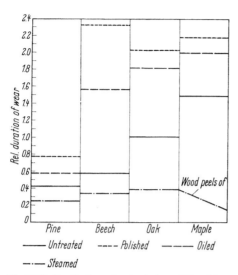

Fig. 7.183. Results from abrasion tests on different treated wood beams. From SACHSENBERG (1929)

Other questions to be considered are: species of wood, rift or slash-sawn, moisture content, condition of surface, curly grain and artificial protection against abrasion.

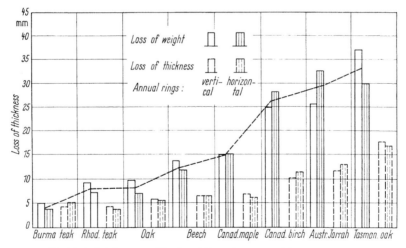

Fig. 7.184. Relative wear of hardwood. From CHAPLIN and ARMSTRONG (1936)

KOLLMANN (1961) has described critically most known abraders and has comparatively tested the most important ones (1961, 1963). His own abrader imitating the steps of the feet, the slipping and the wear by abrasives is shown schematically in Fig. 7.181. A few results showing the different behavior of some woods and wood-based materials are illustrated in Fig. 7.182. SACHSENBERG (1929) investigated the effect of steaming, waxing, and oiling. It is much more pronounced on hardwoods than on softwood (Fig. 7.183). In Princes Risborough the testing machine incorporated two tools: a stamper and an abrasion tool. The panel under test was fixed to a moving bed which slid backwards and forwards under the tools, while the tools themselves moved slowly from side to side. The relative wear of hardwood determined in a first series, is shown in Fig. 7.184. In Fig. 7.185 the mean percentage in thickness is plotted against the mean specific gravity. With the exception of Burma teak and of Australian jarrah the points lie closely to smooth hyperbolic curves. The divergence of Burma teak from the curve is the apparent effect of the natural oil in this wood as a retardant to wear. On the other hand, the higher rate of wear of Australian jarrah may be accounted for by the content of brittle resinous substances accelerating the wear in a manner similar to the presence of grit during the test (CHAPLIN and ARMSTRONG, 1936).

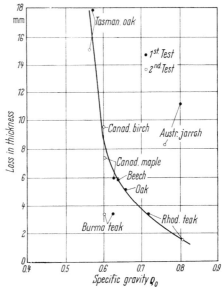

Fig. 7.185. Relationship between wear and specific gravity. From CHAPLIN and ARMSTRONG (1936)

If one takes not the loss in weight or the loss in thickness as mean values for abrasion but the reciprocals as rather arbitrary "wear-resistance values $A$", one obtains the following linear function:

$$A = \beta \cdot \varrho_0 + \alpha. \tag{7.95}$$

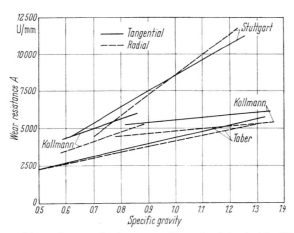

Fig. 7.186. Wear resistance of common beech versus specific gravity determined by different test methods. From KOLLMANN (1963)

This Eq. (7.95) is identical with Eq. (7.94) for hardness. Evaluating the results presented by CHAPLIN and ARMSTRONG (1936) in Fig. 7.185 and choosing as the dimension of $A$ [inch$^{-1}$] the following constants are valid: $\beta = 356$ and $\alpha = -186$.

Increasing moisture content in the hygroscopic range increases the wear, but due to the duration of the abrasion tests and due to the friction which produces heat the relationships are not as clear as for the hardness (cf. Fig. 7.180). In summary,

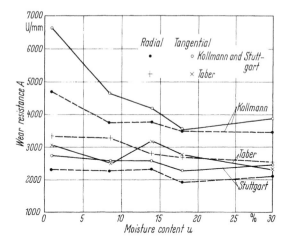

Fig. 7.187. Wear resistance of common beech versus moisture content, determined by different test methods. From KOLLMANN (1963)

the conclusion is justified that hardness and relative resistance to wear are properties which are closely related to crushing strength and shearing strength but which are more variable due to the influence of structure and moisture content. This statement may be illustrated by two diagrams. Fig. 7.186 shows that the slopes of the functions of wear-resistance versus specific gravity determined by different test methods are not identical. Fig. 7.187 compares the dependence of wear-resistance on moisture content. Using the Kollmann-abrader the results correspond with the results of Brinell hardness tests but the Taber and the Stuggart abrader hide the influence of moisture content, probably due to friction. With the Kollmann-apparatus and with the Taber-abrader linear proportionality between wear-resistance and density could be proved experimentally.

### 7.7.4 Some Aspects of Nondestructive Testing of Wood and Timber Grading

By far the largest number of mechanical tests on wood quoted in the literature are destructive tests on small clear specimens which are free of all defects. Structural members are used in much larger sizes and they contain irregularities and defects which may lower their strength. In designing timber structures working stresses must be known which provide the highest economical utilization of wood with adequate safety. Since timber has to meet ever increasing competition from other construction materials such as steel, concrete and plastics, more precise and lower-cost grading is wanted. The conventional visual timbergrading shows a regrettable degree of inefficiency. Improvements in stress grading may be obtained in different ways:

1. By eliminating the human factor in judging and classifying wood by visual inspection and simple measurement of strength reducing characteristics (e.g. ratio of knot diameter to the width of the face).

2. By nondestructive mechanical testing by means of machines which are able to stress-grade timber automatically and continuously.

Nondestructive testing means to determine properties of a material without altering these or any other properties of the material during the test. Nondestructive testing offers some remarkable advantages.

1. Any specimens including structural members, building parts, standing trees or poles may be examined.

2. A single specimen may be tested several times and the various influences on the physical and mechanical properties of wood such as moisture content, temperature and under certain circumstances, decay or duration of load, may be investigated.

3. Different properties tested on a single specimen may be correlated. Amongst the different methods of nondestructive tests for wood there are some which are of high practical importance, e.g. detection of internal discontinuities, decay, grain direction, gross density variations and moisture content by radiographic methods (cf. section 6.1.1) and determination of elastic constants by vibration methods (cf. section 6.7.0 and 7.1.5.2).

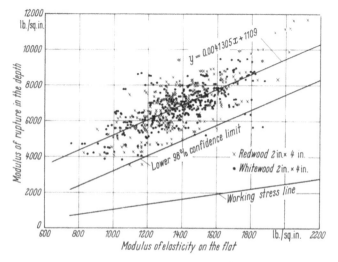

Fig. 7.188. Relationship modulus of rupture and modulus of elasticity. From SUNLEY (1965)

A nondestructive strength test of a material must be based on a known relationship between its strength and some other property that can be evaluated without damaging the material. For wood such convenient relationship is the significant correlation between the modulus of elasticity and the ultimate strength in bending. SUNLEY (1965) determined the relation between modulus of elasticity and bending strength for Baltic redwood and Whitewood (Fig. 7.188). The results of these two species follow the same general law which mathematically can be represented by a regression line. The variability of the material is now taken into consideration by calculation of the lower confidence line at a vertical distance away from the regression line of $-2.33$ times the standard error of estimate of the bending strength. It may be expected that 99% of the results lie above this line. In a second operation a safety factor of 3 is applied. Thus the working stress line is obtained and the basis for mechanical grading is created.

Some consequences of the statistical nature of the relationship between the modulus of elasticity and the bending strength have been pointed out by MILLER (1962, 1964). In particular, individual pieces of timber can be misgraded in two ways, a) timber which is too weak can be accepted, and b) timber which is of adequate strength can be rejected. MILLER defines the selection efficiency of a

## 414 Literature Cited

nondestructive test as "the percentage of the material of acceptable strength which is actually recovered after rejection of the low strength pieces has been ensured". In his second paper MILLER (1964) wrote "... in the nondestructive testing of wood, a compromise must be reached between selection efficiency and tolerance (defined as the percentage of material passed that is actually below acceptable strength). The tolerance should be made as high as is compatible with the consequences of failure in order to achieve a selection efficiency that is economically acceptable."

## Literature Cited

BACH, C., and BAUMANN, R., (1924) Elastizität und Festigkeit, 9. Aufl., Berlin.

BARKAS, W. W., (1949) The swelling of wood under stress. Dep. Sci. Ind. Res., Forestry Prod. Res., London.

BAUMANN, R., (1922) Die bisherigen Ergebnisse der Holzprüfungen in der Materialprüfungsanstalt an der Tech. Hochschule Stuttgart, Forsch. Gebiete Ingenieurw., H. 231, Berlin.

BAUSCHINGER, J., (1883) Mitt. Mech. Tech. Lab. Tech. Hochschule München, H. 9.

BETTS, H. S., (1919) Timber. its strength, seasoning and grading. New York, p. 31.

BRINELL, J. A. (1900), Ein Verfahren zur Härtebestimmung nebst einigen Anwendungen desselben. Gießlers Baumaterialienkunde 5: pp. 276, 280, 294, 297, 317—320, 364, 367, 392, 412—415.

BÜSGEN, M., (1904) Z. Forst.-Jagdwes. 36: 453.

CARRINGTON, H., (1921) The moduli of rigidity for spruce. Phil. Mag. 41: 206, 848—860.

—, (1922a) Young's Modulus and Poisson's Ratio for spruce. Phil. Mag. 43: 871—878, corr. 44: 288.

—, (1922b) The elastic constants of spruce as affected by moisture content. Aeron. J., 26: 462.

—, (1923) The elastic constants of spruce. Phil. Mag. 45: 1055—1057.

CASATI, E., (1932) Essais comparés sur éprouvettes de dimensions différentes de quelques essences de bois. Kongreßbuch Zürich, Intern. Verb. Materialprüfung. p. 21.

CHAPLIN, C. J., and ARMSTRONG, F. H., (1936) Wood, 1, Dec.

ČIŽEK, L., (1932) Diskussionsbeitrag zum Bericht Monnin. Intern. Kongr. Materialprüfung, p. 178, Zürich.

CLARKE, S. H., (1935) Forestry (London) 9: 132.
compare: Dep. Sci. Ind. Res.: Report of the Forest Products Research Board for the year 1935, p. 7, London, 1936.

CLARKE, S. H., (1939) Recent work on the growth, structure, and properties of wood. For. Prod. Res., Spec. Rep., No. 5, London.

—, CHAPLIN, C. J., and ARMSTRONG, F. H., Princes Risborough, Dep. Sci. Ind. Res. For. Prod. Lab., Project 18, Progr. Rep. 3, May.

COKER, E. S., and COLEMAN, G. P., (1930) Proc. Roy. Soc. A., 128: 418.

DOYLE, D. V., DROW, J. T., and MAC BURNEY, R. S., (1945/46) The elastic properties of wood. U.S. Dep. Agr., For. Prod. Lab., Mimeo 1528 and supplements A to H, Madison, Wisc.

DRAFFIN, J. O., and MÜHLENBRUCH, C. W., (1937) ASTM Rep. 96.

EGNER, K., (1941) Ermittlung der Festigkeitseigenschaften der Hölzer. in: Siebel, E., Handbuch der Werkstoffprüfung, Berlin, Vol. 3, p. 101.

—, and ROTHNUND, A., (1944) Zusammenfassender Bericht über Dauerzugversuche mit Hölzern. Materialprüfungsanstalt Tech. Hochschule Stuttgart.

EHRMANN, W., (1937) Über die Scherfestigkeit von Fichten- und Kiefernholz. Forschungsber. Holz H. 4, Berlin.

ELMENDORF, A., (1916) J. Franklin-Inst. 182: 771.

—, (1922) Proc. Am. Soc. Test. Mat., Symposium on Impact Testing of Materials, Philadelphia, p. 87.

FÖPPL, A., (1904) Die Druckfestigkeit des Holzes in der Richtung quer zur Faser. Mitt. Mech. Tech. Lab. Tech. Hochschule München, H. 129.

FORSAITH, C. C., (1921) J. Forestry 19: 237.

GABER, E., (1929) Z. VDI, 73: 932.

—, (1935) Zentr. Bauverw. Verein. Z. Bauwes. 55: 85.

—, (1937) Ein Gütevergleich zwischen deutschen und ausländischen Weichhölzern. Versuchsanst. Holz, Stein, Eisen. H. 5, Karlsruhe.

—, (1937) Zentr. Bauverw. 57: 296.

—, (1940) Druckversuche quer zur Faser an Nadel- und Laubhölzern. Holz als Roh- und Werkstoff, 3: 22-226.

# Literature Cited

GEISTER, C. H., (1926) J. Am. Ceram. Soc. **9**.

GERRY, E., (1915) Science (U. S.) **41**: 179.

GHELMEZIU, N., (1937/38) Untersuchungen über die Schlagfestigkeit von Bauhölzern. Holz als Roh- und Werkstoff **1**: 585—601.

GOENS, E., (1931) Ann. Phys. 5. Folge, **11** (6): 647—678.

GRAF, O., (1921) Bauing. **2**: 498.

—, (1928) Bautechnik **6**: 438.

—, (1929) Masch.-Bau **8**: 642.

—, (1929) Die Dauerfestigkeit der Werkstoffe und der Konstruktionselemente. Berlin.

—, (1930) Masch.-Bau **9**: 375.

—, (1937/38) Wie können die Eigenschaften der Bauhölzer mehr als bisher nutzbar gemacht werden? Welche Aufgaben entspringen aus dieser Frage für die Forschung? Holz als Roh- und Werkstoff **1**: 13—16.

—, (1938) Tragfähigkeit der Bauhölzer und der Holzverbindungen. Mitt. Fachaussch. Holzfragen, H. **20**, Berlin, vgl. auch: —, (1939) Mitt. Fachaussch. Holzfragen, Berlin **23**: 17.

—, (1940) Aus neueren Versuchen mit Bauholz. Mitt. Fachaussch. Holzfragen, Berlin **26**: 1.

—, (1944) Silvae Orbis, Berlin-Wannsee **15**: 41.

HAUBER, O., (1942) Normung von Kiefernholz im Flugzeugbau. Holz Roh Werkstoff **5**: 105—108.

HEARMON, R. F. S., (1948) The elasticity of wood and plywood, For. Prod. Res. Spec. Rep. No. 7, London.

—, (1953) The elastic and plastic properties of natural wood. In: R. Meredith, Mechanical properties of wood and paper. Amsterdam, 1953, p. 19—47.

—, (1959) Forest Products Research 1958. H. M. Stat. Off., London.

HELMHOLTZ, H., (1902) Vorlesungen über theoretische Physik, Vol. II, p. 103, Leipzig.

HÖRIG, H., (1931) Zur Elastizität des Fichtenholzes. Z. tech. Physik **12**: 369—379.

—, (1933) Über die rechnerische Auswertung von Verdrehungsmessungen an Holzstäben. Ing.-Arch. **4**: 570—576.

HÖRIG, H., (1935) Anwendung der Elastizitätstheorie anisotroper Körper auf Messungen an Holz. Ing.-Arch. **6**: 8—14.

—, (1936) Berechnung der Gleitzahlen $s_{44}$, $s_{55}$, $s_{66}$ aus den Verdrehungsmessungen von Stäben mit rechteckigen Querschnitten bei rhombischer Symmetrie. Ing.-Arch. **7**: 165—170.

—, (1937) Über Mittelwertskörper elastisch anisotroper Systeme und deren Anwendung auf Holz. Ing.-Arch. **8**: 174—182.

—, (1943) Über die unmittelbare Messung der Gleitzahlen $s_{41}$, $s_{55}$, $s_{66}$ bei Stoffen von rhombischer Symmetrie und geringer Starrheit. Ann. Phys. **43**: 285—295.

—, (1944) Kritische Bemerkungen zu den auf dem DIN 51190 angegebenen Formeln für die Berechnung des Drillungsmoduls und der sog. Verdrehungsfestigkeit von Holzstäben. Silvae orbis Nr. 15, p. 90—99, Berlin-Wannsee.

HUBER, K., (1928) Z. VDI **72**: 501

—, (1929) Z. Flugtechn. Motorluftsch. **20**: 250.

—, (1937/38) Die Prüfung der Hölzer auf Kugeldruckhärte. Holz als Roh- und Werkstoff **1**: 254—259.

IFJU, G., (1964) Tensile strength behavior as a function of cellulose in wood. For. Prod. J. **14**: 366—372.

IVANOV, G. M., (1938) Acad. Sci. URSS **19** (6—7): 549.

JAMES, W. L., (1962) Dynamic strength and elastic properties of wood. For. Prod. J. **12**: 253—260.

JANKA, G., and HADEK, A., (1900) Untersuchungen über die Elastizität und Festigkeit der österreichischen Bauhölzer. I. Fichte, Südtirol. Mitt. Forstl. Versuchswes. Österr. H. 25, Wien.

—, (1906) Cbl. ges. Forstwes. **9** pp. 193, 241.

—, (1908) Cbl. ges. Forstwes. **11**: 443.

—, (1915) Die Härte der Hölzer. Mitt. Forstl. Versuchswes. Österr., H. 39, Wien.

JAYNE, B. A., (1959) Mechanical properties of wood fiber. Tech. Assoc. Pulp and Paper Ind. **42** (6): 461—467.

—, (1960) Some mechanical properties of wood fibers in tension. For. Prod. J. **10**: 316—322.

JENKIN, C. F., (1920) Materials used in the construction of aircraft and aircraft engines. H. M. Stat. Off., London.

—, (1926) Report on materials of construction used in aircrafts H. M. Stat. Off., p. 98, London.

v. KÀRMÀN, TH., (1910) Untersuchungen über Knickfestigkeit. Forsch. Ingenieurw., H. 81, Berlin.

—, (1927) Grundlagen der Balkentheorie. Abhandl. Aerodyn. Inst. Tech. Hochschule Aachen, No. 7, p. 3—10.

KELLOGG, R.M., (1960) Effect of repeated loading on tensile properties of wood. For. Prod. J. **10**: 586—594.

# Literature Cited

KEYLWERTH, R., (1944/45) Spalten, Spaltbeanspruchung und Querfestigkeit des Holzes. Holz als Roh- und Werkstoff, **7**: 72—78.

—, (1945) Unveröffentlichte Messungen der Reichsanstalt für Holzforschung. Eberswalde.

—, (1951) Die anisotrope Elastizität des Holzes und der Lagenhölzer. VDI-Forschungsheft 430. Deutscher Ingenieur-Verlag GmbH., Düsseldorf.

KING, E. G., JR., (1961) Time-dependent strain behavior of wood. For. Prod. J. **11**: 156—165.

KOEHLER, A., (1933) Causes of brashness in wood. U. S. Dep. Agr. Tech. Bull. No. 342, Washington, D. C.

KOLLMANN, F., (1934) Untersuchungen an Kiefern- und Fichtenholz aus der Rheinpfalz. Forstwiss. Cbl. **56** (6): 181—189.

—, (1937) Eine neue Abnutzungsprüfmaschine. Holz als Roh- und Werkstoff **1**: 87—89.

—, (1940) Über die Schlag- und Dauerfestigkeit der Hölzer. Mitt. Fachaussch. Holzfragen **17**: 17—30, VDI-Verlag, Berlin.

—, (1940) Die mechanischen Eigenschaften verschieden feuchter Hölzer im Temperaturbereich von −100 bis +200 °C. VDI-Forschungsheft No. 403, VDI-Verlag, Berlin.

—, (1941) Die Esche und ihr Holz, Berlin.

—, (1942) Über das Gefrieren und den Einfluß tiefer Temperaturen auf die Festigkeit der Hölzer. Mitt. Akad. Dtsch. Forstwiss. **2** (1): 317—336.

—, (1944) Internationale Normung der Prüfvorschriften für Holz. Silvae Orbis, Berlin-Wannsee (15): 75—86.

—, (1951) Technologie des Holzes und der Holzwerkstoffe, Vol. I, 2nd ed., Springer-Verlag, Berlin—Göttingen—Heidelberg.

—, (1952) Über die Abhängigkeit einiger mechanischer Eigenschaften der Hölzer von der Zeit, von Kerben und von der Temperatur. Holz als Roh- und Werkstoff, **5**: 187—197.

—, (1957) Über Unterschiede im rheologischen Verhalten von Holz und Holzwerkstoffen bei Querdruckbelastung. Forsch. Ingenieurw. **23** (1/2): 49—54.

—, (1961) Rheologie und Struktur-Festigkeit von Holz. Holz als Roh- und Werkstoff **19** (3): 73—80.

—, (1962) Über das rheologische Verhalten von Buchenholz verschiedener Feuchtigkeit bei Druckbeanspruchung längs der Faser. Materialprüfung **4** (9): 313—319.

—, (1963) Untersuchungen über den Abnutzungswiderstand von Holz, Holzwerkstoffen und Fußbodenbelägen. Holz als Roh- und Werkstoff **21** (7): 245—256.

—, (1963) Das Verhalten von Holz bei allseitiger Druckeinwirkung. Materialprüfung **5** (6): 233—239.

—, (1964) Über die Beziehungen zwischen rheologischen und Sorptions-Eigenschaften (am Beispiel von Holz). Rheologica Acta, Bd. **3** (4): 260—270.

—, and KRECH, H., (1960) Dynamische Messungen der elastischen Holzeigenschaften und der Dämpfung. Holz als Roh- und Werkstoff **18**: 41—54.

—, and SCHULZ, F., (1944) Versuche über den Einfluß der Temperatur auf die Festigkeitswerte von Flugzeugbaustoffen, 2. Teilbericht, Reichsanstalt für Holzforschung, Ber. **131**.

KRAEMER, O., (1930) Dauerbiegeversuche mit Hölzern. DVL-Jb., p. 411.

—, (1933) Kunstharzstoffe und ihre Entwicklung zum Flugzeugbaustoff DVL-Jb. p. VI 69.

KRECH, H., (1960) Größe und zeitlicher Ablauf von Kraft und Durchbiegung beim Schlagbiegeversuch an Holz und ihr Zusammenhang mit der Bruchschlagarbeit.

KRÜGER, F., and ROHLOFF, E., (1938) Z. Physik **110**: 58.

KÜCH, W., (1937) Luftwissen **4**: 254.

—, (1943) Der Einfluß des Feuchtigkeitsgehalts auf die Festigkeit von Voll- und Schichtholz. Holz als Roh- und Werkstoff **6**: 157—161.

—, and TELSCHOW, G., (1942) Der Einfluß des Feuchtigkeitsgehalts auf die Festigkeit von Voll- und Schichtholz. DVL-Bericht Kf 300/IX. Berlin-Adlershof.

KUFNER, M., (1963) Über die Spannungsverteilung in hölzernen Zugstäben. Holz als Roh- und Werkstoff **8**: 300—305.

KUNTZE, W., (1933) Z. VDI, **77**: 49.

—, (1934) Z. Metallkd. **26**: 106.

—, (1935) Stahlbau (Beilage zu „Die Bautechnik"), H. 2.

LANGLANDS, J., (1936) J. C. Sci. Ind. Res. (Australia) **9**: 88.

v. LORENZ, M., (1909) Analytische Untersuchung des Begriffs der Holzhärte. Mitt. Forstl. Versuchswes. Österr., Wien, p. 33.

MARK, H., (1952) Cellulose: physical evidence regarding its constitution. Chap. 6, Vol. I Wood, Chemistry, eds, L. E. Wise and E. C. Jahn, Reinhold Publ. Corp., New York.

MARKWARDT, L. J., (1930) Aircraft Woods: Their properties, selection and characteristics. Nat. Advisory Comm. Aeron., Rep. No. 354.

MARKWARDT, L. J. and WILSON, T. R. C., (1935) Strength and related properties of woods grown in the United States. U. S. Dep. Agr. Tech. Bull. Washington, D. C., No. 479, p. 67.

# Literature Cited

MAYER-WEGELIN, H. and BRUNN, G., (1932) Raumgewichte und Druckfestigkeit von Pitch-pine, Oregon pine und deutschem Kiefernholz. Mitt. Fachausschuß Holzfragen, H. 4, Berlin.

MEREDITH, R., (1953) Mechanical properties of wood and paper. North-Holland Publ. Co., Amsterdam.

MEYER, K. H. and MARK, H., (1930) Der Aufbau der hochpolymeren organischen Naturstoffe. Leipzig.

MILLER, D. G., (1962) Selection efficiencies of non-destructive strength tests. For. Prod. J. 12 (8): 358—263, (1964) 14: 179—183.

MÖRATH, E., (1932) Studien über die hygroskopischen Eigenschaften und die Härte der Hölzer, Hannover.

MONNIN, M., (1919) Essais physiques, statiques et dynamiques des bois. Bull. Sect. Tech. L'Aeronaut. Paris.

—, (1932) L'essai des bois. Kongreßbuch Zürich Int. Verb. Materialprüfung. 85 pp.

NATALIS, F., (1918) Druck- und Knickfestigkeit. Tech. Ber. 3: 1—7.

—, (1919) Dinglers Polytech. J., p. 69.

NEWLIN, J. A. and WILSON, T. R. C., (1919) The relation of the shrinkage and strength properties of wood to its specific gravity. U. S. Dep. Agr. Bull. No. 676, Washington, D.C.

—, and TRAYER, G. W., (1924) Form factors of beams subjected to transverse loading only. Nat. Advisory Comm. Aeron. Rep. 181.

—, and GAHAGAN, J. M. (1930) Tests of large timber columns and presentation of the Forest Products Laboratory column formula. U. S. Dep. Agr. Tech. Bull. 167, Washington, D. C.

NÖRDLINGER, H., (1881) Anatomische Merkmale in wichtigsten deutschen Wald- und Gartenholzarten. Stuttgart.

PALLAY, N., (1937/38) Über die Holzhärteprüfung. Holz als Roh- und Werkstoff, 1: 126—130.

—, (1939) Ergänzende Angaben zum Holzhärte-Prüfverfahren nach Krippel. Holz als Roh- und Werkstoff 2: 413—416.

PENTONEY, R. E., (1955) Comp. Wood 2: 43-57.

PEREM, E., (1950) Inverkan av belastningsökningen mot böjning hos svenskt furuvirke, Sv. Träforskningsinst., Trätekn. Avd. Medd. No. 20, Stockholm.

PETERMANN, H., (1941) Schubversuche mit Kiefernholz. Holz als Roh- und Werkstoff, 4: 141—150.

PETTIFOR, C. B., (1936) Toughness of Ash. For. Prod. Res. Lab., Princes Risborough.

—, (1937) Relation between toughness and Izod impact values For. Prod. Res. Labor. Princes Risborough.

PEVZOFF, A., (1935) Die Kugelschlaghärte des Holzes. Moskau.

POULIGNIER, J., (1942) Contribution à l'étude de l'élasticité du bois. Publ. Sci. et tech. du Secr. d'état à l'aviation No. 179, Paris.

PRICE, A. T., (1928) A mathematical discussion on the structure of wood in relation to its elastic properties. Phil. Trans. A 228: 1.

REIN, W., (1943) Kerbempfindlichkeit von Holz. Lilienthal-Ges. Ber. 157, p. 28, Berlin.

REISSNER, E., (1938) Flight, Suppl. Aircraft Eng., 27 January.

RIECHERS, K., (1938) Z. VDI 82: 665.

ROŠ, M., (1925) SIA-Normen für Holzbauten. Disk.-Ber. EMPA, Zürich.

—, (1936) Das Holz als Baustoff. I. Schweiz. Kongr. Förderung Holzverw. Bern, p. 274.

—, and BRUNNER, J., (1932) Die Knickfestigkeit der Bauhölzer. Kongr.-Ber. Zürich IVM, p. 157, Zürich.

ROTH, PH., (1935) Dauerbeanspruchung von Eichenholz- und von Tannenholzprismen in Faserrichtung durch konstante und durch wechselnde Druckkräfte und Dauerbiegebeanspruchung von Tannenholzbalken. Diss. Tech. Hochschule Karlsruhe.

RUDELOFF, M., (1899) Der heutige Stand der Holzuntersuchungen und die Vereinheitlichung der Prüfungsverfahren. Mitt. Mech. Tech. Vers.-Anst. Erg.h. III, Berlin.

RYSKA, K., (1932) Einige Fragen aus dem Gebiete der technischen Prüfungsmethoden für Hölzer. In: Kongreßbuch Zürich des IVM, Prüfung der Hölzer, Zürich.

SACHSENBERG, E., (1929) Holzbearbeitungsmaschine 5: 553.

SALAMON, M., (1963) Quality and strength properties of Douglas fir dried at high temperatures. For. Prod. J. 13 (8): 339—344.

SAVART, F., (1829) Ann. Chim. Physique, 40: 5, 113.

SCHLÜTER, R., (1932) Elastische Messungen an Fichtenholz. Samml. Vieweg No. 107, Braunschweig.

SCHLYTER, R., (1927) The strength of Swedish redwood timber (Pine) and its dependence on moisture content and apparent specific gravity. Congrès international pour l'essai des matériaux. Amsterdam.

418 Literature Cited

SCHLYTER, R., (1932) Researches into durability and properties of Swedish coniferous timbers. Kongreßbuch Zürich IVM, Zürich.

—, and WINBERG, G., (1929) Svenskt furuvirkes hållfasthetsegenskaper och deras beroende av fuktighetshalt och volymvikt. Stat. Provningsanst. Medd. 42, Stockholm.

SCHMIDT, A. X. and MARLIES, C. A., (1948) Principles of high-polymer theory and practice. McGraw-Hill Book Co. Inc., New York, Toronto, London.

SCHNEIDER, H., (1966) Untersuchungen über das Verhalten von Holzwerkstoff-Platten bei Stoßbeanspruchung sowie über ihren dynamischen Elastizitäts- und Schubmodul. Holz als Roh- und Werkstoff 24: 41—52.

SCHOLTEN, J. A., (1935) Wood Handbook, U.S. Dep. Agr. Washington D.C. p. 152.

VON SCHWARZ, M. and BUES, K., (1929) Masch.-Bau 8: 403.

SEEGER, R., (1937) Untersuchungen über den Gütevergleich von Holz nach der Druckfestigkeit in Faserrichtung und nach der Schlagfestigkeit. Forschungsber. Holz, Berlin, H. 4.

SEEWALD, F., (1927) Spannungen und Formänderungen von Balken mit rechteckigem Querschnitt. Abhandl. Aerodyn. Inst. Tech. Hochschule Aachen, No. 7, p. 11—13.

SIIMES, F. E., (1944) Mitteilungen über die Untersuchung von Festigkeitseigenschaften der finnischen Schnittwaren. Silvae Orbis No. 15, p. 60, Berlin-Wannsee.

STAMER, J., (1935) Elastizitätsuntersuchungen an Hölzern. Ing.-Arch. 6: 1—8.

STAMER, J., (1929) Z. VDI 73: 215.

—, and SIEGLERSCHMIDT, H., (1933) Elastische Formänderungen der Hölzer Z. VDI 77: 503—505.

SUENSON, E., (1937/38) Zulässiger Druck auf Querholz. Holz als Roh- und Werkstoff 1: 213—216.

—, (1941) Die Lage der Nullinie in gebogenen Holzbalken. Holz als Roh- und Werkstoff 4: 305—314.

SULZBERGER, P. H., (1943).

SULZBERGER, P. H., (1948) The effect of temperature on the strength properties of wood, plywood and glued joints at various moisture contents C.S.I.R.O., Div. For. Prod. (Australia), South Melbourne.

—, (1953) The effect of temperature on the strength of wood. Aeron. Res. Cons. Comm. Rep. ACA-46, Melbourne.

SUNLEY, J. G., (1965) Working stresses for structural timbers. Dep. Sci. Ind. Res., For. Prod. Res. Bull., No. 47, 2nd ed. London, H. M. Stat. Off.

TETMAJER, L., (1884) Methoden und Resultate der Prüfung der schweizerischen Bauhölzer. Mitt. Prüf. Baumat. Eidgen. Polytechnikum Zürich, H. 2, Zürich.

—, (1903) Die Gesetze der Knickungs- und der zusammengesetzten Druckfestigkeit. Leipzig u. Wien.

—, (1905) Die angewandte Elastizitäts- und Festigkeitslehre. Wien.

TIEBE, H., (1940) Tharandt, Forstl. Jb. 91 (1—3).

TIEMANN, H. D., (1944) Wood Technology, 2nd ed., New York.

THUNELL, B., (1939) Ing. Vet. Akad. Stockholm, H. 3.

—, (1940) Ing. Vet. Akad. Stockholm, H. 4.

—, (1941) Über die Elastizität schwedischen Kiefernholzes. Holz als Roh- und Werkstoff 1: 15—18.

—, (1941) Hällfasthetsegenskaper hos svenskt furuvirke utan kvistar och defekter. Ing. Vet. Akad. Handl. No. 161, p. 40. Stockholm.

—, (1942) Kvalitet och hållfosthet hos virke. Statens Provningsanst. Medd. 89, p. 4, Stockholm.

—, (1943) Inverkan av visse kvalitetsbestämmande faktorer på hållfostheten mot tryck hos svenskt furuvirke. Diss. Tech. Hochschule Stockholm.

TRAYER, G. W., and MARCH, H. W., (1930) The torsion of members having sections common in aircraft construction Nat. Advisory Comm. Aeron. autics Rep. No. 334. Washington D. C.

TRENDELENBURG, R., (1932) Allg. Forst.- und Jagd-Z. 108: 1.

—, (1933) Forstarch. 9: 37.

—, (1939) Das Holz als Rohstoff. München.

UGRENOVIĆ, A., (1941) Untersuchungen über die Spaltfestigkeit und ihren Zusammenhang mit dem Bau der Markstrahlen. Holz als Roh- und Werkstoff, 4: 26—31.

WALLDÉN, P., (1939) Acta forest Fenn 39.

WANGAARD, F. F., (1950) The mechanical properties of wood. J. Wiley and Sons, Inc., New York, Chapman and Hall, Ltd., London.

WEGELIUS, E., (1933) Suomalaisen männyn Väsytyslujuus, Eripainos Teknillisestä Aikakanslehdestä, Nr. 09, Helsinki.

Literature Cited

WILSON, T. R. C., (1921) The effect of spiral grain on the strength of wood. J. Forestry, Washington **19**: 747.
—, (1922) Proc. Amer. Soc. Test. Mat. **22** (II): 55.
—, (1932) U. S. Dep. of Agr., Bull. No. 282, Washington.
WINTER, H., (1944) Richtlinien für den Holzflugzeugbau. Beiträge B IIa 2, B II b, B II c, B II d, B III b. Berlin.
Wood Handbook (1955) No. 72. U. S. Dep. Agr. For. Serv., For. Prod. Lab.
WUOLIJOKI, J., (1947) Über die gleichzeitige Bestimmung des Elastizitäts- und Schubmoduls an Hand der Obertöne eines in Biegeschwingungen stehenden Stabes. Diss. Tech. Hochschule Helsinki.
YLINEN, A., (1938) Die Knickfestigkeit eines zentrisch gedrückten geraden Stabes im elastischen und unelastischen Bereich. Diss. Tech. Hochschule Helsinki.
—, (1942) Dehnmessungen in der Umgebung von Ästen beim Kiefernholz. Holz als Roh- und Werkstoff, **5**: 337—338.
—, (1942) Ein neues Meßverfahren für die Bestimmung der Schubmoduln von Holz. Holz als Roh- und Werkstoff **5**: 375—376.
—, (1942) Über den Einfluß des Spätholzanteiles und der Rohwichte auf die Festigkeits- und elastischen Eigenschaften des Nadelholzes. Acta forest. Fenn. **5**: 1.
—, (1943) Über den Einfluß der Rohwichte und des Spätholzanteils auf die Brinellhärte des Holzes. Holz als Roh- und Werkstoff **6**: 125—127.
—, (1943) Über die Bestimmung der Spannungen und Formänderungen von Holzbalken mit rechteckigem Querschnitt. Helsinki.
—, (1944) Begründung der Abänderungsvorschläge der Prüfnormen für Holz. Silvae Orbis, Berlin-Wannsee **15**: 99.

# 8. STEAMING AND SEASONING OF WOOD

## 8.0 General Considerations

One of the most important and effective measures for wood protection is seasoning. Seasoning starts immediately after the cutting of green trees even in an atmosphere highly saturated with water vapor. At first liquid or free water (in practice commonly called "sap") is removed and eventually the fiber saturation point is reached. As was pointed out in section 6.2.6.2, slight shrinkage may occur even above the fiber saturation point, e.g. at about 50% moisture content. Shrinkage and swelling due to changes in moisture content are undesirable properties of solid wood for most uses. However, the disadvantage of shrinkage in the course of seasoning is more than offset by a remarkable number of advantages: increase in resistance to blue stain and wood-destroying fungi; reduction in susceptibility to attack by some types of insects; reduction of warping, twisting, splitting, checking, and honeycombing; reduction in weight (thus reducing shipping costs); increase in stiffness, mechanical strength, hardness and nail-holding power with decreasing moisture content below the fiber saturation point (cf. Section 6.2.1); improvement of paintability (wet wood should not be painted; below the fiber saturation point, however, the moisture content is of minor importance). Except in a few treating methods (Boucherie process for the treatment of green, unpeeled poles, widely used in Europe; diffusion processes with green or wet wood) any timber to be treated with preservatives should be carefully peeled and either seasoned or otherwise conditioned. Seasoning is therefore of the utmost importance for the economical utilization of wood, though statistical data to support this are not available. In any case the assumption is justified that the majority of wood products are air-dried at least for some time prior to kiln-drying. Air-drying alone is not sufficient for wood items which are used in heated interior rooms. As a rule, timber should be seasoned as near as possible to the average moisture content at which it will be maintained under service conditions (cf. Section 6.2.4). Items suited for interior finish must be kiln-dried.

Among the advantages of kiln-drying over air-drying are the following: even greater reduction in weight; possibility of drying to any desired moisture content; higher drying speed (cf. Section 6.3.2) and therefore much shorter drying times than those required in air-drying; the killing of staining or wood-destroying fungi or insects that may have attacked the wood; and low degrade, if adequate kiln schedules are followed.

## 8.1 Air-drying

### 8.1.1 Moisture Content of Green Wood

Green sapwood contains more moisture than heartwood, as is shown in Table 8.1, and the wood of the lower part of the stem less than the top of the tree which contains more sapwood (cf. Fig. 8.2). The variations in individual trees may

Ref. p. 471]                          8.1 Air-drying                                    421

Table 8.1. *Average Moisture Content, Based on Oven-Dry Weight, of Green Wood, by Species*

| Species | Heart-wood % | Sapwood % | Mixed heart- and sapwood % | Species | Heart-wood % | Sap-wood % | Mixed heart- and sapwood % |
|---|---|---|---|---|---|---|---|
| *Softwoods* | | | | *Hardwoods* | | | |
| Baldcypress | 121 | 171 | — | Alder, red | — | 97 | — |
| Cedar, western | | | | Ash, white | 46 | 44 | — |
| red- | 58 | 249 | — | Aspen | 95 | 113 | — |
| Douglas-fir, | | | | Basswood, | | | |
| coast-type | 37 | 115 | — | American | 81 | 133 | — |
| | | | | Beech, | | | |
| Fir: | | | | American | 55 | 72 | — |
| Noble | 34 | 115 | — | Birch, yellow | 74 | 72 | — |
| White | 98 | 160 | — | Cherry, black | 58 | — | — |
| Hemlock, | | | | Cottonwood, | | | |
| western | 85 | 170 | — | black | 162 | 146 | — |
| Larch, western | 54 | 119 | — | Elm, American | 95 | 92 | — |
| | | | | Hickory | 68···97 | 49···62 | — |
| Pine: | | | | Holly, | | | |
| Eastern | | | | American | — | — | 82 |
| white | — | — | 68 | Locust, black | — | — | 40 |
| Ponderosa | 40 | 148 | — | Maple, sugar | | | |
| Southern | | | | hard | 65 | 72 | — |
| yellow | 30···32 | 106···122 | — | | | | |
| Sugar | 98 | 219 | — | Oak: | | | |
| Western | | | | Northern red | 80 | 69 | — |
| white | 62 | 148 | — | White | 64 | 78 | — |
| Redwood, old- | | | | Persimmon, | | | |
| growth | 86 | 210 | — | common | — | — | 58 |
| Spruce, sitka | 41 | 142 | — | Sweetgum | 79 | 137 | — |
| | | | | Sycamore, | | | |
| | | | | American | 114 | 130 | — |
| | | | | Tupelo, black | 87 | 115 | — |
| | | | | Walnut, black | 90 | 73 | — |
| | | | | Willow, black | — | — | 139 |
| | | | | Yellow-poplar | 83 | 106 | — |

be considerable while the changes with the seasons are, as a rule, relatively small and their causes not quite clear. In spruce heartwood, GÄUMANN (1928) found a nearly constant moisture content, between 33.4 and 34.9% at a height of 6 m above the ground, over the entire year. The sapwood exhibited remarkable variations from stem to stem but the lowest moisture content values (on the average 154.1 $\pm$ 5.7%) appeared during the winter months (December, January, February) in contrast to the remaining nine months with 187.1 $\pm$ 5.2%. Fir heartwood contained an average of 45% moisture content from June until January. From January on, the moisture content decreased and reached a minimum in April; subsequently it reached the normal state in June. In fir sapwood the conditions were more complicated; the moisture content amounted to a maximum of 211% in April followed by a low range between 170% and 186% during the early summer months. Following was a second peak higher than 200% in August and September and then a continuous decrease to a minimum of 134% in March.

KNUCHEL (1930) published the curves shown in Fig. 8.1 for the seasonal moisture content of spruce and fir sapwood and heartwood. The samples were taken at a height of 4 to 8 m above the ground. The seasonal variations in moisture content of beech wood are illustrated by Fig. 8.2 from KNUCHEL (1935). The highest moisture content has been found throughout the year in the outer layers of the stem below the bark; the lowest moisture content (69.2%) near the pith. The geographical range of growth may be of great influence. For examples, the moisture content of Western yellow pine (*Pinus ponderosa* Laws.) may vary between 80 and 118% in the eastern part of its range, while in California 100 to 185% is typical.

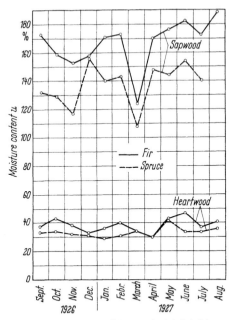

Fig. 8.1. Seasonal moisture content (related to oven-dry weight) of green spruce and fir. From KNUCHEL (1935)

### 8.1.2 Course of Air-drying

As long as liquid water is present in wood a steep moisture gradient exists. Its magnitude may be estimated approximately as follows

$$\frac{2(u_i - u_{he})}{s}$$

where $u_i$ is the moisture content in the core of the wood, $u_{he}$ is the moisture content corresponding with the surrounding atmosphere according to the hygroscopic equilibrium (cf. Section 6.2.3), and $s$ is the thickness of the wood. The rate of drying is directly proportional to this gradient. It may be concluded that the course of air-drying, as that of kiln-drying, depends on temperature, relative

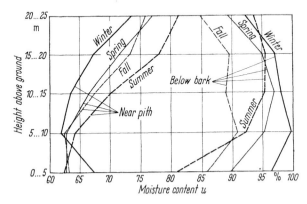

Fig. 8.2. Seasonal variations in moisture content of green beech at different stem heights. From KNUCHEL (1935)

humidity, and speed of circulation of the air surrounding the wood. Climate and geographical site are important factors as are elevation above sea level, prevalence of dry or wet winds, tendency for fog (e.g. near rivers, lakes or swamps), average amount of precipitation, and average duration of sunshine. All these conditions may vary considerably not only seasonally, but also from year to year and from place to place, e.g. in a valley or a mountain slope or near a lake or swamp.

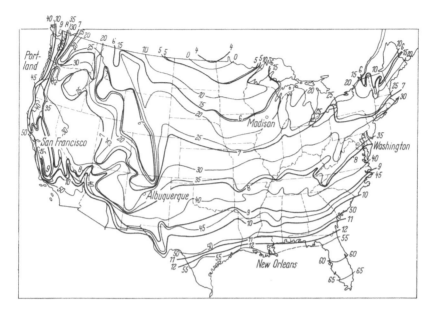

Fig. 8.3. Average external temperature (thin lines, values in °F) and approximate hygroscopic equilibrium values (thick lines, in %) for interior finishing woodwork in heated apartments in the USA during January. From PECK (1932)

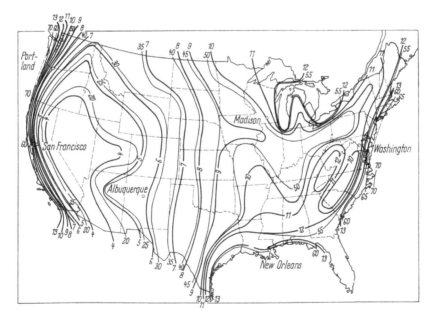

Fig. 8.4. Average minima of relative air humidity (thin lines, in %) and approximate hygroscopic equilibrium values (thick lines, in %) for interior finishing woodwork in heated apartments in the USA during July. From PECK (1932)

Nevertheless, for larger geographical regions the average seasonal course of the hygroscopic equilibrium can be deduced. This leads to the listing of recommended moisture content values for different uses of wood in different countries (cf. Section 6.2.4).

PECK (1932) worked out maps showing lines of equal hygroscopic equilibrium for wood in the USA during January (Fig. 8.3) and July (Fig. 8.4). The rather simple map showing recommended moisture content averages for interior finishing woodwork for use in various parts of the United States (cf. Fig. 6.44) is based on these investigations. Similar investigations were carried out by C.S.I.R.O. (1934), in Australia by REHMAN (1942) in India and by MÖRATH (1933) in Germany.

More important than all the theoretical considerations are the practical measurements made either in the open air or in closed rooms. Fig. 6.45 shows the measured moisture content of yard-piled white pine boards, 1 in. (25.4 mm) thick in eastern Canada (FELLOWS, 1937). The normal yearly variation of hygroscopic equilibrium varied between 6 and 7%. Similar results were obtained for Germany (Fig. 8.5).

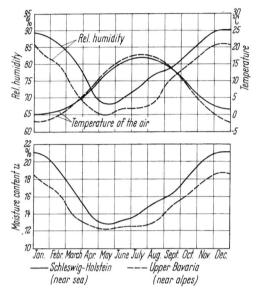

Fig. 8.5. Climate curves and hygroscopic equilibrium for wood in Schleswig-Holstein and Upper Bavaria (Germany). From KOLLMANN (1951)

The sensitivity of wooden parts in taking up moisture depends on their dimensions. Small, and especially very thin pieces may follow the equilibrium moisture content, changing their moisture content rather quickly. Pocket-size wood-hygrometers are based on this fact (cf. Section 6.2.2.4). If, on the other hand, boards are tightly piled so that no air may circulate over their surfaces the moisture content of the wood may remain rather constant. JENKINS and GUERNSEY (1937) proved this for seasoned lumber during ocean shipment. Considerable drying occurs after glazing and heating if rather wet lumber is used in constructing new buildings. An example is given in Fig. 6.46 (Forest Products Research Laboratory, Princes Risborough, 1933). The recommended moisture content values (Table 6.10 and 6.11) therefore should not be neglected.

### 8.1.3 Yard Seasoning

**8.1.3.1 Lumberyard Layout.** Though air seasoning is dependent mainly on the climate and the seasons the yard layout is a very important factor for the success of natural drying. The piles should be arranged in such a way in the lumber yard that an excellent circulation of the air is secured. The piling methods have an effect on the circulation. The air should have uniform access to all parts of the pile which should be protected as well as possible against rain, snow, and the sun's rays. Intensive, direct sun radiation is even more noxious than rain for the drying wood. Sun radiation can elevate the drying rate to such an extent that case-hardening, checking, warping and, in the long run blue-staining of the wood is caused. Under unfavorable circumstances rather large amounts of dust may be deposited on the drying wood. In laying out a lumber yard first the direction of prevailing winds is to be considered. Normally the wide alleys should run parallel to the prevailing wind, but if enough space is available the wide alleys should run both along and across thus facilitating good circulation through the yard.

The following principles should be observed:

1. The yard should be spacious and have free flow of air;
2. Long narrow yards facilitate a better air circulation than square yards;
3. Yards located near lakes or rivers as a rule are well ventilated since air movement over water is usually frequent and powerful;
4. The ground of the yard should be flat, but a slight fall is useful;
5. The best soil is gravel, the worst swamp;
6. Growing grass or weeds must be removed by means of suitable chemicals (e.g. sodium chlorate);
7. For wet grounds a covered water drainage is necessary; surface water is carried away in open ditches, or better, in tile drains;
8. Pile spacing has to be considered carefully. If the ground space is expensive or if only hardwood will be dried, the piles should be located close together and should be built high. Low piles are advantageous if rapid and uniform drying is wanted.
9. Moderate speed of drying may be desired for thick hardwoods, such as oak, ash, maple, beech; in this case the piles should be erected close together but adequate air circulation must be allowed and one must recognize that the danger of stain and rot is linked with close piling.

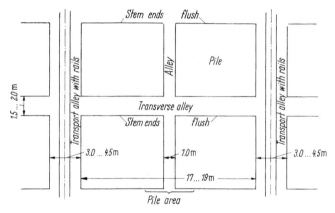

Fig. 8.6. Alley arrangement in a lumberyard

10. A suitable alley arrangement in a lumber yard is shown in Fig. 8.6. The width of the alleys depends on the methods of transportation and handling. If monorails are used a minimum width of 3 m ($\sim$10 feet) is required. In very large lumber yards (area more than 5000 m$^2$) vacant strips of about 30 m or more are left at intervals as fire breaks. These fire breaks also improve air circulation.

11. The height of the piles is determined by economical considerations (costs of handling) and by stability (height of the piles = 3 $\times$ width of the pile). Low piles are typical for small mills; they are cheaper to erect, but they need more ground space than higher piles.

12. The foundations for the piles must be solid and strong. If lumber is used, only durable species should be taken or it should be impregnated with coal tar

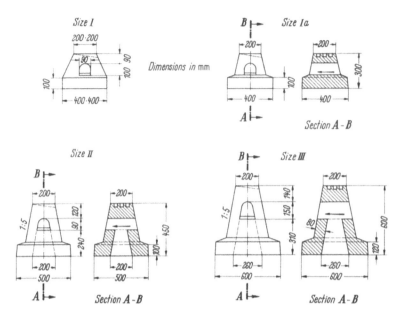

Fig. 8.7. Types of concrete footings. From DEUTSCHE GESELLSCHAFT FÜR HOLZFORSCHUNG

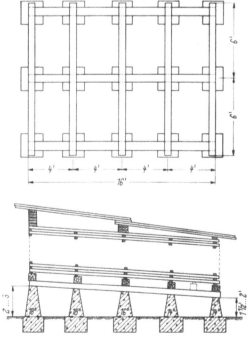

Fig. 8.8. Plan and elevation of a lumber pile foundation. From HENDERSON (1951)

creosote or other suitable preservatives. The lumber should be air-seasoned before treatment. Better and more permanent than wood are concrete footings. Fig. 8.7 shows some types: The foundations must be high enough (a minimum of 40 cm) to secure adequate air circulation beneath the lumber pile. The concrete footings should go down below frost level.

13. Plan and elevation of a lumber pile foundation is illustrated in Fig. 8.8 according to HENDERSON (1951). One can see that the lumber in the pile has a slight slope (pitch) of approximately 2 cm per 1 m length in order to improve the run-off of rain or melting snow.

14. The drying boards are separated by stickers which must be uniform in thickness. The width varies between 1 and 3 in. (25 and 75 mm). The stickers should

be 3 to 5 ft. (1 to 1.5 m) apart and arranged in straight vertical rows to minimize warping (Fig. 8.9).

15. Chimneys in the piles facilitate the movement of air.

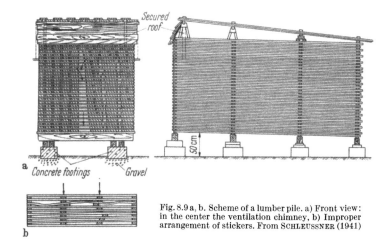

Fig. 8.9 a, b. Scheme of a lumber pile. a) Front view: in the center the ventilation chimney, b) Improper arrangement of stickers. From SCHLEUSSNER (1941)

16. Roofs should protect the piles against rain and sunshine. The roofs are secured against winds by means of wires fastened from the roof crossers to sticks pushed between rows of boards (cf. Fig. 8.9).

17. High-grade hardwoods are dried slowly in sheds (Fig. 8.10).

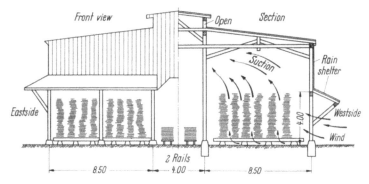

Fig. 8.10. Drying shed. From DEUTSCHE GESELLSCHAFT FÜR HOLZFORSCHUNG

**8.1.3.2 Seasoning Periods.** The great variation in moisture content of green wood has been mentioned in Section 8.1.1. In general, green wood is cut into lumber under favorable conditions for logging and marketing. In mountain areas the logs usually are stored over longer periods so that they lose moisture prior to sawing.

The main purpose of air seasoning in sawmills is a reduction of the moisture content of the lumber to about 25 to 30%; in this condition the lumber is ready for transport and for many building purposes. The fiber saturation point in boards sawn from slightly seasoned round logs will be reached within two or three weeks during the summer months.

The lowest moisture content attainable by air seasoning may vary between 12 and 15%. In the same season and at the same place, the drying speed, if a dry

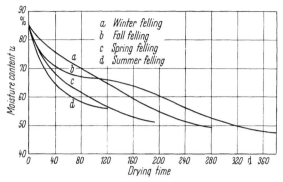

Fig. 8.11. Course of open-air drying of chestnut (*Castanea dentata* Borkh) poles. From MATHEWSON (1930)

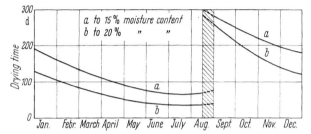

Fig. 8.12. Average drying times for softwood in Upper Bavaria (Germany). From KOLLMANN (1955)

wind blows, may be four times higher than that under the influence of cool and wet weather. The course of the air seasoning depends to a very large extent on the time of felling. Fig. 8.11, for example, shows that chestnut poles in the USA dried from an initial moisture content of 85% after felling in the winter season to about 55% in about 200 days; the same moisture content can be reached after felling in the summer season in 120 days. Felling in the fall season is always unfavorable since during the following winter season practically no air seasoning occurs (cf. Fig. 8.12). It may be mentioned that in practice the drying periods differ widely; in all cases a proper yard layout and careful piling increases the drying rate.

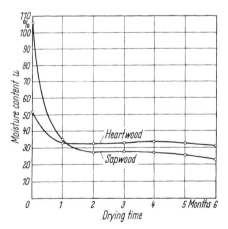

Fig. 8.13. Air-seasoning-curves for pine (*Pinus taeda* L.), cross-section 90 mm/115 mm, length 3 m. From MATHEWSON (1930)

The air seasoning (especially of heartwood) to moisture content values below the fiber saturation point is always lengthy, as shown in Fig. 8.13 according to MATHEWSON (1930). The influence of board thickness is illustrated in Fig. 8.14.

BATESON (1952) published "as a rough guide to assist in planning air-seasoning operations" some approximate drying rates. "Softwood 1 inch thick, if piled in spring, should dry to about 20% moisture content in 2 to 3 months; 2-inch-thick softwoods will require 3 to 4 months under similar conditions. Hardwoods 1 inch

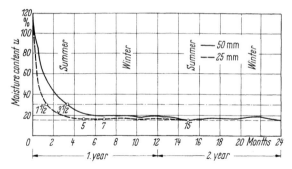

Fig. 8.14. Course of open-air drying of pine 25 mm and 50 mm thick. From SCHLEUSSNER (1941)

thick, if piled in the autumn, should dry to about 20% by the following summer, whereas under favorable conditions, 2 inch hardwoods should dry to about the same figure in a year if piled in October or November." Air-dried wood subsequently stored under roofs changes its moisture content with the different seasons hardly more than 2%.

### 8.1.4 Accelerated Air-drying, Predrying

**8.1.4.1 Fan Air-drying.** The first attempts to accelerate the drying of lumber in the open air by means of fans, such as ventilating the piles, were made by some large mills in the USA at the beginning of the century. These experiments failed for economical reasons, but the idea was not dead and PFEIFFER (1958) described the application of big axial fans to forced air-drying in the Pacific Northwest of the USA. After World War II forced air-drying moved from the experimental or pilot stage and became economically important due to the trend toward higher mechanization and productivity (KOLLMANN and SCHNEIDER, 1965). Several types of fans were developed for this purpose to meet the drying requirements.

After an initial experiment with two 60 in. (~1520 mm) two-blade propeller type fans PFEIFFER (1958) applied two 72 in. (~1830 mm) fans. Each fan was driven by a 7.5 HP electric motor. The lumber charge of about 75,000 board feet (177 m³) of 2 in. (51 mm) western coast hemlock dimension lumber in 6-, 8-, and 12-in. width was stacked in parallel piles. After bad experience with polyethylene sheeting, portable wooden frames covered with aluminium roofing were placed over the piles on each side of the fans to protect against rain. The central alley way between the two piles was covered with planks and plywood. Air velocity through the lumber piles was rather uniform. Fig. 8.15 shows the results obtained for charge 6. Hygrothermograph data reflect the drying conditions. Pulling rather than pushing air through the piles increased the uniformity of air flow. An economic analysis clearly indicated the advantage of forced-air drying. During some months of the year lumber may be dried to an average moisture content of 19% and may be sold in this condition on the basis of dry lumber prices. Humidity-controlled switches on the fan motors reduced power costs since little or no drying occurs at relative humidities above 90%.

HUFFMAN and POST (1960) studied forced-air drying of gum and oak cross ties. The movement of air across 2 parallel ties was accomplished by placing a fan housing which contained a 4.18 in. (425 mm) fan powered by $2^1/_3$ HP electric motors over the plenum chamber. Based on the results of this study it could be concluded that gum and oak cross ties may be seasoned by forced-air drying to a proper moisture content for treating in approximately one-third the time usually required to air season these species. The authors listed the advantages of the forced-air drying methods as follows:

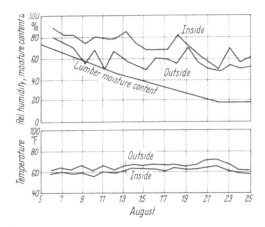

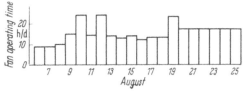

Fig. 8.15. Drying conditions and fan operating time for forced-air drying of West Coast hemlock lumber. From PFEIFFER (1958)

1. Low capital investment
2. Low operating cost
3. Flexible operation
4. Less requirement of space and time than for air seasoning
5. Easy determination of drying costs
6. Lower insurance rates
7. Reduced inventory
8. Decay-free ties

COBLER (1964) pointed out that several fan-air drying systems have been developed and have proven economical and successful. Fan-air drying affords an economic advantage especially for the small operator. He used 2.60 in. (1180 mm) desk fans powered by a 5 HP electric motor, located in a bi-sected plenum chamber. End baffles prevented wetting of the lumber by rain. Adjustable

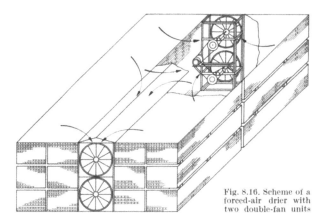

Fig. 8.16. Scheme of a forced-air drier with two double-fan units

wing walls and a canvas cover promoted efficient air movement through the lumber stacks. An automatic humidity controller stopped fans during periods of extremely high relative humidity. The main benefits from fan air drying as compared with yard air drying are more uniform moisture content, reduced kiln

Ref. p. 471]                              8.1 Air-drying                              431

time, lower unit drying costs, higher quality of lumber and decreased yard inventory. The scheme of a forced-air drier with two double-fan units is shown in Fig. 8.16. The drier is designed for lumber loads on rails and unit packages of 5,000 board feet (about 12 m³ and up). Another fan-air system (ANONYMOUS, 1959) consists basically of a battery of fans which move the air through the stacked green lumber. An electric control device permits the operator to make use of the most advantageous atmospheric conditions. The most essential parts of this device are a humidity indicator together with high and low humidity limit controls. In a permanent-type drier two sides are closed, one side is covered by the battery of fans and the side opposite the fans is open for the entry of the lumber. The length of time required for fan-air drying depends on outside weather conditions, but the lumber is usually ready for kiln drying within 5 to 10 days. The reduction in moisture content is shown in Table 8.2.

Table 8.2. *Reduction in Moisture Content after Seven Days Forced-air Seasoning*
(ANONYMOUS, 1959)

| Species | Cross section dimensions | Moisture content % | |
|---|---|---|---|
| | | Initial | After drying |
| Poplar | $^4/_4{}''$ | 95.0 | 19.4 |
| Elm | $^4/_4{}''$ | 22.5 | 26.2 |
| Hackberry | $^4/_4{}''$ | 45.2 | 29.4 |
| Cottonwood | $^5/_4{}''$ | 69.5 | 23.5 |
| Sap gum | $^5/_4{}''$ | 82.5 | 28.7 |

For middle-European weather conditions HERMA (1960) investigated forced-air-drying for different wood species and thicknesses. The results were comparable to those obtained by the U.S.-Forest Products Laboratory, Madison, Wisc. Especially useful were axial ventilators put between two wood piles in such a way that the air from the one pile was sucked and pushed through the second one again in the open air. Periodical change of the sense of revolutions of the fans caused fluctuation of the air and increased equality of the moisture contents. Even under relatively unfavorable weather conditions forced-air-drying can be economical. The following example is instructive: a pile of 5 m³ poplar was dried in the month of November from an initial moisture content of 110% to 32% with a power consumption of only 221 kWh. The weight of the water removed was 1757 kg. Therefore, for 1 kg water removed, no more than 0.127 kWh were necessary.

Experiments carried out by RUCKER and SMITH (1961) indicate that air velocities of about 2.5 to 3.0 m/sec. (500 to 600 ft. per min.) are optimum from the point of view of the drying rate balanced against fan operation costs. Without intermittent reversal of fans, air travel across lumber should be limited to 3 m (~ 10 ft.) to maintain fast and uniform drying in all parts of the lumber pile.

HILDEBRAND (1962) investigated the forced-air drying of oakwood cants with a cross section 90 mm × 100 mm. Fig. 8.17 shows the results. Starting the experiments at the end of March the cants dried from an initial moisture content of 60% down to 30% in a period of 48 days. Below the fiber saturation point the drying was retarded and one can see clearly the influence of rain or very high temperatures on the drying rate. The INSTITUT FÜR HOLZFORSCHUNG UND HOLZTECHNIK at the University of Munich compared natural air-drying and forced-air drying. Each pile of lumber was ventilated by a pair of propeller fans with 1800 mm diameter and 4 kW power consumption. The fans produced an air speed of about 3.1 m/s between supports. The following results were obtained: the length of the air path through the pile should not exceed 2.5 m, if the fans are acting only in one direction. If the fans may be reversed periodically the length can be increased to 5 m. The drying rate is practically independent of the psychrometric difference of the air above the fiber saturation point. With respect to the economy of the method the fans should be stopped if the relative humidity is too high and the temperature of the air is too low.

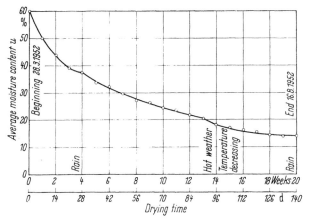

Fig. 8.17 Curve of forced-air drying of oak wood cants, cross-section 90 mm/100 mm. From HILDEBRAND (1962)

**8.1.4.2 Air-drying by Means of Swings or Centrifuges.** In 1937, GRAU, in Chemnitz, Germany, constructed a heavy swing for enforcing accelerated air circulation. The boards—in an amount corresponding to a total weight of 20 Mp (= about 20 ltn)—were piled vertically, the top-end showing downward. The

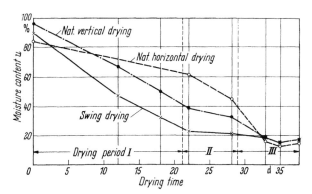

Fig. 8.18. Drying curves for boards 12 mm thick during swing drying compared with natural drying of horizontal and vertical piles. From EGNER and SINN (1941)

electric drive was similar to that of a bell-ringing apparatus; that is electric current was supplied to the motor only when the amplitude of the swing decreased. The consumption of energy was kept very low by this feeding system. EGNER and SINN (1941) proved that swing-drying actually increases the drying rate especially compared with normal air-drying of piles with horizontal boards. The differences compared with end-piles were smaller (Fig. 8.18). PIEST (1941, 1950) obtained similar results for smaller (workshop-)swings having a capacity of 2 to 2.5 m³ of

Ref. p. 471]                                8.1 Air-drying                                    433

boards. The distribution of moisture within the piles after the swinging was very uniform but it was determined that the drying rate, using a swing, is not remarkably higher than that of a normal end-pile. This is the reason why the drying-swings were abandoned. Up to date, drying centrifuges, first built by KASTMARK (1943) in Sweden and later improved technically by EISENMANN (1950) in West Germany (cf. also the rather fine test results published by FESSEL, 1952), had the same fate.

### 8.1.4.3 Air-drying by Solar Heat.

Early in 1934, SCHWALBE and BARTELS proposed to improve the heat economy of air-drying round logs. They stored the logs vertically on frames about 1 m above ground level between two parallel wooden walls; the wall towards the sun is either closed or periodically open, so that an opening of about 1.5 m in height over ground is created.

In this way the sun radiation was led into the lower, dark-painted part of the plant. The temperature of the air inside the plant became a few degrees higher than that of the open air and the drying was accelerated. A plant similar in principle utilizing solar heat and facilitating the drying process by the absorption capacity for water vapor of wood chips filled into the hollow side walls of the plant was described by ALTENKIRCH (1938). In such a plant it was possible to dry softwood boards to 15% moisture content as compared with 22% in an open shed during a wet and cloudy winter season.

Another way of utilizing solar heat in the drying of lumber is to place a transparent sheet between the sun and the lumber to be heated. The transparent sheet offers practically no resistance to the passage of sunlight but does resist the passage of heat waves that are radiated from the lumber. This principle is incorporated into a drying structure, called a predryer, that has been designed at the U.S. Forest Products Laboratory, Madison, Wisc. A series of experiments was carried out using such a predryer with 25 mm (4/4″) red-oak flooring stock in 1958. PECK made another series of similar experiments in 1960/61. He pointed out that the drying time to a moisture content of 20% can be reduced to about one half of that required for drying the same lumber in an open yard. It seems probable that this drying-time ratio can be maintained on an average throughout all of the months of the year even at geographical locations where only a moderate amount of solar energy is obtainable.

### 8.1.4.4 Predriers.

The period of ventilated air-drying and the lowest moisture content obtainable depends on the weather conditions. It has been mentioned that the fans should be switched off for economical reasons during wet and cold weather seasons; in order to eliminate these disadvantages during recent years, especially in the USA and in Australia, ventilated air driers were combined with simple heating plants. These so-called predriers can be operated throughout the year and are used mainly in predrying lumber prior to kiln drying. The capacity of the predriers is often very high. Green lumber should be dried to about 30 to 25% moisture content. The temperature in predriers normally lies in the range between 20 and 40 °C (GABY, 1961; RUCKER and SMITH, 1961; GATSLICK, 1962). For such low temperatures the problem of heat insulation is not critical. An increase of temperature by 6 to 12 °C as compared with the open air temperature allows drying down to 12 to 14% moisture content of wood even if the humidity is high. Predrying is far better than uncontrolled air drying because of reduced drying time, improved uniformity of drying and elimination of many drying defects. During adverse weather conditions a constant drying rate can be maintained in the predrier and the lumber does not suffer degrade from stain.

28   Kollmann/Côté, Solid Wood

434 8. Steaming and Seasoning of Wood [Ref. p. 471

WRIGHT (1962) used predriers for beams and ties before and after pressure impregnation. GABY (1961) could dry 25 mm thick boards of southern pine within 3 to 6 days from an initial moisture content of 110% to a final moisture content of 80%; the temperature was 27 °C. For 51 mm thick pine boards the drying period was about doubled.

## 8.2 Steaming

### 8.2.1 Reasons for Steaming

Steaming preliminary to air seasoning is employed occasionally to sterilize the green lumber for killing fungi, mold or insects. Darkening of the wood of some species (red beech, black walnut, butternut, yellow birch, red gum, pear) is developed by prolonged steam treatment. Preliminary steaming at about 95% relative humidity is the usual practice at the start of kiln drying. This process heats the lumber and relieves it of any stresses set up by casehardening during air-seasoning. Steaming under atmospheric pressure does not change the chemical constituents of the wood. Development of casehardening during kiln drying is a common phenomenon and can easily be removed by intermittent steaming. According to British investigations in 1940, steaming does not remarkably alter the quality of beechwood such as its drying properties, its workability, or its dimensional changes with changes in the hygroscopic moisture content. CAMPBELL (1961) established that a short preliminary steaming treatment of 2 to 4 hours' duration is of considerable value for "difficult" material of some of the ash-type eucalypts growing in the southern parts of Australia. The main effect is to reduce the drying time by increating the drying-rate. Reports from seasoning-plants that have accepted the practice of preliminary steaming indicate that the treatment resulted in improved recovery from collapse, less degrade, and better machining qualities.

### 8.2.2 Methods of Steaming, and Heat Consumption

Steaming can be performed either directly or indirectly. Direct steaming means that mostly unoiled, exhaust steam is led into chambers or cylinders in which the wood is piled. The arrangement with respect to tubes, valves, etc. is simple but the control is difficult, defects in the steamed wood due to local overheating are unavoidable, and the thermal economy is unfavorable since no condensate can be regained. In plants for indirect steaming the exhaust steam is led into coils which are arranged in a basin. The water in the basin is evaporated, the saturated steam mildly treats the wood. The steam may contain oil, its condensate can be regained and therefore the thermal economy is good. Nevertheless, steaming, especially in steaming pits prior to veneer peeling, is characterized by a very low economical efficiency. Fig. 8.19 shows the heat balances for different temperatures in the steaming pits and for varying diameters of Gaboon wood. The percentage of effective heat is very low (approximately 5 to 18%) and it became evident that the heat lost to the outside air due to loose-fitting steaming-pit covers causes by far the most important losses (KUHLMANN, 1962). It is not reasonable to base the amount of steam consumption on the volume of the steamed round logs since the consumption of heat is not at all proportional to the capacity of the steaming pit. Only the figures for the consumption of heat in thermal units (kcal or BTU) per steaming process under consideration of species, log diameter, and average steam temperature are useful. For approximate comparison only, a few figures are listed in Table 8.3.

Table 8.3. *Consumption of Steam (643 kcal/kg) in kg per Hour, Directly Steaming Gaboon Wood in a Pit of 97 m³ Void Volume* (From KUHLMANN, 1962)

| Diameters of logs | cm | 60 | 70 | 80 | 90 | 110 |
|---|---|---|---|---|---|---|
| Average temperature of steam °C | 60 | 102.0 | 960 | 94.0 | 93.0[1] | — |
| | 80 | 168.0 | 155.0 | 146.5 | 143.8 | 140.3 |
| | 99 | — | 268.0 | 262.0 | 240.0 | 232.0 |

[1] value for 15 cm log diameter

FESSEL (1955) arrived at about corresponding values, but DOFFINÉ (1956) found much higher figures. The justified conclusion is that steaming, especially of round logs in steaming pits, represents an uneconomical heating process. Constructive measures are desirable to bring about a reduction of steam consumption and an improvement of the steaming effect. With this aim a short discussion of the heat balance in steaming-pits may be useful. Based on KUHLMANN's (1962) investigations one can say the following:

1. As mentioned previously, the percentage of effective heat varied, under the conditions of the test, only between 5 and 18%, influenced by the ratio of log volume to void volume of the pit, moisture content of the logs and period of steaming.

2. The loss of heat into the open air due to the leakage around the pit covers, amounting

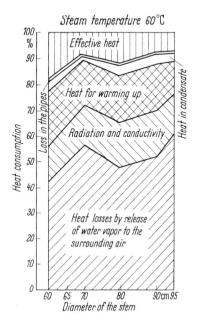

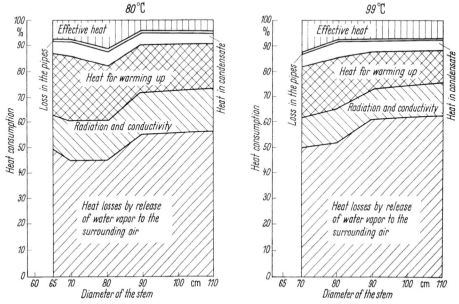

Fig. 8.19. Heat balances for different steam temperatures in the pits and for varying diameters of Gaboon. From KUHLMANN (1962)

to 41%···63% of the total heat applied, is by far the highest sum in the balance. It is not difficult to reduce much of this heat loss by better tightening of the steaming-pit covers.

3. The losses of heat by radiation and conduction varied between 12 and 19%. They may be reduced by reduction of the diffusion of water vapor through the pit walls by means of suitable paints, for example teflon.

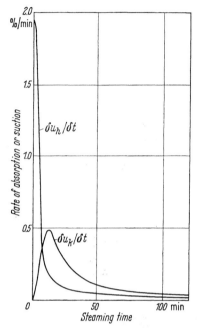

Fig. 8.20. Dependence of the rate of absorption from hygroscopic water causing swelling $\frac{\partial u}{\partial t} h$ and the rate of suction of the capillary water $\frac{\partial u}{\partial t} k$ upon the steaming time for ash wood stock. From KOLLMANN (1939)

4. The heating of the pit walls amounts to 13% to 26%, the second largest term in the heat balance. This term also may be reduced by preventing water vapor diffusion into the pit walls. The explanation is easy: density and specific heat of concrete increase with increasing moisture content.

5. The heat content of the condensate in the pit, varying between 2% and 6%, is relatively low.

6. A low value of 0.3% was estimated for the heat losses in the tubes.

### 8.2.3 Effects of Steaming on Wood

At the beginning of the steaming an exchange of heat between the cold wood and the hot steam occurs. As shown in Section 6.5.4. the phenomena may be analyzed and graphically reproduced using Fourier's theory. Of course one has to know the values for thermal conductivity, the specific heat, and the diffusivity (cf. Section 6.5.4). Figures determined experimentally and those calculated theoretically are in a good agreement.

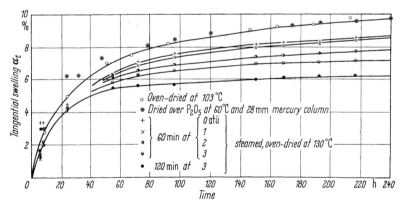

Fig. 8.21. Tangential swelling of beech specimens after different periods of steaming, compared with unsteamed wood. From KOLLMANN (1939)

If green or very moist wood is steamed at atmospheric pressure drying takes place at first. The liquid water evaporates under the influence of the temperature of the steam near the boiling point. On the other hand very dry wood absorbs moisture from the steam. The problem is complicated due to condensation phenomena. At the surface of freshly steamed wood a water skin is always produced. If wood is steamed in a closed vat where condensed water is accumulated, an equilibrium moisture content between 40 and 50% will be reached (KOLLMANN,

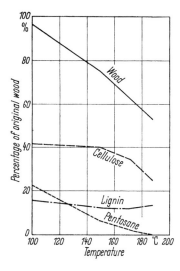

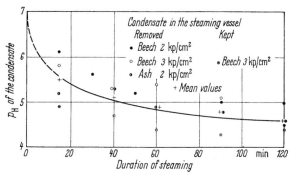

Fig. 8.22. Residues of European aspen wood and its constituents after hot water-extraction. From ARONOWSKY and GORTNER (1930)

Fig. 8.23. Effect of the duration of steaming beech and ash wood on the pH-value of the condensate. From KOLLMANN (1939)

1939). In streaming saturated steam the equilibrium moisture content is between 22 and 25%. If the condensed water is continuously removed from the vat, cylinder or pit, the wood-water relationships in an atmosphere of saturated steam becomes clear: Fig. 8.20 shows that the rate of absorption of hygroscopic

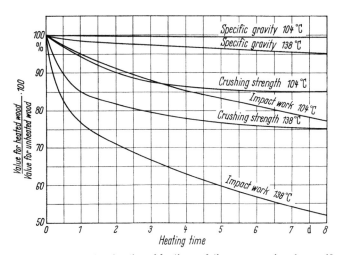

Fig. 8.24. Relationship between the duration of heating and the average values for specific gravity, crushing strength parallel to the grain, and toughness of Sitka spruce expressed in % of the values for unheated control specimens. From KOEHLER and PILLOW (1925)

water causing swelling drops very quickly from a maximum at the beginning of the steaming. The rate of absorption of free liquid water starts with the zero value, and with increasing wetting and formation of filled capillaries quite near the fiber saturation point reaches a maximum from which it also decreases rapidly.

438                    8. Steaming and Seasoning of Wood                    [Ref. p. 471

The capacity of the wood to swell can be reduced by steaming. SCHWALBE and ENDER (1934) found such a reduction for the sapwood, but not for the heartwood of beech. KOLLMANN (1939) came to similar but more illustrative results since he systematically varied the steam pressure and the periods of steaming (Fig. 8.21). Steaming over longer periods and/or with higher pressure always causes a loss of substance. An unavoidable hydrolysis reduces the relative amount of cellulose and hemicelluloses (Fig. 8.22) (ARONOWSKY and GORTNER, 1930). Acetic acid and formic acid are developed and the pH-values quickly decrease (KOLLMANN, 1939, Fig. 8.23.) It is only logical that density is also reduced and that, more importantly, the mechanical properties are influenced adversely (Fig. 8.24, KOEHLER and PILLOW, 1925).

## 8.3 Kiln Drying

### 8.3.0 General Considerations

The necessity and advantages of kiln drying are explained in Section 8.0. For many purposes kiln drying is almost indispensable. Green lumber should be dried at the sawmill. In particular higher grades of lumber should be dried in the kiln without the delay and dangers of air-seasoning. According to HENDERSON (1951) the advantages that accrue from the viewpoint of the sawmill operator are listed as follows:

1. Availability of very cheap heat and power for drying, being waste products from the steam-engines of the mill;
2. Saving repeated handling, carting etc. that usually accompany the air-seasoning process;
3. Much faster turnover; therefore, requirement of less capital and quick adaptation to market conditions;
4. Production of brighter stock, free from decay, stain and insect attacks and with a minimum of warping, checking and splitting.
5. Possibility to meet the consumers' demand for kiln-dried lumber.

As compared to air seasoning one-inch lumber (25 mm) from green to 20% one can save by kiln drying from green to 6% about 75 to 94%, and from 20% to 6% about 90% to 99% of the time. If lumber is kiln-dried at the wood working factory, principally the same advantages are gained and the following additional ones:

1. Possibility of reaching moisture content values prohibitive in air seasoning but necessary to bring various items made of the wood later in an equilibrium with the surrounding atmosphere in service (cf. Section 6.2.4. and Table 6.10);
2. Improvement of workability especially in machining with high cutting speeds;
3. Improvement of glueability;
4. Creation of good conditions for surfacing (painting, varnishing, lacquering).

### 8.3.1 Fundamental Drying Factors

The physical laws governing the rate of diffusion as the predominant factor in air-seasoning as well as in kiln drying of wood are dealt with in Section 6.3. In this connection methods and formulas for an approximate calculation of the drying periods are given.

Generally flat piling is used in dry kilns and, as a rule, the boards are piled lengthwise. The stickers can be made of any kind of dried lumber. They should be uniform and planed to a uniform thickness. In the USA $^7/_8$ in. (= about 22 mm) is the usual thickness; in Germany the thickness of the stickers is adapted to the thickness of the boards (Fig. 8.25). All stickers should be $1^1/_2$ times as wide as they are thick. The stickers should be spaced carefully to prevent warping, cupping and splitting. With thin stock (25 mm or less) or rather green stock (up to 75 mm) the stickers should be spaced 45 cm to 50 cm (18 in. to 20 in.) apart. With air-seasoned stock 50 mm and up in thickness spacing 60 cm (25 in.) to 75 cm (30 in.), in some cases 90 cm to 120 cm (36 in. to 48 in.), is usual. Only boards of the same thickness and of one species should be dried together. If different thicknesses and species are piled in a kiln the schedule used must be adjusted to the slowest drying lumber. The columns of stickers should be vertical. Uneven stickering causes warping. Stickers must be placed flush at both ends of the drying boards. Free ends of the boards will likely warp and check. To increase the circulation of the drying medium through the pile thicker stickers should be used and a central tapered chimney (200 to 400 mm at the base, 100 mm at the top) should be left. Vertical piling is used in some large sawmills of the South and West Coast regions of the USA. This system permits the application of mechanical stickers but it facilitates warping and splitting. The kilns must be designed in such a way that the air-steam mixture is exclusively directed vertically through the pile.

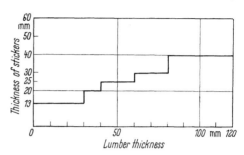

Fig. 8.25. Recommended thickness of stickers in relation to the lumber thickness

Proper piling or stickering of the lumber either on kiln trucks or as load units transferred to the kiln by mean of lift trucks, is the first fundamental factor governing kiln drying. It is also essential with respect to ease of loading and unloading. The second prerequisite is proper control of drying conditions, such as temperature, relative humidity and circulation. Other factors are the required drying time and the power consumption (steam and electricity). After the piles are pushed into the kiln, and after its doors are closed, preliminary steaming is the usual practice. The relative humidity ranges between 80 and 100%; during this period of a few hours green or very wet wood will mostly lose a few percent of moisture whereas air-dry stock will adsorb 5 to 8% moisture. The desired effect of preliminary steaming is the heating of the kiln with its load and the relief of stresses set up in the lumber by casehardening during air seasoning. Some dry kiln manufacturers and operators are of the opinion that preliminary steaming is not necessary. They point out that at the beginning of normal drying conditions condensation on the surfaces of the cold lumber is inevitable and contributes to relief of strains. Nevertheless, preliminary humidity treatment is preferred. The fans blow from the beginning of the spraying process. After the preliminary steaming has been finished, drying is begun. The steam sprays are closed and steam is led into the heating coils. The control of temperature is accomplished either by hand operating valves for the steam supply or by motorvalves in conjunction with a thermostat. The dampers are closed during steaming periods and opened by degrees as the drying proceeds.

The temperatures required for conventional kiln drying range from 40 to 100 °C (104 to 212 °F). The necessary heat is supplied predominantly by steam (either live steam from the boiler or exhaust steam from engines or turbines). Sometimes gas or crude oil burners are used to generate steam. Electricity is used for small dry kilns and for high temperature kilns. With respect to heat economics and dimensional stabilization the temperatures in kiln drying should be as high as possible. Nevertheless, the quality of the lumber should not be reduced by staining, checking, or exudation of resin. Table 8.4 contains ten steps of temperature (dry bulb) for kiln drying of lumber (thickness up to 30 mm) according to KEYLWERTH (1951).

Table 8.4. *Temperatures for the Kiln Drying of Lumber; Thickness up to 30 mm* (From KEYLWERTH, 1951)

| Species | Schedule | Initial temperature °C | Maximum temperature below FSP °C |
|---|---|---|---|
| Oak | T 1 | 40 | 50 |
| Oak, Box-tree, Eucalypt | T 2 | 40 | 60 |
| Oak | T 3 | 40 | 80 |
| Parana pine | T 4 | 50 | 70 |
| Walnut | T 5 | 50 | 80 |
| Beech, Maple, Locust (Robinia), Elm, Hickory | T 6 | 60 | 80 |
| Birch, Larch, White pine, pine | T 7 | 70 | 80 |
| Douglas-fir (Oregon pine), Scots pine | T 8 | 70 | 90 |
| Fir, Spruce, Scots pine | T 10 | 100 | 120 |

The humidity is controlled by regulating the steam valves, steam sprays, and dampers. When the dampers are open the humidity goes down, when they are closed the humidity increases. It is possible to regulate the dampers automatically by means of air operated kiln controllers. It is a difficult task for the kiln operator to find the proper ventilation in a running kiln. He must work by trial and error

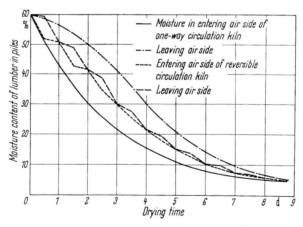

Fig. 8.26. Comparative drying curves for lumber in a one-way circulation kiln and a reversible circulation kiln, using same schedule. From HENDERSON (1951)

and he must know the wet bulb temperature for a desired relative humidity. For practical purposes in control of dry kiln or storage room conditions, tables or nomograms are available showing not only the relative humidity for different dry-bulb temperatures and wet-bulb depressions, but also the equilibrium moisture contents of wood (Table 8.5).

In all modern kilns adequate circulation is secured by means of fans or blowers. Gravity circulation, widely used in the past, is not sufficient. Short circuits of the air, that is movement of the air under and/or over and around the pile but not through the pile impedes the drying and increases the waste of heat. Gravity movement is also slow and cannot be properly controlled. Canvas curtains as air baffles are effective in kilns ventilated by means of fans but not in kilns with gravity circulation. In a number of drying systems cold water condensers, water sprays or steam jet blowers were used to increase the speed of the drying agent.

All these systems are characterized on the one hand by a greater volume of circulation and on the other hand by some undesirable features (e.g. high consumption of expensive water, difficult control of humidity). There is the possibility of placing the fans either outside or inside of the kiln (cf. Section 8.3.2). Alternated or reversed circulation aids uniform drying of the lumber in the pile. Fig. 8.26 according to HENDERSON (1951, p. 170), shows clearly the good results of reversed drying (alternating the air current twice each day) as compared with one-way circulation.

Drying schedules for various species of wood, thicknesses and initial moisture contents are available from different sources. The manufacturer of the kiln normally furnishes such tables. The U.S. Forest Products Laboratory in Madison, Wisc., published (1943) suggestions and instructions for kiln operators (drying aircraft lumber). Reference should also be made to kiln drying schedules worked out at the British Forest Products Research Laboratory (1938) and applicable to various European and overseas commercial timbers. These schedules guarantee very careful drying, preservation of quality, and freedom from casehardening and checking, but the drying periods are relatively long. In the last three or four decades the time necessary to properly dry lumber (even hardwood from the green condition) has been noticeably reduced. Due to improvements in kiln construction and drying schedules not only the drying periods, but also the costs and the defects were reduced. HENDERSON (1951) prepared drying schedules for modern kilns after careful experiments and long experience. Fig. 8.27 shows the recommended schedule for the kiln drying of green one-inch softwood lumber. This schedule and a set of similar ones are characterized by the fact that the wet bulb temperature is kept constant—"a material aid to operation" (HENDERSON, 1951, p. 219). Drying curves for various softwood species with various thicknesses are shown in Fig. 8.28 (B. SCHILDE according to KOLLMANN, 1955, p. 319); these

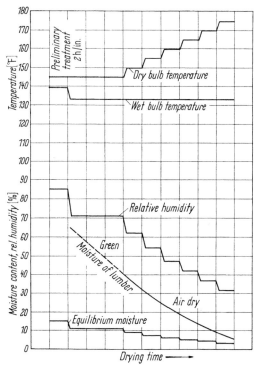

Fig. 8.27. Drying schedule for one-inch softwood lumber. From HENDERSON (1951)

442　　　　8. Steaming and Seasoning of Wood　　　　[Ref. p. 471

Table 8.5 *Relative Humidity*[1] *and Equilibrium Moisture Content*[2]

| Temperature, dry-bulb (°F) | Wet-bulb | | | | | | | | | | | | | | | | |
|---|---|---|---|---|---|---|---|---|---|---|---|---|---|---|---|---|---|
| | 1 | 2 | 3 | 4 | 5 | 6 | 7 | 8 | 9 | 10 | 11 | 12 | 13 | 14 | 15 | 16 | 17 |
| 30 | 89 | 78 | 67 | 57 | 46 | 36 | 27 | 17 | 6 | | | | | | | | |
| | | *15.9* | *12.9* | *10.8* | *9.0* | *7.4* | *5.7* | *3.9* | *1.6* | | | | | | | | |
| 35 | 90 | 81 | 72 | 63 | 54 | 45 | 37 | 28 | 19 | 11 | 3 | | | | | | |
| | | *16.8* | *13.9* | *11.9* | *10.3* | *8.8* | *7.4* | *6.0* | *4.5* | *2.9* | *0.8* | | | | | | |
| 40 | 92 | 83 | 75 | 68 | 60 | 52 | 45 | 37 | 29 | 22 | 15 | 8 | | | | | |
| | | *17.6* | *14.8* | *12.9* | *11.2* | *9.9* | *8.6* | *7.4* | *6.2* | *5.0* | *3.5* | *1.9* | | | | | |
| 45 | 93 | 85 | 78 | 72 | 64 | 58 | 51 | 44 | 37 | 31 | 25 | 19 | 12 | 6 | | | |
| | | *18.3* | *15.6* | *13.7* | *12.0* | *10.7* | *9.5* | *8.5* | *7.5* | *6.5* | *5.3* | *4.2* | *2.9* | *1.5* | | | |
| 50 | 93 | 86 | 80 | 74 | 68 | 62 | 56 | 50 | 44 | 38 | 32 | 27 | 21 | 16 | 10 | 5 | |
| | | *19.0* | *16.3* | *14.4* | *12.7* | *11.5* | *10.3* | *9.4* | *8.5* | *7.6* | *6.7* | *5.7* | *4.8* | *3.9* | *2.8* | *1.5* | |
| 55 | 94 | 88 | 82 | 76 | 70 | 65 | 60 | 54 | 49 | 44 | 39 | 34 | 28 | 24 | 19 | 14 | 9 |
| | | *19.5* | *16.9* | *15.1* | *13.4* | *12.2* | *11.0* | *10.1* | *9.3* | *8.4* | *7.6* | *6.8* | *6.0* | *5.3* | *4.5* | *3.6* | *2.5* |
| 60 | 94 | 89 | 83 | 78 | 73 | 68 | 63 | 58 | 53 | 48 | 43 | 39 | 34 | 30 | 26 | 21 | 17 |
| | | *19.9* | *17.4* | *15.6* | *13.9* | *12.7* | *11.6* | *10.7* | *9.9* | *9.1* | *8.3* | *7.6* | *6.9* | *6.3* | *5.6* | *4.9* | *4.1* |
| 65 | 95 | 90 | 84 | 80 | 75 | 70 | 66 | 61 | 56 | 52 | 48 | 44 | 39 | 36 | 32 | 27 | 24 |
| | | *20.3* | *17.8* | *16.1* | *14.4* | *13.3* | *12.1* | *11.2* | *10.4* | *9.7* | *8.9* | *8.3* | *7.7* | *7.1* | *6.5* | *5.8* | *5.2* |
| 70 | 95 | 90 | 86 | 81 | 77 | 72 | 68 | 64 | 59 | 55 | 51 | 48 | 44 | 40 | 36 | 33 | 29 |
| | | *20.6* | *18.2* | *16.5* | *14.9* | *13.7* | *12.5* | *11.6* | *10.9* | *10.1* | *9.4* | *8.8* | *8.3* | *7.7* | *7.2* | *6.6* | *6.0* |
| 75 | 95 | 91 | 86 | 82 | 78 | 74 | 70 | 66 | 62 | 58 | 54 | 51 | 47 | 44 | 41 | 37 | 34 |
| | | *20.9* | *18.5* | *16.8* | *15.2* | *14.0* | *12.9* | *12.0* | *11.2* | *10.5* | *9.8* | *9.3* | *8.7* | *8.2* | *7.7* | *7.2* | *6.7* |
| 80 | 96 | 91 | 87 | 83 | 79 | 75 | 72 | 68 | 64 | 61 | 57 | 54 | 50 | 47 | 44 | 41 | 38 |
| | | *21.0* | *18.7* | *17.0* | *15.5* | *14.3* | *13.2* | *12.3* | *11.5* | *10.9* | *10.1* | *9.7* | *9.1* | *8.6* | *8.1* | *7.7* | *7.2* |
| 85 | 96 | 92 | 88 | 84 | 80 | 76 | 73 | 70 | 66 | 63 | 59 | 56 | 53 | 50 | 47 | 44 | 41 |
| | | *21.2* | *18.8* | *17.2* | *15.7* | *14.5* | *13.5* | *12.5* | *11.8* | *11.2* | *10.5* | *10.0* | *9.5* | *9.0* | *8.5* | *8.1* | *7.6* |
| 90 | 96 | 92 | 89 | 85 | 81 | 78 | 74 | 71 | 68 | 65 | 61 | 58 | 55 | 52 | 49 | 47 | 44 |
| | | *21.3* | *18.9* | *17.3* | *15.9* | *14.7* | *13.7* | *12.8* | *12.0* | *11.4* | *10.7* | *10.2* | *9.7* | *9.3* | *8.8* | *8.4* | *8.0* |
| 95 | 96 | 92 | 89 | 85 | 82 | 79 | 75 | 72 | 69 | 66 | 63 | 60 | 57 | 55 | 52 | 49 | 46 |
| | | *21.3* | *19.0* | *17.4* | *16.1* | *14.9* | *13.9* | *12.9* | *12.2* | *11.6* | *11.0* | *10.5* | *10.0* | *9.5* | *9.1* | *8.7* | *8.2* |
| 100 | 96 | 93 | 89 | 86 | 83 | 80 | 77 | 73 | 70 | 68 | 65 | 62 | 59 | 56 | 54 | 51 | 49 |
| | | *21.3* | *19.0* | *17.5* | *16.1* | *15.0* | *13.9* | *13.1* | *12.4* | *11.8* | *11.2* | *10.6* | *10.1* | *9.6* | *9.2* | *8.9* | *8.5* |
| 105 | 96 | 93 | 90 | 87 | 83 | 80 | 77 | 74 | 71 | 69 | 66 | 63 | 60 | 58 | 55 | 53 | 50 |
| | | *21.4* | *19.0* | *17.5* | *16.2* | *15.1* | *14.0* | *13.2* | *12.6* | *11.9* | *11.3* | *10.8* | *10.3* | *9.8* | *9.4* | *9.0* | *8.7* |
| 110 | 97 | 93 | 90 | 87 | 84 | 81 | 78 | 75 | 73 | 70 | 67 | 65 | 62 | 60 | 57 | 55 | 52 |
| | | *21.4* | *19.0* | *17.5* | *16.2* | *15.1* | *14.1* | *13.3* | *12.6* | *12.0* | *11.4* | *10.8* | *10.4* | *9.9* | *9.5* | *9.2* | *8.8* |
| 115 | 97 | 93 | 90 | 88 | 85 | 82 | 79 | 76 | 74 | 71 | 68 | 66 | 63 | 61 | 58 | 56 | 54 |
| | | *21.4* | *19.0* | *17.5* | *16.2* | *15.1* | *14.1* | *13.4* | *12.7* | *12.1* | *11.5* | *10.9* | *10.4* | *10.0* | *9.6* | *9.3* | *8.9* |
| 120 | 97 | 94 | 91 | 88 | 85 | 82 | 80 | 77 | 74 | 72 | 69 | 67 | 65 | 62 | 60 | 58 | 55 |
| | | *21.3* | *19.0* | *17.4* | *16.2* | *15.1* | *14.1* | *13.4* | *12.7* | *12.1* | *11.5* | *11.0* | *10.5* | *10.0* | *9.7* | *9.4* | *9.0* |
| 125 | 97 | 95 | 91 | 88 | 86 | 83 | 80 | 77 | 75 | 73 | 70 | 68 | 65 | 63 | 61 | 59 | 57 |
| | | *21.2* | *18.9* | *17.3* | *16.1* | *15.0* | *14.0* | *13.4* | *12.7* | *12.1* | *11.5* | *11.0* | *10.5* | *10.0* | *9.7* | *9.4* | *9.0* |
| 130 | 97 | 94 | 91 | 89 | 86 | 83 | 81 | 78 | 76 | 73 | 71 | 69 | 67 | 64 | 62 | 60 | 58 |
| | | *21.0* | *18.8* | *17.2* | *16.0* | *14.9* | *14.0* | *13.4* | *12.7* | *12.1* | *11.5* | *11.0* | *10.5* | *10.0* | *9.7* | *9.4* | *9.0* |

[1] Relative humidity values in roman type.
[2] Equilibrium moisture content values in italic type.

Ref. p. 471]  8.3 Kiln Drying

*Table for Use with Dry-Bulb Temperatures and Wet-Bulb Depressions*

depression (°F)

| 18 | 19 | 20 | 21 | 22 | 23 | 24 | 25 | 26 | 27 | 28 | 29 | 30 | 32 | 34 | 36 | 38 | 40 | 45 | 50 |
|---|---|---|---|---|---|---|---|---|---|---|---|---|---|---|---|---|---|---|---|
| 5<br>1.3 | | | | | | | | | | | | | | | | | | | |
| 13<br>3.2 | 9<br>2.3 | 5<br>1.3 | 1<br>0.2 | | | | | | | | | | | | | | | | |
| 20<br>4.5 | 16<br>3.8 | 13<br>3.0 | 8<br>2.3 | 6<br>1.4 | 2<br>0.4 | | | | | | | | | | | | | | |
| 25<br>5.5 | 22<br>4.9 | 19<br>4.3 | 15<br>3.7 | 12<br>2.9 | 9<br>2.3 | 6<br>1.5 | 3<br>0.7 | | | | | | | | | | | | |
| 31<br>6.2 | 28<br>5.6 | 24<br>5.1 | 21<br>4.7 | 18<br>4.1 | 15<br>3.5 | 12<br>2.9 | 10<br>2.3 | 7<br>1.7 | 4<br>0.9 | 1<br>0.2 | | | | | | | | | |
| 35<br>6.8 | 32<br>6.3 | 29<br>5.8 | 26<br>5.4 | 23<br>5.0 | 20<br>4.5 | 18<br>4.0 | 15<br>3.5 | 12<br>3.0 | 10<br>2.4 | 7<br>1.8 | 5<br>1.1 | 3<br>0.3 | | | | | | | |
| 38<br>7.2 | 36<br>6.7 | 33<br>6.3 | 30<br>6.0 | 28<br>5.6 | 25<br>5.2 | 23<br>4.8 | 20<br>4.3 | 18<br>3.9 | 15<br>3.4 | 13<br>3.0 | 11<br>2.4 | 9<br>1.7 | 4<br>0.9 | | | | | | |
| 41<br>7.6 | 39<br>7.2 | 36<br>6.8 | 34<br>6.5 | 31<br>6.1 | 29<br>5.7 | 26<br>5.3 | 24<br>4.9 | 22<br>4.6 | 19<br>4.2 | 17<br>3.8 | 15<br>3.3 | 13<br>2.8 | 9<br>2.1 | 5<br>1.3 | 1<br>0.4 | | | | |
| 44<br>7.9 | 42<br>7.5 | 39<br>7.1 | 37<br>6.8 | 34<br>6.4 | 32<br>6.1 | 30<br>5.7 | 28<br>5.3 | 26<br>5.1 | 23<br>4.8 | 22<br>4.4 | 20<br>4.0 | 17<br>3.6 | 14<br>3.0 | 10<br>2.3 | 6<br>1.5 | 2<br>0.6 | | | |
| 46<br>8.1 | 44<br>7.8 | 41<br>7.4 | 39<br>7.0 | 37<br>6.7 | 35<br>6.4 | 33<br>6.1 | 30<br>5.7 | 28<br>5.4 | 26<br>5.2 | 24<br>4.9 | 22<br>4.6 | 21<br>4.2 | 17<br>3.6 | 13<br>3.1 | 10<br>2.4 | 7<br>1.6 | 4<br>0.7 | | |
| 48<br>8.3 | 46<br>7.9 | 44<br>7.6 | 42<br>7.3 | 40<br>6.9 | 37<br>6.7 | 35<br>6.4 | 34<br>6.1 | 31<br>5.7 | 29<br>5.4 | 28<br>5.2 | 26<br>4.8 | 24<br>4.6 | 20<br>4.2 | 17<br>3.6 | 14<br>3.1 | 11<br>2.4 | 8<br>1.8 | | |
| 50<br>8.4 | 48<br>8.1 | 46<br>7.7 | 44<br>7.5 | 42<br>7.2 | 40<br>6.8 | 38<br>6.6 | 36<br>6.3 | 34<br>6.0 | 32<br>5.7 | 30<br>5.4 | 28<br>5.2 | 26<br>4.8 | 23<br>4.5 | 20<br>4.0 | 17<br>3.5 | 14<br>3.0 | 11<br>2.5 | 4<br>1.1 | |
| 52<br>8.6 | 50<br>8.2 | 48<br>7.8 | 45<br>7.6 | 43<br>7.3 | 41<br>7.0 | 40<br>6.7 | 38<br>6.5 | 36<br>6.2 | 34<br>5.9 | 32<br>5.6 | 31<br>5.4 | 29<br>5.2 | 25<br>.47 | 23<br>4.3 | 20<br>3.9 | 17<br>3.4 | 14<br>2.9 | 8<br>1.7 | 2<br>0.4 |
| 53<br>8.7 | 51<br>8.3 | 49<br>7.9 | 47<br>7.7 | 45<br>7.4 | 43<br>7.2 | 41<br>6.8 | 40<br>6.6 | 38<br>6.3 | 36<br>6.1 | 34<br>5.8 | 33<br>5.6 | 31<br>5.4 | 28<br>5.0 | 25<br>4.6 | 22<br>4.2 | 19<br>3.7 | 17<br>3.3 | 17<br>2.3 | 10<br>1.1 |
| 55<br>8.7 | 53<br>8.3 | 51<br>8.0 | 48<br>7.7 | 47<br>7.5 | 45<br>7.2 | 43<br>7.0 | 41<br>6.7 | 39<br>6.5 | 38<br>6.2 | 36<br>6.0 | 35<br>5.8 | 33<br>5.5 | 30<br>5.2 | 27<br>4.8 | 24<br>4.4 | 22<br>4.0 | 19<br>3.6 | 13<br>2.7 | 8<br>1.6 |
| 56<br>8.7 | 54<br>8.3 | 52<br>8.0 | 50<br>7.8 | 48<br>7.6 | 47<br>7.3 | 45<br>7.0 | 43<br>6.8 | 41<br>6.6 | 40<br>6.4 | 38<br>6.1 | 37<br>5.9 | 35<br>5.6 | 32<br>5.3 | 29<br>4.9 | 26<br>4.6 | 24<br>4.2 | 21<br>3.8 | 15<br>3.0 | 10<br>2.0 |

Table 8.5

| Temperature, dry-bulb (°F) | 1 | 2 | 3 | 4 | 5 | 6 | 7 | 8 | 9 | 10 | 11 | 12 | 13 | 14 | 15 | 16 | 17 |
|---|---|---|---|---|---|---|---|---|---|---|---|---|---|---|---|---|---|
| 140 | 97 | 95<br>20.7 | 92<br>18.6 | 89<br>16.9 | 87<br>15.8 | 84<br>14.8 | 82<br>13.8 | 79<br>13.2 | 77<br>12.5 | 75<br>11.9 | 73<br>11.4 | 70<br>10.9 | 68<br>10.4 | 66<br>10.0 | 64<br>9.6 | 62<br>9.4 | 60<br>9.0 |
| 150 | 98 | 95<br>20.2 | 92<br>18.4 | 90<br>16.6 | 87<br>15.4 | 85<br>14.5 | 82<br>13.7 | 80<br>13.0 | 78<br>12.4 | 76<br>11.8 | 74<br>11.2 | 72<br>10.8 | 70<br>10.3 | 68<br>9.9 | 66<br>9.5 | 64<br>9.2 | 62<br>8.9 |
| 160 | 98 | 95<br>19.8 | 93<br>18.1 | 90<br>16.2 | 88<br>15.2 | 86<br>14.2 | 83<br>13.4 | 81<br>12.7 | 79<br>12.1 | 77<br>11.5 | 75<br>11.0 | 73<br>10.6 | 71<br>10.1 | 69<br>9.7 | 67<br>9.4 | 65<br>9.1 | 64<br>8.8 |
| 170 | 98 | 95<br>19.4 | 93<br>17.7 | 91<br>15.8 | 89<br>14.8 | 86<br>13.9 | 84<br>13.2 | 82<br>12.4 | 80<br>11.8 | 78<br>11.3 | 76<br>10.8 | 74<br>10.4 | 72<br>9.9 | 70<br>9.6 | 69<br>9.2 | 67<br>9.0 | 65<br>8.6 |
| 180 | 98 | 96<br>18.9 | 94<br>17.3 | 91<br>15.5 | 89<br>14.5 | 87<br>13.7 | 85<br>12.9 | 83<br>12.2 | 81<br>11.6 | 79<br>11.1 | 77<br>10.6 | 75<br>10.1 | 73<br>9.7 | 72<br>9.4 | 70<br>9.0 | 68<br>8.8 | 67<br>8.4 |
| 190 | 98 | 96<br>18.5 | 94<br>16.9 | 92<br>15.2 | 90<br>14.2 | 88<br>13.4 | 85<br>12.7 | 84<br>12.0 | 82<br>11.4 | 80<br>10.9 | 78<br>10.5 | 76<br>10.0 | 75<br>9.6 | 73<br>9.2 | 71<br>8.9 | 69<br>8.6 | 68<br>8.2 |
| 200 | 98 | 69<br>18.1 | 94<br>16.4 | 92<br>14.9 | 90<br>14.0 | 88<br>13.2 | 86<br>12.4 | 84<br>11.8 | 82<br>11.2 | 80<br>10.8 | 79<br>10.3 | 77<br>9.8 | 75<br>9.4 | 74<br>9.1 | 72<br>8.8 | 70<br>8.4 | 69<br>8.1 |
| 210 | 98 | 96<br>17.7 | 94<br>16.0 | 92<br>14.6 | 90<br>13.8 | 88<br>13.0 | 86<br>12.2 | 85<br>11.7 | 83<br>11.1 | 81<br>10.6 | 79<br>10.0 | 78<br>9.7 | 76<br>9.2 | 75<br>9.0 | 73<br>8.7 | 71<br>8.3 | 70<br>8.0 |

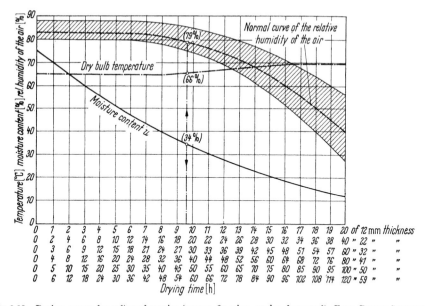

Fig. 8.28. Drying curves for softwood species (spruce, fir, pine, poplar, basswood). From BENNO SCHILDE Maschinenbau AG., Hersfeld, Germany

curves also indicate the drying times to be expected. A schedule used in the U.S.A. for 2 in. (51 mm) tupelowood is reproduced in Fig. 8.29.

Moisture content at each day during the run must be determined by the operator. The sample board method is most convenient. The sample board is about 2 ft. long and cut not closer than 2 ft. from the end of a board. Usually the

(Continued)

depression (°F)

| 18 | 19 | 20 | 21 | 22 | 23 | 24 | 25 | 26 | 27 | 28 | 29 | 30 | 32 | 34 | 36 | 38 | 40 | 45 | 50 |
|---|---|---|---|---|---|---|---|---|---|---|---|---|---|---|---|---|---|---|---|
| 58 | 56 | 54 | 53 | 51 | 49 | 47 | 46 | 44 | 43 | 41 | 40 | 38 | 35 | 32 | 30 | 27 | 25 | 19 | 14 |
| 8.7 | 8.4 | 8.0 | 7.8 | 7.6 | 7.3 | 7.1 | 6.9 | 6.6 | 6.4 | 6.2 | 6.0 | 5.8 | 5.4 | 5.1 | 4.8 | 4.4 | 4.1 | 3.4 | 2.6 |
| 60 | 58 | 57 | 55 | 53 | 51 | 49 | 48 | 46 | 45 | 43 | 42 | 41 | 38 | 36 | 33 | 30 | 28 | 23 | 18 |
| 8.6 | 8.3 | 8.0 | 7.8 | 7.5 | 7.3 | 7.1 | 6.9 | 6.7 | 6.4 | 6.2 | 6.0 | 5.8 | 5.4 | 5.2 | 4.9 | 4.5 | 4.2 | 3.6 | 2.9 |
| 62 | 60 | 58 | 57 | 55 | 53 | 52 | 50 | 49 | 47 | 46 | 44 | 43 | 41 | 38 | 35 | 33 | 31 | 25 | 21 |
| 8.5 | 8.2 | 7.9 | 7.7 | 7.4 | 7.2 | 7.0 | 6.8 | 6.7 | 6.4 | 6.2 | 6.0 | 5.8 | 5.5 | 5.2 | 4.9 | 4.6 | 4.3 | 3.7 | 3.2 |
| 63 | 62 | 60 | 59 | 57 | 55 | 53 | 52 | 51 | 49 | 48 | 47 | 45 | 43 | 40 | 38 | 35 | 33 | 28 | 24 |
| 8.4 | 8.0 | 7.8 | 7.6 | 7.3 | 7.2 | 6.9 | 6.7 | 6.6 | 6.4 | 6.2 | 6.0 | 5.7 | 5.5 | 5.2 | 4.9 | 4.6 | 4.4 | 3.7 | 3.2 |
| 65 | 63 | 62 | 60 | 58 | 57 | 55 | 54 | 52 | 51 | 50 | 48 | 47 | 45 | 42 | 40 | 38 | 35 | 30 | 26 |
| 8.1 | 7.8 | 7.6 | 7.4 | 7.2 | 7.0 | 6.8 | 6.5 | 6.4 | 6.2 | 6.0 | 5.8 | 5.7 | 5.4 | 5.2 | 4.8 | 4.6 | 4.4 | 3.8 | 3.3 |
| 66 | 65 | 63 | 62 | 60 | 58 | 57 | 56 | 54 | 53 | 51 | 50 | 49 | 46 | 44 | 42 | 39 | 37 | 32 | 28 |
| 7.9 | 7.7 | 7.4 | 7.2 | 7.0 | 6.8 | 6.6 | 6.4 | 6.2 | 6.0 | 5.9 | 5.7 | 5.5 | 5.3 | 5.0 | 4.8 | 4.5 | 4.4 | 3.8 | 3.3 |
| 67 | 66 | 64 | 63 | 61 | 60 | 58 | 57 | 55 | 54 | 53 | 52 | 51 | 48 | 46 | 43 | 41 | 39 | 34 | 30 |
| 7.7 | 7.5 | 7.2 | 7.0 | 6.9 | 6.6 | 6.4 | 6.2 | 6.0 | 5.9 | 5.7 | 5.6 | 5.4 | 5.2 | 4.9 | 4.7 | 4.5 | 4.3 | 3.8 | 3.3 |
| 68 | 67 | 65 | 64 | 63 | 61 | 60 | 59 | 57 | 56 | 54 | 53 | 52 | 50 | 47 | 45 | 43 | 41 | 36 | 32 |
| 7.6 | 7.4 | 7.1 | 6.9 | 6.8 | 6.5 | 6.3 | 6.1 | 5.9 | 5.8 | 5.5 | 5.4 | 5.3 | 5.1 | 4.8 | 4.6 | 4.4 | 4.2 | 3.7 | 3.2 |

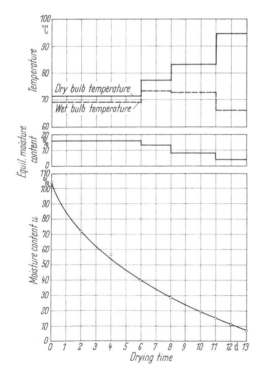

Fig. 8.29. Drying schedule for 51 mm black tupelo (*Nyssa aquatica* L.). From KIMBALL (1947)

same board is used on which the moisture content was determined before drying was started. Fig. 8.30 illustrates where test piece and sample board are taken. Both ends of the sample board are coated with a moisture proof paint and its weight is determined and marked. From the small test piece the moisture content is known and on this basis the oven-dry weight of the sample board may be calculated. Also this figure is marked on the sample board. Sample boards should be

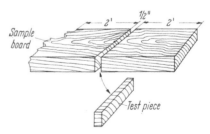

Fig. 8.30. Sample board and test piece for moisture content test. From HENDERSON (1951)

inserted in a lumber pile at positions where rapid, medium, and slow drying is expected. The sample boards are removed periodically from the kiln, weighed and subsequently put back into the pile. The moisture content, based on oven-dry weight, is calculated in the usual manner (cf. Section 6.2.2).

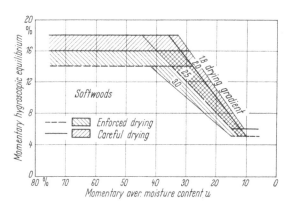

Fig. 8.31. Momentary hygroscopic equilibrium values to be set in the kiln for various drying gradients of softwoods. From KEYLWERTH (1951)

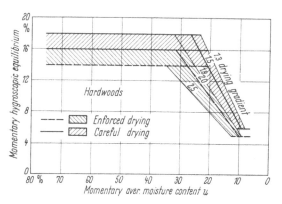

Fig. 8.32. Momentary hygroscopic equilibrium values to be set in the kiln for various drying gradients of hardwoods. From KEYLWERTH (1951)

Many drying schedules, still in use and recommended, are not in accordance with the recent results of drying physics. The drying process is controlled by the "drying-potential" or "drying gradient" expressed by the ratio of the momentary hygroscopic equilibrium $u_h$. This hygroscopic equilibrium for the temperature and the relative humidity of the steam-air mixture in the kiln may be taken from Fig. 6.33. If the drying gradient is too steep the differences of the moisture content in the shell and in the core become too great and severe stresses appear.

According to KEYLWERTH (1951) one can assume that for any drying process at any time the drying gradient is constant

$$\frac{u}{u_h} = \text{constant} \qquad (8.1)$$

A statistical evaluation of tried practical drying schedules has proved this fact. Of course, one has to distinguish between careful and euforced drying of softwoods and hardwoods. The diagrams in Fig. 8.31 and 8.32 may be used. Based on these diagrams, Table 8.6, for the control of the relative humidity of the air in kiln drying of lumber (up to 30 mm thickness) has been developed. For thicker boards a smaller drying gradient (1.6 to 2.0) should be chosen.

For the development of a complete drying schedule the course of the dry bulb temperatures must be known. In Table 8.6 appropriate values are listed. During the drying process it is necessary to determine the moisture content gradients and to check whether casehardening is present. If the moisture gradient is too steep the surface dries too rapidly; this is always the beginning of casehardening, checking and deformations. Average moisture, moisture distribution, and casehardening are tested on samples cut out of an average board from the lumber in the kiln; reference may be made to HENDERSON (1951) p. 142/143 and his scheme, Fig. 8.30. To find the moisture distribution sample $b$ in Fig. 8.30 should be cut as shown and the moisture content of shell and core should be de-

Table 8.6. *Recommended Values for the Control of the Relative Humidity in Kiln Drying Lumber up to 30 mm Thickness* (From KEYLWERTH, 1951)

| Step | Green wood Moisture content u % | | | |
|---|---|---|---|---|
| 1 | Beginning to 60 | Beginning to 50 | Beginning to 40 | Beginning to 35 |
| 2 | 60···50 | 50···40 | 40···35 | 35···30 |
| 3 | 50···40 | 40···35 | 35···30 | 30···25 |
| 4 | 40···35 | 35···30 | 30···25 | 25···20 |
| 5 | 35···30 | 30···25 | 25···20 | 20···15 |
| 6 | 30 up to the end | 25 up to the end | 20 up to the end | 15 up to the end |
| Species | Poplar, Willow, Aspen | Basswood, Walnut Alder, Locust (Robinia) | Oak, Beech, Maple, Birch, Elm | Larch, Douglas-fir, White pine, Scots pine, Spruce, Ash |
| Scale | A | B | C | D |

Air-dry wood

| Step | Moisture content u % | Psychrometric difference [wet bulb depression $(\vartheta_d - \vartheta_u)$ corresponding to drying gradient] | | | | | |
|---|---|---|---|---|---|---|---|
| | | 1.6 | 1.8 | 2.0 | 2.5 | 3.0 | 3.5 |
| 1 | Beginning to 30 | 2 | 2 | 2 | 4 | 6 | 8 |
| 2 | 30···25 | 2 | 3 | 4 | 6 | 8 | 11 |
| 3 | 25···20 | 3 | 4 | 6 | 9 | 12 | 15 |
| 4 | 20···15 | 6 | 8 | 10 | 13 | 17 | 20 |
| 5 | 15···10 | 12 | 14 | 16 | 20 | 25 | 25 |
| 6 | 10 up to the end | 25 | 25 | 25 | 25 | 25 | 25 |
| Species | Air-dry wood | Oak, Boxwood, Eucalypt | Oak, Beech, Eucalypt, Hickory, Ash | Hickory, Ash, Maple, Teak, Parana pine | Birch, Walnut, Mahogany, Locust (Robinia), Elm, Douglas-fir, Larch, White pine | Alder, Basswood, Poplar Willow, Fir, Scots pine, Spruce | Scots pine, Spruce |
| Scale | E | 1 | 2 | 3 | 4 | 5 | 6 |

termined separately. For casehardening tests so-called "fork-samples" (Fig. 8.33) are used. If the prongs remain straight no stress is present (Fig. 8.33a, c, and f). If the prongs turn out as shown in Fig. 8.33b, the surface (at the beginning of the drying process) is dry and tensile stresses are present, the core is still wet and did not shrink. The typical picture of casehardened wood is illustrated in Fig. 8.33d. If the lumber has been steamed too much (with the aim of removing casehardening) the prongs tend to turn out as in Fig. 8.33e.

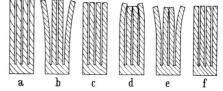

Fig. 8.33a—f. "Fork-samples" for case-hardening tests: a) green wood, b) beginning of the drying process (tensile stresses outward), c) stress equilibrium due to internal shrinkage, d) typical casehardening, e) stress reversion after steaming. f) free of stress dry wood

Casehardening during a drying process is a rather common event and can be easily relieved by steaming. The relative humidity should be drastically elevated. A good example is Henderson's kiln drying schedule for green two-inch-and-thicker beech (Fig. 8.34). Using this schedule, surface checking and splitting are eliminated and no honeycombing or internal stresses are developed. Many operators steam every day for a short period (30 to 60 min.). This intermittent steaming prevents casehardening.

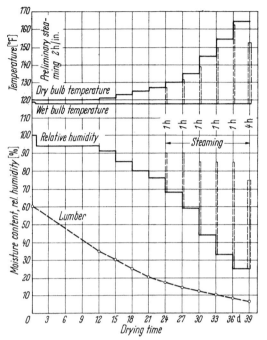

Fig. 8.34. Kiln drying schedule for green two-inch and thicker beech. From HENDERSON (1951)

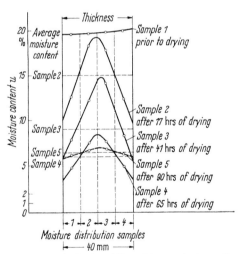

Fig. 8.35. Distribution of the moisture content over the cross-section of 40 mm alder after different drying times and after conditioning period. From SCHLÜTER and FESSEL (1939)

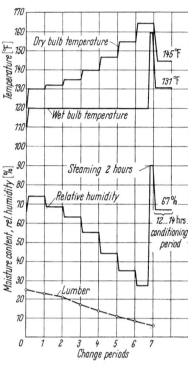

Fig. 8.36. Drying schedule for one-inch and two-inch hardwood with final steaming and conditioning periods. From HENDERSON (1951)

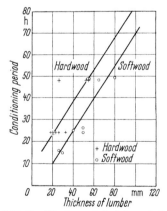

Fig. 8.37. Conditioning period at the end of the drying time in relation to the thickness of lumber. From SCHLÜTER and FESSEL (1939)

At the end of the drying process very frequently the moisture gradients are too steep or the lumber is dried to too low a moisture content (Fig. 8.35). Then final steaming and/or conditioning by high humidity treatment is necessary

Ref. p. 471]                    8.3 Kiln Drying                    449

to equalize the moisture content and to balance it on a reasonable level. An example is given in Fig. 8.36. SCHLÜTER and FESSEL (1939) found for the kiln drying of 20 mm thick hardwood boards in the hygroscopic range that for the equalizing of moisture content a period of about $^1/_4$ to $^1/_3$ of the pure drying time must be spent. From their measurement they also deduced a linear relationship between equalizing period and thickness of board (Fig. 8.37).

Why the whole drying process may be treated as a diffusion phenomenon is explained in Section 6.3.3 and, in Section 6.3.3.2, a method for an approximate calculation of the drying time is described. In detail, the following factors influence the drying time: wood species, initial moisture content, final moisture content, thickness of lumber, dry bulb temperature, wet bulb depression, air velocity, quality specifications, type of kiln.

Eq. (6.4.4) may be applied. For practical use Figs. 6.78 and 6.79 were developed. The curves are valid for a dry bulb temperature $\vartheta_d$ of 65 °C, for a thickness of lumber $s = = 25$ mm and for a normal fan-ventilated compartment kiln running 12 hours per day. As is indicated in the legend, Fig. 6.78 is applicable to softwood with an oven-dry density $\varrho_0 = 0.450$ g/cm³, whereas Fig. 6.79 is valid for hardwood with $\varrho_0 = 0.670$ g/cm³. For other drying conditions a correction factor (cf. Section 5.3.3.2) must be introduced. For convenience they are listed in Table 8.7:

Table 8.7. *Correction Factors for Determination of Drying Time* Softwood $\varrho_0 = 0.450$ g/cm³ (see Fig. 6.78)

| Density g/cm³ | Correction factor $f$ | Density g/cm³ | Correction factor $f$ |
|---|---|---|---|
| 0.350 | 0.686 | 0.700 | 1.940 |
| 0.375 | 0.761 | 0.725 | 2.045 |
| 0.400 | 0.838 | 0.750 | 2.151 |
| 0.425 | 0.918 | 0.775 | 2.260 |
| 0.450 | 0.918 | 0.825 | 2.482 |
| 0.475 | 1.804 | 0.850 | 2.596 |
| 0.500 | 1.172 | 0.875 | 2.711 |
| 0.525 | 1.261 | 0.900 | 2.828 |
| 0.550 | 1.351 | 0.950 | 3.067 |
| 0.575 | 1.444 | 1.000 | 3.312 |
| 0.600 | 1.539 | 1.100 | 3.822 |
| 0.625 | 1.636 | 1.200 | 4.354 |
| 0.650 | 1.735 | 1.300 | 4.910 |
| 0.675 | 1.837 | | |

Hardwood $\varrho_0 = 0.670$ g/cm³ (see Fig. 6.79)

| Density g/cm³ | Correction factor $f$ | Density g/cm³ | Correction factor $f$ |
|---|---|---|---|
| 0.350 | 0.395 | 0.700 | 1.118 |
| 0.375 | 0.438 | 0.725 | 1.178 |
| 0.400 | 0.482 | 0.750 | 1.240 |
| 0.425 | 0.529 | 0.775 | 1.302 |
| 0.450 | 0.576 | 0.800 | 1.365 |
| 0.475 | 0.624 | 0.825 | 1.430 |
| 0.500 | 0.675 | 0.850 | 1.496 |
| 0.525 | 0.726 | 0.875 | 1.563 |
| 0.550 | 0.778 | 0.900 | 1.630 |
| 0.575 | 0.832 | 0.950 | 1.767 |
| 0.600 | 0.887 | 1.000 | 1.912 |
| 0.625 | 0.942 | 1.100 | 2.200 |
| 0.650 | 1.000 | 1.200 | 2.500 |
| 0.675 | 1.058 | 1.300 | 2.830 |

| Temperature | | Thickness | | |
|---|---|---|---|---|
| °C | $f$ | mm | $f_1 = \left(\dfrac{s}{25}\right)^{1,5}$ | $f_2 = \left(\dfrac{s}{25}\right)^{1,25}$ |
| 50 | 1.300 | 15 | 0.465 | 0.528 |
| 55 | 1.180 | 20 | 0.716 | 0.757 |
| 60 | 1.083 | 25 | 1.000 | 1.000 |
| 65 | 1.000 | 30 | 1.315 | 1.256 |
| 70 | 0.928 | 35 | 1.656 | 1.523 |
| 75 | 0.867 | 40 | 2.023 | 1.799 |
| 80 | 0.812 | 50 | 2.828 | 2.379 |
| 85 | 0.765 | 60 | 3.718 | 2.987 |
| 90 | 0.722 | 70 | 4.685 | 3.623 |
| 95 | 0.684 | 80 | 5.720 | 4.280 |
| 100 | 0.650 | 90 | 6.830 | 4.959 |
| | | 100 | 8.000 | 5.657 |
| | | 120 | 10.500 | 7.103 |
| | | 140 | 13.300 | 8.615 |

Table 8.7. (Continued)

| Circulation | f | Operation time hours per day | Correction factor for calculating a daily operation time of 12 hrs |
|---|---|---|---|
| By gravity | 2.00 | 8 | 1.50 |
| by means of water sprayers | 1.80 | 10 | 1.20 |
| by means of normal fans or blowers | 1.20···1.50 | 12 | 1.00 |
| by means of internal fans of highest |  | 16 | 0.75 |
| efficiency and in the case of correct piling | 1.00 | 24 | 0.50 |

KEYLWERTH (1949) has traced a nomogram for a wide range of $\alpha$-values in Eq. (6.44), and for wide ranges of moisture content, board thickness and drybulb temperature (Fig. 8.38). The dotted line shows its use. As a rule the results of the approximation satisfy the needs of practice. Therefore, other statements by PECK, GRIFFITH and RAO (1952) and by KEYLWERTH (1953) are only mentioned.

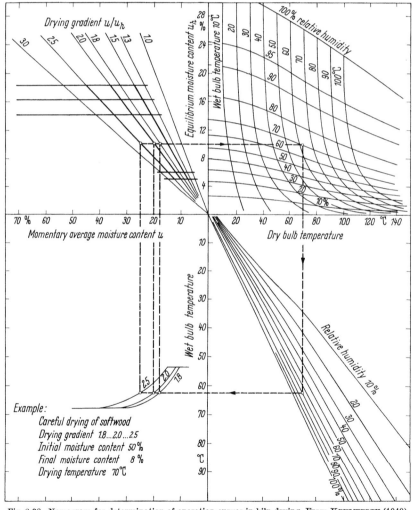

Fig. 8.38. Nomogram for determination of operation curves in kiln drying. From KEYLWERTH (1949)

## 8.3.2 Defects in Wood due to Kiln Drying

**8.3.2.0 General Considerations.** Often defects in wood occur even during air seasoning and they may be worse than those produced in kiln drying. Nevertheless, the types of seasoning and drying defects are all the same and, therefore, they are dealt with exclusively in this section. In modern dry kilns, properly operated, defects are nearly avoidable. The losses in value due to checking, deformations, and staining should not exceed 5%. One must know, of course, how defects occur, how they may be prevented beforehand, and how to remedy them after occurrence. Sections 3.2.2.1 through 3.2.2.7 should be referred to for a discussion of the features that characterize defects due to seasoning and for an explanation of their structural relationships.

**8.3.2.1 Staining.** Blue stain occurs frequently when certain coniferous species are air-seasoned slowly in a humid warm atmosphere. It may be prevented by dipping in chemicals or by kiln drying immediately after sawing. Initial steaming at a minimum temperature of 70 °C is advisable. After drying the lumber should be cooled down in the kiln with fans running. New investigations made by ELLWOOD and ERICKSON (1962) showed that presteaming of stain-susceptible green redwood at 100 °C is the most effective of the methods tried for reducing the amount of subsequent seasoning stain and the exudation of extractives. The recommended duration of presteaming is 4 hours.

Mildew and mold may occur if the circulation is poor or if the temperature is too low and the relative humidity is too high. The air speed should be increased and/or the drying schedule must be improved. The occurrence of brown stain in kiln drying is related to excessively high temperatures. KOLLMANN and MALMQUIST (1952) measured the "relative whiteness" of pine lumber dried with various

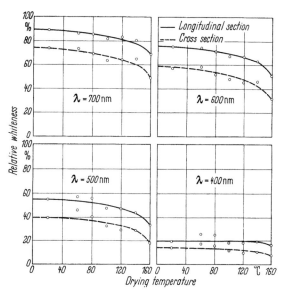

Fig. 8.39. Relative whiteness of pine heartwood after 5 hrs. drying with various temperatures measured by means of a photospectrograph with monochromatic light of the indicated wave lengths. From KOLLMANN and MALMQUIST (1952)

dry-bulb temperatures by means of a photospectrograph. Fig. 8.39 shows some results. The "relative whiteness" of surfaces along the grain is always much higher than that of surfaces across the grain. The influence of moisture content is also important. It may be neglected only between zero and 15%, it is remarkable between 15 and 40%, and it is increased above 40% moisture content. KOLLMANN, KEYLWERTH and KÜBLER (1951) designed an instructive diagram for the loss in brightness of hardwoods treated with various temperatures and various relative humidities over a period of 20 hours (Fig. 8.40). It becomes evident that the influence of the heat is much more pronounced than that of the moisture. Dark-blue stain occurs if water contaminated with iron-salts drips on the surface of wood species which contain a high amount of tannin. Leaky steam coils may be the reason.

**8.3.2.2 Deformations (Warping, Twisting, Cupping).** Woods with spiral and interlocking grain (e.g. spruce, beech, and elms) have a natural tendency to warp and cup. The deformations result from anisotropic shrinkage and are intensified by casehardening and improper piling. Plain sawn boards shrink more in width than quarter-sawn boards. Boards cut from near the pith of the log warp more than side-boards (HENDERSON, 1951, p. 125).

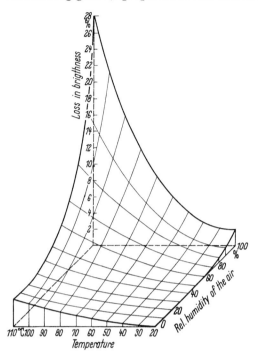

Fig. 8.40. Average loss in brightness of air-dry hardwoods treated with various temperatures and changing relative humidities over a period of 20 hours. From KOLLMANN, KEYLWERTH and KÜBLER (1951)

To prevent warping, twisting, and cupping, careful piling (uniform stickers, adequate in size and number, and evenly spaced) is necessary. Casehardening should be avoided or relieved by steam treatment. Experiments have been made by STEVENS (1961) to find out to what extent deformations can be reduced by imposing restraint on the wood as it is drying. The results show that, if a sample is held flat and straight during drying, some distortion will occur immediately upon removing the restraint, and for some time thereafter a gradual increase in the distortion will take place before a steady shape is finally reached. The ultimate distortion is, however, much smaller than that occurring on matched material dried without any restraint. It was found that restraint was more effective on green than on partially dry wood, and more effective at high temperatures than at low ones. It also appeared that cross-grain distortion could be reduced to a greater extent than longitudinal distortion.

**8.3.2.3 Casehardening.** The causes of casehardening are too rapid or uneven drying. Too rapid surface drying occurs if, in the kiln, the temperature is too high, the relative humidity is too low or the fluctuations of both are too large. In this case schedules with lower temperature and higher humidities must be applied. Temperature and humidity should be controlled carefully in order to keep them uniform. Another remedy against uneven drying is more rapid and even circulation. Frequent tests for casehardening (cf. Fig. 8.33) should be made and intermittent steaming is recommended.

Casehardening is accompanied by surface checking, warping and twisting. Casehardening occurs especially in drying very wet and very dense woods. If a board has an initial moisture content of 50% and is put into a dry kiln where the wood soon reaches an equilibrium moisture content of only 10%, the outer shell is quickly overdried and shrunk before the core can lose moisture to any significant extent. Consequently severe tension stresses are set up in the shell.

If a board in this condition is resawn, cupping as shown in Fig. 8.41a, reflects an early state of casehardening. If the drying schedule is not changed, the core slowly approaches the fiber saturation point and the moisture content soon decreases below it. Then the core begins to shrink. The result is that the tension stresses in the shell are lowered and after some time a stress-free state is reached. From this point on the main danger of defects starts. The overdried shell is stiff and brash and can no longer follow the normal laws of shrinkage. The result is that the shrinking core produces compressive stresses in the shell (Fig. 8.41b). Resawn boards behave like the two parts in Fig. 8.41b on the right side. This phenomenon is very severe since the drying process is interrupted, defects occur and reconditioning is not easy. Surface checks give evidence of casehardening. Casehardening, if not relieved by steaming, may lead to honeycombing. Hard and refractory species of high density with distinct rays, such as oak, hickory, and locust, tend to honeycomb (Fig. 8.42). KOLLMANN (1950) showed that there is a linear relationship between the relative portion of internal checks — determined by a microscopic linear estimation — and the dry-bulb temperature (Fig. 8.43).

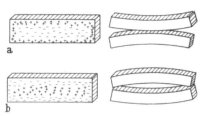

Fig. 8.41 a, b. Casehardening specimens: a) early drying state (outside tension stresses, inside compressive stresses; b) begin of casehardening (outside compressive stresses, inside tension stresses)

Fig. 8.42. Cross-section of a badly dried oakwood post with great internal checks (honeycomb)

**8.3.2.4 Collapse.** In kiln drying certain woods from the green condition the phenomenon of collapse occurs. The heartwood tends to collapse more than the sapwood. Too rapid surface drying during the early stages of the drying, the application of too high drying temperatures, severe casehardening, excessive warping and honeycombing normally accompany collapse. The sinking-in of spots is typical. The surfaces of the boards may appear corrugated. The heartwood of oak may be severely damaged by col-

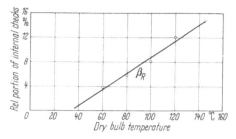

Fig. 8.43. Relationship between the relative portion of internal checks and the dry bulb temperature for oak
From KOLLMANN (1950)

lapse when kiln-dried at temperatures of 50 °C and more, but it will remain free of this defect when air-dried in an open shed. Unlike shrinkage, collapse occurs much above the fiber saturation point. Apparently the cells must be completely filled with water. It is theorized (though not yet proved by experiments) that under some circumstances (impervious cell walls and water free from air bubbles) tension stresses may be set up by capillary forces running up to hundreds of atmospheres (TIEMANN, 1951, p. 176), exceeding the crushing strength of

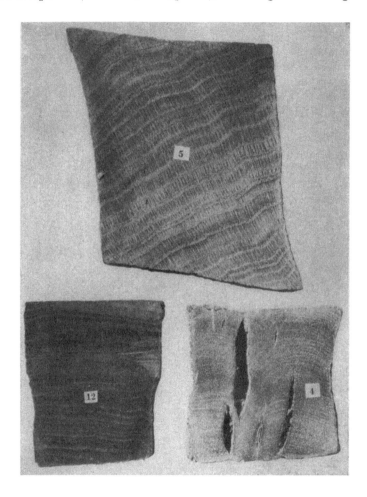

Fig. 8.44. Cross-sections through three squares of eucalyptus. Sample 5 is badly deformed by collapse without interior checks; sample 4 shows honeycomb and collapse; sample 12 shows in the lower part of the cross-section, consisting of heartwood, a severe collapse, whereas the upper part, consisting of sapwood, has normally shrunk (Phot. U. S. Forest Products Laboratory, Madison, Wisc.)

the cells. It is possible that two or more systems of internal forces produce collapse. Considerations on the liquid tension collapse theory suggest that lowering the surface tension of the drying liquid in wood will reduce the collapse. Recent experiments made by ELLWOOD and co-workers (1963) according to which water in wood was replaced by various organic liquids and then dried proved that collapse can be prevented by certain liquids (ethylene glycol monoethyl ether, methanol, ethanol, n-propyl alcohol) even under very severe drying conditions. Organic liquids, however, which dissociate sufficiently to cause chemical deterioration of wood by acidic or basic reactions cause abnormally high collapse, irrespective of surface tension.

As a rule collapse is rare in sapwood since its cell walls are not impervious; they have larger pores or interstices than those in heartwood. In larger pores the tension required for collapse cannot be produced [cf. Eq (6.22)]. In Fig. 8.44

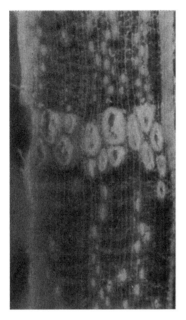
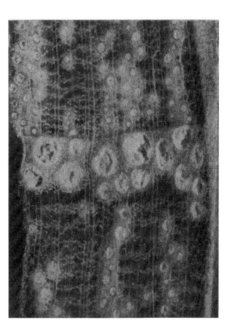

Fig. 8.45. Photomicrograph of a collapsed cross-section of eucalyptus (Phot. U.S. Forest Products Laboratory, Madison, Wisc.)

Fig. 8.46. Photomicrograph of the same piece of eucalyptus, as in Fig. 8.45., after steaming (reconditiong). (Phot. U.S. Forest Products Laboratory, Madison, Wisc.)

cross sections through three squares of eucalyptus are illustrated. Sample 5 is badly deformed without interior checks. Sample 4 shows honeycomb and collapse. In sample 12 the lower part of the cross section, consisting of heartwood, exhibits severe collapse, whereas the upper part, consisting of sapwood, shrank normally. The behavior of individual cells in collapsing varies. Fig. 8.45 shows a photomicrograph of a collapsed cross section of eucalyptus. The portions of the wood fibers appear as continuous black spots evidencing complete collapse. The oval white spots are the vessels hardly affected by the collapse. Fig. 8.46 is a photomicrograph of the same piece of eucalyptus wood after restoration (reconditioning).

The effect of the capillary tension forces is aggravated by an increased plasticity of cell walls. It is generally known that some species tend more to plasticity than others. The content of tannins and of substances formed in heartwood may affect plasticity. During collapse the diameters of the cell cavities become smaller whereas they remain nearly constant in shrinking. The reduction of the cell cavities causes unusual high areal shrinkage. KEYLWERTH (1951) observed for normal beechwood, air-dried and then oven-dried an areal shrinkage of 15.7%, for collapsed beechwood 31.0%, and for collapsed oakwood 36.2% (Fig. 8.47).

A complete recovery from collapse may be impossible if the cells are too greatly distorted. This also is the case when honeycombing has occurred. But the majority of collapsed wood items may be restored to their original uncollapsed condition by steaming at temperatures above 85 °C and preferably at the boiling point of water over periods ranging from 4 to 8 hours. The wood adsorbs moisture by this treatment and its cell walls swell. This swelling restores the normal cross sec-

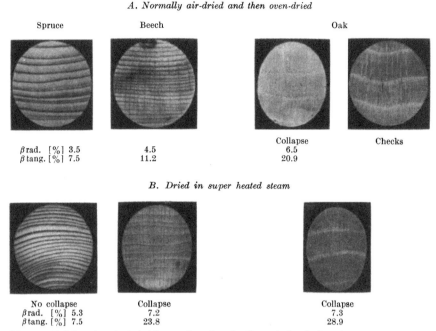

Fig. 8.47. Shrinkage (maximal shrinkage values $\beta$) and collapse during drying of green wood samples. From KEYLWERTH (1951)

tional shapes of the cells. Since the moisture content of the wood increases about 2 to 6%, subsequent to the steaming a careful re-drying at lower temperatures and higher relative humidities must be practiced. A second collapse during the re-drying is not possible for two reasons: 1. Wood cell walls once dried are more pervious; 2. The cells do not become completely refilled with liquid water by the steaming (TIEMANN, 1951, p. 180).

### 8.3.3 Types of Kilns and Instruments

A book devoted to the principles of wood science and technology cannot include descriptions of kilns, machines and other technical equipment in detail. However, at least a short survey of this subject should be given. According to the manner of handling the charge, there are two main types of kilns in use: compartment and progressive.

A compartment kiln (also termed box kiln or charge kiln) is a building which is completely filled with lumber at one operation. After closing the doors the drying process starts. The lumber remains stationary. Temperature and relative humidity in the kiln are varied according to the schedule. In a compartment kiln the same drying conditions are held simultaneously in every part of the kiln. When the desired moisture content is reached, the kiln is turned off, the stock allowed to cool, and then the entire charge is removed. Compartment kilns are built in different sizes and are almost universally used at sawmills and woodworking industries. The main advantages of compartment kilns are as follows:

1. Proper control of drying conditions;
2. Usable for day-time drying only, if operation is impossible at night;
3. Possibility of steaming at any time;
4. Low losses of heat during the drying since the kiln doors are kept closed, throughout the process;
5. As a rule, fast and uniform drying.
6. Compartment kilns are usually more expensive than progressive kilns and need intelligent operation.

In a progressive or tunnel-type kiln, the lumber to be dried enters at the cool, humid "green" or "loading" end and is moved progressively each day along the kiln. Each pile on a track is subjected at each new position to more severe drying conditions and finally reaches the hot, dry unloading end of the kiln. Progressive kilns are usually long (up to 75 m); they are mostly used at sawmills cutting softwoods. Their main advantages are as follows:

1. Cheap construction and equipment;
2. Easy operation;
3. Steady output.

There are the following distinct disadvantages:

1. Suitable only for large production;
2. Inflexibility in use;
3. Necessity to run them day and night;
4. Severe influence of outside weather conditions;
5. Waste of heat during loading and unloading.
6. Difficulties in relieving casehardening.

A survey of the more common types of kilns with particular reference to compartment kilns is given in Table 8.8 (VAROSSIEAU, 1954; as source of data ANG-21 is quoted; kiln certification, 1946). Most modern kilns are constructed of hollow tile. Air circulation in a compartment kiln may be produced by one of the four general mechanisms: natural draft, external-blowers, internal-blowers, and internal-fans. In the USA condenser and water spray kilns were once used. Circulation was produced by the cooling effect of the condenser combined with the heating effect of the steam coils or by the water sprays. These types of kilns have mostly been replaced by compartment kilns. The drying in a natural draft or natural circulation kiln is markedly slower than in a forced-circulation kiln. The experts agree that the most suitable type of kiln is a compartment kiln of the internal fan type. It secures faster and more uniform drying than other types.

It is important to save time in kiln drying. From this viewpoint it is more economical to pile the lumber outside of the kiln on trucks or pallets and to push them into the kiln than to pile the lumber in the kiln. In this connection straddle carriers or lift trucks permit mechanical handling.

The fans or mechanical blowers are mostly operated directly by electric motors. In empty blower kilns the air is changed 100 to 400 times an hour, whereas in empty fan kilns the corresponding figure is between 600 and 900 changes per hour. The amount of power $P$ required to drive the fans can be calculated with the following formula

$$P = \frac{Q \cdot p_t}{75\eta} \ [\text{hp}] \qquad (8.2)$$

Table 8.8. *Conspectus of the More Common Types of Kilns and Dryers with Particular Reference to Compartment Kilns. The Classification is Based on the Construction and Operation of Kilns* (From VAROSSIEAU, 1954)

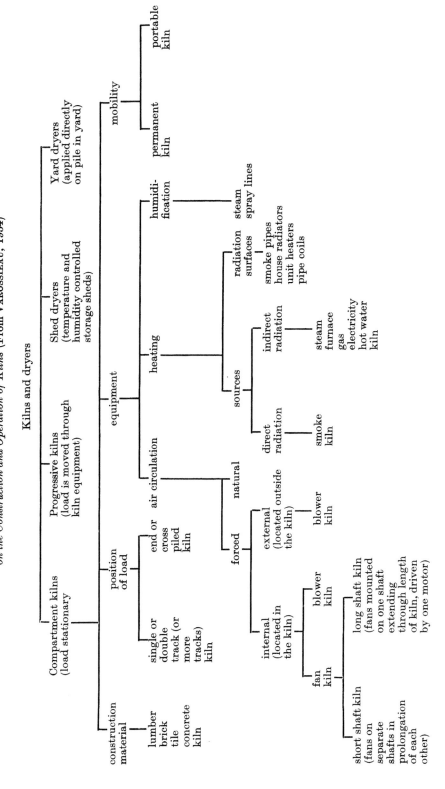

where $Q$ = volume of air circulation in m³/s, $p_t$ = total pressure (= difference of pressure between entry and exit nozzle of the blower) in mm water column = kp/m², and $\eta$ = efficiency (between 0.5 and 0.8).

The volume of air circulation may be measured using anemometers. The quantity of steam consumed is very important where live steam is used. The total consumption of heat consists of the following terms:

1. Amount of heat required to preheat walls, doors, trucks, stickers, and dry lumber;
2. Heating of air;
3. Evaporation (during winter season preheating and melting of ice in the wood;
4. Heat losses.

Table 8.9 lists a few figures for steam consumption in dry kilns under various conditions.

Table 8.9. *Steam Consumption in Dry Kilns*
(From KOLLMANN, 1935)

| Kind of lumber | Oak planks | | Spruce boards | |
|---|---|---|---|---|
| Season | Winter | Summer | Winter | Summer |
| Consumption of steam: kg per kg water evaporated | 2.15 | 1.66 | 1.98 | 1.59 |
| per 1 h of operation | 108.80 | 83.50 | 242.50 | 194.20 |
| Thermal efficiency % | 0.53 | 0.69 | 0.58 | 0.72 |

The thermal efficiency becomes lower with decreasing initial moisture content. The heat economy may be improved considerably by the installation of heat regeneration plants. LÖFGREN (1947) in Sweden calculated the steam consumption

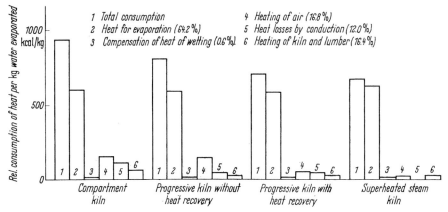

Fig. 8.48. Relative consumption of heat per kg water evaporated during summer season for various kiln types. From ANDERSSON (1952)

in drying softwood lumber without heat regeneration at 1.62 kg per kg water evaporated, with heat regeneration at 1.14 kg per kg water evaporated. ANDERSSON (1952) in Sweden calculated the relative consumption of heat for the drying of 25 mm thick pine boards from 90% to 12% moisture content in summer and winter season, respectively, and for various kiln types. His results are illustrated by Figs. 8.48 and 8.49. If electricity is the source of heat the consumption of energy per kg water evaporated lies between 0.8 and 2 kWh for coniferous species, and between 1 and 4 kWh for broad-leaved species.

Either indicating or recording thermometers are employed for the measurements of temperatures in dry kilns. Thermostatic control of temperature has come more and more into general use. There are three types of regulators: the air-operated instruments, the self-contained, and the electrically controlled. The air-operated type is most commonly used; it consists of a sensitive tube system which opens or closes a diaphragm valve in the steam line by means of compressed air. The system operates satisfactorily, but does not react instantaneously to

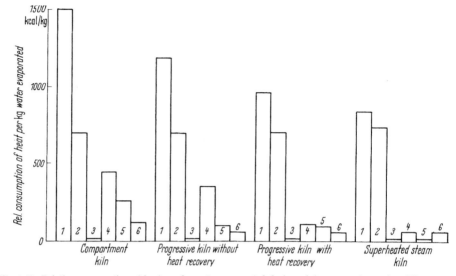

Fig. 8.49. Relative consumption of heat per kg water evaporated during winter season for various kiln types. From ANDERSSON (1952)

temperature changes (VAROSSIEAU, 1954, p. 108). If humidity is controlled, another system with a wet-bulb thermometer operates similarly on the steam spray line.

The self-contained temperature regulator does not need auxiliary power. The regulator opens and closes the diaphragm valve by the pressure exerted in the sensitive tube system. These instruments are less sensitive than the air operated types, but for many purposes they provide adequate control.

Electric instruments record and control dry and wet-bulb temperatures by operating motor valves on the steam supply lines to the heating coils and steam sprays (HENDERSON, 1951, p. 203). Reference may be made to KIMBALL (1947).

## 8.4 Special Seasoning Methods[1]

### 8.4.1 High Temperature Drying

Lumber can be dried very quickly in an atmosphere of superheated steam. As early as 1867 the U.S. Patent No. 64398, "Apparatus for Drying and Seasoning Lumber by Superheated Steam" was granted to ALLEN and CAMPBELL. In 1908 UPHUS and SHAPMAN applied for a patent to dry and treat structural lumber with temperatures up to 163 °C, and TIEMANN used the principle of high temperature drying in a patented superheated-steam kiln in which circulation was

---

[1] Kiln drying of veneers is dealt with in volume II: Wood-base materials.

forced by means of four pairs of steam-spray lines and could be reversed. It is reported that during World War II several superheated-steam kilns were used on the West Coast of the U.S.A. and that very rapid drying rates were achieved (GUERNSEY, 1957; U.S. Forest Products Laboratory, 1957). The reduction in drying time was quite noticeable (25 mm green softwoods were dried to 10% moisture content in 24 hours at drying temperatures of 110 °C (230 °F) in steam atmosphere). However, the severe drying conditions, on the one hand, caused

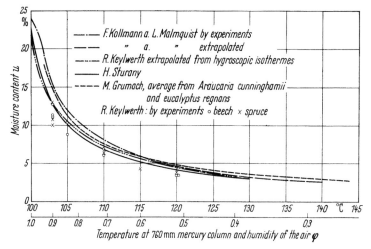

Fig. 8.50. Hygroscopic equilibrium of wood in superheated steam at atmospheric pressure

collapse in some softwoods and in most hardwoods and, on the other hand, the kilns in normal construction deteriorated rapidly. For these reasons the use of superheated-steam kilns was discontinued in the United States.

In Germany, shortly prior to World War II, ZILLNER (1938), and SCHNEIDER (1938), studied the drying of lumber in electrically heated small kilns with temperatures up to 130 °C. Since 1950 considerable research and development work has been devoted to high temperature drying.

Basically one has to distinguish between the following possibilities: if the dry-bulb temperature in a kiln is above 100 °C (212 °F) "high-temperature" drying is always applied. If the wet-bulb temperature in the kiln is below 100 °C (212 °F) the process may be called "superheated-vapor" drying, whereas if the wet-bulb temperature is 100 °C the kiln may be called a "superheated-steam" kiln. In a superheated steam-kiln no air is present. The most essential prerequisite for high temperature drying was the determination of the equilibrium moisture content of wood exposed at atmospheric pressure to superheated steam at temperatures up to 145 °C (293 °F). Fig. 8.50 shows experimental results, obtained by KEYLWERTH (1949), GRUMACH (1951), KOLLMANN and MALMQUIST (1952), and STURANY (1952). EISENMANN (1950) has extrapolated values for pressures up to 3.5 atmospheres (51.4 lb/sq. in. absolute). The U.S. Forest Products Laboratory states (1957) that they prepared in 1926 by mathematical extrapolation, the curves of the hygroscopic equilibrium of wood in superheated steam at atmospheric pressure and that their estimations for temperatures at least up to 120 °C (248 °F) were in good agreement with the values published later on.

Fig. 8.50 is the only tool which an operator of a high-temperature kiln needs. He can control the process by merely regulating the temperature. KOLLMANN

(1952) found that drying by superheated steam is suitable for softwoods and most air-dry hardwoods except oak, which has the inherent tendency to honeycomb and to collapse at temperatures above 65 °C (149 °F). KOLLMANN (1952) and EGNER (1951) reported that a temperature range of 110 °C to 115 °C (230 °F to 239 °F) is most expedient.

STEVENS and PRATT (1954), LADELL (1954, 1956, 1957), and SALAMON (1960) suggested kiln schedules for the high temperature drying of various species and sizes of lumber. They emphasized the importance of careful control of the drying conditions.

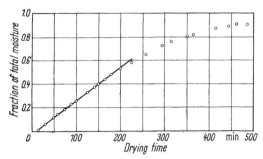

Fig. 8.51. Typical drying rate curve for yellow-poplar specimen dried at 150 °C. Solid line shows that first stage is a linear function of time. From HANN (1964)

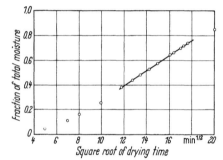

Fig. 8.52. Data in Fig. 8.51. plotted over the square root of drying time. Solid line shows second stage as a linear function. From HANN (1964)

CZEPEK (1952) discussed theory and practice of high-temperature drying and FLEISCHER (1953) investigated the drying rates of thin sections of wood at high temperatures. CALVERT (1958) presented results from the practical high-temperature drying of North American species. The physical fundamentals of high-temperature drying were carefully examined by STURANY (1952), KEYLWERTH and KÜBLER (1954), KOLLMANN and SCHNEIDER (1961), and recently by HANN (1964).

Fig. 8.51 shows a typical drying rate curve for yellow-poplar species dried by HANN (1964) at 150 °C. The linear portion of the drying curve is considered as the constant-rate drying condition. FLEISCHER (1953) and KOLLMANN and SCHNEIDER (1961) pointed out that during this period the drying rate is controlled by the rate of heat transfer to the wood surface. During this first stage of drying the temperature of the entire piece of wood remains at 160 °C, indicating that the wood is drying by vaporization of liquid water at the surface. The steady state drying constant $K_1$ can be calculated from the slope of the linear portion of the drying curve as follows:

$$K_1 = \frac{dw}{dt}\frac{1}{F_a} \qquad (8.3)$$

where $w$ = weight of water lost in g, $t$ = drying time in min., $F_a$ = drying surface area in cm².

When the drying of the free water has ended, the second stage drying moves from the surfaces to the center. The temperature generally remains at or near 100 °C in the center of the wood. The drying rate is a linear function of the square root of time (Fig. 8.52) indicating that the drying rate is controlled by the rate of heat transfer through the wood. At the end of the second stage in HANN's (1964) experiments the moisture content of the specimen ranged from around 6%

at $232\,°C$ to around $13\%$ at $121\,°C$. Heat transfer and not moisture movement is the limiting factor in this process. In the third stage of drying the wood in the center dries from the fiber saturation point down to the equilibrium moisture content for the drying conditions. The drying becomes slower and slower, the temperature gradients decrease, finally the drying stops and the field of temperatures is uniform. Since the rates of drying are higher than those attainable in conventional kilns, rapid transport of moisture away from the pile must be secured. Therefore the rates of circulation in high-temperature drying are higher (from 2 to 10 m/s corresponding to 394 to 1,970 ft. per minute) than in the usual kilns. The fans run throughout the drying process (normally up to about 24 hours). Only the dry-bulb temperature is (e.g. thermostatically) controlled and recorded.

Modern high-temperature kilns have a welded construction consisting of an aluminum lining, a 125 mm (5 in.) thick insulating layer (glass wool or rock wool), and a sheet-steel shell. Such kilns are vaportight and provide protection against high heat losses and corrosion. As a rule, they have a capacity of 1 to 8 m³. The pressure in the kiln is equivalent to the barometric pressure. Due to the lower heat losses, the lower heat capacity of the construction materials, and the shorter drying time the energy required with a superheated vapor kiln ranges only from 0.8 to 1.6 kWh per kg of water evaporated; with low temperature drying the corresponding range is 2 to 4 kWh/kg. The drying time $t$ (without heating, conditioning, and cooling) may be computed with the following formula (KEYL-WERTH and KÜBLER, 1952):

$$t = \frac{k \cdot s^{1.6}}{(\vartheta - 100)} \log \frac{u_b}{u_e} \tag{8.4}$$

where $s$ = thickness of the lumber in mm, $\vartheta$ = average temperature of the superheated steam, $u_b$ = initial moisture content, $u_e$ = final moisture content.

We can summarize the advantages of high-temperature drying and its effects on wood properties as follows:

1. Very short drying times and, therefore, very high output. On the average a high-temperature kiln delivers five-to-ten-times as much as a conventional kiln of the same capacity:

2. Small demand of space for the prefabricated kiln, no building activity at the woodworking plant;

3. Movability of the kilns;

4. Excellent heat economics;

5. Easy operation by far-reaching automation;

6. Reduced hygroscopic capacity (cf. Fig. 6.42) and reduced shrinkage values (70 to 80% as much as air-dry woods);

7. Slight increase of modulus of elasticity, maximum crushing strength, and modulus of rupture.

The disadvantages are less important:

1. Slight chemical brown discoloration (cf. Fig. 8.39 and 8.40);
2. Exudation of resin;
3. Loosening of knots;
4. Somewhat lower values in tension perpendicular to the grain, and in impact bending;
5. Relatively steep moisture gradients at the end of the drying process;
6. Early deterioration of kilns of unsuitable construction (of steel or of masonry).

### 8.4.2 Drying by Boiling in Oily Liquids

Wood can be dried rapidly by dipping into a water-repelling liquid which has a boiling point considerably above that of water and which is maintained at a temperature high enough to vaporize the water. Probably the first process using this method was the Boulton process patented in England in 1879, and in the U.S.A. in 1881. Water removal in hot oil is speeded up by application of a vacuum. Another patent on timber drying as the first step in treating it with hot preservatives was granted to CURTIS and ISAACS in 1895. A German patent for rapid wood drying by submerging in water-repelling liquids or molten substances was granted to ZUHLSDORFF in 1908. The drying conditions of boiling in oil are very severe; therefore, casehardening and checking are frequent phenomena. The process is limited, except in conditioning lumber for preservation.

The water in the wood migrates to the surface, is turned into steam and a thin film of nearly saturated water vapor protects the surface of the wood. When all free water is removed, diffusion starts transporting the water contained in the cell

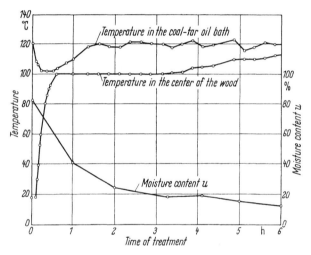

Fig. 8.53. Curves of temperature and moisture content of wood during drying in an oil bath. From KEYLWERTH (1950)

walls to the outside. The movement of water becomes very slow and the outer vapor layer becomes very thin and gradually loses its protective effect. Then the surface of the wood is subjected to the hot liquid and begins to check; also honeycombing may occur, especially if the woods are refractory.

Lumber dried by boiling in oily liquids is always casehardened. It is impossible to relieve the stresses set up by the casehardening during the boiling process. Therefore, subsequent conditioning treatment in a kiln is necessary. The uniformity of drying is impeded by the following factors: 1. Variations in moisture of the green wood; 2. Uneven heating of the oil; 3. Hydrostatic pressure on the lower parts of the pile in the tank, retarding the evaporation of the water.

The U.S. Forest Products Laboratory (1956) reports that 4 in. × 8 in. green southern pine lumber was dried by the boiling-in-oil-process from an average moisture content of 77% to 22% in 16 hours. The final moisture contents ranged from 11 to 34%. Casehardening was severe. The wood retained oil equivalent to 4.1% of the oven-dry weight of the wood.

Ref. p. 471]                    8.4 Special Seasoning Methods                    465

The physical events in drying in hot oil are similar to those in high temperature drying. KEYLWERTH (1950) published characteristic curves for the temperatures in the oil bath and in the center of the wood (Fig. 8.53). There are three stages: 1. Period of heating up, 2. Period of evaporation of the free water ($\vartheta \approx 100\,^{\circ}$C) and, 3. Period of increasing evaporation of the bound water in the cell walls.

Most commercial woods of the U.S.A. are not able to withstand the severe drying conditions without checking or warping unless a vacuum is used. A variation of the boiling-in-oil method is the McDonald process. Perchlorethylene is used as the oily liquid. When the liquid is heated, a vapor mixture or azeotrope of water and the solvent boils off. The liquids are not miscible so they can be separated in a condenser. The solvent is returned to the kiln or to storage. The azeotropic mixture boils slightly below the boiling point of water, thus creating less severe drying conditions.

Many oily liquids suitable for the boiling methods are inflammable. Some of them with low flash points should be avoided entirely. Fumes should be removed with wet steam or an inert gas, not with air. Pervious woods that can be dried without defects by boiling in oil can also be dried rapidly by conventional means.

### 8.4.3 Solvent Seasoning

Solvent seasoning is related to the drying by boiling in oil, to the McDonald process (described in Section 8.4.2) and to distillation methods of determining the moisture content of wood (cf. Section 6.2.2.2). Wood can be dried by heating in a hot solvent, maintained at a temperature above the boiling point of water. Solvents that are miscible with water, even used cold or at temperatures below the boiling point of water, extract water as well as extractives from the wet wood.

This principle is used by the Western Pine Association Research Laboratory according to a patent granted to STAMM (1933). STAMM originally used molten paraffin but the Western Pine Association continuously sprayed hot acetone on the lumber which stood on end in a closed extractor. The acetone, containing both water and the pitch from the wood, recirculates to the sprayer. When the mixture contains too much water it is recovered by distillation. The final step in the drying process is as follows: 1. Shutting off the spray; 2. Ventilation of the lumber by hot air to complete the drying and to remove the acetone; 3. Cooling by introduction of an inert gas.

The Western Pine Association estimated that ponderosa pine can be solvent seasoned in 0.25 to 0.50 the time required for kiln drying. The value of extracted "pitchy" pine is increased and the process yields extractives (e.g. about 60 lb. per 1000 bd.ft., equivalent to 12 kg/m³). Laboratory experiments carried out by ANDERSON and FEARING (1960) showed that 25 mm (4/4'') tanoak sapwood can be dried to 10% moisture content in less than 30 hours — in contrast to the several months required by other seasoning methods — and the lumber was relatively free from seasoning defects.

New investigations by SETH and co-workers (1965) indicated that the drying schedule for 25 mm (4/4'') heavy segregated redwood could be reduced to less than 113 hours by using methanol. Preliminary economic appraisals appear to be promising. The solvent drying process render available, for the first time, various silvichemicals from lumber as a by-product. The yield of extractives recovered from redwood varied from 100 to 300 lb. per 1000 board feet (20 to 60 kg/m³). The extract consists mainly of tannin, other polyphenols, and cyclitols.

30  Kollmann/Côté, Solid Wood

### 8.4.4 Vapor Drying

Wood can be dried by exposing it to the vapors of a boiling organic liquid. ROBBINS (1865), some other American inventors (CRESSOW, VORHEES and CUSTIS, and FIELDER), BESEMFELDER (1910), and WILLIAMSON (1937) used this principle without success for drying and preservation of fibrous materials or for pitch elimination. About 1942, HUDSON, of the Taylor-Colquitt Co., developed a practical vapor-drying process for ties and poles. HUDSON (1947) and Boracure Ltd. in New Zealand (1949) published detailed information.

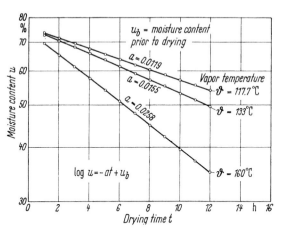

Fig. 8.54. Relationship between moisture content and drying time for vapor drying. From TAYLOR-COLQUITT Co.

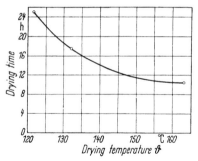

Fig. 8.55. Relationship between drying time and drying temperature for vapor drying. From TAYLOR-COLQUITT Co.

The wood is heated to approximately 150 °C (300 °F) in an ordinary pressure-treating cylinder with hot vapors of an organic chemical, such as xylene. There are many different organic liquids usable in this process and listed in U.S. Patent 2,273,039 (HUDSON, 1942). The mixed vapors of solvent and water are continuously drawn off and condensed. Since the solvent and water are not miscible the water in the condenser precipitates by gravity, is metered and removed, and the solvent is recirculated to the drying cylinder. At the end of the drying process a vacuum is drawn. This removes chemicals adsorbed by the wood and completes the drying as well.

Oak and gum crossties can be dried from the green condition to 40% moisture content in 12 to 16 hours; southern pine poles (8 in. diameter at midlength) from 90 to 35% in 10 hours. The drying is caused by an atmosphere of practically 0% relative humidity and the elevated temperature. Nevertheless, the wood is not attacked, while the rate of heat transfer is optimal. Only very small checks are developed at the surface of the wood.

The relationships between moisture content and drying time for various vapor temperatures are shown in Fig. 8.54. Less clear but, to some extent clarified, is the influence of drying temperature on drying time (Fig. 8.55). The compressive strength perpendicular to the grain of vapor-dried red oak crossties was reduced about 15% as compared with unseasoned control-ties. Vapor-dried and creosoted pine poles were stronger in bending tests than normal steam-conditioned and creosoted poles. A typical cycle for vapor drying followed by an empty-cell treatment is shown in Fig. 8.56. The drying times reached in drying green wood to a final moisture content of 15% using various chemicals were as follows: 30 h with perchloroethylene, 20 h with m-xylene, 16 h with coal tar-naphta.

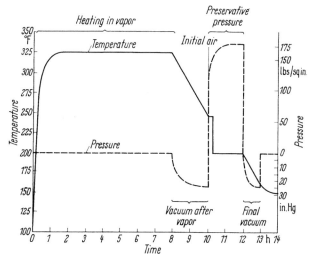

Fig. 8.56. Operation cycle for vapor drying followed by an empty cell-treatment. From TAYLOR-COLQUITT Co.

### 8.4.5 Vacuum Drying

Many different processes have been proposed for using a vacuum to dry goods sensitive to higher temperatures. HOWARD was granted a patent for a vacuum drying process of lumber in 1893. The use of a vacuum, that is of a reduced air

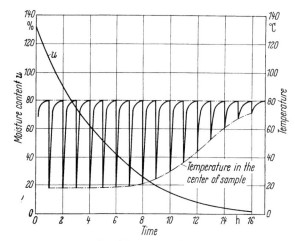

Fig. 8.57. Curves of temperature and moisture content in wood for discontinuous vacuum drying. From KRISCHER and SCHAUSS (1940)

pressure, alone is not effective for rapid wood drying. In drying solid bodies two processes are involved:

1. The movement of water from the interior of the body to the surface;
2. Evaporation and removal of the water at the surface.

The vacuum has very little influence on the rate of moisture diffusion through the wood. The removal of the water at the surface is combined with a cooling effect. This effect slows down the diffusion. Therefore the wood must be heated. This can be done before the vacuum is applied. A large pile of lumber cannot be heated uniformly by usual means during a vacuum period because conduction and convection are minimized. Radiant heat acts only on the outer boards of the pile. Therefore, one heating and vacuum cycle permits only a limited amount of drying. Surface checks are frequent.

In order to avoid the inefficiency of the continuous vacuum drying processes discontinuous methods were developed. A number of alternating cycles of heating and vacuum —the latter not longer than 2 h—must be used (Fig. 8.57). Summarizing, it may be said that there are limitations in the control of vacuum drying and that the equipment is too expensive for commercial utilization. Woods which dry rapidly in a vacuum also can be dried in a normal dry kiln.

### 8.4.6 Chemical Seasoning

Since about 1930 the U.S. Forest Products Laboratory has investigated the question whether, and eventually under which conditions, seasoning or kiln drying of lumber may be improved by the use of chemicals. The basic consideration was as follows:

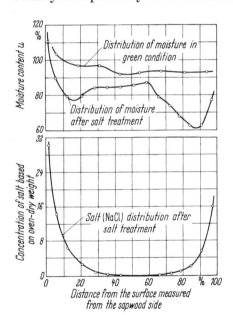

Fig. 8.58. Moisture distribution in green-wood and in wood treated with salt, as well as distribution of salt (sodium chloride) over the cross-section after salt treatment. From LOUGHBOROUGH (1937)

If green wood is treated with a hygroscopic chemical, such as common salt, invert sugar, molasses and other low grade sugars, urea and a urea-formaldehyde mixture or diethylene glycol, the outer zone of the lumber will be impregnated to a depth of about one-tenth of the thickness, with the highest concentration at or near the surface (Fig. 8.58). The maintenance of higher moisture content in this outer zone during drying that occurs with untreated wood results from this condition. The consequence is a reduction of shrinkage and surface checking. Some chemicals produce a permanent antishrink effect. The presence of the chemicals does not impede the drying. The water molecules from the untreated core pass through the impregnated shell and evaporate at the surface. Normally the chemical treatment does not reduce the drying time.

Treating methods employed are soaking, dipping, spraying, and brushing. The most efficient treating method is the soaking method. However, this requires high investment costs for tanks, additional equipment and chemicals. The cheapest method is dry spreading and it has been used extensively. The lumber and the salt are piled in alternate layers. When the chemical has been adsorbed, the lumber moves to the seasoning or kiln-drying stage. HAYGREEN (1962), investigating the factors which determine the optimum kiln-drying conditions for 38 mm (6/4″) red oak treated with sodium chloride, found that a very significant acceleration of drying be safely obtained when only 90 pounds of salt per 1000 sq.ft. of lumber (0.44 kg/m²) are applied.

Some chemicals reduce the strength of the wood and can disturb later gluing and finishing operations. Tests have proved that red oak treated with either sodium chloride or buffered sodium chloride, partially air-dried and then kiln-dried, had a loss in toughness of 30%. Bending strength was reduced 10 to 15%. The reduction in mechanical properties could be avoided when the temperature in kiln drying was reduced from about 74°C to a maximum of 45°C. The hygroscopic and antishrink values of chemicals in relation to chemical seasoning of wood were investigated by PECK (1941). The effect of various chemicals on the volumetric shrinkage is illustrated by Fig. 8.59.

Salt-treated woods may be glued with the usual animal and thermoplastic glues. The joints seem to be poorer than those obtained with normally seasoned wood. Electronic gluing is not advisable due to arcing. The finishing properties of salt-seasoned wood are not badly influenced, especially since the surface layers, containing the relatively highest amount of salt, are removed by planing operations. Where the salt-treated wood will be used in atmospheres of high relative humidities, corrosion of metals in contact with the wood may become severe. The electrical conductivity of salt-treated wood is much higher than that of normal wood. Chemical seasoning is useless for items that can be satisfactorily dried without such a treatment. A more recent and successful method of chemical seasoning is the application of polyethylene glycol (PEG). Green wood turn-

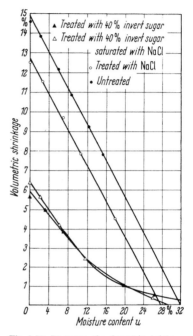

Fig. 8.59. Decrease of volumetric shrinkage of wood treated with common salt, invert sugar, or a mixture of invert sugar with common salt. From PECK (1941)

ings and carvings as well as green veneers were soaked in or pressure-impregnated with solutions of polyethylene glycol. In Sweden a continuous process for the impregnation of green veneer (MORÉN, 1965) using concentrated PEG exists. Due to this impregnation, not only are the properties of the veneer improved but also an increased wood yield is obtained. The treatment of freshly cut green wood in such a way prevents subsequent splitting, checking, warping, and other seasoning degrade (MITCHELL and IVERSEN, 1961). It is possible to impregnate wood with PEG only or to combine the treatment with the application of wood preservatives. Dimensional stability and gloss of PEG-treated wood are improved in comparison to untreated control samples.

### 8.4.7 Drying by Direct Application of Electricity

**8.4.7.1 Drying by Joule's Heat.** Several proposals are known drying wood using its electric resistance and Joule's heat developed under the effect of a high electric current. Since the electric resistance very rapidly increases with decreasing moisture content these proposals are not practicable. VOIGT, KRISCHER and SCHAUSS (1940) investigated the problem scientifically. They found that the lowest final average moisture content is about 60% and that very uneven distributions of moisture content cannot be avoided. Using alternating current with a frequency of 50 c/s arcing was frequent and after the drying the moisture content in both outer layers was near 100% whereas the moisture content in the center reached about 40%. Applying continuous current electrolytic phenomena appeared, the moisture content of the wood surface in touch with the cathode decreased approximately to the fiber saturation point; in the wood layer opposite the anode the moisture was concentrated and reached more than 100%. More recently in the U.S.A. experiments were made to heat and plasticize round logs prior to veneer peeling.

**8.4.7.2 High-frequency Dielectric Drying.** (Fig. 8.60) When wood containing moisture is placed in an alternating field of high frequency (above one million c/s) it is heated due to molecular friction caused by oscillations of the molecules (cf. Section 6.6.2.2). Dielectric properties, specific heat of the material in the field and the electric power available determine the rate of heating. The idea of using this phenomenon for kiln-drying of wood apparently was first patented in 1929 (WHITNEY according to TIEMANN, 1944, p. 302). Lateron, in Germany (KOLLMANN, 1936) and in the Soviet Union investigations in this field were carried out. A survey of the literature is given by KOLLMANN (1955).

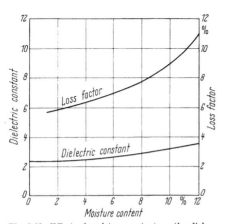

Fig. 8.60. Effect of moisture content on the dielectric constant and the loss factor for spruce at a frequency of 45 MHz. From BROWN, HOYLER and BIERWIRTH (1947)

Any wet material is rapidly and uniformly heated throughout its cross-section. The rate of increase of temperature may range for various wood species between 5 and 20 °C/min. At the exposed surfaces losses of heat and cooling due to evaporation occur. Therefore, the distribution of temperatures is typical. The highest temperatures are produced in the interior of the wood, the lowest prevail in the outer zones. Such a temperature distribution is quite the contrary to that produced by normal convection drying.

In permeable woods the temperature after the heating period levels off at about the boiling point; below the fiber saturation point the temperature rises. Even relatively low temperatures may weaken and split the wood. The danger is considerable for wood species with a very dense structure.

After early high expectations and subsequent disappointments the status of technology in high-frequency drying of wood is as follows:

The schedule of drying must be carefully investigated for each species of wood. The distances between the electrodes or between the electrodes and the wood item must be computed (HAFNER and SCHWOERER, 1965). The frequency of the generator must be adapted to a desired increase in voltage. High frequency drying should be restricted to steam-permeable species of wood; only for such species may very short drying times between 2 and 4 hours be reached. Good experiences were had with red beech, hornbeam, maple, pear tree, birch, alder, linden, spruce, fir, pine, Douglas fir (Oregon pine), poplar, limba, abura, abachi, afrormosia, keruing, tola and zebrano. Continuously working high-frequency kilns are suited for items such as lasts, legs of tables and of chairs, gun stocks, wood for pattern makers, and special lumber for glued trusses. High-frequency drying is economical

8.4 Special Seasoning Methods

only for thick lumber. It is advisable to apply high-frequency drying mainly in the lower ranges of moisture content. The lumber to be dried should be pre-cut or pre-molded prior to drying. The quality of the wood dried in a high-frequency field is high. The moisture content in the center of the high-frequency dried wood is mostly 1...2% lower than near the surfaces. The stresses are very low. The coefficient of shrinkage may be higher than that for wood dried the usual way. In some cases the hygroscopic capacity is reduced. The wood is completely sterilized.

### 8.4.8 Drying by Infra-Red Radiation

Infra-red radiation penetrates wood only to a slight degree (according to Prát (1937), walnut, 0.5 mm; oak, 2 mm; beech and hornbeam, 3 mm; maple and poplar, 4 mm; pine, 3 to 4 mm; spruce, 6 mm). Rawling (1935) and Clark (1939) published the following list for the infra-red penetration for 3-mm-thick wood sheets:

| | |
|---|---|
| *Pinus cembra* L. (Siberia) | |
| *Pinus palustris* Mill. | |
| *Acer pseudoplatanus* L. | very permeable |
| *Boxus sempervirens* L. | |
| *Fagus sempervirens* L. | |
| *Pinus ponderosa* Douglas | moderately permeable |
| *Eucalyptus sp.* | |
| *Tectona grandis* L. f. | |
| *Swietenia macrophylla* King | nearly impermeable |
| *Eucalyptus marginata* J. Smith | |
| *Quercus sp.* | |
| *Endiandra palmerstoni* C. T. White | |
| *Lovoa klaineana* Pierre | |
| *Juglans nigra* L. | impermeable |
| *Terminalia bialata* Steudel | |
| *Chlorophora excelsa* Benth. et Hook. | |
| *Diospyros sp.* | |

Delay and Lecompte (1943) obtained similar results. Experiments by Keylwerth (1951) drying lumber by means of infra-red radiators were not successful. Infra-red radiation is not an advantageous source of heat for drying wood.

### Literature Cited

Allen, C. F. and Campbell, L. W., (1867) U.S. Pat. No. 64 398 "Apparatus for Drying and Seasoning Lumber by Superheated Steam".

Altenkirch, E., (1938) Z. VDI 82: 1347.

Anderson, A. B., and Fearing, jr., W. B., (1960) Solvent Seasoning of Tanoak. For. Prod. J. **10**: 234—238.

Andersson, R., (1952) Paperi Puu **34**: 319.

Anonymous (1959) Air-Seasoning of lumber speeded by new system. South. Lumberman **199**: (2484): 36, 38 and 42.

Aronowsky, S. J. and Gortner, R. A., (1930) The cooking process. Ind. Eng. Chem. **22**: 264—274.

Bateson, R. G., (1952) Timber drying and the behaviour of seasoned timber in use, 3rd ed., Crosby, Lockwood and Son Ltd., London.

472 Literature Cited

BESEMFELDER, R. R., (1910) Verfahren zum Trocknen von Holz and anderen feuchten Gegenständen. German Pat. No. 261 240, April 1, 1910. (see also British Pat. No. 7 975, June 2, 1913).

Boracure Ltd. (New Zealand) (1949) The vapor drying and preservative impregnation of structural timber. Australian. Timber J. **15** (2): 94—102, 152.

BROWN, G. H., HOYLER, C. N., and BIERWITRH, R. A., (1947) Theory and application of radio-frequency heating, New York.

CALVERT, W. W., (1958) High temperature kiln drying of lumber. For. Prod. J. 8: 200—204.

CAMPBELL, G. S., (1961) Value of Presteaming. For. Prod. J. **11**: 343—347.

CLARK, W., (1939) Photography of infrared. Chapman and Hall Ltd., London.

COBLER, W. E., (1964) How the small mill operator can set up fan air-drying. Forest Ind. **91** (5): 106—108.

C.S.I.R.O., Div. For. Prod. The working of wood. Trade Circ. No. 24, Melbourne, Commonw. of Australia.

CZEPEK, R., (1952) Theorie und Praxis der Hochtemperatur-Holztrocknung. Holz als Roh- und Werkstoff **10**: 1—6.

DELAY, A., and LECOMPTE, J., (1943) Sci. Ind. Phot. **14**: 314.

Dep. Sci. Ind. Res., For. Prod. Res. Lab., (1938) Seasoning Series No. 5. Kiln Drying Schedules. Rec. No. 26, London.

Dep. Sci. Ind. Res., For. Prod. Res. Lab., (1940) The steaming and seasoning of English beech. Leaflet No. 16.

DOFFINÉ, E., (1956) Dämpfgruben für die Furniererzeugung. Norddeutsche Holzwirtsch. **4**: 6—7.

EGNER, K., (1951) Zur Trocknung von Hölzern bei Temperaturen über 100 °C. Holz als Roh- und Werkstoff **9** (3): 84—97.

—, and SINN, H., (1941) Über die Holztrocknung mit Großschaukeln. Holz als Roh- und Werkstoff **4**: 386—399.

EISENMANN, E., (1950) Mechanische Holztrocknung mit Trockenzentrifugen. Holz-Zentr. **76**: 106.

ELLWOOD, E. L., ECKLUND, B. A., and ZAVARIN, E., (1963) Collapse and shrinkage of wood. For. Prod. J. **13**: 401—404.

—, and ERICKSON, R. W., (1962) Effect of presteaming and seasoning stain on drying rate of redwood. For. Prod. J. **12**: 328.

FELLOWS, E. S., (1937) The change in moisture content of yard-piled softwood lumber in Eastern Canada. Dep. Mines and Resources Can., Dom. For. Circ. No. 52, Ottawa.

FESSEL, F., (1952) Hartholztrocknung in der Zentrifuge. Holz als Roh- und Werkstoff **10** (10): 391—394.

—, (1955) Bau und Betrieb von Dämpfanlagen für Holz. Holz-Zentr. **81**: 853—854.

FLEISCHER, H. O., (1953) Drying rates of thin sections of wood at high temperature. Yale School of Forestry, Bull. No. 59.

GABY, L. I., (1961) Forced Air-Drying of Southern Pine Lumber. U.S. Forest Serv. Southeast Forest Exp. Station, Paper No. 121.

GÄUMANN, E., (1928) Flora (NF) 23:344.

GATSLICK, H. B., (1962) The potential of the forced-air drying of Northern hardwoods. For. Prod. J. **12**: 385—388.

GRAU, G., (1937) DRP 674644 from 3. 6. 1937, issued 18. 4. 1939.

GRUMACH, M., (1951) The equilibrium moisture content of wood in superheated steam. For. Prod. Lab. South Melbourne, Project S. 17, Progr. Rep. No. 5.

GUERNSEY, F. W., (1957) High-temperature drying of British Columbia Softwood. For. Prod. J. 7: 368—371.

HAFNER, TH., and SCHWOERER, B., (1965) Hochfrequenz-Holztrocknung. Holzwirtschaftliches Jb. No. 15. DRW-Verlags G.m.b.H., Stuttgart.

HANN, R. A., (1964) Drying yellow-poplar at temperatures above 100 °C. For. Prod. J. **14**: 215—220.

HAYGREEN, J. G., (1962) A study of the kiln drying of chemically seasoned lumber. For. Prod. J. **12**: 11—16.

HENDERSON, H. L., (1951) The air seasoning and kiln drying of wood. 5th ed., Albany, New York.

HERMA, W., (1960) A–B–C der Holztrocknung. Oberboihingen: R. Hildebrand G.m.b.H.

HILDEBRAND, R., (1962) Die Schnittholztrocknung. Oberboihingen: Robert Hildebrand G.m.b.H.

HUDSON, M. S., (1942) Treating wood and wood products. U.S. Pat. No. 2273039, Feb. 17, 1942. (See also M. S. HUDSON, U.S. Pat. Nos. 2435218; 2435219; 2535925).

—, (1947) The vapor drying process. Bull. 18, Northeastern Wood Utilization Council, New Haven, Conn.

# Literature Cited

HUFFMAN, J. B., and POST, D. M., (1960) Forced air drying of gum and oak cross ties. South. Lumberman **200** (2500): 33—37.

JENKINS, J. H. and GUERNSEY, F. W., (1937) Ocean shipment of seasoned lumber. Dep. Mines and Resources Can., Dom. For. Serv. Circ. No. 49, Ottawa.

KASTMARK, C. F., (1943) Eine neue Verfahrensweise beim Trocknen von Holzwaren. Trävaru-industrien (9): 134.

KEYLWERTH, R., (1949) Die Ermittlung der Trockenzeit bei künstlicher Holztrocknung. Holz-Zentr. **75**: 735.

—, (1949) Grundlagen der Hochtemperaturtrocknung des Holzes. Holz-Zentr. **75**: 953—954.

—, (1950) Holz-Zentr. **76**: 175—176.

—, (1950) Die gleichzeitige Trocknung und Tränkung in heißen Ölen. Holz-Zentr. **76**: 771—772.

—, (1951) Die Kammertrocknung von Schnittholz. Betriebsblatt 1. Holz als Roh- und Werkstoff, **9**: 289.

—, (1951) Infrarotstrahler in der Holzindustrie. Holz als Roh- und Werkstoff **9**: 224—231.

—, (1953) Furnier-Trocknungs-Versuche. Holz als Roh- und Werkstoff **11**: 11—17.

—, and KÜBLER, H., (1952) Holztrocknung in heißem Teeröl. Holz-Zentr. **78**: 135—136.

—, and —, (1954) Maximum operating temperatures and corresponding surface and interior temperatures in thin softwood lumber during drying. Deutsche Holzwirtschaft **8**: 1—2.

KIMBALL, K. E., (1947) Temperature measuring and controlling devices in lumber dry kilns. U.S. Dep. Agr. For. Serv., For. Prod. Lab., Rep. No. R 1654, Madison, Wisc.

KNUCHEL, H., (1930) Der Einfluß der Fällzeit auf einige physikalische und gewerbliche Eigen-schaften des Holzes. Beih. No. 5, Schweiz. Forstver., Bern.

—, (1935) Mitt. Schweiz. Anst. Forstl. Vers.-Wes., **19** (1).

KOEHLER, A., and PILLOW, M. Y., (1925) Effect of high temperatures on the mode of fracture of a softwood. South. Lumberman **121**: 219—221.

KOLLMANN, F., (1935) Arch. Wärmewirtsch. **16**: 329.

—, (1936) Technologie des Holzes. 1. Aufl., Bd. I/II. Springer-Verlag, Berlin.

—, (1939) Vorgänge und Änderungen von Holzeigenschaften beim Dämpfen. Holz als Roh- und Werkstoff, **2**: 1—11.

—, (1950) Untersuchungen über die Ursachen von Schäden bei der Trocknung von grünem Eschenholz. Svenska Träforskninginstitutet, Trätekn. avd., Medd. 21, Stockholm.

—, (1952) Investigations of the drying of sawn pine timber at elevated temperatures. Svenska Träforskninginstitutet. Trätekniska Avd. Sweden Medd. 23, 40 pp.

—, (1955) Technologie des Holzes und der Holzwerkstoffe. Bd. II, Springer-Verlag, Berlin, Göttingen, Heidelberg.

—, KEYLWERTH, R. and KÜBLER, H., (1951) Verfärbungen des Vollholzes und der Furniere bei der künstlichen Holztrocknung. Holz als Roh- und Werkstoff **9**: 382—391.

—, and MALMQUIST, L., (1952) Untersuchungen über die Trocknung von Kiefernschnittholz mit erhöhten Temperaturen. Svenska Träforskningsinstitutet. Trätekn. adv., Medd. 23. Stockholm.

—, and SCHNEIDER, A., (1961) Der Einfluß der Strömungsgeschwindigkeit auf die Heißdampf-trocknung von Schnittholz. Holz als Roh- und Werkstoff **19**: 461—478.

—, and —, (1965) Freilufttrocknung und beschleunigte Freilufttrocknung. DRW Verl. Stuttgart.

KRISCHER, O., and SCHAUSS, H., (1940) Wissensch. Veröffentl. Tech. Hochschule Darmstadt, **1** (2), 3. Beitr.

KUHLMANN, A., (1962) Wärmebilanzen beim Dämpfen von Gaboon. Holz als Roh- und Werkstoff **6**: 224.

LADELL, J. J., (1954) The outlook for high-temperature seasoning in Canada. For. Prod. J. **4** (5): 260—263.

—, (1956) High-temperature drying of yellow birch. For. Prod. J. **6** (11): 469—475.

—, (1957) High-temperature kiln-drying of Eastern Canadian softwoods: Drying guide and tentative schedules. Tech. Note No. 2, For. Prod. Lab. Canada, Ottawa.

LÖFGREN, B., (1947) Moderna Sågverkstorkar. Fläkten, No. 1.

LOUGHBOROUGH, W. K., (1937) Amer. Lumberman **63**: 66-67.

MATHEWSON, J. S., (1930) The air seasoning of wood. U.S. Dep. Agr. Tech. Bull. No. 174, Washington, D. C. (1961).

MITCHELL, H. L., and IVERSEN, E. S., (1961) Seasoning green-wood carvings with poly-ethyleneglycol-1000. For. Prod. J. **1**: 6—7.

MÖRATH, E., (1933) Holzind. **13**: 73.

MORÉN, R., (1965) Die Polyäthylenglykol-Imprägnierung von Holz und ihre Auswirkungen bei Holztrocknung und Holzbearbeitung. Holz als Roh- und Werkstoff **23** (4): 142.

PECK, E. C., (1932) Moisture content of wood in dwellings. U.S. Dep. Agr. Circ. No. 239, Washington, D. C.

—, (1941) The hygroscopic and antishrink values of chemicals in relation to chemical seasoning of wood. U.S. For. Prod. Lab. Rep. No. R 1270, Madison, Wisc.

474 Literature Cited

PECK, E. C., GRIFFITH, R. T., and RAO, N. K., (1952) Relative magnitudes of surface and internal resistance in drying. Ind. Eng. Chem. 44 (3): 664—669.

PFEIFFER, J. R., (1958) Forced air drying pays dividends. For. Prod. J. 8: 22 A.

PIEST, H., (1941) Kammertrocknung im Handwerksbetrieb, Werkstattschaukeln, künstlich belüftete Hänge. Mitt. Fachaussch. Holzfragen Berlin (29): 166.

—, (1950) DGfH-Nachrichten 1: 124.

PRÁT, S., (1937) Ber. dtsch. Bot. G. 55: 165.

RAWLING, S. O., (1935) Infrared photography. 2nd. ed., London.

REHMAN, M. A., (1942) A survey of the seasonal variation of the moisture content of Indian woods. Ind. For. Res. Utilization (New Serie) 2 (10), Delhi.

ROBBINS, L. S., (1865) Improved process for preserving wood. U.S. Pat. No. 47132. April 4, 1865.

RUCKER, T. W., and SMITH, W. R., (1961) Forced air drying of lumber. Research and experimental. For. Prod. J. 11: 390—394.

SALAMON, M., (1960) Kiln drying of unbundled shingles at high temperatures. Timber of Canada 9.

SCHLEUSSNER, J., (1941) Zweckmäßige Stapelung für die natürliche Holztrocknung. Mitt. Fachaussch. Holzfragen H. 29, S. 146, Berlin.

SCHLÜTER, R., and FESSEL, F., (1939) Neue praktische Erfahrungen bei der künstlichen Holztrocknung. Trockentechnik. Holz als Roh- und Werkstoff, 2: 169—193.

SCHNEIDER, W., (1938) Elektrizitätswirtschaft 37: 435.

SCHWALBE, C. G., and BARTELS, J., (1934) Z. Forst- und Jagdwes. 75: 1.

—, and ENDER, W., (1934) Forstarch. 10: 33.

SETH, K. K., ANDERSON, A. B., and WILKE, C. R., (1965) Solvent drying of California redwood with methanol. For. Prod. J. 15: 297.

STAMM, A. J., (1933) Method for simultaneously seasoning and treating water-swollen fibrous materials. U.S. Pat. No. 2060902 issued Nov. 17, 1936.

STEVENS, W. C., (1961) Drying with and without restraint. For. Prod. J. 11: 348—356.

—, and PRATT, G. H., (1954) Results of high-temperature drying studies on lumber. Timber Technology and Wood Working 62 (2186).

STURANY, H., (1952) Trocknung im reinen Heißdampf. Holz als Roh- und Werkstoff 10: 358—362.

SULZBERGER, P. H., (1948) The effect of temperature on the strength properties of wood, plywood and glued joints at various moisture content. C. S. I. R., Div. For. Prod. (Australia), South Melbourne.

TIEMANN, H. D., (1944) Wood Technology. 2nd ed., New York and Chicago.

—, (1951) Wood Technology. 3rd ed., Pitman Publ. Corp., New York, Toronto, London.

UPHUS, A. J. and SHAPMAN, N. Y., (1908). U.S. Pat. No. 903635 from 10. 11. 1908.

U.S. For. Prod. Lab., Madison, Wisc., (1956) Boiling in oily liquids. Rep. No. 1665 (revised).

U.S. For. Prod. Lab., Madison, Wisc., (1957) High Temperature Drying: Its Application to the Drying of Lumber. Rept. No. 1665-1 (revised).

VAROSSIEAU, W. W., (1954) Forest Products Research and Industries in the United States.

VOIGT, K., KRISCHER, O., and SCHAUSS, H., (1940) Sonderverfahren der Holztrocknung. Holz als Roh- und Werkstoff 3: 364—375.

WILLIAMSON, R. V., (1937) Pitch elimination. West Coast Lumberman 64 (5): 42—43.

WRIGHT, G. W., (1962) Factors affecting seasoning economics. Australian Timb. J. 28: 124—139.

ZILLNER, W., (1938) Elektrowärme 8: 46.

# 9. WOOD MACHINING

## 9.1 Introduction

Until about 50 years ago woodworking was more a handicraft than a scientifically established technology. Rising material and labor costs as well as new wood base materials, severe competition of other materials, and new wood machining processes have led to many studies of wood machining. Many goals have been sought: higher machining accuracy, kerf-loss reduction, increased safety, noise reduction, lower energy consumption, longer blade life. Nevertheless wood machining and especially wood sawing is still, and will remain, a rather conservative technique. Sawing is the most important and most frequent process in woodworking.

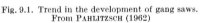

Fig. 9.1. Trend in the development of gang saws. From PAHLITZSCH (1962)

## 9.2 Technology of Sawing

### 9.2.1 Sash Gang Sawing

#### 9.2.1.1 Cutting Velocity.

The reciprocating sash gang saw, originally used mainly in Europe, has been increasingly applied in other industrilized countries. During World War II doubtless the machines have been considerably improved in design, but Fig. 9.1 shows that for gang saws, due to kinematic and dynamic reasons, further essential improvements hardly can be expected (PAHLITZSCH, 1962). The angular velocity of the crank has reached 400 rpm, but the average cutting velocity will not exceed about 6 m/s. There are close interrelationships between stroke and width of sash, rpm of crank, average cutting velocity, and power consumption (Fig. 9.2).

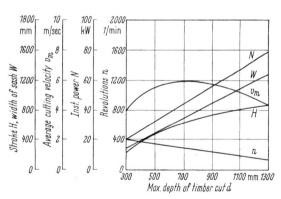

Fig. 9.2. Characteristics of vertical gang saws. From PAHLITZSCH (1962)

The momentary velocity $c$ of the sash can be expressed as:

$$c = r \cdot \omega \left( \sin \gamma + \frac{\lambda}{2} \sin 2\gamma \right) \qquad (9.1\,\text{a})$$

where $r$ = radius of the crank [m], $\omega = \dfrac{n\pi}{30}$ = angular velocity of the crank $\left(\dfrac{1}{\text{sec.}}\right)$, $n$ = rpm of crank, $\gamma$ = angle of the crank (corresponding to $c$), $\lambda = \dfrac{r}{l}$ = connecting rod ratio, $l$ = length of the connecting rod, $H = 2r$ = stroke. If we assume $l = \infty$ we obtain $\lambda = 0$. The velocity $c$ of the sash (or any tooth) then theoretically follows the law

$$c = r \cdot \omega \cdot \sin \gamma \tag{9.1 b}$$

and we obtain the average sash velocity $c_m$ (if we measure the angle as arc $x$)

$$c_m = \dfrac{r \cdot \omega_0 \int_0^\pi \sin x \, dx}{\pi} = \dfrac{rn}{15} \; [\text{m/s}]. \tag{9.2}$$

Since gang saws cut on the down-stroke only, the effective average cutting speed $v_m$ is

$$v_m = \dfrac{c_m}{2} = \dfrac{rn}{30} \; [\text{m/s}]. \tag{9.3}$$

**9.2.1.2 Chip Thickness and Average Cutting Resistance.** Gang saws do not produce chips of equal thickness during the stroke; Fig. 9.3 shows that the sinus

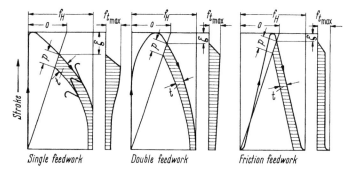

Fig. 9.3. Typical diagrams for usual feedworks of gang saws. From PAHLITZSCH (1962)

law (Eq. 9.1 b) for the tooth speed, the type of feed-work, and the amount of overhang 0 determine the chip thickness $t$. Friction feedworks which move the logs during the up-stroke of the sash and during the down-stroke with constant velocity produce chips with only small changes in thickness. THUNELL (1951) has recorded the feed way of the log as a function of time by means of a special apparatus; he obtained typical discontinuous way-time-diagrams (Fig. 9.4) for the log caused by stops in the feedwork. In sawing, as in planing wood, feed per tooth (or per knife) is one of the principal factors which determines the thickness of the chip. In turn, the force required to remove the chips depends mainly on its thickness. There are, of course, many additional factors, such as wood species, density, moisture, content, grain orientation, geometry and material of tool, cutting speed etc., which may influence the power consumption. If all these factors are held constant and if the mean chip thickness is small then the average force $k_s$ required to cut a chip varies according to a power law

$$k_s = K \cdot t_a^m, \tag{9.4}$$

where $k_s$ = average cutting resistance (force) in the direction of motion of the cutting edge, $K$ = a constant, $t_a$ = mean chip thickness and $m$ = a constant between 0 and 1 (according to BERSHADSKIJ (1953) $m \approx 0.36$) depending on grain direction.

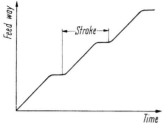

Fig. 9.4. Way-time diagram for the feed way of the log in a gang saw. From THUNELL (1951)

Fig. 9.5. Average cutting force versus chip thickness. (Dotted line from HARRIS (1954); full line from BERSHADSKY (1953)

The power law may be represented by a curve like Fig. 9.5. Away from the region $OP$ it is possible to approximate the curve by a straight line function, such as HARRIS (1954) has stated

$$k_s = K_1 + K_2 \cdot t_a, \tag{9.5}$$

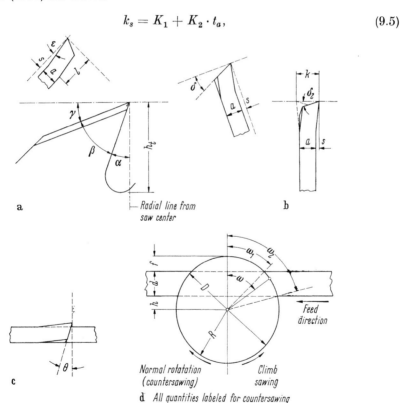

Fig. 9.6a–d. Survey of definitions and symbols used in circular saw blade geometry. a) side view of tooth: $\alpha$ hook angle; $\beta$ lip angle; $\gamma$ clearance angle; $a$ thickness of saw blade; $l$ length of tooth affected by set; $s$ set; $\varepsilon$ inclination of the tooth to the plane of the sawblade, b) front view of tooth: $\delta$ and $\delta_2$ top bevel angle; $k$ kerf width, c) projection along rake plane: $\Theta$ front bevel angle, d) cutting action of circular saw: $R$ blade radius; $D$ blade diameter; $\omega$ instantaneous position angle of tooth under consideration; $\omega_1$ angle through which tooth edge has rotated from reference line to entry in wood; $\omega_2$ angle through which tooth edge has rotated from reference line to exit from wood; $h_t$ distance between work piece and axis of saw rotation; $d$ depth of cut; $f$ saw projection beyond the work piece. From KOCH (1964)

where the constants $K_1$ and $K_2$, depending on wood species and grain orientation, have to be chosen appropriately. The factor $K_1$ may be influenced also by the wear of the cutting edge (cutting edge sharpness) and the factor $K_2$ by the rake or hook angle $\alpha$. Unfortunately there is some discrepancy in the international literature as far as blade angles are concerned. A survey of definitions and symbols may, therefore, be helpful (Fig. 9.6), anticipating circular sawblade geometry which does not differ from gang sawblade geometry as far as definitions of teeth are concerned.

| *American* | *British* | *German* |
|---|---|---|
| rake or hook angle $\alpha$ | cutting or hook angle | Spanwinkel $\gamma$ |
| tooth or wedge or lip angle $\beta$ | sharpness angle | Keilwinkel $\beta$ |
| clearance angle $\gamma$ | clearance angle | Freiwinkel $\alpha$ |
| top or back bevel angle $\delta$ | | Anschrägwinkel $\varepsilon$ |

"The sharper the saw tooth, the more nearly is cutting resistance directly proportional to mean chip thickness $t_a$, according to Eq. 9.5" (LUBKIN, 1957, p. 28).

**9.2.1.3 Consumption of Energy.** REINEKE (1950), who also found a linear law similar to that of Eq. (9.5), explained the consumption of energy in sawing by the following steps: In the first step work must be done to sever the wood fibers; this fiber severance energy is independent of the thickness of the chip to be severed and depends mainly on tool sharpness. Resharpening reduces fiber severance thus

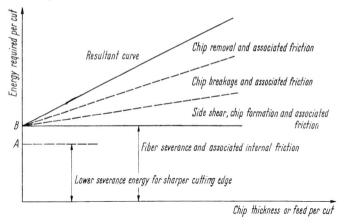

Fig. 9.7. Analysis of energy required by cut in dependence on chip thickness or feed per cut. From REINEKE (1950)

translating curve $B$ in Fig. 9.7 downward to position $A$. In the second step the chip is separated by shearing from the kerf walls. In the third step the loosened chip is broken into small particles (sawdust). In a fourth action the sawdust is removed from the kerf. In these last three actions the energy consumption is increased by frictional losses between workpiece, tool and chips. Later on, REINEKE (1956) added to his analysis a fifth step: the energy required for chip acceleration, as postulated by KOCH (1954, 1955, 1956).

For gang saws and band saws the main cutting force $P_s$, i.e. the predominant component of the cutting force below one tooth in the direction of tooth travel, may be expressed as follows (PAHLITZSCH, 1951):

$$P_s = k_s \cdot q = k_s \cdot \frac{k \cdot f \cdot d}{60 \cdot v_m} \text{ [kp]}, \tag{9.6}$$

where $k_s$ = average cutting resistance in kp/mm², $q$ = chip cross-section in mm², $k$ = kerf width in mm, $f$ = feed speed in m/min, $d$ = depth of timber cut in mm, and $v_m$ = average cutting speed in m/s. The values of $k_s$ vary greatly due to different conditions for chip formation and for friction losses; they must therefore be determined for each case. From the basic study by BUES (1928) an average value of $k_s = 4$ kp/mm² can be deduced.

According to VOIGT (1925, 1949) the value for $P_s$ may be lowered or increased by multiplying by a coefficient $\psi$ taking into consideration the moisture content $u$ of the timber:

$$\psi = 0.48 \cdots 0.60 \text{ for floated wood } (u > 100\%)$$
$$\psi = 0.60 \cdots 0.75 \text{ for green wood } (u = 45 \cdots 100\%)$$
$$\psi = 0.72 \cdots 0.95 \text{ for semidry wood } (u = 20 \cdots 45\%)$$
$$\psi = 0.85 \cdots 1.05 \text{ for airdry wood } (u = 10 \cdots 20\%)$$
$$\psi = 0.95 \cdots 1.15 \text{ for kiln dried wood } (u < 10\%)$$

The cutting power will be analyzed in more detail in Section 9.2.2.

The total power consumption $N$ of any woodworking machine can be expressed by:

$$N = N_i + N_s + N_f \tag{9.7}$$

where $N_i$ = idling power, $N_s$ = cutting power, and $N_f$ = power required for the feedworks.

According to the ESTERER machine-works, Altötting, Bavaria (1930), one can approximate

for spruce
$$N_s + N_f = 0.1 \cdot F_s \text{ [hp]} \tag{9.8a}$$
for oak
$$N_s + N_f = 0.3 \cdot F_s \text{ [hp]} \tag{9.8b}$$

where in both cases $F_s$ is the surface cut per hour in m²/h, unilaterally measured[1].

The power consumption of gang saws increases with the time of service due to dulling of the sawblades (cf. Table 9.1).

Table 9.1. *Increase of Power Consumption of Gang Saws with the Time of Service* (From VOIGT, 1949)

| Time of service after sharpening h | 1 | 2 | 3 |
|---|---|---|---|
| | Increase of power consumption % | | |
| Softwoods | 18 | 33 | 42 |
| Hardwoods | 23 | 52 | — |

### 9.2.1.4 Effects of Tooth Geometry, Tooth Height and Pitch.

Fig. 9.8 shows rake, tooth, and clearance angles, recommended by DOMINICUS, a manufacturer of gang-saw blades in the Federal Republic of Germany. According to the German standard specification DIN 8803 the values of Table 9.2 are standardized.

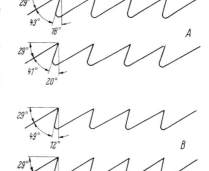

Fig. 9.8. Recommended types of teeth for gang saws. From DOMINICUS. *A*. for softwood; *B*. for hardwood

---

[1] In the USA productivity in sawing is measured as total linear footage of timber sawn per hour in ft. If the depth $d$ of the workpiece in inches is known, one can deduce $F_s$ from $L$ as follows $F_s = 0.0774 \cdot d \cdot L$.

In sawing green wood the German AWF (Ausschuß für wirtschaftliche Fertigung = Committee for economic production) recommends the following rake angles:

|  | rake angle |
|---|---|
| softwood | 8°···12° |
| hardwood | 5°··· 8° |

from which clearance angles follow between 40° and 45°.

Table 9.2. *Dimensions and Tooth Angles for Gang Saw Blades*
(normal thicknesses 1.6···2.0 mm)

| Tooth pitch $p$ mm | Tooth height $h_t$ mm (admissible tolerance ±1 mm) | Rake angle $\alpha$ | Tooth angle $\beta$ |
|---|---|---|---|
| 18 | 9 | for softwood and hardwood 18° | 40° |
| 22 | 12 | | |
| 25 | 14.5 | for hardwood 12° | 44° |
| 28 | 17 | | |

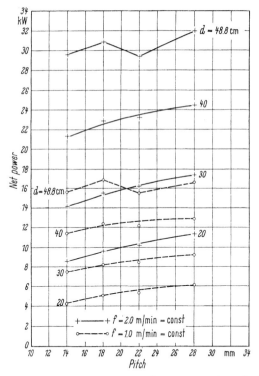
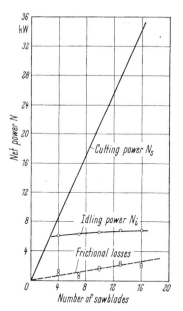

Fig. 9.9. Relationship between cutting power requirement and tooth pitch. From BUES (1928)

Fig. 9.10. Dependence of cutting power, idling power and frictional losses upon the number of saw blades. From BUES (1928)

The power consumption increases with tooth pitch in approximately linear fashion (BUES, 1928). The cutting power requirements (Fig. 9.9) and the frictional losses appear to be directly proportional to the number of sawblades (Fig. 9.10). Cutting power and feed force increase with sawblade thickness (or width of kerf) in a linear manner (KAIUKOVA and KONJUKHOV, 1934; 1935).

### 9.2.1.5 Influence of Setting.

The saw blade needs side clearance, otherwise contact and resulting friction occurs between the saw blade and the kerf wall. Heat will be generated and will affect the saw tension, diminishes hardness and strength of the saw teeth, burns the kerf walls, and increases the power necessary to drive the saw. For gangsaw blades there are two methods of providing this clearance, illustrated in Fig. 9.11. Spring-set clearance is effectuated by bending the tips of each successive tooth in alternate directions as shown in Fig. 9.11a to establish a set $s$. Swage-set clearance is provided by upsetting the steel at the tooth tip as shown in Fig. 9.11b. All teeth are treated identically. The set for gangsaw blades should generally

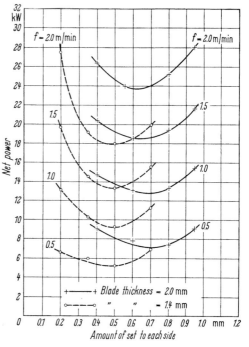

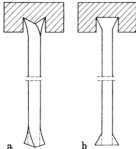

Fig. 9.11a, b. Saw blade sets: a) springset clearance; b) swage set clearance

Fig. 9.12. Influence of the amount of set to each side to the saw blade on the power consumption. From BUES (1928)

amount to 0.4 to 0.6 mm corresponding to the saw blade thickness. Thinner blades must have the greater set. For wet wood, a set of 0.7 to 0.8 mm is advisable. For hardwood and frozen wood the set may be reduced by 0.1 to 0.2 mm.

The power consumption is influenced markedly by the width of set. BUES (1928) observed, for different blade thicknesses and feed speeds, typical curves as shown in Fig. 9.12. The minima lay in the range between 0.5 and 0.7 mm for the set. MEYER (1926) had obtained similar results for circular saws and already explained the concave course of the curves. For small sets (i.e. also for small widths of kerf) the cutting energy is small, but the frictional losses become very important. With increasing set, the cutting energy rises approximately proportional to the width of kerf whereas the friction losses decrease and approach a constant value. Therefore the curve must run through a minimum. Spring-setting should be done according to Fig. 9.13a as "curved-set" since this type of set is more durable and delivers smoother cut surfaces.

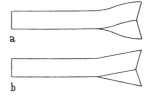

Fig. 9.13a, b. Types of sets: a) curved-set b) bent-set

THUNELL (1951) compared the results of gang sawing with spring-set and swage-set blades. He could not find differences in power consumption though, theoretically, spring-set saws should require about 20% less energy (PAHLITZSCH,

1962). The temperature of swage-set saw blades increases in cutting less than that of spring-set saw blades. The width of kerf produced by swage-set saws decreases somewhat in service. Swage-set saws deliver (about 9.5%) more sawdust than spring-set saws. The yield of lumber is practically independent of the method of setting.

**9.2.1.6 Strain and Stresses in Gang Saw Blades; Thermal Effects.** Smooth surfaces may be produced only when the saw blades in the sash exhibit an adequate tension stress even at working temperatures. The saw must receive a "strain"

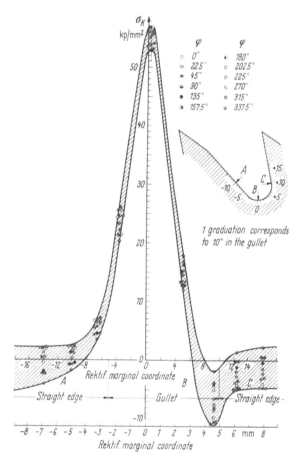

Fig. 9.14. Stress distribution along the gullet of the tooth interspace. From THUNELL and HILTSCHER (1951)

by means of mechanical devices, or by hydraulic mechanisms. The latter offer the advantage of maintaining constant "strain" on all blades, regardless of occasional changes in blade load and temperature. BIERMANN (1942) measured tension stresses between 12 and 15 kp/mm² in gang saw blades; based on experiments by THUNELL (1951) stresses between 23 and 26 kp/mm² could be calculated. In reality "strain" must be adapted to the respective cutting process. Photoelastic investigations by THUNELL and HILTSCHER (1951) on transparent phenol-resin models of gang saw blades clarified the stress distribution in the blades. The results are shown in Fig. 9.14. Stresses vary within the hatched region during the period of a stroke. The maximum with the value $\sigma_K = +56.8 \text{ kp/mm}^2$ lies practically in the ground of the gullet. The radius of the gullet-furrow may

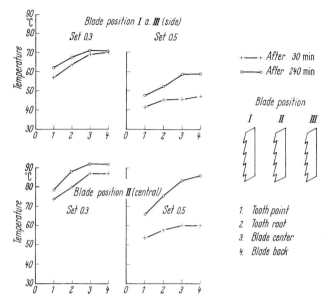

Fig. 9.15. Rise of temperature in gang saw blades, cutting pine. From THUNELL (1951)

greatly influence this maximum stress. Mistakes in sharpening or infractions of the edges in the ground of the gullet may lead to fractures of the saw blade. Statistical investigations have proved this fact.

With increasing cutting power required more friction-losses are caused. The elevation of temperatures in gang saw blades as influenced by the position of the blades in the sash, by the position in the tooth, by the width of spring-set, and by the working time are illustrated in Fig. 9.15 (THUNELL, 1951). A thermal expansion is associated with the heating of saw blades. MARSCHNER (1942) computed a reduction of the tension-stress in gang saw blades of about 0.24 kp/mm² for each degree (centigrade) increase in temperature.

**9.2.1.7 Surface Quality.** Plane and smooth surfaces are essential for marketability of sawn products. Rough or

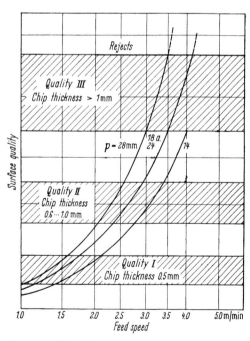

Fig. 9.16. Relationship between surface quality, tooth pitch and feed speed. From BUES (1928)

fuzzy surfaces are a potential source of waste since they require thicker chips in subsequent milling and since they are more easily attacked by wood destroying fungi.

Uneven, trough-shaped boards are frequent in badly managed sawmills; the reasons are disproportionate setting, unsuitable tensioning of saw blades, too high feed speed or defects in the machinery.

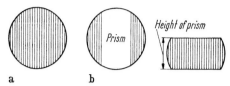

Fig. 9.17a, b. Cutting forms in cutting logs on gang saws. (a) Through-and-trough cutting; (b) prism-cutting

The following rules are valid:

1. The surface quality is improved by increasing average cutting velocity;

2. Increasing feed speed reduces the sharpness of teeth so that the surfaces become markedly rougher (Fig. 9.16);

3. The smaller the tooth pitch the better is the surface quality;

4. The roughness will be increased with the width of set, assuming that pitch and feed speed remain constant, "curved-set" leads, as has been mentioned already, to higher surface quality than conventional set;

5. Unobjectionable sharpness of all teeth is a prerequisite for adequate surface quality. Dull teeth do not shear but tear the wood fibers;

6. The "cutting angle"—according to the German and Russian definition $\Delta = 90° - \alpha = \gamma + \beta$, not to be confused with the British cutting or hook angle, should range between 78° and 82° for softwoods and between 82° and 85° for hardwoods;

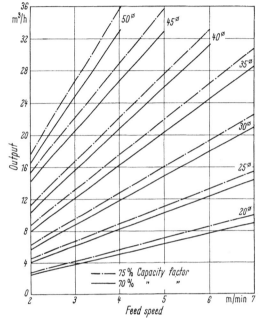

Fig. 9.18. Output of a heavy duty gang saw with one connecting rod as a function of feed speed. From BRAUNSHIRN (1929)

7. The surface quality depends also on the wood species and becomes better as denser, harder and dryer wood is used.

**9.2.1.8 Yield.** Most modern sawmills (as in Sweden) use tandem gangsaws. The first saw slabs the sides of the log and removes the sideboards so that a cant or prism of desired thickness is produced; then the cant is turned and the second saw converts it into boards (Fig. 9.17).

The average yield or volume of recovery depends on type of cutting, width of kerf, log diameter, kind of lumber produced, etc.

BRAUNSHIRN (1929) reports the following mean values:

|  | Yield % |
|---|---|
| Production of parallel-edged lumber | 56···68 |
| Production of conically edged lumber | 60···70 |

Of the 30 to 44% waste, 15 to 17% are slabs and edgings; the rest is sawdust. Table 9.3 compares yield values of gang saw mills with that of circular saw mills.

Ref. p. 551]                    9.2 Technology of Sawing                    485

Table 9.3. *Yield and Sawdust-Refuse in Sawing Logs of Different Diameters in Northern Sweden*

(From G. KINNMAN, without year)

|  |  | Log diameter at the top cm | | | | | | | | |
|---|---|---|---|---|---|---|---|---|---|---|
|  |  | 12.7 | 15.2 | 17.8 | 20.3 | 22.9 | 25.4 | 27.9 | 30.5 | 33.0 |
| *Yield %* | | | | | | | | | | |
| Gang saw | based on the top measure | 66 | 66 | 67 | 68 | 68 | 68 | 68 | 68 | 68 |
|  | based on middle measure | 42 | 46 | 47 | 49 | 51 | 52 | 52 | 53 | 53 |
| *Sawdust-refuse %* | | | | | | | | | | |
| Gang saw | based on log volume | 13 | 12 | 11.5 | 10.5 | 10 | 9 | 8.5 | 8 | 7.5 |
| *Yield %* | | | | | | | | | | |
| Circular saw | based on top measure | 58 | 60 | 62 | 62 | 62 | 62 | 62 | 63 | 63 |
|  | based on middle measure | 37 | 40 | 44 | 45 | 46 | 47 | 47 | 49 | 49 |
| *Sawdust-refuse %* | | | | | | | | | | |
| Circular saw | based on log volume | 17 | 15 | 14 | 13 | 12 | 11.5 | 11 | 10.5 | 10 |

JUŠTŠUK (1951) published the following yield values for sawing round logs with a top-diameter of 20 cm in Sweden:

|  |  | Gang saw mills | Simple circular saw mills |
|---|---|---|---|
| lumber | % | 56 | 44 |
| sawdust | % | 11 | 19 |
| fuelwood, blocks | % | 3 | 3 |
| shrinkage | % | 3 | 3 |
| chippable waste or edgings, laths, slabs | % | 27 | 31 |

JONES (1956) reports that Swedish sawmills, operating tandem gang saws with spring-set blades (width of kerf 3.2 mm), obtain 57% lumber, 29% chippable waste, 3% end trimmings, and 11% sawdust. KOCH (1964) states that, over a period of years, in cutting 8-foot eastern white pine logs into 1-inch boards he obtained: 49% lumber, 38% chippable waste, 2% trim ends, and 11% sawdust. KERSTEN (1944) could prove that in spite of all irregularities in growth there are some general guides which allow one to estimate the yield of sawmills in advance.

The output of gangsaws may be reported in volume cut or total lineal length of lumber produced (m or feet). If the log diameter and the number of cutting blades are not added, these figures do not give clear evidence of the gang saw productivity. Fig. 9.18 shows the output of a heavy duty gang saw, with one connecting rod, as a function of feed speed.

### 9.2.2 Band Sawing

**9.2.2.1 General Considerations, Saw Blade Dimensions.** Bandsaws of different types and sizes are employed from primary log conversion to furniture manufacture. Their main advantages are the very high cutting speed and the low kerf waste. Log bandsaws have a very high output. The trend in the development of bandsaws is shown in Fig. 9.19 according to PAHLITZSCH (1962). Already since about 1920 the curve for the cutting speed exhibits a decreasing rate.

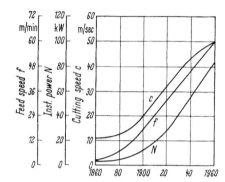

Fig. 9.19. Trend in the development of band saws. From PAHLITZSCH (1962)

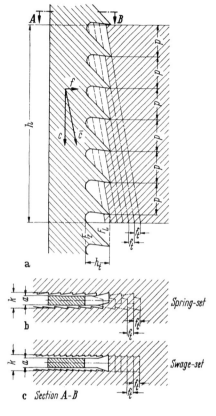

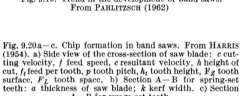

Fig. 9.20a–c. Chip formation in band saws. From HARRIS (1954). a) Side view of the cross-section of saw blade: $c$ cutting velocity, $f$ feed speed, $\bar{c}$ resultant velocity, $h$ height of cut, $f_t$ feed per tooth, $p$ tooth pitch, $h_t$ tooth height, $F_Z$ tooth surface, $F_L$ tooth space, b) Section A–B for spring-set teeth: $a$ thickness of saw blade; $k$ kerf width. c) Section A–B for swage-set teeth

Bandsaws theoretically produce chips with constant cross section. The way of the saw teeth through a workpiece with the height $h$, when $c$ is the cutting velocity in m/s and $f$ is the feed speed in m/min (vertically to the direction of saw movement) shown in Fig. 9.20, designed by PAHLITZSCH and DZIOBEK (1959) is in agreement with HARRIS (1954). The velocity $\bar{c}$ of the tooth relative to the workpiece is as follows

$$\bar{c} = \sqrt{c^2 + \left(\frac{f}{60}\right)^2} \quad [\text{m/s}]. \tag{9.9}$$

In practice $\bar{c} \approx c$ since $f \ll c$. The relationships between cutting force and chip thickness are dealt with in Section 9.2.1.2.

Bandsaws with spring-set teeth consume less power, but their maintenance requires more labor. This is the reason why, in many parts of the world, especially in the U.S.A. and Canada, swage-set teeth are used nearly exclusively. The thickness or gauge of the saw blade depends on the diameter of the band saw wheels. Table 9.4 shows the relationships between important log band saw characteristics in the U.S.A.

Table 9.4. *Relationships between Wheel Diameter, Saw Gauge, Saw Width, and Power Consumption for Swage-Set Log Band Saws in USA*

| Wheel diameter | | Saw gauge | | Saw width | | Average power consumption | |
|---|---|---|---|---|---|---|---|
| inches | mm | BWG | mm | inches | mm | HP | kW |
| 60 | 1,524 | 17 | 1.47 | 5···9 | 127···229 | 62.5 | 46 |
| 66 | 1,676 | 16 | 1.65 | 7···11 | 178···279 | 100 | 73.5 |
| 72 | 1,829 | 15 | 1.82 | 8···13 | 203···330 | 125 | 92 |
| 84 | 2,134 | 14 | 2.10 | 10···14 | 254···356 | 150 | 110 |
| 96 | 2,438 | 13 | 2.41 | 12···16 | 305···406 | 175 | 129 |
| 108 | 2,743 | 12 | 2.77 | 14···16 | 356···406 | 225 | 165 |
| 120 | 3,048 | 11 | 3.05 | 14···18 | 356···457 | 275 | 292 |

One can see that an appropriate thickness $a$ of the blade may be chosen; when $D$ is the wheel diameter (both in mm) $a = 0.001 D$. In Germany tooth styles used for log band saws are shown in Fig. 9.21. A standard band-saw tooth shape

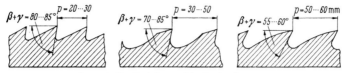

Fig. 9.21. Tooth styles used for log band saws

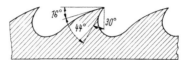

Fig. 9.22. Standard band-saw tooth shape, adopted in the United States

adopted in the United States is illustrated in Fig. 9.22. The average force, parallel to blade motion, and the average power demanded may be computed for spring-set and swage-set band saws either using the linear or the power law (cf. Eq. 9.4 and 9.5) in English units according to P. KOCH (1964, p. 185 to 187).

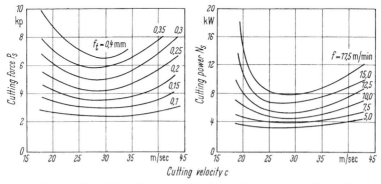

Fig. 9.23. Influence of the cutting velocity on cutting force and power consumption for band saws. From PAHLITZSCH and DZIOBEK (1959)

**9.2.2.2 Cutting Velocity and Cutting Resistance.** As shown in Fig. 9.19, band saws have already reached cutting speeds of 50 m/s ($\approx$10,000 fpm). Such high cutting speeds are suitable for softwoods, such as fir, spruce, pine and basswood.

Harder woods, such as beech, oak, etc. require lower saw speeds between 40 and 46 m/s (about 8,000 to 9,000 fpm). For extremely hard woods and frozen hardwoods, saw speeds of 30 to 35 m/s (about 6,000 to 7,000 fpm) are frequently used.

The influence of the cutting velocity on cutting force or resistance and power consumption has been investigated on a horizontal log band saw by PAHLITZSCH and DZIOBEK (1959). For cutting force, as well as for power consumption, a minimum has been found in the range of about $c = 30$ m/s (Fig. 9.23).

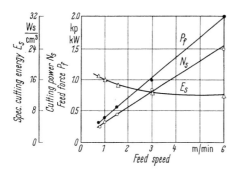

Fig. 9.24. Effect of feed speed on feed force, cutting power, and specific cutting energy in bandsawing Sugi. From SUGIHARA (1953)

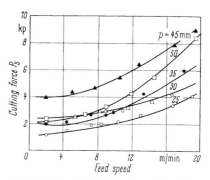

Fig. 9.25. Effect of feed speed on cutting force. From PAHLITZSCH and DZIOBEK (1959) (p - pitch)

**9.2.2.3 Influence of Feed Speed.** SUGIHARA (1953) studied the effect of feed speed on cutting force, cutting power, and specific cutting energy in bandsawing Sugi. He observed that feed force $P_f$ and cutting power $N_s$ increase in a linear manner with the feed speed (Fig. 9.24). Quite similar results were obtained for circular saws by other authors (MEYER, 1926; KAIUKOVA and KONJUKHOV, 1934; LOTTE and KELLER, 1949; McMILLIN and LUBKIN, 1959).

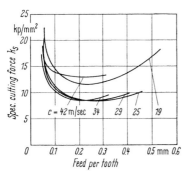

Fig. 9.26. Influence of the feed per tooth on the specific cutting force. From PAHLITZSCH and DZIOBEK (1959)

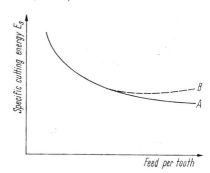

Fig. 9.27. Schematic representation of two possible types of specific cutting energy curve. From LUBKIN (1957)

PAHLITZSCH and DZIOBEK (1959) obtained slightly progressive curves (Fig. 9.25) and gave the following explanation: For low feed speeds the cutting forces may be relatively too high, probably due to friction forces in the kerf. In the middle part the curves approach proportionality, to some extent, but again become more progressive for high feed speeds apparently under the influence of the increasing filling of the gullets with sawdust. The feed of one tooth $f_t$ may be calculated as follows:

$$f_t \approx \frac{f \cdot p}{c} \;\; [\text{mm}]. \tag{9.10}$$

PAHLITZSCH and DZIOBEK (1959) presented the curves shown in Fig. 9.26 for the specific cutting force as a function of the feed per tooth $f_t$; there always appeared an optimum for $f_t = 0.2 \cdots 0.3$ mm. It is known that the specific cutting force decreases with increasing cross section of chips. For a tooth feed $f_t \leq 0.1$ mm, the process is hardly a cutting but a scraping with a rapid dulling of cutting edges and therefore a rapidly increasing specific cutting force. The rise after the optimum for higher tooth feeds is caused by an overloading of the gullets.

Theoretical considerations and the results of many experiments lead to hyperbolic specific sawing energy curves with an asymptotic decrease with increased feed per tooth. The conclusion for the user is the application of the largest possible feed per tooth if minimum energy consumption is desired. This case is schematically represented in Fig. 9.27 by curve $A$. There are occasionally limitations for feed per tooth because the cutting energy curve shows an increase following a minimum as indicated by curve $B$. This phenomenon may be traced to saw "overloading" which is, of course, a relative term.

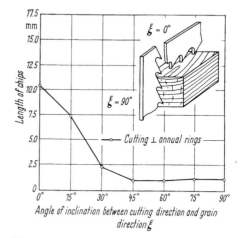

Fig. 9.28. Influence of grain orientation on the length of chips. From SAITO and co-workers (1957)

**9.2.2.4 Effect of Depth of Timber Cut and of Grain Orientation.** For all types of saws, cutting force and power consumption increase linearly with the depth or height $\bar{d}$ of cutting. Also the gradient of temperature rises with $\bar{d}$ as well as with feed of tooth. It is generally assumed that cutting perpendicular to the grain consumes more power than along the grain; in Section 9.2.3.5 evidence is given that in a special cases of circular sawing the contrary holds true. SAITO and co-workers (1957) investigated the influence of the grain orientation on the length of the chips (Fig. 9.28). There is a rapid reduction of the length with growing angle of inclination $\xi$. For $\xi = 45°$, the length of chips is only one tenth of that for $\xi = 0°$. Between $\xi = 45°$ and $\xi = 90°$ there is apparently no more change. A large amount of very short chips ("crumbly chips") increases the friction losses.

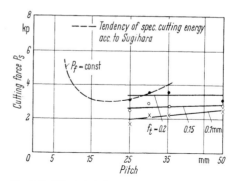

Fig. 9.29. Effect of pitch of the teeth on cutting force. From PAHLITZSCH and DZIOBEK (1959)

**9.2.2.5 Effect of Tooth Geometry and Pitch.** Investigations on the effects of tooth angles, gullet shape and area on band saw performance are apparently not yet available. Therefore, the many studies on circular saws may be referred to Section 9.2.3.7. The results of these studies may permit some conclusions also for the user of bandsaws. The effect of the pitch of the teeth on the cutting force is shown in Fig. 9.29 (PAHLITZSCH and DZIOBEK, 1959) for band sawing with constant feed per tooth. For this case it should theoretically be expected that the cutting force is independent on the pitch. In reality, a slight linear increase seems to exist, perhaps due to poorer "guidance" of the blades with large pitch; if there are fewer teeth working over the depth of the workpiece, the blade can move more in the lateral direction and the kerf may become somewhat larger.

**9.2.2.6 Band Tension and Stability.** The band tension depends, in general, upon the band axial velocity and pulley mounting systems. These two factors should be considered when chosing the initial tension so that an optimum is attained. The optimum running tension is the highest tension the band can withstand without failure during operation.

The band saw natural frequencies decrease continuously with increasing velocity; the band can be considered less stable with increasing velocity. The condition of impending instability, i.e., zero fundamental frequency, always occurs at a specific critical velocity, but this velocity depends upon the pulley mounting system, initial tension, band geometry, and material. "If the band initial tension is applied by a dead weight and lever mechanism support, both the running tension and natural frequencies attain the maximum values for all possible pulley mounting systems. The band can be considered "most stable" for the given initial tension. If, on the other hand, a fixed pulley mounting system is used, the resultant tension remains constant over the band velocity range, and the band natural frequencies attain minimum values. The band is "least stable" for the given initial tension" [MOTE, Jr. 1964].

### 9.2.3 Circular Sawing

**9.2.3.1 Introduction, Saw Blade Geometry, Kinematics.** Circular saws are the most useful tools for general woodworking. They are used from primary lumber production to cabinet shop. Circular saws are very versatile and may be used for

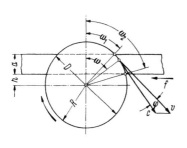 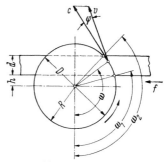

Fig. 9.30. Saw blade geometry: exit and entry angles and velocity vectors for normal countersawing. From LUBKIN (1957)

Fig. 9.31. Saw blade geometry: exit and entry angles and velocity vectors for climb-sawing. From LUBKIN (1957)

ripping and edging, for crosscutting of round logs and lumber, for the production of boards, planks, beams, laths, etc. Fig. 9.6 illustrated saw blade geometry, Fig. 9.30 exit and entry angles and velocity vectors for normal countersawing (up-sawing) and Fig. 9.31 shows for climb-sawing (down-sawing). Definitions and nomenclature are as introduced and used earlier, with the following supplements:

$\Theta$ = front bevel angle
$R$ = blade radius in mm
$i$ = number of teeth in sawblade
$h$ = distance between workpiece and axis of saw rotation in mm
$\omega$ = instantaneous position angle of tooth under consideration
$\omega_1$ = angle through which tooth edge has rotated from the vertical reference line to entry into wood
$\omega_2$ = angle through which tooth edge has rotated from the vertical reference line to exit from wood

Based on the analysis of LUBKIN (1957) these angles are given by the following formulas:

$$\omega_1 = \text{arc cos}\left(\frac{d+h}{R}\right) \tag{9.11}$$

in countersawing

$$\omega_2 = \text{arc cos}\frac{h}{R} \tag{9.12}$$

or

$$\omega_1 = \text{arc cos}\frac{h}{R} \tag{9.13}$$

in climb-sawing

$$\omega_2 = \text{arc cos}\left(\frac{d+h}{R}\right) \tag{9.14}$$

The tooth pitch $p$, measured along the circle of cutting tips for uniformly spaced teeth of equal height can be computed:

$$p = \frac{2\pi R}{i} = \frac{\pi D}{i}. \tag{9.15}$$

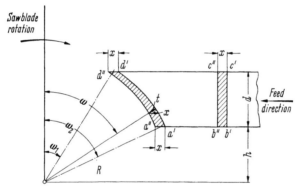

Fig. 9.32. Path of circular saw tooth: $x$ movement of wood corresponding to a tooth movement equal to the pitch; $t$ instantaneous chip thickness. From LUBKIN (1957)

The feed per revolution of the saw blade is

$$f_r = \frac{f}{n}. \tag{9.16}$$

The feed per tooth or "bite" per tooth is defined as:

$$f_t = \frac{f_r}{i} = \frac{f}{n \cdot i}. \tag{9.17}$$

The following units are generally used

$$f_t = 10^3 \frac{f}{n \cdot i} \text{ mm/tooth with } f \text{ in m/min}, n \text{ in rpm} \tag{9.17a}$$

and

$$f_t = 12 \frac{f}{n \cdot i} \text{ in./tooth with } f \text{ in fpm}, n \text{ in rpm}. \tag{9.17b}$$

The chip separated by a sawtooth has a varying thickness along the length of the tooth path (Fig. 9.32). As HARRIS (1954) has shown, the instan-

492                              9. Wood Machining                         [Ref. p. 551

taneous chip thickness $t$ for swage-set circular saw teeth can be approximated as
follows:

$$t = f_t \cdot \sin \omega = \frac{f}{n \cdot i} \sin \omega. \tag{9.18}$$

Eq. (9.18) shows that the chip thickness in normal sawing is smallest at tooth
entry, largest at tooth exit.

The cutting velocity $c$ (or peripheral velocity of the saw teeth) can be expressed as:

$$c = \pi D n \tag{9.19}$$

or

$$c = \frac{\pi D n}{6 \cdot 10^4}, \text{ m/s with } D \text{ in mm, } n \text{ in rpm} \tag{9.19a}$$

and

$$c = \frac{\pi D n}{12} \text{ fpm with } D \text{ in in., } n \text{ in rpm}. \tag{9.19b}$$

One can now relate chip thickness $t$ to cutting speed $c$ and feed speed $f$, using
Eq. (9.18) and eliminating $n$ with Eq. (9.19) and $i$ with Eq. (9.15). For a swage-set
saw one obtains:

$$t = \frac{pf}{c} \cdot \sin \omega \tag{9.20}$$

$$f_t = \frac{pf}{c}.$$

For a spring-set saw HARRIS (1960) gives:

$$t = \frac{2pf}{c} \sin \omega \tag{9.21}$$

$$f_t = \frac{2pf}{c}.$$

**9.2.3.2 Effect of Cutting Velocity on the Cutting Resistance.** HARRIS (1946)
recommends as a suitable cutting velocity for general conversion and ripsawing
with circular saws 10,000 fpm ($\approx 51$ m/s), and up to 13,000 fpm ($\approx 66$ m/s) for swage
saws, ground-off saws, and thin plate saws. FOYSTER (1953) notes that circular
saws for hardwoods may have speeds between 13,000 and 14,500 fpm (66 to
74 m/s). Fig. 9.33 shows the impressive trend of development of circular saws
within the last century. High speed sawing is doubtless in progress. In the
Soviet Union, LAPIN (1954) advocates velocities from 80 to 120 m/s for ripsawing
and 80 to 100 m/s for cross cutting.

In his classical paper, MEYER (1926) showed a diagram of cutting power $N_s$
and feed force $P_f$ as functions of cutting speed (Fig. 9.34). We observe that
higher speed is associated with less efficient utilization of sawing energy, but with
smaller feed forces which is an advantage in manual feeding. A further advantage
is the increased production rate by high-speed sawing. The trend found by
MEYER (1926) has been confirmed to some extent by the results of experiments
carried out by KAIUKOVA and KONJUKHOV (1935) and by DOWER and OAKEY
(1957) (Fig. 9.35). TELFORD (1949) notes: "Presumably at most sawmills the more
efficient use of horsepower results from lower as contrasted with higher saw
speeds."

LOTTE and KELLER (1949) published Fig. 9.36, but they added that in the case of automatic feeding a cutting velocity of 42 m/s guarantees the highest economy in power consumption. In Section 9.2.2.2 we have seen that for bandsaws the curves for the cutting force versus cutting speed exhibit minima.

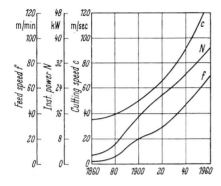

Fig. 9.33. Trend in the development of circular saws. From PAHLITZSCH (1962)

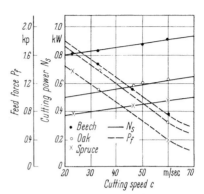

Fig. 9.34. Cutting power and feed force as functions of cutting speed for circular saws. From MEYER (1926)

BERSHADSKII (1956) introduced in his empirical formulas a coefficient $\alpha_v$ characterizing the dependence of power consumption on cutting velocity (Table 9.5).

Table 9.5. *Velocity Coefficients in Ripsawing*

| Cutting velocity c m/s | 1 | 5 | 10 | 15 | 20 | 30 | 40 | 50 | 60 | 70 | 80 | 90 | 100 |
|---|---|---|---|---|---|---|---|---|---|---|---|---|---|
| coefficient $\alpha_v$ | 1.0 | 1.2 | 1.3 | 1.4 | 1.45 | 1.30 | 1.2 | 1.2 | 1.3 | 1.4 | 1.5 | 1.6 | 1.7 |

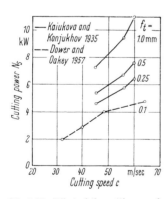

Fig. 9.35. Effect of the cutting speed on the cutting power for circular saws. From KAIUKOVA and KONJUKHOV (1935) and DOWER and OAKEY (1957)

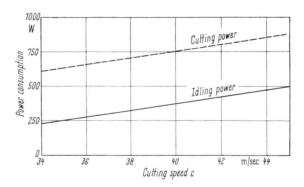

Fig. 9.36. Effect of the cutting speed on the power consumption for sawing Corsican black pine. From LOTTE and KELLER (1949)

If we disregard the very low speeds below 20 m/s, then the coefficients of BERSHADSKII form a curve with a minimum in the range between 40 and 50 m/s; for higher speeds the coefficients are directly proportional to the speed.

LUBKIN (1957, p. 21, 65) mentions that some results of TELFORD (1949) show an influence of cutting velocity in the range of 19.2 to 38.3 m/s very much like

494                                9. Wood Machining                        [Ref. p. 551

BERSHADSKII's coefficients, indicating that the mean cutting force at any given bite decreases slightly with increasing cutting speed.

### 9.2.3.3 Cutting Force and Cutting Power, Effect of Feed Rate or Feed per Tooth.
BERSHADSKII (1956) has developed empirical formulas for cutting force $P_s$ and cutting power $N_s$ based on experiments of saw behavior. Using metric units these equations may be written:

$$P_s = \frac{K_1 \cdot k \cdot d \cdot f^{1-m}}{(p \cdot \sin \omega_a)^m \left(\dfrac{k}{a}\right)^m (D\pi n/1{,}000)^{1-m}} \quad [\text{kp}] \tag{9.22}$$

$$N_s = \frac{K_1 \cdot k \cdot d \cdot f^{1-m} (D\pi n/1{,}000)^m}{6{,}120 (p \cdot \sin \omega_a)^m \left(\dfrac{k}{a}\right)^m} \quad [\text{kW}] \tag{9.23}$$

in which (partly repeated):

$K_1$     = specific cutting energy in mkp/cm³, at unit chip thickness (1 mm)
$k$       = kerf width in mm
$d$       = depth of timber cut in mm
$f$       = feed velocity in m/min
$p$       = tooth pitch in mm
$m$       = empirical constant $\approx 0.36$
$\sin \omega_a = d/b =$ sin of mean position angle (identical with the angle between the cutting velocity vector and the feed velocity vector)
$b$       = arc length of engagement of saw in mm
          = $R(\omega_2 - \omega_1)$ in counter sawing          (9.24)
          = $R(\omega_1 - \omega_2)$ in climb-sawing           (9.25)
$D$       = blade diameter in mm
$a$       = saw blade thickness in mm
$n$       = blade rotative speed in rpm

LUBKIN (1957, p. 39) has given these equations also, using English units. HARRIS (1954), based on his linear law [Eq. (9.5)] for the cutting force, derived the following equations for cutting force $P_s$ and cutting power $N_s$ for spring-set teeth:

$$P_s = k \left( K_1 \cdot \frac{i_t}{2} + \frac{K_2 f d}{D\pi n/1{,}000} \right) [\text{kp}] \tag{9.26}$$

and

$$N_s = \frac{k}{6{,}120} \left( K_1 \cdot \frac{i_t}{2} \cdot \frac{D\pi n}{1{,}000} + K_2 f d \right) \quad [\text{kW}] \tag{9.27}$$

where

$K_1$ and $K_2$ are constants,
$i_t =$ average number of cutting teeth.

MEYER (1926) and LOTTE and KELLER (1949) observed that cutting power and feed force increase linearly with feed rate (Fig. 9.37 and 9.38). It has been mentioned in Section 9.2.2.3 that SUGIHARA (1953) came to the same result for bandsaws (cf. Fig. 9.24). LAURENT (1965) has pointed out that the introduction of lateral vibrations in a saw tooth results in a reduction of over 50% in the cutting force in some cases. The experiments undertaken at the OTTAWA LABORATORY were conducted on dry, end-grain, hard maple samples. Best results were obtained within a frequency range of 50 to 200 HZ and double amplitude values of 0.25 to 0.38 mm.

The effect of feed per tooth on the cutting power has been studied by KAIUKOVA and KONJUKHOV (1934, 1935) and by SAITO and co-workers (1957). Figs. 9.39 and

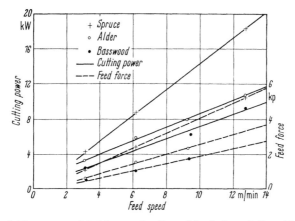

Fig. 9.37. Cutting power and feed force as functions of the feed speed. From MEYER (1926)

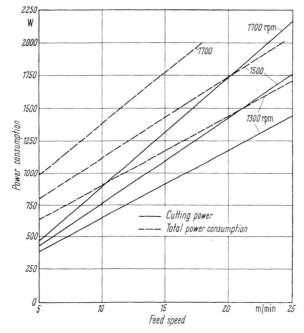

Fig. 9.38. Influence of feed speed on the power consumption of circular sawing Corsican black pine. From LOTTE and KELLER (1949)

9.40 confirm an approximately linearly-increasing relation between cutting power and feed rate or feed per tooth. Other authors (ANDREWS and BELL, 1953; ANDREWS 1954, 1955; ENDERSBY, 1953; ENGLESSON, HVAMB and THUNELL, 1954; SAITO and co-workers, 1956) also found linear variation of cutting power and feed force with feed speed. The latter authors obtained with Japanese oak species (Nara) typical, "overload"-power curves (Fig. 9.41).

Considering the equations for cutting force and cutting power by BERSHADSKII and HARRIS [Eqs. (9.22), (9.23), (9.26), (9.27)] it may be mentioned that all the

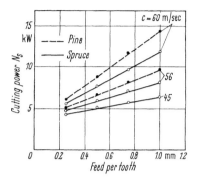

Fig. 9.39. Relationship between cutting power and feed per tooth for various cutting velocities. From KAIUKOVA and KONJUKHOV (1935)

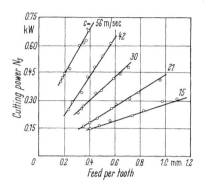

Fig. 9.40. Relationship between cutting power and feed per tooth for various cutting velocities. From SAITO and co-workers (1957)

straight-line functions represented may be approximated parts of power-law curves. In agreement with the cutting-force law the character of the curves should be a degressive one. Such curves were published by DOWER and OAKEY (1957) for ripsawing ponderosa pine planks with carbide-tipped circular saws (Fig. 9.42). PAHLITZSCH (1962) points out that even a progressive trend of the curves for higher feed per tooth would not be surprising due to larger amounts of chips to be detached.

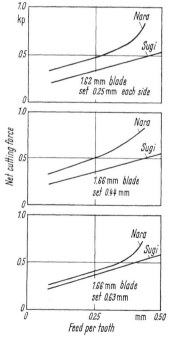

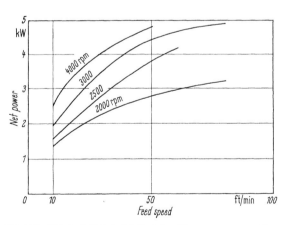

Fig. 9.41. Effect of feed per tooth on net cutting force. From SAITO and co-workers (1956)

Fig. 9.42. Effect of feed speed on net cutting power consumption. From DOWER and OAKEY (1957)

### 9.2.3.4 Specific Cutting Energy.

The specific cutting energy $E_s$ is, together with the specific cutting force, a very instructive figure for comparing and evaluating the efficiency of cutting tools. In sawing, specific cutting energy $E_s$ is usually defined as (LUBKIN, 1957, p. 33):

$$E_s = \frac{\text{net cutting energy consumption}}{\text{volume of kerf removed}} \qquad (9.28)$$

LUBKIN (1957, p. 34, 35) simply writes:

$$E_s = B(1 + A/Bt_a) \qquad (9.29)$$

or applying the power-law:

$$E_s = E_{s1}/t_a^m \qquad (9.30)$$

where

$$m = 1 - m_0 \text{ lies between 0 and 1 } (\approx 0.36 \text{ in sawing}). \qquad (9.31)$$

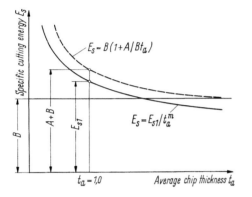

Fig. 9.43. Specific cutting energy curves as influenced by average chip thickness. From LUBKIN (1957)

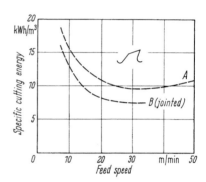

Fig. 9.44. Specific cutting energy for a circular ripsaw effected by feed speed. From AZZOLA (1954)

Typical metric units for $E_s$ are Ws/cm², kWh/m³ or mkp/cm³. An example for an $E_s$ versus $f$-curve in bandsawing is given in Fig. 9.24. Eqs. (9.29) and (9.30) are plotted schematically in Fig. 9.43. Both show that it consumes less energy or power to remove a given volume of wood with a thick chip than with a thin one. The reason is that the strength properties of solid bodies (especially of wood) become smaller with increasing size due to more frequent internal weak spots (e.g. defects) which are statistically distributed.

It should be remembered (Section 9.2.2.3) that the hyperbolic curves, as shown in Fig. 9.43, may be changed by "overloading" (cf. Fig. 9.27) so that they pass a minimum with a subsequent rise. AZZOLA (1954) published some typical examples of specific energy curves which exhibit a minimum. In Fig. 9.44 curve $A$ was obtained with a conventional ripsaw ($D = 436$ mm, $i = 32$ teeth, $h_t = 11{,}5$ mm, $p = 42.5$ mm, $\alpha = 43°$, $\beta = 40°$, $\gamma = 7°$, $a = 2.8$ mm, $s = 0.5$ mm). The second curve $B$ shows the favorable influence of jointing[1] the same saw blade. The specific cutting energies are reduced by some 20%.

### 9.2.3.5 Effect of Depth of Timber Cut and of Grain Orientation.
It has already been stated in section 9.2.2.5 that, for all types of saws, cutting force and cutting

---

[1] The term "jointing" unfortunately has different meanings in wood technology, e.g. smoothing of one surface of a board, uniting parts by glue or other joints, or as used here, equalizing a saw by bringing all cutting edges to a common cutting circle at operating rpm (KOCH, 1964, p. 150, 285).

power increase proportional to the depth of cut [cf. also Eq. (9.6)], of course under the assumption that the gullets are not clogged.

The theory is generally confirmed by tests of different technologists (MEYER, 1926; KAIUKOVA and KONJUKHOV, 1934, 1935; ENDERSBY, 1953; ANDREWS and BELL, 1953; ANDREWS, 1955). Table 9.6 shows the results of the latter.

Table 9.6. *Cutting Power as a Function of Depth of Cut*
(From ANDREWS, 1955)

Depth of cut $\dfrac{\text{in.}}{\text{mm}}$

| Species | $\dfrac{2}{50.8}$ | $\dfrac{4}{101.6}$ | $\dfrac{6}{152.4}$ | $\dfrac{8}{203.2}$ | $\dfrac{10}{254}$ | $\dfrac{12}{304.8}$ | $\dfrac{14}{355.6}$ |
|---|---|---|---|---|---|---|---|
| White pine | 10 | 20 | 29 | 39 | 49 | 59 | 68 |
| White birch | 14 | 29 | 44 | 59 | 73 | 88 | 102 |
| Yellow birch | 16 | 33 | 48 | 65 | 79 | 95 | 111 |
| Sugar maple | 18 | 37 | 54 | 71 | 89 | 106 | 123 |

The trend of these figures is clear.

TELFORD (1949) found that cutting torque is a linear function of depth of cut, but is not proportional to it. LUBKIN (1957) criticizes, with good reason, that the linear function leads to negative cutting-force values at small depths of cut, which is physically impossible.

A further problem in circular sawing is the influence of blade protrusion (= extension of sawblade beyond workpiece in mm) on the performance of the

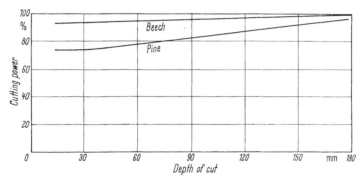

Fig. 9.45. Effect of depth of cut as a measure of blade protrusion on net cutting power, for a given blade diameter. From LOTTE and KELLER (1949)

saw. Though general rules may not be given, as far as power consumption is concerned, the results of studies agree very well. LUBKIN (1957) summarizes: "At fixed feed rate and depth of cut, the effect of increasing the blade protrusion beyond the minimum possible value is to cause a small increase in power consumption, usually of negligible interest from the economic point of view". The curves observed by LOTTE and KELLER (1949) and shown in Fig. 9.45 verify this statement. The problem of optimum blade protrusion is much less a question of energy consumption than a matter of safe saw operation. Too large protrusion may increase the risk of a "kickback" of the workpiece which is a very dangerous event for the worker.

Systematic experiments on the influence of grain orientation on cutting forces are still scarce, but McMILLIN and LUBKIN (1959) made a very interesting contribution to this problem using an overcutting radial arm saw (diameter 16 inches = 406 mm). The workpiece was flat-sawn, knot-free hard maple at 11% moisture content. The angle $\Theta$ referred to in this report is the angle between the direction of the movement of the saw blade and the crosscutting direction, i.e.

$\Theta = 0$ degrees when crosscutting,
$\Theta = 45$ degrees on a miter cut,
$\Theta = 90$ degrees when ripsawing.

For climb-sawing the net cutting power $N_s$ can be related to grain orientation by the following equation:

$$N_s = d(C_1 + C_2 f)(1 + C_3 \sin^2 \Theta) \ [\text{HP}] \qquad (9.32)$$

where, for the particular conditions of the experiment, the constants had the following empirically determined values: $C_1 = 0.284$, $C_2 = 0.0548$, $C_3 = 1.343$; the unit of the feed speed $f$ is in this case fpm.

Fig. 9.46 shows the experimentally observed points and the curves calculated with Eq. (9.32) using the foregoing constants. The fit of the equation is remarkably good. It is evident, as opposed to the prevalent opinion, that ripping (in this particular case) requires more cutting power than crosscutting.

### 9.2.3.6 Effect of Blade Diameter and Blade Thickness.
LUBKIN (1957) has discussed, in his excellent status report on circular sawing, the effect of blade diameter. We refer to his quotation of KAIUKOVA and KONJUKHOV (1934, 1935): "With an increase in the diameter of the saw, all the other conditions being the same, the effort of feeding decreases, while the cutting effort increases. MEYER's (1926) conclusion that a larger value of the ratio $D/h$ (diameter of the blade to distance between workpiece and axis of saw rotation) should be taken, is of practical importance when work is being done on benches where the material is fed by hand." LUBKIN (1957) then examined the effects of varying sawblade diameters by means of BERSHADSKII's equations (9.22 and 9.23). He concludes that no great change in net cutting power should be expected but a very slight increase with diameter. The thickness of circular sawblades must be adequate for avoiding vibrations. If saw thickness is inadequate vibrations occur and the surface quality becomes poor. There are some empirical formulas relating to sawblade thickness $a$ and diameter $D$. According to VOIGT (1949) one can assume:

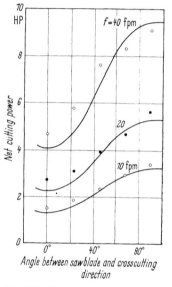

Fig. 9.46. Effect of the angle between sawblade and crosscutting direction on net cutting power. From KOCH (1964)

$$a = G\sqrt{D}, \ \text{mm}. \qquad (9.33)$$

To provide sufficient lateral stiffness, the constant $G$ is normally 0.1, but may vary from 0.075 to 0.142. The heavier the load and the higher the feed rate, the thicker should be the blade.

The German AWF (cf. Section 9.2.1.4) recommends a straight-line function:

$$a = A + D/B \text{ mm}. \tag{9.34}$$

As a rule one can insert the following constants in Eq. (9.34):

$$A = 0.45, \quad B = 260.$$

If the blade is supported by extra-large collars, or otherwise, the thickness may be lowered by using the constants:

$$A = 0.65, \quad B = 460.$$

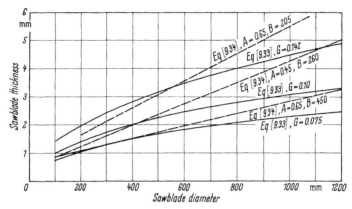

Fig. 9.47. Relationship between sawblade thickness, cutting depth and sawblade diameter

If cutting deep, in hardwoods, with automatic feed, extra thick blades, having a diameter of at least 200 mm should be used; in this special case the constants are as follows:

$$A = 0.65, \quad B = 205.$$

Fig. 9.48. Different customary circular sawtooth forms

Fig. 9.47 shows the straight-lines and curves calculated with Eqs. (9.33) and (9.34) using the mentioned constants. The equations do not apply to blades with variable thickness.

**9.2.3.7 Effect of Tooth Geometry and Pitch.** As mentioned in section 8.2.3.1, circular saws are very versatile and have a broad field of application. Therefore there are many types of suitable teeth. However, a general rule is, that all blades should be truely round and exactly set, both dimensions being held to the nearest 0.1 mm (0.004 in.). The blades must be properly tensioned.

The AWF recommends 40 to 50 teeth on all ripsaw blades up to 400 mm diameter, to obtain a surface of high quality. Different customary sawtooth forms are shown in Fig. 9.48. In general, mean tooth pitch $p$ may be calculated from blade diameter $D$ and number of teeth $i$ by using Eq. (9.15). The tooth height $h_t$ of ripsaws normally conforms to

$$h_t = 0.5p + 1 \cdots 3, \text{ [mm]} \tag{9.35}$$

where the tooth height $h_t$ is measured along a radial line through the tooth tip.

Ref. p. 551]                9.2 Technology of Sawing                501

The smaller the pitch, the better the quality of cut, assuming that the feed rate is small enough. Pitches of less than 15 mm are uncommon for ripsaws. In cross-cutting, smaller pitches must be employed to get good surface quality.

Recommended tooth angles are listed in Table 9.7.

Table 9.7. *Tooth Angle Recommendations* (From KOLLMANN 1955)

| Angle degress | Ripsaws | | Crosscut Saws | |
|---|---|---|---|---|
| | Softwoods (Sp & W) | Hardwoods (W) | Sp | W |
| Clearance $\gamma$ | Sp 33 (AWF) <br> Sp 35 (Br) <br> W 20 (FPRL) <br> W 16 (INB) | 20···24 (AWF) <br> 24···28 (Br) <br> 15 (FPRL) | 37···38 (Br) <br> 45···60 (FPRL) | 20···24 (AWF) <br> 30 (Br) <br> 25 (FPRL) |
| Sharpness or Wedge $\beta$ | Sp 39 (AWF) <br> Sp 37 (Br) <br> W 40···45 (FPRL) <br><br> W 34···39 (INB) | 38···42 (AWF) <br> 40 (Br) <br> Dry wood 55···65 <br> (FPRL) <br> Green wood 50···65 <br> (FPRL) | 45···49 (Br) <br> 50···60 (FPRL) | 55···60 (AWF) <br> 44 (Br) <br> 70 (FPRL) |
| Rake $\alpha$ | Sp 18 (AWF, Br) <br> W Dry wood, 25 <br> (FPRL) <br> W Green wood, <br> 25···30 (FPRL) <br> W 35···40 (INB) | 28 (AWF) <br> 20···26 (Br) <br><br> Dry wood 10···20 <br> (FPRL) <br> Green wood 10···25 <br> (FPRL) | 5···15 (AWF) <br> 3···8 (Br) <br><br> −10···−30 <br> (FPRL) | 5···15 (AWF) <br> 16 (Br) <br><br> −5···−10 <br> (FPRL) |
| Cutting $\varDelta = \gamma + \beta$ | Sp 72 (AWF, Br) <br> Sp 80···85 (H) <br> W 60···65 (FPRL) <br> W 50···55 (INB) | 58···66 (AWF) <br> 64···68 (Br) <br> 60 (H) <br> Dry wood 70···80 <br> (FPRL) <br> Green wood 65···80 <br> (FPRL) | 75···85 (AWF) <br> 80···85 (H) <br> 95···120 <br> (FPRL) | 75···85 (AWF) <br> 95 (FPRL) |

*Symbols and Abbreviations*: Sp = „Spitzwinkelzahn"; W = „Wolfszahn"; AWF = (German) Ausschuß für wirtschaftliche Fertigung; FPRL = (British) Forest Products Research Lab., Princes Risborough; INB = (French) Institut National du Bois, Paris (now Centre Technique du Bois); Br = (German) Braunshirn; H = HÜTTE, Des Ingenieurs Taschenbuch, Vol. II, 27th Ed. (a German Handbook).

Reference may also be made to HARRIS' (1947) recommendations in his original paper, or in LUBKIN's (1957) status report.

The effects of tooth angles have been very intensively studied. MEYER (1926) obtained, for the effect of clearance angle on power demand, the curves given in Fig. 9.49. All species tested demanded minimum power in the range between 14° and 16° clearance angle. MEYER explained the power increase with clearance angle decrease as follows: In the course of chip formation the wood fibers under the cutting edge are compressed until the chip is detached; at this instant the ends of the compressed fibers are relieved and spring upwards, rubbing on the back of the tooth. The smaller the clearance angle, the larger is the rubbing surface and there-fore the greater the friction loss and power demand. KAIUKOVA and KONJUKHOV (1934, 1935) examined the relationship between clearance angle, cutting power, and feed force. Their results, showing a miminum for net power near 15° to 16° clearance angle are in good agreement with MEYER's results.

SAITO and co-workers (1955) obtained the curves shown in Fig. 9.50. The unusually large clearance angles (25° to 40°) for the point of minimum power are noteworthy. SKOGLUND and HVAMB (1953) observed that the clearance angle had only a small effect on cutting force. LUBKIN (1957) has discussed the problem and pointed out, in agreement with KOLLMANN (1955), that the sharpness of the cutting edge has a decisive influence on the results of various investigators. He concludes that the results of MEYER (1926), KAIUKOVA and KONJUKHOV (1934, 1935), SAITO and co-workers (1955) might be characterized as typical "normal-edge" sawing

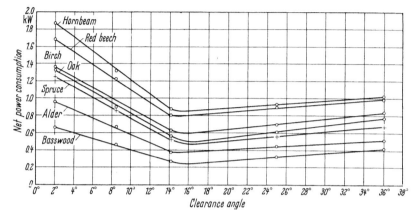

Fig. 9.49. Effect of clearance angle on net power consumption. From MEYER (1926)

results. Those of SKOGLUND and HVAMB (1953) and KIVIMAA (1950), in planing, are typical "sharp-edge" results. Therefore, only the results with "normal-edges" are representative for normal circular sawing.

The tooth angles are not only of great importance for saw performance but rake angle, tooth angle and clearance angle may not be varied independently of each

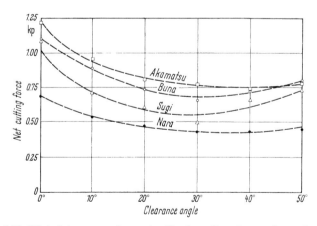

Fig. 9.50. Effect of clearance angle on net cutting force. From SAITO and co-workers (1955)

other. Any variation of one tooth angle causes a change of another angle or of both of them. MEYER (1926) examined the effect of varying rake angle. He found a linear decrease in power demand as the rake (or hook) angle is increased. All wood species tests (moisture content 9.5 to 14.3%) exhibited the same general behavior. By contrast, LOTTE and KELLER (1949) found that the influence of the rake angle on power consumption in sawing beech hardwood was rather small but in sawing pine it was very pronounced and showed about the same tendency as observed by MEYER (Fig. 9.51). KAIUKOV and KONJUKHOV (1934, 1935) and ENDERSBY (1953) also reported a slight decrease with increase in rake angle. Assuming

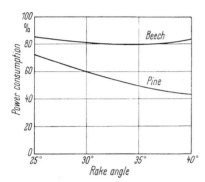

Fig. 9.51. Effect of rake angle on power consumption in sawing beech and pine. From LOTTE and KELLER (1949)

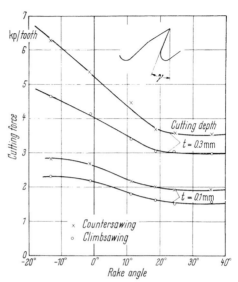

Fig. 9.52. Effect of rake angle on cutting force per tooth in countersawing and climbsawing of pine. From SKOGLUND and HVAMB (1953)

that the power required $N_s$, in the range of angles tested, is roughly a linear function of rake angle $\alpha$ and also of depth of cut $d$, he gives the equation

$$N_s = (-m \cdot \alpha + C) d \qquad (9.36)$$

where $m$ and $C$ are suitable constants.

ENDERSBY comments on his findings as follows: "In the test, a change of hook from 15° to 30° caused the power to fall by about 12%, which is equivalent to 1.0 and 0.4 hp for depths of cut of 6 and 2 in., respectively. Thus the use of a little more hook gives a worthwhile reduction in power if the depth of cut is substantial." HARRIS, 1947 (rev. 1955) came to similar conclusions. SCHIMMING (1952/53) points out that the variation with rake angle is not linear, but exhibits a concave-downward trend in the range from −11 to +17°. SKOGLUND and HVAMB (1953), using a pendulum dynamometer which allows very exact measurement of the energy consumption but which restricts the cutting speed to rather low values, e.g. to 6.3 m/s, as opposed to 40 to 60 m/s in normal circular sawing operations, obtained the results plotted in Fig. 9.52. Cutting forces decrease rather rapidly with increasing rake angle in the range from −15° to +15°. The influence is more pronounced for thick chips (0.3 mm) than for thin chips (0.1 mm). SKOGLUND and HVAMB state that 20° to 30° is the optimum rake angle for general rip sawing. Most of the investigators dealing with rip sawing pointed out that too large rake angles are disadvantageous; for this reason they limited their tests to about 40° of rake. In fact, there is proof that power consumption tends to increase more or less rapidly with rake angles going beyond about 40°. SAITO and co-workers (1957) noticed that the minimum power required occurs at 20° to 25° rake angle with a rapid increase beyond the maximum (Fig. 9.53). The shape of their curves correspond well with curves obtained by FOYSTER (1953) and completely reinterpreted by LUBKIN (1957, p. 151).

In manual feeding of ripsaws, and in the case of crosscutting, a slight inclination or bevel of the top of the tooth is usual. As a rule, power consumption is reduced as tooth bevel angle is increased. A linear relationship has been found by MEYER (1926) (Fig. 9.54) and KAIUKOVA and KONJUKHOV (1934, 1935). According to EMMERICH (no date) a bevel-ground saw cuts better than a square-ground saw at the beginning of its operation period. The power consumption is lower for the first 30 to 35 min, about equal for the next 30 to 35 min, then progressively greater thereafter.

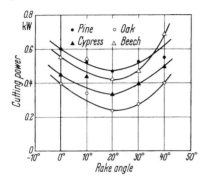

Fig. 9.53. Effect of rake angle on cutting power. From SAITO and co-workers (1957)

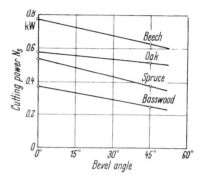

Fig. 9.54. Effect of bevel angle on cutting power. From MEYER (1926)

Another reference (Anonymous, 1954) states that top bevel has been gradually replaced by square-ground teeth, since the latter provide better kerf quality. SAITO and co-workers (1957) present curves showing a minimum of cutting power required for a bevel angle between 20 and 25°. LOTTE and KELLER (1949) recommend a top bevel of 25° in ripsawing which gives remarkably improved surfaces.

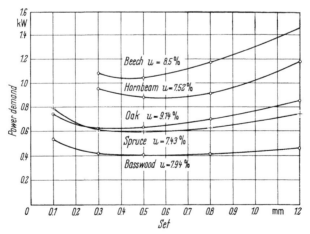

Fig. 9.55. Effect of set on power requirement. From MEYER (1926)

SKOGLUND and HVAMB (1953) used a pendulum dynamometer in investigating the effect of top bevel on cutting force. Their statement that a top bevel angle (in the range of 0 and 25°) gives only a small reduction in cutting force is contradictory to the opinions expressed by other authors. LUBKIN (1957), discussing this discrepancy, came to the conclusion that the results of SKOGLUND and HVAMB may not be representative for sawing. Full agreement exists with respect to the effect of the width of set. There is a minimum of power consumption for a set $s = 0.5$ mm to each side of saw blades (Fig. 9.55). The reasons for this phenomenon are explained in Section 9.2.1.5.

PAHLITZSCH (1962) points out that the effects of tooth geometry are principally the same for all sawing methods with the exception that for gang saw blades the rake angle may lie below $\alpha = 10°$ (cf. Section 9.2.1.4) and the width of set on each side may reach about 0.75 mm.

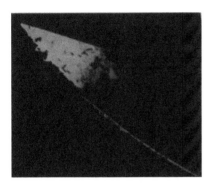

Fig. 9.56. Photograph of chip formation in a circular saw blade with 350 mm diameter, 8 teeth, 2.3 mm thickness, 137 mm pitch, 2.8° clearance angle, 56.8° lip angle and 30.4° rake angle; swaging 0.5 mm. From THUNELL (1951)

Fig. 9.57. Photograph of chip formation in a circular saw blade with 400 mm diameter, 110 teeth, 2.01 mm thickness, 11 mm pitch, 39.0° clearance angle, 47.6° lip angle and 3.4° rake angle; set 0.15 mm. From THUNELL (1951)

**9.2.3.8 Chip Formation.** Formation and movement of chips in the gullet is of great importance for the whole sawing process, for the stresses in the blade, and for surface quality. The formation of chips can be studied using a light source capable of a 1 μs flash duration. THUNELL (1951) published and discussed a series of photographs taken in such a way. Fig. 9.56 shows that the amount of chips in the gullet for circular blades, with only a few teeth, is very much larger than for normal teeth (Fig. 9.57). The last figure also elucidates how unequally the different tooth-edges cut, though the blade was jointed after tensioning. Therefore, the stresses in the different teeth must vary over a wide range.

KOCH (1954), studying peripheral milling, described three distinct types

Fig. 9.58. Type I chip. Photograph by FRANZ (1958)

of chips. Subsequently FRANZ (1958) confirmed the occurrence of these three basic chip types when machining wood parallel to the grain. FRANZ classifies them as follows:

Type I  Formed when cutting conditions are such that the wood splits ahead of the tool by cleavage until failure in bending as a cantilever beam occurs, as shown in Fig. 9.58.

Type II  Occurs when wood failure in the chip is along a line extending from the cutting edge to the work surface, as indicated in Fig. 9.59a.

Type III  Results when tool forces cause compression and shear failures in the wood ahead of the cutting edge, as shown in Fig. 9.59b.

A more or less abrupt or regular transition from one chip type to another may occur.

Cutting perpendicular to the grain (with the cutting edge either parallel to a tangent on the annual rings, or parallel to the rays or in an angle between these orientations) chip formation and failure type are different.

Fig. 9.59 a,b. a) Type II chip, b) Type III chip. Photographs by FRANZ (1958)

**9.2.3.9 Thermal Effects, Stresses, and Stability of Circular Saw Blades.** Sawing is always combined with the development of heat by friction. The largest proportion of the heat-loss is carried away with the chips or is radiated to the surrounding air. A part of the friction-heat increases the temperature of the sawblade. Normally the heat is produced by the cutting processes at the tooth-tips. From there the heat is conducted to the center of the blade. Usually the highest temperatures occur at the tooth-tips. The uneven distribution of temperatures causes

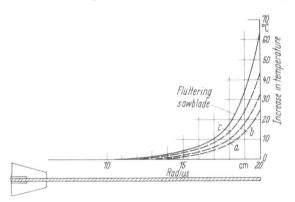

Fig. 9.60. Temperature distribution over the blade radius in a circular saw blade (400 mm diameter, 2 mm thickness, 2.400 rpm) heated by friction at its edge. From SKOGLUND (1950)

stresses in the saw blade. Skoglund (1950), heating a circular sawblade without teeth by friction at its edge, observed the temperature-curves shown in Fig. 9.60. One can see how rapidly the temperature decreases due to air-cooling towards the center of the blade. The largest part of the blade remains comparatively "cold" and impedes free extension of the heated parts. The consequences are high compressive stresses at the outer edge of the blade. Fig. 9.61 shows the thermal stresses developed in a circular sawblade (400 mm diameter, 2,400 rpm, temperature gradient over the blade radius 30 °C). For the conditions mentioned the maximum compressive stress at the outer edge of the blade would reach about 630 kp/cm². Stresses of this magnitude may lead to buckling or at least large amplitude transverse vibrations of the blades. This results in still further heating

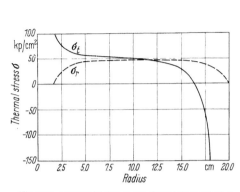

Fig. 9.61. Tangential ($\sigma_t$) and radial ($\sigma_r$) thermal stresses developed in a circular sawblade (400 mm diameter, 2.400 rpm) for a temperature gradient over the blade radius 30°C. From SKOGLUND (1950)

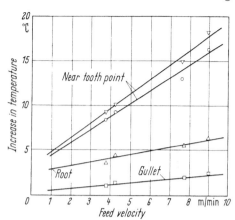

Fig. 9.62. Increase in temperature at various points of a circular sawblade in relation to feed velocity (400 mm diameter, 2.0 mm thickness, 2.400 rpm). From SKOGLUND (1950)

of the blades, "blue spots", increased power consumption, and a loss of dimensional tolerance of the work piece (MOTE, Jr., 1964). The heating of sawblades also depends on external factors such as feed speed (Fig. 9.62), depth of timber cut (Fig. 9.63), and time of sawing (Fig. 9.64). The oscillations of the curves in Fig. 9.64 may be

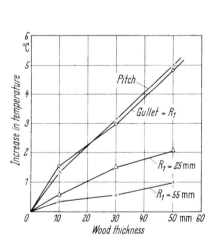

Fig. 9.63. Increase in temperature at various points of a circular sawblade in relation to wood thickness (400 mm diameter, 2.400 rpm, feed velocity 4.3 m/min). From SKOGLUND (1950)

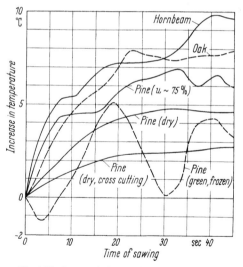

Fig. 9.64. Increase in temperature at the tooth root of a circular sawblade in relation to the time of sawing (40 mm wood thickness, 4.3 m/min. feed velocity). From SKOGLUND (1950)

partly explained by the fact that in sawing over a longer period parts of wood of varying properties (e.g. density or moisture content) must be cut. The following conclusions may be drawn: 1. The rise of temperature is proportional to the specific gravity of the wood; 2. Pinewood of high moisture content is heated more than that having low moisture content; 3. The elevation of temperatures is higher in sawing along than across the grain.

Temperature distribution, rotational speed, blade thickness, etc., determine circular saw stability "Saw blade vibrations are composed of a combination of specific amplitude distributions and corresponding frequencies of vibration. These frequencies are designated as the natural frequencies of vibration of the system. In many instances, external forces acting on a system will oscillate. For example, external forces acting on a saw blade result from intermittent cutting at the rim and from transverse shaft vibration at the hub. Resonance occurs when the frequency of oscillation of an external force is equal to one of the structural natural frequencies (usually the lowest). The resulting large amplitude oscillations cause structural failure, decreased fatigue life, and many other obvious circular saw problems. Resonance usually precedes actual buckling, and may, indirectly, account for the actual buckling itself." (MOTE, Jr., 1964). Resonance is more frequent in thin than in thick blades. Saw blade stability is improved by tensioning. Tensioning is effected by hammering of the saw. An analysis of the blade temperature and stress distributions (LINDHOLM, 1950; BARZ, 1953, 1957; SAITO and NIGA, 1954; SKOGLUND, 1949; SUGIHARA and SUMIYA, 1955) and of the deformations (strains) is necessary in order to predict the blade stability. A blade is optimally tensioned if the fundamental frequency of oscillation is as large as possible (MOTE, Jr., 1964).

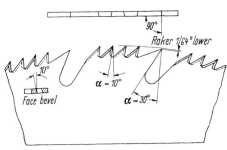

Fig. 9.65. Tooth style for hollow-ground combination planer saw. From KOCH (1964)

There are other methods of avoiding thermal buckling such as cooling the blade rim, heating the interior of blade with friction packings, expansion slits, etc.

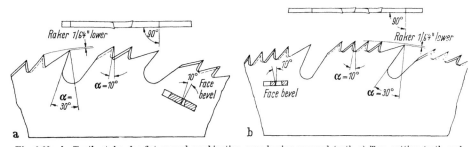

Fig. 9.66a, b. Tooth styles for flat-ground combination saws having grouped teeth a) Two cutting teeth and one raker tooth per section, b) Four cutting teeth and one raker tooth per section. From KOCH (1964)

**9.2.3.10 Special Types of Circular Saw Blades.** Hollow-ground combination planer saws having grouped teeth with a minimum of clearance generate excellent surfaces. Feed per tooth is limited. Tooth style is shown in Fig. 9.65. The raker teeth remove the kerf material. Flat-ground combination saws, also with grouped teeth, have cutting teeth springset for clearance (Fig. 9.66). These saws require more power but can withstand higher feed speeds and will rip, miter and crosscut.

Dado heads are assemblies of two hollow-ground outside saws in combination with one or more inside cutters. They are designed to cut smooth-surfaced grooves from 3 to 100 mm wide in any grain direction (KOCH, 1964, p. 271). "Kickback-free" safety blades possess only a few teeth, e.g. 6 to 10 teeth for a blade with 350 mm diameter. Fig. 9.67 illustrates the tooth style; side clearance may be obtained either by spring-set teeth or by swage-set teeth. It may be recalled that in sawing net cutting power can be expressed by a linear law [Eq (9.29)] or by a power law [Eq. (9.30)]. KOCH (1964), based on an exhaustive analysis of cutting-force, power, and energy relationships (expressed as functions of average chip thickness), came to the conclusion that, if all elements of the power equations are held constant except number of teeth $i$, and pitch $p$ (in other words, saw diameter $D$, saw rpm $n$, kerf width $k$, depth of cut $d$, cutting velocity $v$, and length of arc of

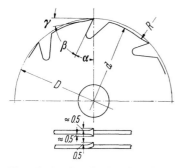

Fig. 9.67. Tooth style of a "Wigo" circular sawblade. $D$ maximal saw diameter, $d$ saw diameter without the tooth limitation, $R$ radius of the curved sector, $\alpha$ hook angle, $\beta$ sharpness angle, $\gamma$ clearance angle

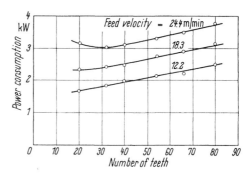

Fig. 9.68. Effect of the number of teeth on the power consumption for cutting speed 51 m/s. From ENDERSBY (1953)

tooth engagement with work piece $b$ are all held constant) then Eqs. (9.27) and (9.29) for net cutting power $N_s$ can be rewritten as

$$N_s = \frac{C_1}{p} + C_2 \qquad \text{linear law} \quad (9.37)$$

and Eq. (9.30) can be rewritten as:

$$N_s = \frac{C_3}{p^{1-m_0}} \qquad \text{power law} \quad (9.38)$$

where $C_1$, $C_2$, and $C_3$ are all suitable new constants and $(1 - m_0)$ is on the order of 0.36 as previously mentioned.

Considering that pitch $p$ is defined as $\pi D/i$, the net cutting power can be expressed as a function of the number of teeth $i$ by formulas as follows:

$$N_s = C_4{}^i + C_5 \qquad \text{linear law} \quad (9.39)$$

or

$$N_s = C_6{}^{i^{1-m_0}} \qquad \text{power law} \quad (9.40)$$

where $C_4$, $C_5$, and $C_6$ are all suitable new constants and $1 - m_0$ is about 0.36.

Experiments by ENDERSBY (1953) confirm this theoretical deduction (Fig. 9.68). The following advantages accompany the use of saws with few teeth, i.e. saws with long pitch:

1. No kickback, therefore safe operation;
2. Reduced (to 65 to 70%) energy consumption;
3. Reduced feed force; LINDNER (1949) determined a reduction between 30 and 40%;
4. Longer operational periods between sharpening (DOSKER, 1950);
5. Easy maintenance and sharpening;
6. Quiet operation.

REINEKE (1950) at the U.S. Forest Products Laboratory designed the Duo-Kerf Ripsaw. Shape and action are illustrated by Fig. 9.69. The gullets between chipper teeth (Ch) and side-dresser teeth (SD) may be alike or unlike, as indicated by solid and broken-line profiles. Side-dresser teeth have parallel sides and concave face and are lower than chipper teeth by $d$ which is related to feed speed. The side sketch shows how kerf material is removed in sequences of cutting actions by chipper and side-dresser teeth. The first chipper tooth starts the cut with a chip $k_i$ wide. Next, the first side-dresser tooth cuts material from both sides of the kerf to a width $k_f$ and to a depth slightly off the bottom of the first chipper-tooth cut. Following this kerf widening, the second chipper tooth advances the narrow kerf, ejecting the sawdust into the widened kerf behind it. The second side-dresser tooth then widens most of this new chipper kerf along with the remnant left by the first side dresser preparing the kerf for easy removal of the chip produced by the third chipper tooth.

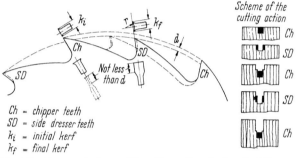 

Fig. 9.69. Shape and scheme of the cutting action of a Duo-Kerf Ripsaw. From REINEKE (1950)  Fig. 9.70. Schematical top view of a Lundberg circular saw

Duo-Kerf principles may be applied also to inserted tooth circular saws. Advantages are power saving (20 to 25%) and surface improvement. The smooth surfaces allow reduction in side clearance which means less kerf. In Sweden the LUNDBERG circular saw has been developed. By special set sharpening according to Fig. 9.70, Duo-Kerf is also produced. The tips $A$ of the saw blade first cut a small kerf, then the tips $B$ cut on both sides as dressers. The LUNDBERG circular saw is said to have the following advantages:

1. Smooth cutting even in wood with cross grain. Uniform loading of tooth tips.
2. Constancy of set.
3. Reduction of set and therefore reduction of energy consumption thus producing higher yield.
4. Suitability for ripping as well as for crosscutting.
5. Application for sawing dry or wet wood.

### 9.2.4. Chain sawing

**9.2.4.1 Introduction.** Chain saws were first utilized in the USA about 1915. Their design and construction were markedly improved after 1930 and since that time they have found increasing applications in the forest — for felling and related operations —, in mills, and on wood constructions activities. A chain saw consists of a motor-driven cutting plate. The chain is a combination of right hand cutters, left hand cutters, right hand rakers, left hand rakers, tie straps, and rivets.

Fixed saws may be driven by any power source. In the USA heavy (30 HP) chain saws on a hydraulically moved carriage are used for bucking large logs up to 2.5 m (~ 8′) in diameter. The chains have a 19 mm pitch and a kerf width of about 14 mm (9/16″). Fixed or semiportable, air or hydraulically fed machines can be used for the following operations:

1. Log and veneer bolt bucking in the pond;
2. Crosscutting of bundled small logs, timbers, or flitches;
3. Log ripping.

Portable saws are powered by light gasoline engines or, occasionally, by electric motors. There are chain saws with direct drive (chain speeds 12 to 19 m/s) or with gear drive (chain speeds 5 to 11 m/s).

**9.2.4.2 Machine Types.** There are two principal machine types:

1. Chain saws equipped with a straight bar. One-man portable machines have bar lengths from 35 to 150 cm (about 14 to 60 inches), power consumption ranges from 2 to 6 HP.

Two-man portable machines have bar lengths from 90 to 365 cm (about 36 to 144 inches) with power consumption from 5 to 10 HP.

Kerf width for the smaller machines is 6 to 8 mm and for the larger ones (with gear drive) up to 12 mm.

2. Chain saws equipped with bow. These curved rim chain saws are used primarily to

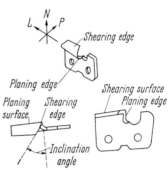

Fig. 9.71. Detail of chain saw tooth and definition of the three principal directions: $L$ longitudinal, $P$ perpendicular, and $N$ normal. From GAMBRELL Jr. and BYARS (1966)

avoid pinching the saw blade when bucking small round wood on the ground. They enable the operator to make plunging cuts and thus avoid stooping unnecessarily (KOCH, 1964, p. 277).

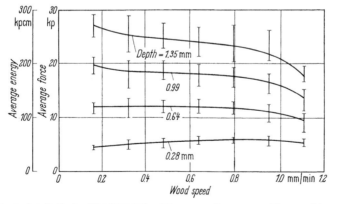

Fig. 9.72. Average longitudinal cutting force and energy consumption versus cutting speed for various depths of cut. Curves show both average values and range of variation. From GAMBRELL Jr. and BYARS (1966)

**9.2.4.3 Chip Formation, Power Requirement.** An initial study of the effects of several major parameters on cutting forces, energy requirements, and chip formation when cutting wood with chainsaw teeth was carried out by GAMBRELL JR. and BYARS (1966). Cutting forces and energy were measured over a speed range from 2.5 to 18.5 m/sec, at depths of cut of about 0.25 to 1.5 mm (0.010 to

0.060 in.), when cutting green red oak with chipper-type saw teeth (Fig. 9.71). The data plotted in Fig. 9.72 indicate that cutting force and power requirements in the longitudinal direction, relative to a chain-saw tooth, are dependent on depth of cut and relatively independent of cutting speed. Fig. 9.73 shows that the energy consumed by cutting in the longitudinal directions is relatively independent from cutting angle of the saw tooth.

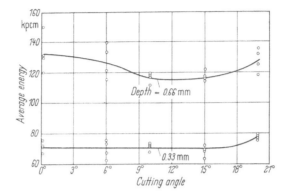

Fig. 9.73. Energy consumption in longitudinal direction versus cutting angle for two depths of cut. From GAMBREL Jr. and BYARS (1966)

"The mechanics of chip formation with chain-saw teeth can be broken into two basic actions resulting from the cutting characteristics of the shearing and planing edges. The shearing edge, having an inclination angle, separates the cellular structure of the wood by wedge action and directs the resulting chip towards the cutter bar at angle to the plane of the cutter bar. The planing edge,

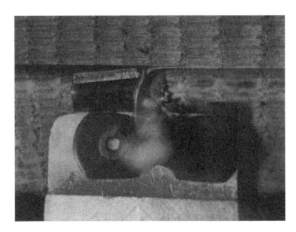

Fig. 9.74. High-speed photograph of chip formation. From GAMBRELL Jr. and BYARS (1966)

being essentially orthogonal to the cutting velocity vector, severs or crushes the wood cell walls and directs the resulting chip in a plane perpendicular to the cutter bar. This action produces a chip whose characteristics depend upon the depth of cut but not directly upon cutting speed" (GAMBRELL JR. and BYARS, 1966). A high-speed photograph of chip formation is shown in Fig. 9.74.

## 9.3 Proposed Methods of Chipless Wood-Cutting

### 9.3.1 Peeling and Slicing

Rotary peeling and slicing are well-known techniques for the production of veneers but not yet suited for the conversion of round logs into boards, planks and similar kinds of lumber. Nevertheless, there are already some machines for the manufacture of thin (up to 10 mm) small boards for cheap boxes.

One operates with a sturdy knife clamped into a sliding sash which is moved like a reciprocating gang saw by means of a crank shaft and a connecting rod. Another machine uses a disc with knives inserted in radial slits. Prior to the slicing operation the wood in form of a square block is plasticized by steaming.

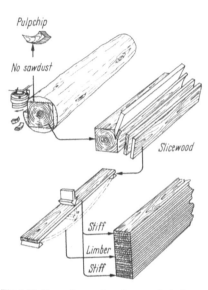

Fig. 9.75. Processing system for manufacturing laminated beams. From KOCH (1964)

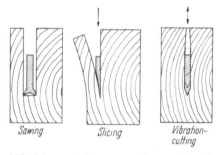

Fig. 9.76. Scheme of the processes of saw-cutting, slicing and vibration cutting. From KÜBLER (1960)

KOCH (1964) suggests the conversion of round logs by means of special tools, not producing saw dust but flake-like chips usable as raw material for pulp and paper as well as for fiberboard and particleboard. Square blocks could be produced from such material cut on special machines into lamella-like boards, and finally laminated into beams of high and uniform stiffness and strength (Fig. 9.75).

### 9.3.2 Cutting with Vibration Cutters

Vibration cutters are cutter or wedge-type tools oscillating back and forth in feed direction, at knife level, while the wood is advanced towards the tool cutter. The vibrating effects the separation of the wood through minor forces, so that very thin tools may be used. Fig. 9.76 schematically confronts the process of saw cutting, of slicing and of vibration-cutting. Vibration seems to counteract the splitting of the wood before the tool cutter, so that the tool floats less and the surfaces cut become smoother.

Even prior to the end of World War II, H. FLEMMING, Dresden was the first to voice the idea of cutting wood by means of vibration cutters. Since 1953 Russian scientists have dealt with this problem. For the time being, they are accounting for the favorable cutting results by the fact that the periodically acting cutting pressure breaks down the wood structure. Also, the small cutting forces are seen in relation to the low friction of oscillating surfaces.

In his tests, BAKIEV (1959) used a knife of 1···5.5 mm in thickness and 30 mm in width with a lip angle 40°. An eccentric drove the cutter at 24, 39 or 74 oscillations per second. Solely perpendicular to the grain, the cutting surface was good. The feed forces rose substantially less than proportionally with increasing feed velocity. With advancing oscillation width and oscillation rate, the feed force diminished according to expectation (Fig. 9.77 and 9.78). SMIRNOV (1957) used an electro-dynamic vibrator for his tests. His reflections were based on the idea that the percussions of the cutter blade against the material are

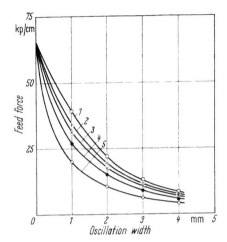

Fig. 9.77. Relationship between feed force and oscillation width for green spruce, 2.4 mm blade thickness and various feed velocities: 1...15 cm/s; 2...10 cm/s; 3...8 cm/s; 4...5 cm/s; 5...2 cm/s. From BAKIEV (1959)

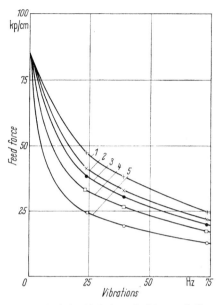

Fig. 9.78. Relationship between feed force and vibration frequency for green spruce, 1.2 mm blade thickness, and various feed velocities: 1...15 cm/s; 2...10 cm/s; 3...8 cm/s; 4...5 cm/s; 5...2 cm/s. From BAKIEV (1959)

of decisive importance. Therefore, he also put the wood into oscillation, i.e. with the same oscillation rate as the cutter but dephased by 1/2 oscillation period, the wood thus vibrating towards the blade. In this way, very little current was needed for the cutting.

MANZOS (1955) obtained good results with an electro-magnetic vibrator, although the noise of oscillations was disturbing. Moreover, in the Russian literature, tests with a series of disc cutters are mentioned, rotations being easier to produce than oscillation movement. These investigations, however, would not have yielded any serviceable result. Based on the present publications, in 1959 vibration cutting had not yet left the experimental stage.

### 9.3.3 Cutting with High-Energy Jets

Liquid jets have been used to extract coal or ores from mines, to debark roundwood, and even to separate hard materials (SCHWACHA, 1961). In 1960 in the Soviet Union YUGOV and OSIPOV published information concerning jet-cutting of various materials. BRYAN (1963) has studied liquid cutting of wood. He developed the following formula for the force $F$ [kp] exerted on the workpiece by the jet acting normal to the surface of impact:

$$F = 0.7076 \, d^2 \, p \tag{9.41}$$

where $d=$ diameter of the nozzle opening expressed in cm, and $p=$ pressure in kp/cm².

Figs. 9.79, 9.80 and 9.81 show graphically the most important relationships obtained experimentally by BRYAN. The coincidence between the theoretical and the experimental lines in Fig. 9.79 is not too bad considering that the theory

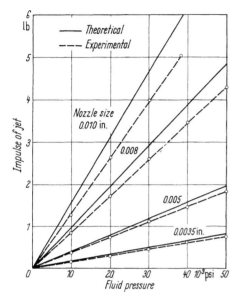

Fig. 9.79. Relationship between fluid pressure and jet force for four nozzle sizes. From BRYAN (1963)

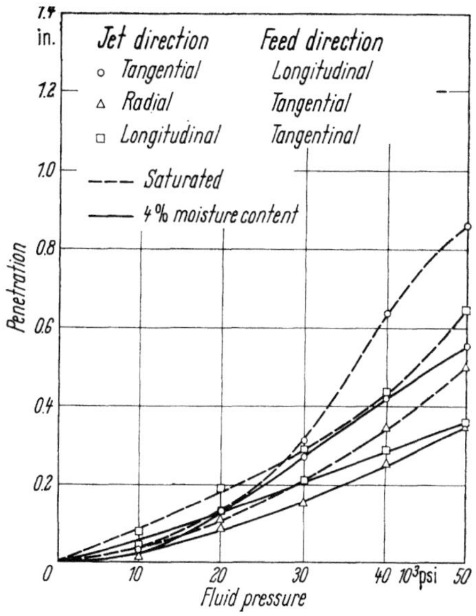

Fig. 9.80. Relationship between fluid pressure and penetration for three grain orientations and two moisture contents. From BRYAN (1963)

is based on the assumption of nonviscous, incompressible, and frictionless flow. Wet wood was penetrated more easily than dry wood for two reasons:

1. Wood strength and therefore cohesion decreases, within the hygroscopic moisture content range, with increasing moisture content. In section 9.2.1.2 it is reported how the average cutting force in gang-sawing may be reduced for wet wood according to VOIGT (1925, 1949).

2. In wet wood the kerf formed will not be closed by swelling as it is done in dry wood. The moisture effect is least when the jet cuts across the fiber. In this case the tendency for the kerf to close is negligible.

In experiments where the jet direction corresponded to the fiber direction, cell destruction was not complete and the effectiveness of the jet was reduced. The influence of three grain orientations is shown in Fig. 9.80. For a given nozzle size the penetration is a curvilinear function of jet velocity which depends on fluid pressure. The points of inflection in the curves of Fig. 9.81 show that above about 22,000 psi (1540 kp/cm²) cutting efficiency decreases. There is an approximately linear dependence of penetration on the square of nozzle diameter. Fig. 9.82 permits the conclusion that penetration is not markedly influenced by wood species. Nevertheless the surfaces produced differed substantially. Dense woods showed smoother surfaces than wood of low specific gravity. Fig. 9.83 shows that penetration decreases as the angle of incidence on the workpiece is increased. Penetration also decreases as the distance from the nozzle is increased. It can be assumed that the optimum distance in any case will be small.

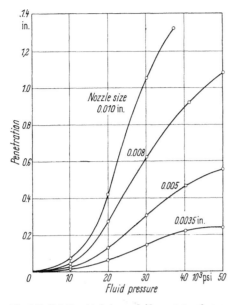

Fig. 9.81. Relationship between fluid pressure and penetration for four nozzle sizes. From BRYAN (1963)

Fig. 9.82. Relationship between fluid pressure and penetration for five wood species. From BRYAN (1963)

Efficiency increases with feed rate. This increase is pronounced up to approximately 250 cm (about 100 inches) per minute, and continues to increase in a somewhat linear fashion as higher feed rates are applied.

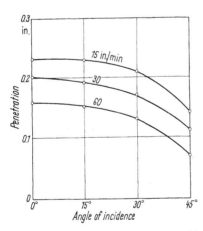

Fig. 9.83. Effect of angle of incidence on jet penetration for three specimen feed rates. From BRYAN (1963)

In general, the width of the kerf is approximately equal to the diameter of the nozzle opening. The quality of the surface generated by a high energy liquid jet is related to the jet diameter and also to the grain direction. Very smooth surfaces can be attained up to the point where full penetration is approached. Rough and torn surfaces, however, appear when soft wood is cut at high pressure levels.

The material removed can be described as "defibrated wood fibers". Microscopic investigations of the surfaces produced indicated that these fibers were separated by shear and/or tensile failures from the solid material.

Under the conditions of the study carried out by BRYAN (1963) the cutting efficiency in terms of area generated per unit of energy input was, at best, 50 times less than that normally obtained with conventional power saws. At higher feed rates this difference would become smaller, but jet-cutting very probably never will be highly efficient with respect to energy consumption. BRYAN (1963) writes: "Whether or not losses in power efficiency can be justified by savings in material and other advantages will depend mostly on the type of application ... The use of high-energy liquid jets for wood machining and processing offers many advantages in regard to equipment maintenance and complete utilization of material. It is felt that these factors will more than outweigh the low energy efficiency in many instances, and will justify applications of the process."

### 9.3.4 Cutting with the Laser

Kerfless sawing should be realized by highly concentrated sources of energy projected to a given plane of a solid workpiece and thus cutting it. The development of the "laser" (Light Amplification by Stimulated Emission of Radiation) created such a tool. The laser consists of a coherent beam of highly concentrated monochromatic light. Focussed optically to a minimum diameter the power density can exceed one million watts per cm$^2$.

The laser is applied in such areas as medicine, photochemistry, metal cutting, welding, etc. For the lasering process synthetic pink ruby is the most powerful material. It is used as a slender rod with one end fully reflective and the other semireflective. Under the influence of a flash light there results an intense burst of red light from the semireflective ruby rod end. BRYAN (1963) carried out experiments using a ruby laser with a maximum energy output of 3.0 joules per pulse. Woods of various densities were subjected to repeated laser pulses. The material in the path of the beam was vaporized. Holes of about 0.75 mm diameter were produced with depths from 0.8 to 1.6 mm per pulse.

The tests have demonstrated that wood can be cut by means of a laser beam. However, a continuous beam of high energy must be available. Producing a kerf rather than a hole the dispersion of the beam by interfering vapors probably could be avoided. At present laser equipment is expensive and inefficient. The future will show whether machining wood with a high power continuous laser can find commercial application.

## 9.4 Technology of Jointing, Planing, Moulding and Shaping

### 9.4.1 General Considerations

In this chapter "jointing" is defined as the smoothing of one suface of a board of lumber by means of a single peripheral milling head. "Planing" "refers to the peripheral milling of wood to smooth one or more surfaces of the workpiece and at the same time bringing the workpiece to some predetermined dimension in thickness, width, or profile pattern" (KOCH, 1964, p. 290). "Moulding" also is a peripheral milling process with the aim of machining pieces of lumber into form having different cross-sectional shapes. "Shaping" means cutting an edge profile or edge pattern on the side, end or periphery of a workpiece.

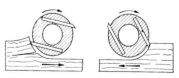

Fig. 9.84. Scheme of up-milling and down-milling technique

Shaping also belongs to the peripheral milling processes, in which wood is removed from lumber in the form of single chips. These chips are produced by the intermittent engagement of knives clamped on the periphery of a rotating cutterhead or cylinder.

The finished surface consists of a series of individual knife traces.. The engaged knives can move counter to the direction of the workpiece or in its direction (Fig. 9.84). The first technique, called "up-milling" is predominant in milling, whereas the latter technique, called "down-milling" is rarely used though it delivers very smooth surfaces.

## 9.4.2 Geometry of Cutterhead-knives

In jointing, planing, moulding, and shaping the movement of the tool relative to the work piece is combined by a rotation with the revolutions $n$ per min and a linear translation $m$ (m/min). The edges of the knives relative to the wood surface trace cycloides. Fig. 9.85 shows the geometrical conditions. The equations for these types of prolonged cycloides are

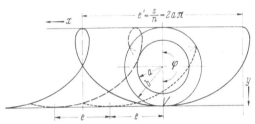

Fig. 9.85. Geometrical conditions for shaping and planing with cutterhead knives

$$x = a \pm b \cdot \sin \varphi$$

$$y = a - b \cos \varphi \qquad (9.42)$$

In these equations $a = \dfrac{f}{2n} =$ radius of the rolling circle of the cycloide, $f =$ feed speed (m/min), $n =$ revolutions per min, $b =$ distance of the cutter head edge from the center of the rolling circle and $\varphi =$ angle of rotation.

One can replace practically without important error the cycloide by a circle. KÖBERLE (1935) in Switzerland has given an instructive figure how chips in planing are formed and how smoothness of the planed surface can be evaluated. Fig. 9.86 reproduces his scheme. The decisive factor is the feed or step $e$ by one knife-cut

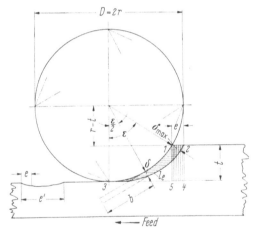

Fig. 9.86. Scheme of chip formation in planing. From KÖBERLE (1935). $\delta$ chip thickness, $e$ step, $\varepsilon$ arch angle corresponding to the chip length, $l_e$ arch corresponding to the chip length, $t$ depth of cut, $b$ width of cut, $D$ diameter of the cutterhead

$$e = \frac{1000 f}{n \cdot i} \qquad (9.43)$$

where $f =$ feed speed m/min, $n =$ revolutions of the cutterhead per min, and $i =$ number of cutting edges.

The formation of sickle-like chips is also shown in Fig. 9.86. The average thickness of the chip can be calculated after some approximation to

$$d = e \frac{t}{D} \text{ (mm)} \qquad (9.44)$$

The values for $d$, $t$ and $D$ can be taken from Fig. 9.86.

KÖBERLE (1935) proposed the following classes for average chip-thickness:

| Surface | Chip-thickness mm |
|---|---|
| very smooth | 0.014···0.04 |
| smooth | 0.041···0.16 |
| rough | 0.161···0.4 |

Geometry and kinematics of peripheral milling parallel to grain have been explored by MARTELLOTTI in greater detail (1941, 1945) and discussed by KOCH (1964, p. 113).

### 9.4.3 Cutting Velocity and Cutting Force

**9.4.3.1 Effect of Cutting Velocity on the Cutting Force.** According to KÖBERLE (1935) the cutting velocities in wood machining range as follows:

| | Cutting velocity m/s |
|---|---|
| I. low range | 5 ··· 20 |
| II. mean range | 21 ··· 60 |
| III. upper range | 61 ··· 100 |

Cutting velocities between 20 and 40 m/s seem to be optimal, higher ones should, if possible, be avoided. KIVIMAA (1952) found that there is no influence of cutting

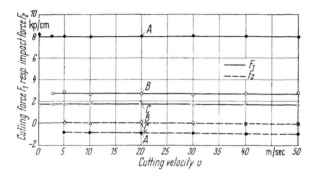

Fig. 9.87. Dependence of cutting force and impact force upon cutting velocity for various cutting directions: A cut surface perpendicular to the grain, B cutsurface and cutter motion parallel to the grain, C cut surface parallel to the grain, cutter motion perpendicular to the grain. Chip thickness 0.1 mm, birch wood, specific gravity 0.61 g/cm³, moisture content 11%, knive: $\alpha = 10°$, $\beta = 45°$, $\gamma = 35°$. From KIVIMAA (1952)

velocity on cutting force (Fig. 9.87). To some extent similar results were published by THUNELL (1950, cf. KOLLMANN 1955, p. 701) but more for freshly sharpened than for dull cutting edges.

**9.4.3.2 Effect of Cutting-Circle Diameter, Feed Speed, and Number of Knives.** Fig. 9.88 shows data obtained by KOCH (1964, p. 144). It is seen that for cutter heads carrying the same number of knives the larger one demands more power (about 9% in the example of the figure) than the small one. Larger cutterheads develop lesser wave heights than smaller ones. This, and perhaps also the higher cutting velocities of larger heads, contribute to a superior surface quality. The net cutterhead power consumption is more or less directly proportional to the feed speed.

If feed per knife is held constant, then a direct proportionality exists between number of jointed knives cutting and net power demand. In the general case in which the feed speed is increased without regard to feed per knife, things are more complicated. At high feed speeds (about 300 ft/min = 90 m/min) and great depths of cut (1/8′′′ 3.2 mm) a 2-knife head requires more power than a 12-knife head. KOCH (1964, p. 146/147) states summarizing "that with low feed speeds and ligth cuts, the horsepower demand increased with number of jointer knives cutting

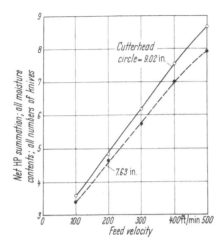

Fig. 9.88. Effect of feed speed and cutter head diameter on net cutterhead horsepower. From KOCH (1964)

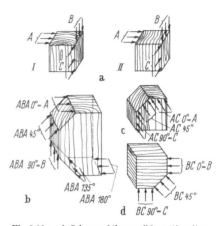

Fig. 9.89a–d. Scheme of the possible cutting directions. a) Main cutting directions referred to the annual rings, b) Cutting directions between the main directions $A$ and $B$, c) Cutting directions between the main directions $A$ and $C$, d) Cutting directions between the main directions $B$ and $D$. From KIVIMAA (1952)

... "When machining saturated stock, a more marked proportionality exists between the number of knives cutting and the horsepower demand at any particular feed speed."

**9.4.3.3 Effect of Grain Orientation, Inclination of the Cutting Edge, and Chip Thickness.** The cutting force depends on the shearing strength and probably on some other mechanical properties of the wood. All mechanical properties are more

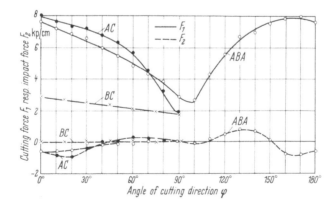

Fig. 9.90. Dependence of cutting force and impact force upon the cutting directions. Chip thickness 0.1 mm, birch, specific gravity 0.62 g/cm³, moisture content 11%, knive: $\alpha = 10°$, $\beta = 45°$, $\gamma = 35°$. From KIVIMAA (1952)

or less affected by the grain orientation (cf. Sections 7.3.3.1 and 7.4.2.1). Therefore, analogous relationships between cutting force and cutting direction with respect to the orientation of fibers and annual rings are to be expected. KIVIMAA (1952) could prove this fact by moulding experiments. He developed a scheme for various possible cutting directions (Fig. 9.89). The figure is self-explanatory. The results of a test-series with a normal knife cutting a chip of 0.1 mm thickness are shown in Fig. 9.90. The author stated that the effect of the angle between cutting direc-

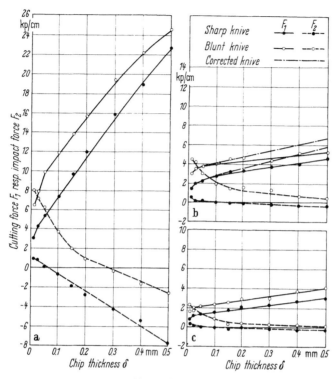

Fig. 9.91 a—c. Dependence of cutting force and impact force upon chip thickness and knive sharpness in the three main cutting directions a) in direction A, b) in direction B, c) in direction C. From KIVIMAA (1952)

tion and grain direction increases with increasing thickness of chips. The impetus force or acceleration force was, at least for the conditions of the KIVIMAA's experiments, independent of the chip thickness for various angles of the cutting force. Variations in chip formation, cutting force, and cutting energy were studied by McKENZIE and FRANZ (1964) for inclined cutting (cutting edge inclined to its direction of motion) achieved by feeding a work-piece radially to the edge of a rotating disk and cutting in a plane perpendicular to the fiber direction. In cutting wood of moderate density with rake angles and edge states in the practical range, the greatest effects of inclination were observed in the range of inclination angles from 0° to 79°. Resultant cutting force was reduced by 50% or more, and surface quality was greatly improved, even with a very blunt edge. Efficiency was increased, if at all, only in the lower part of this range, above which it decreased steeply.

The thickness of the chips is one of the principal factors in wood cutting. KIVIMAA (1952) published Fig. 9.91. In the range of the thicknesses which are most important in a practical sense, (from about 0.025 until 0.3 mm) the curves for the main cutting force can be substituted by straight lines. This means that the cutting forces reduced to the unit of cutting width are directly proportional to the chip thickness. KIVIMAA explained that in wood cutting two types of forces are necessary, one for the real cutting process, influenced by shape and sharpness of the knife but not affected by the chip thickness, the other probably for the deformation of the chips. This part of the cutting force is nearly proportional to

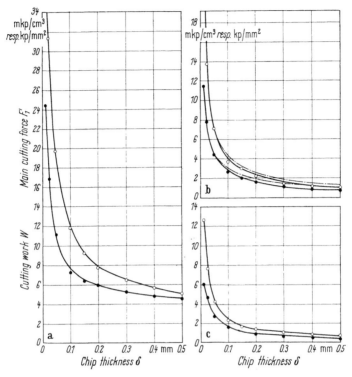

Fig. 9.92a—c. Dependence of cutting work (resp. main cutting force) upon the chip thickness in the three main directions, a) in direction A, b) in direction B, c) in direction C. From KIVIMAA (1952)

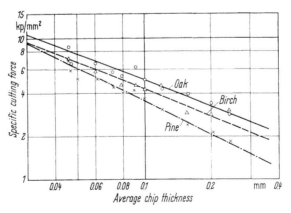

Fig. 9.93. Relationship between specific cutting force and average chip thickness. From THUNELL (1950)

the chip thickness. Calculating the specific cutting force (= cutting per unit of chip cross-section, kp/mm²) KIVIMAA (1952) obtained the curves in Fig. 9.92 from the values plotted in Fig. 9.91. The numerical values of the specific cutting forces are identical to those of the mechanical work $W$, necessary to convert the unit volume wood into chips (kp/cm³).

One can see that the chip thickness $d = 0.1$ mm, which is very common in practice, takes a singular position insofar as with further decreasing chip thickness

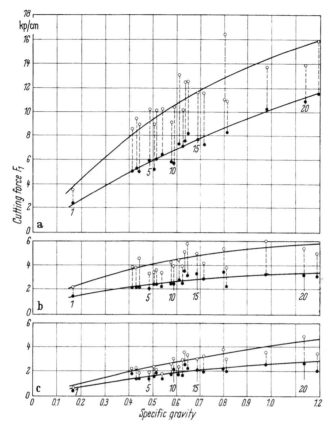

Fig. 9.94 a – c. Relationship between cutting force and specific gravity in the three main directions. Lower and higher points obtained with sharp, resp. with dull knives. Chip thickness 0.1 mm; moisture content 10.5 to 12%; kinves: $\alpha = 10°$, $\beta = 45°$, $\gamma = 35°$; wood species: 1 Balsa, 2 Spruce, 3 Aspen, 4 Poplar, 5 Fir, 6 Elm, 7 Alder, 8 Pitch pine, 9 Oak, 10 Mahogany, 11 Birch, 12 Teak, 13 Maple, 14 Hornbeam, 15 Beech, 16 Ash, 17 Hickory, 18 Oregon pine, 19 Ekki, 20 Lignum vitae, 21 Ebony (a) in direction A, (b) in diretion B, (c) in direction C. From KIVIMAA (1952)

the specific power consumption rises very rapidly. Plotted in a double-logarithmic net the curves of Fig. 9.91 appear as straight lines. THUNELL (1950) found quite similar results for three wood species (Fig. 9.93). He also expressed the relationship between the energy required ($W'$) to machine the workpiece and the feed ($f'$) per cut by the following parabolic equation:

$$W' = k \cdot f'^n \quad \text{kpcm/cm} \tag{9.45}$$

where

$k = f(d)$ ($d =$ depth of cut)

$f' =$ feed in mm/cut

$n =$ a constant.

Again plotted in a double-logarithmic net the functions are approximately represented by straight lines. For a rake angle = 30° THUNELL (1950) stated

$$W' = (0.8\,d + 1.6)\,(10f')^{0.25} \quad \text{kpcm/cm} \tag{9.46}$$

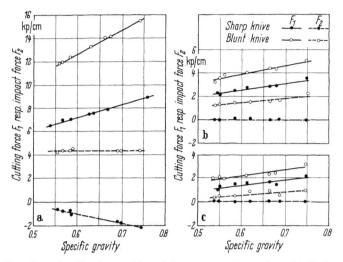

Fig. 9.95a—c. Dependence of cutting force and impact force upon specific gravity and knife sharpness in the three main cutting directions, according to Fig. 9.89: a) A, b) B and c) C. Chip thickness 0.1 mm, birch, moisture content 11%, knife: $\alpha = 10°$, $\beta = 45°$, $\gamma = 35°$. From KIVIMAA (1952)

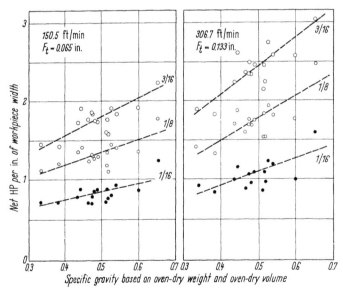

Fig. 9.96. Effect of specific gravity on net cutterhead horsepower requirement. From KOCH (1964)

**9.4.3.4 Effect of Wood Species, Moisture Content and Temperature.** KIVIMAA (1952) observed an increase of power demand with specific gravity. Fig. 9.94 shows that the higher points were obtained with dull knives, the lower points with sharp knives. The picture became still clearer when the tests were restricted to one single wood species. In this case the cutting force as well as the impact force showed a linear increase in power demand with specific gravity (Fig. 9.95). Koch (1964, p. 138) observed about the same effect of specific gravity on net cutterhead horsepower requirement under specified test conditions (Fig. 9.96). He pointed out that the increase in power demand with increasing specific gravity is mainly a result of increased quantities of extraneous substances in the wood.

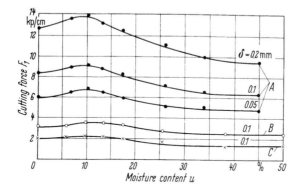

Fig. 9.97. Dependence of cutting force upon moisture content and chip thickness in the three main cutting directions, birch, specific gravity 0.70 g/cm³, knife: $\alpha = 10°$, $\beta = 45°$, $\gamma = 35°$. From Kivimaa (1952)

The effect of moisture content on the cutting force is somewhat the same as on the shearing strength (Kivimaa, 1952, Fig. 9.97). Even marked variations in temperature have no influence on the main cutting force in machining air-dry wood; in machining very wet wood the main cutting force decreases slightly with rising temperatures (Kivimaa, 1952). This is important from the viewpoint of the manufacture of veneers. The power consumption in peeling hot logs is less than that in peeling cold logs.

**9.4.3.5 Effect of Cutter Materials.** Pahlitzsch [1966] pointed out that the question where and when to apply hard metals for planing and moulding tools is mainly an economical problem. It is possible by the application of hard metal tools to increase the cutting speed, but the costs of investment and maintenance of tools are also increased. According to Pahlitzsch (1966) there is, not only for wood but in the same sense for metals an optimum cutting speed from the economical point of view. He points out that one has to consider that higher cutting speeds need higher revolutions per minute of the tool spindles and the machines must be constructed better and more rigidly. He mentioned quite validly that these relationships are not always considered in the proper way, but there is no doubt that the application of hard metals for tools in woodworking is a consequence of the increasing application of plywood and laminated products on the one hand and fiberboards, chipboards, especially with an overlay, on the otherhand. The differences in the behavior of different materials for planing and moulding tools were investigated by several researchers. Grube and Alekseev (1961) published results which are illustrated in Fig. 9.98. Reference may be made also to investigations by Pahlitzsch and Jostmeier (1965).

**9.4.3.6 Effect of Cutting Depth.** There does not exist an international terminology for woodworking or even for metalworking, but the quality of any planed or moulded wood surface depends on the step $e$ by one knife cut according to equation

9.43. In this connection Fig. 9.99 according to DEHNE und REHN (1960) shows the influence of feed speed, on the one hand, and of the number of cutting edges on the other. It is well-known that absolutely accurate adjustment of several cutting edges to the same cutting circle is practically impossible and Fig. 9.99 proves this fact since there are remarkable deviations from the theoretical curves of the measured points.

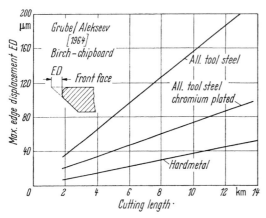

Fig. 9.98. Relationship between maximum edge displacement and cutting length. From GRUBE and ALEKSEEV (1964)

### 9.4.3.7 The Blunting of Cutter Head Knives.

Blunting is a problem more important with the cutting of wood base materials which are densified or which contain hardened artificial resins. In this book, devoted to solid wood, only a few remarks will be made on this point. PAHLITZSCH and JOSTMEIER (1964a) investigated the wear properties of hard metals and high speed steel tools in moulding two *different parts* of particleboard. It appeared that the wood species of which the particleboards were made influenced the blunting effect. The same authors (1964b) subsequently showed that ap-

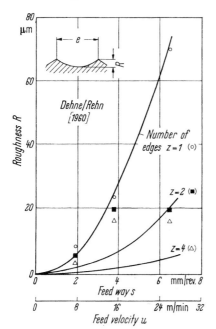

Fig. 9.99. Effect of feed way or feed velocity on the roughness of planed wood surfaces of cutting edges. From DEHNE and REHN (1960)

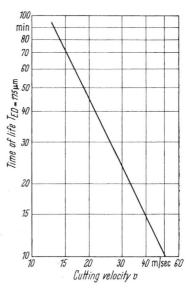

Fig. 9.100. Effect of cutting velocity on time of life of a wood-working tool with maximal edge displacement $ED = 175\,\mu m$. From PAHLITZSCH and JOSTMEIER (1964b)

parently tool wear depends on the density of the different layers in the boards. Silica impurities cause a very unfavorable effect. Once more it could be proved that there does exist a minimum for the cutting force as a function of the cutting

Ref. p. 551]    9.4 Technology of Jointing, Planing, Moulding and Shaping    527

speed (cf. 9.4.3.1). This tendency has been reported by several investigators, e.g. HEMMING and KNOSPE (1958) and by WEBER (1961).

Fig. 9.100 from PAHLITZSCH and JOSTMEIER (1964b) illustrates the effect of cutting speed on tool life. Similar relationships are known from metal machining. According to PAHLITZSCH and JOSTMEIER (1965), in the moulding of overlayed sandwich boards, cutting speeds between 36 and 38 cm/s proved to be best with respect to the cutting force. Cut abrasion profiles, plotted through measurement of the blunted edge after various lengths of feed clearly showed the influence of the structure of the sandwich boards.

### 9.4.4 Formation of Chips through Knife-Cutting.

**9.4.4.1 Influence of Wood Moisture Content on Chip Formation.** The results of PAHLITZSCH and MEHRDORF (1962a) show that it is possible with practical chipping thicknesses to produce chips of corresponding good quality with respect to small surface roughness, uniform chip thickness, sufficient chip size, etc. The test results, for pine as well as for beech wood, remain nearly independent of moisture content as long as this value lies below the fiber saturation point. If, however, the moisture content is increased beyond this point the chipping forces and energies, the formation of splits occurring during cutting, the roughness of the chip surface, the coefficient of variation of chip thickness, the binding of the chips and the portion of refined materials are increased. Though these results were obtained with chipping machines for particleboards they are undoubtedly applicable also for planing and moulding.

**9.4.4.2 Influence of Knife Geometry on Chip Formation.** The last remarks of the foregoing section are valid also for the following statements based on investigations of PAHLITZSCH and MEHRDORF (1962b) on producing flakes with a cutter disc chipping machine for particleboards. The cutting speed within the tested range between 10 and 40 m/s hardly influences the cutting forces, the cutting output, the quality of the chip surfaces and the uniformity of the chip thicknesses. A change of the hook angle $\alpha$ at a constant clearance angle influences the cutting forces and the properties of the chips produced to a small extent within the region of $\alpha = = 45...55°$ and a little more within $\alpha = 35...45°$. By increasing the hook angle the chip deformation occurring with chipping is diminished; therefore, with increasing hook angle the cutting force, feed force, depth of splits (determining the roughness of the chip surfaces) and chip bending become smaller.

**9.4.4.3 Other Cutting Factors and their Effect on Chip Formation and Quality.** A few years ago chips produced by planing or moulding were treated as wood waste and in most cases were burned in a refuse burner or in the boiler house. But the development of particleboard industries changed these conditions. Now we speak of industrial wood residues instead of wood waste and the quality of the chips produced in planing and moulding gain more and more importance and led directly to the development of special tools with the aim of obtaining chips of a high quality. Especially undesirable are checks and extreme roughness of the chip surfaces. The greater the thickness of the chips the greater becomes the depth of checks or splits and the distance between the splits. Bending of the chips is also an important property. As Fig. 9.101 shows, the behavior of beechwood and pinewood in this respect is quite different. The influence of chip thickness below a critical thickness of approximately 0.3 mm for pinewood is noteworthy. The condition of the knife edge, sharp or blunt, increases not only the cutting force and the feed force, but also the roughness of the chip surfaces. The formation of splits in the chips as well as the bending of the chips remain approximately constant.

A few words may be devoted to the problem of the effect of the cut-surface angle. The cut-surface angle is the angle between the wood fiber and the plane of the cutting edge. It is measured in a plane parallel to the wood fiber and perpendicular to the plane of the cutting edge. As Fig. 9.102 illustrates, the cutting surface angle is positive if the cutting is effected along or "with" the grain and negative if cutting is done against the fiber direction. With respect to the production of a good

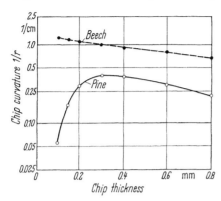

Fig. 9.101. Relationship between chip thickness and chip curvature for beechwood and pinewood. From PAHLITZSCH and MEHRDORF (1962)

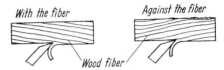

Fig. 9.102. Cutting effected "with" and "against" the wood fiber. From PAHLITZSCH and MEHRDORF (1962)

surface and to a uniform thickness of the chips according to PAHLITZSCH and MEHRDORF (1962) a cut-surface angle between $+3$ and $+6°$ is regarded as particularly favorable. Various wood species apparently behave in different ways, but the values mentioned above were found satisfactory not only for chipping pine, but also for poplar. The moisture content of the wood was above 30% (PAHLITZSCH and MEHRDORF, 1963).

## 9.5 Sanding

### 9.5.1 General Considerations

Quality and smoothness of all machined wood articles which will be stained, polished or lacquered depends on the sanding operation. Planing, moulding, and routing with solid cutters or cutterheads nearly always tears fibers and produces rhythmic regular waves, rather close together on the wood surfaces. Softer woods are particularly susceptible. There are remedies such as the increase of the cutting angle or the decrease of the number of cuts but they are not completely satisfactory. Therefore abrasion techniques must be applied. The smoothness of a sanded surface depends mainly on the size, shape, and quality of the particles of grit and on the speed of sander and workpiece. To prevent scratching, mechanical sanders should have an oscillatory movement except in the case of moulding sanders (LISTER, 1948].

### 9.5.2 Abrasives

The task of any abrasive is to cut and, therefore, the particles require sharp and durable cutting edges. Table 9.8 lists the common abrasives for wood. The first abrasive was natural sand.

According to SUTTER (1935) the abrasive particles should be compact and strong, with rough surfaces and sharp edges. From this point of view, mainly garnet and brown aluminium oxide are recommended for the sanding of hardwoods. The raw materials for the abrasive particles are ground in mills, cleaned by washing and dried, separated from iron by means of magnets and classified according to grain sizes. Average and small grain sizes are chosen for the finest sanding work.

Experiments prove that a high surface quality also may be obtained with greater grain sizes if the sanding speed is high enough. Another relevant point is the density of distribution of the abrasive particles on the paper or cloth.

The efficiency of sanding can be expressed by the loss of weight of sanded material per unit time. Fig. 9.103 illustrates that with different types of abrasive particles coated papers show different behavior in use. According to LISTER (1948) close coated paper is used for hardwoods and free coated paper for softwoods. According to SACHSENBERG and BALZER (1930) the specific sanding efficiency (expressed in cm³/min and cm² sanded surface) reaches a maximum after a short period of sanding (in the experiments mentioned, after approximately 40 s) and then decreases slowly. The amount of material sanded also depends on the density of the wood species and is higher for softwoods for example, than for hardwoods under otherwise similar conditions of use.

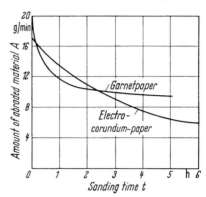

Fig. 9.103. Effect of sanding time on the amount of abraded material for garnet paper and Electro-corundum-paper No. 60. Beech, specific sanding pressure 70 g/cm². From SUTTER (1935)

Table 9.8. *Principal Properties of Abrasives for Wood*
(From Sutter, 1935)

| Abrasive | Origin | Color | Type of fracture | Toughness | Hardness | Applicability on paper or cloth |
|---|---|---|---|---|---|---|
| 1. Glass (green bottle-glass) | artificial | green | sharp edges, pointed | brash, splintering | 4 to 6 | hand sanding |
| 2. Flint | natural | grey | pointed | brash | 5 to 6 | hand and machine sanding |
| 3. Garnet | natural | reddish yellow-brown | mussel-shaped, sharp edges | rather brash | 7 | machine sanding of all wood species |
| 4. Aluminium oxide | artificial | light grey until brown | grain or mussel shaped, sharp edges | brash, but also tough | 8 to 9 | machine sanding of hardwood and decorative veneers |
| 5. Silicon carbide | artificial | green black to black | sharp edges and tips | brash | 9 | for special purposes (sanding of very hard woods) |

### 9.5.3 Technology of Sanding Process

Besides the few scientific papers mentioned in the previous section more recent investigations about sanding have been published by PAHLITZSCH and DZIOBEK (1959, 1961a, 1961b, 1962).

Belt sanding was particularly investigated. As test materials spruce, pine, poplar, alder, red beech, oak and teak wood have been used. In the first study the following summarized results were obtained:

1. The volume of abrasion as the measure of the efficiency of the whole process is influenced by structure, strength and wood extractives.

2. The volume of abrasion increases with sanding perpendicular to the grain.

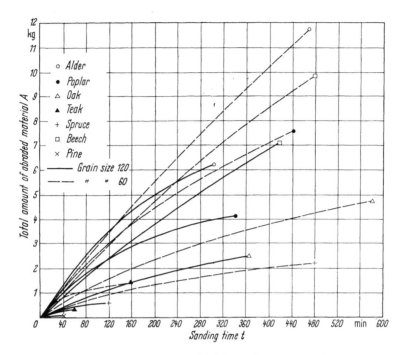

Fig. 9.104. Effect of sanding time on total amount of abraded material in sanding various wood species. Average moisture content 10%; grain size 60 and 120; sanding area 150 cm$^2$; working conditions: contact force 5.1 kp, band velocity 16 m/s, sanding direction along the grain. From PAHLITZSCH and DZIOBEK (1959)

3. The specific volume of abrasion increases with constant pressure of the sanding belt on the wood surface to be sanded, proportional with the reduction of the sanding area.

4. The sanding time has very marked influence on the total amount of abraded wood. Fig. 9.104 proves this fact for different species and shows furthermore the influence of size of the abrasive particles. Fig. 9.105 is also revealing. One can see, of course, that the total volume increases continuously with the sanding time, but that the specific sanding efficiency, expressed in weight per unit time, after a very small decrease remains nearly constant. The specific sanding work, that is consumption of energy per unit volume abrased, increases at first, then remains to some extent nearly constant, but increases again after a critical period.

5. The optimum belt speed as determined by the specific quantities of abrasion is 30 m/s for particle size 60 and slightly less than 30 m/s for particle size 120. PAHLIZTSCH and DZIOBEK (1959) stated that with increasing belt speed the volume of abraded wood per unit time also is increased.

6. Smaller abrasive particles produce smaller volumes of abrasion and are more sensitive to high pressures.

## 9.5 Sanding

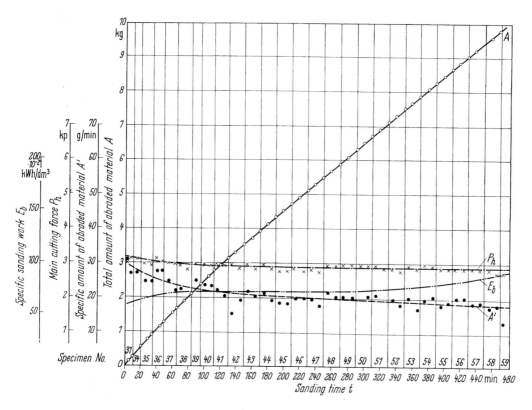

Fig. 9.105. Effect of sanding time on amount of abraded material, main cutting force and specific sanding work in sanding beech. Working conditions as in Fig. 9.104. From PAHLITZSCH and DZIOBEK (1959)

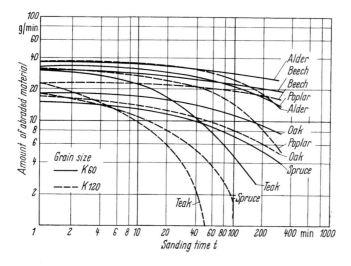

Fig. 9.106. Effect of sanding time on the amount of abraded material for various wood species and grain sizes. Working conditions as in Fig. 9.104. From PAHLITZSCH and DZIOBEK (1961a)

34*

PAHLITZSCH and DZIOBEK (1961a) reported the test results again using an edge sander. Several wood species were tested for constant contact pressure and constant sanding volume per min. The principal conclusions of these investigations are as follows:

7. Two different stages of blunting are apparent. The blunting can be seen from the curves of the abrasive volume or abrasive weight per unit time. Fig. 9.107 gives the instructive survey. The authors of the paper mentioned above stated in the first stage a strongly decreasing development of blunting, caused mainly by first rapid and later slow breaking off or out of abrasive particles and in the second stage a progressive falling trend, caused very probably by continuously increasing deposits of very fine wood particles in the belt.

8. The roughness of the workpiece sanded increases with progressing blunting.

9. Very high contact pressure, especially within the time of initial sharpness of the belt, is of greatest advantage. The same is valid for the sanding speed; approximately 30 m/s is very favorable.

10. The temperature of belt and work piece are of particular interest. Preliminary tests show that, probably, increasing contact pressure and belt speed after a short initial stage result in a nearly proportional rise of temperature (Fig. 9.107).

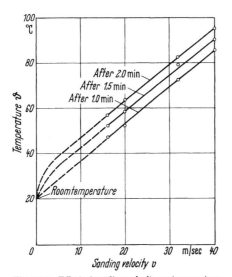

Fig. 9.107. Effect of sanding velocity on temperature of the workpiece in sanding beech. Working conditions as in Fig. 9.104. From PAHLITZSCH and DZIOBEK (1961a)

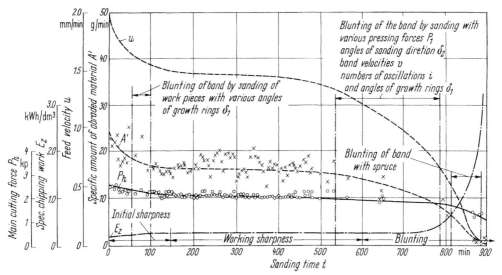

Fig. 9.108. Interrelationship between sanding time and specific amount of abraded material, feed velocity, main cutting force and specific chipping work in sanding beech. Working conditions as in Fig. 9.104. From PAHLITZSCH and DZIOBEK (1962)

PALITZSCH and DZIOBEK (1961b) also dealt with the methods of classifying the surface quality of belt sanded woods. As an introduction they summarized the literature dealing with this problem. After an examination of the suitability of light-and-shadow microscopes and of instruments making use of a stylotracer,

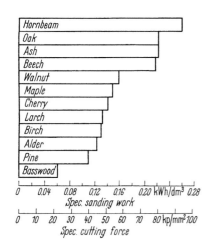

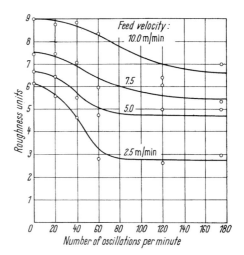

Fig. 9.109. Specific sanding work, resp. specific cutting force for drum sanding of various wood species. From KRATZ (1932)

Fig. 9.110. Effect of number of oscillations per minute and feed velocity on roughness of wood surfaces in drum sanding. From KRATZ (1932)

the latter method was preferred. Five different wood species were investigated with respect to the maximum roughness depth. The results were evaluated statistically. Frequency curves allow conclusions to be drawn on the average profile form. In this way also the maximum roughness depth can be limited.

Another publication by PAHLITZSCH and DZIOBEK (1962) is devoted to the determination of the quality of sanded wood surfaces and especially describes the effect of working conditions. In this connection wetting before and after sanding was investigated.

The following results may briefly be given:

11. The relative and the absolute increases in roughness for a dry surface towards the end of the initial sharpness of the belt show a low value until the end of the working sharpness and a further decrease with the blunting (Fig. 9.108).

12. With respect to the angle between fiber and sanding direction the roughness is greatest with small angles (0 ... 30°).

13. Sanding with oscillation of the belt, the mean values of roughness decreased 15 to 25%: a maximum roughness occurred at the belt velocity of 20 m/s.

In concluding the section it may be said that the principal observations for belt sanders can be applied also, at least to some extent, to bobbin sanders, disc sanders and drum sanders.

KRATZ (1932) published some contributions to the influence of wood species on the cutting force (Fig. 9.109) and of the number of oscillations per unit time in drum sanding (Fig. 9.110). The last graph clearly shows that it is practically useless to increase the number of oscillations above about 120 p. min.

## 9.6 Turning

### 9.6.1 General Considerations

Turning of wood as a technological process is quite similar to the turning of metals. The cutting forces, of course, are much smaller and in general turning of wood from the point of view of industrial woodworking is of less interest.

### 9.6.2 Effects on the Turning of Wood

WARLIMONT (1932) published the first scientific investigations on the turning of wood. The most important figure is the main cutting force. From the geometrical point of view the conditions should be well considered. Fig. 9.111 shows, based on German standard specifications, the interrelationship of cutting forces and feed speed.

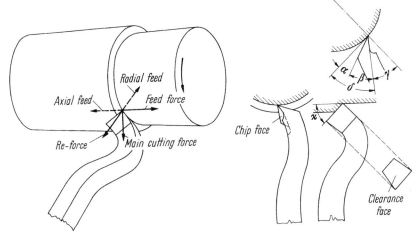

Fig. 9.111. Interrelationship of cutting forces and feed speed in turning wood, $\alpha$ clearance angle, $\beta$ wedge angle, $\gamma$ rake angle, $\delta$ cutting angle, $\varkappa$ inclination angle of the edge. (Based on German Standard Specifications)

WARLIMONT (1932) stated that the consumption of energy ($N_s$) in turning wood can be expressed by the same formula as introduced by KLOPSTOCK (1926) for metals. The consumption of energy ($N_s$) can be calculated with the following formula under the assumption that $P_g$ is the main cutting force and $v$ is the cutting velocity expressed in m/s:

$$N_s = \frac{M_d \cdot n}{716 \cdot 1.36} = \frac{P_g \cdot v}{75 \cdot 1.36} \text{ (kW)} \tag{9.47}$$

The main cutting force $P_g$ is obtained by multiplying the specific cutting force $k_s$ by the chip cross section $q$ (mm²). Another consideration is the efficiency of the turning lathe (on the average 0.7). This value has to be introduced in the denominator of Eq. (9.47). The specific cutting force for laminated wood products depends only to a very small extent on either the cutting speed or the depth of cutting (Fig. 9.112).

The relatively small influence of the cutting depth on the cutting force has been established theoretically by WARLIMONT (1932) and for planing by BRÜNE (1930). Much more pronounced is the effect of the feed speed, especially below a value of 0.3 mm p.r. The relationships between feed speed and the different components of forces in turning on the one hand, specific cutting pressure on the other and feed speed are illustrated by the curves in Fig. 9.113. One can assume that the cutting forces in turning, similar to the conditions in moulding, increase proportionally with increasing density of wood. WARLIMONT (1932) found, for example, that the values of the specific cutting forces in turning pine and beech were in proportion as 1 to 0.5. The average values of density for these wood species are in proportion as 1 to 1.4.

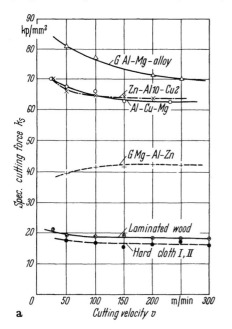

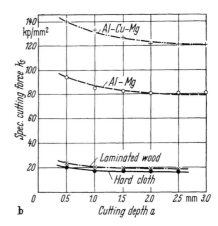

Fig. 9.112a, b. Dependence of the specific cutting force (a) upon cutting velocity, (b) upon cutting depth for various materials. From SCHALLBROCH and DODERER (1943)

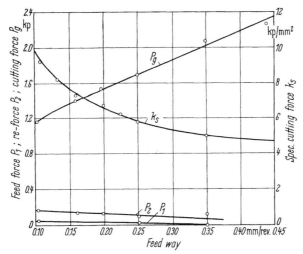

Fig. 9.113. Effect of feed way on the components of the cutting force and the specific cutting force. From WARLIMONT (1932)

Grain orientation also has a strong influence on the consumption of energy in turning. Under the assumption of constant cross section of chips and constant feed speed for the specific cutting force it has been found that adjusting the angle of the cutting tool $\varkappa$ between 45 and 60 degree is preferable. There is another important influence to be considered; that is, the width of the chip as a function of the angle $\varkappa$. BRÜNE (1930) found the following equation for the dependence of the specific cutting force in turning on the adjusting angle

$$k_s = k_s \parallel \cos^2 \varkappa + k_s \perp \sin^2 \varkappa \tag{9.48}$$

where:

$k_s \parallel$ is specific cutting force for turning with the cutting edge parallel to the fiber and feed perpendicular to the fiber and

$k_s \perp$ is the cutting force with the cutting edge perpendicular to the fiber and feed perpendicular to the fiber.

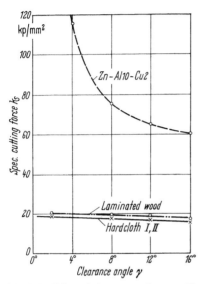

Fig. 9.114. Effect of clearance angle on specific cutting force for various materials. From SCHALLBROCH and DODERER (1943)

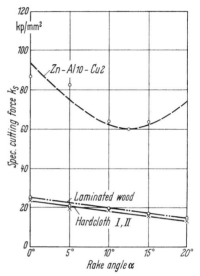

Fig. 9.115. Effect of rake angle on specific cutting force for various materials. From SCHALLBROCH and DODERER (1943)

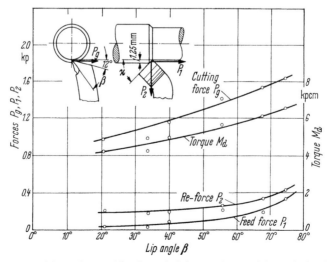

Fig. 9.116. Effect of lip angle on cutting force, feed force, re-force and torque in turning beech. From WARLIMONT (1932)

The shape of the turning tool also has an effect on the cutting force (BETHMANN, 1940; SCHALLBROCH and DODERER, 1943). Figs. 9.114 and 9.115 illustrate the effect of the clearance angle and the hook angle on the specific cutting force $k_s$. One can see that in comparison to metals the effect is very small. From practical experience and the investigations of WARLIMONT (1932), a clearance angle between 12 and 18 degrees offers an optimum cutting condition. Another interesting problem is the lip angle $\beta$. WARLIMONT in his experiments found the curves represented in Fig. 9.116. In practice, a wedge angle between 20 and 30° is recommended, but there is no doubt that such small wedge angles are only justified for turning softwoods. Taking into consideration all aspects WARLIMONT recommends lip angles between 50 and 60° though the energy consumption will be increased for the rough turning of hardwoods.

### 9.6.3 Quality of Turned Surfaces

For the evaluation and application of most turned wood articles the quality of the surface is of the utmost importance, e.g. for bobbins, tools, handles, etc. The experimental values with respect to this problem are small. As Fig. 9.117 accord-

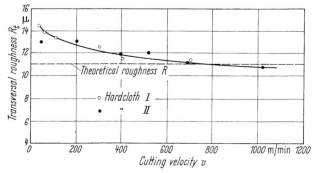

Fig. 9.117. Effect of cutting velocity on transversal roughness of a turned hardcloth-surface. From SCHALLBROCH and DODERER (1943)

ing to SCHALLBROCH and DODERER (1943) illustrates, the roughness decreases with increasing cutting speed. The roughness perpendicular to the grain increases in a parabolic manner with the feed speed. There is no doubt that also the specific pressure of the tool on the turned surface has a remarkable influence on the roughness. In practice there is a distinction made between "cutting" and "pressing".

## 9.7 Tenoning, Mortising and Boring

Tenoning, mortising and boring have the goal of producing holes with various shapes and sizes and various depths in wooden parts. The shape of the hole may be round, oval or more or less rectangular.

Mortise and tenon joints are basic to good joinery. Formerly these joints were made by hand, but now machines are used. First the circular saw may be mentioned as successful in some cases. In other cases the spindle moulder is a very fine substitute.

The large proportion of the tenons is not more than 75 mm long. Special tools are the so-called shoulder cutters. They make it possible to work at one stroke a number of sash-bars, narrow rails, etc. (HUDSON, 1953).

An example of a corner lock sash joint which shows the type of woodworking to be done is given in Fig. 9.118. In summary, the following technological operations are possible: to use square chisels, borers, also the type of automatic multiple spindle borers, double end borers, chain mortisers of the single and of the multiple type, combined mortisers and borers. Jigs are widely used to hold the piece to be worked in the correct position. A jig can be very simple or it can be quite elaborate (LISTER, 1949).

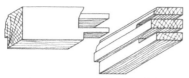

Fig. 9.118. Corner lock sash joint. From HUDSON (1953)

There are various types of tools which can be used on tenoning machines. Fig. 9.119 shows examples of the components of the so-called outside and inside cutters of Dado heads. These various components make a complete head.

Fig. 9.119. Outside and inside cutters of Dado heads

In modern woodworking factories the double-end tenoning machine is one of the most important time savers. Both ends of the pieces are tenoned and if necessary, scribed and cut off to dead length simultaneously (HAYCOCK, 1949). Feed

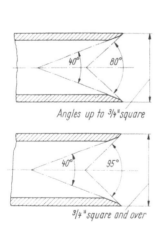

Fig. 9.120. Geometry of a square chisel. From LISTER (1949)

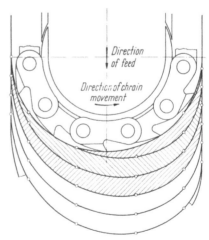

Fig. 9.121. Chip formation in chain mortising. From WALTER (1931)

can be done by hydraulic or mechanical means. The mechanical feed is involved with the danger of wear slackness and shuttering. The hydraulic feed is noiseless and smooth (LISTER, 1949).

## 9.7 Tenoning, Mortising and Boring

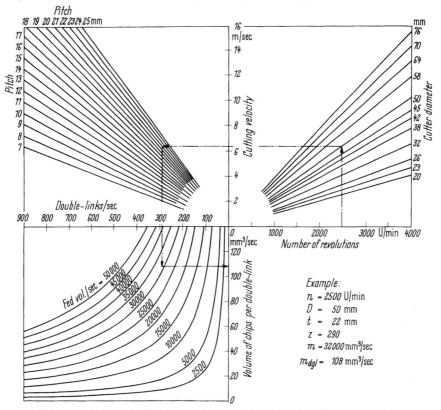

Fig. 9.122. Volume of chips per double-link of chain mortiser in relation to cutting velocity, cutter diameter and pitch. From KULLMANN (1930)

Example:
$n = 2500$ U/min
$D = 50$ mm
$t = 22$ mm
$z = 290$
$m = 32000$ mm³/sec
$m_{dgl} = 108$ mm³/sec

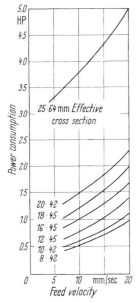

Fig. 9.123. Effect of effective cross-section and feed velocity on power consumption in chain mortising fir. From KULLMANN (1930)

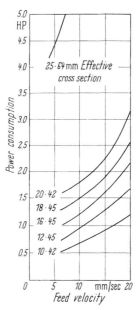

Fig. 9.124. Effect of effective cross-section and feed velocity on power consumption in chain mortising beech. From KULLMANN (1930)

Mortising can be done with a chain, with a square chisel or with a revolving oscillating bit. The square chisel gang mortisers with a feed speed varying between 10 and 20 strokes p.m., for example make four holes in the same action. The machine is important insofar as the chain mortiser produces round buttons in the

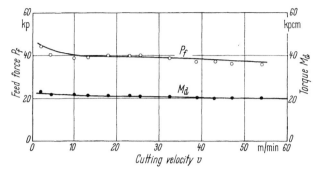

Fig. 9.125. Effect of cutting velocity on feed force and torque in boring hardcloth. From SCHALLBROCH and DODERER (1943)

holes. Square chisel gang mortisers can be fitted with air clamps, automatic motions and hydraulic feed mechanisms. Fig. 9.120 shows the geometry of a square chisel in section.

To produce square holes two operations are combined: first a circular hole is bored and then the corners are squared out using the chisel. It is possible to run an borer in the chisel and so carry out both operations at once.

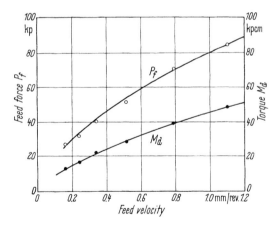

Fig. 9.126. Effect of feed velocity on feed force and torque in boring hardcloth. From SCHALLBROCH and DODERER (1943)

Scientific investigations on the technology of turning, mortising and boring are rather scarce. However, there are some experiments with chain mortising. Chip formation in chain mortising is illustrated by Fig. 9.121 according to WALTER (1926). The shape of the chips is asymmetrical since the incising teeth are accelerated to some extent by the feed speed whereas on the opposite side they are retarded. KULLMANN (1930) designed Fig. 9.122 which allows the determination of chip-volume per double-link of chain mortiser in relation to chain speed, chain diameter, and pitch. There are always small differences in the heights of the teeth and therefore the real values of the chip-volume may vary markedly. Power consumption increases about parabolically with feed speed, with working cross section, and with the density of the wood (Fig. 9.123 and 9.124, from KULLMANN, 1930). WALTER, showed that power consumption reaches a minimum value for a chain speed of about 8 m/s (corresponding to 3000 rpm). For higher chain speeds the higher friction causes higher power consumption and more rapid wear. The hook angle of chain-teeth normally should not exceed 20°.

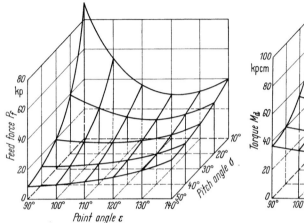 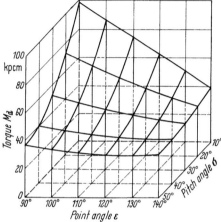

Fig. 9.127. Effect of point angle and pitch angle on feed force in boring hardcloth. From SCHALLBROCH and DODERER (1943)

Fig. 9,128. Effect of point angle and pitch angle on torque in boring hardcloth. From SCHALLBROCH and DODERER (1943)

For boring a few remarkable results are available (OSENBERG, 1927, HETZEL 1928, BÜTTNER, 1930 and SCHALLBROCH and DODERER, 1943).

A difficulty exists also insofar as the shapes of boring or drilling tools are quite different (KOLLMANN, 1955). The following results are interesting:

1. Feed force $P_f$ and torque $M_t$ are nearly independent on the cutting force (Fig. 9.125);

2. Feed force $P_f$ and torque $M_t$ increase with feed speed (Fig. 9.126);

3. The angles of the tool (sharpness angle or wedge angle and twist angle) have an important influence on either the feed force (Fig. 9.127) or the torque (Fig. 9.128).

## 9.8 Bending of Solid Wood

### 9.8.1 General Considerations

Bending or moulding of wood in forms is a technique that offers the following advantages:

1. There are no losses through wood waste;

2. The shaping as a rule, is simpler and can be quicker than by means of the usual woodworking machines;

3. The investment costs for bending machines are compartively low;

4. Power consumption in bending is low;

5. Strength and stiffness of bent parts is higher than the same properties of parts shaped by sawing or spindle moulding.

6. The surfaces of properly bent parts are smoother than those of sawn parts;

7. In some cases, e.g. in the manufacture of various sporting goods, of crooks for walking sticks and umbrellas, of parts of chairs, barrel staves, wood rings, etc., bending is the only economical technique.

### 9.8.2 Strains and Stresses in Wood Bending

It is difficult to bend air-dry wood without plasticizing beforehand. This is done mostly by steaming or boiling. For several special purposes after boiling, a combination of bending and compression is applied at the end of the parts to be

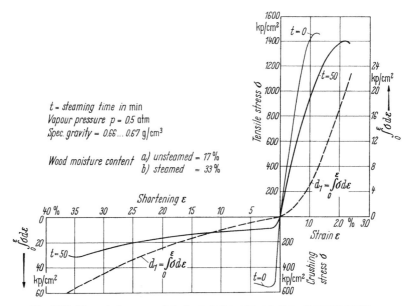

Fig. 9.129. Stress-strain curves of air-dry and steamed beech. From PRODEHL (1921)

bent, such as staves for beer barrels. Other special processes: plasticizing by chemical agents, e.g. urea, liquid ammonia or by high-frequency-fields which were successful to a limited extent only. The reason for the difficulties in bending is the low extension of wood in tension failure. For bending a wood stick in a circular shape a nondestructive moulding is only possible if the following condition is secured:

$$\varepsilon_t \geqq \frac{a}{2R}$$

where $\varepsilon_t$ = extension at tension failure, $a$ = thickness of the bent part, and $R$ = radius of the bent part, measured up to its neutral axis.

To air dry beech wood one obtains values $\varepsilon_t = 0.75$ until $1.0\%$ and thus a ratio $\frac{a}{R} = \frac{1}{67}$ until $\frac{1}{50}$. Steaming or boiling makes the wood more plastic, the values for $\varepsilon_t$ will be increased to 1.5 or 2.0%. In these cases we find $\frac{a}{R} \leqq \frac{1}{33}$ to $\frac{1}{25}$. PRODEHL (1921), STEVENS and TURNER (1948) and FESSEL (1952) published stress-strain-curves of air-dry and steamed beechwood and oakwood in tension and compression tests (Fig. 9.129 and 9.130). TEICHGRÄBER (1953) showed that the compressive stress-strain diagrams are straight lines in the elastic range and to some extent also in the plastic range (Fig. 9.131). The distribution of the compressive and the tensile stresses over the cross-section of bent oakwood sticks is illustrated in Fig. 9.132 according to FESSEL (1951). ,,a" shows the distribution of stresses and strains. The stresses in wood particles during bending are less dangerous if the wood has a high capacity for changing its shape under stress. Broad-leaved species are better suited for bending than all coniferous species. High density as is prevalent in beech, ash, hickory and elm is not at all disadvantageous. In coniferous species the abrupt change of the mechanical properties between earlywood and latewood is probably the main cause

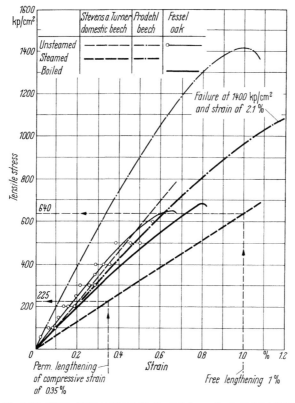

Fig. 9.130. Stress-strain curves of air-dry and steamed or boiled beech and oak

for failures in bending. One has to keep in mind that rather high compressive strains are imposed upon the fibers in the whole structure of the wood (STEVENS and TURNER, 1948).

Any point of weakness, e.g. caused by defects such as irregular fibers, knots, checks, fractures, pitch pockets or ingrown bark will facilitate these fractures. On this basis only clear wood should be chosen for bending. If defects cannot be completely avoided they should lie near the convex face or, better yet, in the immediate neighborhood of the neutral axis.

In selecting species of timber for bending the following prerequisites should be taken into consideration: availability of material, strength properties and bending properties. A few hints for the bending properties of a number of timbers according to the research at the Forest Products Laboratory in Princes Risborough are given in Table 9.9. The authors remark that the data in the table refer to good quality, air-dry material, 25 mm thick and steamed at atmospheric pressure.

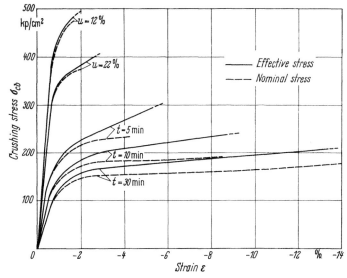

Fig. 9.131. Crusing stress-strain-curves for beech at two moisture contents and after various steaming times. From TEICHGRÄBER (1953)

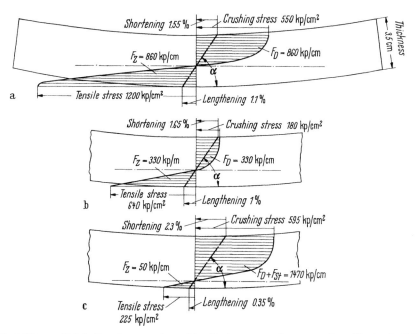

Fig. 9.132a—c. Distribution of compressive and tensile stresses over the cross-section of bent oak sticks. a) unsteamed stick, b) steamed stick, c) steamed stick edges (under pressure). From FESSEL (1951)

It may be seen that there is no remarkable difference in the bending properties of ash, beech and oak, but that mahogany, teak and spruce are inferior. The influence of supporting straps is evident. A severe problem is the proper choice of

[Ref. p. 551] 9.8 Bending of Solid Wood 545

whole trees to select for bending. STEVENS and TURNER (1948) discussed this question and remarked that few bending firms seem to be entirely in agreement. In this respect factors such as age, width of annual rings, and soil characteristics have an influence, but they are of secondary importance assuming that the material is clear and straight grown. Very old trees do not yield excellent material and any rot makes wood worthless for bending.

Most timber can be bent in the green condition immediately after felling, though some woods, if bent to small radii of curvature, "are liable to rupture as result of hydraulic pressures induced within the moisture-laden cells." (STEVENS and TURNER, 1948).

Table 9.9. *Approximate Radius of Curvature (in mm) at which Breakages during Bending should not Exceed 5 per cent*

| Species | When supporting straps | |
|---|---|---|
| | are used | are not used |
| Ash, American (*Fraxinus* sp.) | 110 | 330 |
| Ash, European (*Fraxinus excelsior* L.) | 64 | 300 |
| Vanak (*Virola merendonis* Pittier) | 760 | 1220 |
| Beech European (*Fagus sylvatica* L.) imported | 100 | 370 |
| Beech European (*Fagus sylvatic.t* L.) home-grown in England | 38 | 330 |
| Birch, Canadian yellow (*Betula lutea* Michx.) | 76 | 430 |
| Blackbutt (*Eucalyptus pilularis* Smith) | 610 | 1220 |
| Camphorwood, East African (*Ocotea usambarensis* Engl.) | 250 | 410 |
| Elm, Dutch (*Ulmus hollandica* Mill. var. *major* Rehd.) home grown in England | > 10 | 240 |
| Greenheart (*Ocotea rodiaei* Mez) | 300 | 1420 |
| Gurjun (*Dipterocarpus* sp.), Indian | 760 | — |
| Idigbo (*Terminalia ivorensis* A. Chev.) | 760 | 1020 |
| Iroko (*Chlorophora excelsa* Benth. et Hook. f.) | 410 | — |
| Karrik, South African (*Eucalyptus diversicolor* F. Muell.) | 200 | 320 |
| Mahogany, African (*Khaya ivorensis* A. Chev.) | 760 | 810 |
| Mahogany, Central American (*Swietenia macrophylla* King) | 300 | 710 |
| Oak, American white (*Quercus* sp.) | 25 | 330 |
| Oak, European (*Quercus robur*) home-grown | 50 | 330 |
| Obeche (*Triplochiton scleroxylon* K. Schum.) | 300 | 710 |
| Odoko (*Scotellia coriacea* A. Chev.) | 760 | 1520 |
| Pine, British Honduras pitch (*Pinus caribaea* Morelet) | 360 | 710 |
| Robinia (*Robinia pseudoacacia* L.) | 38 | 280 |
| Spruce, European [*Picea abies* (L.) Karst] | 760 | — |
| "Tasmanian Oak" (*Eucalyptus obliqua* L'Herit.) | 400 | 610 |
| Teak (*Tectona grandis* L. f.) | 400 | 710 |

As a rule air dry wood with a moisture content between 17 and 25% is most suitable for bending for the following reasons:

1. Failures after removing the supporting strap are rare;
2. The duration of drying after the bending is short;
3. Deformations during the drying after the bending are small.

It may be mentioned that of course the consumption of energy is higher and that the danger of surface checks is apparent.

### 9.8.3 Pretreatment of the Wood Prior to Bending

The plasticity of the wood as a fundamental property prior to bending mostly is effected by steaming; usually saturated steam of approximately 100 °C is applied. One steams approximately 1 hour per 25 mm thickness of wood. The application

35 Kollmann/Côté, Solid Wood

of higher steam pressures is possible; it increases the velocity of softening to the third power of the pressure (KOLLMANN, 1939), but above $2\ \text{kp/cm}^2$ the risk of deterioration of the wood is too high. Instead of steaming, boiling is possible, but the swelling of the wood is not so uniform and the absorption of moisture is too high.

### 9.8.4 Methods and Machines for Wood Bending

Bending by hand is rare as is the bending of wood in the cold condition. Wood, not pretreated and air-dry, generally can

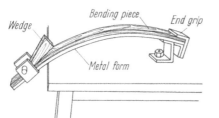

Fig. 9.133. Bending over a metal form. From STEVENS and TURNER (1948)

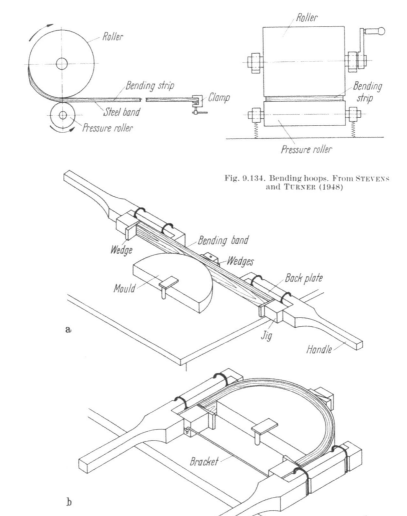

Fig. 9.134. Bending hoops. From STEVENS and TURNER (1948)

Fig. 9.135a, b. a) Initial set-up for making a simple U-shaped bend. b) Completion of bending operation with bracket attached. From STEVENS and TURNER (1948)

be bent only if the ratio of the radius of curvature $r$ to the thickness of the wood $d$ is $\frac{r}{d} \leq \frac{1}{50}$. Wood parts bent in the cold condition also have the tendency to spring back to a nearly flat state. For bends of smaller radius the wood needs to be plasticized as mentioned above.

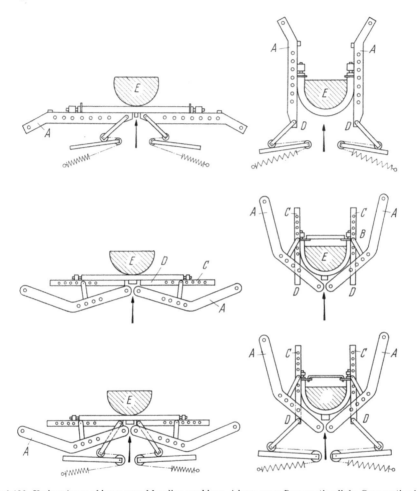

Fig. 9.136. Various types of lever armed bending machines. *A* lever arm; *B* connecting link; *C* supporting lever; *D* ends of the supporting lever; *E* bending form

A steamed, but unsupported material, if its thickness is small, may be bent over a metal form as shown in Fig. 9.133. Sieves, cattle drums, barrels, and so on can be bent by a hand-operated machine, consisting essentially of two rollers as illustrated in Fig. 9.134.

If the ratio of the radius of curvature to the thickness of wood is lower than approximately 30, frequent fractures are unavoidable. It was MICHAEL THONET, a cabinet maker, who detected at the end of the 19th century that supporting straps on the convex side prevent fractures. Normally these straps are manufactured of steel. The thickness of the supporting strap must be adapted to the thickness of the piece of wood to be bent and to the radius of curvature. It lies between 0.2 and 2 mm and averages from 0.75 to 0.88 mm. Fig. 9.135 shows a rather simple set-up for making a U-shaped bend at an initial stage (above) and after completion of the operation (below). For machine bending the mechanical, physical and technological prerequisites are approximately the same as for hand bending. In this book where only the scientific principles of wood bending are dealt with,

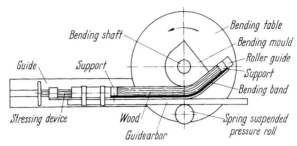

Fig. 9.137. Revolving table bending machine. From KIVIMAA (1948)

just a few remarks may be given about the construction and operation of such bending machines. The consumption of energy is rather low and amounts to 1.5 to 3 kW. Winch and bell presses are applied for bending barrel staves. Fig. 9.136 shows schematically different types of lever-armed bending machines. Another type is the revolving table machine (Fig. 9.137).

### 9.8.5 Properties of Bent Wood

**9.8.5.1 Sorption Properties.** The sorption capacity of cellulose controls the hygroscopical behavior of wood, though there are still some differences in the opinions of the experts about some not yet absolutely clarified facts. We know that hydroxyl

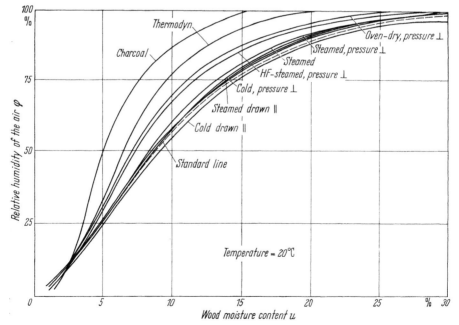

Fig. 9.138. Hygroscopic isotherms of beech after different steaming or mechanical treatment in comparison to charcoal and Thermodyn (a material, highly compressed under heat consisting of sawdust). From TEICHGRÄBER (1953)

groups are of greatest importance especially in the coordination of coupling between adjacent chain molecules. The structure of wood is changed markedly during bending and the setting that follows. Considering this fact one can expect that the hygroscopic isotherms also will be changed either by cold treatment, e.g. application of tension or by high frequency treatment or by steaming and bending.

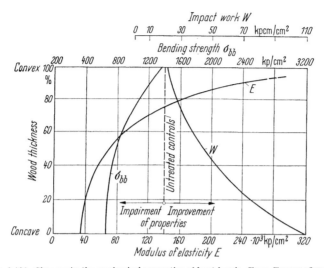

Fig. 9.139. Changes in the mechanical properties of bent beech. From TEICHGRÄBER (1953)

Fig. 9.138 shows the influence briefly mentioned above and also shows the much greater effect of a very intensive thermal treatment as it is used in producing a material on the basis of wood waste under the effect of high pressure and exclusion of air or by charring wood.

**9.8.5.2 Mechanical Properties.** Any bending procedure is a complex phenomenon to a degree depending on wood species, density, moisture content, ratio of thickness, bending radius and perhaps some chemical properties. TEICHGRÄBER (1953) has clearly proved that the mechanical properties depend in an interesting manner upon the ratio of wood thickness to bending radius and may induce either a decrease in mechanical quality or a remarkable increase. Fig. 9.139 shows a summary of the results. The diagram needs a rather intensive study. It shows, for example, that the crushing strength is practically always lessened whereas in comparison to untreated solid wood the modulus of elasticity and the shock resistance become higher, especially if the thickness of the bent parts is smaller and the measurements approach the outer zones. Thes rather different behavior as far as the modulus of elasticity or the shock resistance ability are concerned is remarkable.

## 9.9 Laminated Bending

It is a well known fact that veneers or other thin strips of wood may be bent easily without defect, even to a small radius of curvature. Unfortunately due to their elastic properties, they will try to resume the original shape upon removal of the bending forces (STEVENS and TURNER, 1948, p. 35). For this reason they must be secured and held to their shape in an appropriate manner. The best method is to bend a series of veneers concentrically combined with the gluing operation. This is the method of laminated bending. It is important that the individual laminae are of uniform thickness, that they are free of defects and decay, and that they are not brittle. The moisture content should be less than approximately 20%.

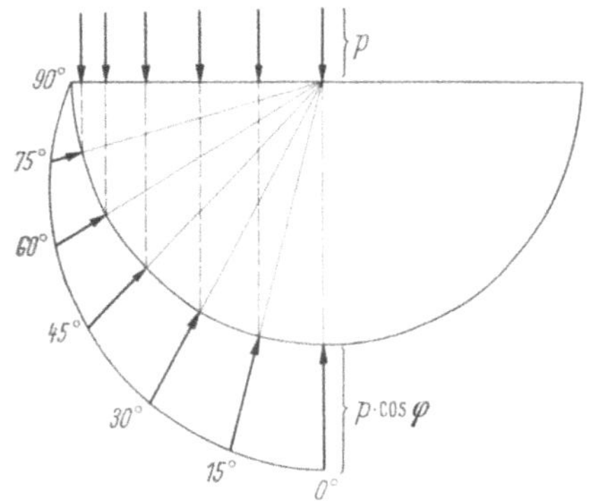

Normally veneers from hardwoods can be bent to smaller radii than comparable laminae from softwoods. Ten years ago nearly all types of glue were considered as suitable for this operation. Currently, practically only urea formaldehyde or phenolic types of glues are used.

The method of pressing laminae between male and female forms is limited. The reason is shown in Fig. 9.140. One can see that the pressure applied follows a cosine function and therefore it would

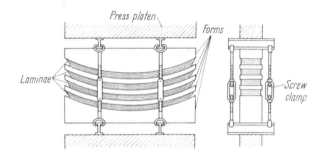

Fig. 9.140. Diagram of the pressure applied on a hemispherical form

Fig. 9.141. Simultaneous pressing of a number of bent members

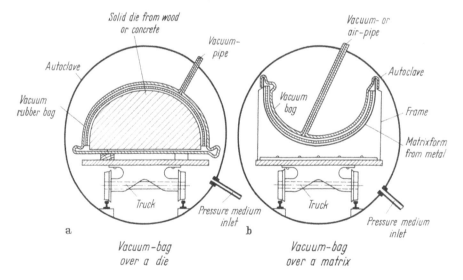

Fig. 9.142a, b. Bending of laminae by means of vacuum rubber bags in autoclaves

be impossible e.g. to produce hemispherical shapes in such a way. Only flatter forms like those shown in Fig. 9.141, are producible, but in this case simultaneous pressing of a number of bent members is possible.

The problem can be solved, of course, by the application of metal tension bands or by clamps. From the economical point of view these arrangements are not satisfactory. By far the best method of applying a uniform pressure over the total

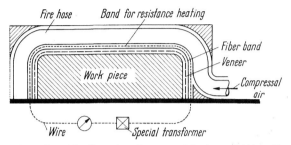

Fig. 9.143. Veneering of furniture pieces by means of fire hoses and low voltage strips

surface area of a bend is by means of hydraulic pressure. Flexible rubber hoses or tubes may be used. Another method is the application of rubber sheet or bags (Fig. 9.142).

The veneers should have a moisture content of approximately 15%. Still more important than the type of glue and the correct amount of glue spread is the pressure applied across the surfaces. The setting of laminated bends depends mainly on the type of glue used. As a rule, heat is applied to the laminated products, either by means of steam, hot water or electricity. The latter method includes the use of electrically heated blankets or low voltage strips. Fig. 9.143 gives an example of such an arrangement. A fluid-pressure molding of plywood is described in detail by HEEBINK (1946).

## Literature Cited

ANDREWS, G. W., (1954) Power at the headsaw. Timber of Canada (4).
—, (1955) Sawing wood with circular headsaws. For. Prod. J. 5: 186—192.
—, and BELL, G. E., (1953) Fundamental Sawmill Research: Rep. No. 1. Ottawa Lab., FPL Div. Canada.
ANONYMOUS, (1954) Technische Neuheiten: Eine neue Sägetechnik. Holz-Zentr. 80 (136): 1602.
AZZOLA, F. K., (1954) Beziehungen zwischen Zahnform, Vorschubgeschwindigkeit und spezifischer Zerspanungsarbeit beim Kreissägenlängsschnitt. Holz-Zentr. 80 (108): 1266.
BAKIEV, R., (1959) Some aspects of the vibration cutting of wood (in Russian) Izv. Vys. Ucheb. Zaved. Lesn. J. 2 (2): 144—149.
BARKAS, W. W., (1949) The swelling of wood under stress. Dep. Sci. Ind. Res., For. Prod. Res., London.
BARZ, E., (1953) Schwingungs- und Arbeitsverhalten von Kreissägeblättern für Holz. Forschungsberichte des Wirtschafts- und Verkehrsministeriums Nordrhein-Westfalen, No. 61, Westdeutscher Verlag, Köln und Opladen.
—, (1957) Fertigungsverfahren und Spannungsverlauf bei Kreissägeblättern für Holz. Forschungsberichte des Wirtschafts- und Verkehrsministeriums Nordrhein-Westfalen No. 360, Westdeutscher Verlag, Köln und Opladen.
BERSHADSKIJ, A. L., (1953) Complex solution of problems in high-speed sawing (in Russian). Lesnaja Promyshlenost (Timber Industry) No. 5, 26—29 and No. 6, 22—25.
—, (1956) Calculation of the working conditions during wood cutting. Derev. Prom. 5 (5): 6—10.
BETHMANN, H., (1940) Wesensart und kritische Beurteilung der Kurzprüfverfahren zur Bestimmung der Zerspanbarkeit von Werkstoffen. Diss., Tech. Hochschule München.

# Literature Cited

BIERMANN, O., (1942) Betriebserfahrungen mit dem Vollgatter und ihre Auswertung. Holz als Roh- und Werkstoff, Vol. **5**: 390—397.

BRAUNSHIRN, F., (1929) Das Sägewerk. Wien.

BRÜNE, H. B., (1930) Untersuchungen über den Einfluß der Faserrichtung auf die zur Holzbearbeitung erforderliche Zerspanungsarbeit. Diss., Tech. Hochschule Dresden.

BRYAN, E. L., (1963) Machining wood with light. For. Prod. J. **13** (1): 14.

—, (1963) High energy jets as a new concept in wood machining. For. Prod. J. **13** (8): 305—312.

BUES, K., (1928) Versuche an Gattersägen. Diss. Tech. Hochschule München.

BÜTTNER, R., (1930) Untersuchungen über den Holzzerspanungsvorgang mittels zwangläufigen Vorschubs unter besonderer Berücksichtigung der Schneidhaltigkeit der Holzbohrer. Diss., Tech. Hochschule Dresden.

DEHNE, E., and REHN, E., (1960) Einfluß der Schneidenzahl auf die Oberflächengüte beim Fräsen. Holztechnologie **1**: 146—150.

DOSKER, C. D., (1950) Wood (USA) (3): 15.

DOWER, E. J., and OAKEY, W., (1957) Selection of carbide-tipped saw blades for optimum performance. Paper No. 57-WDI-9, ASME Wood Industries Conference, Winston-Salem.

EMMERICH, H., (no date) Der Sägewerker und sein Wirkungskreis. Berlin o. J., p. 142.

ENDERSBY, H. J., (1953) The performance of circular plate ripsaws. Gt. Britain, Dep. Sci. Ind. Res., For. Prod. Res. Bull. No. 27, London. H. M. S. O.

ENGLESSON, T., HVAMB, G., and THUNELL, B., (1954) Noen resultater fra undersøkelser over saging med og mot fibrene. Norsk Treteknisk Institutt, Rep. No. 7, Oslo.

FESSEL, F., (1951) Probleme beim Holzbiegen. Holz als Roh- und Werkstoff **9**: 56—62.

FLEMMING, H., and KNOSPE, L., (1958) Der Einfluß der Schnittgeschwindigkeit auf die spezifische Arbeit bei der spangebenden Holzbearbeitung. Fertigungstechnik **8**: 565—566.

FOYSTER, R., (1953) Modern mechanical saw practice. Crosby Lockwood Son London.

FRANZ, N. C., (1958) An analysis of the wood-cutting process. Ph. D. thesis, University of Michigan Press, Ann Arbor.

GAMBRELL JR., S. C. and BYARS, E. F., (1966) Cutting characteristics of chain saw teeth. For. Prod. J. **16** (1): 63—71.

GRUBE, A., and ALEKSEEV, A. V. (1961) Specific cutting work in planing and moulding of chipboards (in Russian). Derev. prom. **10** (2): 7—8.

HARRIS, P., (1947) A Handbook of Wood Cutting, Dep. Sci. Ind. Res., For. Prod. Res., H. M. S. O., London.

—, (1954) Mechanics of sawing: Band and circular saws. For. Prod. Res. Bull. No. 30, H.M.S.O., London.

HAYCOCK, A. H., (1949) Tenoning, mortising and boring. London, The Technical Press Ltd.

HEEBINK, B. G., (1946) Fluid-pressure molding of plywood. U.S. For. Prod. Lab. Rep. No. R 1624, Madison, Wisc.

HETZEL, F., (1928) Über die Bearbeitbarkeit von Spanholzplatten und Sperrholzplatten (Bohren und Stanzen). Diss., Tech. Hochschule Dresden.

HUDSON, H. R., (1953) Woodworking machinery. London, George Newnes' Ltd.

JONES, D. S., (1956) Swedish frame saws. The Australian Timber J. **22** (5): 484—498.

JUŠTŠUK, G., (1951) Avfallmänder inom sågverksindustrien. Svenska Träforskningsinstitutet, Trätekniska avdelningen. Medd. 20 B, Stockholm.

KAIUKOVA, M. V., and KONJUKHOV, D. N., (1934, 1935) Research on circular saws and the cutting process when they are used (in Russian). Lesnaja Prom. (11): 36—55; 1935 (1): 29—39.

KERSTEN, G., (1944) Ausnutzung beim Einschnitt von Nadelholz. Mitt. Dtsch. Ges. f. Holzforschg., H. 35, Berlin.

KIVIMAA, E., (1948) Koivuu taivutuksesta (On the bending of birch). Valtion teknillinen tutkimuslaitos No. 54, Helsinki.

—, (1950) The cutting force in woodworking. Publ. No. 18, The State Inst. for Techn. Research, Helsinki, Finland.

—, Die Schnittkraft in der Holzbearbeitung. Holz als Roh- und Werkstoff **10**: 94—108.

KLOPSTOCK, V., (1962) Die Untersuchung der Dreharbeit. Ber. Tech. Hochschule Berlin H. 8, Berlin.

KOCH, P., (1954) An analysis of the lumber planing process, Unpubl. Ph. D. thesis, Univ. of Washington; available on microfilm from the Univ. of Michigan.

—, (1955, 1956) An analysis of the lumber planing process, Part I, For. Prod. J. Vol. 5, No. 4, 255—264; Part II, For. Prod. J. Vol. 6, No. 10, 393—402.

—, (1964) Square cants from round bolts without slabs or sawdust. For. Prod. J. **14**: 332—336.

—, (1964) Beams from boltwood: A feasibility study. For. Prod. J. **14** (11): 497—500.

—, (1964) Wood machining processes. Ronald Press Co., New York.

KÖBERLE, J., (1935) Richtlinien für die Wahl wirtschaftlicher Tourenzahlen an Kehl- und Fräsmaschinen. Tech. Rdsch. (Bern) **27**, No. 44, 45, 47, 48.

# Literature Cited

KOLLMANN, F., (1939) Vorgänge und Änderung von Holzeigenschaften beim Dämpfen. Holz als Roh- und Werkstoff **2**: 1—11.

—, (1955) Technologie des Holzes und der Holzwerkstoffe. Bd. II 2nd ed. Berlin/Göttingen/ Heidelberg.

KRATZ, E., (1939) Untersuchungen über das Holzschleifen an Trommelschleifmaschinen. Diss., Tech. Hochschule Dresden.

KÜBLER, H., (1960) Das Schneiden von Holz mit Vibrationsmessern. Holz-Zentr. **86**: 1605-1606.

KULLMANN, H., (1930) Holzbearb.-Masch. **6**: 17.

LAPIN, P. J. (1954) High-speed cutting of wood (in Russian). Derev. Prom. **3** (3): 3—8.

LAURENT, A. St., (1965) Effect of induced lateral vibration of a saw tooth on the cutting of wood. For. Prod. J. **15**: 113—116.

LINDHOLM, E., (1950) Cirkelsagars Buckling vid Symmetrisk Temperaturfördelning. Teknisk Tidskrift, **80** (11): 243—247.

LINDNER, J., (1949) Rückschlagsichere Kreissägeblätter. Holztechnik **29**: 171—179.

LISTER, S., (1948) Sanders and Sanding. Lomax Erskine and company Ltd., London.

—, (1949) Boring, mortising and tenoning. Lomax, Erskine and Company Ltd., London.

LOTTE, M., and KELLER, M., (1949) La scie circulaire. Serv. Centr. d'Essais des Bois et Lab. d'Essais de L'Institut Nat. du Bois. Broch. Tech. No. 8, Paris.

LUBKIN, J. L., (1957) A Status Report on Research in the Circular Sawing of Wood, Vol. 1, Centr. Res. Lab., American Machine and Foundry, Greenwich, Conn.

MANZOS, F., (1955) Chipless cutting of wood using vibration (in Russian). Lesn. Prom. (110): 4.

MARSCHNER H., (1942) Die Berechnung der Einspannkraft von Gattersägen. Holz als Roh- und Werkstoff **5**: 427—434.

MARTELLOTTI, M. E., (1941, 1945) An analysis of the milling process. Trans. of ASME **63**: 677—700; also: Part II, Down Milling Trans. of ASME **67**: 233—251.

McKENZIE, W. M. and FRANZ, N. C., (1964) Basic aspects of inclined or oblique wood cutting. For. Prod. J. **14** (12): 555—566.

McMILLIN, C. W., and LUBKIN, J. L., (1959) Circular sawing experiments on a radial arm saw. For. Prod. J. **9** (10): 361—367.

MEYER, M., (1926) Untersuchungen über die den Zerspanungsvorgang mittels Holzkreissägen beeinflussenden Faktoren. Ausgewählte Arbeiten des Lehrstuhls für Betriebswissenschaften in Dresden **3**: 90, Berlin.

MOTE, Jr., C. D., (1964) Circular saw stability — a theoretical approach. For. Prod. J. **14** (6): 244—250.

OSENBERG, W., (1926) Untersuchungen über den Zerspanungsvorgang mittels Holzbohrern beeinflussenden Faktoren. Diss., Tech. Hochschule Dresden.

PAHLITZSCH, G., (1962) Stand der Forschung auf dem Gebiet des Sägens. Holz als Roh- und Werkstoff **20**: 381—392.

—, (1951) Spangebendes Formen, in „Hütte", Taschenbuch für Betriebsingenieure (Betriebs-hütte) 4th ed., Part 1, Verlag Wilhelm Ernst u. Sohn, Berlin, p. 393.

—, (1966) Internationaler Stand der Forschung auf dem Gebiet des Hobelns und Fräsens von Holz und Holzwerkstoffen. Holz als Roh- und Werkstoff, **24**: 579—593.

—, and DZIOBEK, K., (1959) Untersuchungen über das Bandschleifen von Holz mit gerad-liniger Schnittbewegung. Holz als Roh- und Werkstoff **17**: 121—134.

—, and —, (1959) Untersuchungen an einer horizontalen Blockbandsäge. Holz als Roh- und Werkstoff **17**: 364—376.

—, and —, (1961a) Über das Wesen der Abstumpfung von Schleifbändern beim Band-schleifen von Holz. Holz als Roh- und Werkstoff **19**: 136—149.

—, and —, (1961b) Beitrag zur Bestimmung der Oberflächengüte spanend bearbeiteter Hölzer — Erste Mitteilung. Holz als Roh- und Werkstoff **19**: 403—417.

—, and —, (1962) Beitrag zur Bestimmung der Oberflächengüte spanend bearbeiteter Hölzer — Zweite Mitteilung. Holz als Roh- und Werkstoff **20**: 125—137.

—, and JOSTMEIER, H., (1964a) Beobachtungen über das Abstumpfungsverhalten beim Fräsen von Spanplatten. Holz als Roh- und Werkstoff **22**: 139—146.

—, and —, (1964b) Weitere Beobachtungen über das Abstumpfverhalten und den Einfluß der Schnittgeschwindigkeit beim Fräsen von Spanplatten. Holz als Roh- und Werkstoff **22**: 424—429.

—, and —, (1965) Beobachtungen über das Abstumpfungsverhalten beim Fräsen von Schicht-stoff-Verbundplatten. Holz als Roh- und Werkstoff **23**: 121—125.

—, and MEHRDORF, F., (1962a) Herstellen von Schneidspänen mit Flachscheiben-Spanern. Erste Mitteilung. Einfluß von Spannungsdicke und Holzfeuchtigkeit auf die Erzeugung von Holzspänen. Holz als Roh- und Werkstoff **20**: 314—322.

—, and —, (1962b) Herstellen von Schneidspänen mit Flachscheiben-Spanern. Zweite Mit-teilung. Einfluß von Schnittgeschwindigkeit, Spanwinkel und Schnittflächenwinkel. Holz als Roh- und Werkstoff **20**: 408—418.

554                 Literature Cited

PAHLITZSCH, G., (1962) and —, (1963) Herstellen von Schneidspänen mit Flachscheiben-Spanern. Vierte Mitteilung. Spanen von Pappelholz im Vergleich zum Spanen von Kiefernholz. Holz als Roh- und Werkstoff **21**: 144—149.

PRODEHL, A., (1921) Untersuchungen über das Biegen gedämpfter Rotbuche. Diss. Tech. Hochschule Dresden. Comp.: Z. VDI 75: 1217.

REINEKE, L. H., (1950) Sawteeth in action, Proc. For. Prod. Res. Soc., 4: 36—51.

—, (1956) Motion energy of wood particles. For. Prod. J. **6**: 507—509.

SACHSENBERG, E., and BALZER (1930) Holzbearb. Masch. **6**: 29.

SAITO, J., EDAMATSU, N., and OHIRA, J., (1955) The influence of tooth angle and set on ripsawing. J. Japan Wood Res. Soc. **1** (1): 42—45.

—, —, and —, (1956) The effect of tooth bite on the cutting forces in ripsawing. J. Japan Wood Res. Soc. **2**: 65—68.

—, —, and —, (1957) Holzzerspanungsforschung mit mechanischen Sägezähnen. Govt. For. Exper. Stat. Bull. (97): 33—60.

SAITO, J., and NIGA, T., (1954) The buckling of the circular saw. Wood Ind. (Tokyo) 9 (6 and 7).

SCHALLBROCH, H., and DODERER P. v., (1943) Zerspanbarkeitsuntersuchungen an geschichteten Kunstharzpreßstoffen. Berichte über betriebswissenschaftliche Arbeiten **15**, Berlin.

SCHIMMING, H., (1952/53) Untersuchungen zum Problem der Kreissägeblätter für Holzbearbeitung unter besonderer Berücksichtigung von Hobelsägeblättern. Wiss. Z. Tech. Hochschule Dresden **2** (1): 65—88.

SCHWACHA, B. G., (1961) Liquid cutting of hard materials. U.S Pat. No. 2, 985.050. Issued May 23, 1961. Assigned to North American Aviation, Inc.

SKOGLUND, C., (1949) Temperaturmätnigar pa sagklingor. Föreningen Svenska Sagverksmän, Sjunde Årsmotet, Stockholm, pp. 50—54.

—, (1950) Spänningar och svängningar i roterande sägklingor. Svenska Träforskningsinstitutet, Trätekniska avd., Medd. 12 B, Stockholm.

—, and HVAMB, G., (1953) Tannvinklenes innvirkning på kraftforbruket ved saging med og mot fibrene. Reg. No. 4, Norsk Treteknisk Institutt, Oslo.

SMIRNOV, V., (1957) About the problem of the formation of a theory of vibration cutting of wood (in Russian). Nauch. Trudy Lvov. Lesotechn. Inst. **3**: 88—92.

STEVENS, W. C., and TURNER, N., (1948) Solid and Laminated Wood Bending. London, H. M. S. O.

SUGIHARA, H., (1953) Untersuchungen über das Bandsägen von Holz, 1. Bericht. Wirkungen der Vorschubkraft und der Zahnungsbreite. Wood Res. Bull. of the Wood Res. Inst. Kyoto **10**: 1—16.

—, (1956) Studies on the sawing with bandsaw blade, IV: Clearance angle and pitch. J. Japan Wood Res. Soc., **2** (1): 1—4.

—, and SUMIYA, K., (1955) Theoretical study on temperature distribution of a circular sawblade. Wood Res. Bull. No. 15 of the Kyoto University, Kyoto, Japan, pp. 60—74.

SUTTER, A., (1935) Masch.-Bau Betrieb **14**: 325.

TEICHGRÄBER, R., (1953) Über die Spannungszustände bei der Verformung von Holz und die dadurch geänderten Holzeigenschaften. Diss., Univ. Hamburg.

TELFORD, C. J., (1949) Energy requirements for insert point circular headsaws. For. Prod. Res. Soc., Preprint No. 49.

THUNELL, B., (1950) Neuere Ergebnisse der holztechnologischen Forschung in Schweden. „Holztagung 1949", Schriftenreihe der Österr. Ges. f. Holzforschung **2**: 8.

THUNELL, B., (1951) Fortschritte bei der Zerspanungsforschung von Holz. Holz als Roh- und Werkstoff **9**: 11—20.

—, (1951) Provsågning vid korsnäsverkan våren 1951. Svenska Träforskningsinstitutet, Träteknikska avdelningen, Medd. 21 B, Stockholm.

THUNELL, B., and HILTSCHER, R., (1951) Die Beanspruchung von Gattersägeblättern im Betrieb. Holz als Roh- und Werkstoff **9**: 232—242.

VOIGT, F., (1925) Holzmarkt **42**: 188—189.

—, (1949) in: Hütte, Des Ingenieurs Taschenbuch, Vol. 2, 27th ed., Berlin, Wilhelm Ernst u. Sohn, p. 862.

WALTER, J., (1931) Untersuchungen an einer Kettenfräsmaschine. Diss. Tech. Hochschule Dresden.

WARLIMONT, J., (1932) Zerspanungsuntersuchungen beim Drehen von Holz. Diss. Tech. Hochschule Dresden.

WEBER, A., (1961) Messung der Schnittkräfte beim Fräsen von Holz mit Hilfe eines magnetostriktiven Meßverfahrens. Diss. Tech. Hochschule Stuttgart.

YUGOV, V. G., and OSIPOV, A. J. (1960) The use of high-speed water jets in wood cutting and processing (in Russian). The transactions of the Central Scientific Research Institute of Mechanization and Energy Requirements of the Forest Industries of USSR **15** (6).

# AUTHOR INDEX

Adams and Douglas 64
Adler 67, 69, 71
Alexander and Mitchell 58
Allen and Campbell 460
Altenkirch 433
Amati 274
Amy 151
Andersson 90, 459, 460
Andersson and Fearing 465
Andrews 495, 498
Andrews and Bell 495, 498
Anonymous 150, 431, 504
Argue and Maass 201, 203
Armstrong and Savory 112
Aro 180
Aronowsky and Gortner
  347, 348
Aspinall and Ross 63
Asunmaa and Lange 73
AWF 500, 501
A. W. P. A. Manual of Re-
  commended Practice 143,
  149
Azzola 497

Bach and Baumann 292
Baechler and Roth 143
Bailey 35, 73, 137
Bailey and Vestal 104
Bakiev 514
Barefoot 48
Barkas, 193, 315, 316
Barz 408
Bateman, Hohf and Stamm
  220
Bateson 428
Baumann 301, 304, 305, 306,
  313, 326, 327, 336, 341,
  342, 361, 365, 367, 371,
  396, 406
Bauschinger 342, 397
Bazhenov 271, 272, 273
Becker 113, 119, 120, 124,
  128, 129, 130, 131
Becker and Gersonde 131
Bershadskij 477, 493, 494, 496
Besemfelder 466
Bethmann 537
Betts 368, 369
Biermann 482
Björkman 67, 70
Björkman and Person 67
Bland 74
Blew et al. 146

Boltzmann 279
Boracure Ltd. (New Zealand)
  466
Bouveng 64
Bouveng and Lindberg 61
Bouveng and Meier 63
Brake and Schütze 268, 270
Brauns 67, 155
Brauns and Brauns 56
Braunshirn 484, 501
Breuil 161
Brillié 279
British Forest Products
  Research Laboratory 441
British Standards Air Raid
  Precaution Series 39, 154
Brotero 398
Brown et al. 141, 259
Brown, Hoyler and Bierwirth
  470
Brown, Panshin and Forsaith
  168, 268, 275, 340, 356,
  381, 402
Browne 151
Browning 56
Brunauer, Emmet and Teller
  194
Brüne 534, 536
Bryan 93, 515, 516, 517
Bucher 75
Burr and Stamm 228, 234
Bues 479, 480, 481, 483
Büsgen 403

Calvert 462
Campbell 101, 110, 111, 434
Carrington 293, 294, 297,
  298, 302, 303, 309, 311,
  396
Cartwright and Findlay 105,
  107, 108, 111
Casati 380
Chaplin and Armstrong 410,
  411
Chattaway 16, 17
Chevandier and Wertheim 160
Choong 221
Christensen and Kelsey 190,
  191, 192
Čižek 370, 371
Clark 471
Clark and Williams 258, 261
Clarke 354, 355, 392, 393

Clarke et al. 346, 347
Cobler 430
Cochrane 109
Cockrell 208
Coker and Coleman 330, 331,
  332, 399
Colvin 60
Commonwealth of Australia,
  Counc. Sci. Ind. Res. Div.
  For. Prod. 197
Core 81, 83
Core et al. 6, 44, 92
Côté 34, 38, 137
Côté and Day, 35, 36, 49,
  50, 51, 74
Côté and Krahmer 137, 139
Côté and Marton 18
Côté and Timell 73
Cowling 98, 99, 102, 104, 111
Craighead 128
Cressow, Vorhees Curtis and
  Fielder 466
Croon et al. 63, 74
Crossley 111
C. S. I. R. O. Division of
  Forest Products, Australia
  187, 188, 424
Curtis and Isaacs 464
Czepek 462

Dadswell and Wardrop 48,
  49, 83, 84
Dadswell, Wardrop and Wat-
  son 47
Davidson 163
De Bruyne 217, 218
Dehne and Rehn 526
Delay and Lecompte 471
Delmhorst Instrument Co.,
  Boonton, N. J. 186
De Luca, Campbell and Maass
  270
Dept. Sci. Ind. Res., London
  194
Doffiné 435
Dominicus 479
Dosker 509
Dower and Oakey 492, 493, 496
Doyle, Drow and McBurney,
  294, 295, 296
Draffin and Mühlenbruch 308
Duncan 104, 107
Dunlap, 184, 201, 245, 246
Dunlap and Bell 185

# Author Index

Eberius 183
Egner 224, 229, 230, 406, 462
Egner, Klauditz, Kollmann and Noack 155
Egner and Sinn 432
Egner and Rothmund 335
Ehrmann 398, 400, 401
Eisenmann 433, 461
Elbein, Barber and Hassid 71
Ellwood and Erickson 451
Elmendorf 383
Emmerich (no date) 504
Endersby 495, 498, 503, 509
Engesser 359
Englesson, Hvamb and Thunell 495
Entwistle et al. 60
Erdtman 70
Erickson, Schmitz and Gortner 240
Esterer Machine-works, Altötting, Bavaria 479

Fellows 197, 198,
Fessel 433, 435, 542
Findlay and Savory 104
Fischer 183
Fischbein 41
Fisher 116
Flatscher 180
Fleischer 150, 462
Flemming 513
Fletcher and Munson 281, 282
Föppl 341
Forest Products Research Laboratory, Princes Risborough 199, 333, 501, 543
Forsaith 393
Foyster 492, 503
Franco 37
Franz 505, 506
Frei et al. 20, 22
Frei and Preston 60
French 12
Freudenberg 71
Freundlich 194
Frey-Wyssling 20, 56, 208
Frey-Wyssling et al. 29, 40, 137
Frey-Wyssling and Bosshard 4
Fukada 269, 271, 272
Fukada et al. 274

Gaber 327, 339, 340, 373, 398, 406
Gaby 433, 434
Galligan and Bertholf 271, 272, 273
Galligan and Jayne 271, 272
Gambrell, Jr. and Byars 511, 512
Gardner 70

Gatslick 433
Gäumann 108, 421
Geister 410
George et al. 114, 115
Gerby 16
Gerry 393
Gewecke 147
Ghelmeziu 380, 385, 387, 388, 391
Gierer 67
Glatzel 241
Goens 302
Goldschmid and Perlin 63
Goodwin and Hug 147
Goring 67
Goring and Timell 58
Graf 80, 327, 328, 329, 340, 341, 352, 353, 355, 373, 376
Grau 432
Greathouse 71
Greenhill 205
Greenwood 64
Griffiths and Kaye 246
Grube and Alekseev 525, 526
Grumach 461
Guarneri 274
Guernsey 461
Gustafsson et al. 57

Hafner and Schwoerer 470
Hägglund 56
Hale and Prince 168, 179
Hamilton et al. 63
Hamilton and Quimby 61
Hann 462
Hansen 279
Harada 22, 26, 40, 137
Harada and Miyazaki 40
Harada and Wakashima 22, 47
Harada and Wardrop 24
Harris 70, 477, 486, 492, 494, 496, 501, 503
Hart 224
Hartig 171
Hartmann 82, 153
Hasselblatt 217, 218, 258
Hauber 342
Hathway 70
Hawley 223, 237, 239
Haycock, 538
Haygreen 468
Hearmon 263, 275, 294, 299, 301, 309, 313
Hearle 259
Heebink 551
v. Helmholtz 293
Hendershot 241
Henderson 196, 426, 438, 440, 445, 448, 460
Henley 60
Henriksson 146
Herma 431

Hermans 58, 60, 191, 193
Hesehus 276
Hildebrand 431
Hillis 56, 73
Hirst and Jones 64
Hiruma 258, 260
Honda and Konno 279
Honeyman 58
Hörig 279, 294, 297, 301, 395
Howard 467
Howard and Hulett 163
Howes 18
Howick 127
Howsmon 60
Huber 396, 403
Hudo and Yoshida 152
Hudson 466, 537
Huffman and Post 430
Hunt and Garratt 133, 137, 142

I. A. W. A. (International Association of Wood Anatomists) 6, 15, 27, 48
Ifju 323
Inokuty 398
Institut National du Bois, Paris 501
Ivanov 334

Jaccard 48
Jalava 176, 180
James 279, 381
Janka 169, 342, 403, 406
Janka and Hadek 347
Jayme and Fengel 22
Jayme and Krause 161
Jayne 323
Jenkin 297
Jenkins and Guernsey 197, 424
Jennison 109
Johansson 174
Johnston and Maass 238, 239
Jones 485
Jones et al. 62
Jorgensen 58
Jurd 70
Jürges 232
Juštšuk 485
Juon 151, 152
Jutte 50

Kaiukova and Konjukhov 480, 488, 492, 493, 495, 496, 498, 499, 501, 502, 503, 504
Kalninš 346
Kármán 301, 359
Kastmark 433
Katz 201, 217
Katz and Miller 259
Keith 15
Kellog 318
Kelsey and Clarke 202
Kemp 94

## Author Index

Kerr 20
Kersten 485
Keylwerth 166, 190, 191, 205, 208, 209, 211, 212, 213, 214, 216, 297, 303, 304, 305, 331, 332, 333, 334, 400, 401, 402, 440, 446, 450, 455, 456, 461, 464, 465, 471
Keylwerth and Kübler 462, 463
Keylwerth and Noack 188, 261
Kimball 445, 460
King 316, 318
Kinnman 485
Kivimaa 502, 519, 520, 521, 522, 523, 524, 525, 548
Klauditz et al. 321
Klem 176
Klemm 347
Klopstock 534
Knight and Newall 214, 215
Knight and Pratt 191, 193
Knuchel 421, 422
Kobayashi and Utsumi 40
Köberle 518, 519
Koch 9, 44, 45, 57, 477, 478, 485, 487, 497, 499, 505, 508, 509, 511, 513, 517, 519, 520, 524, 525
Koehler 95, 213, 387, 393
Koehler and Pillow 437, 438
Kohara 273, 274
Kohlmeyer et al. 132
Kollmann 151, 152, 155, 164, 167, 176, 177, 178, 179, 194, 198, 199, 201, 202, 212, 221, 224, 225, 226, 227, 229, 231, 232, 246, 247, 250, 255, 284, 299, 303, 304, 306, 307, 316, 317, 318, 321, 326, 335, 336, 337, 341, 342, 343, 344, 346, 347, 348, 349, 350, 351, 352, 354, 355, 364, 377, 378, 380, 382, 384, 386, 387, 391, 392, 393, 394, 396, 399, 408, 410, 411, 412, 424, 428, 436, 437, 438, 453, 459, 461, 462, 470, 501, 502, 519, 541, 546
Kollmann and Dosoudil 335
Kollmann and Höckele 182, 183
Kollmann and Keylwerth 399, 402
Kollmann, Keylwerth and Kübler 451, 452
Kollmann and Krech 277, 278, 301, 302, 307, 308, 309, 310, 311
Kollmann and Malmquist 247, 248, 249, 250, 256, 257, 451, 461

Kollmann and Schneider 194, 195, 232, 462
Kollmann and Schulz 328
Koppers Co. 157
Koran 16, 37
Krahmer and Côté 28, 29, 30, 31, 38, 73, 138
Kraemer 377, 378, 396
Kratz 533
Kratzl 71
Kratzl and Billek 56, 71
Krech 379, 380, 388, 394
Kribs 12, 14
Krischer 221, 222, 231, 232
Krischer and Schauss 467
Kröll 220
Kröner 267, 269, 270
Krüger and Rohloff 301
Kübler 261, 513
Küch 327, 328 367, 368, 374, 375
Küch and Telschow 347
Kufner 324, 325
Kuhlmann 434, 435
Kullmann 539, 540, 541
Kuntze 337, 339
Kurth 70
Kurth and Sherrard 173
Kutscha 104

Ladell 462
Lang 398
Lange 73
Langlands 375
Langmuir 194
Lantican 30
Lapin 492
Laurent 494
Levon 180
Lewis 232
Liang et al. 62
Liese 24, 30, 31, 40, 41, 42, 137
Liese and Hartmann-Fahnen-brock 40
Liese and Johann 40
Liese and Schubert 237
Liese and Schmid 106
Lilly and Barnett 109
Limbach 90
Lin 257, 258, 259, 260, 261, 262
Lindberg 70
Lindgren 70, 97
Lindholm 508
Lindner 509
Lister 528, 529, 538
Löfgren 232, 459
Loos 161
Lorenz 88
v. Lorenz 406
Lotte and Keller 488, 493, 494, 495, 498, 502, 503, 504
Loughborough 221, 468

Loughborough and Hawley 190
Lübecke 276
Lubkin 478, 488, 490, 491, 493, 494, 497, 498, 499, 501, 502, 503, 504
Ludwig 235, 236
Lundberg 510

MacLean 145, 238, 239, 240, 244, 245, 246, 247, 251, 253, 254
Maku 247
Malcolm 85
Mallis 115, 123
Malmquist 194
Manley 60
Mann and Marrinan 60
Manners 64
Manzos 514
March 244
Marchessault et al. 62
Marchessault and Liang 62
Mark 59, 321
Markwardt 375, 376
Markwardt, Bruce and Freas 155
Markwardt and Wilson 205, 213, 385, 388
Marschner 483
Martellotti 519
Martley 224
Mathewson 212, 428
Mayer-Wegelin and Brunn 343, 344
Mayr 173
McIntosh 208
McKenzie and Franz 521
McMillin and Lubkin 488, 498
Meier 18, 22, 42, 57, 60, 62, 63, 73, 74, 102
Meier and Wilkie 74
Meier and Yllner 75
Menzel 241
Menzies 130
Menzies and Turner 130, 131
Meredith 292, 293, 299
Merewether 70
Merrick 148
Metz 149, 151, 152, 153, 154
Metzger 48
Meyer 282, 283, 284, 481, 488, 492, 493, 494, 495, 498, 499, 501, 502, 503, 504
Meyer and Mark 59, 321
Meyer and Misch 59
Meyer and Rees 258
Miller 413
Mitchell and Iversen 469
Mitchell and Wahlgren 156
Moll 226
Mombächer 167, 168
Monnin 343, 344, 371, 378, 379, 384, 387

Mörath 165, 166, 197, 199, 207, 208, 212, 213, 404, 406, 408, 424
Morén 469
Mote, Jr. 490, 507, 508
Mühlethaler 19, 20, 60
Müller-Stoll 174
Mutton 70

Narayanamurti 228, 259
Narayanamurti and Ranganathan 246
Natalis 358
Neish 71
Nelson and Schuerch 61
Neufeld and Hassid 71
Newlin 205, 206, 359
Newlin and Gahagan 356
Newlin and Trayer 372
Newlin and Wilson 344, 386, 401, 406
Newsletter 128
Niethammer 160
Nilakantan 271
Nord and Schubert 71
Nördlinger 331, 403
Northcote 56
Noskowiak 84
Nusser 258, 259

Oberg and Hossfeld 217
Onaka 48
Osenberg 541
Ott, Spurlin and Grafflin 58
Ottawa Laboratory 494

Pahlitzsch 475, 476, 478, 481, 486, 493, 496, 505, 525
Pahlitzsch and Dziobek 486, 487, 488, 489, 529, 530, 531, 532, 533
Pahlitzsch and Jostmeier 525, 526, 527
Pahlitzsch and Mehrdorf 527, 528
Pallay 404, 405, 408
Pascheles 207
Paul 160, 171, 174, 175
Paul and Marts 168, 169
Peck 197, 205, 423, 424, 433, 469
Peck, Griffith and Rao 450
Pence 113
Pentoney 208, 211
Perem 337
Perilä 12, 73
Perilä and Seppa 12, 73
Perilä and Heitto 12, 73
Perkitny 215
Petermann 398
Pettifor 383, 385
Pevzoff 406
Pfeiffer 429, 430
Piest 432
Pillow 84

Pillow and Luxford 83, 174
Pinder 70
Poulignier 301
Prát 471
Price 297
Preston and Ranganathan 50
Preusser, Dietrichs and Gottwald 26
Proctor 101
Prodehl 542, 543
Prütz 173

Rasmussen 91
Rawling 471
Ray 128, 131, 132, 133
Ray and Stuntz 131
Record and Hess 18
Rehman 197, 424
Reimer 60
Rein 329, 330, 385, 389, 390, 391
Reineke 478, 510
Reischel and Wedekind 271
Reissner 397
Reuleaux 323
Riechers 334
Risi and Arseneau 219
Ritter and Fleck 261
Rochester 176, 177, 178, 179
Roe, Hochman and Holden 133
Roelofsen 56
Roš 361, 365, 372
Roš and Brunner 358, 369
Rosenbohm 201
Roth 88, 334, 355, 376
Rother 184
Rucker and Smith 431, 433
Rudeloff 397
Runkel and Lüthgens 191
Ryska 336

Sabine 284
Sachs, Clark and Pew 18, 73
Sachsenberg 410, 411
Sachsenberg and Balzer 529
Saito, Edamatsu and Ohira 489, 495, 496, 501, 502, 503, 504
Saito and Niga 508
Salamon 313, 462
Sargent 138
Savart 293
Savkov 176
Savory 104
Savory and Pinion 105
Schallbroch and Doderer 535, 536, 537, 540, 543
Schilde Maschinenbau AG., Germany, 441, 444
Schimming 503
Schleussner 427
Schlüter 297, 301
Schlüter and Fessel 226, 229, 231, 232, 248, 249

Schlyter 327, 343, 367, 376, 377, 378
Schlyter and Winberg 300, 306, 327, 328, 347, 368, 400, 401
Schmid 27
Schmid and Machado 35, 36
Schmidt 180, 257
Schmidt and Marlies 315, 319, 320
Schneider 301, 302, 379, 461
Schniewind 211
Scholten 373
Schwacha 514
Schwalbe and Bartels 433
Schwalbe and Becker 261
Schwalbe and Berling 154
Schwalbe and Ender 438
Schwankl 331
Schwappach 168, 169, 344
v. Schwarz and Bues 406
Schwerin 57
Scurfield and Wardrop 48
Seborg 192
Sedziak 150
Seeger 385, 387
Seekamp 155
Seewald 301
Sepall and Mason 60
Seth et al. 465
Sharpe and O'Kane 268, 270
Sheppard and Newsome 218
Sherwood 221
Shubnikov 271
Siemens & Halske AG 186, 187, 188
Siimes 80, 174, 372, 373
Simeone 116, 118, 120, 124, 125, 126
Simonsen 70
Sinnott 81
Siu 109
Skaar 262, 266, 267, 270
Skoglund 506, 507, 508
Skoglund and Hvamb 502, 503, 504
Smirnov 514
Snyder 112, 114, 115, 122
Sonnleithner 226
Stamer 294, 295, 297, 299, 304, 349, 404
Stamer and Sieglerschmidt 294, 295, 297, 298
Stamm 156, 157, 163, 198, 199, 200, 205, 217, 218, 219, 220, 221, 223, 224, 232, 233, 234, 235, 238, 257, 258, 259, 260, 465
Stamm and Baechler 156
Stamm and Hansen 163, 164, 203
Stamm and Harris 189, 191
Stamm and Hauser 200
Stamm and Loughborough 191, 202, 203, 205

# Author Index

Stamm and Nelson 234
Stamm and Woodruff 192
Stevens 205, 207, 208, 452
Stevens and Pratt 462
Stevens and Turner 542, 543, 545, 546, 548
Stewart et al. 73
Stillwell 224
Stolley 84
Stoops 270
Stradivarius 274
Struve 241
Sturany 461, 462
Suenson 340, 341, 362
Sugihara 488, 494
Sugihara and Sumiya 508
Suits and Dunlap 259
Sulzberger 312, 313, 349, 350, 352, 353, 369, 370, 389, 390
Sunley 412, 413
Sutter 528, 529

Tag und Heppenstall 258
Takechi and Inose 258
Tarkow and Stamm 157
Taylor-Colquitt Co. 466, 467
Technical Association of the Pulp and Paper Industry (TAPPI) 56
Teichgräber 542, 544, 548, 549
Telford 492, 493, 498
Tetmajer 357
Thompson 150, 204
Thonet 547
Thornber and Northcote 73
Thunell 301, 309, 311, 312, 313, 361, 362, 368, 369, 370, 373, 387, 389, 476, 477, 481, 482, 483, 508, 519, 522, 523, 524
Thunell and Hiltscher 482
Thunell and Lundquist 246
Tiebe 355
Tiemann 92, 94, 338, 339, 379, 434, 454, 456, 460, 470
Tillyard 127
Timell 58, 61, 62, 63, 65
Timell and Côté 47
Tjaden 398

Tolman and Stearn 216
Tooke 116, 118, 120
Torgeson 232
Trayer and March 394, 395, 397
Treiber 56
Trendelenburg 83, 162, 166, 167, 169, 170, 171, 174, 176, 177, 178, 199, 327, 354, 355, 406
Trendelenburg and Mayer-Wegelin, 168
Truax and Harrison 150
Tuomola 226, 227, 228, 229, 230, 231
Turner 128
Tuttle 221, 224, 225

Ugrenovič 331
Uphus and Shapman 460
Urquhart 192
Urquhart and Williams 192, 203
U.S. Forest Products Laboratory Madison 184 196, 214, 242, 327, 332, 374, 378, 464, 468, 510 431, 433, 441, 454, 455, 461

Van Bemmelen 191
Van Groenou et al. 136, 148
Van Kleeck 153, 154
Van Kleeck and Martin 155
Varossieau 457, 458, 460
Verrall 142
Villari 241
Vind et al. 133
Vintila 174
Voigt 279, 297, 302, 479, 499, 515
Voigt, Krischer and Schauss 222, 469
Volbehr 201, 202
Volkert 169, 172
Vorreiter 180

Wahlberg 174
Wald et al. 67
Walldén 368
Walter 538, 541

Wandt 170
Wangaard 173, 246, 364, 365, 409
Wangaard and Zumwalt 171
Wardrop 18, 21, 22, 24, 29, 34, 37, 42, 47, 48, 49, 51
Wardrop and Davies 47
Wardrop and Dadswell 22, 26, 45, 49, 50, 73
Wardrop and Harada 21, 35
Warlimont 534, 535, 536, 537
Weatherwax and Stamm 242, 243, 244, 258
Weber 527
Wegelius 378
Weichert 194, 195
Weinberg 221
Wergin and Casperson 37, 47
Western Pine Association 465
Whelan 64, 71
Whitney 470
Whistler and BeMiller 60
Whistler and Smart 64
Wiesner 173
Wigglesworth 113, 115, 119
Williamson 466
Wilson 309, 310, 311, 365, 366, 381, 382, 383
Winter 330, 358, 369, 365, 366
Wirka 147
Wise and Jahn 56
Wise, Murphy and D'Addieco 58
Wright 434
Wuolijoki 301

Ylinen 175, 176, 301, 308, 309, 314, 326, 327, 336, 346, 359, 367, 371, 372, 384, 407
Yugov and Osipov 514

Zabel and St. George 97
Zaitseva 73
Zillner 461
Zimmermann 56
Zsigmondy 192
Zuhlsdorff 464

# SUBJECT INDEX

Abachi, high frequency drying 470
—, weight 168
*Abies balsamea* Mill., specific gravity 177
— —, specific gravity and summerwood 176
*Abies pectinata* DC., specific gravity 177
Abietic acid 70
Abrasion resistance 409
— test machine 409
— tests on different treated wood beams 410
Abrasives 528
— for wood, properties 529
Abrupt transition from earlywood to latewood 5, 6
Absorption rate of hygroscopic water, causing swelling 436
Abura, high frequency drying 470
—, weight 168
Acacia, damping of sound radiation and sound wave resistance 280
Accessibility of cellulose 60
*Acer pseudoplatanus* L., infrared drying 471
*Acer rubrum* L., specific gravity 179
*Acer* sp., density 174
Acetic acid, formation 73
Acetone, use for solvent seasoning 465
—, swelling agent 218
Acetyl (wood analysis) 56
Acetylated wood 157
Acetylating 218
Acicular and styloid crystals 17
Acid copper chromate 149
Acidity of wood 73
Acoustical properties of wood 274
Acoustics of buildings 281
Adsorption and volumetric swelling vs. molecular weight of alcohols 219
Adsorption-desorption hysteresis curves 192, 195
— —, isotherms for untreated and heat-treated wood, 195
Adult *Lyctus* beetle 116
AFNOR 330
Afrormosia, high frequency drying 470
Afzelia, weight 168
Air-circulation in compartment kiln 457
Air driers, ventilated 433
Air-drying 420
—, accelerated 429
— by means of swings or centrifuges 432
Air-drying by solar heat 433
—, course 422
Airplane cabin, sound intensity 281
— engine, sound intensity 281

Air-seasoning curves for pine (*Pinus taeda* L.) 428
Air, sound wave resistance 279
— spaces in dry wood 162
— speed and the interspace between the single boards, effect on drying time 232
Air speeds in kiln drying 232
Akamatsu, effect of clearance angle on net cutting force during circular sawing 502
Alaska yellow cedar (*Chamaecyparis nootkatensis* Sudw.), electrical resistivity 260
Alate or winged reproductive 114
Albumin concentration 120
Alcohol, decreasing swelling 218
Alder, average moisture content 421
—, cutting force and specific gravity 523
—, cutting power, feed speed and feed force 495
—, effect of clearance angle on net power consumption during circular sawing 502
—, fiber saturation point 199
—, hardness and density 407
—, high frequency drying 470
—, mechanical properties across the grain 333
—, relative humidity during kiln drying 447
—, specific gravity variation 170
—, specific sanding work 533
Algae 60
Aliform axial parenchyma 7
— paratracheal parenchyma 6, 7
Alkali silicates 154
— solubility of decayed wood 98
— — — soft-rotted wood 105
$\alpha$-, $\beta$-, and $\gamma$-thujaplicin 71
$\alpha$-pinene 70
Alpine maple, elastic constants 295
*Alstonia* spp., density 162
*Alternaria tenuis* 112
Alternating-current conductivity 265
Alternating-current properties of wood 262
Aluminium oxide as abrasive for wood 529
American and English Standard Specifications 331
Ammoniacal copper arsenate 149
Amorphous regions of cellulose 59
Amorphous zones, regions in the elementary fibrils 20
Amount of set, influence on power consumption 481
Amphipod, Chelura 132
Amphipoda 128
Amylopectin 64
Amylose 64
Angiosperms 1, 57, 70

# Subject Index

Angle between sawblade and crosscutting direction, effect on net cutting power 499
Anisotropic nature of wood 297
Anisotropic shrinkage, drying deformation 452
Anisotropy 271
Annual ring 5
Anobiidae 115
*Anobium punctatum* damage 119
Anschrägwinkel 478
Antenna masts 271
Anthocyanins 70
Apotracheal-banded axial parenchyma 7
Apotracheal-diffuse axial parenchyma 7
Apotracheal parenchyma 6
Appressorium against tracheid wall 106
Apparatus for the titration method according to Fischer 183
Arabino-4-O-methylglucurono-xylan 63
Arabinogalactan 55
— in epithelial cells 73
— in larch 73
— in ray cells 73
Arabino-glucurono-xylan in $S_3$ 75
Arabinose 64
— (wood analysis) 56
Araucaria, swelling 209
—, hygroscopic equilibrium in superheated steam 461
Areal shrinkage 455
Aromatic oils 18
Arrangement of the fibers in wood 248
Artifact, cutting 48
Artificially pruned branch (spike knot) 82
Ascomycetes 98, 104
Ash, American (*Fraxinus* sp.), bending radius 545
—, average moisture content 421
—, Baumann's ratio, 395
—, bending strength and load angle 365, 366
—, compliance and specific gravity 306
— content of wood 71
—, crushing strength and specific gravity 342
—, cutting force and specific gravity 523
—, damping of sound radiation and sound wave resistance 280
—, elastic constants 295
—, European (*Fraxinus excelsior* L.), bending radius 545
—, fiber saturation point 200
—, fatigue resistance 378
— hardness and density 407
— knots and notches 313
—, linear thermal expansion 241
—, mechanical properties across the grain 333
Ash, modulus of elasticity 307
—, notch factor 330
—, Poisson's ratio 298
—, relative humidity during kiln drying 447
—, shearing strength 402
—, sound velocity 276
—, specific sanding work 533
—, stress curve 358
—, swelling gradient 209
—, tensile strength and specific gravity 327

Ash, torsional strength 396
— trees, (*Fraxinus pennsylvanica* var. *lanceolata* Sarg.), variations in wood density 169
— wood analysis 56
Aspen, average moisture content 421
—, cutting force and specific gravity 523
—, hardness and density 407
—, relative humidity during kiln drying 447
—, volumetric swelling 165
Aspirated pits 31
Aspiration of bordered pits in heartwood 73
ASTM Designation: D 143—52, 324, 330, 331, 336, 339, 363
Auger or shot-hole borers 121
Australian jarrah, wear and specific gravity 411
Automatic feed in circular sawing 500
Average cutting resistance, in sawing 479
— cutting resistance in the direction of motions of the cutting edge, in sawing 477
— moisture content of green wood 421
— molecular weight 58
AWF (Ausschuß für wirtschaftliche Fertigung) 480
Axes and planes in wood 293
Axial parenchyma 6
— — in hardwoods 7
— — in softwoods 6, 8
— ventilators 430

Baldcypress, average moisture content 421
Balsa (*Ochroma lagopus* Swartz), density 162
—, linear thermal expansion 243
—, compressibility and density 299
—, cutting force and specific gravity 523
—, elastic constants 295
—, electrical conductivity 261
—, modulus of elasticity and density 308
—, crushing strength 350
—, swelling 209
—, weight 168
—, Young's modulus and density 308
Balsam fir, average density 168
Bamboo 1
—, tensile strength and breaking length 323
Banded apotracheal parenchyma 6
Band saw blades, dulling of cutting edges 487, 489
— sawing 486, 488
Band saws, feed speed 486, 490
Band tension and stability in sawing 490
*Bankia* (Teredinidae) 128
Bark, modified 48
—, content of solid matter 179
— pockets 87
Basidiomycetes 98
Baskets and fruit packages, recommended moisture content 197
Basswood, average moisture content 421
—, bending strength and load angle 365
—, cutting power, feed speed and feed force 495
—, cutting velocity and resistance 487

36 Kollmann/Coté, Solid Wood

## 562 Subject Index

Basswood, damping of sound radiation and sound wave resistance 280
—, drying curves 44
—, effect of bevel angle on cutting power 504
—, effect of clearance angle on net power consumption 502
—, effect of set on power requirement 504
—, hardness and density 407
—, linear thermal expansion 241
—, modulus of elasticity 304
—, relative humidity during kiln drying 447
—, specific sanding work 533
—, stress curve 358
— tensile strength and direction of the grain 326
—, volumetric swelling 165
Baumann's ratio on pine, hemlock and ash 395
Beech (*Fagus sylvatica* L.), air-dry crushing strength 350
—, American, average moisture content 421
—, areal shrinkage 455
—, average sound velocity 277
—, Brinell side hardness 408
—, chain mortising 539
—, chip formation 528
—, chip thickness and curvature 528
—, compressibility and density 299
—, compression-load diagram 339
—, conductivity and resistivity 259
—, crushing and tensile strength ratio and moisture content 336
—, crushing strength and grain angle to load direction ratio 342
—, crushing strength and moisture content 351
—, cutting force and specific gravity 523, 535
—, cutting velocity and resistance 488
—, damping of sound radiation and sound wave resistance 280
—, decrease in toughness 385
—, density 171, 174
—, dependence of the absorption coefficients on the temperature 256
—, drying deformations 452
—, drying temperatures 440
—, effect of bevel angle on cutting power 504
—, effect of rake angle on cutting power 504
—, effect of rake angle on power consumption 503
—, effect of set on power requirements 504
—, effects of various forces during turning 536
— European 545
Beech, fiber saturation point 199
—, frequency curves for specific gravity 168
—, frozen 351
—, hardness and density 407
—, hygroscopic equilibrium in superheated steam 461
—, impact work and moisture content 388
—, infrared drying 471
—, mechanical properties across the grain 333
—, mine timber, conversion factors for dry matter content 179

Beech, modulus of elasticity 303
—, moisture content and temperature curves in drying boards 231
—, packing density 162
—, Poisson's ratio 298
—, power factor 268
— props 353
—, relative humidity during kiln drying 447
—, sanding time and abraded material 529, 531
—, sanding velocity and surface temperature 532
—, seasonal variations in moisture content 422
—, shipping weight 166
—, shearing strength 402
—, shrinkage 208
—, shrinkage and collapse 456
—, slenderness ratio and impact work 382
—, sound velocity 276
—, sound wave resistance 279
—, specific electrical conductivity and dielectric constant 263
—, specific gravity variation 170
—, specific sanding work 533
—, steaming 434
—, stress-strain curves 542
—, surface of torsion 305
—, swelling gradients 210
—, tangential swelling and radial shrinkage 245
—, toughness and width of annual rings 391
—, unfrozen 351
—, volumetric swelling 165, 209
—, wear and specific gravity 411
—-wood, average density 168
— —, change of dielectric constant 267
— — electrical conductivity 261
Bell presses for bending 548
Belt sander 533
— sanding 530
— speed optimum 530
Bending and crushing of pine 360
Bending broad-leaved species 543
Bending coniferous species 543
Bending hoops, 546
Bending machine, revolving table 548
Bending machines, lever armed 547
Bending of laminae 550
— of solid wood 541
—, radius of curvature 545
—, pretreatment of the wood 545
Bending strength 326, 333, 359, 365, 367, 376, 413
Bending test, beams under static center loading 363
Bending test, load deflection diagram 364, 378
— —, set-up of notched beams 374
Bent wood, properties 548
Benzene, content of moisture and volatiles 183
— solvent for the distillation method 182
Benzyl alcohol group 67
— ether group 67
Bershadskii's equation 499

## Subject Index

$\beta$-pinene 70
Beta-rays 161
Bethell (full-cell) process 145
Bevel angle, effect on cutting power in circular sawing 504
Biosynthesis of wood constituents 71
Birch (*Betula lutea* Michx.), specific gravity 179
— *papyrifera* Marsh., specific gravity 179
— (*Betula pubescens* Ehrhardt), fatigue resistance 378
— (*Betula verrucosa* Ehrhardt), fatigue resistance 378
— Canadian yellow (*Betula lutea* Michx.), bending radius 545
— bending strength and moisture content 368
— chipboard 526
— compressed, power factor 268
—, cutting force, chip thickness and moisture content 525
— cutting force and specific gravity 523
—, crushing strength and speed of testing 337
—, dependence of the absorption coefficient on the moisture content 256
—, drying temperatures 440
—, effect of clearance angle on net power consumption during circular sawing 502
—, elastic constants 295
—, fiber saturation point 199
—, hardness and density 407
—, high frequency drying 470
— normal, power factor 268
—, Poisson's ratio 298
—, relative humidity during kiln drying 447
—, specific cutting force and average chip thickness 522
—, specific electrical conductivity and dielectric constant 264
—, specific sanding work 533
—, torsional strength 396
—, volumetric swelling 165
—, yellow, average moisture content 421
Birefringence 20
Bite per tooth of the circular saw blade 491
Björkman lignin or milled wood lignin 67
Blackbutt (*Eucalyptus pilularis* Smith), bending radius 545
Black carpenter ants (*Camponotus pennsylvanicus* (DeG.)) 125
Black check 87
Black locust, fiber saturation point 200
— —, swelling gradients 209
Black walnut, fiber saturation point 200
Black —, darkening during steaming 434
Blade diamter, effect on cutting power in circular sawing 499
— life 475
— protrusion 498
— stability 508
— thickness, effect on cutting power in circular sawing 499
Blinds, recommended moisture content 197
Blocks, yields from gang saws 485
Blue spots 507

Blue stain, during kiln drying 451
Board thickness, influence on rate of diffusion 229
Bobbin sander 533
Bollywood, modulus of elasticity and temperature 312
Bongossi, swelling gradients 209
—, weight 168
Bone, piezoelectricity 272
Bordered pit, hardwood 31
— — membranes of conifers 28
Bordered pit pair membranes in hardwood 33
Bordered pit pairs, intervessel 34
Bore holes 102
Boring 537, 540
Borings, frass and flight (exit) holes of horntails, *Sirex gigas* L. 127
Bornyl acetate 70
Bostrichidae 115, 121
Boucherie process 140, 146
Boulton boiling-under-vacuum process 145
Boulton process 464
Bound-water vapor movement mechanisms in the wood 221
Bow, special warp form 92
Box kiln 456
Boxes and crates, recommended moisture content 197
—, boards for — in chipless wood cutting 513
Boxwood, (*Boxus sempervirens* L.) compressibility and density 299
—, infrared drying 471
—, hardness and density 407
—, linear thermal expansion 241
—, relative humidity during kiln drying 447
—, swelling gradients 210
—, volumetric swelling 165
Brashness 341
— in wood 391
Brass, sound wave resistance 279
Braun's native lignin 67
Brazilian pine, compressibility and density 299
— —, swelling gradients 210
Breaking length 321, 323, 344
Breaking strength 323
Brightness loss of air-dry hardwoods 452
Brinell hardness 404
— —, effect of moisture content on pine 408
— —, effect of specific gravity 404
— — test for wood 403
Brinell end-hardness, effect of moisture content 199
Brinell side hardness, frequency distribution of pine and beech 408
— — —, effect of moisture content 199
Broadleaved wood, crushing length and lignin content 354
— —, tensile strength and breaking length 323
Broadleaved species, fatigue resistance 378
Bronze, sound wave resistance 279
*Brosimum aubletii* Poepp., density 162
Brown kiln stain 88
Brown rot 97, 98

## 564

Subject Index

Brown stain in kiln drying 451
Brush handles and backs, recommended moisture content 197
Brushwood, content of solid matter 179
Buckling of the cells 339
Buckling tests with beams 357
Bulk density of the wood 162
Bulk modulus 299
— — for broad-leaved species 299
— — for coniferous species 299
Bulk Poisson's ratio 299
Bulking the fibers 218
Buna, effect of clearance angle on net cutting force during circular sawing 502
Burma teak, wear and specific gravity 411
Burnett process 145
Burst checks and shakes 245
Buttswell of tree 85
Butternut, darkening during steaming 434
Buttresses of ash trees 169

Cadmium triethylenediamine (cadoxen) 60
Calcium 71.
— oxalate 17
California red fir, tangential swelling and radial shrinkage 245
Calorific value of wood 150
Cambial tissue, chemical composition 73
— zone 72
Cambium 3
Camphor 70
Camphorwood, East African (*Ocotea usambarensis* Engl.), bending radius 545
Canadian birch, wear and specific gravity 411
— maple, wear and specific gravity 411
*Canarium euphyllum* Kurz, electrical resistance 259
Canvas curtains 441
Capacity and radio-frequency power-loss types of moisture meters 185
Capillaries, menisci 221
Capillary condensation 194
— condensed water 193
— forces 220, 454
— movement and diffusion in wood 219
— — of water in wood 221
— walls 220
Car decking, recommended moisture content 197
Carbide-tipped circular saws 496
Carbohydrate depletion in soft-rotted wood 105
— metabolism in Limnoria 131
Carbon disulphide, swelling and dielectric constant 218
Carbonates 71
Carbonyl groups 67
Careful drying of softwood 446
Carpenter ants (*Camponotus* spp.) 112, 124
— ant nest 125
— bees, Hymenoptera 126
Casehardening 92, 452, 464
— during air seasoning 439
— test 447

Caskets and coffins, recommended moisture content 197
Cast iron, Hooke's law 292
— —, sound wave resistance 279
— zinc, Hooke's law 292
Caste system in termites 113
*Castanea dentata*, bending strength and casehardening 368
*Cavanillesia platanifolia* H. B. K., elastic constants 296
Cavities in the cell wall 209
Cedar, average moisture content 421
—, sound velocity 276
Cell inclusions 15
— sorting and arrangement
— wall, chemical composition 72
— — depolymerization of the cellulosic components 111
— — layering of gelatinous fibers 49
— — organization, diagram of 21
— — sculpturing 27
— — thickness 233
"Cellon Process" 147
Cellophane, power factor 270
Cellular composition 2
Cellulase 119, 128
Cellulolytic enzymes 111
Cellulose 57
— content of mature tracheid primary wall 22
— -digesting protozoa 113
— formation 73
— in tensionwood 74
— in wood cell walls 55
— nitrate 58
— of teakwood, magnetic susceptibility 271
—, power factor and dielectric constant 269, 270
Cerambycidae 115, 119
Chain sawing 510
— mortiser, volume of chips per double-link in relation to cutting velocity, cutter diameter and pitch 539
— —, chip formation 538
— —, effect of effective cross-section and feed velocity on power consumption 539
— saws equipped with bow 511
— — equipped with a straight bar 511
Chairs and chair stock, recommended moisture content 197
Characteristics of preservatives 147
— of wood-destroying fungi 98
— of wood staining fungi 105
Charge kiln 456
Checks in logs 90
— in wood 91
*Chelura* 130
— *terebrans* Philippi 132
Chemical composition of compressionwood 47
— — of decayed wood 98
— — of residual material following decay 104
— — of tensionwood 52
— — of wood 56
— constituents of wood and their determination 56

Chemical phenomena in combustion of wood 151
— seasoning 468
— staining 88
Cherry, fiber saturation point 200
—, black, average moisture content 421
—, conductivity and resistivity 259
—, sound velocity 276
—, specific sanding work 533
Chestnut, fiber saturation point 200
—, conductivity and resistivity 259
—, linear thermal expansion 241
—, volumetric swelling 209, 211
Chip acceleration in sawing 478
— breakage and associated friction in sawing 478
Chip cross-section in sawing 479
— formation and fiber saturation point 527
— — in a circular sawblade 505
— — in chain sawing 511, 512
— — in circular sawing 506
— — in planing 518
— —, influence of knife geometry 527
— —, influence of wood moisture content 527
— — through knife-cutting 527
Chippable waste, yield 485
Chipless wood-cutting 513
Chip removal and associated friction in sawing 478
Chips from band sawing, cross-section 489
Chip surface roughness 527
— thickness 483, 520, 527
— — and chip curvature 528
— — and sharpness of a saw tooth 478
— — for circular saw teeth 492
— — in planing 519
— — in wood cutting 521
— —, power law 477
Chip types in circular sawing 505
Chlordane as insecticide 123, 126
Chlorine-ethanolamine 58
Chlorite 58
Chloroform, content of moisture and volatiles 183
—, solvent for the distillation method 182
Chloroform, swelling agent 218
*Chlorophora excelsa* Benth. et Hook, infrared drying 471
Chromated zinc arsenate 149
— — chloride 149
— — — for impregnating wood 145
Circular sawblade 490, 506
Circular sawing 490, 498
Circular saws 481, 485, 488, 491
Circulation by gravity in kilns 441
Clausius-Clapeyron equation for heat effects 202
Clay, damping of sound radiation and sound wave resistance 280
Clearance angle (American) of saw blade 478
— —, circular saw blade 477
— —, effect on net power consumption 502
— —, effect on power demand 501
— —, influence on chip formation 527
— — of saws 501

Clearance angles for gang saw blades 480
Cleavage 330, 333
— of fastenings or joints 323
— specimens 331
Climate curves and hygroscopic equilibrium for wood 424
Climb sawing 477, 499
— in circular sawing 490, 491, 503
Clogged gullets of sawblades 498
Coachwood 353
— (*Ceratopetalum apetalum* D. Don), modulus of elasticity and temperature 312
Coal-tar creosote 133, 238
Coast Douglas-fir, tangential swelling and radial shrinkage 245
Cobalt-chloride paper-strip for determining the moisture content of wood 184
Coefficients of linear thermal expansion of various woods 241
Cold soaking process 141
Cold water condensers in kilns 441
Collapse 93, 205, 453
Collapsed cross-section of eucalyptus 455
— redwood board after kiln drying 93
Collapse in heartwood 455
— in oak 462
— in sapwood 455
— in tension wood 84
— recovery 434
—, shrinkage and collapse 456
— theory liquid tension 454
College classroom, quiet house, sound intensity 281
Combustibility of wood 149
Common Horsechestnut, swelling gradients 210
Comparison of dimensional stability of decayed wood blocks 99
— of tension wood and compression wood characteristics 43
Compartment kilns 456, 458
Compliance and specific gravity 306
Composition of normal and compression wood 57
Compreg 329
—, notch-factor 330
— resin-impregnated wood product 156
Compressed wood, ratio $\sigma_\perp/\sigma_\parallel$ 341
Compressibility (bulk modulus) 299
— and density 299
Compression failure 89, 339
— of the adsorbed water on the internal surface of the wood 163
— parallel to grain 336
— perpendicular to grain 339
—, dependence upon angle between specimen axis and grain direction 326
— —, effect of moisture content 199
—, stress-strain curve 292
— wood 43, 55, 58, 73, 81, 83, 171, 213
— — density, 174
— — galactan 63
Compressive and tensile stresses distribution over the cross-section of bent oakwood sticks 544

compressive strain $\varepsilon$ of oven-dry red beech 316
— — $\varepsilon$ of red beech wood in compression test parallel to the grain 317
Compressive strength 328
— — perpendicular to the grain of vapordried red oak crossties 466
— tests on wood-props 353
Concentration of the treating solution in the Burnett process 146
Concrete, Hooke's law 292
Condensed guaiacyl units 67
Condensed tannins 70
Condenser and water spray in a kiln 457
Conditioning period at the end of the drying time 448
Conditioning procedure for relieving stresses of case-hardening 92
Conduction or diffusion in the fiber direction 233
Conductivity 259, 261
Conductivity increase and salt content 260
Confluent axial parenchyma 7
Confluent paratracheal parenchyma 6, 7
Conformation of xylan backbone 62
Conidendrin 70
Coniferin 71
Conifer tracheid 10
Conifers, coefficient of shrinkage 212
Coniferous trees 213
Coniferous woods 178
— —, crushing length and lignin content 354
— —, fatigue resistance 378
— —, tensile strength and breaking length 323
— — with distinctly colored heartwood 200
— — without colored heartwood 200
Constant bending moment, stress distribution 371
Constriction of hypha 101
— phenomenon of hypha in *Fomes annosus* 102
Construction methods as decay prevention control of wood in service 107
Consumption of energy in sawing 478
— of steam per kg of evaporated water 459
Continuity equation 236
Contraction of volume for the swelling 202
Control measures against wood-boring insects 121
Control of termites 115
Conversion factors for different assortments of wood 179
— factors from phons into decibels 282
Convoluted G-layers 49
Convolutions (stages of cell wall formation) 48
Cooperage, recommended moisture content 197
Copper, damping of sound radiation and sound wave resistance 280
—, Hooke's law 292
Copperized chromated zinc arsenate 149
— — — chloride 149
— — — — for wood impregnating 145
Copper naphthenate 149
—, sound velocity 276
— sulfate as wood preservative 147

Copper wire, hard drawn, tensile strength and breaking length 323
Cores, recommended moisture content 197
Cork, sound velocity 276
Cornel, hardness and density 407
Corner lock sash joint 538
Correlation between mechanical properties, hardness, abrasion resistance 160
Cos $\varphi$ 263
Cost, important factor in the application of wood preservatives 148
Cotton, integral heat of swelling 201
— fibers 58
— linters, density 164
— linters, pressure of volume contraction 203
—, tensile strength and breaking length 323
Cottonwood, black, average moisture content 421
—, forced air-drying 431
— (*Populus deltoides* Marsh.), linear thermal expansion 243
Countersawing 477, 490
— in circular sawing 503
Course of wood charring 152
Creep and creep recovery 317
— phenomena 334
— recovery 334
Creosote 148
Cribriform (sieve-like) pit membrane 35
Crook of tree 85
—, special warp form 92
Crossband veneers, recommended moisture content 197
Crosscut saws 501
Cross field pitting in conifers 39
Cross fracture 92
Cross grain 84, 323
Cross-linking in the cellulose chains of the fibers 218
Cross-section, radial section or tangential section of hardwood stem 3
Cross-ties, recommended moisture content 197.
Crude oil burners to generate steam 440
Crumble chips in bandsawing 489
Crushing length 344, 349
— — in relation to lignin content 354
Crushing strength 336, 341, 346, 544
Crustacean borers 128, 130
Crystal lattice 271
— sand 17
Crystalline and amorphous areas 321
Crystalline regions 60
Crystallites or micelles, zones of elementary fibrils 20
Crystallization, additional of wood substance 321
Crystallized region of cellulose 59
Crystals in wood cells 16
Cubic compressibility 299
Cubical pattern of cross-checks in brown-rotted wood 99
Cucumber Magnolia, (*Magnolia acuminata* L.), linear thermal expansion 241
Cup, special warp form 92

Cupping 452
Cuprammonium hydroxide 60
Cupressales 70
Cupressoid pits 37, 39
Cupriethylenediamine 60
Current in a system of electrodes and wood 265
Curved set of sawblades 484
Cut-surface angle 528
Cutter disc chipping machine 527
Cutterhead knives 518, 526
Cutterhead, power consumtion directly proportional to the feed speed, in planing 519
Cutter materials 525
Cutting angle in sawing 484
— — of sawblades 591
Cutting artifact 48
Cutting-circle diameter, effect in planing 519
Cutting depth, effects 525
Cutting directions 520
Cutting edge, effect of the inclination 520
Cutting effects with and against the wood fiber 528
Cutting force 479, 489, 503, 511, 519, 523, 533
Cutting forms in cutting logs on gang saws 484
Cutting or hook angle (Brit.) of saw blade 478
Cutting perpendicular to the grain in band-sawing 489
Cutting power 479, 495, 498
Cutting resistance of saw blades 478
Cutting speed 479, 493, 525, 527
Cutting torque as a linear function of depth of cut in sawing 498
Cutting velocity 475, 519, 526
Cutting vibrator, electro-magnetic 514
Cutting work, dependence upon the chip thickness 522
Cyclitols, extract 465
Cypress earlywood fibers, stress-strain properties 322
—, effect of rake angle on cutting power 504
— latewood fibers, stress-strain properties 322
— longitudinal shrinkage 213
*Cytospora pini* 112

**D**ado heads 538
Dampers 439
Damping capacity 275
— of sound radiation 279
Dark-blue stain 451
DDT as insecticide 123, 126
Dead load 334
Death-watch beetle (*Xestobium rufovillosum* DeG.) 118
Debye's theory of polar molecules 270
Decay of free vibrations in a wood member 275
— prevention methods 107
Decayed knot 82
Decibels 281
Defects due to processing 90, 95
— in wood due to kiln drying 451

Deflection tests, effect of grain orientation to the compliance 304
Deformation and strength of twisted wood members, computation 394
— of an elastic-plastic body as a function of time 316
— planes in drilling beech and pine wood 297
Deformations drying defects 452
Degree of toxicity toward fungi, insects and marine borers 148
Delignification of wood 47
$\Delta_3$-carene 70
Density 160, 210, 250
— and compliance 305
— and modulus of elasticity 306, 307
— defined as mass per unit volume 160
— determination by immersion in mercury 161
— effect on shear modulus of spruce 308
— influenced by moisture content 160
— of cellulose 164
— of compressed adsorbed water 163
— of coniferous species, average 168
— of lignin 164
— of pulp wood 160
— of resin 173
— of solid wood substance 161
— of springwood 173
— of spruce 178
— of summerwood 173
— of surface-bound water, 193
— of wet cell wall substance 250
— relationship to moisture content 164
Dentate ray tracheids 42
Department store, sound intensity 281
Depolymerization of the cellulosic components in the cell wall 110
Depth of cut as a measure of blade protrusion on net cutting power in circular sawing 498
— — — of saw blade 477
Depth of timber cut 489, 507
— — — — in sawing 479
Desorption-adsorption hysteresis curves 193
Detecting compression failures in wood 90
Deterioration of wood slats in cooling towers 104
— of the cell wall 103
Determination of elastic constants 300
Deuterium exchange 60
D-galactose 61
D-glucuronic acid 61
Diagonal grain 84
Diamagnetic susceptibility 271
Diamond, special warp form 92
Dicotyledons 1
Dielectric constant of wood species 263, 264
— — and moisture content of wood below and above fiber saturation point 267
— and specific gravity 266
— — effect of moisture content on — and the loss factor for spruce 470
— — for waterfree pure cellulose 270
— — of commercial woods 266
Dieldrin as insecticide 126

568    Subject Index

Differential heat of wetting 202, 203
— longitudinal shrinkage in compression and normal wood 83
— swelling 211
Differentiating phloem 73
— xylem 73
Diffuse apotracheal parenchyma 6, 7
— axial parenchyma in softwoods 8
— parenchyma in softwoods 6, 8
— porous hardwood 5, 6
— porous hardwoods, density 174
— porous woods, specific gravity 179
Diffusion coefficient 224, 232
— — free 234
— — hindered 234
— — of spruce wood vs. drying temperature 229
— method 143, 144
— — non-pressure process 141
— of bound water 234
— coefficient of water vapor 234
— of enzymes 108
— of water in wood 224
— — through wood, tangential 234
— — — vapor below about 25% of moisture content controls the drying process 230
Diffusion phenomenon 220
— resistance 222
—, wood preservation processes 140
Diffusivity 250
—, influence of density and moisture content 251
— of wood 250
— perpendicular to the grain as influenced by density and moisture content 251
Dimensional stability of wood 218
— — of decayed wood 98
Dimensional stabilization 155, 157
— — of wood 218
DIN 8803 479
DIN 52185, 336, 341, 348
DIN 52186, 363
*Dinoderus minutus* Fab., distribution 121
Diphenyl linkage 67
Dipping non-pressure process 141
Direct-current electrical conductivity of wood 258
Disc sander 533
Distillation apparatus for the determination of the moisture content of wood 182
— method of determining the moisture content of wood 181, 465
Distribution of chemical constituents in wood 72, 75
— of lignin in cell wall 73
— of preservative as indicator of the efficiency of wood treatment 136, 137
D-mannose 61
Dogwood, hardness and density 407
Doors, recommended moisture content 197
Double diffusion method 143
Double-end tenoning machine 538
Douglas-fir, average moisture content 421
— — damping of sound radiation and sound wave resistance 280

Douglas-fir, density 171
— — depolymerization 323
— — — earlywood fibers stress-strain properties 322
— — elastic constants 294
— — electrical resistivity and moisture content 260
— — fiber saturation point 200
— — fluid pressure and penetration 516
— — internal friction 279
— — latewood fibers, stress-strain properties 322
— — middle lamella 73
— — Poisson's ratio 298
— — quality factor 343
— — quality zones 344
— — relative humidity during kiln drying 447
— — spiral and diagonal grained lumber 366
— — spring and summerwood 173
— — strength properties 313
— — volumetric swelling 209
— — (Oregon pine) drying temperatures 440
— — (Oregon pine), high frequency drying 470
— — (*Pseudotsuga taxifolia* (Poir). Britt.), fatigue resistance 378
— — (*Pseudotsuga taxifolia* Britt.), linear thermal expansion 243
Drum sander 533
— sanding, effect of number of oscillations per minute and feed velocity on roughness of wood surfaces 533
Druses, crystals in wood 17, 19
Dry kilns, steam consumption 459
— -mass production of forest 165
— -volume of wood substance 233
— -wood termites 112
Drying and seasoning lumber by superheated steam 460
— by boiling in oily liquids 464
— — direct application of electricity 469
— — infrared radiation 471
— — Joule's heat 469
— centrifuges 433
— conditions 430
— cracks 87
— curves for lumber in a one-way circulation kiln and a reversible circulation kiln 440
— curves for softwood species 444
— deformations, cross-sections through three squares of eucalyptus 454
— diffusion coefficients for softwoods vs. the swollen volume specific gravity of the wood 235
—, first stage 462
— gradient 446
— — relationship between — — of pine sapwood and air temperature 230
— in an oil bath, curves of temperature and moisture content 464
— of green softwood lumber 441
— — wood as a diffusion problem 225
— potential 446
— rate curve for yellowpoplar specimen 462

Drying rate influence of temperature and psychrometric difference 230
— rates, ratio of the longitudinal and the radial — — vs. average moisture content 228
— schedule for green two-inch and thicker beech 448
— — for one-inch and two-inch hardwood with final steaming and conditioning periods 448
— — for 51 mm Black tupelo (*Nyssa aquatica* L.) 445
— — of Henderson 447
—, second stage 462
— shed 427
—, third stage 463
— time 463
— —, approximated calculation 226
— — correction factors 449
— —, diagram for the determination of — — of 1″ hardwood board 227
— — diagram for the determination of — — of 1″ softwood boards 227
— — relationship between — — and board thickness 229
— —, relationship between — — and drying temperature for vapor drying 466
— — ratio for tangential and radial diffusion vs. temperature 228
— water-saturated Sitka spruce 220
du Pont de Nemour and Co. abrader 410
Duo-Kerf Ripsaw 510
— — —, shape and scheme of cutting action 510
Duration of heating, relationship to average values for specific gravity, crushing strength parallel to the grain and toughness of Sitka spruce 437
Dynamic quality factor 387
— tests 301
*Dyospyros* sp. infrared drying 471
D-xylose 61

Earlywood 327, 341
— rheological models of — and latewood zone in the annual rings 318
Eastern hemlock, penetration depth of preservatives 240
East-Prussian pine, quality zones 344
Eastern red cedar, volumetric swelling and shrinking 205
— white pine, yield 485
Ebonite, magnetic susceptibility 271
— sound velocity 276
Ebony, cutting force and specific gravity 523
—, hardness and density 407
—, linear thermal expansion 241
Eccentric growth of tensionwood 83
— loading of columns 359
— —, effect of — — and moisture content on the stresses in pine columns 359
Eccentricity of stem cross section 43
Ecology of wood-deteriorating fungi 107
Edge displacement and cutting length 526
— notches 330
— sander 532
Edgings 484

Edgings, yields from circular saws 485
—, yield from gang saws 485
— and trimmings, content of solid matter 180
— from peeled and sliced veneers, content of solid matter 180
Effects and properties of fire retardants 152
— of extractives on the durability of heartwood 110
— of stacking time on penetration depth in osmose process 144
— of temperature on the growth of *Polyporus vaporarius* 108
— of wood structure on treatment 137, 140
Ekki, cutting force and specific gravity 523
Elastic and viscous elements acting in series 319
— constants 307
— —, variation with moisture content 311
— — of wood 301
— deformation, disappearance of a plane of shear through transition 337
— limit for the compressive strength perpendicular to the grain of green softwoods 223
— properties of wood, influences 302
Elasticity, plasticity and creep 293
Elastomat, 301
Electric radiating panel test 155
Electrical conductance of green and oven-dry wood 199
— conductivity 257
— — analogy applicable to the drying of wood 235
— moisture meters 184
— properties of wood 257
— resistance 257
— — and moisture content 259
— — of wood at fiber saturation point 257
— resistivity below fiber saturation point 258
— — of redwood and moisture content 257
— — of woods as affected by grain orientation 260
Electricity for small dry kilns and for high temperature kilns 440
Electron microscopy 50, 60
— — for studies on cellulose microfibrils 18, 19, 20
Electronic gluing 468
Elementary fibril 60
— fibrils of cellulose 19, 20
Ellagic acid 70
Ellagtannins 70
Elm, American, average moisture content 421
—, cutting force and specific gravity 523
—, damping of sound radiation and sound wave resistance 280
—, drying deformations 452
— —, temperatures 440
—, Dutch (*Ulmus hollandica* Mill. var. major Rehd.), bending radius 545
—, forced air-drying 431
—, hardness and density 407
—, heartwood, swelling gradients 209

Elm, linear thermal expansion 241
—, relative humidity during kiln drying 447
Embrittlement of wood 218
Embrittling effect of phenol-formaldehyde treatment 157
Emissivity 256
— of wood above fiber saturation point 257
— of wood at room temperature 257
Empty-cell process 145
Encased or loose knot 79, 81
Encrustation of pit membranes in heartwood 73
Encrusting materials of the wood cell wall 18, 31
End checks in wood 91
*Endiandra palmerstoni* C. T. White, infrared drying 471
Endurance load 335
Energy consumption in longitudinal direction versus cutting angle for two depths of cut in chain sawing 512
— — in wood machining 475
— for cut depending on chip thickness or feed per cut in sawing 478
— required and feed per cut in wood machining 523
Enforced drying, of softwood 446
Engelmann spruce, electrical resistivity and moisture content 260
Enzymatic mechanism of decay 98
Equilibrium moisture content 189
— — — and relative humidity 442
Eradication of marine animals from wood in service 133
*Ernobius mollis* L., distribution 118
*Erythrina* sp., density 162
Ether, swelling and dielectric constant 218
Ethyl alcohol, swelling and dielectric constant 218
Eucalypt, drying temperatures 440
—, relative humidity during kiln drying 447
—, steaming 434
*Eucalyptus marginata* J. Smith, infrared drying 471
— *regnans*, hygroscopic equilibrium in superheated steam 461
— —, hygroscopic isotherms 190
— sp., infrared drying 471
Euler curve 356
Euler and FPL curves interrelation 356
European house borer, *Hylotrupes bajulus* L. 119
— old-house borer 121
Excessive taper of tree 85
Expansion of yellow birch in the three structural directions 242
Exterior trim, recommended moisture content 196
External closing membrane of a pit 27, 31
— volumetric shrinkage and moisture content 205
— — swelling of thin cross-sections of white pine in distilled water vs. equilibrium pH of water 217
Extra lignin 47

Extractives (wood analysis) 56
Exudation of extractives during kiln drying 451
Face veneers, recommended moisture content 197
Factor of completeness 365
*Fagus grandifolia* Ehrh., specific gravity 179
— *sempervirens* L., infrared drying 471
Failures in compression, according to ASTM designation D 143-52 337
Fan-air drying 429
Fan operating time 430
Fans, power 457
Fatigue in bending 376
— — tension parallel to grain 334
— resistance values for wood species 378
Fatty acids 70
Feed force and oscillation width for green spruce 514
— — and vibration frequency 514
— per revolution of the circular saw blade 491
— per tooth and net cutting force in circular sawing 496
Feed per tooth of the circular saw blade 491
— speed and cutter head diameter, effect on net cutterhead horsepower 520
— — and net cutting power consumption in circular sawing 496
— — in planing 519
— — of circular saws 507
— — of sawblade 479
— way or feed velocity, effect on the roughness of planed wood surfaces 526
Feeding of ripsaws 504
Fenestriform (Window like) pit 37, 38, 39
Ferromagnetic properties of wood ashes 271
Fibrillar alignment 44
Fibrils, cellulose 19
Fiberboard, sound absorption 285
Fiber length of a softwood 233
— saturation point 90, 107, 108, 180, 198, 200, 205, 213, 220
— — — and dielectric constant 267
— — — and case hardening 453
— — — and collapse 454
— — — at third stage of drying 463
— severance and associated internal friction in sawing 478
— tracheids 11
Fick's law of diffusion 224
Filter paper, integral heat of wetting 201
Fine bore hole through the cell wall 106
Finnish pine, crushing strength and specific gravity 346
— —, specific gravity and summerwood 175
— —, tensile strength 347
Fir, average moisture content 421
—, average sound velocity 277
—, bending strength and load angle 365
—, buckling 357
—, chain mortising 539
—, compression wood, sound velocity 276
—, conductivity and resistivity 259
—, cutting force and specific gravity 523

Fir, cutting velocity and resistance 487
—, damping of sound radiation and sound wave resistance 280
—, drying curves 444
—, drying temperatures 440
—, fatigure in compression 355
—, fiber saturation point 200
—, hardness and density 407
—, high frequency drying 470
—, linear thermal expansion 241
—, mine timber, conversion factors for dry matter content 179
—, shipping weight 166
—, sound velocity 276
—, sound wave resistance 279
—, swelling gradients 210
Fire injury 89
— resistance of wood preservatives 148
— retardants acting mechanically 153
— —, properties 153
— — treatment 149, 155
— tube test 155
Flake-like chips 513
Flavonols 70
Flax, Irish, tensile strength and breaking length 323
Flint as abrasive for wood 529
Flooring, recommended moisture content 196
Flow curve of an elastic and viscous element 320
— in sapwood 240
— velocity of elastic and plastic elements 319
Fluid motion through a cross-sectional reduction, schematic representation 237
— pressure and penetration for four nozzle sizes 516
— — — — for three grain orientations and two moisture contents 515
— — — — for five wood species 516
Fluor chrome arsenate phenol 149
Flush doors, recommended moisture content 196
Fluttering sawblade in circular sawing 506
Foam-forming chemicals 153
— — organic compounds 154
*Fomes annosus*, construction phenomenon 102
Food preference as difference between *Anobium punctatum* and Lyctus beetle 118
Footage of timber sawn per hour 479
Footings, types 426
Forced air drier with two double fan units 430
— — drying of oak cants 432
— — —, reduction in moisture content 431
Forest Products Laboratory fourth power formula 356
Fork samples 447
Formaldehyde treatment 157
Formation of heartwood 138
Formicidae 124
4-O-methyl-D-glucuronic acid 61
Fourier. partial differential equation 251
Fourier's analysis for heat conduction 225
F. P. L. toughness-testing machine 383

Framework substance of the wood cell wall 18
Framing, recommended moisture content 196
*Fraxinus americana* L., longitudinal shrinkage 213
— *excelsior*, density 174
— — L., specific gravity 177
Frass from work of Anobiidae 120
— — — of *Lyctus* 120
Freiwinkel 478
French Standard Specification 331
Frequency and dielectric constant 267
Friction forces in the kerf in band sawing 488
Frictional losses and number of sawblades 480
— — in sawing 481
Fringe micellar theory 60
Fromage de Hollande, (*Bombax malabaricum* D. C.), density 162
Frost crack 86
— rings 87
Frozen hardwood, cutting velocity and resistance 488
Fructose 71
Fuelwood, content of solid matter 179
— from sawing round logs, yield values 485
Full-cell process 145, 236
— — in wood preservative treatment 133
Fungi causing wood deterioration 97, 112
— Imperfecti 98, 104
Furniture beetle (*Anobium punctatum* DeG.) 118
—, recommended moisture content 196, 197
Fusiform rays 8, 14

Gaboon, steam consumption during steaming 435
Galactan 47, 55
Galactan, new type in tensionwood 52
Galactoglucomannan 47, 57, 63
Galactose 56, 64
— residues 58
Galacturonic acid 64
Gallic acid 70
Gallotannins 70
Gamma-radiation 161, 323
Garnet as abrasive for wood 529
Gas burners to generate steam 440
Gas constant 202, 204
Gang saws 475, 485
Gang saw blades 477, 480, 482
Gang sawing 476
Gelatinous fiber of angiosperm reactionwood 48
— fibers in tensionwood 43
— layer (G-layer) in tensionwood 55, 73
Generalized structure and terminology of cell wall 20
German pine, tensile strength and temperature 328
German Standard Specification DIN 4074 328
Glass, sound velocity 276
—, damping of sound radiation and sound wave resistance 280
Glucomannan 57, 61

## Subject Index

Glucomannan, distribution in cell wall 75
Glucose 56, 71
— in sapwood-heartwood transition 73
Glucosidic uninterrupted chains 321
Glycerine, swelling and dielectric constant 218
Glycosides 70
Gold, damping of sound radiation and sound wave resistance 280
G-layer, cellulose 74
— of tensionwood fiber 43, 48, 50
Gradual transition from earlywood to latewood 5
Grain angle 302
— — influence on tensile strength 326
Grain orientation and cutting force in circular sawing, hard maple 499
— — and length of chips 489
— — and penetration depth of NaF as a function of time 237
Grain orientation, effect on cutting force 520
— —, effect on bandsawing 489
— — in circular sawing 497
Granite, Hooke's law 292
Green bottle-glass as abrasive for wood 529
Greenheart, mineral constituents 173
— (Ocotea rodiaei Mez.) bending radius 545
Green sawn boards, woolly 43
Green wood, density 165
— — heated in steam 244
— — treated with a hygroscopic chemical 468
Gross structure of wood 1, 9
Growth increment 4
— increments in softwood and hardwood structure 13
Guaiacyl units 67
Guaiacylglycerol-ß-coniferyl 67
Guajacum officinale, density 162, 173
Gullet of circular saw tooth 507
— shape effect on bandsaw performance 489
Gum crossties, drying 466
— ducts 9
Gums 18, 161
— and resins 18
Gurjun (Dipterocarpus sp.) Indian, bending radius 545
Gymnosperms 1

Hackberry, forced air-drying 431
Hadrobrogmus carinatus (Say), distribution 118
Hagen-Poiseuille's law for capillary movement 220
Half-bordered pit pairs 36
Half-value time in drying 232
Handles, recommended moisture content 197
Hankinson's formula 326, 341
Hard maple, tangential swelling and radial shrinkage 245
— —, vibrations during sawing 495
Hard metals for planing tools 525
Hardness and abrasion resistance 403
— figure 406
—, relationship to crushing strength 403
—, relationship to impression depth for common beech 403

Hardness, relationship to density 407
— tests 403
— tester and test block 402
Hardwood bordered pits 31
Hardwoods, careful drying 446
—, coefficient of shrinkage 212
—, enforced drying 446
—, rake angle 480
Hatt-Turner machine 382
— — test 379, 381
Hearing and feeling, threshold and curves of equal loudness as function of frequency 281
Heart checks 86
— shake 86
Heartwood 4, 72
—, integral heat of wetting 201
— physical properties 55
Heat and power for drying 438
— balances for different steam temperatures in the pits and for varying diameters of Gaboon 435
— bridges in wood 248
— conduction of wood 242
— consumption per kg water evaporated 459
— of evaporation 150
— of wetting 199, 202
— transfer 463
— transmission coefficient 232
— value of wood 160
Heating of saw blades 507
Heavy traffic, pneumatic, sound intensity 281
Helical checks 43
Helical checking of compressionwood tracheids 45
Helical thickenings in cells (spiral thickenings) 37, 40
Hemicellulose 57, 61, 73
— of softwoods 63
Hemicelluloses in wood cell walls 55
Hemispherical menisci in capillaries, surface tension 204
Hemlock, Baumann's ratio 395
Hemlock, western, average moisture content 421
Hemp ropes, Hooke's law 292
Hemp, tensile strength and breaking length 323
Heterocellular or heterogeneous rays in conifers 14
Hickory, average moisture content 421
—, cutting force and specific gravity 523
—, damping of sound radiation and sound wave resistance 280
—, drying temperatures 440
—, relative humidity during kiln drying 447
—, tangential swelling and radial shrinkage 245
Hicoria ovata Britt., specific gravity 178
High-frequency dielectric drying 470
High net retention in the full-cell process 146
High-temperature drying 313, 460, 463
High-temperature kilns 463
Hinoki cypress 273
Hollow-horning or honeycomb 94

# Subject Index

Holly, average moisture content 421
Holocellulose 58, 63, 190
Homocellular or homogeneous rays in conifers 14
Honeycomb 94
— during casehardening 453
Honeycombing 448
—, areal shrinkage 455
— in badly dried oakwood post with great internal checks 453
Hook angle, circular saw blade 477
— —, influence on chip formation 527
Hooke's law 292
— —, modulus of elasticity 292
Hoop pine, modulus of elasticity and temperature 312
Hornbeam, cutting force and specific gravity 523
—, damping of sound radiation and sound wave resistance 280
—, effect of clearance angle on net power consumption during circular sawing 502
—, effect of set on power requirement 504
—, fiber saturation point 199
—, hardness and density 407
—, high frequency drying 470
—, infrared drying 471
—, linear thermal expansion 241
—, mechanical properties across the grain 333
—, specific sanding work 533
—, swelling gradients 209
Horntails (Siricidae) 126
Host-wood preference 102
Hot-and-cold bath 141
Hot box test 155
Hot water extraction, residues of European aspen and its constituents 437
Hydrates, formation 193
Hydrogen ion concentration for fungal growth 109
Hydrolyzing exoenzymes 110
Hydroxyl groups 192
Hygrometric methods for determining the moisture content of wood 183
Hygroscopic bond, energy to break 202
— equilibrium moisture content 225
— — of wood in superheated steam at atmospheric pressure 461
— — values for interior finishing woodworks 423
— — values to be set in the kiln for various drying gradients of softwoods 446
— isotherms 190
— — for different temperatures 190
— — for Klason-lignin and methanol-lignin 192
Hygroscopicity of the wood 218
*Hylotrupes bajulus* L., European house borer 119
Hymenoptera, carpenter bees 126

Idigbo (*Terminalia ivorensis* A. Chev.) bending radius 545

Idling power of woodworking machines 479
Ilomba, weight 168
Immersion vessel for determination of density 160
Impact bending 379, 393
— —, temporal course of the force on support 392
— — test assembly 379
— —, types of rupture in ash specimens 392
Impact deflection and impact work, effect of moisture content, in beech 388
Impact test, single blow 380
— —, test sample of ash 392
Impact work, dependence upon the depth of notch for pine 385
— —, effect of moisture content for pine, spruce and beech 388
— —, effect of slenderness ratio l/h for pine and beech 382
— —, effect of the specific gravity on air-dry ash 386
— —, effect of specific gravity for pine, spruce, beech and oak 387
— — for notched specimens, effect of moisture content, for pine and fir 389
— — for notched specimens of pine, effect of temperature 391
— — for pine with different density in the air-dry and nearly green condition, effect of temperature 389
Impreg, resin-impregnated wood product 156, 218
Impregnation, physical aspects 235
Inaccurate edging, as source of wane 90
Incipient decay of wood 105
Incising of timber 140, 236
Inclination of the tooth to the plane of the sawblade 477
Inclined cutting 521
— panel test 155
Included bark 88
Incombustibility 150
Increase in the solubility of wood 110
Index of refraction of warts 41
Industrial wood residues 527
Infestation of wood 122
Inflammability of wood 150
Influence of decay on mechanical properties 111
Infrared radiation, drying 471
— spectroscopy 60
Initial pit border 34
Injuries caused by fire or logging 89
Inorganic constituents of wood resins 70
Insecticides 123
Inside box car material, recommended moisture content 197
Integral heat of wetting 201
Intensity of sound 281
Intercellular canals 8
Intercellular layer 20, 22, 34
— — in pit membrane 30
Intercellular spaces 43, 73
Intercellular substance, middle lamella 20, 22, 34

Intergrown or tight knots 79
Interlocked grain 85
Interlocking grain, drying defects 452
Intermittent steaming 452
Internal checks, relationship to the dry bulb temperature for oak 453
Internal friction 279
Internal notches 330
Internal viscosity 320
Internal weak spots in wood 497
Interior-finish wood-work 196
Interior woodwork, recommended moisture content 196
Inter-tracheid pit pairs in hardwood 32
Intervessel bordered pits 34
Ionic conduction 257
Iroko (*Chlorophora excelsa* Benth. et Hook. f.), bending radius 545
—, weight 168
Iron, damping of sound radiation and sound wave resistance 280
— salt, drops during kiln drying 451
—, sound velocity 276
— wire, hard drawn, tensile strength and breaking length 323
Irradiation with $\gamma$-rays and sorption capacity 194
Irregularity in growth 323
Isolation fiberboard, sound absorption 285
Isopod marine wood borers 130
Isopoda 128
Isoptera 112
Isothermal contours in a timber cross-section 255
Ivory nuts 60
Izod impact test, specimen and apparatus 383, 384
Izod-values for different species with varying moisture content 385

Japanese cypress (*Chamaecyparis obtusa* Endl.) 273
Japanese oak, cutting power curves 496
Jet force, relationship on fluid pressure 515
Jet penetration, effect of angle of incidence 516
Jets, cutting with high-energy jets 514
Jig 538
Jointed knives, number and net power demand 519
Jointing, defined as smoothing one surface of a board by means of a single peripheral milling head 517
Jointing, different meanings in wood terminology 497
Jointing, equalizing a saw 497
Joists, recommended moisture content 197
*Juglans nigra* L., infrared drying 471

Kalotermes flavicollis, distribution 112
Kalotermitidae 112
Karrik, South African (*Eucalyptus diversicolor* F. Muell.), bending radius 545
Keilwinkel 478
Kerf loss reduction 475

Kerf quality of circular saw teeth 504
— width of saw blade 477, 479
Keruing, high frequency drying 470
Khaya, elastic constants 295
— sp., specific electrical conductivity and dielectric constant 264
—, weight 168
Kickback in circular sawing 498
Kickback-free safety blades 508
Kickback of preservative 145
Kiln dried wood 479
— drying 438
— — instruments 456
— — of lumber, temperatures 440
— —, operation curves nomogram 450
— — process control 230
— — temperatures 439
— trucks 439
Kilns and dryers, common types 458
— brown stain 94
Kirchhoff's law for conduction 233
Knot hole 81
— size and crushing strength for large specimens 352
Knots 79, 323
— and notches 313
—, influence on tensile and compressive strength 328
"Knotty pine" 79
Kollmann-abrader 410
Kyanizing 142

Lamellation of the ray cell walls 26
— within G-layer 51
Laminated beams, processing system for manufacturing 513
Laminated beechwood 330, 350
Laminated bending 549
Laminated fabric, static endurance load 334
Laminated products, recommended moisture content 196
Laminated wood 329, 535
Laminating 218
L-arabinose 61
Larch arabinogalactan 64
—, average density 168
—, buckling 357
—, damping of sound radiation and sound wave resistance 280
—, drying temperatures 440
—, fiber saturation point 200
—, mine timber, conversion factors for dry matter content 179
—, relative humidity during kiln drying 447
—, specific gravity 172
—, specific gravity variation 170
—, mechanical properties across the grain 333
—, specific sanding work 533
—, volumetric swelling 165
—, western, average moisture content 421
Lariciresinol 70
*Larix*, chemical composition 55, 64
— *europaea* D. C., density 174, 176, 178

*Larix leptolepis* Gord., density 174
— *occidentalis* Nutt., hemicelluloses 64
Larvae and adults of Anobiidae 118
— of Lyctus beetle 116
Laser cutting 517
Lateral bonding between microfibrils 51
Lateral stiffness of circular sawblades 499
Lateral strains measurement 301
Lateral vibrations in a saw tooth 494
Latewood 327, 341
—, rheological models of earlywood and —
zones in the annual rings 318
Laths, yield from circular saws 485
—, yield from gang saws 485
Lauraceae 18
Lead, damping of sound radiation and sound
wave resistance 280
—, sound velocity 276
—, sound wave resistance 279
Leather, Hook's law 292
*Leitneria* spp., density 162
Length of saw tooth affected by set 477
L-fucose 61
Libriform fiber 11
Life cycle of anobiids 119
Lift trucks 439
Light scattering 58
Light weight construction 345
Lignans 70
Lignification of cell walls 18, 55
Lignin 64
Lignin deposit in compression wood tra-
cheids 46
—, extra 47
— from maple and spruce, density 164
— in wood cell walls 55, 74
—, magnetic susceptibility 271
—, power factor 269, 270
— in wood analysis 56
Lignosulfonic acids in lignin 67
*Lignum vitae*, ratio $\sigma_\perp/\sigma_\parallel$ 341
— —, cutting force and specific gravity
523
— —, damping of sound radiation and sound
wave resistance 280
— —, hardness and density 407
— —, weight 168
Lilac, hardness and density 407
Lime tree, fiber saturation point 199
Limba, high frequency drying 470
— (*Terminalia superba* Engl. & Diels) 116
—, weight 168
Limonene 70
Lime, average sound velocity 277
—, compressibility and density 229
—, mechanical properties across the grain
333
*Limnoria*, protection 132
— (Isopoda) 128
— sp., female 130
Lindane solution as insecticide 123
Linden, high frequency drying 470
Linear shrinkage of beechwood coefficient
214
Linear thermal expansion 243

Linear swelling coefficient 217
Lip angle, circular saw blade 477
Liquefied gas 147
Liquid cutting of wood 514
*Liriodendron tulipifera* L., power factor 268
— —, specific electrical conductivity and
dielectric constant 264
Load compression diagrams for compression
perpendicular to the grain 339
Load deflection diagram for bending tests 364
Load deformation curve for compression
perpendicular to the grain 339
Loblolly pine, external volumetric shrinkage
and moisture content relationship 205
— —, spring and summerwood 175
Local dissolution of the cell wall 102
Location of knots 80
Locust, black, average moisture content 421
— (Robinia), drying temperatures 440
— (Robinia), relative humidity during kiln
drying 447
Log band saws 486
— — —, tooth styles 487
Logarithm of conductivity and moisture
content 199
— of resistivity of oven-dry birch wood 261
— of resistivity vs. reciprocal value of absolute temperature for various moisture
contents of Engelmann spruce 262
— of the elctrical conductivity and moisture
content 258
— of the electrical resistivity and moisture
content 260
Logarithmic blending rule for dielectric
constant 267
Logarithmic decrement, damping capacity
of wood 276, 278, 302
Logging injury 89
Long columns 356
Long-horned beetles 119
Longitudinal parenchyma 6
Longitudinal resonant vibrations 275
Longitudinal shrinkage of different woods as
affected by moisture content 213
— — of tensionwood 24, 214
Longitudinal strains measurement 301
Longleaf pine, spring and summerwood 175
Loose or encased knot 79, 81
Loosened grain 95
Loss factor, effect of moisture content on the
dielectric constant and the — — for spruce
470
Loss of weight at the combustion of various
wood species 151
*Lovoa klaineana* Pierre, infrared drying 471
Lowry process 146
L-rhamnose 61
Lumber, recommended moisture content 197
— yield values 485
— piles 427
Lumberyard alley arrangement 425
— layout 424
Lundberg circular saw 510
Lyctidae 115
Lyctus attack, susceptibility to 121

Lyctus brunneus 116
— cavicollis Lec. 116
— linearis Goeze 116
—, parallelopipedus Melsh, 116
— planicollis Lec. 116
— powderpost work 117
Lysigenous canals 9

Machining accuracy 475
Macrofibrils 20
Magnesium 71
Magnetic properties of wood and wood
   constituents 271
Magnetic susceptibility 271
Major cell types 10
Mahogany, African (Khaya ivorensis A. Chev.),
   bending radius 545
—, Central American (Swietenia macrophylla
   King), bending radius 545
—, conductivity and resistivity 259
—, cutting force and specific gravity 523
—, damping of sound radiation and sound
   wave resistance 280
—, elastic constants 296
—, linear thermal expansion 241
—, relative humidity during kiln drying 447
—, specific electrical conductivity and dielectric constant 264
Main cutting force in sawing 478
Maintaining dimensional stability by coating methods 156
Makoré, weight 168
Malus pumila 62
Mannose in sapwood—heartwood transition
   73
Mannose (wood analysis) 56
Manufacturing defects 90
Maple, average sound velocity 277
—, compressibility and density 299
—, conductivity and resistivity 259
—, cutting force and specific gravity 523
—, damping of sound radiation and sound
   wave resistance 280
—, drying temperatures 440
—, high frequency drying 470
—, infrared drying 471
—, relative humidity during kiln drying 447
—, linear thermal expansion 241
—, Poisson's ratio 298
—, specific sanding work 533
—, sugar, hard, average moisture content 421
—, swelling gradients 209
—, volumetric swelling 209
— syrup 88
Marble, Hook's law 292
Margo, supporting membrane 30
Marine borers 128, 136
Marine fungi, nutritional aids for Limnoria
   131
Martesia 129
Masonite, sound absorption 285
Matrix substances of the wood cell wall 18,
   31
Mäule reaction 70
Maximum longitudinal shrinkage, 214

Maximum moisture content of wood 198, 200
Maximum retention of preservative 145
Maximum shrinkage for compressionwood 214
Maximum volumetric shrinkage and swell-
   ing 204
Maximum volumetric swelling 165
McDonald process 465
Mean modulus of rigidity 394
Mechanical properties of bent beech 549
Mechanical properties of some commercial
   woods across the grain 333
Mechanics 292
Mechanism of wood decay 110
Melt-forming chemicals 153
Mercuric chloride in water as solution for
   non-pressure processes 142
Metal-impregnated wood 234
Methacrylate embedding 50
Methoxyl 66
Methyl bromide for the eradication of horn-
   tail larvae 128
Micelles or crystallites, zones of elementary
   fibrils 20
Microcapillary system in wood cell walls 18
Microfibrils 18, 58, 60
Microfibrillar impression in intercellular
   layer 34
Microfibrillar orientation 21
— — in S-layers of cell wall 74
Micronutrients of inorganic nature for fungal
   growth 109
Micro-section perpendicular to the fibers 392
Middle lamella 55
— —, chemical composition 72
— —, intercellular substance 20, 22
Mildew, during kiln drying 451
Milled wood lignin or Björkman lignin 67
Mine timbers, content of solid matter 179
Mineral constituents in wood 173
Mineral streak or mineral stain 88, 89
Minerals 161
Minimum of cutting power required in circular
   sawing 503, 504.
Minute compression failures in fiber walls
   49
Miscellaneous natural defects 87
Mixed residues, content of solid matter 180
Modulus of buckling resistance 358
Modulus of elasticity 292
— — — along the grain in spruce and oak
   302
— — —, average percentage of six species
   plotted against temperature 313
— — —, determination 302
— — —, effect of moisture content 199
— — —, effect of temperature on various
   wood species 312
— — — for Baltic redwood and Whitewood
   413
— — — for Swedish pine 312
— — — in compression and bending, and
   density of balsa 308
— — — in tension 324
— — — (tension) vs. grain angle 304
— — rigidity 275, 301

## Subject Index

Modulus of rigidity, determination 302

— — rupture, dependence upon stress at the proportional limit and upon logarithm of load duration for Sitka-spruce 375

— — —, effect of moisture content 199

— — —, effect of size 371

— — —, effect of temperature 370

— — — for unnotched and notched beams of pine 375

— — — reduction by the effect of square-cornered notches at the ends of beams 375

Moisture conductivity 222

Moisture content 211, 213, 309

— — and grain orientation at different frequencies and specific high-frequency resistivity of wood 262

— — at the fiber saturation point 162

— — changes of yard-piled softwood lumber in Eastern Canada 198

— —, definition 180

— —, determination 181

— — distribution at different times during the drying of a piece of hardwood 186

— — distribution over the cross-section of 40 mm alder 448

— —, effect on elastic constants of Sitka spruce 209

— —, effect on mechanical propterties 199

— —, effect on tensile strength 327

— —, effect on the dielectric constant and the loss factor for spruce 470

— —, effect on the modulus of elasticity parallel to the grain of oak wood 310

— —, effect on the modulus of elasticity parallel to the grain of spruce wood 310

— —, equalizing 449

— — relationship between — — and the temperature of wood 261

— —, relationship to drying time for vapor drying 466

— —, relative humidity desorption-isotherms 189

Moisture content for Douglas-fir veneer of various thicknesses 188

— — of green wood 420

— — of wood below and above fiber-saturation point and dielectric constant 267

— — values, recommended for different uses of wood in the USA 196

— — variations with the seasons in spruce and fir 421

Moisture curves 231

Moisture distribution in a drying spruce board 226

— — in drying wood 223

Moisture gradient set up in tangential drying 220

—, influence on the elasticity of wood 309

— movement in the capillaries of drying wood 223

— requirements of fungi 108

Mold, during kiln drying 451

Molecules of water in a small capillary 220

Molluscan borers 128

Momentary hygroscopic equilibrium values to be set in the kiln for various drying gradients of hardwoods 446

Monoclinic unit cell 59

Monocotyledons 1, 70

Mora, compressibility and density 299

Mortising 537

Moulding 517

— as peripheral milling with the aim of machining lumber into different cross-sectional shapes 517

—, non-destructive 542

— of overlayed sandwich boards 527

— sanders 528

Mountain ash, bending strength, temperature and moisture content 369

— —, modulus of elasticity and temperature 312

*Musanga Smithii*, packing density 161

Musical instruments 274, 279

— —, recommended moisture content 197

Nara, effect of clearance angle on net cutting force during circular sawing 502

—, power curves 496

Natalis equation 358

National Bureau of Standards abrader 410

Native cellulose 58

Natural defects 79

Natural durability of heartwood 110

— — of wood 136

Natural frequency 301

— —, elastomat instrument for measurement 301

Natural frequencies of wood 275

Naturally pruned branch 82

*Nausitora* 128, 129

Negative contrast preparation technique for electron microscopy 20

Net cutting power of special circular saw blades 509

Network of water-filled capillaries of wood 233

Neutrons and sorption capacity 194

Newtonian liquid 319

Niangon, weight 168

Nitration 58

Nitrobenzene 66

Nitrogen, essential for growth of fungi in wood 109

Nitrogen requirements of marine borers 128

Noble fir, average moisture content 421

Noise reduction in buildings with wood and wood- base materials, 274

Noise reduction in wood machining 475

Nomogram for correcting readings obtained with the Siemens moisture meter for varying temperatures 188

Non-corrosiveness to metals, as property of wood preservatives 148

Non-damaging of wood, as property of wood preservatives 148

37  Kollmann/Côté, Solid Wood

Non-destructive measurements of wood properties 161
Non-destructive testing 273, 275, 413
Non-leachable chemical wood preservatives 107
Non-pressure processes in wood preservation 140, 141, 235
Northern red oak, (*Quercus borealis* Michx.) linear thermal expansion 241
Notched beams 373
Notched compression specimen of isotropic material, stress distribution over the cross-section 354
Notch factors 330
Notches 329
Number of knives in planing 519
Number of teeth, effect on the power consumption 509
Nutritional requirements of wood-rotting and wood-staining fungi 109
Nymphal stage (termites) 113

**0**-Acetyl-4-0-methylglucuronoxylan 61
Oak, American white (*Quercus* sp.), bending radius 545
—, average sound velocity 277
—, buckling 357
—, conductivity and resistivity 259
—, crossties drying 466
—, crushing strength and speed of testing 337
—, cutting force and specific gravity 523
—, cutting velocity and resistance 487
—, damping of sound radiation and sound wave resistance 280
—, density 171, 178
—, drying temperatures 440
—, effect of bevel angle on cutting power 504
Oak, effect of clearance angle on net power consumption during circular sawing 502
—, effect of rake angle on cutting power 504
—, effect of set on power requirement 504
—, elastic constants 295
—, emissivity 257
—, European (*Quercus robur*) home-grown, bending radius, 545
—, fatigue in compression 355
—, fiber saturation point 200
—, hardness and density 407
—, infrared drying 471
—, internal checks and honeycombing 453
—, linear thermal expansion 241
—, logarithmic decrement 278
—, mechanical propterties across the grain 333
—, mine timber, conversion factors for dry matter content 179
—, modulus of elasticity and density 302, 307, 310
—, Northern red, average moisture content 421
—, Poisson's ratio 298
—, power factor 268
—, power for cutting and speed 479
—, plastic deformation 334
— (*Quercus alba* L.), fatigue resistance 378

Oak, ratio $\sigma_\perp/\sigma_\parallel$ 341
—, relative humidity during kiln drying 447
—, shipping weight 166
—, shrinkage and collapse 456, 462
—, sound velocity 276
—, specific cutting force and average chip thickness 522
—, specific gravity variation 170
—, specific sanding work 533
—, steam consumption during drying 459
—, stress-strain curves 542
—, swelling gradients 209
—, torsional strength 396
—, toughness and width of annual rings 391
—, volumetric swelling 165
—, wear and specific gravity 411
—, white, average moisture content 421
Obeche (*Triplochiton scleroxylon* K. Schum.), bending radius 545
*Ochroma lagopus* Sw., longitudinal shrinkage 213
Odoko (*Scottelia coriacea* A. Chev.) bending radius 545
Off-cuts from bobbins, content of solid matter 180
Ohm's law 269
Oil cells 17
Okoumé, Baumann's ratio 395
—, swelling gradients 209
—, weight 168
Old-house borer 120
Open-air drying of chestnut (*Castanea dentata* Borkh) poles 428
— — of pine 429
*Ophiostoma piceae* 106
Oregon pine, cutting force and specific gravity 523
—, swelling gradients 210
Organ pipes 274
Organization of the cell wall 18, 42
Orientation of cellulose microfibrils in cell wall 74
Orthochlorobenzene as insecticide 123
Oscillating vacuum-pressure treatment 146
Osmose process 143
Osmosis in green pine sapwood, distribution of NaF 236
Osmotic procedure 235
Output of gangsaws 485
Oven-dry method of determining the moisture content of wood 181
Oven-dry vulcanized fiber, power factor 270
Overloading 497
— of the gullets 489
Oxygen, essential for growth of fungi 107

Packing density 161
Paper, power factor 268, 270
Paradichlorobenzene as insecticide 123
Paraffin, magnetic susceptibility 271
Parana pine, *Araucaria angustifolia* 82
— —, drying temperatures 440
— —, relative humidity during kiln drying 447
— —, volumetric swelling 209

Subject Index

Paratracheal parenchyma 6,7
Parenchyma 2, 10, 55
—, chemical composition 72
Parquetry, recommended moisture content 196
Partially decayed wood, view of cross-section 35
Particle structure of wood cells 40
Parts of aircraft, recommended moisture content 196
Pear, darkening during steaming 434
—, hardness and density 407
— tree, high frequency drying 470
Pecky cypress 79
Pectic material 55, 57
— — in cell wall 74
Pectin in wood 64
Peeling and slicing in chipless wood-cutting 513
—, power consumption 525
PEG 219, 469
Pendulum dynamometer 503
Pendulum hammer 380
Penetrability of softwood 137
Penetration depth and time of impregnation 239
—, influence of · moisture content, grain orientation, and structure 239
— of preservative as indicator of the efficiency of wood treatment 136
— of the wood cell walls by hyphae 102
Penetrating ability of preservatives 148
Pentachlorophenol 123, 148
Perchlorethylene, special seasoning methods 465
Periodate lignin 67
Permanence of preservatives 148
Permanent pit membrane pores 233
Permanent set in elastoplastic bodies 315
Permeability of microscopic and submicroscopic structure 219
Persimmon, average moisture content 421
Petrol, swelling and dielectric constant 218
Petroleum oil and creosote, viscosity and temperature 238
Phenolic group 67
Phenylcoumaran dimers 67
Phenylpropane units 64
Phloem 2
Pholadidae 128
Phosphates 71
Photoelastic stress pattern and boundary stresses on laminated models 325
Photoelastic tests 324
p-hydroxybenzaldehyde 66, 70
Physics of wood 160
Physiological requirements of fungi — relationship to wood preservation 110
— — of wood-destroying and wood-inhabiting fungi 107
Piano cases, recommended moisture content 197
*Picea abies* constituents of ray cells 73
— *canadensis*, specific gravity and summerwood 176

*Picea, excelsa* Link., density 174
—, specific gravity 177
— *glauca* Voss., width of annual rings 179
— *mariana* B. S. P., width of annual rings 179
— *sitchensis*, Wöhler curve 376
Piceoid pits 37, 39
Piezoelectric behavior and change of compressive strength with age in old timbers 273
— modulus, crushing strength and dynamic Young's modulus 274
— properties of wood 271
— stress constant and modulus of elasticity 273
— texture 272
— voltage generation, average $\Delta A$ max. and static modules of elasticity 273
— — — vs. density for Douglas-fir 273
— voltage profiles 273
Piezoelectricity and degree of crystallinity of cellulose 274
—, intensity in wood 272
Piles, foundations 426
Piling, improper 452
— in dry kilns 439
— methods 424
— vertical 439
Pin knots 82
Pine, average density 168
—, average sound velocity 277
—, Baumann's ratio 395
— beams, load deflection diagram 364
—, bending strength and load angle 365
— bending strength and moisture content 368
—, bending strength ratio 371
—, breaking stress 362
—, Brinell side hardness 408
—, British Honduras pitch (*Pinus caribaea* Morelet), bending radius 545
—, buckling 357
—, chip formation 527
—, chip thickness and curvature 528
—, compliance and specific gravity 306
—, compression-load diagram 339
—, conductivity and resistivity 259
—, creep 334
—, crushing strength and angle of grain 341
—, crushing strength and speed of testing 337
—, cutting forces for turning 535
—, cutting power and feed per tooth 496
—, cutting velocity and resistance 487
—, damping of sound radiation and sound wave resistance 280
—, dependence of the absorption coefficient on the moisture content 256
—, dependence of the absorption coefficients on the temperature for oven-dry wood 256
—, drying curves 444
—, drying temperatures 440
—, Eastern, average moisture content 421
—, effect of knots on structural lumber 353
—, effect of rake angle on cutting power 504
—, effect of rake angle on power consumption 502

37*

580 Subject Index

Pine, elastic compliance in a knot 314
—, elastic constants 294
—, fiber saturation point 200
—, Finnish, fatigue resistance 378
—, frequency curves for specific gravity 169
—, hardness and density 407
—, high frequency drying 470
—, impact work and moisture content 388
—, infrared drying 471
—, knots and notches 313
—, longitudinal swelling 199
—, Mark Brandenburg, quality zones 344
—, mechanical properties across the grain 333
—, mine timber, conversion factors for dry matter content 179
—, notch factor 330
—, Poisson's ratio 298
Pine, props 353
—, quality factor 343
—, radial swelling 199
—, shearing strength 402
—, shearing strength and moisture content 401
—, shipping weight 166
—, slenderness ratio and impact work 382
—, sound absorption 284
—, specific cutting force and average chip thickness 522
—, specific gravity 172
—, specific gravity and crushing strength 355
—, specific gravity variation 170
—, specific sanding work 533
—, static endurance load 334
—, stress curve 358
—, surface of torsion 305
—, Swedish, fatigue resistance 378
—, swelling gradients 207, 209
—, tangential swelling 199
—, tensile strength and specific gravity 327
—, torsional strength 396
—, tension tests with knots 329
—, toughness and width of annual rings 391
—, volumetric swelling 165, 199
—, Wöhler curve 376
Pinoid pits 37, 39
Pinoresinol 67, 70
Pinosylvin 70
*Pinus banksiana* Lamb., density 174, 178
— *caribaea* Morelet, density 174
— *cembra* L. 80
— — L. (Siberia), infrared drying 471
— *echinata* Mill., density 174
— — —, longitudinal shrinkage 213
— *excelsa*, specific gravity and summerwood 176
— *mercusii* Jungh, fibers 323
— *palustris* Mill., density 173, 174
— — —, infrared drying 471
— — —, longitudinal shrinkage 213
— *pinea* L., density 174
— *ponderosa* Douglas, infrared drying 471
— — Laws., longitudinal shrinkage 213
— *resinosa* Ait., specific gravity 178
— — —, specific gravity and summerwood 176

*Pinus strobus* L. 80
— — —, specific gravity and summerwood 176, 178
— *sylvestris* L. 73
— — —, density 174, 176, 178
— — —, mechanical properties 199
— — —, specific electrical conductivity and dielectric constant 263, 264
— *taeda* L., density 174
Pit, aspirated 31
— aspiration 137
—, blind 27
Pit border in secondary wall formation 27
— —, initial 34
— —, intervessel 35
—, bordered, in hardwood 31
Pits, cupressoid 37, 39
Pit cavity (pit chamber) 27
Pit membrane permeability 235
— — pore, diameter 238
— — thickness 233, 238
— openings 233
— pair, bordered 27, 31
— —, connecting pits 27
— —, half-bordered 27
— —, simple 36
Pits, piceoid 37, 39
—, pinoid 37, 39
Pit, recess in the secondary wall of the cell 27
—, simple 21
— structure in the secondary cell wall 27
Pits, taxodioid 37
—, vestured 35
Pitch, of circular saw tooth 507
—, of gang saw blades 479
— of the teeth, effect on cutting force 489
— pine, cutting force and specific gravity 523
— —, power factor 268
— —, quality factor 343
— —, quality zones 344
— pockets 87
— streaks 87
Pith 3
Planer shavings, content of solid matter 180
Planes of wood 3
Planing 517
Plasmalemma 42
Plasmodesmata 33, 36
Plastic deformations 334
Plasticity and creep 315
— as a cause of sorption hysteresis 193
— of cell walls 455
Plasticizing by chemical agents 542
Platinum, damping of sound radiation and sound wave resistance 280
Plum, hardness and density 407
Plywood, recommended moisture content 196
—, sound absorption 284
Poiseuille's equation and law 221, 236, 237, 238
Poisson's ratios 297
Polarity of a liquid 217
Polarization and stress-strain relationships 271
— microscopy for studies on cellulose in wood 18

Subject Index

Polarized light photomicrograph 23, 45
— — microscopy for microfibrillar orientation in layers of wood cell walls 21
Poles, recommended moisture content 197
Polymerization of methacrylate 50
Polyphenolic substances in wood cells 55
Polyphenols 70
—, extract 465
— in heartwood 72
Polyethylene glycols 156, 219, 469
*Polyporus versicolor* L. ex Fries 100, 102, 104, 111
Polysaccharides 57
Ponderosa pine 421, 465, 496
Poplar, compressibility and density 299
—, compression-load diagram 339
—, cut surface angle 528
—, cutting force and specific gravity 523
—, damping of sound radiation and sound wave resistance 280
—, drying curves 444
—, fiber saturation point 199
—, forced air-drying 431
—, hardness and density 407
—, high frequency drying 470
—, infrared drying 471
—, linear thermal expansion 241
—, relative humidity during kiln drying 447
*Populus balsamifera* L., specific gravity 179
*Populus tremuloides* L., specific gravity 179
Pores 5, 10, 14
Pore volume and specific gravity of wood 162
— — of a wood specimen 163
*Poria monticola* 100, 111
Porosity 160
— of wood constituents and of wood substance 160
Potassium 71
Powder-post beetles 115, 124
Powder-post-type of damage 112
Power consumption and tooth bevel angle 504
— —, for sawblade sets 481
— — of gang saws, increase with the time of service 479
— — increase with tooth pitch 480
Power demand in planing, effect of moisture content 524
— — — —, effect of temperature 524
— — — —, effect of wood species 524
— —, linear increase with specific gravity in planing 525
Power factor 263, 267, 269
— — and dielectric constant of spruce, cellulose, lignin and resin as a function of frequency 269
— — and moisture content 270
— —, dependence on frequency 270
— — of different wood species and frequencies 268
Power loss 268
Power loss factor 265
Power required as a function of the rake angle and depth of cut 503
— required for the feedworks 479
— requirement in chain sawing 511
Predryer 433

Predrying 429
Preferred host-woods 98
Preliminary vacuum in full-cell process 145
Preservative flow, kinetics 239
— materials toxic to insects, fungi and marine borers 148
— temperature in Bethell process 145
— treatment of wood 122
—, viscosity 238
Pressure and melting temperature of ice 352
— and vacuum impregnation in wood preservation processes 140
— in laminated bending 550
— processes 144, 236
— treatment of wood, theory 236
Pre-treatment for wood preservation treatment 145
Primary cell wall 20, 55
Primary creep 318
Primary wall of the microfibrils 74
— — structure 25
Principal wood constituents 58, 70
Prism-cutting in gang saw mills 484
Privet, hardness and density 407
Productivity in sawing 479
Progressive kilns 456, 458, 459
Progressive or tunnel-type kiln 457
Prong test for case-hardening defects 92
Proportions of cell types in hardwoods 12
Propyl alcohol, swelling and dielectric constant 218
Prosenchyma 2, 10
Prosenchymatous cells, chemical composition 72
Protection against marine wood borers 132
Protection of marine structures 133
Protein 161
Protein substances, piezoelectricity 272
*Prunus serotina* 62
*Pseudotsuga taxifolia* Britt., electrical resistance 259
— — —, longitudinal shrinkage 213
— — —, specific gravity, 174, 177
*Ptilinus pectinicornis* Geof. 118
*Ptilinus ruficornis* (Say) 118
Pulp, integral heat of wetting 201
Pulpwood 179
—, content of solid matter 179
— piles, content of solid matter 180

Quality factor 343, 378
Quality figures and properties of wood 344
Quality zones of air-dry pine wood 344
Quartz, damping of sound radiation and sound wave resistance 280
*Quercus borealis*, longitudinal shrinkage 213
*Quercus borealis* Michx., specific gravity 178
*Quercus pedunculata* Ehrh., specific electrical conductivity and dielectric constant 263
— — —, specific gravity 178
— —, infrared drying 471
— —, specific electrical conductivity and dielectric constant 264
Quinones 70
Quipo, elastic constants 296

582 Subject Index

Radial drying 235
Radial section of a hardwood stem 3
Radial shrinkage 91
— — and tangential swelling in softwoods and hardwoods under different heating conditions 245
— — coefficient $\beta_r$ 212
Radiation methods for measurement of material quality 161
— of heat with respect to wood 256
Radios, recommended moisture content 197
Raised grain 94
Rake angle 524
— — and depth of cut, linear function 503
— — effect on cutting force per tooth in countersawing and climbsawing 503
— —, effect on cutting power 504
— —, effect on power consumption in sawing 503
— — for gang saw blades 480
Rake angles of sawblades, 478, 501
Raphides (crystals in wood) 17
Rate of combustion of wood 149
— of diffusion in longitudinal direction and across the grain 228
— of suction of the capillary water 436
Ratio $\varepsilon$ of tangential shrinkage to radial shrinkage or swelling 212
Ray cell contents 71, 73
— parenchyma 24
— tracheid 2, 10
— tracheids, dentate 42
Rayon, tensile strength and breaking length 323
Reactance 265
Reaction wood 55, 80, 84
— — anatomy and ultrastructure 43
Reconditioning 455
—, after casehardening 453
— of collapsed lumber 94
Red beech, adsorption, desorption 211
— —, dashening during steaming 434
— —, effect of clearance angle on net power consumption during circular sawing 502
— —, elastic constants 295
— —, high frequency drying 470
— —, ice lattice 352
— —, swelling 207, 214
Red gum, dashening during steaming 434
Red oak, fluid pressure and penetration 516
— —, predrying 433
— —, tangential swelling and radial shrinkage 245
— — treated with sodium chloride 468
Red pine, revistivity 259
Red tulip oak (*Tarrietia argyrodendron* var. *peralata*), effect of notches on bending strength 375
Reduction in toughness of decaying wood 101
Redwood, average moisture content 421
—, earlywood fibers, stress-strain properties 322
—, fluid pressure and penetration 516
— latewood fibers, stress-strain properties 322

Redwood, stain susceptibility 451
— (*Sequoia sempervirens* Endl.), linear thermal expansion 243
Relationship between drying rate and psychrometric difference 230
Relative air humidity, average minima 423
Relative crystallinity of cellulose 60
Relative humidity and equilibrium moisture content 442
— — in kiln drying lumber, recommended values 447
Relative percentage of polysaccharides in the different layers of the cell wall 74
Relative strength properties and angle between specimen axis and grain direction 326
Relative sugar composition of normal and tension wood 57
Relative vapor pressure, radius of capillaries and swelling pressure 204
Relative whiteness of pine heartwood after 5 hrs. drying 451
Relaxation time 319
Residues from plywood factories, content of solid matter 180
— from sawmilling, content of solid matter 180
— from woodworking, content of solid matter 180
Resin acids 70
— canals 138
— — in coniferous wood 8, 13
— exudation 440
—, power factor and dielectric constant 268, 270
Resins 161
Resistance of wood against compression perpendicular to the grain, effect of moisture content 349
Resistance-type moisture meter 185, 187
Resistivity, electrical 259, 261, 270
Retardation of ignition of wood 149
Retarded elastic deformation 320
Retention of preservative as indicator of the efficiency of wood treatment 136
*Reticulitermes flavipes* 112
*Reticulitermes hesperus* 112
Revolving table bending machine 548
Rheological models of earlywood and latewood zone in the annual rings, 317, 318
Rheology of wood 292
Rhinotermitidae 112
Rhodesian teak, wear and specific gravity 411
Rhombic symmetry of wood 293, 303
Rhombic system, Poisson's ratio 298
Rift cracks 86
Ring-porous broadleaved woods with distinctly colored heartwood 200
Ring-porous hardwood 5
— —, density 174, 179
— width and density 179
Ring shake 86
Ripsawing, velocity coefficients 493
Ripsaws 501
Robinia, hardness and density 407

## Subject Index

Robinia (*Robinia pseudoaccacia* L.), bending radius 545
Root wood 200
Rosewood, linear thermal expansion 241
Rotary peeling in chipless wood-cutting 513
Rough boxes, recommended moisture content 197
Rough lumber construction, recommended moisture content 197
Roughness in sawing 484
Round-headed borers 119
Rounded tracheids 43
Round-wood or logs, content of solid matter 179
Rowan tree, hardness and density 407
Rubber, damping of sound radiation and sound wave resistance 280
—, Hooke's law 292
—, sound wave resistance 279
Rüping process 146
Rupture in the middle lamellae 392
Rustle of leaves, sound intensity 281

$S_1$ (outer layer) of secondary cell wall 22, 25
$S_2$ (middle layer) of secondary cell wall 22, 24
$S_3$ (inner layer) of secondary cell wall 22, 24
$S_4$ layer lining the cell lumen 26
Safety in handling as property of the wood preservatives 148
— in wood machining 475
Salt and sugar treated wood, decrease of volumetric shrinkage 469
—-seasoned wood, finishing properties 469
—-treated wood, eletrical conductivity 469
— treatment, moisture distribution in green wood and in treated wood over the cross-section 468
Samba, weight 168
*Sambucus callicarpa* Greene, longitudinal shrinkage 213
Sample board and test piece for moisture content 445
— — method, for moisture content test 444
Sanded surfaces, roughness 533
Sanding 528
—, blunting stages 532
— of hardwood 529
— process, technology 529
— speed 532
— time, effect on the amount of abraded material 529
— —, effect on amount of abraded material, main cutting force and specific sanding work 531
— —, effect on the amount of abraded material for various wood species and grain sizes 530, 531
— —, related on specific amount of abraded material, feed velocity, main cutting force and specific chipping work in 532
— velocity, effect on temperature of work-piece 532
— with oscillation 533
— work for drum sanding 533

Sandstone, Hooke's law 292
Sapelli, weight 168
Sap gum, forced air-drying 431
— replacement in wood preservation processes 140
Sap-staining fungi 97, 105
Sapwood 4, 72
—, integral heat of wetting 201
— of coniferous species with distinctly colored heartwood 200
— of ring-porous hardwoods, fiber saturation point 199
— of semi-ring-porous hardwoods, fiber saturation point 199
— physical properties 55
Saw blades, cutting power 483
— — dimension 486
— —, fractures 483
— — geometry, survey of definitions and symbols 477
— —, gullet ground 483
— —, sharpness of teeth 484
— — sets, springset clearance, swage set clearance 481
— —, tension stresses 482
— — thickness, cutting depth and sawblade diameter 500
Saw cutting, slicing and vibration cutting, scheme of the processes 513
Sawdust 484
—, content of solid matter 180
—, yield values 485
Sawing, average cutting velocity 484
—, energy curves 489
—, feed speed 484
—, friction losses 483
—, power law 476, 477
—, sash gang 475
—, technology 475
—, thermal effect 482
Sawmills, advance estimation of yield 485
Sawtooth forms of circular saws 499
Scanty paratracheal parenchyma 6
Schedules for the high temperature drying 462
Schizogenous canals 9
Scots pine, drying temperatures 440
— —, relative humidity during kiln drying 447
Seasonal moisture content 422
— variations in moisture content 422
Seasoning, special methods 460
— defects 90
Secondary cell wall layers 20, 22, 392
— creep 318
— reproductive of termite 113
— thickening 73
Sedimentation-diffusion 58
Semi-diffuse porous hardwood 6
Semi-ring-porous broadleaved woods with distinctly colored heartwood 200
— — — hardwood 5
*Sequoia sempervirens* Endl. longitudinal shrinkage 213
Set, effect on power requirement for circular sawing 504
— of saw blades 477, 481

584                                  Subject Index

Shake 86
Shakes in logs 90
Shaping 517
— and planing with cutterhead knives geometrical conditions 518
— as cutting an edge profile or edge pattern
   on the side, end or periphery 517
Sharpness angle (British) of saw blade 478
— angle of saws 501
Shear-in-one-plane specimens and shear test
   devices 397
Shear-in-two-planes specimens and shear
   test devices 398
Shear specimens, special shaped 398
— — using bolts 399
Shear strain between the fibers 272
— strains, determination 301
— stress 301
— —, distribution in a bent beam 362
Shearing, orientations between plane of —
   and cross force to fiber direction on the
   one side and layers direction 400
— strength along the grain 323
Shearing, effect of moisture content for pine
   and beech 401, 402
— strength for Swedish pine 400
— —, parallel to grain, determination 397
— —, radial, and specific gravity for ash 399
— — values for beech and ash 402
Sheating, recommended moisture content 196
Shed dryer 458
Shellac, magnetic susceptibility 271
Shingles, recommended moisture content 197
Ship and boat lumber, recommended moisture
   content 197
Shipping weight of commercial timbers in
   green condition 166
— — of green wood 167
Shipworms 128
— damage in wood 129
Shock resistance 333
— — or toughness 379
Shoe heels, recommended moisture content 197
Short columns, stresses 356
Shortleaf pine, damping of sound radiation
   and sound wave resistance 280
— —, spring and summerwood 175
Shot-hole or auger borers 121
Shrinkage 204
—, abnormally high 205
— and distortion of flats, squares, and
   rounds as affected by the direction of the
   annual rings 214
— and swelling, anisotropy 205
— during wood decay 98
— maximal values $\beta$ and collapse during
   drying of green wood samples 456
—, radial and tangential 208
Shuttles and bobbins, recommended moisture
   content 197
Siberian Larch, swelling gradients 209
Sickle-like chips 518
Side clearance of saw blade 481
Side shear, chip formation and associated
   friction in sawing 478

Sidings, recommended moisture content 196
Siemens moisture meter 261
Significance of capillary size 140
— of rays in the penetration of wood 138
Silicates 71
Silicon carbide as abrasive for wood 529
Silk, tensile strength and breaking length 323
Silver, damping of sound radiation and
   sound wave resistance 280
— Quandong, modulus of elasticity and
   temperature 312
Silvery sheen of tensionwood 83
Simple pit pairs 36
Single crystals 60
*Sirex gigas* L. 127
— *juvencus* (L.) 127
— *noctilio* Fab. 127
Sitka spruce, bending strength, effect of
   temperature and moisture content 369
— —, diffusion of water 224
— —, drying 220
— —, earlywood fibers, stress-strain properties 322
— —, latewood fibers, stress-strain properties 322
— —, modulus of elasticity and temperature
   312
— —, moisture content-relative humidity
   desorption-isotherms 189
— —, elastic constants 294
— —, electrical resistivity 260
— —, fatigue resistance 378
— —, fiber saturation point 199
— —, linear thermal expansion 243
— —, spiral and diagonal grained lumber 366
— —, tangential swelling and radial shrinkage 245
Slab chimney test 155
Slabs 484
—, content of solid matter 180
—, yield from circular and gang saws 485
Slack staves and heading, recommended
   moisture content 197
Slake-gas-developing chemicals 152
Slash pine, earlywood fibers, stress-strain
   properties 322
— —, latewood fibers, stress-strain properties
   322
— —, spring and summerwood 175
Slenderness ratio of adsorbed energy W to
   volume V as a function of — — for beech
   and pine 384
Slicing in chipless wood-cutting 513
Slip planes in fiber walls 49
Soda-boiled cotton, differential heat, free-
   energy and entropy changes 203
Soft rot fungi 98, 104
Soft-rotted wood 104
Softwood, careful drying 446
—, enforced drying 446
—, rake angle 480
Soldier (termites) 113
Solid matter content in piles of wood
   179
Solid volume of wood 162

## Subject Index

Solutions of creosote-pentachlorophenol 148
Solvent seasoning 465
Solvents for the distillation method 182
Sorbed molecules 220
Sorption 189
— isotherms, composition 194
— — of cellulose 193
— properties of bent wood 548
Sound 274
Sound absorption 284
— — of various building materials 285
Sound energy 281
— figures 280
—-radiating properties of vibrating wood 274
— transmission in wood 276
— — loss and frequency of a plywood and a brick wall 282
— — — for plywood walls 283
— — — for various constructions 282
— — — through single walls and weight of the wall 282
— velocity in woods and other materials 276, 277
— — parallel to fibers and density 277
— — parallel to the grain and moisture content 278
— wave resistance 279
Sounding boards in pianos or violins 274, 279
Source of nitrogen for termites 113
Southern pine, average moisture content 421
— — poles, drying 466
— —, tangential swelling and radial shrinkage 245
Spaces, intercellular 43
Spanwinkel 478
Specific cutting energy curve, possible types 488
— — — for circular sawing 496
— — — in sawing 497
— — — curves influenced by average chip thickness 497
— — — for a circular ripsaw effected by feed speed 497
— — force and average chip thickness 522
— electrical conductivity 263
— — conductivity and dielectric constant for various woods, impregnated and not impregnated 264
— — resistance and moisture content 258
Specific gravity 160, 209
— — and crushing strength 355
— — and summerwood percentage for ashwood 177
— — and summerwood percentage for coniferous species 176
— — and summerwood percentage, theoretical relationship 176
— — and width of annual rings 177
— —, dependence of linear thermal expansion 243
— —, effect on net cutterhead horsepower requirement 524
— —, effect of wide annual rings 179

Specific gravity, frequency distribution in temperate and subtropical zones, tropical zones and other sources 173
— — and moisture content diagram 167
— —, moisture content of wood and dielectric constant 266
— — of Douglas-fir, effect of position in the tree 171
— — of pine wood, multinodal frequency curve 170
— — of springwood and summerwood, variability curves 175
— — of spruce and pine, frequency curves 168
— — of the wood substance 233
— gravity of water adsorbed in cell walls of wood as a function of moisture content 200
— —, variation with height in the tree 170
— heat 250
— — and diffusivity 251
— — of wood 245
— volume of the wood, average 234
Speed of combustion of wood 150
Speed of the drying medium, influence 232
*Sphaeroma* 130
Spike knot 80, 82
Spiral and diagonal grained lumber, deficiency in strenght properties with respect to straigth-grained material 366
— grain, drying defects 452
— —, effect on the strength 365
— — lumber 84
— thickenings in cells (helical thickenings) 37, 40
Spitzwinkelzahn 501
Splits in logs 90
Spools, recommended moisture content 197
Spraying process 439
Spring-set clearance for gang saw blades 481
Springwood 56, 72
— cell walls 73
—, $S_2$ layer in cells 75
Spruce, seasonal moisture content 422
—, average density 168
—, average sound velocity 277
—, Baumann's ratio 395
— beams, load deflection diagram 364
—, buckling 357
—, compliance and specific gravity 306
—, conductivity and resistivity 259
—, crushing and tensile strength ratio and moisture content 336
—, cutting force and specific gravity 523
—, — power and feed per tooth 496
—, — power, feed speed and feed force 495
—, — velocity and resistance 487
—, damping of sound radiation and sound wave resistance 280
—, dependence of the absorption coefficient on the moisture content 256
—, drying deformations 452
—, drying temperatures 440
—, drying curves 444
—, effect of bevel angle on cutting power 504

# Subject Index

Spruce, effect of clearance angle on net consumption during circular sawing power 502
—, effect of knots on structural lumber 353
—, effect of set on power requirement in sawing 504
—, elastic constants 294, 303, 307
—, European (*Picea abies* (L.). Karst), bending radius 545
—, fiber saturation point 200
—, frequency curves for specific gravity 169
—, hardness and density 407
—, heat of wetting and contraction of volume 202
—, high frequency drying 470
—, hygroscopic equilibrium in superheated steam 461
—, impact work and moisture content 388
—, infrared drying 471
—, integral heat of wetting 201
—, linear thermal expansion 241
—, logarithmic decrement 278
—, mechanical properties across the grain 333
—, mine timber, conversion factors for dry matter content 179
—, modulus of elasticity 302, 307, 310
—, modulus of rupture and wane 373
— (*Picea rubra* Link) 338
— (*Picea excelsa* Link.), fatigue resistance 378
—, plastic deformation 334
—, Poisson's ratio 298
—, power factor and dielectric constant 269
—, power for cutting and feed in sawing 479
—, pressure at volume contraction 203
—, rate of diffusion in drying 230
—, relative humidity during kiln drying 447
—, shipping weight 166
—, shrinkage and collapse 456
—, sitka, average moisture content 421
—, specific gravity 172
—, specific gravity variation 170
—, steam consumption during drying 459
—, stress curve 358
—, swelling gradients 209
—, tensile strength of fibers 321
— tension tests with knots 329
—, torsional strength 396
—, toughness and width of annual rings 391
—, volumetric swelling 165
—, Wöhler curve 376
Square chisel geometry 538
Square columns, stresses 357
Stability of circular saw blades 506
Staining during kiln drying 451
Stains in wood 105
Stamm's theoretical drying diffusion co-efficients 232
Standard C 10 of the A. W. P. A. 143
Standard-cotton-cellulose, integral heat of wetting 201
Star checks 245
Starch 57, 64, 71, 73
— content of wood and *Lyctus* attack 115

Static bending test, set-up of a beam 363
— bending by a central load 300
— endurance load in tension of pine wood and laminated fabric 334
Staves for beer barrels 542
—, recommended moisture content 196
Staypak, thermally stabilized wood product 157
Steady state drying constant 462
Steam consumption in a pit 435
— jet blowers in kilns **441**
— treatment 452
Steaming 434
— against staining 451
— and seasoning of wood 420
— and vacuum in the preservation process 145
— duration, effect on the pH-value of condensate from beech and ash 437
— effects 436
— in chipless wood-cutting 513
—, intermittent 434
— period until equilibrium and thickness of wood 255
— pits, heat balances 434
— plants 434
— preliminary for relief of stresses 439
— — to air-seasoning 434
— prior to veneer peeling 434
—, swelling of beech after different periods of steaming 436
Steam, spray 440
Steel, tensile strength and breaking length 323
—, sound wave resistance 279
— wire, maximum tensile strength and breaking length 323
Steeping and cold soaking (non-pressure process) 142
— process 141
Stefan's and Boltzmann's law 256
Step *R* by one knife-cut 518
Sterilization of green lumber 434
— of wood products 123
Sticker marking 94
Sticker, mechanical 439
Stickers 426, 439, 452
—, recommended thickness in relation to the lumber thickness 439
Stilbenes 70
Stone wall, sound absorption 285
Strain distribution in the cross-section of cleavage specimens 332
— stresses in wood bending 542
Strength of endless cellulose molecules 321
— of primary valence bonds 321
— properties of decayed wood 98
— — of wood 326
Stress-compression diagrams for compression tests on wood columns loaded perpendicular to the grain 340
— condition in drying wood 93
— curves of ash, pine, spruce and basswood 358
— diagram for the cross-section of a bent beam 362

Stress-diagramm for trapezoid-like distribution of the stress 361
— distribution, at a constant bending moment 371
Stress distribution in a bent wooden beam 361
— — in a symmetrically loaded cube 399
— — in a tension bar of pine in the vicinity of a knot 314
— — in double-cleavage specimens 331
— — in the vicinity of a knot 314
— — over the medium cross-section of the double-cleavage specimen 330
Stresses on circular sawblades 506
Stress-strain behavior 315
— — curves of air-dry and steamed beech 542
— — curves of air-dry and steamed or boiled wood 543
— — curves of soft vulcanized Hevea rubber 315
— — curves for wood 292
— — curves for wood and other elastic-plastic materials 315
— — cycles for repeated loading and unloading with increasing approach to the ideal elastic behavior 316
— — cycles for repeated loading and unloading with increasing plasticity 315
— — diagram for tensile and compressive tests 361
— — plane for spruce wood, intersection 302
— — properties of wood fibers 322
String instruments 274
Structural formula for larch arabinogalactan 64
— — of arabino 4-O-Methylglucuronoxylan 63
— — of cellulose 59
— — of hardwood glucomannan 62
— — of O-Acetylgalacto-glucomannan 63
— fractures in spruce (*Picea rubra* Link) 338
Structure of 4-O-Methylglucuronoxylan 61
Studs, recommended moisture content 197
Stump wood, content of solid matter 179
Sub-flooring, recommended moisture content 197
Subterranean termites 113
Sucrose 71
Sugar maple, (*Acer saccharum* Marsh.), linear thermal expansion 241, 243
— —, cutting power and depth of cut 498
— —, fluid pressure and penetration 516
— pine, average moisture content 421
Sugars 70
Sugi, cutting force and feed per tooth 496
—, effect of clearance angle on net cutting force during circular sawing 502
Sulfates (ash constituents) 71
Sulphite pulp from spruce, density 164
— —, pressure at volume contraction 203
— pulpwood, bleached, integral heat of wetting 201
Summary of lignin structure 69
Summative analyses of wood 56

Summerwood 56, 72
— cell walls 73
— $S_r$ layer in cells 75
Superheated steam 456, 460
— — kiln 459
— — —, energy requirement 463
Surface-bound water 193
— checks 91
— — during casehardening 453
— forces in capillaries 221
— molds on wood 105
— quality of sawn products 483
— replicas for electron microscopy 47
— tension, 222, 305
— — of the water 221
—, woolly, of tension wood 83
Swage-set clearance, for gang saw blades 481
— saw blade, temperature 482
— saws, width of kerf 482
Swedish pine, Baumann's ratio 395
— — knots and knotches 313
— —, tensile strength and specific gravity 327
— —, tensile strength and temperature 328
— —, shearing strength 400
— —, Young's modulus 306
Sweep of tree 85
Sweet chestnut, hardness and density 407
Sweet gum, elastic constants 296
—, average moisture content 421
Swelling 204
— and moisture content 199
— and shrinkage of temporarily clamped and of free control samples 215
— capacity of wood reduced by steaming 438
— components, superposition 214
— compressive stress and the modulus of elasticity perpendicular to the grain 216
— gradient 209
— in aqueous solutions and organic liquids 216
— in Bromides 217
— in Chlorides 217
— in Iodides 217
—, longitudinal 207
— of beech after different periods of steaming 436
— phenomenon of methacrylate polymerization 51
— pressure 204
—, radial 207
—, restrained 214
—, tangential 207
*Swietenia macrophylla* King, infrared drying 471
— sp., specific electrical conductivity and dielectric constant 264
Swing drying, drying curves for boards 432
Swollen volume density 205
— — specific gravity of the wood 233
Sycamore, American, average moisture content 421
Syringaldehyde 66
Syrigin 71

Taber-abrader 410
Tandem gang saws, yield 485
Tandem gangways in sawmills 484
Tangential drying 235
— section of a hardwood stem 3
— shrinkage 91
— shrinkage coefficient $\beta_t$ 212
— swelling and radial shrinkage in soft-woods and hardwoods under different heating conditions 245
— swelling of birch wood an the dielectric constant of the swelling medium 218
Tanks and silos, recommended moisture content 197
Tannins 70
Tannin, extract 465
—, influence of drying heat 452,
Tasmanian Oak (*Eucalyptus oblique* L'Herit.), bending radius 545
— —, wear and specific gravity 411
Taxodioid pits 37
—, bending radius 545
Teak, cutting force and specific gravity 523
—, damping of sound radiation and sound wave resistance 280
—, infrared drying 471
—, knots and knotches influence on modulus of electicity 313
—, relative humidity during kiln drying 447
—, sound wave resistance 279
—, weight 168
Teeth types of gang saws 479
Temperature change in heated wood 250
— coefficients for crushing strength 350
— curves and moisture content in beech 231
— distribution at different distances in timbers 254
—, effect on the crushing strength 350
—, external in the USA 423
— for optimum growth of fungi 107
—, thermostatic control in kilns 460
Temperatures at the centers of green round Southern pine timbers when steamed 253
— at the centers of green sawed Southern pine timbers of different dimensions when steamed 254
— in the interior of green pine lumber after various steaming periods 253
Tenoning 537
— machine tools 538
Tensile strength 321, 327, 330, 333, 335
— — and breaking length of some materials 323
— — along the grain, determination 324
— — across the grain in turned specimens 333
— — and fiber saturation point 327
— — and moisture content 327
— — and temperature 328
— — and the crushing strength of various wood species 335
— — -density relationship 326
— — of compressionwood 327
— — of earlywood 323

Tensile strength of cellulose molecules 321
— — of latewood 323
— — of separated wood fibers 321
— — of single fibers of coniferous species 323
— — of wood parallel to the grain 321
— — parallel to the grain and moisture content 328
— — parallel to the grain and specific gravity 327
— — perpendicular to the grain, determination 330
— tests perpendicular to the grain, specimen 330
Tension and compression test specimen 335
— parallel to grain 324
—, strength and angle between specimen-axis and grain direction 326
—, stress-strain curve 292
— tests of specimens with knots of different sizes 329
Tensionwood 43, 48, 52, 55, 58, 73, 83
Tensionwood galactan 62
Tensioning of saw blades 484
*Teredo* (Teredinidae) 128, 131
Terminal apotracheal parenchyma 6
— Terminal axial parenchyma 7
— — — in softwoods 8
— lamella 55
— — of tension wood fiber 51
*Terminalia bialata* Steudel, infrared drying 471
Termite castes of *Reticulitermes lucifugus* 113
Termites 112
Termite work in oak flooring 114
*Terminalia superba* Engl. & Diels 116
Termitidae 112
Terpenes 70
Terrestrial Basidiomycetes 132
Test for weakened fiber 107
Testing of fire retardants 155
Tetmajer straight line 357
Tetrachlorethane, content of moisture and volatiles 183
Tetrachloroethane, distillation method 182
Tetrachlorethylene, solvent for the distillation method 182
Thermal conductivity 246, 249
— effects in circular sawing 506
— efficiency of kilns 459
— expansion 240
— modification of wood 157
— phenomena, effect on tensile strength 327
— properties of wood 240
Thermodynamics of sorption 201
Thickenings, dentate ray tracheid 42, 44
Thickness of saw blade 477
— variation of lumber 90
Thick-walled epithelial cells of normal canals 8
Thin-walled epithelium of normal canals 9
Thomson's law and equation 198, 220, 222
Threshold of hearing, sound intensity 281
Threshold of pain, sound intensity 281
Through-and-through cutting in gangsaw mills 484

Tight or intergrown knots 79, 81
*Tilia* 49
*Tilia* sp., density 174
Timber grading 413
— preparation 140
Timbers, recommended moisture content 197
Time of sawing 507
Tight staves and heading, recommended moisture content 197
Titration according to Fischer and Eberius, for determining the moisture content of wood 183
Tola, high frequency drying 470
Toluene, content of moisture and volatiles 183
—, solvent for the distillation method 182
Tonoplast 42
Tooth angles of bandsaw blades 489
— — recommendations for ripsaws and cross-cut saws 501
Tooth geometry and pitch at circular saws 500
— — and pitch in bandsawing 489
— —, effects on sawing 479
Tooth height of saws 479
— — of ripsaws 500
— or wedge or lip angle (American) of saw blade 478
— pitch, of saw blades 484
— style for hollow-ground combination planer saw 508
— — of a "Wigo" circular sawblade 509
— styles for flat-ground combination saws having ground teeth 508
Top level angle of saw blade 477
Torsion 301, 304
Torsional properties and shear strenght 394
— — of Sitka spruce, effect of moisture content 394
— resonant vibrations 275
— strength, determination 395
— — $\tau_{tb}$ for square cross-section 395
— — of wood 396
— vibrations in a wood member 275
Torus, pit 29
Total moisture and square root of drying time 462
Total parenchyma volume 12
Total power consumption of any wood-working machine 479
Toughness, decrease with increase of the angle between specimen axis and fiber direction 385
— machine 380
— of coachwood, effect of moisture content and temperature 390
— of decayed wood 98
— of hoop pine, effect of moisture content and temperature 390
— of pine, spruce, beech, oak, effect of the width of annual rings 391
— of spruce, effect of moisture content and
— temperature 390
of wood 219
Tracheid, coniferous 10
Tracheids, rounded 43

*Trametes serialis* 101
Transition lamellae of the cell wall 22
Transmission loss of sound for typical constructions 283
Transverse resonant vibrations of flexural vibrations 275
Traumatic resin canals 9
Trichloroethylene content of moisture and volatiles 183
Trichloroethylene, solvent for the distillation method 182
Trim ends, yield 485
Tritium exchange 60
Tropical hardwoods 57
— woods 173
Tropolones 70
Trough-shaped boards in sawmills 484
Trunks and valises, recommended moisture content 197
Tunnel-type kiln 457
Tupelo, black, average moisture content 421
Turned surfaces, quality 537
Turning 534, 541
Turning tools 537, 541
Turpentine, swelling and dielectric constant 218
Twisting 452
Twist, special warp form 92
Tyloses in hardwood vessels 15
— in heartwood 73
Tylosoids in softwood 16

Ulmaceae 12
*Ulmus americana* 62
Ultra-thin cross-sections for electron microscopy 47
Ultraviolet absoprtion spectrum of lignin 67
— photomicrograph 101
Uncondensed guaiacyl units 67
Unit cell cellulose I 59
— — of crystal of cellulose 269
Upmilling and downmilling technique 517
Uronic anhydride (wood analysis) 56
U-shaped bend 546
Utile, swelling gradients 209
UV-absorption 50

Vacuum drying, temperature and moisture content for discontinuous — 467
Vanak (*Virola merendonis* Pittier) bending radius 545
van der Walls forces 59
Vanillin 66
Vapor-dried and creosoted pine poles in bending tests 466
Vapor drying 466
— — in the preservation process 145
— — operation cycle followed by an empty cell-treatment 467
Variations in density 168
Vascular cambium 4
— tracheids 11
Vasicentric paratracheal parenchyma 6
Velocity of sound, determination 302

590 Subject Index

Veneer cores, content of solid matter 180
Veneer production in chipless wood-cutting 513
Veneering of furniture pieces by means of fire hoses and low voltage strips 551
Veneers, recommended moisture content 197
Vessels 10
— arrangement 12
— elements 10
Vestured pits 35
Vibration cutters 513
Vibratory resonant motion 275
Violins manufactured by Italian masters 274
Virola, swelling gradients 209
Viscosity 319, 321
— and temperature of different preservatives 238
— coefficient 221
— of different preservatives and penetration depth into air-dried Eastern hemlock thresholds 240
Vitamin $B_1$ 109
Void volume 233
— — of wood containing moisture 163
Volume contraction of the water within chemisorption 209
— per 1 g dry wood substance and 1% m.c., increase 211
— susceptibility 271
— swelling of wood proportional to the density 209
Volumetric swelling 205, 207
— — as a function of moisture 165
— — per 1% increase in moisture content 166, 209
Volumetric shrinkage 205
— — of different wood species vs. specific gravity 206

Wagons, recommended moisture content 196
Wane 90, 373
Walnut, black, average moisture content 421
—, drying temperatures 440
—, elastic constants 296
—, fatigue resistance 378
—, hardness and density 407
—, infrared drying 471
—, linear thermal expansion 241
—, mechanical properties across the grain 333
—, relative humidity during kiln drying 447
—, sound velocity 276
—, specific sanding work 533
—, swelling gradients 209
—, torsional strength 396
—, volumetric swelling 165
Warp 92, 452
Wart structure of wood cells 39
Warts in compression wood tracheids 46
—, index of refraction 41
—, size and shape 42
Warty layer 106
— — of wood cells 41
— — on vessel wall 36
— membrane 55

Washboarding 94
Waste in sawing 484
Water adsorbed in a monomolecular layer 193
— adsorbed in multimolecular layers 193
— movement in wood above and below fiber saturation point 219
— of constitution 193
— -resistant surface and internal coatings 218
— -soluble extractives 260
— — molecular fragments of decayed wood 110
— — phenol formaldehyde 156
— — salts 153
—, sound wave resistance 279
— sprays in kilns 441
—, swelling and dielectric constant 218
Wawa, weight 168
Wavy (quilted or blistered) grain 85
Wear lines of various woods and wood-base materials 409
— of hardwood 410
— of the cutting edge of saw blade 478
—, relationship to specific gravity 411
— resistance of common beech versus moisture content 412
— — of common beech versus specific gravity 411
— — values 411
Wedge angle of saws 501
Weight loss test 155
Weights of tropical timbers 168
West Coast hemlock, drying conditions 430
Western fir, movement of moisture 221
Western hemlock earlywood fibers, stress-strain properties 322
— — latewood fibers, stress-strain properties 322
Western larch, tangential swelling and radial shrinkage 245
Western red cedar (*Thuja plicata* D. Don), earlywood fibers stress-strain properties 322
— — — electrical resistivity 260
— — — latewood fibers, stress-strain properties 322
— — — resistivity 259
Western white pine, average moisture content 421
— — —, earlywood fibers, stress-strain properties 322
— — —, latewood fibers stress-strain properties 322
Western yellow pine, (*Pinus pondersosa* Laws.), longitudinal shrinkage 214
— — — moisture content 422
Wet bulb depression corresponding to drying gradient 447
Wetwood zones 94
Wheel, recommended moisture content 196
— spokes, recommended moisture content 196
Whisper, sound intensity 281
White ants (termites) 112

## Subject Index

White ash, (*Fraxinus americana L.*) linear thermal expansion 241
— —, spiral and diagonal grained lumber 366
White birch, cutting power and depth of cut 498
White fir (*Abies concolor* Lindl. et Gordon), linear thermal expansion 243
— —, average moisture content 421
— —, earlywood fibers, stress-strain properties 322
— —, latewood fibers, stress-strain properties 322
— —, modulus of elasticity 304
— —, specific gravity 172
— —, tensile strength and direction of the grain 326
— —, volumetric swelling 165
White pine (*Pinus strobus L.*), average moisture content 421
— —, cutting power and depth of cut 498
— —, fiber saturation point 200
— —, drying temperatures 440
— —, electrical resistivity and moisture content 260
— —, hardness and density 407
— —, linear thermal expansion 241
— —, relative humidity during kiln drying 447
— —, specific gravity 172
— —, volumetric swelling 217
White rot fungi 97
White spruce, density 164
— —, earlywood fibers, stress-strain properties 322
— —, latewood fibers, stress-strain properties 322
White thorn, hardness and density 407
Whitewood, elastic constants 296
Width of annual rings 178
— of growth rings of coniferous woods 178
— of kerf of circular saws 481
— of set, effect on power consumption in circular sawing 505
— of set in sawing 484
Willow, black, average moisture content 421
—, damping of sound radiation and sound wave resistance 280
—, hardness and density 407
—, fiber saturation point 199
—, relative humidity during kiln drying 447
—, swelling gradients 209
Winch presses for bending 548
Window frames, recommended moisture content 196
Winged reproductive or alate (termite) 114
Wöhler curves 376
Wolman salts 147
Wood as a building material 160
Wood bending, consumption of energy 548
— — machines and methods 546
— — over metal form 546
— boring insects 112, 136
—-charring preservatives 153
— columns, stresses 356
—-destroying fungi 97, 114, 136

Wood extractives 70
—-feeding Anobiidae 118
— flour, integral heat of wetting 201
— for sounding boards, damping of sound radiation and sound wave resistance 280
— hydrolysis 99
— hygrometers, pocket-size 184
— in service, recommended moisture content 195
—, magnetic susceptibility 271
— machining 475
— liquid relations 180
— preservation processes 140
— preservative treatment for marine structures 133
— residues, content of solid matter per unit volume 179, 180
— rays 2
— — multiseriate and uniseriate 14
Wooden mine sweepers 271
— wool board, sound absorption 285
Wool, piezoelectricity 272
Woolliness of green sawn boards 43
— of tensionwood surfaces 83
Woolly surface of board of mahogany 84
Woolly surfaces of tensionwood 83
Worker termite 113
Work of the European old-house borer in softwood 124
— to ultimate load, effect of temperature 369
Wormy chestnut 79

*Xeris spectrum* (L.) 121
*Xestobium rufovillosum* DeG. (death-watch beetle) 118
X-rays 161
X-ray diagram of cellulose 58
— diffraction 50
— diffraction for studies on cellulose in wood 18
Xylan 56, 58
—, in cell wall 75
— in differentiated cell 73
—, proportion in tensionwood 52
Xylem, 2, 72
Xylene, content of moisture and volatiles 183
— mixture of isomers, solvent for the distillation method 182
Xylophones 274
Xylose. wood analysis 56

Yard dryer 458
— seasoning 424
Yellow birch (*Betula lutea* Mich.), linear thermal expansion 243
— —, cutting power and depth of cut 498
— —, dashening during steaming 434
— —, elastic constants 295
— —, electrical resistivity and moisture
— —, content 260
— —, linear thermal expansion 241
Yellow pine, average moisture content 421
Yellowpoplar (*Liriodendron tulipifera* L.), and paper, power factor 268, 270

## Subject Index

Yellowpoplar, average moisture content 421
— —, fluid pressure and penetration 516
— —, linear thermal expansion 243
— —, tangential swelling and radial shrinkage 245
Yew, hardness and density 407
Yield and quality of pulp products from decayed wood 98
— and sawdust-refuse in sawing logs 485
— in sawmills 484
— of extractives recovered from redwood 465
—, production of conically-edged parallel-edged and lumber 484

Yield, values for gang saw mills 485
— — for simple circular saw mills 485
Young's modulus 292, 300, 322
— — and density, linear proportion 308

Zebrano, high frequency drying 470
Zinc chloride for impregnating wood 145
— — solution, viscosity and temperature 238
Zonate axial parenchyma in softwoods 6, 8
Zone lines, present with incipient decay 107
Zootermopsis 112